Quantum Mechanics with Basic Field Theory

Students and instructors alike will find this organized and detailed approach to quantum mechanics ideal for a two-semester graduate course on the subject.

This textbook covers, step-by-step, important topics in quantum mechanics, from traditional subjects like bound states, perturbation theory and scattering, to more current topics such as coherent states, quantum Hall effect, spontaneous symmetry breaking, superconductivity, and basic quantum electrodynamics with radiative corrections. The large number of diverse topics are covered in concise, highly focused chapters, and are explained in simple but mathematically rigorous ways. Derivations of results and formulas are carried out from beginning to end, without leaving students to complete them.

With over 200 exercises to aid understanding of the subject, this textbook provides a thorough grounding for students planning to enter research in physics. Several exercises are solved in the text, and password-protected solutions for remaining exercises are available to instructors at www.cambridge.org/9780521877602.

Bipin R. Desai is a Professor of Physics at the University of California, Riverside, where he does research in elementary particle theory. He obtained his Ph.D. in Physics from the University of California, Berkeley. He was a visiting Fellow at Clare Hall, Cambridge University, UK, and has held research positions at CERN, Geneva, Switzerland, and CEN Saclay, France. He is a Fellow of the American Physical Society.

Quantum Mechanics with Basic Field Theory

Bipin R. Desai

University of California at Riverside

CAMBRIDGE UNIVERSITY PRESS
Cambridge, New York, Melbourne, Madrid, Cape Town, Singapore,
São Paulo, Delhi, Dubai, Tokyo, Mexico City

Cambridge University Press
The Edinburgh Building, Cambridge CB2 8RU, UK

Published in the United States of America by Cambridge University Press, New York

www.cambridge.org
Information on this title: www.cambridge.org/9780521877602

First published 2010
Reprinted with corrections 2010

Printed in the United States of America

A catalog record for this publication is available from the British Library

ISBN 978-0-521-87760-2 Hardback

To Ba, Bapuji, and Blaire

Contents

Preface

While writing this book I was reminded at times of what Professor Francis Low used to say when I took his class on undergraduate electromagnetism at the University of Illinois, Urbana-Champaign. "Be sure to understand the subject thoroughly," he said, "otherwise, your only other chance will be when you have to teach it." Knowing now what I know by having written this book, I would add that, if at that point one still does not understand the subject, there will be yet another opportunity when writing a book on it. That was certainly the case with me and this book.

For the last twenty years or so I have taught a one-year graduate course in quantum mechanics at the University of California, Riverside. I have used several books, including the text by Schiff which also happens to be the text I used when I was taking my graduate courses at the University of California, Berkeley (along with my class notes from Professor Eyvind Wichmann who taught the quantum electrodynamics course). However, it became clear to me that I would need to expand the subject matter considerably if I wanted the book not only to be as thorough and up-to-date as possible but also organized so that one subject followed the other in a logical sequence. I hope I have succeeded.

Traditionally, books on graduate quantum mechanics go up to relativity and in some cases even cover the Dirac equation. But relativistic equations lead to the troublesome negative-energy solutions. It would be unsatisfactory then to just stop there and not go to second quantization, to show how the negative-energy states are reinterpreted as positive-energy states of antiparticles. It was, therefore, logical to cover elementary second quantization, which in a sense is many-body quantum mechanics with quantization conditions. And once this topic was addressed it would be unfair not to cover the great successes of many-body systems in condensed matter, in particular, superconductivity and Bose–Einstein condensation. A logical concurrent step was to include also full relativistic quantum field theory, at least basic quantum electrodynamics (QED) and then finish on a triumphant note describing the stunning success of QED in explaining the anomalous magnetic moment and the Lamb shift. With the vast acreage that I wanted to cover, it seemed only appropriate to include as well the modern subject of spontaneous symmetry breaking, which has its applications both in condensed matter physics and in particle physics. This then was the rationale behind this book's content and organization.

I have organized the book with small chapters in what I believe to be a logical order. One can think of the layout of the chapters in terms of the following blocks, each with a common thread, with chapters arranged in an increasing degree of complexity within each block

Chs. 1, 2, 3	Basic Formalism
Chs. 4, 5, 6, 7	Free Particles
Chs. 8, 9, 10, 11, 12	Exactly Solvable Bound State Problems
Chs. 13, 14, 15	Two-Level Problems
Chs. 16, 17, 18	Perturbation Theory
Ch. 24	New approximation methods
Ch. 25	Lagrangian and Feynman integral formalisms
Chs. 19, 20, 21, 22, 23	Scattering Theory
Chs. 26, 27, 28, 29, 30	Symmetry, Rotations, and Angular Momentum
Chs. 31, 32, 33, 34, 35, 36	Relativistic theory with Klein–Gordon, Dirac, and Maxwell's equations
Chs. 37, 38, 39, 40	Second Quantization, Condensed Matter Problems
Chs. 41, 42	Classical Fields and Spontaneous Symmetry Breaking
Chs. 43, 44, 45	Quantum Electrodynamics and Radiative Corrections

In the chapters on scattering theory, one may find an extra coverage in this book on the properties of the S-matrix especially with reference to its analytical properties. This is thanks to my thesis advisor at Berkeley, Professor Geoffrey Chew who emphasized the importance of these properties to his students.

I believe it is feasible to complete the first 32 chapters in one year (two semesters or three quarters). The remaining chapters beginning with the Dirac equation could well be taught in the first semester or first quarter of an advanced quantum mechanics course. Since these topics cover quantum field theory applied to both particle physics and condensed matter physics, it could be taken by students specializing in either subject.

Except at the beginning of each chapter, this book does not have as much narrative or as many long descriptive paragraphs as one normally finds in other textbooks. I have instead spent extra space on deriving and solving the relevant equations. I feel that the extra narrative can always be supplemented by the person teaching the course.

There are an adequate number of problems in this book. They are fairly straightforward. I suppose I still have scars left from the days when I took graduate quantum mechanics from Professor Edward Teller at Berkeley, who gave very inspiring lectures full of interesting and topical episodes while on the blackboard he usually wrote down just the basic formulas. But then he turned around and gave, as homework, a huge number of some of the toughest problems this side of the Atlantic! Those assignments routinely took care of our entire weekends.

I have many people to thank, beginning with Dustin Urbaniec and Omar Moreno who did a good bit of the typing for me, and Barbara Simandl who did all the figures. I am also grateful to a number of graduate students from my Quantum Mechanics course for pointing out errors in my write-up; in particular, I am thankful to Eric Barbagiovanni, for suggesting a number of improvements. I must also thank Dr. Steve Foulkes, a former graduate student at UC Riverside, who read a number of chapters and, following my instructions not to show any mercy in criticizing what he read, did exactly that! I also wish to thank my colleagues who critically read parts of the manuscript: Professors Robert Clare (who also directed me to Cambridge University Press), Leonid Pryadkov, G. Rajasekaran and Utpal Sarkar.

At Cambridge University Press, my special thanks to Simon Mitton, with whom I corresponded in the early years, for his kind support and encouragement; to John Fowler and Lindsay Barnes for their constant help and, more importantly, for their patience with this long project.

There is one individual, Alex Vaucher, whom I must single out, without whose help this book would neither have been started nor completed. After finishing my graduate course on Quantum Mechanics at UC Riverside some years ago, he strongly encouraged me to write this book. He supplied the necessary software and, knowing how computer-ignorant I was, continued to provide me with technical instructions during all phases of this project. Initially the two of us were planning to collaborate on this book but, because of his full time position with the Physics and Astronomy department at the University of California, Los Angeles, he was not able to participate. My deepest gratitude to him.

Physical constants

Planck's constant	$\hbar$	6.581×10^{-16} eV s
Velocity of light in vacuum	c	2.9979×10^{10} cm/s
Fine structure constant	$\alpha = e^2/\hbar c$	$1/137.04$
Rest mass of the electron	mc^2	0.511 MeV
Mass of the proton	Mc^2	938.28 MeV
Bohr radius	$\hbar^2/me^2$	5.2918×10^{-9} cm
Bohr magneton	$e\hbar/2mc$	0.58×10^{-8} eV/gauss
Boltzmann constant	k	8.62×10^{-5} eV/K
1 eV		1.6×10^{-12} erg

1 Basic formalism

We summarize below some of the postulates and definitions basic to our formalism, and present some important results based on these postulates. The formalism is purely mathematical in nature with very little physics input, but it provides the structure within which the physical concepts that will be discussed in the later chapters will be framed.

1.1 State vectors

It is important to realize that the Quantum Theory is a linear theory in which the physical state of a system is described by a vector in a complex, linear vector space. This vector may represent a free particle or a particle bound in an atom or a particle interacting with other particles or with external fields. It is much like a vector in ordinary three-dimensional space, following many of the same rules, except that it describes a very complicated physical system. We will be elaborating further on this in the following.

The mathematical structure of a quantum mechanical system will be presented in terms of the notations developed by Dirac.

A physical state in this notation is described by a "ket" vector, $|\rangle$, designated variously as $|\alpha\rangle$ or $|\psi\rangle$ or a ket with other appropriate symbols depending on the specific problem at hand. The kets can be complex. Their complex conjugates, $|\rangle^*$, are designated by $\langle|$ which are called "bra" vectors. Thus, corresponding to every ket vector there is a bra vector. These vectors are abstract quantities whose physical interpretation is derived through their so-called "representatives" in the coordinate or momentum space or in a space appropriate to the problem under consideration.

The dimensionality of the vector space is left open for the moment. It can be finite, as will be the case when we encounter spin, which has a finite number of components along a preferred direction, or it can be infinite, as is the case of the discrete bound states of the hydrogen atom. Or, the dimensionality could be continuous (indenumerable) infinity, as for a free particle with momentum that takes continuous values. A complex vector space with these properties is called a Hilbert space.

The kets have the same properties as a vector in a linear vector space. Some of the most important of these properties are given below:

(i) $|\alpha\rangle$ and $c\,|\alpha\rangle$, where c is a complex number, describe the same state.

(ii) The bra vector corresponding to $c\,|\alpha\rangle$ will be $c^*\,\langle\alpha|$.

(iii) The kets follow a linear superposition principle

$$a\,|\alpha\rangle + b\,|\beta\rangle = c\,|\gamma\rangle \tag{1.1}$$

where a, b, and c are complex numbers. That is, a linear combination of states in a vector space is also a state in the same space.

(iv) The "scalar product" or "inner product" of two states $|\alpha\rangle$ and $|\beta\rangle$ is defined as

$$\langle\beta|\,\alpha\rangle. \tag{1.2}$$

It is a complex number and not a vector. Its complex conjugate is given by

$$\langle\beta|\,\alpha\rangle^* = \langle\alpha|\,\beta\rangle. \tag{1.3}$$

Hence $\langle\alpha|\,\alpha\rangle$ is a real number.

(v) Two states $|\alpha\rangle$ and $|\beta\rangle$ are orthogonal if

$$\langle\beta|\,\alpha\rangle = 0. \tag{1.4}$$

(vi) It is postulated that $\langle\alpha|\,\alpha\rangle \geq 0$. One calls $\sqrt{\langle\alpha|\,\alpha\rangle}$ the "norm" of the state $|\alpha\rangle$. If a state vector is normalized to unity then

$$\langle\alpha|\,\alpha\rangle = 1. \tag{1.5}$$

If the norm vanishes, then $|\alpha\rangle = 0$, in which case $|\alpha\rangle$ is called a null vector.

(vii) The states $|\alpha_n\rangle$ with $n = 1, 2, \ldots$, depending on the dimensionality, are called a set of basis kets or basis states if they span the linear vector space. That is, any arbitrary state in this space can be expressed as a linear combination (superposition) of them. The basis states are often taken to be of unit norm and orthogonal, in which case they are called orthonormal states. Hence an arbitrary state $|\phi\rangle$ can be expressed in terms of the basis states $|\alpha_n\rangle$ as

$$|\phi\rangle = \sum_n a_n\,|\alpha_n\rangle \tag{1.6}$$

where, as stated earlier, the values taken by the index n depends on whether the space is finite- or infinite-dimensional or continuous. In the latter case the summation is replaced by an integral. If the $|\alpha_n\rangle$'s are orthonormal then $a_n = \langle\alpha_n\,|\phi\rangle$. It is then postulated that $|a_n|^2$ is the probability that the state $|\phi\rangle$ will be in state $|\alpha_n\rangle$.

(viii) A state vector may depend on time, in which case one writes it as $|\alpha(t)\rangle$, $|\psi(t)\rangle$, etc. In the following, except when necessary, we will suppress the possible dependence on time.

(ix) The product $|\alpha\rangle\,|\beta\rangle$, has no meaning unless it refers to two different vector spaces, e.g., one corresponding to spin, the other to momentum; or, if a state consists of two particles described by $|\alpha\rangle$ and $|\beta\rangle$ respectively.

(x) Since bra vectors are obtained through complex conjugation of the ket vectors, the above properties can be easily carried over to the bra vectors.

1.2 Operators and physical observables

A physical observable, like energy or momentum, is described by a linear operator that has the following properties:

(i) If A is an operator and $|\alpha\rangle$ is a ket vector then

$$A\,|\alpha\rangle = \text{another ket vector.} \tag{1.7}$$

Similarly, for an operator B,

$$\langle\alpha|\,B = \text{another bra vector} \tag{1.8}$$

where B operates to the left

(ii) An operator A is linear if, for example,

$$A\left[\lambda\,|\alpha\rangle + \mu\,|\beta\rangle\right] = \lambda A\,|\alpha\rangle + \mu A\,|\beta\rangle \tag{1.9}$$

where λ and μ are complex numbers. Typical examples of linear operators are derivatives, matrices, etc. There is one exception to this rule, which we will come across in Chapter 27 which involves the so called time reversal operator where the coefficients on the right-hand side are replaced by their complex conjugates. In this case it is called an antilinear operator.

If an operator acting on a function gives rise to the square of that function, for example, then it is called a nonlinear operator. In this book we will be not be dealing with such operators.

(iii) A is a called a unit operator if, for any $|\alpha\rangle$,

$$A\,|\alpha\rangle = |\alpha\rangle, \tag{1.10}$$

in which case one writes

$$A = \mathbf{1}. \tag{1.11}$$

(iv) A product of two operators is also an operator. In other words, if A and B are operators then AB as well as BA are operators. However, they do not necessarily commute under multiplication, that is,

$$AB \neq BA \tag{1.12}$$

in general. The operators commute under addition, i.e., if A and B are two operators then

$$A + B = B + A. \tag{1.13}$$

They also exhibit associativity, i.e., if A, B, and C are three operators then

$$A + (B + C) = (A + B) + C. \tag{1.14}$$

Similarly $A(BC) = (AB)C$.

(v) B is called an inverse of the operator A if

$$AB = BA = \mathbf{1}, \tag{1.15}$$

in which case one writes

$$B = A^{-1}. \tag{1.16}$$

(vi) The quantity $|\alpha\rangle\langle\beta|$ is called the "outer product" between states $|\alpha\rangle$ and $|\beta\rangle$. By multiplying it with a state $|\gamma\rangle$ one obtains

$$[|\alpha\rangle\langle\beta|]\gamma\rangle = [\langle\beta|\gamma\rangle]|\alpha\rangle \tag{1.17}$$

where on the right-hand side the order of the terms is reversed since $\langle\beta|\gamma\rangle$ is a number. The above relation implies that when $|\alpha\rangle\langle\beta|$ multiplies with a state vector it gives another state vector. A similar result holds for the bra vectors:

$$\langle\gamma[|\alpha\rangle\langle\beta|] = [\langle\gamma|\alpha\rangle]\langle\beta|. \tag{1.18}$$

Thus $|\alpha\rangle\langle\beta|$ acts as an operator.

(vii) The "expectation" value, $\langle A\rangle$, of an operator A in the state $|\alpha\rangle$ is defined as

$$\langle A\rangle = \langle\alpha|A|\alpha\rangle. \tag{1.19}$$

1.3 Eigenstates

(i) If the operation $A|\alpha\rangle$ gives rise to the same state vector, i.e., if

$$A|\alpha\rangle = (\text{constant}) \times |\alpha\rangle \tag{1.20}$$

then we call $|\alpha\rangle$ an "eigenstate" of the operator A, and the constant is called the "eigenvalue" of A. If $|\alpha\rangle$'s are eigenstates of A with eigenvalues a_n, assumed for convenience to be discrete, then these states are generally designated as $|a_n\rangle$. They satisfy the equation

$$A|a_n\rangle = a_n|a_n\rangle \tag{1.21}$$

with $n = 1, 2, \ldots$ depending on the dimensionality of the system. In this case one may also call A an eigenoperator.

(ii) If $|\alpha_n\rangle$ is an eigenstate of both operators A and B, such that

$$A|\alpha_n\rangle = a_n|\alpha_n\rangle, \quad \text{and} \quad B|\alpha_n\rangle = b_n|\alpha_n\rangle \tag{1.22}$$

then we have the results

$$AB|\alpha_n\rangle = b_n A|\alpha_n\rangle = b_n a_n|\alpha_n\rangle, \tag{1.23}$$

$$BA|\alpha_n\rangle = a_n B|\alpha_n\rangle = a_n b_n|\alpha_n\rangle. \tag{1.24}$$

If the above two relations hold for all values of n then

$$AB = BA. \tag{1.25}$$

Thus under the special conditions just outlined the two operators will commute.

1.4 Hermitian conjugation and Hermitian operators

We now define the "Hermitian conjugate" $A^\dagger$, of an operator A and discuss a particular class of operators called "Hermitian" operators which play a central role in quantum mechanics.

(i) In the same manner as we defined the complex conjugate operation for the state vectors, we define $A^\dagger$ through the following complex conjugation

$$[A|\alpha\rangle]^* = \langle\alpha|A^\dagger \tag{1.26}$$

and

$$\langle\beta|A|\alpha\rangle^* = \langle\alpha|A^\dagger|\beta\rangle. \tag{1.27}$$

If on the left-hand side of (1.26), $|\alpha\rangle$ is replaced by $c|\alpha\rangle$ where c is a complex constant, then on the right-hand side one must include a factor c^*.

(ii) From (1.26) and (1.27) it follows that if

$$A = |\alpha\rangle\langle\beta| \tag{1.28}$$

then

$$A^\dagger = |\beta\rangle\langle\alpha|. \tag{1.29}$$

At this stage it is important to emphasize that $|\alpha\rangle$, $\langle\beta|\alpha\rangle$, $|\alpha\rangle\langle\beta|$, and $|\alpha\rangle|\beta\rangle$ are four totally different mathematical quantities which should not be mixed up: the first is a state vector, the second is an ordinary number, the third is an operator, and the fourth describes a product of two states.

(iii) The Hermitian conjugate of the product operator AB is found to be

$$(AB)^\dagger = B^\dagger A^\dagger. \tag{1.30}$$

This can be proved by first noting from (1.27) that for an arbitrary state $|\alpha\rangle$

$$[(AB)|\alpha\rangle]^* = \langle\alpha|(AB)^\dagger. \tag{1.31}$$

If we take

$$B|\alpha\rangle = |\beta\rangle \tag{1.32}$$

where $|\beta\rangle$ is another state vector, then the left-hand side of (1.31) can be written as

$$[(AB)|\alpha\rangle]^* = [A|\beta\rangle]^*. \tag{1.33}$$

From the definition given in (1.26) we obtain

$$[A|\beta\rangle]^* = \langle\beta| A^\dagger = \left[\langle\alpha| B^\dagger\right] A^\dagger = \langle\alpha| B^\dagger A^\dagger \tag{1.34}$$

where we have used the fact that $\langle\beta| = [B|\alpha\rangle]^* = \langle\alpha| B^\dagger$. Since $|\alpha\rangle$ is an arbitrary vector, comparing (1.31) and (1.34), we obtain (1.30).

(iv) Finally, if

$$A = A^\dagger \tag{1.35}$$

then the operator A is called "Hermitian."

1.5 Hermitian operators: their eigenstates and eigenvalues

Hermitian operators play a central role in quantum mechanics. We show below that the eigenstates of Hermitian operators are orthogonal and have real eigenvalues.

Consider the eigenstates $|a_n\rangle$ of an operator A,

$$A|a_n\rangle = a_n|a_n\rangle \tag{1.36}$$

where $|a_n\rangle$'s have a unit norm. By multiplying both sides of (1.36) by $\langle a_n|$ we obtain

$$a_n = \langle a_n| A |a_n\rangle. \tag{1.37}$$

Taking the complex conjugate of both sides we find

$$a_n^* = \langle a_n| A |a_n\rangle^* = \langle a_n| A^\dagger |a_n\rangle = \langle a_n| A |a_n\rangle. \tag{1.38}$$

The last equality follows from the fact that A is Hermitian ($A^\dagger = A$). Equating (1.37) and (1.38) we conclude that $a_n^* = a_n$. Therefore, the eigenvalues of a Hermitian operator must be real.

An important postulate based on this result says that since physically observable quantities are expected to be real, the operators representing these observables must be Hermitian.

We now show that the eigenstates are orthogonal. We consider two eigenstates $|a_n\rangle$ and $|a_m\rangle$ of A,

$$A\,|a_n\rangle = a_n\,|a_n\rangle\,, \tag{1.39}$$

$$A\,|a_m\rangle = a_m\,|a_m\rangle\,. \tag{1.40}$$

Taking the complex conjugate of the second equation we have

$$\langle a_m|\,A = a_m\,\langle a_m| \tag{1.41}$$

where we have used the Hermitian property of A, and the fact that the eigenvalue a_m is real. Multiplying (1.39) on the left by $\langle a_m|$ and (1.41) on the right by $|a_n\rangle$ and subtracting, we obtain

$$(a_m - a_n)\,\langle a_m|\,a_n\rangle = 0. \tag{1.42}$$

Thus, if the eigenvalues α_m and α_n are different we have

$$\langle a_m|\,a_n\rangle = 0, \tag{1.43}$$

which shows that the two eigenstates are orthogonal. Using the fact that the ket vectors are normalized, we can write the general orthonormality relation between them as

$$\langle a_m|\,a_n\rangle = \delta_{mn} \tag{1.44}$$

where δ_{mn} is called the Kronecker delta, which has the property

$$\begin{aligned} \delta_{mn} &= 1 \ \text{ for } \ m = n \\ &= 0 \ \text{ for } \ m \neq n. \end{aligned} \tag{1.45}$$

For those cases where there is a degeneracy in the eigenvalues, i.e., if two different states have the same eigenvalue, the treatment is slightly different and will be deferred until later chapters.

We note that the operators need not be Hermitian in order to have eigenvalues. However, in these cases none of the above results will hold. For example, the eigenvalues will not necessarily be real. Unless otherwise stated, we will assume the eigenvalues to be real.

1.6 Superposition principle

A basic theorem in quantum mechanics based on linear vector algebra is that an arbitrary vector in a given vector space can be expressed as a linear combination – a superposition – of a complete set of eigenstates of any operator in that space. A complete set is defined to be the set of all possible eigenstates of an operator. Expressing this result for an arbitrary state vector $|\phi\rangle$ in terms of the eigenstates $|a_n\rangle$ of the operator A, we have

$$|\phi\rangle = \sum_n c_n |a_n\rangle \tag{1.46}$$

where the summation index n goes over all the eigenstates with $n = 1, 2, \ldots$. If we multiply (1.46) by $\langle a_m|$ then the orthonormality relation (1.44) between the $|a_n\rangle$'s yields

$$c_m = \langle a_m|\,\phi\rangle. \tag{1.47}$$

It is then postulated that $|c_m|^2$ is the probability that $|\phi\rangle$ contains $|a_m\rangle$. That is, $|c_m|^2$ is the probability that $|\phi\rangle$ has the eigenvalue a_m. If $|\phi\rangle$ is normalized to unity, $\langle\phi|\,\phi\rangle = 1$, then

$$\sum_n |c_n|^2 = 1. \tag{1.48}$$

That is, the probability of finding $|\phi\rangle$ in state $|a_n\rangle$, summed over all possible values of n, is one.

Since (1.46) is true for any arbitrary state we can express another state $|\psi\rangle$ as

$$|\psi\rangle = \sum_n c_n' |a_n\rangle. \tag{1.49}$$

The scalar product $\langle\psi\,|\phi\rangle$ can then be written, using the orthonormality property of the eigenstates, as

$$\langle\psi\,|\phi\rangle = \sum_m c_m'^* c_m \tag{1.50}$$

with $c_m' = \langle a_m|\,\psi\rangle$ and $c_m = \langle a_m|\,\phi\rangle$.

The above relations express the fact that the state vectors can be expanded in terms of the eigenstates of an operator A. The eigenstates $|a_n\rangle$ are then natural candidates to form a set of basis states.

1.7 Completeness relation

We consider now the operators $|a_n\rangle\langle a_n|$, where the $|a_n\rangle$'s are the eigenstates of an operator A, with eigenvalues a_n. A very important result in quantum mechanics involving the sum of the operators $|a_n\rangle\langle a_n|$ over the possibly infinite number of eigenvalues states that

$$\sum_n |a_n\rangle\langle a_n| = \mathbf{1} \tag{1.51}$$

where the **1** on the right-hand side is a unit operator. This is the so called "completeness relation".

To prove this relation we first multiply the sum on the left hand of the above equality by an arbitrary eigenvector $|a_m\rangle$ to obtain

$$\left[\sum_n |a_n\rangle \langle a_n|\right] |a_m\rangle = \sum_n |a_n\rangle \langle a_n| a_m\rangle = \sum_n |a_n\rangle \delta_{nm} = |a_m\rangle \tag{1.52}$$

where we have used the orthonormality of the eigenvectors. Since this relation holds for any arbitrary state $|a_m\rangle$, the operator in the square bracket on the left-hand side acts as a unit operator, thus reproducing the completeness relation.

If we designate

$$P_n = |a_n\rangle \langle a_n| \tag{1.53}$$

then

$$P_n |a_m\rangle = \delta_{mn} |a_m\rangle \tag{1.54}$$

Thus P_n projects out the state $|a_n\rangle$ whenever it operates on an arbitrary state. For this reason P_n is called the projection operator, in terms of which one can write the completeness relation as

$$\sum_n P_n = \mathbf{1}. \tag{1.55}$$

One can utilize the completeness relation to simplify the scalar product $\langle \psi | \phi \rangle$, where $|\phi\rangle$ and $|\psi\rangle$ are given above, if we write, using (1.51)

$$\langle \psi | \phi \rangle = \langle \psi | \mathbf{1} |\phi\rangle = \langle \psi | \left[\sum_n |a_n\rangle \langle a_n|\right] |\phi\rangle = \sum_n \langle \psi | a_n\rangle \langle a_n | \phi\rangle = \sum_n c_n'^* c_n. \tag{1.56}$$

This is the same result as the one we derived previously as (1.50).

1.8 Unitary operators

If two state vectors $|\alpha\rangle$ and $|\alpha'\rangle$ have the same norm then

$$\langle \alpha | \alpha \rangle = \langle \alpha' | \alpha' \rangle. \tag{1.57}$$

Expressing each of these states in terms of a complete set of eigenstates $|a_n\rangle$ we obtain

$$|\alpha\rangle = \sum_n c_n |a_n\rangle \quad \text{and} \quad |\alpha'\rangle = \sum_n c_n' |a_n\rangle. \tag{1.58}$$

The equality in (1.57) leads to the relation

$$\sum_n |c_n|^2 = \sum_n |c_n'|^2, \tag{1.59}$$

which signifies that, even though c_n may be different from c'_n, the total sum of the probabilities remains the same.

Consider now an operator A such that

$$A|\alpha\rangle = |\alpha'\rangle. \tag{1.60}$$

If $|\alpha\rangle$ and $|\alpha'\rangle$ have the same norm, then

$$\langle\alpha|\alpha\rangle = \langle\alpha'|\alpha'\rangle = \langle\alpha|A^\dagger A|\alpha\rangle. \tag{1.61}$$

This implies that

$$A^\dagger A = \mathbf{1}. \tag{1.62}$$

The operator A is then said to be "unitary." From relation (1.60) it is clear that A can change the basis from one set to another. Unitary operators play a fundamental role in quantum mechanics.

1.9 Unitary operators as transformation operators

Let us define the following operator in term of the eigenstates $|a_n\rangle$ of operator A and eigenstates $|b_n\rangle$ of operator B,

$$U = \sum_n |b_n\rangle\langle a_n|. \tag{1.63}$$

This is a classic example of a unitary operator as we show below. We first obtain the Hermitian conjugate of U,

$$U^\dagger = \sum_n |a_n\rangle\langle b_n|. \tag{1.64}$$

Therefore,

$$UU^\dagger = \left[\sum_n |b_n\rangle\langle a_n|\right]\sum_m |a_m\rangle\langle b_m| = \sum\sum |b_n\rangle\langle a_n|a_m\rangle\langle b_m| = \sum |b_n\rangle\langle b_n| = \mathbf{1} \tag{1.65}$$

where we have used the orthonormality relation $\langle a_n|a_m\rangle = \delta_{nm}$ and the completeness relation for the state vectors $|b_n\rangle$ discussed in the previous section. Hence U is unitary.

We note in passing that

$$\sum_n |a_n\rangle\langle a_n| \tag{1.66}$$

is a unit operator which is a special case of a unitary operator when $\langle b_n| = \langle a_n|$.

We also note that

$$U |a_m\rangle = \sum_n |b_n\rangle \langle a_n| a_m\rangle = \sum_n |b_n\rangle \delta_{nm} = |b_m\rangle . \tag{1.67}$$

Hence U transforms the eigenstate $|a_m\rangle$ into $|b_m\rangle$. In other words, if we use $|a_n\rangle$'s as the basis for the expansion of a state vector then U will convert this basis set to a new basis formed by the $|b_n\rangle$'s. Thus U allows one to transform from the "old basis" given by $|a_n\rangle$'s to the "new basis" given by the $|b_n\rangle$'s. One can do the conversion in the reverse order by multiplying both sides of (1.67) by $U^\dagger$ on the left:

$$U^\dagger U |a_m\rangle = U^\dagger |b_m\rangle . \tag{1.68}$$

Hence, from the unitary property of U, we find

$$U^\dagger |b_m\rangle = |a_m\rangle . \tag{1.69}$$

Furthermore, the matrix element of an operator, A, in the old basis set can be related to its matrix elements in the new basis as follows

$$\langle b_n| A |b_m\rangle = \langle b_n| UU^\dagger AUU^\dagger |b_m\rangle = \langle a_n| U^\dagger AU |a_m\rangle \tag{1.70}$$

where we have used the property $U^\dagger U = \mathbf{1}$ and the relations (1.69). This relation will be true for all possible values of $|a_n\rangle$'s and $|b_n\rangle$'s. Therefore, it can be expressed as an operator relation in terms of the "transformed" operator A^T. We then write

$$A^T = U^\dagger AU. \tag{1.71}$$

Finally, if $|a_n\rangle$'s are the eigenstates of the operator A,

$$A |a_n\rangle = a_n |a_n\rangle \tag{1.72}$$

and relation (1.67) connecting $|a_n\rangle$ and $|b_n\rangle$ holds, where $|b_n\rangle$'s are the eigenstates of an operator B, then we can multiply (1.72) by U on both sides to obtain

$$UA |a_n\rangle = a_n U |a_n\rangle \tag{1.73}$$

and write

$$UAU^\dagger [U |a_n\rangle] = a_n [U |a_n\rangle] . \tag{1.74}$$

Hence

$$UAU^\dagger |b_n\rangle = a_n |b_n\rangle . \tag{1.75}$$

Thus $|b_n\rangle$ is the eigenstate of $UAU^\dagger$ with the same eigenvalues as the eigenvalues of A. However, since $|b_n\rangle$'s are eigenstates of the operator B we find that $UAU^\dagger$ and B are in some sense equivalent.

1.10 Matrix formalism

We define the "matrix element" of an operator A between states $|\alpha\rangle$ and $|\beta\rangle$ as

$$\langle\beta\,|A|\,\alpha\rangle\,, \tag{1.76}$$

which is a complex number. To understand the meaning of this matrix element we note that when A operates on $|\alpha\rangle$ it gives another ket vector. This state when multiplied on the left by $\langle\beta|$ gives a number. When there are many such $|\alpha\rangle$'s and $\langle\beta|$'s then we have a whole array of numbers that can be put into the form of a matrix. Specifically, the matrix formed by the matrix elements of A between the basis states $|b_n\rangle$, with $n = 1, 2, \ldots, N$ depending on the dimensionality of the space, is then a square matrix $\langle b_m\,|A|\,b_n\rangle$ written as follows:

$$\{A\} = \begin{bmatrix} A_{11} & A_{12} & A_{13} & . & A_{1N} \\ A_{21} & A_{22} & A_{23} & . & . \\ A_{31} & A_{32} & A_{33} & . & . \\ . & . & . & . & . \\ A_{N1} & . & . & . & A_{NN} \end{bmatrix} \tag{1.77}$$

where

$$A_{mn} = \langle b_m\,|A|\,b_n\rangle\,. \tag{1.78}$$

The matrix (1.77) is then called the matrix representation of A in terms of the states $|b_n\rangle$. It gives, in a sense, a profile of the operator A and describes what is an abstract object in terms of a matrix of complex numbers. The matrix representation will look different if basis sets formed by eigenstates of some other operator are used.

The matrices follow the normal rules of matrix algebra. Some of the important properties are given below, particularly as they relate to the Hermitian and unitary operators.

(i) The relation between the matrix elements of $A^\dagger$ and A is given by

$$\langle\alpha|\,A^\dagger\,|\beta\rangle = \langle\beta|\,A\,|\alpha\rangle^*\,. \tag{1.79}$$

Thus the matrix representation of $A^\dagger$ can be written as

$$\{A^\dagger\} = \begin{bmatrix} A^*_{11} & A^*_{21} & A^*_{31} & . & A^*_{N1} \\ A^*_{12} & A^*_{22} & A^*_{23} & . & . \\ A^*_{13} & A^*_{23} & A^*_{33} & . & . \\ . & . & . & . & . \\ A^*_{1N} & . & . & . & A^*_{NN} \end{bmatrix} \tag{1.80}$$

where A is represented by the matrix (1.77).

(ii) If the operator A is Hermitian then

$$\langle\beta|\,A\,|\alpha\rangle = \langle\beta|\,A^\dagger\,|\alpha\rangle\,. \tag{1.81}$$

Using the property (1.79) we find

$$\langle\beta|\, A\, |\alpha\rangle = \langle\alpha|\, A\, |\beta\rangle^* . \tag{1.82}$$

In particular, the matrix elements with respect to the eigenstates $|b_n\rangle$ satisfy

$$\langle b_m\, |A|\, b_n\rangle = \langle b_n\, |A|\, b_m\rangle^* . \tag{1.83}$$

A Hermitian operator will, therefore, have the following matrix representation:

$$\{A\} = \begin{bmatrix} A_{11} & A_{12} & A_{13} & . & A_{1N} \\ A_{12}^* & A_{22} & A_{23} & . & . \\ A_{13}^* & A_{23}^* & A_{33} & . & . \\ . & . & . & . & . \\ A_{1N}^* & . & . & . & A_{NN} \end{bmatrix}. \tag{1.84}$$

We note that the same result is obtained by equating the matrices A and $A^\dagger$ given by (1.77) and (1.80) respectively. We also note that the diagonal elements $A_{11}, A_{22}, \ldots$ of a Hermitian operator are real since the matrix elements satisfy the relation $A_{mn}^* = A_{nm}$.

(iii) The matrix representation of an operator A in terms of its eigenstates is a diagonal matrix because of the orthonormality property (1.44). It can, therefore, be written as

$$\{A\} = \begin{bmatrix} A_{11} & 0 & 0 & . & 0 \\ 0 & A_{22} & 0 & . & . \\ 0 & 0 & A_{33} & . & . \\ . & . & . & . & . \\ 0 & . & . & . & A_{NN} \end{bmatrix} \tag{1.85}$$

where $A_{mm} = a_m$ where a_m's are the eigenvalues of A. The matrix representation of A in terms of eigenstates $|b_n\rangle$ of an operator B that is different from A and that does not share the same eigenstates is then written as

$$\{A\} = \begin{bmatrix} A_{11} & A_{12} & A_{13} & . & A_{1N} \\ A_{21} & A_{22} & A_{23} & . & . \\ A_{31} & A_{32} & A_{33} & . & . \\ . & . & . & . & . \\ A_{N1} & . & . & . & A_{NN} \end{bmatrix} \tag{1.86}$$

where $A_{mn} = \langle b_m\, |A|\, b_n\rangle$. If the operator B has the same eigenstates as A then the above matrix will, once again, be diagonal.

(iv) We now illustrate the usefulness of the completeness relation by considering several operator relations. First let us consider the matrix representation of the product of two operators $\{AB\}$ by writing

$$\langle b_m\, |AB|\, b_n\rangle = \langle b_m\, |A\mathbf{1}B|\, b_n\rangle = \sum_p \langle b_m\, \big|A\left[|b_p\rangle\langle b_p|\right] B\big|\, b_n\rangle. \tag{1.87}$$

In the second equality we have inserted a unit operator between the operators A and B and then replaced it by the sum of the complete set of states. Hence the matrix representation of AB is simply the product of two matrices:

$$\{AB\} = \begin{bmatrix} A_{11} & A_{12} & . & . & A_{1N} \\ A_{21} & A_{22} & . & . & . \\ . & . & . & . & . \\ . & . & . & . & . \\ A_{N1} & . & . & . & A_{NN} \end{bmatrix} \begin{bmatrix} B_{11} & B_{12} & . & . & B_{1N} \\ B_{21} & B_{22} & . & . & . \\ . & . & . & . & . \\ . & . & . & . & . \\ B_{N1} & . & . & . & B_{NN} \end{bmatrix}. \tag{1.88}$$

Next we consider the operator relation

$$A\,|\alpha\rangle = |\beta\rangle\,. \tag{1.89}$$

It can be written as a matrix relation if we multiply both sides of the equation by the eigenstates $\langle b_m|$ of an operator B and then insert a complete set of states $|b_p\rangle$ with $p = 1, 2, \ldots$:

$$\langle b_m|\,A\left[\sum_p |b_p\rangle\,\langle b_p|\right]|\alpha\rangle = \sum_p \langle b_m|\,A\,|b_p\rangle\,\langle b_p\,|\alpha\rangle = \langle b_m\,|\beta\rangle\,. \tag{1.90}$$

This is a matrix equation in which A is represented by the matrix in (1.86), and $|\alpha\rangle$ and $|\beta\rangle$ are represented by the column matrices

$$\begin{bmatrix} \langle b_1\,|\alpha\rangle \\ \langle b_2\,|\alpha\rangle \\ . \\ \langle b_N\,|\alpha\rangle \end{bmatrix} \quad \text{and} \quad \begin{bmatrix} \langle b_1\,|\beta\rangle \\ \langle b_2\,|\beta\rangle \\ . \\ \langle b_N\,|\beta\rangle \end{bmatrix} \tag{1.91}$$

respectively, and hence the above relation can be written in terms of matrices as

$$\begin{bmatrix} A_{11} & A_{12} & . & . & A_{1N} \\ A_{21} & A_{22} & . & . & . \\ . & . & . & . & . \\ . & . & . & . & . \\ A_{N1} & . & . & . & A_{NN} \end{bmatrix} \begin{pmatrix} \langle b_1\,|\alpha\rangle \\ \langle b_2\,|\alpha\rangle \\ . \\ . \\ \langle b_N\,|\alpha\rangle \end{pmatrix} = \begin{pmatrix} \langle b_1\,|\beta\rangle \\ \langle b_2\,|\beta\rangle \\ . \\ . \\ \langle b_N\,|\beta\rangle \end{pmatrix}. \tag{1.92}$$

We can now follow the rules of matrix multiplications and write down N simultaneous equations.

(v) A matrix element such as $\langle\psi\,|\,A\,|\phi\rangle$ can itself be expressed in terms of a matrix relation by using the completeness relation

$$\begin{aligned} \langle\psi|\,A\,|\phi\rangle &= \langle\psi|\left[\sum_m |b_m\rangle\,\langle b_m|\right] A \left[\sum_n |b_n\rangle\,\langle b_n|\right]|\phi\rangle \\ &= \sum_m \sum_n \langle\psi\,|b_m\rangle\,\langle b_m|\,A\,|b_n\rangle\,\langle b_n|\,\phi\rangle. \end{aligned} \tag{1.93}$$

The right-hand side in the matrix form is then

$$\left(\langle\psi\,|b_1\rangle \quad \langle\psi\,|b_2\rangle \quad \ldots \quad \langle\psi\,|b_N\rangle\right)\begin{bmatrix} A_{11} & A_{12} & . & . & A_{1N} \\ A_{21} & A_{22} & . & . & . \\ . & . & A_{33} & . & . \\ . & . & . & . & . \\ A_{N1} & . & . & . & A_{NN} \end{bmatrix}\begin{pmatrix} \langle b_1\,|\phi\rangle \\ \langle b_2\,|\phi\rangle \\ . \\ . \\ \langle b_N\,|\phi\rangle \end{pmatrix}, \tag{1.94}$$

which after a long multiplication reduces to a single number which is, of course, the single matrix element $\langle\psi|\,A\,|\phi\rangle$.

Often it is more convenient to write the above product by utilizing the relation $\langle\psi\,|b_m\rangle = \langle b_m\,|\psi\rangle^*$,

$$\left(\langle b_1\,|\psi\rangle^* \quad \langle b_2\,|\psi\rangle^* \quad \ldots \quad \langle b_N\,|\psi\rangle^*\right)\begin{bmatrix} A_{11} & A_{12} & A_{13} & . & A_{1N} \\ A_{21} & A_{22} & A_{23} & . & . \\ A_{31} & A_{32} & A_{33} & . & . \\ . & . & . & . & . \\ A_{N1} & . & . & . & A_{NN} \end{bmatrix}\begin{pmatrix} \langle b_1\,|\phi\rangle \\ \langle b_2\,|\phi\rangle \\ . \\ . \\ \langle b_N\,|\phi\rangle \end{pmatrix}. \tag{1.95}$$

(vi) The trace of a matrix A is defined as a sum of its diagonal elements,

$$\mathrm{Tr}(A) = \sum_n \langle a_n|\,A\,|a_n\rangle = \sum_n A_{nn}, \tag{1.96}$$

where $|a_n\rangle$'s form an orthonormal basis set. An important property of a trace is

$$\mathrm{Tr}(AB) = \mathrm{Tr}(BA) \tag{1.97}$$

This can be proved by noting that

$$\mathrm{Tr}(AB) = \sum_n \langle a_n|\,AB\,|a_n\rangle = \sum_n\sum_m \langle a_n|\,A\,|a_m\rangle\,\langle a_m|\,B\,|a_n\rangle \tag{1.98}$$

where we have used the completeness relation for the basis sets $|a_m\rangle$. Since the matrix elements are numbers and no longer operators, they can be switched. Hence using completeness for the $|a_m\rangle$'s we have

$$\mathrm{Tr}(AB) = \sum_m\sum_n \langle a_m|\,B\,|a_n\rangle\,\langle a_n|\,A\,|a_m\rangle = \sum_m \langle a_m|\,BA\,|a_m\rangle = \mathrm{Tr}(BA), \tag{1.99}$$

which is the desired result. This leads to the generalization involving a product of an arbitrary number of operators that

$$\mathrm{Tr}(ABC\ldots) = \text{ invariant} \tag{1.100}$$

under a cyclic permutation of the product $ABC\ldots$.

1.11 Eigenstates and diagonalization of matrices

Consider the case where we know the matrix elements of A with respect to the basis given by $|a_n\rangle$. That is, we know $\langle a_m| A |a_n\rangle$ where $|a_n\rangle$'s are not necessarily the eigenstates of A. We call $|a_n\rangle$'s the old basis. We now wish to obtain the eigenstates and eigenvalues of A. Let $|b_n\rangle$'s be the eigenstates of A, we will call them the new basis, which satisfy

$$A |b_n\rangle = b_n |b_n\rangle . \tag{1.101}$$

We proceed in a manner very similar to the previous problem. We multiply both sides of (1.101) by $\langle a_m|$ and insert a complete set of states $|a_p\rangle$,

$$\sum_p \langle a_m| A |a_p\rangle\langle a_p| b_n\rangle = b_n \langle a_m| b_n\rangle. \tag{1.102}$$

The above relation can be written as a matrix relation by taking $m = 1, 2, \ldots, N$ for a fixed value of n

$$\begin{bmatrix} A_{11} & A_{12} & . & . & A_{1N} \\ A_{21} & A_{22} & . & . & . \\ . & . & . & . & . \\ . & . & . & . & . \\ A_{N1} & . & . & . & A_{NN} \end{bmatrix} \begin{pmatrix} \langle a_1 |b_n\rangle \\ \langle a_2 |b_n\rangle \\ . \\ . \\ \langle a_N |b_n\rangle \end{pmatrix} = b_n \begin{pmatrix} \langle a_1 |b_n\rangle \\ \langle a_2 |b_n\rangle \\ . \\ . \\ \langle a_N |b_n\rangle \end{pmatrix} \tag{1.103}$$

where, as stated earlier, the matrix elements $A_{mp} = \langle a_m| A |a_p\rangle$ are known. The relation (1.103) can be written as

$$\begin{bmatrix} A_{11} - b_n & A_{12} & . & . & A_{1N} \\ A_{21} & A_{22} - b_n & . & . & . \\ . & . & . & . & . \\ . & . & . & . & . \\ A_{N1} & . & . & . & A_{NN} - b_n \end{bmatrix} \begin{pmatrix} \langle a_1 |b_n\rangle \\ \langle a_2 |b_n\rangle \\ . \\ . \\ \langle a_N |b_n\rangle \end{pmatrix} = 0, \tag{1.104}$$

which is valid for each value of n. Thus, effectively, the above relation corresponds to "diagonalization" of the matrix formed by the A_{mn}.

A solution of equation (1.104) is possible only if the determinant of the $N \times N$ matrix vanishes for each value of n. Hence the eigenvalues b_n are the roots of the determinant equation

$$\det [A - \lambda I] = 0 \tag{1.105}$$

where I is a unit matrix. The different values of λ correspond to the different eigenvalues b_n. The corresponding elements $\langle a_m | b_n \rangle$ can then be determined by solving N simultaneous equations in (1.104).

1.11.1 Diagonalization through unitary operators

Let us bring in the unitary operators, which will shed further light on determining eigenstates and eigenvalues of A. Let U be the unitary operator that transforms the old basis to the new basis,

$$|b_n\rangle = U |a_n\rangle . \tag{1.106}$$

This transformation preserves the norms of the two basis sets since U is unitary. Equation (1.101) can be expressed as

$$AU |a_n\rangle = b_n U |a_n\rangle . \tag{1.107}$$

We now multiply both sides on the left by $U^\dagger$; then

$$U^\dagger AU |a_n\rangle = b_n U^\dagger U |a_n\rangle = b_n |a_n\rangle \tag{1.108}$$

where we have used the unitary property $U^\dagger U = \mathbf{1}$. Once again we multiply on the left, this time by $\langle a_m|$. We find

$$\langle a_m| U^\dagger AU |a_n\rangle = b_n \langle a_m |a_n\rangle = b_n \delta_{mn}. \tag{1.109}$$

The right-hand side corresponds to a diagonal matrix. Thus the operator U must be such that $U^\dagger AU$ is a diagonal matrix and we write

$$A_D = U^\dagger AU. \tag{1.110}$$

Once we find U then we can immediately obtain the eigenstates $|b_n\rangle$ from (1.106) and eigenvalues b_n from (1.109).

Taking the trace of both sides of (1.10) we obtain

$$\mathrm{Tr}(A_D) = \mathrm{Tr}(U^\dagger AU). \tag{1.111}$$

Since A_D is a diagonal operator with matrix elements given by the eigenvalues of A, the trace on the left of the above equation is simply the sum of the eigenvalues. For the right-hand side we note, using the invariance of a trace under cyclic permutation, that

$$\mathrm{Tr}(U^\dagger AU) = \mathrm{Tr}(AUU^\dagger) = \mathrm{Tr}(A). \tag{1.112}$$

Thus

$$\text{Tr}(A) = \text{sum of the eigenvalues of the operator } A. \tag{1.113}$$

This result holds even though A itself is not diagonal in the basis set $|a_n\rangle$.

In a two-channel system, i.e., in a system that consists of only two eigenstates, U is relatively easy to express. One writes

$$U = \begin{bmatrix} \cos\theta & \sin\theta \\ -\sin\theta & \cos\theta \end{bmatrix}, \tag{1.114}$$

which is manifestly unitary. One then imposes the condition

$$U^\dagger AU = \text{diagonal matrix}. \tag{1.115}$$

This is accomplished by taking the off-diagonal elements to be zero. From this relation one can determine the angle θ and, therefore, the matrix U. The diagonal elements of the matrix (1.116) give the eigenvalues, while the eigenstates $|b_n\rangle$ are obtained in terms of U and $|a_n\rangle$ through (1.107). We will return to this formalism in Chapter 13 when we focus on two-channel phenomena.

1.12 Density operator

The expectation value $\langle\alpha|A|\alpha\rangle$, taken with respect to a state $|\alpha\rangle$, of an operator A was defined earlier. It was defined with respect to a single state, often referred to as a pure quantum state. Instead of a pure quantum state one may, however, have a collection of states, called an ensemble of states. If each state in this ensemble is described by the same ket $|\alpha\rangle$ then one refers to it as a pure ensemble.

When all the states in an ensemble are not the same then it is called a mixed ensemble. If one is considering a mixed ensemble where w_α describes the probability that a state $|\alpha\rangle$ is present in the ensemble, w_β describes the probability that a state $|\beta\rangle$ is present, and so on, then, instead of the expectation value, $\langle\alpha|A|\alpha\rangle$, the relevant quantity is the ensemble average, which is defined as

$$\langle A\rangle_{av} = \sum_\alpha w_\alpha \langle\alpha|A|\alpha\rangle \tag{1.116}$$

where the sum runs over all the states in the ensemble. Naturally, if $w_\alpha = 1$, with all other w_i's zero, then only the state $|\alpha\rangle$ contributes in the sum, in which case we have a pure ensemble and the ensemble average is then the same as the expectation value. We note that a mixed ensemble is also referred to as a statistical mixture.

Below we outline some important properties of $\langle\alpha|A|\alpha\rangle$ and $\langle A\rangle_{av}$.

Inserting a complete set of states $|b_n\rangle$ with $n = 1, 2, \ldots$ in the expression for the expectation value $\langle\alpha|A|\alpha\rangle$ we can write

$$\langle\alpha|A|\alpha\rangle = \sum_n\sum_m \langle\alpha|b_n\rangle\langle b_n|A|b_m\rangle\langle b_m|\alpha\rangle \tag{1.117}$$

$$= \sum_n\sum_m \langle b_n|A|b_m\rangle\langle b_m|\alpha\rangle\langle\alpha|b_n\rangle \tag{1.118}$$

where we have made an interchange of some of the matrix elements, which is allowed because they are numbers and no longer operators. We now introduce a "projection operator" defined by

$$P_\alpha = |\alpha\rangle\langle\alpha|. \tag{1.119}$$

It has the property

$$P_\alpha|\beta\rangle = |\alpha\rangle\langle\alpha|\beta\rangle = \delta_{\alpha\beta}|\alpha\rangle \tag{1.120}$$

where we have taken the states $|\alpha\rangle$, $|\beta\rangle\ldots$ as orthonormal. Thus P_α projects out the state $|\alpha\rangle$ when operating on any state. Furthermore,

$$P_\alpha^2 = |\alpha\rangle\langle\alpha|\alpha\rangle\langle\alpha| = |\alpha\rangle\langle\alpha| = P_\alpha, \tag{1.121}$$

$$P_\alpha^\dagger = (\langle\alpha|)(|\alpha\rangle) = |\alpha\rangle\langle\alpha| = P_\alpha. \tag{1.122}$$

The completeness theorem gives

$$\sum_\alpha P_\alpha = \sum_\alpha |\alpha\rangle\langle\alpha| = \mathbf{1}. \tag{1.123}$$

From (1.119) and (1.120) we can write $\langle\alpha|A|\alpha\rangle$ in terms of P_α by noting that

$$\langle b_m|\alpha\rangle\langle\alpha|b_n\rangle = \langle b_m|P_\alpha|b_n\rangle \tag{1.124}$$

where $|b_n\rangle$'s are eigenstates of an operator. Therefore,

$$\langle\alpha|A|\alpha\rangle = \sum_n\sum_m \langle b_n|A|b_m\rangle\langle b_m|P_\alpha|b_n\rangle \tag{1.125}$$

$$= \sum_n \langle b_n|AP_\alpha|b_n\rangle \tag{1.126}$$

where we have used the completeness relation $\sum_m |b_m\rangle\langle b_m| = \mathbf{1}$ for the eigenstates $|b_m\rangle$. Thus,

$$\langle\alpha|A|\alpha\rangle = \mathrm{Tr}\,(AP_\alpha) \tag{1.127}$$

where "Tr" indicates trace. If we take

$$A = \mathbf{1} \tag{1.128}$$

then

$$\mathrm{Tr}\,(P_\alpha) = 1. \tag{1.129}$$

We consider now a mixed ensemble that contains the states $|\alpha\rangle, |\beta\rangle, \ldots$ etc. with probabilities $w_\alpha, w_\beta \ldots$, respectively. We define a density operator ρ as

$$\rho = \sum_\alpha w_\alpha P_\alpha = \sum_\alpha w_\alpha |\alpha\rangle\langle\alpha|. \tag{1.130}$$

Since w_α, being the probability, is real and P_α is Hermitian, ρ is, therefore, Hermitian,

$$\rho^\dagger = \rho. \tag{1.131}$$

From (1.117) and (1.128) the ensemble average is

$$\langle A\rangle_{av} = \sum_\alpha w_\alpha \mathrm{Tr}\,(AP_\alpha)\,. \tag{1.132}$$

We note that since w_α is a number and not a matrix, and, at the same time, since the operator A is independent of α, and thus does not participate in the summation over α, we can reshuffle the terms in (1.133) to obtain

$$\langle A\rangle_{av} = \mathrm{Tr}\left(A\sum_\alpha w_\alpha P_\alpha\right) = \mathrm{Tr}\,(A\rho) = \mathrm{Tr}\,(\rho A)\,. \tag{1.133}$$

The last equality follows from the property that the trace of a product of two matrices is invariant under interchange of the matrices.

From (1.117) we find, by taking $A = \mathbf{1}$, that

$$\langle \mathbf{1}\rangle_{av} = \sum_\alpha \omega_\alpha \langle\alpha\,|\mathbf{1}|\,\alpha\rangle = \sum_\alpha \omega_\alpha = \mathbf{1}. \tag{1.134}$$

Therefore, from (1.134), we get

$$\langle \mathbf{1}\rangle_{av} = \mathrm{Tr}\,(\rho)\,. \tag{1.135}$$

Finally, (1.135) and (1.136) imply

$$\mathrm{Tr}\,(\rho) = 1. \tag{1.136}$$

We will discuss the properties of ρ in Chapter 14 for the specific case of the spin ½ particles, and again in the chapter on two-level problems.

1.13 Measurement

When a measurement of a dynamical variable is made on a system and a specific, real, value for a physical observable is found, then that value is an eigenvalue of the operator representing the observable. In other words, irrespective of the state, the act of measurement itself kicks the state into an eigenstate of the operator in question.

From the superposition principle, the state of a system, say $|\phi\rangle$, can always be expressed as a superposition of basis states, which we take to be normalized kets. One can choose these states to be the eigenstates of the operator, e.g., $|a_n\rangle$. Thus one can write the superposition as

$$|\phi\rangle = \sum_n c_n |a_n\rangle . \tag{1.137}$$

When a single measurement is made, one of the eigenstates of this operator in this superposition will be observed and the corresponding eigenvalue will be measured. In other words, the state $|\phi\rangle$ will "collapse" to a state $|a_n\rangle$. The probability that a particular eigenvalue, e.g., a_n will be measured in a single measurement is given by $|c_n|^2$. But a second measurement of an identical system may yield another eigenstate with a different eigenvalue and a different probability. Repeated measurements on identically prepared systems, will then give the probability distribution of the eigenvalues and yield information on the nature of the system. In practice, one prefers to make measurements on a large number of identical systems, which gives the same probability distribution.

Similar arguments follow if a measurement is made to determine whether the system is in state $|\psi\rangle$. In this case the original state $|\phi\rangle$ will "jump" into the state $|\psi\rangle$ with the probability $|\langle\psi\,|\phi\rangle|^2$.

The concept of measurement and the probability interpretation in quantum mechanics is a complex issue that we will come back to when we discuss Stern–Gerlach experiments in Chapter 7 and entangled states in Chapter 30.

1.14 Problems

1. Define the following two state vectors as column matrices:

$$|\alpha_1\rangle = \begin{bmatrix} 1 \\ 0 \end{bmatrix} \quad \text{and} \quad |\alpha_2\rangle = \begin{bmatrix} 0 \\ 1 \end{bmatrix}$$

with their Hermitian conjugates given by

$$\langle\alpha_1| = \begin{bmatrix} 1 & 0 \end{bmatrix} \quad \text{and} \quad \langle\alpha_2| = \begin{bmatrix} 0 & 1 \end{bmatrix}$$

respectively. Show the following for $i, j = 1, 2$:

(i) The $|\alpha_i\rangle$'s are orthonormal.

(ii) Any column matrix

$$\begin{bmatrix} a \\ b \end{bmatrix}$$

can be written as a linear combination of the $|\alpha_i\rangle$'s.

(iii) The outer products $|\alpha_i\rangle\langle\alpha_j|$ form 2×2 matrices which can serve as operators.

(iv) The $|\alpha_i\rangle$'s satisfy completeness relation

$$\sum_i |\alpha_i\rangle\langle\alpha_i| = \mathbf{1}$$

where $\mathbf{1}$ represents a unit 2×2 matrix.

(v) Write

$$A = \begin{bmatrix} a & b \\ c & d \end{bmatrix}$$

as a linear combination of the four matrices formed by $|\alpha_i\rangle\langle\alpha_j|$.

(vi) Determine the matrix elements of A such that $|\alpha_1\rangle$ and $|\alpha_2\rangle$ are simultaneously the eigenstates of A satisfying the relations

$$A|\alpha_1\rangle = +|\alpha_1\rangle \quad \text{and} \quad A|\alpha_2\rangle = -|\alpha_2\rangle.$$

(The above properties signify that the $|\alpha_i\rangle$'s span a Hilbert space. These abstract representations of the state vectors actually have a profound significance. They represent the states of particles with spin ½. We will discuss this in detail in Chapter 5.)

2. Show that if an operator A is a function of λ then

$$\frac{dA^{-1}}{d\lambda} = -A^{-1}\frac{dA}{d\lambda}A^{-1}.$$

3. Show that a unitary operator U can be written as

$$U = \frac{\mathbf{1} + iK}{\mathbf{1} - iK}$$

where K is a Hermitian operator. Show that one can also write

$$U = e^{iC}$$

where C is a Hermitian operator. If

$$U = A + iB.$$

Show that A and B commute. Express these matrices in terms of C. You can assume that

$$e^M = 1 + M + \frac{M^2}{2!} + \cdots$$

where M is an arbitrary matrix.

4. Show that

$$\det\left(e^A\right) = e^{\mathrm{Tr}(A)}.$$

5. For two arbitrary state vectors $|\alpha\rangle$ and $|\beta\rangle$ show that

$$\mathrm{Tr}\left[|\alpha\rangle\langle\beta|\right] = \langle\beta|\alpha\rangle.$$

6. Consider a two-dimensional space spanned by two orthonormal state vectors $|\alpha\rangle$ and $|\beta\rangle$. An operator is expressed in terms of these vectors as

$$A = |\alpha\rangle\langle\alpha| + \lambda\,|\beta\rangle\langle\alpha| + \lambda^*\,|\alpha\rangle\langle\beta| + \mu\,|\beta\rangle\langle\beta|.$$

Determine the eigenstates of A for the case where (i) $\lambda = 1, \mu = \pm 1$, (ii) $\lambda = i, \mu = \pm 1$. Do this problem also by expressing A as a 2×2 matrix with eigenstates as the column matrices.

2 Fundamental commutator and time evolution of state vectors and operators

In the previous chapter we outlined the basic mathematical structure essential for our studies. We are now ready to make contact with physics. This means introducing the fundamental constant $\hbar$, the Planck constant, which controls the quantum phenomena. Our first step will be to discuss the so-called fundamental commutator, also known as the canonical commutator, which is proportional to $\hbar$ and which essentially dictates how the quantum processes are described. We then examine how time enters the formalism and thus set the stage for writing equations of motion for a physical system.

2.1 Continuous variables: *X* and *P* operators

Eigenvalues need not always be discrete as we stated earlier. For example, consider a one-dimensional, continuous (indenumerable) infinite-dimensional position space, the x-space. One could have an eigenstate $|x'\rangle$ of a continuous operator X,

$$X\,|x'\rangle = x'\,|x'\rangle \tag{2.1}$$

where x'corresponds to the value of the x-variable.

The ket $|x'\rangle$ has all the properties of the kets $|\alpha\rangle$ and of the eigenstates $|a_n\rangle$ that were outlined in the previous chapter. The exceptions are those cases where the fact that x is a continuous variable makes an essential difference.

For continuous variables the Kronecker delta defined in (1.45) must be replaced by Dirac δ-function $\delta(x - x')$, which has the following properties:

$$\delta(x - x') = 0 \text{ for } x \neq x', \tag{2.2}$$

$$\int_{-\infty}^{\infty} dx\,\delta(x - x') = 1. \tag{2.3}$$

From these two definitions it follows that

$$\int_{-\infty}^{\infty} dx f(x)\,\delta(x - x') = f(x'). \tag{2.4}$$

The properties of the delta function are discussed in considerable detail in Appendix 2.9.

The orthogonality condition involving the Kronecker δ-function that we used for the discrete variables is now replaced by

$$\langle x | x' \rangle = \delta(x - x') \tag{2.5}$$

The completeness relation is then expressed as

$$\int_{-\infty}^{\infty} dx \; |x\rangle \langle x| = \mathbf{1} \tag{2.6}$$

where the summation in the discrete case is replaced by an integral. Just as in the discrete case one can prove this relation quite simply by multiplying the two sides of (2.6) by the ket $|x'\rangle$. We obtain

$$\left[\int_{-\infty}^{\infty} dx \; |x\rangle \langle x|\right] |x'\rangle = \int_{-\infty}^{\infty} dx \; |x\rangle \langle x | x' \rangle = |x'\rangle . \tag{2.7}$$

Thus the left-hand side of (2.6), indeed, acts as a unit operator.

We also note that

$$\langle x | X | x' \rangle = x \, \delta(x - x'). \tag{2.8}$$

It does not matter if we have the factor x or x' on the right-hand side multiplying the δ-function since the δ-function vanishes unless $x = x'$. The matrix element $\langle x | X | x' \rangle$, in this continuous space, is called the "representative" of the operator X in the x-space.

A "wavefunction" in the x-space, $\phi(x)$, corresponding to a state vector $|\phi\rangle$ is defined as

$$\phi(x) = \langle x | \phi \rangle \tag{2.9}$$

and from the properties of the bra and ket vectors outlined earlier,

$$\phi^*(x) = \langle \phi | x \rangle . \tag{2.10}$$

The function $\phi(x)$ is then a manifestation of an abstract vector $|\phi\rangle$ in the x-space which we also call the "representative" of $|\phi\rangle$ in the x-space. The linear superposition principle stated in (1.1) can be written in the following form by taking the representatives in the x-space,

$$a\alpha(x) + b\beta(x) = c\gamma(x) \tag{2.11}$$

where $\alpha(x) = \langle x | \alpha \rangle$ etc.

The product $\langle \psi | \phi \rangle$ can be expressed in the x-space by inserting a complete set of intermediate states:

$$\langle \psi | \phi \rangle = \langle \psi | \mathbf{1} | \phi \rangle = \int_{-\infty}^{\infty} dx \, \langle \psi | x \rangle \langle x | \phi \rangle = \int_{-\infty}^{\infty} dx \, \psi^*(x) \phi(x) \tag{2.12}$$

In particular, if the states $|\phi\rangle$ are normalized, $\langle\phi|\phi\rangle = 1$, then replacing $\langle\psi|$ by $\langle\phi|$ in the above relation we obtain the normalization condition in the x-space for the wavefunction $\phi(x)$,

$$\int_{-\infty}^{\infty} dx\, \phi^*(x)\phi(x) = 1. \tag{2.13}$$

For the integrals to be finite, it is important to note that $\phi(x)$ and $\psi(x)$ must vanish as $x \to \pm\infty$.

As with the x-variables, one can also consider the momentum variable, p, in one dimension (p-space) which can be continuous. If P is the momentum operator with eigenstates $|p'\rangle$ then

$$P|p'\rangle = p'|p'\rangle \tag{2.14}$$

with

$$\langle p|p'\rangle = \delta(p - p') \tag{2.15}$$

and

$$\langle p|P|p'\rangle = p\delta(p - p'). \tag{2.16}$$

A wavefunction in the p-space, $f(p)$, for an abstract vector $|f\rangle$ is defined as

$$f(p) = \langle p|f\rangle \tag{2.17}$$

with the normalization condition

$$\int_{-\infty}^{\infty} dp f^*(p)f(p) = 1. \tag{2.18}$$

If the state vectors depend on time then the above conditions can be expressed in terms of $\phi(x, t)$ and $f(p, t)$.

2.2 Canonical commutator $[X, P]$

The commutator between two operators B and C is defined as

$$[B, C] = BC - CB \tag{2.19}$$

It is itself an operator. As we saw earlier, it will vanish if B and C have a common eigenstate.

The commutator between the canonical variables X and P plays a fundamental role in quantum mechanics, which in the one-dimensional case is given by

$$[X, P] = i\hbar\mathbf{1} \tag{2.20}$$

where the right-hand side has the Planck constant, $\hbar$, multiplied by a unit operator. The appearance of $\hbar$ signals that we are now in the regime of quantum mechanics.

The relation (2.20) shows that $[X, P]$ does not vanish, and, therefore, X and P cannot have a common eigenstate, i.e., they cannot be measured simultaneously. The commutator, however, allows us to connect two basic measurements: one involving the x-variable and the other involving the p-variable. If the right-hand side were zero then it would correspond to classical physics, where the two measurements can be carried out simultaneously

To obtain the representatives of the momentum operator in the x-space, we proceed by taking the matrix elements of both sides of (2.20),

$$\langle x|(XP - PX)|x'\rangle = i\hbar\delta(x - x'). \tag{2.21}$$

The left-hand side is simplified by inserting a complete set of intermediate states,

$$\langle x|(XP - PX)|x'\rangle = \int_{-\infty}^{\infty} dx'' \langle x|X|x''\rangle\langle x''|P|x'\rangle - \int_{-\infty}^{\infty} dx'' \langle x|P|x''\rangle\langle x''|X|x'\rangle. \tag{2.22}$$

For the right-hand side of the above equation we use the relation (2.8) and replace the matrix elements of X by Dirac δ-functions. The integrals are further simplified by using (2.4) and (2.8). We thus obtain

$$x\langle x|P|x'\rangle - \langle x|P|x'\rangle x' = i\hbar\delta(x - x'). \tag{2.23}$$

We notice that when $x \neq x'$ the right-hand side of the above equation vanishes. The left-hand side, however, does not vanish when $x \neq x'$ unless $\langle x| P \left|x'\right\rangle$ itself has a δ-function in it. Writing it as

$$\langle x|P|x'\rangle = \delta(x - x')P(x) \tag{2.24}$$

where $P(x)$ is an operator expressed in terms of the x-variable, we obtain the following from (2.23), and (2.24) after cancelling the δ-function from both sides,

$$xP(x) - P(x)x = i\hbar. \tag{2.25}$$

This is an operator relation in which the operators are expressed as functions of x. This relation becomes meaningful only upon operating it on a wavefunction. Multiplying both sides by a wavefunction $\phi(x)$ on the right, we obtain

$$[xP(x) - P(x)x]\,\phi(x) = i\hbar\phi(x), \tag{2.26}$$

which we can write as

$$xP(x)\left[\phi(x)\right] - P(x)\left[x\phi(x)\right] = i\hbar\phi(x). \tag{2.27}$$

Considering the left-hand side in (2.26), we note that since $P(x)$ is assumed to be a linear operator that operates on the quantities to the right of it, we will have two types of terms $P(x)\left[\phi(x)\right]$ and $P(x)\left[x\phi(x)\right]$. Furthermore, relation (2.27) suggests that $P(x)$ will be a derivative operator, in which case one can simplify the second term involving the product $\left[x\phi(x)\right]$ by using the product rule for first order derivatives and writing

$$P(x)\left[x\phi(x)\right] = \left[P(x)x\right]\phi(x) + x\left[P(x)\phi(x)\right] \tag{2.28}$$

where the square bracket in $\left[P(x)x\right]\phi(x)$ implies that $P(x)$ operates only on x and not on $\phi(x)$. Thus the left-hand side of (2.27) can be written as

$$x\left[P(x)\phi(x)\right] - \left[P(x)x\right]\phi(x) - x\left[P(x)\phi(x)\right] = -\left[P(x)x\right]\phi(x). \tag{2.29}$$

The equation (2.27) now reads

$$-\left[P(x)x\right]\phi(x) = i\hbar\phi(x). \tag{2.30}$$

Since on the left-hand side of (2.30), $P(x)$ operates only on x and does not operate on $\phi(x)$, we can remove $\phi(x)$ from both sides and obtain the relation

$$-\left[P(x)x\right] = i\hbar. \tag{2.31}$$

It is easy to confirm that the following linear differential operator for $P(x)$ satisfies the above relation:

$$P(x) = -i\hbar\frac{d}{dx}. \tag{2.32}$$

This is then the representation of the operator P in the x-space. In terms of matrix elements in the x-space we can write this relation as

$$\langle x'|P|x\rangle = \delta(x - x')\left(-i\hbar\frac{d}{dx}\right). \tag{2.33}$$

A matrix element that one often comes across is $\langle x'|P|\phi\rangle$ which can be expressed, by inserting a complete set of states, as

$$\langle x'|P|\phi\rangle = \int_{-\infty}^{\infty} dx''\,\langle x'|P|x''\rangle\langle x''|\phi\rangle. \tag{2.34}$$

Using (2.33) we obtain

$$\langle x'|P|\phi\rangle = -i\hbar\frac{d}{dx'}\phi(x'). \tag{2.35}$$

2.3 P as a derivative operator: another way

We describe another way in which the relation (2.35) can be established. Let us start with the fundamental commutator written as

$$[P, X] = -i\hbar. \tag{2.36}$$

Consider

$$[P, X^2] = PX^2 - X^2P. \tag{2.37}$$

This relation can be rewritten as

$$[P, X^2] = [P, X]X + X[P, X] = -2i\hbar X \tag{2.38}$$

where we have used the relation (2.36). Expressing the right-hand side as a derivative we obtain

$$[P, X^2] = -i\hbar\frac{d}{dX}(X^2). \tag{2.39}$$

Proceeding in a similar manner one can show that

$$[P, X^n] = -ni\hbar X^{n-1} = -i\hbar\frac{d}{dX}(X^n). \tag{2.40}$$

Consider now the commutator $[P, f(X)]$ where $f(X)$ is a regular function which can be expanded in a Taylor series around $X = 0$,

$$f(X) = a_0 + a_1X + \cdots + a_nX^n + \cdots. \tag{2.41}$$

Using the result (2.40) we conclude that

$$[P, f(X)] = -i\hbar\frac{df(X)}{dX}. \tag{2.42}$$

We operate the two sides on a state vector $|\phi\rangle$,

$$[P, f(X)]|\phi\rangle = -i\hbar\frac{df(X)}{dX}|\phi\rangle. \tag{2.43}$$

The left-hand side is

$$(Pf - fP)|\phi\rangle = [(Pf)|\phi\rangle + fP|\phi\rangle] - f(P|\phi\rangle) = (Pf)|\phi\rangle. \tag{2.44}$$

In the above relation (Pf) means P operates on f alone. Since $|\phi\rangle$ is any arbitrary state vector one can remove the factor $|\phi\rangle$. Hence

$$(Pf) = -i\hbar\frac{df(X)}{dX}. \tag{2.45}$$

Thus

$$P = -i\hbar\frac{d}{dX}. \tag{2.46}$$

Taking the matrix element of the above, we find

$$\langle x'|P|\phi\rangle = -i\hbar\frac{d}{dx'}\phi(x'), \tag{2.47}$$

confirming our result (2.35).

2.4 *X* and *P* as Hermitian operators

Let us now show that X and P are Hermitian, as they should be since they correspond to physically measurable quantities.

For X it is quite trivial as we demonstrate below. From the definition of Hermitian conjugate we notice that the matrix element of $X^\dagger$ is given by

$$\langle x'|X^\dagger|x\rangle = \langle x|X|x'\rangle^* = x\delta(x-x'). \tag{2.48}$$

where we have taken account of the fact that x' and $\delta(x-x')$ are real. The matrix element of X is given by

$$\langle x'|X|x\rangle = x\delta(x-x'). \tag{2.49}$$

Therefore,

$$X = X^\dagger. \tag{2.50}$$

The Hermitian conjugate of P can be obtained by first taking the matrix element of P between two arbitrary states $|\psi\rangle$ and $|\phi\rangle$ and then evaluating it by inserting a complete set of states. We obtain

$$\langle\psi|P|\phi\rangle = \int_{-\infty}^{\infty} dx' \int_{-\infty}^{\infty} dx''\ \langle\psi|x'\rangle\langle x'|P|x''\rangle\langle x''|\phi\rangle \tag{2.51}$$

$$= \int_{-\infty}^{\infty} dx' \int_{-\infty}^{\infty} dx''\ \psi^*(x')\delta(x'-x'')\left(-i\hbar\frac{\partial}{\partial x''}\right)\phi(x''), \tag{2.52}$$

remembering that P operates on the functions to the right. Therefore,

$$\langle\psi|P|\phi\rangle = (-i\hbar)\int_{-\infty}^{\infty} dx'\ \psi^*(x')\frac{\partial\phi(x')}{\partial x'}. \tag{2.53}$$

Interchanging ψ and ϕ, we obtain

$$\langle\phi|P|\psi\rangle = (-i\hbar)\int_{-\infty}^{\infty} dx'\,\phi^*(x')\frac{\partial\psi(x')}{\partial x'}. \tag{2.54}$$

Integrating the right-hand side above by parts, and also, to ensure the convergence of the integrals, assuming $\psi(x') \to 0, \phi(x') \to 0$ as $x' \to \pm\infty$, we obtain

$$\langle\phi|P|\psi\rangle = (i\hbar)\int_{-\infty}^{\infty} dx'\,\frac{\partial\phi^*(x')}{\partial x'}\psi(x'). \tag{2.55}$$

Next we consider the matrix element of $P^\dagger$, and use its Hermitian conjugation properties,

$$\langle\phi|P^\dagger|\psi\rangle = \langle\psi|P|\phi\rangle^* = (i\hbar)\int_{-\infty}^{\infty} dx'\,\psi(x')\frac{\partial\phi^*(x')}{\partial x'} \tag{2.56}$$

where we have used (2.53). Comparing this result with that of $\langle\phi|P|\psi\rangle$ in (2.55) we conclude that, since $\psi(x')$ and $\phi(x')$ are arbitrary functions, we must have

$$P^\dagger = P \tag{2.57}$$

and thus P is Hermitian.

We can extend the above relations to three dimensions with the commutation relations given by

$$\left[X_i, P_j\right] = i\hbar\delta_{ij}\mathbf{1}, \tag{2.58}$$

$$\left[X_i, X_j\right] = 0, \tag{2.59}$$

$$\left[P_i, P_j\right] = 0 \tag{2.60}$$

where $i,j = 1,2,3$ correspond to the three dimensions. Thus X_1, X_2, X_3 correspond to the operators X, Y, Z respectively, while P_1, P_2, P_3 correspond to operators P_x, P_y, P_z respectively.

Beginning with the relation

$$[X, P_x] = i\hbar\mathbf{1} \tag{2.61}$$

we conclude, following the same steps as before, that the representation of P_x in the x-space will be given by

$$P_x = -i\hbar\frac{\partial}{\partial x}. \tag{2.62}$$

Note that we now have a partial derivative rather than the total derivative that we had earlier for the one-dimensional case. The above operator satisfies the relation

$$[Y, P_x] = 0 = [Z, P_x]. \tag{2.63}$$

Since P_x commutes with Y and Z, it cannot involve any derivative operators involving the y- and z-coordinates. Therefore, the relation (2.61) remains a correct representation of P_x. Similarly, we have

$$P_y = -i\hbar\frac{\partial}{\partial y}, \quad P_z = -i\hbar\frac{\partial}{\partial z}. \tag{2.64}$$

Finally, we note that the operators P_x, P_y, and P_z as written above also satisfy the relations (2.60). All the remaining results we derived for the one-dimensional case stay the same for the three-dimensional operators.

2.5 Uncertainty principle

As we discussed earlier, when two operators commute, i.e., when the commutator between two operators vanishes, then these operators can share the same eigenstates. This implies that the observables associated with these operators can be measured simultaneously. A commutator, therefore, quantifies in some sense the degree to which two observables can be measured at the same time.

In quantum mechanics, Planck's constant, $\hbar$, provides a measure of the commutator between the canonical operators X and P through the fundamental commutator (2.20). The uncertainty relation, one of the most famous results in quantum mechanics, relates the accuracies of two measurements when they are made simultaneously.

The uncertainty of an operator A is defined as the positive square root $\sqrt{\langle(\Delta A)^2\rangle}$ where

$$\Delta A = A - \langle A\rangle \mathbf{1} \tag{2.65}$$

and $\langle A\rangle$ is given by

$$\langle A\rangle = \langle\phi|A|\phi\rangle, \tag{2.66}$$

which is the expectation value of A with respect to an arbitrary state $|\phi\rangle$. We note that

$$\langle\Delta A\rangle = \langle A - \langle A\rangle \mathbf{1}\rangle = \langle A\rangle - \langle A\rangle = 0 \tag{2.67}$$

and

$$\left\langle(\Delta A)^2\right\rangle = \left\langle A^2 - 2A\langle A\rangle + \langle A\rangle^2 \mathbf{1}\right\rangle = \left\langle A^2\right\rangle - \langle A\rangle^2 . \tag{2.68}$$

To derive the uncertainty relation let us consider three Hermitian operators C, D, F that satisfy the commutation relation

$$[C, D] = iF. \tag{2.69}$$

We consider the following relation involving the state vector $|\phi\rangle$,

$$(C + i\lambda D)\,|\phi\rangle = C|\phi\rangle + i\lambda D|\phi\rangle \tag{2.70}$$

where λ is real. We note that

$$|C|\phi\rangle + i\lambda D|\phi\rangle|^2 \geq 0 \tag{2.71}$$

for all values of λ. The left-hand side can be written as

$$\left(\langle\phi|C^\dagger - i\lambda\langle\phi|D^\dagger\right)(C|\phi\rangle + i\lambda D|\phi\rangle) \tag{2.72}$$

$$= \langle\phi|C^\dagger C|\phi\rangle + i\lambda\langle\phi|C^\dagger D|\phi\rangle - i\lambda\langle\phi|D^\dagger C|\phi\rangle + \lambda^2\langle\phi|D^\dagger D|\phi\rangle \tag{2.73}$$

$$= \langle\phi|C^\dagger C|\phi\rangle + i\lambda\langle\phi|\,[C, D]\,|\phi\rangle + \lambda^2\langle\phi|D^\dagger D|\phi\rangle \tag{2.74}$$

$$= |C|\phi\rangle|^2 - \lambda\langle\phi|F|\phi\rangle + \lambda^2|D|\phi\rangle|^2 \tag{2.75}$$

where we have used the Hermitian property of the operators and the commutation relation (2.69).

The relation (2.71) can then be written as

$$|C|\phi\rangle|^2 - \lambda\langle\phi|F|\phi\rangle + \lambda^2|D|\phi\rangle|^2 \geq 0 \tag{2.76}$$

for all values of λ. Let us write

$$x^2 = |D|\phi\rangle|^2, \quad y = \langle\phi|F|\phi\rangle \quad \text{and} \quad z^2 = |C|\phi\rangle|^2. \tag{2.77}$$

The relation (2.76) can be written as a quadratic in λ,

$$\lambda^2 x^2 - \lambda y + z^2 \geq 0 \quad \text{for all values of } \lambda. \tag{2.78}$$

First we confirm that this relation is correct for $\lambda \to \infty$. To determine the sign of the left-hand side of (2.78) for finite values of λ, let us express the relation in terms of its roots in λ:

$$\lambda^2 x^2 - \lambda y + z^2 = x^2\,(\lambda - \lambda_1)\,(\lambda - \lambda_2) \tag{2.79}$$

where λ_1 and λ_2 are the roots, which we write as

$$\lambda_1 = a_0 - \Delta \text{ and } \lambda_2 = a_0 + \Delta \tag{2.80}$$

where

$$a_0 = \frac{y}{2x^2} \text{ and } \Delta = \frac{\sqrt{y^2 - 4x^2z^2}}{2x^2}. \tag{2.81}$$

Since x^2 defined by (2.77) is positive definite, the relation (2.78) implies that

$$(\lambda - \lambda_1)(\lambda - \lambda_2) \geq 0. \tag{2.82}$$

However, the left-hand side is negative for

$$\lambda_1 < \lambda < \lambda_2. \tag{2.83}$$

To solve this apparent contradiction we substitute (2.80) for λ_1 and λ_2 in the product

$$(\lambda - \lambda_1)(\lambda - \lambda_2) = (\lambda - a_0 - \Delta)(\lambda - a_0 + \Delta) = (\lambda - a_0)^2 - \Delta^2. \tag{2.84}$$

In order to satisfy (2.82), the right-hand side above must be positive, which implies that $\Delta^2 < 0$, i.e., Δ is pure imaginary,

$$\Delta = i|\Delta|. \tag{2.85}$$

Hence,

$$(\lambda - \lambda_1)(\lambda - \lambda_2) = (\lambda - a_0)^2 + |\Delta|^2 \tag{2.86}$$

This is a positive definite quantity. We note that since F is Hermitian, a_0 defined by (2.81) is real. For Δ to be pure imaginary we must have, from (2.81) and (2.85),

$$4x^2z^2 \geq y^2. \tag{2.87}$$

In terms of the operators C, D, and F this implies

$$\left[|C|\phi\rangle|^2\right]\left[|D|\phi\rangle|^2\right] \geq \frac{1}{4}|\langle\phi|F|\phi\rangle|^2. \tag{2.88}$$

To derive the uncertainty relation we consider two Hermitian operators A and B, and relate them to the operators C and D as follows:

$$C = \Delta A, \quad D = \Delta B. \tag{2.89}$$

Then from the relations (2.65), (2.69), we have

$$[C, D] = [\Delta A, \Delta B] = [A, B] \tag{2.90}$$

and

$$iF = [A, B]. \tag{2.91}$$

We, therefore, conclude from (2.88) that

$$\left\langle(\Delta A)^2\right\rangle\left\langle(\Delta B)^2\right\rangle \geq \frac{1}{4}\left|\langle[A, B]\rangle\right|^2. \tag{2.92}$$

This is true for any set of Hermitian operators. We choose $A = X$, and $B = P$. Since $[X, P] = i\hbar\mathbf{1}$ we obtain from (2.92)

$$\left\langle (\Delta X)^2 \right\rangle \left\langle (\Delta P)^2 \right\rangle \geq \frac{\hbar^2}{4} \implies \Delta x \Delta p \geq \frac{\hbar}{2} \tag{2.93}$$

where we have taken Δx and Δp as the uncertainties defined by $\sqrt{\left\langle (\Delta X)^2 \right\rangle}$ and $\sqrt{\left\langle (\Delta P)^2 \right\rangle}$ respectively. These are the uncertainties in the measurements of the x and p variables. One often writes the relation (2.93) more simply as

$$\Delta x \Delta p \gtrsim \hbar. \tag{2.94}$$

This is the statement of the uncertainty relation in a quantum-mechanical system where the commutation relations between the canonical variables are governed by the Planck constant, $\hbar$.

2.6 Some interesting applications of uncertainty relations

2.6.1 Size of a particle

By "size" we mean Δx in the sense of the uncertainty relations according to which

$$\Delta x \sim \frac{\hbar}{\Delta p}. \tag{2.95}$$

A typical value of Δp is given by the momentum mv of the particle. The maximum possible value of the velocity according to the theory of relativity is the velocity of light, c. The minimum value of Δx is then

$$(\Delta x)_{\text{min}} = \frac{\hbar}{mc}. \tag{2.96}$$

Quantum-mechanically therefore, one does not think of a particle of mass m as a point particle but rather as an object with a finite size, $\hbar/mc$. This is also called the Compton wavelength characteristic of that particle and it enters into calculations whenever a length scale for the particle appears.

2.6.2 Bohr radius and ground-state energy of the hydrogen atom

Classically, an electron of charge $-e$ orbiting around a proton of charge e would lose energy due to radiation and eventually fall into the proton. This is, of course, in contradiction to the observed fact that the electron executes stable orbits. This is explained quantum-mechanically in the simplest terms through the uncertainty principle.

The total energy E of the electron is a sum of the kinetic energy $\left(\frac{1}{2}\right) mv^2$ and the potential energy, which in the case of hydrogen is just the Coulomb potential. Thus, writing the kinetic energy in terms of $p = mv$, the momentum, we have

$$E = \frac{p^2}{2m} - \frac{e^2}{r} \tag{2.97}$$

where r is the distance between the proton and electron. Taking $\Delta p \sim p$ and $\Delta r \sim r$, the uncertainty relation says that

$$pr \sim \hbar. \tag{2.98}$$

Thus, as r for the electron gets small due to the Coulomb attraction, p becomes large due to the uncertainty principle. In other words, as the attractive Coulomb potential moves the electron toward the proton, the increasing kinetic energy pushes the electron away from the system. The electron will then settle down at a minimum of the total energy. We can obtain this minimum by writing

$$E = \frac{\hbar^2}{2mr^2} - \frac{e^2}{r} \tag{2.99}$$

and then taking

$$\frac{\partial E}{\partial r} = 0. \tag{2.100}$$

This gives

$$-\frac{\hbar^2}{mr^3} + \frac{e^2}{r^2} = 0. \tag{2.101}$$

We find

$$r_{\min} = \frac{\hbar^2}{me^2}. \tag{2.102}$$

This is, indeed, the Bohr radius of the hydrogen atom, which is designated as a_0. Substituting this in the expression (2.99) we obtain the minimum value of E given by

$$E_{\min} = -\frac{e^2}{2a_0} \tag{2.103}$$

which is precisely the ground-state energy of the hydrogen atom.

Thus we find that even a simple application of the uncertainty relations can often give meaningful results.

2.7 Space displacement operator

We consider two unitary operators that are of fundamental importance: space displacement, also called translation, and time evolution. The displacement operator, $D(x)$, involves translation of the space coordinates and, in one dimension, it has the property

$$D(\Delta x')|x'\rangle = |x' + \Delta x'\rangle. \tag{2.104}$$

Hence

$$\langle x'|D^{\dagger}(\Delta x') = \langle x' + \Delta x'| \tag{2.105}$$

where $\Delta x'$ is the displacement. The displacement operator for $\Delta x' = 0$ corresponds to an operator that causes no displacement, hence,

$$D(0) = \mathbf{1}. \tag{2.106}$$

Furthermore, by repeated translations one finds

$$D(\Delta x'')D(\Delta x') = D\left(\Delta x'' + \Delta x'\right) = D(\Delta x')D(\Delta x''). \tag{2.107}$$

This follows from the fact that when all three of the operators that appear in the above equality operate on $\left|x'\right\rangle$, they give the same state, $\left|x' + \Delta x' + \Delta x''\right\rangle$. In particular, taking $\Delta x'' = -\Delta x'$, we find, since $D(0) = \mathbf{1}$,

$$D(\Delta x')D(-\Delta x') = \mathbf{1}. \tag{2.108}$$

Hence

$$D(-\Delta x') = D^{-1}(\Delta x') \tag{2.109}$$

which implies that the inverse of a displacement operator is the same as displacement in the opposite direction.

The set of operators $D(\Delta x')$ for different Δx's form what is called a "group." The precise definition of a group and its detailed properties will be discussed in some of the later chapters, but for now it is important to note that the operator $D(\Delta x')$ varies continuously as $\Delta x'$ varies. The D's then form a "continuous" group.

Since $D(\Delta x')$ merely translates the coordinates of a state vector, we assume the norm of $\left|x'\right\rangle$ to remain unchanged. Therefore,

$$\langle x'|x'\rangle = \langle x' + \Delta x'|x' + \Delta x'\rangle \Longrightarrow \langle x'|x'\rangle = \langle x'|D^{\dagger}(\Delta x')D(\Delta x')|x'\rangle, \tag{2.110}$$

i.e.,

$$D^{\dagger}(\Delta x')D(\Delta x') = \mathbf{1}. \tag{2.111}$$

Hence D is a unitary operator.

If $\Delta x'$ is infinitesimally small then, since $D(\Delta x')$ is a continuous function of $\Delta x'$, one can expand it in a power series as follows, keeping only the leading term in $\Delta x'$,

$$D(\Delta x') = \mathbf{1} - iK\Delta x'. \tag{2.112}$$

The operator K is called the "generator" of an infinitesimal space displacement transformation. Inserting the above expression in (2.111), we have

$$\mathbf{1} + i\left(K^{\dagger} - K\right)\Delta x' + O(\Delta x')^2 = \mathbf{1} \tag{2.113}$$

where $O\left(\Delta x'\right)^2$ means of the order of $\left(\Delta x'\right)^2$. This term can be neglected since $\Delta x'$ is infinitesimal and because we want to keep only the leading terms in $\Delta x'$. The above result then implies that

$$K^{\dagger} = K. \tag{2.114}$$

The operator K must, therefore, be Hermitian.

We will not prove it here but the fact that the generators of infinitesimal unitary transformations are Hermitian is actually a very general property.

To obtain further insight into K, consider the commutator $\left[X, D(\Delta x')\right]$. When the commutator operates on a state vector $\left|x'\right\rangle$ we obtain

$$\left[X, D(\Delta x')\right]|x'\rangle = XD(\Delta x')|x'\rangle - D(\Delta x')X|x'\rangle \tag{2.115}$$

$$= \left(x' + \Delta x'\right)|x' + \Delta x'\rangle - x'|x' + \Delta x'\rangle \tag{2.116}$$

$$= \Delta x'|x' + \Delta x'\rangle. \tag{2.117}$$

Since $\Delta x'$ is infinitesimal, we can make the expansion

$$|x' + \Delta x'\rangle = |x'\rangle + O\left(\Delta x'\right). \tag{2.118}$$

Substituting (2.118) in (2.117) and neglecting terms of order $\left(\Delta x'\right)^2$, we have from (2.114)

$$\left[X, D(\Delta x')\right] = \Delta x'\mathbf{1} \tag{2.119}$$

where we retain the unit operator on the right to preserve the fact that both sides of (2.119) are operators. Substituting the expression for D in terms of its generator K in (2.119) we obtain

$$\left[X, (\mathbf{1} - iK\Delta x')\right] = \Delta x'\mathbf{1}. \tag{2.120}$$

The commutator $[X, \mathbf{1}]$ vanishes, so we have, after canceling $\Delta x'$ from both sides,

$$[X, K] = i. \tag{2.121}$$

Comparing this commutator with the canonical commutator relation $[X, P] = i\hbar\mathbf{1}$, we obtain the important relation

$$K = \frac{P}{\hbar}, \tag{2.122}$$

which leads to the observation that the generator for infinitesimal space displacement transformations of a state vector is proportional to the momentum operator. Thus the space displacement operator can be written as

$$D(\Delta x') = \mathbf{1} - i\frac{P}{\hbar}\Delta x'. \tag{2.123}$$

To confirm that K has the same form as P we proceed by considering the quantity $D(\Delta x)|\phi\rangle$. Inserting a complete set of states we have

$$D(\Delta x)|\phi\rangle = \int_{-\infty}^{\infty} dx\, D(\Delta x)|x\rangle\langle x|\phi\rangle = \int_{-\infty}^{\infty} dx\, |x + \Delta x\rangle\langle x|\phi\rangle. \tag{2.124}$$

We change variables by taking $(x + \Delta x) = x''$, and notice that the limits $\pm\infty$ will remain unchanged. Thus, on the right-hand side above, $\langle x|\phi\rangle$ will be replaced by $\langle (x'' - \Delta x)|\phi\rangle$, which can be expanded in a Taylor series in terms of Δx to give

$$D(\Delta x)|\phi\rangle = \int_{-\infty}^{\infty} dx''\, |x''\rangle\langle x'' - \Delta x|\phi\rangle = \int_{-\infty}^{\infty} dx''|x''\rangle \left[\phi(x'') - \Delta x\frac{d}{dx''}\phi(x'')\right] \tag{2.125}$$

where we have taken $\langle x''|\phi\rangle = \phi(x'')$. Multiplying the two sides above by $\langle x'|$, we have

$$\langle x'|D(\Delta x)|\phi\rangle = \int_{-\infty}^{\infty} dx''\, \langle x'|x''\rangle \left[\phi(x'') - \Delta x\frac{d}{dx''}\phi(x'')\right]. \tag{2.126}$$

Since $\langle x'|x''\rangle = \delta(x' - x'')$, we obtain

$$\langle x'|D(\Delta x)|\phi\rangle = \phi(x') - \Delta x\frac{d}{dx'}\phi(x'). \tag{2.127}$$

Inserting $D(\Delta x) = \mathbf{1} - iK\Delta x$, we find

$$\langle x'|K|\phi\rangle = -i\frac{d}{dx'}\phi(x'). \tag{2.128}$$

Hence K has the same form as P, given by

$$\langle x'|P|\phi\rangle = -i\hbar\frac{d}{dx'}\phi(x'), \tag{2.129}$$

confirming (2.122).

Finally, we can convert D defined in terms of an infinitesimal transformation, to finite transformations. For example, space displacement by a finite amount, x, can be obtained if we divide x, into N equal parts $\Delta x = x/N$ and then take the products of $D(\Delta x)$ N times in the limit $N \to \infty$,

$$D(x) = \lim_{N\to\infty} D\left(\frac{x}{N}\right) D\left(\frac{x}{N}\right) D\left(\frac{x}{N}\right) \cdots = \lim_{N\to\infty} \left[D\left(\frac{x}{N}\right)\right]^N \tag{2.130}$$

$$= \lim_{N\to\infty} \left[\mathbf{1} - i\frac{1}{N}\frac{Px}{\hbar}\right]^N. \tag{2.131}$$

To determine the right-hand side we note that one can write

$$\lim_{N\to\infty} \left[\mathbf{1} + a\frac{x}{N}\right]^N = \lim_{N\to\infty} \exp\left[N \ln\left(\mathbf{1} + a\frac{x}{N}\right)\right]. \tag{2.132}$$

If we expand the exponent containing the logarithmic function, we obtain

$$N \ln\left(\mathbf{1} + a\frac{x}{N}\right) = N\left[a\left(\frac{x}{N}\right) - \frac{1}{2}a^2\left(\frac{x}{N}\right)^2 + \cdots\right] = ax - \frac{1}{2}(ax)^2\frac{1}{N} + \cdots. \tag{2.133}$$

In the limit $N \to \infty$, the right-hand side is simply ax. Hence

$$\lim_{N\to\infty} \left[\mathbf{1} + a\frac{x}{N}\right]^N = e^{ax}. \tag{2.134}$$

Thus $D(x)$ in (2.131) can be written as

$$D(x) = \exp\left(-i\frac{P}{\hbar}x\right) \tag{2.135}$$

where the exponential of the operators are understood in terms of the power series expansion as

$$e^A = 1 + A + \frac{A^2}{2!} + \cdots. \tag{2.136}$$

We can derive the relation (2.133) analytically also by using the definition of the derivative

$$\frac{dD(x)}{dx} = \lim_{\Delta x\to 0} \frac{D(x + \Delta x) - D(x)}{\Delta x}. \tag{2.137}$$

Let us then write

$$D(x + \Delta x) = D(x)\, D(\Delta x) = D(x)\left(\mathbf{1} - i\frac{P}{\hbar}\Delta x\right). \tag{2.138}$$

Inserting this in (2.137) we obtain

$$\frac{dD(x)}{dx} = \lim_{\Delta x\to 0} \frac{D(x)\left(\mathbf{1} - i\frac{P}{\hbar}\Delta x\right) - D(x)}{\Delta x} = -i\frac{P}{\hbar}D(x). \tag{2.139}$$

The solution of this is an exponential given by

$$D(x) = \exp\left(-i\frac{P}{\hbar}x\right), \tag{2.140}$$

confirming our earlier result.

We can extend the above results to three dimensions quite simply through the following steps for the infinitesimal translation of the state $|x, y, z\rangle$ in three mutually independent directions:

$$\left|x' + \Delta x', y' + \Delta y', z' + \Delta z'\right\rangle \tag{2.141}$$

$$= D(\Delta x')\left|x', y' + \Delta y', z' + \Delta z'\right\rangle \tag{2.142}$$

$$= \left[D(\Delta x')\right]\left[D(\Delta y')\right]|x', y', z' + \Delta z'\rangle \tag{2.143}$$

$$= \left[D(\Delta x')\right]\left[D(\Delta y')\right]\left[D(\Delta z')\right]|x', y', z'\rangle. \tag{2.144}$$

We then obtain the relation

$$D(\Delta x_i') = \left(\mathbf{1} - i\frac{P_i}{\hbar}\Delta x_i'\right) \tag{2.145}$$

where $i = 1, 2, 3$ correspond to the x, y, z components. For finite transformation one follows the procedure of N infinitesimal transformations in each direction in the limit $N \to \infty$. We can then write

$$D(x, y, z) = D(x)D(y)D(z) = \exp\left(-i\frac{P_x x + P_y y + P_z z}{\hbar}\right). \tag{2.146}$$

In vector notation the transformation can be written compactly as

$$D(\mathbf{r}) = \exp\left(-i\frac{\mathbf{P}\cdot\mathbf{r}}{\hbar}\right) \tag{2.147}$$

where $\mathbf{r} = (x, y, z)$ and $\mathbf{P} = \left(P_x, P_y, P_z\right)$.

2.8 Time evolution operator

Following the same procedure that we followed for the space displacement operator, we define another unitary operator, the time evolution operator, $U(t, t_0)$, which shifts the time parameter of a state from t_0 to t,

$$|\alpha(t)\rangle = U(t, t_0)\,|\alpha(t_0)\rangle. \tag{2.148}$$

We note that

$$U(t_0, t_0) = \mathbf{1}. \tag{2.149}$$

Furthermore, we can write (2.148) in two steps, e.g., we first write it as

$$|\alpha(t)\rangle = U(t, t_1)\,|\alpha(t_1)\rangle \tag{2.150}$$

and then write

$$|\alpha(t_1)\rangle = U(t_1, t_0)\,|\alpha(t_0)\rangle. \tag{2.151}$$

From (2.148), (2.150), and (2.151) we find, just as with the D-operator,

$$U(t, t_0) = U(t, t_1)\,U(t_1, t_0)\,. \tag{2.152}$$

If the norms of $|\alpha(t_0)\rangle$ and $|\alpha(t)\rangle$ are the same then we have

$$\langle\alpha(t_0)|\alpha(t_0)\rangle = \langle\alpha(t)|\alpha(t)\rangle = \langle\alpha(t_0)|U^\dagger(t, t_0)\,U(t, t_0)\,|\alpha(t_0)\rangle, \tag{2.153}$$

which implies that U is unitary,

$$U^\dagger(t, t_0)\,U(t, t_0) = \mathbf{1}. \tag{2.154}$$

Consider an infinitesimal transformation from t_0 to $t_0 + \Delta t$. As we did for the space displacement operator, D, the operator $U(t_0 + \Delta t, t_0)$ can be expanded in terms of its generator Ω, as follows:

$$U(t_0 + \Delta t, t_0) = \mathbf{1} - i\Omega\Delta t. \tag{2.155}$$

From the unitary property of $U(t_0 + \Delta t, t_0)$ we have

$$U^\dagger(t_0 + \Delta t, t_0)\,U(t_0 + \Delta t, t_0) = \mathbf{1}. \tag{2.156}$$

Substituting (2.155) we find, following the same arguments as for the generator K of D, that Ω is Hermitian,

$$\Omega^\dagger = \Omega. \tag{2.157}$$

As in classical mechanics we identify time evolution with the Hamiltonian, H, of the system and write the following relation that includes the Planck constant $\hbar$:

$$\Omega = \frac{H}{\hbar}. \tag{2.158}$$

An equivalent quantum-mechanical description for the time evolution of a state vector is then given by

$$U(t_0 + \Delta t, t_0) = \mathbf{1} - i\frac{H}{\hbar}\Delta t. \tag{2.159}$$

Thus we can write

$$|\alpha(t + \Delta t)\rangle = U(t + \Delta t, t)\,|\alpha(t)\rangle = \left(\mathbf{1} - i\frac{H}{\hbar}\Delta t\right)|\alpha(t)\rangle. \tag{2.160}$$

We expand $|\alpha(t+\Delta t)\rangle$ for an infinitesimal Δt in a Taylor series in terms Δt, keeping only the first two terms,

$$|\alpha(t+\Delta t)\rangle = |\alpha(t)\rangle + \Delta t \frac{\partial}{\partial t}|\alpha(t)\rangle. \tag{2.161}$$

Comparing this equation with the one above, we obtain the result

$$i\hbar \frac{\partial}{\partial t}|\alpha(t)\rangle = H|\alpha(t)\rangle, \tag{2.162}$$

which is the equation for the time evolution of a state vector in terms of the Hamiltonian H.

As we will discuss further in the next section, this is the time-dependent Schrödinger equation. In comparing space and time transformations, we note that, unlike X which is an operator, t is simply a parameter and not an operator and, therefore, has no commutation properties with H. This is not unexpected since x and t are not on an equal footing in nonrelativistic problems. In the relativistic case, where they are on an equal footing, one finds that instead of elevating t to an operator, x is in fact demoted to being a parameter in the quantum-mechanical treatments, keeping the status of x and t the same.

Let us now obtain an explicit expression for $U(t, t_0)$. We can do this by following the same procedure as we followed for $D(x)$ which is to take the product of N infinitesimal transformations in the limit $N \to \infty$, or to do it analytically. We will follow the analytical path. We first obtain the differential equation for U, by using the definition for the partial derivative

$$\frac{\partial U(t, t_0)}{\partial t} = \lim_{\Delta t \to 0} \frac{U(t+\Delta t, t_0) - U(t, t_0)}{\Delta t}. \tag{2.163}$$

From (2.152) we write $U(t+\Delta t, t_0) = U(t+\Delta t, t)\, U(t, t_0)$ and from (2.159), replacing t_0 by t we write $U(t+\Delta t, t) = \mathbf{1} - iH/\hbar \Delta t$. The above relation then reads

$$\frac{\partial U(t, t_0)}{\partial t} = \lim_{\Delta t \to 0} \frac{\left(\mathbf{1} - i\Delta t \frac{H}{\hbar}\right) U(t, t_0) - U(t, t_0)}{\Delta t}, \tag{2.164}$$

which leads to

$$i\hbar \frac{\partial U(t, t_0)}{\partial t} = HU(t, t_0). \tag{2.165}$$

In other words, as we will elaborate in the next section, U itself satisfies the same time-dependent equation that the ket vector $|\alpha(t)\rangle$ satisfies. We can now integrate the above equation with the condition $U(t_0, t_0) = \mathbf{1}$ and obtain

$$U(t, t_0) = \exp\left[-i\frac{H}{\hbar}(t - t_0)\right] \tag{2.166}$$

where we have assumed that H, is independent of time.

We can also obtain the equation for $U^\dagger$ by taking the Hermitian conjugate of the equation (2.165)

$$-i\hbar\frac{\partial U^\dagger}{\partial t} = U^\dagger H \tag{2.167}$$

where we have made use of the fact that H is Hermitian. The solution of this equation, as expected, directly from (2.167) is

$$U^\dagger(t, t_0) = \exp\left[i\frac{H}{\hbar}(t - t_0)\right]. \tag{2.168}$$

2.9 Appendix to Chapter 2

2.9.1 Dirac delta function

The Dirac delta function, $\delta(x)$, in one dimension, is defined by the following two relations:

$$\delta(x) = 0 \text{ for } x \neq 0, \tag{2.169}$$

$$\int_{-\infty}^{\infty} dx\, \delta(x) = 1. \tag{2.170}$$

The above two relations then also imply that the limits of integration can be different from $\pm\infty$, as long as they include the point $x = 0$. Indeed, one can write

$$\int_b^c dx\, \delta(x) = 1 \text{ for } b < 0 < c. \tag{2.171}$$

One can extend the definition to include a point that is not the origin:

$$\delta(x - a) = 0 \quad \text{for} \quad x \neq a, \tag{2.172}$$

$$\int_{-\infty}^{\infty} dx\, \delta(x - a) = \int_b^c dx\, \delta(x - a) = 1 \text{ for } b < a < c \tag{2.173}$$

where a is a real number.

Properties of the δ-function

Following are some of the interesting properties satisfied by the δ-function.

Property (a)

$$\delta(-x) = \delta(x). \tag{2.174}$$

This can be proved by noting that $\delta(-x)$ satisfies the same properties as $\delta(x)$ given in (2.169) and (2.170), namely, it vanishes for $x \neq 0$, and, through change of variables, $x \to -x$, its integral over $(-\infty, \infty)$ is found to be the same as that of $\delta(x)$. This establishes the fact that $\delta(x)$ is an even function of x.

Property (b)

$$\int_{-\infty}^{\infty} dx f(x)\delta(x-a) = f(a) \tag{2.175}$$

where $f(x)$ is assumed to be regular, without any singularity, along the interval of integration. The proof is quite simple if we write

$$\int_{-\infty}^{\infty} dx f(x)\delta(x-a) = \int_{a-\epsilon}^{a+\epsilon} dx f(x)\delta(x-a) = f(a)\int_{a-\epsilon}^{a+\epsilon} dx\, \delta(x-a) = f(a) \tag{2.176}$$

where ϵ is a small positive quantity.

Property (c)

$$\delta(ax) = \frac{1}{|a|}\delta(x). \tag{2.177}$$

This relation can be derived by changing variables, $ax = y$, and taking account of the even nature of the δ-function.

Property (d)

$$\delta\left(x^2 - a^2\right) = \frac{\delta(x-a) + \delta(x+a)}{2|a|}. \tag{2.178}$$

To prove this we first note that both sides vanish for $x \neq a$. Furthermore, we can write the integral on the left-hand side in the following manner:

$$\int_{-\infty}^{+\infty} \delta\left(x^2 - a^2\right) = \int_{-\infty}^{+\infty} \delta\left[(x-a)(x+a)\right] = \int_{-a-\epsilon}^{-a+\epsilon} \delta\left[(-2a(x+a))\right] + \int_{a-\eta}^{a+\eta} \delta\left[(2a(x-a))\right] \tag{2.179}$$

where ϵ and η are small positive quantities. From (2.177) we recover the right-hand side of (2.178).

Property (e)

$$\delta(f(x)) = \sum_n \frac{\delta(x - x_n)}{\left|\dfrac{df}{dx}\right|_{x=x_n}} \tag{2.180}$$

where $f(x)$ is a regular function, and x_n are the (real) zeros of $f(x)$. This is a generalization of the case (d) where there were only two zeros, $x = \pm a$. To prove this one expands $f(x)$ around each of its zeros, e.g., around $x = x_n$:

$$f(x) = f(x_n) + (x - x_n)\left(\frac{df}{dx}\right)_{x=x_n} + \cdots = (x - x_n)\left(\frac{df}{dx}\right)_{x=x_n} \tag{2.181}$$

where in the last equality we have used the fact that $f(x_n) = 0$, and have kept only the leading term in $(x - x_n)$. Therefore,

$$\delta\left(f(x)\right) = \delta\left[(x - x_n)\left(df/dx\right)_{x=x_n}\right]. \tag{2.182}$$

Using (2.177) we obtain (2.180).

As a strict mathematical function, the relation (2.169) for $\delta(x)$ will imply that the right-hand side of (2.170) must also be zero, because a function that is everywhere zero except at one point (or at a finite number of points) gives an integral that will also be zero. However, it can be represented in terms of limiting functions.

Representations of the δ-function

We consider the following four well-known limiting functions and discuss their role in mathematical problems, particularly in quantum mechanics

(i) The most common example of a δ-function behavior in quantum-mechanical problems, or in Fourier transforms, is the following limiting function:

$$\lim_{L\to\infty} \frac{1}{\pi}\frac{\sin Lx}{x}. \tag{2.183}$$

This function oscillates along the $\pm$ x-axis with zeros at $x = \pm\pi/L, \pm 2\pi/L, \ldots, \pm n\pi/L, \ldots$ where n is an integer. The spacing between two consecutive zeros is given by π/L. Therefore, as $L \to \infty$, this spacing becomes increasingly narrower with the zeros almost overlapping with each other, while at the same time the function $\sin(Lx/x)$ itself goes to zero as $x \to \pm\infty$. Thus, for all practical purposes this function vanishes along the $\pm$ x-axis. That is,

$$\lim_{L\to\infty} \frac{1}{\pi}\frac{\sin Lx}{x} \to 0 \text{ for } x \neq 0. \tag{2.184}$$

The only exceptional point is the origin itself, $x = 0$. At this point $\sin Lx = Lx$ and $\sin(Lx/\pi x) = L/\pi$ which grows as $L \to \infty$. Thus the function goes to infinity at one point, while vanishing everywhere else, which is a classic situation for a δ-function. As far as its integral is considered, one can show, using standard techniques – through the complex variables method – that

$$\frac{1}{\pi}\int_{-\infty}^{\infty} dx \frac{\sin Lx}{x} = 1, \tag{2.185}$$

independent of L.

Thus as a limiting function one can express the δ-function as

$$\delta(x) = \lim_{L\to\infty} \frac{1}{\pi} \frac{\sin Lx}{x}. \tag{2.186}$$

The following equivalent integral form,

$$\frac{1}{2} \int_{-L}^{L} dk\, e^{ikx} = \frac{\sin Lx}{x} \tag{2.187}$$

enables us, through the relation (2.186), to write

$$\delta(x) = \frac{1}{2\pi} \int_{-\infty}^{\infty} dk\, e^{ikx} \tag{2.188}$$

where it is understood that

$$\frac{1}{2\pi} \int_{-\infty}^{\infty} dk\, e^{ikx} = \lim_{L\to\infty} \frac{1}{2\pi} \int_{-L}^{L} dk\, e^{ikx}. \tag{2.189}$$

More generally, one can write

$$\delta(x-a) = \frac{1}{2\pi} \int_{-\infty}^{\infty} dk\, e^{ik(x-a)}. \tag{2.190}$$

(ii) The following Gaussian also mimics a δ-function behavior:

$$\lim_{\beta\to\infty} \sqrt{\frac{\beta}{\pi}} \exp(-\beta x^2). \tag{2.191}$$

We note that for $x \neq 0$ the right-hand side vanishes in the limit $\beta \to \infty$, but at $x = 0$, in the same limit, it goes to infinity like $\sqrt{\beta/\pi}$. This is once again a typical δ-function behavior. From the well-known Gaussian integral

$$\int_{-\infty}^{\infty} dy \exp(-y^2) = \sqrt{\pi} \tag{2.192}$$

we deduce, by a change of variables $\sqrt{\beta}x = y$, that

$$\sqrt{\frac{\beta}{\pi}} \int_{-\infty}^{\infty} dx \exp(-\beta x^2) = 1. \tag{2.193}$$

Thus, one can write

$$\delta(x) = \lim_{\beta\to\infty}\sqrt{\frac{\beta}{\pi}}\exp(-\beta x^2). \tag{2.194}$$

Another way of writing the above result is by taking $\beta = 1/\left[\alpha\left(t-t'\right)\right]$, and writing a very general condition, replacing x by $\left(x-x'\right)$,

$$\delta(x-x') = \lim_{t\to t'}\sqrt{\frac{1}{\alpha\left(t-t'\right)\pi}}\exp\left[-\frac{\left(x-x'\right)^2}{\alpha\left(t-t'\right)}\right]. \tag{2.195}$$

(iii) Here is another δ-function type behavior:

$$\lim_{\epsilon\to 0}\frac{1}{\pi}\frac{\epsilon}{x^2+\epsilon^2}. \tag{2.196}$$

We find that for $x \neq 0$, this function vanishes in the limit $\epsilon \to 0$. However, if $x = 0$, the function behaves like $1/\pi\epsilon$ and goes to infinity in the same limit (the limit has to be taken after the value of x is chosen). Furthermore, one can easily show through standard integration techniques that

$$\frac{1}{\pi}\int_{-\infty}^{\infty} dx\frac{\epsilon}{x^2+\epsilon^2} = 1. \tag{2.197}$$

Once again we can identify

$$\delta(x) = \lim_{\epsilon\to 0}\frac{1}{\pi}\frac{\epsilon}{x^2+\epsilon^2}. \tag{2.198}$$

This form often occurs in the so-called dispersion relations and in Green's function problems, e.g.,

$$\lim_{\epsilon\to 0}\frac{1}{\pi}\int_0^{\infty} dx'\frac{f(x')}{x'-x-i\epsilon} \tag{2.199}$$

where one writes

$$\begin{aligned}\lim_{\epsilon\to 0}\left[\frac{1}{x'-x-i\epsilon}\right] &= \lim_{\epsilon\to 0}\left[\frac{(x'-x)}{(x'-x)^2+\epsilon^2}\right] + \lim_{\epsilon\to 0}\left[\frac{i\epsilon}{(x'-x)^2+\epsilon^2}\right] \\ &= \lim_{\epsilon\to 0}\left[\frac{(x'-x)}{(x'-x)^2+\epsilon^2}\right] + i\pi\delta(x'-x).\end{aligned} \tag{2.200}$$

This relation is often written as

$$\frac{1}{x' - x - i\epsilon} = P\left(\frac{1}{x' - x}\right) + i\pi\delta(x' - x) \tag{2.201}$$

where P in the first term on the right stands for what is called the "principal part." It vanishes at the point $x' = x$ and hence excludes that point when integrated over it. The contribution of the singularity at $x' = x$ is now contained in the second term. Thus,

$$\lim_{\epsilon \to 0} \frac{1}{\pi} \int_0^\infty dx' \frac{f(x')}{x' - x - i\epsilon} = \frac{1}{\pi} P \int_0^\infty dx' \frac{f(x')}{x' - x} + if(x). \tag{2.202}$$

(iv) Finally, an interesting representation is given by the function,

$$\frac{d}{dx}\theta(x) \tag{2.203}$$

where $\theta(x)$ is called the step-function,

$$\theta(x) = 0 \quad \text{for} \quad x \leq 0, \tag{2.204}$$

$$\text{and } \theta(x) = 1 \quad \text{for} \quad x > 0. \tag{2.205}$$

Notice that the derivative on either side of $x = 0$ vanishes, but at the point $x = 0$ it becomes infinite. The integral can be carried out and is given by

$$\int_{-\infty}^{\infty} dx \frac{d}{dx}\theta(x) = \theta(\infty) - \theta(-\infty) = 1. \tag{2.206}$$

We also note that the above result holds if we take the integration limits to be $(-L, +L)$. Thus, one can write

$$\delta(x) = \frac{d}{dx}\theta(x). \tag{2.207}$$

This θ-function representation appears in Green's function calculations, often for finite-dimensional problems.

In summary, we note that for the above representations of $\delta(x)$, as long as $x \neq 0$, each of the functions on the right-hand side vanishes. But when they are integrated over an interval that includes the point $x = 0$, the result is 1, keeping in mind that the limits are to be taken after the integration is carried out. Thus the limiting process is nonuniform.

Three dimensions

We define the δ-function in the Cartesian system as a product of three one-dimensional δ-functions as follows:

$$\delta^{(3)}(\mathbf{r} - \mathbf{r}') = \delta(x - x')\delta(y - y')\delta(z - z'). \tag{2.208}$$

Therefore, it satisfies

$$\delta^{(3)}(\mathbf{r}-\mathbf{r}') = 0 \quad \text{if} \quad x \neq x' \text{ or } y \neq y' \text{ or } z \neq z' \tag{2.209}$$

and

$$\int_{(\infty)} d^3r\,\delta^{(3)}(\mathbf{r}-\mathbf{r}') = 1, \tag{2.210}$$

which implies, as in the one-dimensional case, that

$$\int_{(\infty)} d^3\mathrm{r} f(\mathbf{r})\delta^{(3)}(\mathbf{r}-\mathbf{r}') = f(\mathbf{r}'). \tag{2.211}$$

Since

$$\delta(x-x') = \frac{1}{2\pi}\int_{-\infty}^{\infty} dk\, e^{ik(x-x')}, \tag{2.212}$$

we can express the three-dimensional δ-function as the product

$$\delta^{(3)}(\mathbf{r}-\mathbf{r}') = \left[\frac{1}{2\pi}\int_{-\infty}^{\infty} dk_x\, e^{ik_x(x-x')}\right]\left[\frac{1}{2\pi}\int_{-\infty}^{\infty} dk_y\, e^{ik_y(y-y')}\right]\left[\frac{1}{2\pi}\int_{-\infty}^{\infty} dk_z\, e^{ik_z(z-z')}\right], \tag{2.213}$$

which can be written in a compact form as

$$\delta^{(3)}(\mathbf{r}-\mathbf{r}') = \frac{1}{(2\pi)^3}\int d^3k e^{i\mathbf{k}.(\mathbf{r}-\mathbf{r})} \tag{2.214}$$

where d^3k is the three-dimensional volume element in the k-space given by

$$d^3k = dk_x\, dk_y\, dk_z. \tag{2.215}$$

We have defined the vector $\mathbf{k}$ as a three-dimensional vector with components (k_x, k_y, k_z) so that

$$\mathbf{k}.(\mathbf{r}-\mathbf{r}') = k_x(x-x') + k_y(y-y') + k_z(z-z'). \tag{2.216}$$

In spherical coordinates, (r, θ, ϕ), the δ-function is easily defined, once again, in terms of products of three one-dimensional δ-functions:

$$\delta^{(3)}(\mathbf{r}-\mathbf{r}') = A(r,\theta,\phi)\delta(r-r')\delta(\theta-\theta')\delta(\phi-\phi') \tag{2.217}$$

where the relation between the Cartesian and spherical coordinates is given by

$$x = r\sin\theta\cos\phi, \quad y = r\sin\theta\sin\phi, \quad z = r\cos\theta. \tag{2.218}$$

The parameter A in (2.217) can be determined from the definition (2.214):

$$1 = \int_{(\infty)} d^3r\,\delta^{(3)}(\mathbf{r}-\mathbf{r}') = \int_0^\infty dr \int_0^\pi d\theta \int_0^{2\pi} d\phi\, J\, A(r,\theta,\phi)\delta(r-r')\delta(\theta-\theta')\delta(\phi-\phi') \tag{2.219}$$

where on the right-hand side we have converted d^3r from Cartesian to spherical coordinates through the relation

$$d^3r = dx\,dy\,dz = J\,dr\,d\theta\,d\phi \tag{2.220}$$

with J as the Jacobian, $\dfrac{\partial(x,y,z)}{\partial(r,\theta,\phi)}$, defined as

$$J = \det\begin{bmatrix} \dfrac{\partial x}{\partial r} & \dfrac{\partial x}{\partial \theta} & \dfrac{\partial x}{\partial \phi} \\ \dfrac{\partial y}{\partial r} & \dfrac{\partial y}{\partial \theta} & \dfrac{\partial y}{\partial \phi} \\ \dfrac{\partial z}{\partial r} & \dfrac{\partial z}{\partial \theta} & \dfrac{\partial z}{\partial \phi} \end{bmatrix}. \tag{2.221}$$

We find

$$J = r^2\sin\theta. \tag{2.222}$$

Substituting this in the above relation, we obtain

$$d^3r = r^2\sin\theta\,dr\,d\theta\,d\phi. \tag{2.223}$$

Relation (2.219) gives

$$\int_0^\infty dr \int_0^\pi d\theta \int_0^{2\pi} d\phi \left(r^2\sin\theta\right) A(r,\theta,\phi)\delta(r-r')\delta(\theta-\theta')\delta(\phi-\phi') = 1. \tag{2.224}$$

Hence,

$$A(r,\theta,\phi) = \frac{1}{r^2\sin\theta} \tag{2.225}$$

and

$$\delta^{(3)}(\mathbf{r}-\mathbf{r}') = \frac{1}{r^2\sin\theta}\delta(r-r')\delta(\theta-\theta')\delta(\phi-\phi'). \tag{2.226}$$

It is often more convenient to use $\cos\theta$ as a variable rather than θ so that we can write the integrals more simply

$$\int_0^\infty dr \int_0^\pi d\theta \int_0^{2\pi} d\phi\, r^2 \sin\theta \to \int_0^\infty dr\, r^2 \int_{-1}^1 d\cos\theta \int_0^{2\pi} d\phi, \tag{2.227}$$

in which case we obtain the relation

$$\delta^{(3)}(\mathbf{r}-\mathbf{r}') = \frac{1}{r^2}\delta(r-r')\delta(\cos\theta - \cos\theta')\delta(\phi-\phi'). \tag{2.228}$$

This expression is consistent with the relation (2.180) for converting $\delta(\theta-\theta')$ to $\delta(\cos\theta - \cos\theta')$.

2.10 Problems

1. The state vectors $|\gamma\rangle$, $|\alpha\rangle$ and $|\beta\rangle$ are related as

$$|\gamma\rangle = |\alpha\rangle + \lambda|\beta\rangle$$

where λ is an arbitrary complex constant. By choosing an appropriate λ and the fact that $\langle\gamma|\gamma\rangle \geq 0$, derive the Schwarz inequality relation

$$\langle\alpha|\alpha\rangle\langle\beta|\beta\rangle \geq |\langle\alpha|\beta\rangle|^2.$$

2. For the above problem consider a state $|\phi\rangle$ such that

$$\Delta A|\phi\rangle = |\alpha\rangle, \Delta B|\phi\rangle = |\beta\rangle$$

where A and B are Hermitian operators and ΔA and ΔB are the corresponding uncertainties. Expressing the product $\Delta A\Delta B$ as a sum of a commutator and an anticommutator,

$$\Delta A\Delta B = \frac{1}{2}[\Delta A, \Delta B] + \frac{1}{2}\{\Delta A, \Delta B\},$$

and using the Schwarz inequality relation derived above, show that

$$|\Delta A\Delta B|^2 \geq \frac{1}{4}|[A,B]|^2.$$

From this result show that the uncertainty relation follows

$$\Delta x\Delta p \geq \frac{\hbar}{2}.$$

3. Put $B = H$ in the above relation and show that

$$\Delta A \Delta H \geq \frac{1}{2}\hbar \left| \frac{d}{dt} \langle A \rangle (t) \right|.$$

Defining the time uncertainty Δt as

$$\frac{1}{\Delta t} = \frac{1}{\Delta A} \left| \frac{d}{dt} \langle A \rangle (t) \right|,$$

show that one obtains

$$\Delta E \Delta t \geq \frac{1}{2}\hbar,$$

which is often called the energy–time uncertainty relation.

4. Use uncertainty relations to estimate the bound-state energies corresponding to
   (i) the linear potential

$$V(r) = Kr;$$

   (ii) the Coulomb potential (hydrogen atom)

$$V(r) = -\frac{Ze^2}{r}.$$

5. Show that the operator in spherical coordinates given by $-i\hbar\partial/\partial r$ is not Hermitian. Consider then the operator

$$-i\hbar \left(\frac{\partial}{\partial r} + \frac{a}{r} \right).$$

Determine a so that it is Hermitian.

6. Show that the operator

$$D = \mathbf{p}.\left(\frac{\mathbf{r}}{r}\right) + \left(\frac{\mathbf{r}}{r}\right).\mathbf{p}$$

is Hermitian. Obtain its explicit form in spherical coordinates. Compare your result with that of problem 5.

7. For the operator D defined above, obtain

$$\text{(i) } [D, x_i], \quad \text{(ii) } [D, p_i], \quad \text{and} \quad \text{(iii) } [D, L_i]$$

where $\mathbf{L}\,(= \mathbf{r} \times \mathbf{p})$ is the angular momentum operator. Also show that

$$e^{i\alpha D/\hbar} x_i e^{-i\alpha D/\hbar} = e^{\alpha} x_i.$$

8. Using the fundamental commutator relation, determine $\left[x, p^2\right]$, $\left[x^2, p\right]$ and $\left[x^2, p^2\right]$.

9. Consider the operator which corresponds to finite displacement

$$F(d) = e^{-ipd/\hbar}.$$

Show that

$$[x, F(d)] = dF(d).$$

If for a state $|\alpha\rangle$ we define $|\alpha_d\rangle = F(d)\,|\alpha\rangle$, then show that the expectation values with respect to the two states satisfy

$$\langle x \rangle_d = \langle x \rangle + d.$$

10. For a Hamiltonian given by

$$H = \frac{p^2}{2m} + V(x),$$

evaluate the commutator $[H, x]$ and the double commutator $[[H, x], x]$. From these derive the following identity involving the energy eigenstates and eigenvalues:

$$\sum_k (E_k - E_n) |\langle k|x|n\rangle|^2 = \frac{\hbar^2}{2m}.$$

11. For a Hamiltonian given by

$$H = \frac{\mathbf{p}^2}{2m} + V(\mathbf{r}),$$

use the properties of the double commutator $\left[\left[H, e^{i\mathbf{k}\cdot\mathbf{r}}\right], e^{-i\mathbf{k}\cdot\mathbf{r}}\right]$ to obtain

$$\sum_n (E_n - E_s)\, |\langle n|e^{i\mathbf{k}.\mathbf{r}}|s\rangle|^2.$$

3 Dynamical equations

We are now ready to derive the equation of motion of the state vectors and operators that determine the dynamics of a physical system. These equations are considered within the framework of three commonly used pictures: the Schrödinger, the Heisenberg and the interaction pictures.

3.1 Schrödinger picture

Here the motion of the system is expressed in terms of time- and space-variation of the wavefunctions. Consequently the operators in the coordinate representation are expressed in terms of time and space derivatives. The Hamiltonian operator, as we discussed in the preceding section, is expressed as

$$H \rightarrow i\hbar \frac{\partial}{\partial t}. \tag{3.1}$$

The Hamiltonian operator corresponds to the energy of the particle. In other words, if a state $|a_n(t)\rangle$ is an eigenstate of energy, E_n, then it satisfies the eigenvalue equation

$$H\,|a_n(t)\rangle = E_n\,|a_n(t)\rangle \tag{3.2}$$

where the time dependence of the state vector has been made explicit. Therefore,

$$H\,|a_n(t)\rangle = i\hbar \frac{\partial}{\partial t}\,|a_n(t)\rangle = E_n\,|a_n(t)\rangle\,. \tag{3.3}$$

The solution of this equation is quite simple, given by,

$$|a_n(t)\rangle = |a_n(0)\rangle \exp\left(-\frac{iE_n}{\hbar}t\right) \tag{3.4}$$

where we have normalized the eigenvector to its $t = 0$ value.

Confining to one dimension, the eigenfunction in the x-space is given by

$$\phi_n(x,t) = \langle x|\, a_n(t)\rangle. \tag{3.5}$$

Equation (3.4) then reads

$$\phi_n(x,t) = \phi_n(x,0) \exp\left(-\frac{iE_n}{\hbar}t\right). \tag{3.6}$$

We now discuss the properties of the Hamiltonian. In a dynamical system, the motion of a particle is described in terms of the Hamiltonian, H, which is written as

$$H = T + V \tag{3.7}$$

where T is the kinetic energy and V the potential energy. Thus, for a state vector $|\phi\rangle$ one can write

$$H|\phi\rangle = T|\phi\rangle + V|\phi\rangle . \tag{3.8}$$

In terms of operators X and P, H is given by

$$H = \frac{P^2}{2m} + V(X) \tag{3.9}$$

where V is assumed to be a function of x only, thus $V(X)|x\rangle = V(x)|x\rangle$. Implicit in this statement is the assumption that the interaction is local, i.e., $\langle x'| V(X)|x\rangle = V(x)\delta(x' - x)$. Furthermore, we assume V to be real.

In the coordinate-space the momentum operator, P, is given by

$$P \to -i\hbar \frac{\partial}{\partial x}. \tag{3.10}$$

Writing

$$\langle x|\phi(t)\rangle = \phi(x, t) \tag{3.11}$$

and

$$\langle x|V(X)|\phi(t)\rangle = V(x)\phi(x, t), \tag{3.12}$$

the equation of motion in the Schrödinger picture is then described by the following differential equation for the wavefunction $\phi(x, t)$:

$$i\hbar \frac{\partial \phi(x, t)}{\partial t} = -\frac{\hbar^2}{2m} \frac{\partial^2 \phi(x, t)}{\partial x^2} + V(x)\phi(x, t). \tag{3.13}$$

This is the classic Schrödinger equation.

If $\phi(x, t)$ is an eigenstate of energy E_n, which we previously designated as $\phi_n(x, t)$, then it satisfies the equation

$$-\frac{\hbar^2}{2m} \frac{\partial^2 \phi_n(x, t)}{\partial x^2} + V(x)\phi_n(x, t) = E_n \phi_n(x, t). \tag{3.14}$$

The t-dependence of the energy eigenfunctions is already known from our previous results as

$$\phi_n(x, t) = u_n(x) \exp\left(-\frac{iE_n}{\hbar} t\right) \tag{3.15}$$

where

$$u_n(x) = \phi_n(x, 0). \tag{3.16}$$

The equation for $u_n(x)$ will now be in terms of the total derivative, given by

$$-\frac{\hbar^2}{2m}\frac{d^2u_n(x)}{dx^2} + V(x)u_n(x) = E_n u_n(x). \tag{3.17}$$

This equation can easily be extended to three dimensions by writing

$$-\frac{\hbar^2}{2m}\nabla^2 u_n(\mathbf{r}) + V(\mathbf{r})u_n(\mathbf{r}) = E_n u_n(\mathbf{r}) \tag{3.18}$$

where

$$\nabla^2 = \frac{\partial}{\partial x^2} + \frac{\partial}{\partial y^2} + \frac{\partial}{\partial z^2} \tag{3.19}$$

and

$$u_n(\mathbf{r}) = u_n(x, y, z), \qquad V(\mathbf{r}) = V(x, y, z). \tag{3.20}$$

Solving the above equations for different potentials and boundary conditions will be our task in the coming chapters.

3.2 Heisenberg picture

In the Schrödinger picture the state vectors evolve with time, while the operators are independent of time. In the Heisenberg picture, which we will describe below, the state vectors are fixed in time while the operators evolve as a function of time. The quantities in the two pictures are related to each other, however, due to the fact that the results of any experiment should be the same when described in either of the two pictures.

The Schrödinger picture deals with the equations for the wavefunctions. It is more often used because its framework allows for easier calculations, while the Heisenberg picture effectively involves equations between operators.

We designate the states and the operators in the Schrödinger picture with subscript S, i.e., as $|\alpha_S(t)\rangle$ and A_S respectively. As we have already stated, in this picture the states depend on time, while the operators do not (except for special cases which we ignore). Consider now the matrix element

$$\langle \alpha_S(t) | A_S | \beta_S(t) \rangle . \tag{3.21}$$

From our earlier discussions we can connect $|\alpha_S(t)\rangle$ to $|\alpha_S(0)\rangle$ through the unitary operator $U(t,0) = U(t)$,

$$|\alpha_S(t)\rangle = U(t)\,|\alpha_S(0)\rangle \tag{3.22}$$

where, as we found in the previous section,

$$U(t) = e^{-iHt/\hbar} \tag{3.23}$$

where H is the Hamiltonian, assumed to be independent of time. The matrix element (3.21) can be written as

$$\langle \alpha_S(t)\,|A_S|\,\beta_S(t)\rangle = \left\langle \alpha_S(0)\left|U^\dagger(t)A_S U(t)\right|\beta_S(0)\right\rangle. \tag{3.24}$$

Let $|\alpha_H\rangle$ be the state vector in the Heisenberg picture. It is independent of time and we define it to be the same as the state vector in the Schrödinger picture at $t = 0$. That is,

$$|\alpha_H\rangle = |\alpha_S(0)\rangle. \tag{3.25}$$

As we stated earlier, the measurement of an observable is reflected in the value of the matrix element of the corresponding operator. The matrix elements of an operator in the Schrödinger and Heisenberg pictures must, therefore, be the same, as the result of any measurement must be independent of the type of picture which one uses to describe it. If A_H is the operator in the Heisenberg picture then we must have

$$\langle \alpha_S(t)\,|A_S|\,\beta_S(t)\rangle = \langle \alpha_H\,|A_H|\,\beta_H\rangle. \tag{3.26}$$

From (3.22) and (3.23) we conclude that

$$A_H = U^\dagger(t)A_S U(t) = e^{iHt/\hbar}A_S e^{-iHt/\hbar}. \tag{3.27}$$

In the Heisenberg picture, therefore, it is the operators that change as a function of time.

Taking the time-derivatives on both sides of the equation (3.27) above, we obtain

$$\frac{dA_H}{dt} = (iH/\hbar)\left(e^{iHt/\hbar}A_S e^{-iHt/\hbar}\right) + \left(e^{iHt/\hbar}A_S e^{-iHt/\hbar}\right)(-iH/\hbar) \tag{3.28}$$

where we have assumed that A_H does not have any explicit dependence on time. From (3.27) and (3.28) we obtain

$$i\hbar\frac{dA_H}{dt} = [A_H, H]. \tag{3.29}$$

Thus, in the Heisenberg representation the time dependence is governed by the commutator of the operator with the Hamiltonian.

In the following we will confine our attention entirely to the Heisenberg picture and suppress the index H. The space and momentum operators, X and P respectively, will then satisfy

$$i\hbar\frac{dX}{dt}=[X,H], \tag{3.30}$$

$$i\hbar\frac{dP}{dt}=[P,H]. \tag{3.31}$$

We see, at once, that if an operator commutes with H, i.e.,

$$[A,H]=0 \tag{3.32}$$

then A is an "invariant." It stays constant as a function of time and is called the constant of the motion. Since H is given by

$$H=\frac{P^2}{2m}+V(X), \tag{3.33}$$

the evaluation of the right-hand side above proceeds by the application of the fundamental commutator $[X,P]$ (2.20).

The Heisenberg picture plays an important role in many problems, including the harmonic oscillator problem (Chapter 9) among other topics, and in quantum field theory (Chapter 43).

3.3 Interaction picture

The interaction picture accommodates certain aspects of both the Schrödinger and Heisenberg representations and is used most often when the interaction Hamiltonian, representing the potential, depends on time. Consider the following Hamiltonian:

$$H=H_0+H_I(t) \tag{3.34}$$

where $H_0(=p^2/2m)$ represents the kinetic energy and is independent of time, while H_I corresponds to the interaction potential, which now depends on time. We have thus replaced $V(X)$ which we used in the previous two pictures by $H_I(t)$.

The time evolution operator $U(t,t_0)$ satisfies the equation

$$\frac{\partial U(t,t_0)}{\partial t}=-i\frac{H}{\hbar}U(t,t_0)=-i\frac{H_0}{\hbar}U(t,t_0)-i\frac{H_I}{\hbar}U(t,t_0). \tag{3.35}$$

We introduce an operator $U_I(t,t_0)$, which is defined as

$$U_I(t,t_0)=e^{iH_0/\hbar(t-t_0)}U(t,t_0). \tag{3.36}$$

Since it is a product of two unitary operators, $U_I(t, t_0)$ is also a unitary operator. Taking the time derivative of both sides we obtain

$$\frac{\partial U_I(t, t_0)}{\partial t} = e^{iH_0/\hbar(t-t_0)} \left(i\frac{H_0}{\hbar} \right) U(t, t_0) + e^{i\frac{H_0}{\hbar}(t-t_0)} \frac{\partial U(t, t_0)}{\partial t}. \tag{3.37}$$

Using (3.35) and (3.36) we obtain

$$\frac{\partial U_I(t, t_0)}{\partial t} = -\frac{i}{\hbar} e^{i\frac{H_0(t-t_0)}{\hbar}} H_I(t) e^{-i\frac{H_0(t-t_0)}{\hbar}} U_I(t, t_0). \tag{3.38}$$

We now define

$$H_I'(t) = e^{i\frac{H_0(t-t_0)}{\hbar}} H_I(t) e^{-i\frac{H_0(t-t_0)}{\hbar}}. \tag{3.39}$$

The equation for U_I now reads

$$\frac{\partial U_I(t, t_0)}{\partial t} = -\frac{iH_I'(t)}{\hbar} U_I(t, t_0). \tag{3.40}$$

We note from (3.36) that, since $U(t_0, t_0) = \mathbf{1}$, we must have $U_I(t_0, t_0) = \mathbf{1}$. Integrating both sides of (3.40) we obtain

$$U_I(t, t_0) = \mathbf{1} - \frac{i}{\hbar} \int_{t_0}^{t} dt' \, H_I'(t') \, U_I(t', t_0). \tag{3.41}$$

Through recursion of (3.41) we obtain the following series expansion.

$$U_I(t, t_0) = \mathbf{1} - \frac{i}{\hbar} \int_{t_0}^{t} dt' \, H_I'(t') + \left(-\frac{i}{\hbar} \right)^2 \int_{t_0}^{t} dt' \, H_I(t') \int_{t_0}^{t'} dt'' \, H_I(t'') + \cdots. \tag{3.42}$$

Since the order of integration is unimportant, we can write

$$\int_{t_0}^{t} dt' \int_{t_0}^{t'} dt'' H_I'(t') H_I'(t'') = \int_{t_0}^{t} dt'' \int_{t_0}^{t''} dt' \, H_I'(t'') H_I'(t'). \tag{3.43}$$

Hence

$$\int_{t_0}^{t} dt' \int_{t_0}^{t'} dt'' \, H_I'(t') H_I'(t'')$$
$$= \frac{1}{2} \left[\int_{t_0}^{t} dt' \int_{t_0}^{t'} dt'' \, H_I'(t') H_I'(t'') + \int_{t_0}^{t} dt'' \int_{t_0}^{t''} dt' \, H_I'(t'') H_I'(t') \right]. \tag{3.44}$$

To obtain a simple expression for U_I, let us define the "time-ordered product" of two operators $A(t')$ and $B(t'')$ as

$$T\left[A(t')B(t'')\right] = \theta\left(t' - t''\right) A(t')B(t'') + \theta\left(t'' - t'\right) B(t'')A(t') \tag{3.45}$$

where the θ-function is defined by

$$\theta(t_1 - t_2) = 0, \quad t_1 < t_2 \tag{3.46}$$

$$= 1, \quad t_1 > t_2. \tag{3.47}$$

This function is also called a "step-function." The relation (3.43) can then be written as

$$\int_{t_0}^{t} dt' \int_{t_0}^{t'} dt'' H_I'(t') H_I'(t'') = \frac{1}{2} \int_{t_0}^{t} dt' \int_{t_0}^{t} dt'' T\left(H_I'(t') H_I'(t'')\right) \tag{3.48}$$

where, following the definition (3.45),

$$T\left(H_I'(t') H_I'(t'')\right) = \theta\left(t' - t''\right) H_I'(t') H_I'(t'') + \theta\left(t'' - t'\right) H_I'(t'') H_I'(t'). \tag{3.49}$$

We note that the upper limits in the double integral are now the same, which will make it very convenient to write the series.

Even though it is somewhat complicated, one can define the time-ordered product when a product of more than two H_I's is involved. We will not pursue this matter further. For now we note that (3.42) leads to

$$U_I(t, t_0) = \mathbf{1} + \sum_{n=1}^{\infty} \left(\frac{-i}{\hbar}\right)^n \frac{1}{n!} \int_{t_0}^{t} dt_1 \int_{t_0}^{t} dt_2 \cdots \int_{t_0}^{t} dt_n\, T\left(H_I'(t_1) H_I'(t_2) ... H_I'(t_n)\right). \tag{3.50}$$

One can write this more compactly as

$$U_I(t, t_0) = T\left(\exp\left[-i \int_{t_0}^{t} dt\, H_I'(t')\right]\right) \tag{3.51}$$

since the series in (3.50) is the same as the exponential series. In contrast, in the Schrödinger picture where the Hamiltonian H is independent of time, the time evolution operator is

$$U(t, t_0) = e^{-i\frac{H}{\hbar}(t-t_0)}. \tag{3.52}$$

We write $|\alpha_I(t)\rangle$ as a state vector in the interaction picture, which we normalize by assuming the following relation at $t = 0$ involving the state vectors in the Schrödinger and Heisenberg pictures,

$$|\alpha_I(0)\rangle = |\alpha_H\rangle = |\alpha_S(0)\rangle. \tag{3.53}$$

To obtain the time dependence of $|\alpha_I(t)\rangle$ we employ the unitary operator $U_I(t, t_0)$. We take $t_0 = 0$ to conform to the initial conditions, and define $U(t, 0) = U(t)$ and $U_I(t, 0) = U_I(t)$. From (3.36), and (3.53) we obtain

$$|\alpha_S(t)\rangle = U(t)|\alpha_S(0)\rangle = e^{-i\frac{H_0 t}{\hbar}} U_I(t)|\alpha_S(0)\rangle = e^{-i\frac{H_0 t}{\hbar}} U_I(t)|\alpha_I(0)\rangle. \tag{3.54}$$

Since the matrix elements of an operator in different pictures must be the same, we have

$$\langle \alpha_S(t)|A_S|\beta_S(t)\rangle = \langle \alpha_I(t)|A_I(t)|\beta_I(t)\rangle. \tag{3.55}$$

Using (3.54) we obtain

$$\langle \alpha_S(t)|A_S|\beta_S(t)\rangle = \langle \alpha_I(0)|\, U_I^\dagger e^{i\frac{H_0 t}{\hbar}} A_S e^{-i\frac{H_0 t}{\hbar}} U_I\, |\beta_I(0)\rangle. \tag{3.56}$$

The definitions of $|\alpha_I(t)\rangle$ and $A_I(t)$ follow quite simply as

$$|\alpha_I(t)\rangle = U_I(t)\,|\alpha_I(0)\rangle \tag{3.57}$$

and

$$A_I(t) = U_I^\dagger e^{i\frac{H_0 t}{\hbar}} A_S e^{-i\frac{H_0 t}{\hbar}} U_I. \tag{3.58}$$

The relations (3.58) when substituted in (3.56) reproduce (3.55).

To obtain the time dependence of $|\alpha_I(t)\rangle$ we note from the relation of $|\alpha_S(t)\rangle$ given by (3.54) that

$$|\alpha_S(0)\rangle = U_I^\dagger e^{i\frac{H_0 t}{\hbar}}\,|\alpha_S(t)\rangle. \tag{3.59}$$

Substituting this in (3.53) we obtain

$$|\alpha_I(t)\rangle = e^{i\frac{H_0 t}{\hbar}}\,|\alpha_S(t)\rangle. \tag{3.60}$$

Taking the derivatives of both sides of (3.60) we obtain

$$\begin{aligned}
i\hbar\frac{\partial|\alpha_I(t)\rangle}{\partial t} &= -e^{i\frac{H_0 t}{\hbar}} H_0\,|\alpha_S(t)\rangle + e^{i\frac{H_0 t}{\hbar}}\, i\hbar\frac{\partial}{\partial t}\,|\alpha_S(t)\rangle \\
&= -e^{i\frac{H_0 t}{\hbar}} H_0\,|\alpha_S(t)\rangle + e^{i\frac{H_0 t}{\hbar}}\,(H_0 + H_I(t))\,|\alpha_S(t)\rangle \\
&= e^{iH_0 t/\hbar} H_I(t)\,|\alpha_S(t)\rangle \\
&= e^{iH_0 t/\hbar} H_I(t)\, e^{-iH_0 t/\hbar} e^{iH_0 t/\hbar}\,|\alpha_S(t)\rangle.
\end{aligned} \tag{3.61}$$

Hence

$$i\hbar\frac{\partial|\alpha_I(t)\rangle}{\partial t} = H_I'(t)\,|\alpha_I(t)\rangle. \tag{3.62}$$

We note that $H_I(t)$ and $H_I'(t)$ are operators in the Schrödinger representation:

$$H_I(t) = (H_I(t))_S \quad \text{and} \quad H_I'(t) = (H_I(t))_S. \tag{3.63}$$

The time derivative of A_I can be calculated from (3.58). After certain mathematical steps similar to the case of A_H it is found to be

$$i\hbar\frac{dA_I(t)}{dt} = [A_I, H_0]. \tag{3.64}$$

Thus relations (3.62) and (3.64) show that, in the interaction picture, the time dependence of the state vectors is governed by the interaction Hamiltonian, while the time evolution of the operators is determined by the free Hamiltonian.

3.4 Superposition of time-dependent states and energy–time uncertainty relation

Let us assume that an arbitrary state $|\psi(t)\rangle$ can be expanded as a sum of an infinite number of energy eigenstates $|a_n(t)\rangle$. At $t = 0$ we write

$$|\psi(0)\rangle = \sum_n c_n |a_n(0)\rangle \tag{3.65}$$

where

$$\sum_n |c_n|^2 = 1. \tag{3.66}$$

The eigenstates $|a_n(t)\rangle$ are expressed in terms of their eigenvalues E_n as

$$|a_n(t)\rangle = |a_n(0)\rangle \exp(-iE_n t/\hbar). \tag{3.67}$$

Hence

$$|\psi(t)\rangle = \sum_n c_n |a_n(0)\rangle \exp(-iE_n t/\hbar). \tag{3.68}$$

Let us now consider the product $\langle\psi(0)\,|\psi(t)\rangle$ to determine the evolution of $|\psi(t)\rangle$ with respect to $|\psi(0)\rangle$. We find

$$\langle\psi(0)\,|\psi(t)\rangle = \sum_n |c_n|^2 \exp(-iE_n t/\hbar). \tag{3.69}$$

We consider this sum for the case of continuous or near continuous values of the energy eigenstates. We note that since $\Delta n = 1$ we can convert the above sum as follows:

$$\sum_n \to \sum_n \Delta n = \sum_n \frac{\Delta n}{\Delta E}\Delta E \to \int dE\, \rho(E) \tag{3.70}$$

where $\rho(E)$ $(= dn/dE)$ is the density of states. We obtain

$$\langle\psi(0)\,|\psi(t)\rangle = \int dE\rho(E)\,|c(E)|^2 \exp(-iEt/\hbar) = \int dE\, g(E) \exp(-iEt/\hbar) \tag{3.71}$$

where $c(E)$ replaces c_n in the continuum limit and

$$g(E) = \rho(E)\,|c(E)|^2\,, \tag{3.72}$$

which is normalized as

$$\int dE\, g(E) = 1. \tag{3.73}$$

For the purposes of illustration and to simplify our discussion, we assume $g(E)$ to be peaked at $E = E_0$ and write

$$g(E) = g_0 \exp\left[-(E - E_0)^2 / 4(\Delta E)^2\right] \tag{3.74}$$

where ΔE corresponds to the width of the peak. We then write

$$\langle \psi(0) | \psi(t) \rangle = \exp(-iE_0 t/\hbar) g_0 \int dE \exp\left[-(E - E_0)^2 / 4(\Delta E)^2\right] \exp(-i(E - E_0)t/\hbar). \tag{3.75}$$

To carry out the integration we make the change of variables

$$(E - E_0) = y \tag{3.76}$$

and take

$$A = \frac{1}{4(\Delta E)^2} \quad \text{and} \quad B = -i\frac{t}{2\hbar}. \tag{3.77}$$

Therefore, the term in the exponent is given by

$$-\frac{(E - E_0)^2}{4(\Delta E)^2} - i\frac{(E - E_0)t}{\hbar} = -Ay^2 + 2By \tag{3.78}$$

and the integral can be written as

$$\int dy \exp\left[-Ay^2 + 2By\right] = \exp\left[B^2/A\right] \int du \exp[-Au^2] \tag{3.79}$$

where

$$u = y - \frac{B}{A}. \tag{3.80}$$

The integral can be obtained quite simply. It is given by

$$\int_{-\infty}^{\infty} du \exp[-Au^2] = \sqrt{\frac{\pi}{A}}. \tag{3.81}$$

Hence, using the normalization condition (3.75) we find

$$\langle \psi(0) | \psi(t) \rangle = \exp(-iE_0 t/\hbar) e^{-(\Delta E)^2 t^2/\hbar^2}. \tag{3.82}$$

This expression equals 1 at $t = 0$, but for $t \gtrsim \hbar/\Delta E$ it will start differing appreciably from 1. Thus the state $|\psi(t)\rangle$ will retain its original form only for relatively short times of order

$$\Delta t \sim \hbar/\Delta E, \tag{3.83}$$

beyond which it will deteriorate and eventually go to zero.

The two extremes of this result are interesting to consider. One extreme is when the state is an energy eigenstate,

$$|\psi(t)\rangle = |\psi(0)\rangle\, e^{-iE_0 t/\hbar}. \tag{3.84}$$

This corresponds to taking the limit $\Delta E = 0$ in $g(E)$ in (3.74), which leads to

$$\lim_{\Delta E \to 0} e^{[-(E-E_0)^2/4(\Delta E)^2]} \sim \delta(E - E_0) \tag{3.85}$$

and, therefore, only one term, $E = E_0$, will contribute in the integral. The state vector $|\psi(t)\rangle$ will retain its original form indefinitely (i.e., $\Delta t = \infty$). The other extreme is when $g(E) \approx 1$, which corresponds to $\Delta E = \infty$. Here all the states contribute equally to the integral. From (3.74) it implies that the state $|\psi(t)\rangle$ will disappear immediately after $t = 0$ (i.e., $\Delta t = 0$).

The above results correspond to the time–energy uncertainty relation

$$\Delta t \Delta E \sim \hbar. \tag{3.86}$$

We point out that the origin of this result is very different from the uncertainty relation $\Delta x \Delta p \sim \hbar$, which follows from the fundamental commutator $[X, P] = i\hbar$. No such commutator exists involving time and the Hamiltonian.

3.4.1 Virtual particles

The energy–time relation (3.86) tells us that energy conservation can be violated by an amount ΔE if such a violation occurs over a time

$$\Delta t \sim \frac{\hbar}{\Delta E}. \tag{3.87}$$

Particle A can, therefore, emit a particle B even though the energy–momentum of the initial state (particle A) and the final state (particles A and B) are not the same. Particle B will not act as a free particle but what one calls a "virtual" particle, which appears and disappears within a short time (e.g., A emits B and then re-absorbs it: $A \to A + B \to A$).

This remarkable result led Yukawa to propose that in neutron (n)–proton (p) interactions, called "strong" interactions, the forces between the two particles are due to the exchange of virtual particles. If r_0 is the range of interactions, then assuming the particle travels with the speed of light, c, one estimates $\Delta t \sim r_0/c$ and, therefore, from (3.87),

$$\Delta E \sim \hbar c/r_0\,. \tag{3.88}$$

The experimental value of r_0 is found to be 10^{-13}cm. Putting this value in (3.88) and equating ΔE to the mass of the exchanged particle (using the relativistic formula $E = mc^2$), one finds that the mass of this particle must be 140 MeV. It was called the π-meson and was subsequently discovered. Such exchange processes are nicely described by diagrams called Feynman diagrams. We will return to this subject in Chapter 43 involving relativistic quantum mechanics, where we discuss the particle interpretation of Yukawa and Coulomb potentials in scattering processes.

3.5 Time dependence of the density operator

Having considered the time dependence of the state vectors, let us obtain the time dependence of the density operator. We will confine ourselves to just two states $|\alpha\rangle$ and $|\beta\rangle$ with the corresponding projection operators given by

$$P_\alpha = |\alpha\rangle\langle\alpha| \quad \text{and} \quad P_\beta = |\beta\rangle\langle\beta|. \tag{3.89}$$

The density operator is then

$$\rho = w_\alpha P_\alpha + w_\beta P_\beta. \tag{3.90}$$

For a Hamiltonian H the time evolution equation for the state vectors is given by

$$i\hbar\frac{d|\alpha\rangle}{dt} = H|\alpha\rangle \quad \text{and} \quad i\hbar\frac{d|\beta\rangle}{dt} = H|\beta\rangle. \tag{3.91}$$

Thus,

$$i\hbar\frac{dP_\alpha}{dt} = i\hbar\left[\left(\frac{d|\alpha\rangle}{dt}\right)\langle\alpha| + |\alpha\rangle\left(\frac{d\langle\alpha|}{dt}\right)\right] = [H, P_\alpha] \tag{3.92}$$

where we have used the Hermitian conjugate relation

$$-i\hbar\left(\frac{d\langle\alpha|}{dt}\right) = H\langle\alpha|. \tag{3.93}$$

Similarly,

$$i\hbar\frac{dP_\beta}{dt} = \left[H, P_\beta\right]. \tag{3.94}$$

Hence,

$$i\hbar\frac{d\rho}{dt} = [H, \rho] = -[\rho, H]. \tag{3.95}$$

This relation, we note, has exactly the opposite sign to the Heisenberg relation obtained in this chapter. There is actually no contradiction because ρ is not an Heisenberg operator since it is made up of states that evolve in time according to the Schrödinger picture.

3.6 Probability conservation

Consider the three-dimensional Schrödinger equation for $\phi(\mathbf{r}, \mathbf{t})$, which we express as

$$i\hbar\frac{\partial\phi}{\partial t}+\frac{\hbar^2}{2m}\nabla^2\phi - V\phi = 0. \tag{3.96}$$

Taking the complex conjugate of the above equation we obtain

$$-i\hbar\frac{\partial\phi^*}{\partial t}+\frac{\hbar^2}{2m}\nabla^2\phi^* - V\phi^* = 0, \tag{3.97}$$

where as stated before, V is only a function of the magnitude $r = |\mathbf{r}|$ and is real. Multiply (3.96) by ϕ^* on the left and (3.97) by ϕ on the right and subtract:

$$i\hbar\left[\phi^*\left(\frac{\partial\phi}{\partial t}\right)+\left(\frac{\partial\phi^*}{\partial t}\right)\phi\right]+\frac{\hbar^2}{2m}\left[\phi^*\left(\nabla^2\phi\right)-\left(\nabla^2\phi^*\right)\phi\right]=0. \tag{3.98}$$

If we define

$$\rho(\mathbf{r},t)=\phi^*\phi \quad \text{and} \quad \mathbf{j}(\mathbf{r},t)=\frac{\hbar}{2im}\left[\phi^*(\nabla\phi)-\left(\nabla\phi^*\right)\phi\right] \tag{3.99}$$

where ρ is called the probability density and $\mathbf{j}$ the probability current density, then the first term in (3.98) is proportional to $\partial\rho/\partial t$. The square bracket in the second term in (3.98) can be written as

$$\phi^*\left(\nabla^2\phi\right)-\left(\nabla^2\phi^*\right)\phi=\nabla\cdot\left[\phi^*(\nabla\phi)-\left(\nabla\phi^*\right)\phi\right] \tag{3.100}$$

where we note that the cross terms $\nabla\phi\cdot\nabla\phi^*$ in (3.100) cancel. The term on the right-hand side in (3.100) is proportional to $\nabla.\mathbf{j}$. The relation (3.98) can then be expressed as

$$\frac{\partial\rho}{\partial t}+\nabla\cdot\mathbf{j}=0. \tag{3.101}$$

This is the classic probability conservation relation. We elaborate on this below.

Let the total probability in a volume V be written as

$$Q(t)=\int_V d^3r\,\rho(\mathbf{r},t) \tag{3.102}$$

where the integral is over a volume V. Taking the time derivative of both sides we get

$$\frac{dQ}{dt}=-\int_V d^3r\,\nabla\cdot\mathbf{j} \tag{3.103}$$

where we have used (3.101). Using Gauss's theorem to convert the volume integral to a surface integral, we obtain the following from (3.103):

$$\frac{dQ}{dt}=-\oint\mathbf{j}\cdot d\mathbf{S} \tag{3.104}$$

where $\mathbf{dS}$ is an element of the boundary surface enclosing the volume V and the integral is around the closed surface. Using the language of electromagnetism, with ρ as the charge density and $\mathbf{j}$ the current density, the above relation is simply a statement of an obvious fact: the current leaving the volume by crossing its surface must result in the decrease of the total charge inside the volume. In other words, (3.104) reflects the fact that everything is accounted for, nothing simply disappears. In the context of quantum mechanics, one can substitute the current density by probability current density and charge density by probability density. In spite of the fact that this relation looks obvious, it has profound consequences, among them the so called unitarity of the S-matrix in scattering theory among many other things.

Let us consider a situation in which ϕ_n is an eigenstate of energy given by

$$\phi_n(\mathbf{r},t) = u_n(\mathbf{r})\, e^{-i\frac{E_n t}{\hbar}}. \tag{3.105}$$

This relation implies that $\rho(= \phi_n^*\phi_n)$ is independent of time. The relation (3.101) then says that

$$\nabla \cdot \mathbf{j} = 0, \quad \text{i.e. } \mathbf{j} = \text{continuous}. \tag{3.106}$$

Hence in solving eigenvalue problems the continuity of $\mathbf{j}$ must be an essential input. For example, considering one dimension for the moment, if the potential $V(x)$ changes from one region in the x-space to another, then across the boundary of the two regions $\mathbf{j}$ must be continuous. This can be accomplished by having $d\phi_n/dx$ continuous, or $\phi_n = 0$ on both sides of the boundary. In addition, of course, ϕ_n must be continuous since probability can not be different on two sides of the boundary. Boundary value problems such as these will be discussed in the following chapters.

3.7 Ehrenfest's theorem

At this stage it is of interest to ask whether there are circumstances in which one can recover the classical equations of motion from quantum-mechanical equations. For this purpose let us look at the behavior of the expectation values of the operators.

Consider the following expectation value of the X-operator written with respect to a wavefunction $\phi(\mathbf{r})$,

$$\langle x \rangle = \int d^3r\, \phi^* x \phi. \tag{3.107}$$

Taking the time-derivative we obtain

$$\frac{d}{dt}\langle x \rangle = \int d^3r \left(\frac{\partial \phi^*}{\partial t}\right) x\phi + \int d^3r\, \phi^* x \left(\frac{\partial \phi}{\partial t}\right). \tag{3.108}$$

From the Schrödinger equation we have

$$i\hbar\frac{\partial\phi}{\partial t} = -\frac{\hbar^2}{2m}\nabla^2\phi + V\phi \tag{3.109}$$

and its complex conjugate

$$-i\hbar\frac{\partial\phi^*}{\partial t} = -\frac{\hbar^2}{2m}\nabla^2\phi^* + V\phi^*. \tag{3.110}$$

Substituting these in (3.108) we obtain

$$\frac{d}{dt}\langle x\rangle = \frac{i\hbar}{2m}\int d^3r\left[\phi^* x\left(\nabla^2\phi\right) - \left(\nabla^2\phi^*\right)x\phi\right]. \tag{3.111}$$

By partial integration one can show that

$$\int d^3r\left(\nabla^2\phi^*\right)x\phi = \int d^3r\,\phi^*\left(\nabla^2 x\phi\right) \tag{3.112}$$

where we assume that the wavefunction vanishes at infinity. Hence the right-hand side of (3.111) can be written as

$$\int d^3r\left[\phi^* x\left(\nabla^2\phi\right) - \left(\nabla^2\phi^*\right)x\phi\right] = \int d^3r\left[\phi^* x\left(\nabla^2\phi\right) - \phi^*\left(\nabla^2 x\phi\right)\right]. \tag{3.113}$$

To simplify the right-hand side of the above equation, we note that

$$\nabla^2(x\phi) = 2\frac{\partial\phi}{\partial x} + x\nabla^2\phi. \tag{3.114}$$

Hence, the right-hand side of (3.113) is simply

$$-2\int d^3r\,\phi^*\left(\frac{\partial\phi}{\partial x}\right). \tag{3.115}$$

Combining the results from (3.112), through (3.115) we obtain

$$\frac{d}{dt}\langle x\rangle = \frac{1}{m}\int d^3r\,\phi^*\left(-i\hbar\frac{\partial}{\partial x}\right)\phi = \frac{1}{m}\langle p_x\rangle \tag{3.116}$$

where for the last equality in (3.116) we have used

$$-i\hbar\frac{\partial}{\partial x} = p_x. \tag{3.117}$$

Hence we find

$$\frac{d}{dt}\langle x\rangle = \frac{1}{m}\langle p_x\rangle, \tag{3.118}$$

which is the x-component of the classical equation of motion

$$\frac{d\mathbf{r}}{dt} = \frac{\mathbf{p}}{m} \tag{3.119}$$

where the x-components of $\mathbf{r}$ and $\mathbf{p}$ stand for the expectation values, $\langle x \rangle$ and $\langle p_x \rangle$ respectively. Similarly, we can derive the classical equations for $\langle y \rangle$ and $\langle z \rangle$.

We now consider the time derivative of the expectation value $\langle p_x \rangle$

$$\begin{aligned}\frac{d}{dt}\langle p_x \rangle &= \frac{d}{dt}\int d^3r\phi^* \left(-i\hbar\frac{\partial}{\partial x}\right)\phi \\ &= -i\hbar\int d^3r\left[\frac{\partial\phi^*}{\partial t}\frac{\partial\phi}{\partial x} + \phi^*\frac{\partial}{\partial x}\left(\frac{\partial\phi}{\partial t}\right)\right].\end{aligned} \tag{3.120}$$

Using the Schrödinger equations (3.109) and (3.110) and the relation

$$\int d^3r\left[\phi^*\frac{\partial}{\partial x}\left(\nabla^2\phi\right)\right] = \int d^3r\left(\nabla^2\phi^*\right)\frac{\partial\phi}{\partial x}, \tag{3.121}$$

which is obtained through integration by parts, we obtain

$$\frac{d}{dt}\langle p_x \rangle = -\int d^3r\,\phi^*\left[\frac{\partial}{\partial x}(V\phi) - V\frac{\partial\phi}{\partial x}\right]. \tag{3.122}$$

Hence

$$\frac{d}{dt}\langle p_x \rangle = \int d^3r\,\phi^*\left(-\frac{\partial V}{\partial x}\right)\phi = \left\langle -\frac{\partial V}{\partial x}\right\rangle, \tag{3.123}$$

which is the x-component of the classical equation

$$\frac{d\mathbf{p}}{dt} = \mathbf{F} = -\,\nabla\mathbf{V} \tag{3.124}$$

where the x-components of $\mathbf{p}$ and $\nabla\mathbf{V}$ stand for $\langle p_x \rangle$ and $-\langle \partial V/\partial x \rangle$ respectively. Similar results are found for $\langle p_y \rangle$ and $\langle p_z \rangle$.

We conclude then that even though we are in the quantum domain, the expectation values will still follow classical dynamics.

3.8 Problems

1. Consider the Hamiltonian as a sum of kinetic energy (T) and potential energy (V)

$$H = \frac{\mathbf{p}^2}{2m} + V(\mathbf{r}) = T + V.$$

Assuming $\langle \mathbf{r} \cdot \mathbf{p} \rangle$ to be time-independent show that

$$2\langle T \rangle = \langle \mathbf{r} \cdot \nabla V \rangle .$$

This is the quantum-mechanical version of the virial theorem.

2. From the virial theorem determine the average kinetic energy of a particle moving in a potential given by

$$V(r) = \lambda \ln(r/a).$$

3. Express x in the Schrödinger representation as an operator x_H in the Heisenberg representation for the case of a free particle with the Hamiltonian

$$H = \frac{\mathbf{p}^2}{2m}.$$

Carry out the expansion of the exponentials in terms of the corresponding Hamiltonian. Consider also the case where the potential has the form $V(x) = \lambda x^n$, where n is an integer.

4. If $|\psi(\lambda)\rangle$ is a normalized eigenstate of Hamiltonian $H(\lambda)$ with eigenvalue $E(\lambda)$ where λ is a continuous parameter, then show that

$$\frac{\partial E}{\partial \lambda} = \left\langle \psi(\lambda) \left| \frac{\partial H}{\partial \lambda} \right| \psi(\lambda) \right\rangle .$$

This is called the Feynman–Hellman relation.

5. Let $|\phi_1\rangle$ and $|\phi_2\rangle$ be the eigenstates of the Hamiltonian H with eigenvalues E_1 and E_2 respectively. And let $|\chi_1\rangle$ and $|\chi_2\rangle$ be the eigenstates of an operator A, which does not commute with H, with eigenvalues a_1 and a_2. The two sets of eigenstates are related to each other as follows:

$$|\chi_1\rangle = \frac{1}{\sqrt{2}}(|\phi_1\rangle + |\phi_2\rangle),$$

$$|\chi_2\rangle = \frac{1}{\sqrt{2}}(|\phi_1\rangle - |\phi_2\rangle).$$

If at time t the system is described by a state $|\psi(t)\rangle$ such that

$$|\psi(0)\rangle = |\chi_1\rangle$$

then determine $|\psi(t)\rangle$ at arbitrary times in terms of the eigenstates $|\phi_1\rangle$ and $|\phi_2\rangle$.

6. Consider the Schrödinger equation for a particle of mass m with a Coulomb-like potential $-\lambda/r$ and with angular momentum $l = 0$. If the eigenfunction is given by

$$u(r) = u_0 e^{-\beta r}$$

where u_0 and β are constants, then determine u_0, β and the energy eigenvalue E in terms of λ and m.

7. A particle of charge e is subjected to a uniform electric field $\mathbf{E}_0$. Show that the expectation value of the position operator $\langle \mathbf{r} \rangle$ satisfies

$$m\frac{d^2 \langle \mathbf{r} \rangle}{dt^2} = e\mathbf{E}_0.$$

8. Let $|E_n\rangle$ s denote the eigenstates of the Hamiltonian, H. If $C = AB$ and $A = [B, H]$, then obtain the matrix element, $\langle E_m | C | E_n \rangle$, in terms of the matrix elements of A.
9. The states $|a_i\rangle$ are eigenstates of an operator A and U is a unitary operator that changes the basis from $|a_i\rangle$ to $|b_i\rangle$. Determine, in terms of U and A, the operator for which $|b_i\rangle$ are the eigenstates. Also determine the eigenvalues.
10. Show that

$$\sum_{n \neq 0} \frac{\langle 0 | A | n \rangle \langle n | A | 0 \rangle}{(E_n - E_0)} = \langle 0 | AF | 0 \rangle$$

where $H|n\rangle = E_n |n\rangle$ for $n = 0, 1, 2, \ldots$ and where $A = [H, F]$ and $\langle 0 | F | 0 \rangle = 0$.
11. Show that if $\mathbf{L}$ is the angular momentum operator then $[\mathbf{L}, H] = 0$ where $H = \mathbf{p}^2/2m + V(r)$.

4 Free particles

After discussing the basic definitions and formalism in quantum mechanics, it is now time for dynamical calculations. We will do perhaps the easiest and yet very informative case of free particles. What we learned in the previous chapters will now be put into play. We will obtain the free particle wavefunction in one dimension and in three dimensions. For the latter case we will consider both the Cartesian and spherical coordinate systems and finally, in the spherical system, we will introduce the angular momentum operator and spherical harmonics.

4.1 Free particle in one dimension

In the absence of any forces, that is, with the potential $V = 0$, the Hamiltonian for a particle of mass m is given entirely by the kinetic energy

$$H = \frac{P^2}{2m} \tag{4.1}$$

where P is the momentum operator. In this section we will be discussing the motion of the free particle only in one dimension, the x-direction. Thus P is just the x-component, P_x, given by the operator relation

$$P_x = -i\hbar \frac{\partial}{\partial x}. \tag{4.2}$$

Let $u_p(x)$ be the momentum eigenfunction in the x-space ($= \langle x | p \rangle$). We note that momentum is a vector and, therefore, has direction. The momentum eigenfunction for a particle moving along the positive x-direction with eigenvalue p satisfies the equation

$$-i\hbar \frac{\partial u_p(x)}{\partial x} = p u_p(x). \tag{4.3}$$

The solution of this equation is given by

$$u_p(x) = C_1 \exp\left(\frac{ip}{\hbar} x\right). \tag{4.4}$$

The eigenfunction for a particle moving along the negative direction with momentum p satisfies the equation

$$-i\hbar\frac{\partial u_p(x)}{\partial x} = (-p)\, u_p(x), \tag{4.5}$$

whose solution is

$$u_p(x) = C_2 \exp\left(-\frac{ip}{\hbar}x\right). \tag{4.6}$$

The constants C_1 and C_2 will be determined once we decide how we will normalize the eigenfunctions.

The eigenstates of the Hamiltonian operator are the energy eigenfunctions with eigenvalues, E, that are simply related, through (4.1), to the momentum eigenvalues, p, by

$$E = \frac{p^2}{2m}. \tag{4.7}$$

We note that the energy is positive ($E \geq 0$) since we are considering free particles. The wavefunction $\phi_E(x,t)$ corresponding to the eigenstate of H satisfies the Schrödinger equation

$$H\phi_E(x,t) = i\hbar\frac{\partial}{\partial t}\phi_E(x,t) = E\phi_E(x,t). \tag{4.8}$$

Following the derivations in the Schrödinger picture, the wavefunction can be written as

$$\phi_E(x,t) = u_E(x) \exp\left(\frac{-iEt)}{\hbar}\right) \tag{4.9}$$

where, from (4.1), $u_E(x)$ satisfies the equation

$$-\frac{\hbar^2}{2m}\frac{d^2u_E(x)}{dx^2} = Eu_E(x) = \frac{p^2}{2m}u_E(x) \tag{4.10}$$

and where we have used (4.7) to express E in terms of p. The solution of this equation is given by

$$u_E(x) = A\, \exp\left(\frac{ip}{\hbar}x\right) + B\, \exp\left(-\frac{ip}{\hbar}x\right) \tag{4.11}$$

where $p = \sqrt{2mE}$. Thus the energy eigenfunction $u_E(x)$ given by (4.11) is a linear combination of two momentum eigenfunctions. The (+) sign in the exponent of the first term in (4.11) corresponds to the momentum eigenfunction for a particle moving in the positive x-direction, while the (−) sign signifies that the particle is moving in the opposite direction, in each case with the magnitude of the momentum given by $p = \sqrt{2mE}$. In the following we will, for simplicity, remove the $\pm$ sign in the exponent in (4.11) and write $u_E(x)$ as a single exponent

$$u_E(x) = C \exp\left(\frac{ip}{\hbar}x\right) \tag{4.12}$$

allowing p, however, to take positive and negative values, $\pm\sqrt{2mE}$. This is the so-called "plane wave" solution for one-dimensional motion.

In the following we will not make any distinction between the energy and momentum eigenfunctions and will use the momentum eigenfunctions throughout.

4.2 Normalization

First, to facilitate the derivations we will write

$$p = \hbar k, \tag{4.13}$$

so that

$$\exp\left(\frac{ip}{\hbar}x\right) = \exp(ikx). \tag{4.14}$$

There are two standard prescriptions for normalization. The first corresponds to a finite space. Here one divides up the space into units of L, with $-L/2 < x < L/2$, in which the wavefunction is taken to be periodic,

$$u(x + L) = u(x), \tag{4.15}$$

which implies that

$$\exp[ik\,(x + L)] = \exp(ikx). \tag{4.16}$$

The eigenvalues of k are then discrete, satisfying the relation $Lk_n = 2n\pi$, where n is an integer. Thus we have

$$k_n = \frac{2n\pi}{L}, \quad E_n = \frac{2n^2\pi^2\hbar^2}{mL^2}, \quad n = 0, \pm 1, \pm 2, \ldots. \tag{4.17}$$

The momentum eigenfunction is

$$u_n(x) = C \exp(ik_n x), \quad -\frac{L}{2} < x < \frac{L}{2} \tag{4.18}$$

which is normalized in the traditional manner as follows:

$$\int_{-\frac{L}{2}}^{\frac{L}{2}} dx\, u_m^*(x) u_n(x) = \delta_{mn}. \tag{4.19}$$

The normalization constant is found to be $C = 1/\sqrt{L}$, and, hence,

$$u_n(x) = \frac{1}{\sqrt{L}} \exp(ik_n x), \quad -\frac{L}{2} < x < \frac{L}{2}. \tag{4.20}$$

The energy eigenfunction as a function of space and time is then given by

$$\phi_n(x,t) = u_n(x) \exp\left(\frac{-iE_n t}{\hbar}\right), \quad -\frac{L}{2} < x < \frac{L}{2} \tag{4.21}$$

where E_n is given in (4.17) and $u_n(x)$ is given in (4.20). When we go to three dimensions this type of normalization will be extended to the space divided up into units of cubes. This is then called "box normalization." We will use this term even when normalizing in one or two dimensions. In Chapter 1 it was shown that the eigenstates satisfy the completeness relation. The eigenstates u_n above must, therefore, also do the same. Hence,

$$\sum_n u_n(x) u_n^*(x') = \delta(x - x'). \tag{4.22}$$

The second method of normalization involves the entire infinite space $(-\infty, \infty)$ by taking $L \to \infty$. In order to do this we first start with the completeness relation (4.22) and insert expression (4.20) for $u_n(x)$. The left-hand side of (4.22) is then

$$\frac{1}{L}\sum_n \exp\left[ik_n\left(x - x'\right)\right]. \tag{4.23}$$

First we note that, since n runs over integers, the separation, Δn, between two adjacent values of n, is simply given by $\Delta n = 1$. Hence,

$$\frac{1}{L}\sum_n \exp\left[ik_n\left(x - x'\right)\right] = \frac{1}{L}\sum (\Delta n) \exp\left[ik_n\left(x - x'\right)\right] \tag{4.24}$$

$$= \frac{1}{L}\sum \frac{L\,(\Delta k_n)}{2\pi} \exp\left[k_n(x - x')\right] \tag{4.25}$$

$$= \frac{1}{2\pi}\sum (\Delta k_n) \exp\left[k_n(x - x')\right]. \tag{4.26}$$

With k_n given by (4.17), we find, therefore, that in the limit when L becomes large, the interval extends to $(-\infty, \infty)$ and Δk_n becomes infinitesimal. Therefore, k_n can be replaced by a continuous variable, which we designate as k in the interval $(-\infty, \infty)$. Expression (4.23), in the limit, $L \to \infty$, can then be written as

$$\frac{1}{2\pi}\int_{-\infty}^{\infty} dk e^{ik(x-x')} \tag{4.27}$$

where we have replaced Δk_n by dk and the sum over n by an integral. From the properties of the Dirac δ-function, the integral above, indeed, reproduces $\delta(x - x')$. Thus our results are consistent with the completeness relation (4.22). We will, therefore, write the wavefunction in $(-\infty, \infty)$ as

$$u_k(x) = \frac{1}{\sqrt{2\pi}} e^{ikx} \quad \text{with } p = \hbar k. \tag{4.28}$$

The completeness condition now reads

$$\int_{-\infty}^{\infty} dk\, u_k(x) u_k^*(x') = \delta(x - x'). \tag{4.29}$$

Thus the unit operator that appeared for the discrete case is replaced by a δ-function. The functions $u_k(x)$ also satisfy the orthogonality condition

$$\int_{-\infty}^{\infty} dx\, u_{k'}^*(x) u_k(x) = \frac{1}{2\pi} \int_{-\infty}^{\infty} dx\, e^{i(k-k')x} = \delta(k - k') \tag{4.30}$$

where instead of the Kronecker delta, which appeared for finite dimensions, we have Dirac's δ-function on the right-hand side.

Expression (4.28) for the momentum eigenfunction corresponds to the second way of normalization, which here is referred to as δ-function normalization. We need to point out that in contrast to the finite-dimensional case the wavefunctions here cannot be normalized to unity since for $k = k'$, the right-hand side of (4.30) becomes infinite. This is the price one has to pay by going from finite dimensions to infinite space.

We note also that the two relations (4.29) and (4.30) are, respectively, restatements of the relations we discussed in Chapter 1,

$$\langle x | x' \rangle = \delta(x - x'), \tag{4.31}$$

$$\langle k | k' \rangle = \delta(k - k'). \tag{4.32}$$

Since the momentum eigenfunction in the bra-ket notation is

$$\langle x | k \rangle = u_k(x) = \frac{1}{\sqrt{2\pi}} e^{ikx}, \tag{4.33}$$

the relation (4.29) is derived from (4.31) by inserting a complete set of states $|k\rangle$, while (4.30) is derived from (4.32) by inserting a complete set of states $|x\rangle$.

In summary, a particle normalized with the Dirac δ-function normalization is described by the wavefunction

$$\phi_k(x, t) = u_k(x) \exp\left(\frac{-iE_k t}{\hbar}\right), \quad -\infty < x < \infty \tag{4.34}$$

where $u_k(x)$ is given by (4.33) and $E_k = \hbar^2 k^2 / 2m$.

4.3 Momentum eigenfunctions and Fourier transforms

If $|\phi\rangle$ corresponds to an arbitrary state vector with its x-representation given by $\phi(x) = \langle x|\,\phi\rangle$, then we can write, using completeness,

$$\phi(x) = \int_{-\infty}^{\infty} dk\, \langle x|\,k\rangle\, \langle k|\,\phi\rangle = \int_{-\infty}^{\infty} dk \frac{1}{\sqrt{2\pi}} e^{ikx}\, \langle k|\,\phi\rangle = \frac{1}{\sqrt{2\pi}} \int_{-\infty}^{\infty} dk\, f(k) e^{ikx} \tag{4.35}$$

where $\langle k|\,\phi\rangle = f(k)$. This is just the Fourier transform expression with $f(k)$ being the transform of $\phi(x)$. It tells us that an arbitrary wavefunction can be expanded in terms of the continuous momentum eigenfunctions as the basis. Therefore, the probability of finding a particle, described by $\phi(x)$, with momentum k is $|f(k)|^2$.

One can obtain $f(k)$ through the inverse relation by multiplying both sides of (4.35) by $e^{-ik'x}$ and integrating over x:

$$\begin{aligned}
\int_{-\infty}^{\infty} dx e^{-ik'x} \phi(x) &= \frac{1}{\sqrt{2\pi}} \int_{-\infty}^{\infty} dx e^{-ik'x} \int_{-\infty}^{\infty} dk f(k) e^{ikx} \\
&= \frac{1}{\sqrt{2\pi}} \int_{-\infty}^{\infty} dk\, f(k) \int_{-\infty}^{\infty} dx\, e^{i(k-k')x} \\
&= \frac{1}{\sqrt{2\pi}} \int_{-\infty}^{\infty} dk\, f(k) 2\pi\, \delta(k-k') = \sqrt{2\pi} f(k')
\end{aligned} \tag{4.36}$$

which leads to the well-known inverse Fourier transform result

$$f(k) = \frac{1}{\sqrt{2\pi}} \int_{-\infty}^{\infty} dx \phi(x) e^{-ikx}.$$

Finally, if $\phi(x)$ is normalized,

$$\int_{-\infty}^{\infty} dx \phi^*(x) \phi(x) = 1, \tag{4.37}$$

then we can obtain a relation for $f(k)$ through (4.35). The left-hand side of (4.37) can be written as

$$\frac{1}{2\pi}\int_{-\infty}^{\infty} dx \int_{-\infty}^{\infty} dk f^*(k) e^{-ikx} \int_{-\infty}^{\infty} dk' f(k') e^{ik'x} = \frac{1}{2\pi}\int_{-\infty}^{\infty} dk \int_{-\infty}^{\infty} dk' f^*(k) f(k') \int_{-\infty}^{\infty} dx\, e^{i(k'-k)x}$$

$$= \int_{-\infty}^{\infty} dk \int_{-\infty}^{\infty} dk' f^*(k) f(k') \delta\left(k'-k\right)$$

$$= \int_{-\infty}^{\infty} dk \, |f(k)|^2 . \tag{4.38}$$

Hence, from (4.37),

$$\int_{-\infty}^{\infty} dk \, |f(k)|^2 = 1. \tag{4.39}$$

Thus, as expected, the probabilities add up to unity. Once again, we could derive this result quite simply by expressing the normalization condition (4.39) as $\langle \phi | \phi \rangle = 1$, and writing

$$1 = \langle \phi | \phi \rangle = \int_{-\infty}^{\infty} dk \, \langle \phi | k \rangle \, \langle k | \phi \rangle. \tag{4.40}$$

Since $\langle k | \phi \rangle = f(k)$, we obtain (4.39).

4.4 Minimum uncertainty wave packet

A free particle traveling along the positive x-axis with momentum p_0 $(= \hbar k_0)$ is described by a wavefunction, $\psi(x,t)$, which at $t = 0$ is given by

$$\psi(x,0) = \phi(x) = \frac{1}{\sqrt{2\pi}} e^{ik_0 x}. \tag{4.41}$$

The probability of finding this particle at a point x is

$$|\phi(x)|^2 = \text{constant} \tag{4.42}$$

for all values of x. This is, of course, a reflection of the uncertainty principle, $\Delta x \Delta p \geq \hbar/2$, with $p = p_0$, and $\Delta p = 0$, which says that if one tries to localize the momentum of a particle then its position cannot be localized and vice versa.

It is then interesting to examine a physical situation in which neither the momentum nor the position of a particle is localized very sharply but both vary within a region consistent with the uncertainty principle. A wavefunction describing such a situation, called a *wave packet*, will be closer to reality. Below we will consider some of the properties of such a wave packet consistent with the minimum uncertainty, $\Delta x \Delta p = \hbar/2$.

Let us now consider $\phi(x)$, which is not a momentum eigenfunction but rather a superposition of momentum eigenfunctions written as

$$\phi(x) = \frac{1}{\sqrt{2\pi}} \int_{-\infty}^{\infty} dk\, e^{ikx} f(k). \tag{4.43}$$

Instead of $f(k) = \delta(k - k_0)$, which corresponds to $\phi(x)$ having a definite momentum $\hbar k_0$, we consider another function which, however, is still centered at $k = k_0$ but not as sharply described as a δ-function. A natural choice is a Gaussian function of the form

$$f(k) = N e^{-\alpha(k-k_0)^2}. \tag{4.44}$$

We show below that this profile of $f(k)$ corresponds to a wave packet that is consistent with the minimum uncertainty product

$$(\Delta x)(\Delta k) = \frac{1}{2}. \tag{4.45}$$

First we note that the probability interpretation for $\phi(x)$ implies that

$$\int_{-\infty}^{\infty} dk\ |\phi(x)|^2 = 1. \tag{4.46}$$

As we have already verified, this leads to

$$\int_{-\infty}^{\infty} dk\ |f(k)|^2 = 1. \tag{4.47}$$

In order to do the integrals involved in carrying out our calculations, the following two results will be useful:

$$\int_{-\infty}^{\infty} dy\, e^{-\alpha y^2} = \sqrt{\frac{\pi}{\alpha}} \tag{4.48}$$

and,

$$\int_{-\infty}^{\infty} dy\, y^2 e^{-\alpha y^2} = \frac{1}{2}\frac{\sqrt{\pi}}{\alpha^{3/2}}. \tag{4.49}$$

Relation (4.49) is obtained by taking the derivative of (4.48) with respect to α.

Substituting (4.44) in (4.47) and using (4.48) we find

$$N = \left(\frac{2\alpha}{\pi}\right)^{1/4}. \tag{4.50}$$

Thus, we have

$$f(k) = \left(\frac{2\alpha}{\pi}\right)^{1/4} e^{-\alpha(k-k_0)^2}. \quad (4.51)$$

In the following we will continue using the functional form (4.44) for $f(k)$ without substituting the value of N given by (4.50).

The expectation value $\langle k \rangle$ is given by

$$\langle k \rangle = \int_{-\infty}^{\infty} dk\, k\, |f(k)|^2 = N^2 \int_{-\infty}^{\infty} dk\, k e^{-2k(k-k_0)^2}. \quad (4.52)$$

The integration can easily be performed by changing variables, $k - k_0 = q$. We find

$$\langle k \rangle = k_0. \quad (4.53)$$

Furthermore, by definition,

$$(\Delta k)^2 = \left\langle (k - \langle k \rangle)^2 \right\rangle = \left\langle k^2 \right\rangle - k_0^2 \quad (4.54)$$

where

$$\left\langle k^2 \right\rangle = N^2 \int_{-\infty}^{\infty} dk\, k^2 e^{-2\alpha(k-k_0)^2}. \quad (4.55)$$

Carrying out the integration one finds

$$(\Delta k)^2 = \frac{1}{4\alpha}. \quad (4.56)$$

Thus,

$$\alpha = \frac{1}{4(\Delta k)^2}. \quad (4.57)$$

Substituting this in (4.44) and (4.51) we obtain

$$f(k) = N \exp\left[-\frac{(k-k_0)^2}{4(\Delta k)^2}\right]. \quad (4.58)$$

To simplify things we take $k_0 = 0$ in (4.44) leaving N and α in place without substituting their values,

$$f(k) = N e^{-\alpha k^2}. \quad (4.59)$$

Therefore,

$$\phi(x) = \frac{N}{\sqrt{2\pi}} \int_{-\infty}^{\infty} dk\, e^{ikx} e^{-\alpha k^2}. \quad (4.60)$$

It is then easy to show that

$$\phi(x) = \frac{N}{\sqrt{2\alpha}} \exp\left[-\frac{x^2}{4\alpha}\right]. \tag{4.61}$$

In the same manner as we determined $(\Delta k)^2$ in (4.56), one can determine $(\Delta x)^2$, which is found to be

$$(\Delta x)^2 = \alpha. \tag{4.62}$$

Using the expression for α from (4.57) we have

$$(\Delta x)^2 = \alpha = \frac{1}{4(\Delta k)^2}. \tag{4.63}$$

Therefore,

$$(\Delta x)(\Delta k) = \frac{1}{2}. \tag{4.64}$$

This is the minimum uncertainty relation. Hence the Gaussian form (4.44) we chose is called the minimum uncertainty wave packet.

Including now the time dependence, the wavefunction $\psi(x,t)$ can be written as

$$\psi(x,t) = \int_{-\infty}^{\infty} dk\, e^{ikx} e^{-i\frac{E_k}{\hbar}t} f(k) \tag{4.65}$$

where $E_k = \hbar^2 k^2/2m$. Substituting $f(k)$ from (4.59),

$$\alpha' = \alpha + \frac{i\hbar}{2m}t, \tag{4.66}$$

we obtain

$$\psi(x,t) = \frac{N}{\sqrt{2\alpha'}} \exp - \left[\frac{x^2}{4\left(\alpha + \frac{i\hbar}{2m}t\right)}\right] \tag{4.67}$$

where

$$\alpha' = \alpha + \frac{i\hbar}{2m}t. \tag{4.68}$$

Using (4.63) for α, the exponent can be written in terms of Δx^2 as

$$\frac{x^2}{4(\Delta x)^2 + \frac{2i\hbar}{m}t}. \tag{4.69}$$

To obtain $|\psi(x,t)|^2$ we note that

$$\left|\exp\left[-\left(\frac{1}{a+ib}\right)\right]\right| = \exp\left[-\frac{a}{a^2+b^2}\right]. \tag{4.70}$$

Hence,

$$|\psi(x,t)|^2 \sim \exp\left[-\frac{x^2}{(\Delta x(t))^2}\right] \tag{4.71}$$

where we define $(\Delta x(t))^2$ as

$$(\Delta x(t))^2 = (\Delta x)^2 + \frac{\hbar^2 t^2}{4m^2(\Delta x)^2} = (\Delta x)^2 + \hbar^2(\Delta k)^2\frac{t^2}{m^2}. \tag{4.72}$$

On the right-hand side of (4.72) we have incorporated the result

$$\Delta x \Delta k = \frac{1}{2} \tag{4.73}$$

which was derived in (4.64).

Thus, we conclude that the center of the peak remains at $x = 0$ but its width given by (4.72) spreads with the speed $(\Delta p)/m$ where $\Delta p = \hbar\Delta k$. It travels a distance $(\Delta p)t/m$ in time t. This is the same behavior as that exhibited by a classical particle.

4.5 Group velocity of a superposition of plane waves

Let us consider the following wavefunction expressed as a superposition of energy–momentum eigenfunctions:

$$\psi(x,t) = \int_{-\infty}^{\infty} dk\, e^{ikx}e^{-i\omega t}f(k). \tag{4.74}$$

This is the time-dependent representation of a wave packet, where the individual waves travel with velocity.

$$v_p = \frac{\omega}{k}, \tag{4.75}$$

which is called the phase velocity.

If $f(k)$, however, has a peak at $k = k_0$, and ω is a function of k, then we can make the expansion of ω around the point $k = k_0$:

$$\omega = \omega(k) = \omega_0 + \left(\frac{d\omega}{dk}\right)_{\omega=\omega_0}(k-k_0) = \omega_0 + v_g(k-k_0) \tag{4.76}$$

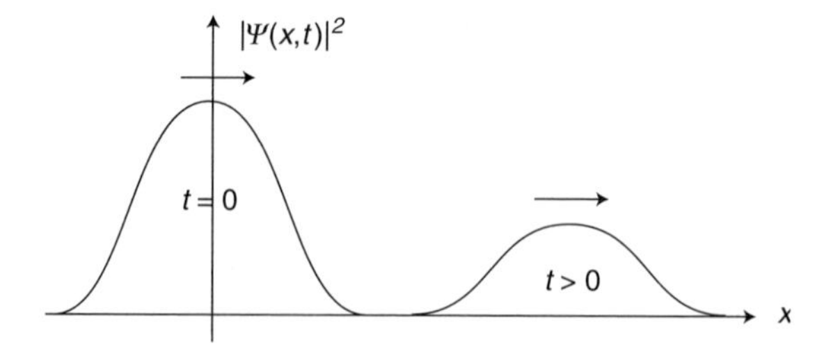

Fig. 4.1

where

$$v_g = \left(\frac{d\omega}{dk}\right)_{\omega=\omega_0}. \tag{4.77}$$

Since dominant contributions to the integral (4.74) come from the region around $k = k_0$, we have kept the expansion in (4.76) only to first order in $(k - k_0)$. We can then write

$$kx - \omega t = kx - \left[\omega_0 + v_g(k - k_0)\right]t. \tag{4.78}$$

Thus

$$\psi(x,t) = e^{i(k_0 v_g - \omega_0)t} \int_{-\infty}^{\infty} dk\, e^{ik(x - v_g t)} f(k), \tag{4.79}$$

which we can write as

$$\psi(x,t) = e^{i(k_0 v_g - \omega_0)t} \psi\left(x - v_g t, 0\right). \tag{4.80}$$

We note that the packet as a whole travels with a unique velocity, v_g, maintaining the same probability density (see Fig. 4.1)

$$|\psi(x,t)|^2 = \left|\psi\left(x - v_g t, 0\right)\right|^2. \tag{4.81}$$

Hence v_g is called the group velocity.

$$v_g = \text{group velocity}. \tag{4.82}$$

4.6 Three dimensions – Cartesian coordinates

Let us now return to the plane wave solutions and consider their properties in three dimensions. We will write the momentum eigenfunction as $u_p(\mathbf{r})$ where $\mathbf{r}$ represents the three Cartesian coordinates (x, y, z). Hence $u_p(\mathbf{r}) = u_p(x, y, z)$. In the following we will suppress the index p in $u_p(\mathbf{r})$, and write the eigenfunction simply as $u(\mathbf{r})$. As in the case of the one-dimensional problems, we will not make any distinction between the momentum and energy eigenfunctions as far as the dependence on the coordinates is concerned.

The Schrödinger equation for $u(\mathbf{r})$ for free particles is then given by

$$-\frac{\hbar^2}{2m}\nabla^2 u(\mathbf{r}) = Eu(\mathbf{r}) \tag{4.83}$$

where

$$E = \frac{\mathbf{p}^2}{2m} \tag{4.84}$$

and $\mathbf{p}$ is the momentum in three dimensions with components $\left(p_x, p_y.p_z\right)$. We express equation (4.83) in Cartesian coordinates as

$$-\frac{\hbar^2}{2m}\left(\frac{\partial^2}{\partial x^2} + \frac{\partial^2}{\partial y^2} + \frac{\partial^2}{\partial z^2}\right)u(x,y,z) = Eu(x,y,z). \tag{4.85}$$

The standard method to solve this equation is to use the so-called separation of variables technique, in which we write $u(x, y, z)$ as a product of three functions each depending on only one of the variables, that is,

$$u(x,y,z) = X(x)Y(y)Z(z). \tag{4.86}$$

Substituting this expression in (4.85), we can write the equation in terms of total derivatives of the individual functions:

$$-\frac{\hbar^2}{2m}\left[YZ\frac{d^2X}{dx^2} + ZX\frac{d^2Y}{dy^2} + XY\frac{d^2Z}{dz^2}\right] = EXYZ. \tag{4.87}$$

Dividing both sides of the above equation by XYZ, we have

$$-\frac{\hbar^2}{2m}\left[\frac{1}{X}\frac{d^2X}{dx^2} + \frac{1}{Y}\frac{d^2Y}{dy^2} + \frac{1}{Z}\frac{d^2Z}{dz^2}\right] = E. \tag{4.88}$$

We note that each of the three terms on the left depends on only one variable. Since these variables are independent of one another, the only way in which they can add up to a constant, E, the eigenvalue, is if each term itself is a constant. We then write

$$-\frac{\hbar^2}{2m}\frac{1}{X}\frac{d^2X}{dx^2} = E_1, \tag{4.89}$$

$$-\frac{\hbar^2}{2m}\frac{1}{Y}\frac{d^2Y}{dy^2} = E_2, \tag{4.90}$$

$$-\frac{\hbar^2}{2m}\frac{1}{Z}\frac{d^2Z}{dz^2} = E_3, \tag{4.91}$$

with

$$E_1 + E_2 + E_3 = E. \tag{4.92}$$

The above three equations are individually one-dimensional equations of the type we have already considered. For example, for the function $X(x)$ the equation is

$$-\frac{\hbar^2}{2m}\frac{d^2X}{dx^2} = E_1 X. \tag{4.93}$$

The solution of this equation was obtained previously as

$$X(x) = C_1 \exp\left(\pm\frac{ip_1}{\hbar}x\right) \quad \text{with } E_1 = \frac{p_1^2}{2m}. \tag{4.94}$$

Similarly, we can obtain the solutions for $Y(y)$ and $Z(z)$. As in the one-dimensional case we will write

$$p_i = \hbar k_i, \quad i = 1, 2, 3. \tag{4.95}$$

Once again we have two types of normalizations. We will first consider periodic boundary conditions within a cube of dimensions

$$-\frac{L_1}{2} < x < \frac{L_1}{2}; \quad -\frac{L_2}{2} < y < \frac{L_2}{2}; \quad -\frac{L_3}{2} < z < \frac{L_3}{2}. \tag{4.96}$$

The product solution can now be written down inside the above cube of volume V $(= L_1L_2L_3)$ as

$$u_{nlm}(x,y,z) = \frac{1}{\sqrt{V}} \exp i(k_{1n}x + k_{2l}y + k_{3m}z) \tag{4.97}$$

where

$$k_{1n} = \frac{2n\pi}{L_1}, \quad k_{2l} = \frac{2l\pi}{L_2}, \quad k_{3m} = \frac{2m\pi}{L_3} \quad \text{with} \quad n, l, m = 0, \pm 1, \pm 2 \ldots. \tag{4.98}$$

The complete wavefunction including the time dependence is

$$\phi_{nlm}(x,y,z,t) = u_{nlm}(x,y,z) \exp\left(\frac{-iE_{nlm}t)}{\hbar}\right), \tag{4.99}$$

with

$$E_{nlm} = \frac{\hbar^2\left(k_{1n}^2 + k_{2l}^2 + k_{3m}^2\right)}{2m} \tag{4.100}$$

and with the normalization relation given by

$$\int_{-\frac{L_1}{2}}^{\frac{L_1}{2}} dx \int_{-\frac{L_2}{2}}^{\frac{L_2}{2}} dy \int_{-\frac{L_3}{2}}^{\frac{L_3}{2}} dz\, u_{n'l'm'}(x,y,z) u_{nlm}(x,y,z) = \delta_{n'n}\delta_{l'l}\delta_{m'm}. \tag{4.101}$$

For the normalization in the infinite space we follow the same procedure as in the one-dimensional case to write the solution as the product $X(x)Y(y)Z(z)$ to obtain the following wavefunction:

$$u_k(\mathbf{r}) = \frac{1}{\left(\sqrt{2\pi}\right)^3} \exp(i\mathbf{k}.\mathbf{r}), \quad \text{with } \mathbf{k}.\mathbf{r} = k_1x + k_2y + k_3z, \tag{4.102}$$

with the normalization given by

$$\int d^3\, r u_k^*(\mathbf{r}) u_{k'}(\vec{\mathbf{r}}) = \delta^{(3)}(\mathbf{k} - \mathbf{k}') \tag{4.103}$$

where the three-dimensional δ-function has already been defined in the Appendix of Chapter 2 as

$$\delta^{(3)}(\mathbf{k} - \mathbf{k}') = \delta\left(k_1 - k_1'\right)\delta\left(k_2 - k_2'\right)\delta\left(k_3 - k_3'\right). \tag{4.104}$$

The completeness relation is then

$$\int d^3k\, u_k(\mathbf{r}) u_k^*(\mathbf{r}') = \delta^{(3)}(\mathbf{r} - \mathbf{r}'). \tag{4.105}$$

The free particle energy eigenfunction, $\phi_k(\mathbf{r})$, in the infinite three-dimensional space, is then given by

$$\phi_k(\mathbf{r}) = u_k(\mathbf{r}) \exp\left(\frac{-iE_k t)}{\hbar}\right) \tag{4.106}$$

with

$$E_k = \frac{\hbar^2\left(k_1^2 + k_2^2 + k_3^2\right)}{2m}. \tag{4.107}$$

4.7 Three dimensions – spherical coordinates

Once again we start with the Schrödinger equation for the free particle,

$$-\frac{\hbar^2}{2m}\nabla^2 u(\mathbf{r}) = Eu(\mathbf{r}), \tag{4.108}$$

and express it in spherical coordinates by making the following transformation:

$$x = r\sin\theta\cos\phi, \tag{4.109}$$

$$y = r\sin\theta\sin\phi, \tag{4.110}$$

$$z = r\cos\theta \tag{4.111}$$

where r is the magnitude of the coordinate vector $\mathbf{r}$, θ is the polar angle and ϕ is the azimuthal angle. The ∇^2 operator given by (4.85) is then found in the spherical coordinates to give

$$\nabla^2 u = \frac{1}{r}\frac{\partial^2 (ru)}{\partial r^2} + \frac{1}{r^2 \sin\theta}\frac{\partial}{\partial\theta}\left(\sin\theta\frac{\partial u}{\partial\theta}\right) + \frac{1}{r^2\sin^2\theta}\frac{\partial^2 u}{\partial\phi^2}. \tag{4.112}$$

The Schrödinger equation is then given by

$$-\frac{\hbar^2}{2m}\left[\frac{1}{r}\frac{\partial^2 (ru)}{\partial r^2} + \frac{1}{r^2 \sin\theta}\frac{\partial}{\partial\theta}\left(\sin\theta\frac{\partial u}{\partial\theta}\right) + \frac{1}{r^2\sin^2\theta}\frac{\partial^2 u}{\partial\phi^2}\right] = Eu. \tag{4.113}$$

We, once again, use the separation of variables technique to solve this equation:

$$u(\mathbf{r}) = R(r)Y(\theta,\phi). \tag{4.114}$$

Substituting this expression in (4.113) and multiplying both sides of the equation by r^2/RY, we obtain

$$-\frac{\hbar^2}{2m}\left[\frac{r}{R}\frac{d^2 (rR)}{dr^2} + \frac{1}{Y \sin\theta}\frac{\partial}{\partial\theta}\left(\sin\theta\frac{\partial Y}{\partial\theta}\right) + \frac{1}{Y\sin^2\theta}\frac{\partial^2 Y}{\partial\phi^2}\right] - r^2 E = 0. \tag{4.115}$$

The above equation is thus split into terms that depend only on r and terms that depend only on the combination (θ,ϕ). We can rewrite it as follows:

$$\frac{r}{R}\frac{d^2 (rR)}{dr^2} + \frac{2mr^2E}{\hbar^2} = -\left[\frac{1}{Y \sin\theta}\frac{\partial}{\partial\theta}\left(\sin\theta\frac{\partial Y}{\partial\theta}\right) + \frac{1}{Y\sin^2\theta}\frac{\partial^2 Y}{\partial\phi^2}\right]. \tag{4.116}$$

Following the arguments presented in the case of Cartesian coordinates, we take each side of the above equation to be a constant. If we designate this constant as λ, then we are led to the following two equations:

$$\frac{r}{R}\frac{d^2 (rR)}{dr^2} + \frac{2mr^2E}{\hbar^2} = \lambda, \tag{4.117}$$

$$\frac{1}{Y \sin\theta}\frac{\partial}{\partial\theta}\left(\sin\theta\frac{\partial Y}{\partial\theta}\right) + \frac{1}{Y\sin^2\theta}\frac{\partial^2 Y}{\partial\phi^2} = -\lambda. \tag{4.118}$$

The equation for Y can be written as

$$\sin\theta\frac{\partial}{\partial\theta}\left(\sin\theta\frac{\partial Y}{\partial\theta}\right) + \frac{\partial^2 Y}{\partial\phi^2} + \lambda\sin^2\theta\, Y = 0. \tag{4.119}$$

This equation can also be solved by the separation of variables technique:

$$Y(\theta,\phi) = P(\theta)Q(\phi). \tag{4.120}$$

Substituting this expression in the equation (4.119) for Y we obtain

$$Q \sin\theta \frac{d}{d\theta}\left(\sin\theta \frac{dP}{d\theta}\right) + P\frac{d^2Q}{d\phi^2} + \lambda PQ \sin^2\theta = 0. \tag{4.121}$$

Dividing the above equation by PQ we obtain equations that individually depend only on a single variable,

$$\frac{d^2Q}{d\phi^2} = -\mu Q \tag{4.122}$$

and

$$\frac{1}{\sin\theta}\frac{d}{d\theta}\left(\sin\theta \frac{dP}{d\theta}\right) + \left(\lambda - \frac{\mu}{\sin^2\theta}\right)P = 0 \tag{4.123}$$

where μ is a constant.

The Q-equation (4.122) has a simple solution,

$$Q(\phi) = Ce^{\pm\sqrt{\mu}\phi}, \tag{4.124}$$

where C is a constant. Since $u(\mathbf{r})$, which corresponds to the probability amplitude, must be single-valued, the value of Q at ϕ and at physically the same point, $\phi + 2\pi$, must be the same,

$$Q(\phi + 2\pi) = Q(\phi), \tag{4.125}$$

which implies

$$\sqrt{\mu} = im \tag{4.126}$$

where m is an integer. Normalizing Q over the ϕ-interval $(0, 2\pi)$, we have

$$\int_0^{2\pi} d\phi Q^*(\phi)Q(\phi) = 1 \tag{4.127}$$

and we obtain the normalization constant $C = 1/\sqrt{2\pi}$. Hence,

$$Q(\phi) = \frac{1}{\sqrt{2\pi}}e^{im\phi} \quad \text{with } m = 0, \pm 1, \pm 2, \ldots. \tag{4.128}$$

The equation for P now becomes

$$\frac{1}{\sin\theta}\frac{d}{d\theta}\left(\sin\theta \frac{dP}{d\theta}\right) + \left(\lambda - \frac{m^2}{\sin^2\theta}\right)P = 0. \tag{4.129}$$

By making the transformation, $w = \cos\theta$, one can express the above equation in terms of w:

$$\frac{d}{dw}\left[\left(1 - w^2\right)\frac{dP}{dw}\right] + \left(\lambda - \frac{m^2}{1 - w^2}\right)P = 0. \tag{4.130}$$

This is an eigenvalue equation that is a second order differential equation and, therefore, has two solutions. The solutions are obtained in standard mathematical text books on special functions. One finds that for the solution that is regular in θ in the interval $(0,\pi)$, i.e., $\cos\theta$ in the interval $(-1, 1)$, the eigenvalues are given by

$$\lambda = l(l+1), \tag{4.131}$$

$$-l \leq m \leq l \tag{4.132}$$

where l is a positive integer. In the next section we will show that the quantum number l is linked to the orbital angular momentum, while m is given by the z-component of the orbital angular momentum. In what follows we will use the terms orbital angular momentum, orbital momentum, and just angular momentum interchangeably.

The function P is well known and is called the associated Legendre function, designated as $P_l^m(\cos\theta)$. The product $Y = PQ$ is now of the form

$$Y_{lm}(\theta,\phi) = \sqrt{\frac{2l+1}{4\pi}\frac{(l-|m|)!}{(l+|m|)!}}(-1)^m e^{im\phi} P_l^m(\cos\theta) \tag{4.133}$$

for $m \geq 0$. For $m < 0$, the same expression holds but without the factor $(-1)^m$. This function is called a spherical harmonic function or simply a spherical harmonic, which we will discuss at considerable length in Section 4.9 in this chapter and also in the chapter on rotations and angular momentum. Substituting the value of λ given by (4.131) in (4.117), we obtain the equation for R, which now depends on l,

$$R = R_l(r). \tag{4.134}$$

The wavefunction $u(\mathbf{r})$ is now written in the product form,

$$u(\mathbf{r}) = R_l(r)Y_{lm}(\theta,\phi). \tag{4.135}$$

The equation for $R_l(r)$, after regrouping some terms in (4.117) and dividing by r^2, is of the form

$$-\frac{\hbar^2}{2m}\frac{1}{r}\frac{d^2(rR_l)}{dr^2} + \frac{\hbar^2 l(l+1)}{2mr^2}R_l = ER_l. \tag{4.136}$$

This is called the radial wave equation for a free particle. The equation for $Y_{lm}(\theta,\phi)$ is given by (4.119)

$$\frac{1}{\sin\theta}\frac{\partial}{\partial\theta}\left(\sin\theta\frac{\partial Y_{lm}}{\partial\theta}\right) + \frac{1}{\sin^2\theta}\frac{\partial^2 Y_{lm}}{\partial\phi^2} = -l(l+1)Y_{lm}. \tag{4.137}$$

In the following sections we individually discuss the radial wavefunctions and the spherical harmonics.

4.8 The radial wave equation

In order to simplify the radial equation we make the following change of variables,

$$\rho = kr \quad \text{and} \quad k = \sqrt{\frac{2mE}{\hbar^2}}, \tag{4.138}$$

and write the equation as

$$\frac{d^2R_l}{d\rho^2} + \frac{2}{\rho}\frac{dR_l}{d\rho} - \frac{l(l+1)R_l}{\rho^2} + R_l = 0. \tag{4.139}$$

The general solutions of the above equation are well known; they are called spherical Bessel and Neumann functions designated by $j_l(\rho)$ and $n_l(\rho)$ respectively with $l = 0, 1, 2 \ldots$.

The following list contains some interesting and useful information about $j_l(\rho)$:

$$j_0(\rho) = \frac{\sin\rho}{\rho}, \tag{4.140}$$

$$j_1(\rho) = \frac{\sin\rho}{\rho^2} - \frac{\cos\rho}{\rho}, \tag{4.141}$$

$$j_l(\rho) \to \frac{\rho^l}{(2l+1)!!}, \quad \text{as } \rho \to 0 \tag{4.142}$$

where $(2l+1)!! = 1 \cdot 3 \cdot 5 \cdot \ldots (2l+1)$,

$$j_l(\rho) \to \frac{1}{\rho}\cos\left[\rho - (l+1)\frac{\pi}{2}\right] \quad \text{as } \rho \to \infty. \tag{4.143}$$

And the following is a similar list for $n_l(\rho)$:

$$n_0(\rho) = -\frac{\cos\rho}{\rho}, \tag{4.144}$$

$$n_1(\rho) = -\frac{\cos\rho}{\rho^2} - \frac{\sin\rho}{\rho}, \tag{4.145}$$

$$n_l(\rho) \to -\frac{(2l-1)!!}{\rho^{l+1}}, \quad \text{as } \rho \to 0. \tag{4.146}$$

$$n_l(\rho) \to \frac{1}{\rho}\sin\left[\rho - (l+1)\frac{\pi}{2}\right], \quad \text{as } \rho \to \infty. \tag{4.147}$$

Linear combinations of $j_l(\rho)$ and $n_l(\rho)$, called Hankel functions of the first and second kind, designated by $h_l^{(1)}(\rho)$ and $h_l^{(2)}(\rho)$, will also be useful in our calculations. Some of the interesting properties of these functions are listed below:

$$h_l^{(1)}(\rho) = j_l(\rho) + in_l(\rho) \to \frac{1}{\rho}e^{i[\rho-(l+1)\frac{\pi}{2}]}, \quad \text{as } \rho \to \infty, \tag{4.148}$$

$$h_l^{(2)}(\rho) = j_l(\rho) - in_l(\rho) \to \frac{1}{\rho}e^{-i[\rho-(l+1)\frac{\pi}{2}]}, \quad \text{as } \rho \to \infty. \tag{4.149}$$

Both functions are infinite as $\rho \to 0$ because of the presence of $n_l(\rho)$.

Returning to the free particle radial wavefunction R_l, the only solution compatible with the probabilistic interpretation of the wavefunction is $j_l(\rho)$ since it is finite at $r = 0$, while the other solution, $n_l(\rho)$, is infinite at that point. Hence we write

$$R_l = A_l j_l(\rho) = A_l j_l(kr). \tag{4.150}$$

4.9 Properties of $Y_{lm}(\theta, \phi)$

In order to determine the properties of the spherical harmonics, $Y_{lm}(\theta, \phi)$, let us first consider $P_l^m(\cos\theta)$. We note that for $m = 0$, $P_l^m(\cos\theta)$ becomes a simple polynomial called the Legendre polynomial, written as $P_l(\cos\theta)$. We summarize the properties of $P_l(\cos\theta)$ first, and then those of $P_l^m(\cos\theta)$ before discussing $Y_{lm}(\theta, \phi)$.

To simplify writing we take

$$x = \cos\theta. \tag{4.151}$$

One then finds that $P_l(x)$ can be expressed as

$$P_l(x) = (-1)^l \frac{1}{2^l l!} \frac{d^l}{dx^l} \left(1 - x^2\right)^l \tag{4.152}$$

and $P_l^m(x)$ can then be expressed in terms of $P_l(x)$ as follows:

$$P_l^m(x) = \left(1 - x^2\right)^{\frac{|m|}{2}} \frac{d^{|m|}}{dx^{|m|}} P_l(x). \tag{4.153}$$

We note that for $m = 0$, $P_l^m(x)$ is identical to $P_l(x)$.

Some of the individual functions can be written as

$$P_0(x) = 1, \tag{4.154}$$

$$P_1(x) = x, \tag{4.155}$$

$$P_2(x) = \frac{1}{2}(3x^2 - 1). \tag{4.156}$$

We also note that

$$P_l(-x) = (-1)^l P_l(x). \tag{4.157}$$

and

$$P_l(0) = 1. \tag{4.158}$$

The normalization relation for the Legendre functions is found to be

$$\int_{-1}^{1} dx\, P_l(x) P_{l'}(x) = \frac{2}{2l+1}\delta_{ll'}. \tag{4.159}$$

We also note that $P_{l'}(x)$ appears as a coefficient of expansion for the following function,

$$\frac{1}{\sqrt{(1-2xs+s^2)}} = \sum_{l=0}^{\infty} P_l(x)s^l. \tag{4.160}$$

Many important properties of $P_l(x)$ can be derived from this relation.

For the associated Legendre functions, $P_l^m(x)$, we have correspondingly the following:

$$P_1^1(x) = \sqrt{1-x^2}, \tag{4.161}$$

$$P_2^1(x) = 3x\sqrt{1-x^2}, \tag{4.162}$$

$$P_2^2(x) = 3(1-x^2) \tag{4.163}$$

with the normalization condition

$$\int_{-1}^{+1} dx\, P_l^m(x) P_{l'}^m(x) = \frac{2}{2l+1}\frac{(l+m)!}{(l-m)!}\delta_{ll'}. \tag{4.164}$$

The spherical harmonic function $Y_{lm}(\theta,\phi)$ is already defined in (4.133). Some typical values of Y_{lm} are given by

$$Y_{00} = \frac{1}{\sqrt{4\pi}}, \tag{4.165}$$

$$Y_{10} = \sqrt{\frac{3}{4\pi}}\cos\theta, \tag{4.166}$$

$$Y_{1\pm1} = \mp\sqrt{\frac{3}{8\pi}}e^{\pm i\phi}\sin\theta, \tag{4.167}$$

$$Y_{20} = \sqrt{\frac{5}{16\pi}}(3\cos^2\theta - 1). \tag{4.168}$$

And the normalization relation is found to be

$$\int_0^{2\pi}\int_0^{\pi} d\theta\, d\phi\, Y_{lm}(\theta,\phi)Y_{lm'}(\theta,\phi)\sin\theta = \delta_{ll'}\delta_{mm'}. \tag{4.169}$$

Finally, let us determine the relation between the free particle wavefunctions in the Cartesian and spherical coordinates. For a free particle traveling in the z-direction, the wavefunction in Cartesian coordinates is given by $\exp(ikz)$, while in spherical coordinates, as we have found, the wavefunction is proportional to the product

$$j_l(kr)P_l^m(\cos\theta)e^{im\phi}. \tag{4.170}$$

Since the z-axis is the axis of reference with $z = r\cos\theta$, there will be no dependence on ϕ in the wavefunction and hence one must take $m = 0$. From the superposition principle, one can write the wavefunction in the Cartesian system in terms of a sum over the eigenstates in the spherical system. The exact relation for this case is found to be

$$e^{ikz} = \sum_{l=0}^{\infty}(2l+1)i^l j_l(kr)P_l(\cos\theta). \tag{4.171}$$

4.10 Angular momentum

We will show below that the quantum number l discussed in the previous section is actually related to the angular momentum of the particle through the eigenvalue equation

$$\mathbf{L}^2 Y_{lm} = l\,(l+1)\,\hbar^2 Y_{lm} \tag{4.172}$$

where the angular momentum operator $\mathbf{L}$ is defined by

$$\mathbf{L} = \mathbf{r} \times \mathbf{p}. \tag{4.173}$$

From the above relation the ith component of $\mathbf{L}$ is given by

$$L_i = \sum_{jk} \epsilon_{ijk} r_j p_k \tag{4.174}$$

where r_i and p_i are the Cartesian components of the operators $\mathbf{r}$ and $\mathbf{p}$, and ϵ_{ijk} is a totally antisymmetric tensor that satisfies the following properties with respect to the indices i, j, k:

$$\epsilon_{123} = 1, \tag{4.175}$$

$$\epsilon_{ijk} = 1 \text{ for even permutations of } (123) \tag{4.176}$$

$$= -1 \text{ for odd permutations of } (123) \tag{4.177}$$

$$= 0 \text{ otherwise.} \tag{4.178}$$

The ϵ_{ijk}'s also satisfy the property

$$\sum_i \epsilon_{ijk}\epsilon_{iab} = \delta_{ja}\delta_{kb} - \delta_{jb}\delta_{ka}. \tag{4.179}$$

We can now compute $\mathbf{L}^2$:

$$\mathbf{L}^2 = \sum_i L_i L_i = \sum_i \sum_{jk} \sum_{ab} \left(\epsilon_{ijk} r_j p_k\right)\left(\epsilon_{iab} r_a p_b\right). \tag{4.180}$$

Using the relation (4.179) we obtain

$$\mathbf{L}^2 = \sum_{jk} \left(r_j p_k r_j p_k - r_j p_k r_k p_j\right) \tag{4.181}$$

where the order of the operators **r** and **p** is maintained since they do not necessarily commute. We now implement the canonical quantization relation

$$[r_\alpha, p_\beta] = i\hbar\delta_{\alpha\beta} \tag{4.182}$$

to simplify (4.181). Considering each term in (4.181) individually we find, for the first term,

$$\sum_{jk} r_j p_k r_j p_k = \sum_{jk} r_j \left[r_j p_k - i\hbar\delta_{jk}\right] p_k \tag{4.183}$$

$$= \mathbf{r}^2\mathbf{p}^2 - i\hbar\mathbf{r}\cdot\mathbf{p} \tag{4.184}$$

where $\mathbf{r}^2 = \sum_j r_j r_j$ and $\mathbf{p}^2 = \sum_k p_k p_k$ and where we have used the relation (4.182). For the second term in (4.181) we obtain

$$\sum_{jk} r_j p_k r_k p_j = \sum_{jk} \left[p_k r_j + i\hbar\delta_{kj}\right] r_k p_j \tag{4.185}$$

$$= \sum_{jk} p_k r_j r_k p_j + i\hbar\mathbf{r}\cdot\mathbf{p} \tag{4.186}$$

$$= \sum_{jk} p_k r_k r_j p_j + i\hbar\mathbf{r}\cdot\mathbf{p} \tag{4.187}$$

where we have taken into account the fact that r_j and r_k commute. To simplify this further, we make use of the canonical commutator (4.182), in which, if we take $\beta = \alpha$ and then sum over α, we find

$$\sum_\alpha (r_\alpha p_\alpha - p_\alpha r_\alpha) = i\hbar\sum_\alpha \delta_{\alpha\alpha} = 3i\hbar. \tag{4.188}$$

Hence the first term in (4.187) gives

$$\sum_{jk} p_k r_k r_j p_j = \sum_k r_k p_k \sum_j r_j p_j - 3i\hbar\sum_j r_j p_j \tag{4.189}$$

$$= (\mathbf{r}\cdot\mathbf{p})^2 - 3i\hbar\mathbf{r}\cdot\mathbf{p}. \tag{4.190}$$

From (4.181), (4.184), (4.187) and (4.190) we obtain

$$\mathbf{L}^2 = \mathbf{r}^2\mathbf{p}^2 - (\mathbf{r}\cdot\mathbf{p})^2 + i\hbar\mathbf{r}\cdot\mathbf{p}, \tag{4.191}$$

which after rearranging the terms and dividing both sides by $2mr^2$ gives

$$\frac{\mathbf{p}^2}{2m} = \frac{(\mathbf{r}\cdot\mathbf{p})^2}{2mr^2} - i\hbar\frac{\mathbf{r}\cdot\mathbf{p}}{2mr^2} + \frac{\mathbf{L}^2}{2mr^2}. \tag{4.192}$$

Since $\mathbf{p} = -\mathbf{i}\hbar\nabla$, we can write

$$\mathbf{r}\cdot\mathbf{p} = -i\hbar r\frac{\partial}{\partial r} \tag{4.193}$$

and

$$(\mathbf{r}\cdot\mathbf{p})^2 = \hbar^2\left(r\frac{\partial}{\partial r}\right)\left(r\frac{\partial}{\partial r}\right) = \hbar^2\left[r\frac{\partial}{\partial r} + \mathbf{r}^2\frac{\partial^2}{\partial r^2}\right]. \tag{4.194}$$

Thus operating on the wavefunction $u(\mathbf{r})$, we have

$$\frac{\mathbf{p}^2}{2m}u = -\frac{\hbar^2}{2m}\cdot\frac{1}{r}\frac{\partial^2}{\partial r^2}(ru) + \frac{\mathbf{L}^2}{2mr^2}u, \tag{4.195}$$

which relates the kinetic energy term on the left to angular momentum, which is the second term on the right.

Substituting $\mathbf{p} = -\mathbf{i}\hbar\nabla$ on the left-hand side of (4.195) we obtain

$$-\frac{\hbar^2}{2m}\nabla^2 u = -\frac{\hbar^2}{2m}\cdot\frac{1}{r}\frac{\partial^2}{\partial r^2}(ru) + \frac{\mathbf{L}^2}{2mr^2}u. \tag{4.196}$$

Comparing (4.196) with (4.112) we find that

$$\frac{\mathbf{L}^2 u}{2mr^2} = -\frac{\hbar^2}{2mr^2}\left[\frac{1}{\sin\theta}\frac{\partial}{\partial\theta}\left(\sin\theta\frac{\partial u}{\partial\theta}\right) + \frac{1}{\sin^2\theta}\frac{\partial^2 u}{\partial\phi^2}\right]. \tag{4.197}$$

Replacing u by $R_l(r)Y_{lm}(\theta,\phi)$ and taking into account equation (4.137) in Section 4.7, we observe that the right-hand side of (4.197) is

$$\frac{\hbar^2}{2mr^2}l(l+1)\,Y_{lm}R_l. \tag{4.198}$$

Hence factoring out $R_l(r)$ we have

$$\mathbf{L}^2 Y_{lm} = \hbar^2 l(l+1)\,Y_{lm}. \tag{4.199}$$

Therefore, the quantum number l is related to the angular momentum operator $\mathbf{L}$, and Y_{lm} is an eigenstate of $\mathbf{L}^2$.

4.10.1 Angular momentum in classical physics

The Hamiltonian for a particle is given by

$$H = T + V \tag{4.200}$$

where T is the kinetic energy and V the potential energy. For a particle carrying out a radial motion,

$$T = \frac{1}{2}m\dot{r}^2 + \frac{1}{2}mr^2\dot{\theta}^2. \tag{4.201}$$

The angular momentum is given by

$$L = mr^2\dot{\theta}. \tag{4.202}$$

The expression for T then becomes

$$T = \frac{1}{2}m\dot{r}^2 + \frac{L^2}{2mr^2}. \tag{4.203}$$

Thus, in classical mechanics, just as we found in quantum mechanics, there is a presence of the angular momentum term in the Hamiltonian. But while classically L is continuous, we find that in quantum mechanics, where angular momentum is an operator, L is quantized.

4.11 Determining L^2 from the angular variables

Let us now reconfirm directly from the definition of $\mathbf{L}$ in (4.173) that the right-hand side in the square bracket of (4.197) is, indeed, related to $\mathbf{L}^2$, that is,

$$\mathbf{L}^2 u = -\hbar^2\left[\frac{1}{\sin\theta}\frac{\partial}{\partial\theta}\left(\sin\theta\frac{\partial u}{\partial\theta}\right) + \frac{1}{\sin^2\theta}\frac{\partial^2 u}{\partial\phi^2}\right]. \tag{4.204}$$

From (4.173) we write

$$\mathbf{L} = -\mathrm{i}\hbar\mathbf{r}\times\nabla. \tag{4.205}$$

The operator ∇ is found in terms of the spherical coordinates to be

$$\nabla = \hat{\mathbf{r}}\frac{\partial}{\partial r} + \hat{\boldsymbol{\theta}}\frac{1}{r}\frac{\partial}{\partial\theta} + \boldsymbol{\phi}\frac{1}{r\sin\theta}\frac{\partial}{\partial\phi} \tag{4.206}$$

where $\hat{\mathbf{r}}, \hat{\boldsymbol{\theta}}$ and $\hat{\boldsymbol{\phi}}$ are unit orthogonal vectors in the respective directions. They can be expressed in terms of the unit Cartesian coordinates $\hat{\imath}$, $\hat{\jmath}$ and $\hat{\mathbf{k}}$ in the x-, y-, and z-directions respectively as follows:

$$\hat{\mathbf{r}} = \sin\theta\cos\phi\hat{\imath} + \sin\theta\sin\phi\hat{\jmath} + \cos\theta\hat{\mathbf{k}}, \tag{4.207}$$

$$\hat{\boldsymbol{\theta}} = \cos\theta\cos\phi\hat{\imath} + \cos\theta\sin\phi\hat{\jmath} - \sin\theta\hat{\mathbf{k}}, \tag{4.208}$$

$$\hat{\boldsymbol{\phi}} = -\sin\phi\hat{\imath} + \cos\phi\hat{\jmath}. \tag{4.209}$$

It is then straightforward to show that

$$\mathbf{L} = \mathbf{r}\times\mathbf{p} = -\,i\hbar r\hat{\mathbf{r}}\times\nabla = -i\hbar\left(\hat{\boldsymbol{\phi}}\frac{\partial}{\partial\theta} - \hat{\boldsymbol{\theta}}\frac{1}{\sin\theta}\frac{\partial}{\partial\phi}\right), \tag{4.210}$$

which leads to

$$L_x = -i\hbar\left(-\sin\phi\frac{\partial}{\partial\theta} - \cos\phi\cot\theta\frac{\partial}{\partial\phi}\right), \tag{4.211}$$

$$L_y = -i\hbar\left(-\cos\phi\frac{\partial}{\partial\theta} - \sin\phi\cot\theta\frac{\partial}{\partial\phi}\right), \tag{4.212}$$

$$L_z = -i\hbar\frac{\partial}{\partial\phi}, \tag{4.213}$$

and hence we obtain, after a somewhat lengthy calculation,

$$\mathbf{L}^2 = L_x^2 + L_y^2 + L_z^2 = -\hbar^2\left[\frac{1}{\sin\theta}\frac{\partial}{\partial\theta}\left(\sin\theta\frac{\partial}{\partial\theta}\right) + \frac{1}{\sin^2\theta}\frac{\partial^2}{\partial\phi^2}\right], \tag{4.214}$$

which reproduces (4.204).

We also note, since Y_{lm} given by (4.133) contains a factor $e^{im\phi}$, that

$$L_z u = -i\hbar R_l\left[\frac{\partial}{\partial\phi}Y_{lm}\right] = m\hbar u. \tag{4.215}$$

Thus the eigenvalue m corresponds to the z-component of $\mathbf{L}$.

We need to point out that, apart from mentioning mathematical textbooks as references and providing, indirect, derivations, we have not established directly that the eigenvalues of $\mathbf{L}^2$ and L_z are $l\,(l+1)\,\hbar^2$ and $m\hbar$ respectively. For that we will have to wait until Chapter 26.

4.12 Commutator $[L_i, L_j]$ and $[\mathbf{L}^2, L_j]$

It is interesting to consider the commutators of the angular momentum operators based on the fundamental commutation relation given in (4.182). Specifically, we consider

$$[L_x, L_y] = L_x L_y - L_y L_x \tag{4.216}$$

where we note that $\mathbf{L} = \mathbf{r} \times \mathbf{p}$. We obtain

$$L_x = yp_z - zp_y = -i\hbar\left(y\frac{\partial}{\partial z} - z\frac{\partial}{\partial y}\right) \tag{4.217}$$

etc.

From (4.174) and the relation $\mathbf{p} = -i\hbar\nabla$ we find

$$L_x L_y = -\hbar^2\left(y\frac{\partial}{\partial z} - z\frac{\partial}{\partial y}\right)\left(z\frac{\partial}{\partial x} - x\frac{\partial}{\partial z}\right)$$

$$= -\hbar^2\left(y\frac{\partial}{\partial x} + yz\frac{\partial^2}{\partial z\partial x} - yx\frac{\partial^2}{\partial z^2} - z^2\frac{\partial^2}{\partial y\partial x} + zx\frac{\partial^2}{\partial y\partial z}\right). \tag{4.218}$$

Similarly,

$$L_yL_x = -\hbar^2\left(zy\frac{\partial^2}{\partial x\partial y} - z^2\frac{\partial^2}{\partial x\partial y} - xy\frac{\partial^2}{\partial z^2} + x\frac{\partial}{\partial y} + xz\frac{\partial^2}{\partial z\partial y}\right). \tag{4.219}$$

Subtracting (4.219) from (4.218) we obtain

$$L_xL_y - L_yL_x = +\hbar^2\left(x\frac{\partial}{\partial y} - y\frac{\partial}{\partial x}\right) = i\hbar L_z \tag{4.220}$$

where L_z is determined by (4.205).

Thus, in summary,

$$\left[L_x, L_y\right] = i\hbar L_z, \tag{4.221}$$

$$\left[L_y, L_z\right] = i\hbar L_x, \tag{4.222}$$

$$\left[L_z, L_x\right] = i\hbar L_y. \tag{4.223}$$

We can write the above relations compactly as

$$\left[L_i, L_j\right] = i\hbar\sum_k \epsilon_{ijk}L_k. \tag{4.224}$$

To determine $\left[\mathbf{L}^2, L_i\right]$, let us consider the specific case of $\left[\mathbf{L}^2, L_x\right]$. We can write it as

$$\left[\mathbf{L}^2, L_x\right] = \left[L_x^2, L_x\right] + \left[L_y^2, L_x\right] + \left[L_z^2, L_x\right] = \left[L_y^2, L_x\right] + \left[L_z^2, L_x\right] \tag{4.225}$$

where we have taken into account the fact that L_x^2 and L_x commute. The two terms on the right together can be written as

$$\begin{aligned}
\left[L_y^2, L_x\right] + \left[L_z^2, L_x\right] &= L_y\left[L_y, L_x\right] + \left[L_y, L_x\right]L_y + L_z\left[L_z, L_x\right] + \left[L_z, L_x\right]L_z \\
&= L_y\left(-i\hbar\right)L_z + \left(-i\hbar\right)L_zL_y + L_z\left(i\hbar\right)L_y + \left(i\hbar\right)L_yL_z \\
&= 0
\end{aligned} \tag{4.226}$$

where we have used the commutation relations (4.224). Therefore, we can write the following general result:

$$\left[\mathbf{L}^2, L_i\right] = 0. \tag{4.227}$$

We also note that L_i will be Hermitian since r_i and p_i are Hermitian and r_i and p_j commute for $i \neq j$. That is,

$$L_i^\dagger = L_i \tag{4.228}$$

where $i = 1, 2, 3$.

Another way, perhaps a more elegant one, of deriving the commutation relations is described in Problem 3.

4.13 Ladder operators

Instead of considering L_x and L_y individually, it is often more instructive to work with the so-called ladder operators, $L_\pm$, defined by

$$L_\pm = L_x \pm iL_y. \tag{4.229}$$

One finds, using (4.211) and (4.212),

$$L_\pm = \hbar e^{\pm i\phi}\left(\pm\frac{\partial}{\partial\theta} + i\cot\theta\frac{\partial}{\partial\phi}\right). \tag{4.230}$$

The $\mathbf{L}^2$ operator can be written in terms of $L_\pm$ as

$$\mathbf{L}^2 = L_+L_- + L_z^2 - \hbar L_z. \tag{4.231}$$

Furthermore, from (4.221), (4.222), and (4.223) one can show that

$$[L_z, L_+] = \hbar L_+, \tag{4.232}$$

$$[L_z, L_-] = -\hbar L_-. \tag{4.233}$$

Using relation (4.231) for $\mathbf{L}^2$ we can also show that

$$\left[\mathbf{L}^2, L_z\right] = 0 = \left[\mathbf{L}^2, L_+\right] = \left[\mathbf{L}^2, L_-\right]. \tag{4.234}$$

Let us consider the state (L_-Y_{lm}). From (4.233) we find that

$$L_z\,(L_-Y_{lm}) = (L_-L_z - \hbar L_-)\,Y_{lm}. \tag{4.235}$$

Since $L_zY_{lm} = m\hbar Y_{lm}$, therefore,

$$L_z\,(L_-Y_{lm}) = (m-1)\,\hbar\,(L_-Y_{lm}). \tag{4.236}$$

Thus (L_-Y_{lm}) is an eigenstate of L_z with eigenvalue $(m-1)\,\hbar$. Furthermore, one finds that $\mathbf{L}^2$ commutes with L_-:

$$\mathbf{L}^2\,(L_-Y_{lm}) = L_-\mathbf{L}^2Y_{lm} = l\,(l+1)\,\hbar^2\,(L_-Y_{lm}). \tag{4.237}$$

Hence one can write

$$(L_-Y_{lm}) = CY_{lm-1}, \tag{4.238}$$

which implies that L_- acts as a "lowering" operator by lowering the eigenvalue m to $m-1$.

Since the spherical harmonic functions are normalized, we use the normalization condition

$$\begin{aligned}&\int_0^{2\pi}\int_0^{\pi} d\theta\, d\phi\, (L_-Y_{lm})^\dagger\,(L_-Y_{lm})\sin\theta\\&=|C|^2\int_0^{2\pi}\int_0^{\pi} d\theta\, d\phi\, Y^*_{lm-1}(\theta,\phi)Y_{lm-1}(\theta,\phi)\sin\theta\\&=|C|^2.\end{aligned} \tag{4.239}$$

The left-hand side corresponds to

$$\int_0^{2\pi}\int_0^{\pi} d\theta\, d\phi\, L_-^\dagger L_- Y^*_{lm}Y_{lm}\sin\theta. \tag{4.240}$$

The product $L_-^\dagger L_-$ can be simplified to

$$L_-^\dagger L_- = L_+L_- = \mathbf{L}^2 - L_z^2 + \hbar L_z \tag{4.241}$$

where we have used the relation $L_-^\dagger = L_+$ and the relation (4.223). Since Y_{lm} are eigenstates of $\mathbf{L}^2$ and L_z with eigenvalues $l\,(l+1)\,\hbar^2$ and $m\hbar$ respectively, and since Y_{lm}'s are normalized, the left-hand side of (4.239) simply gives

$$\left[l(l+1) - m^2 + m\right]\hbar^2 = (l+m)\,(l-m+1)\hbar^2. \tag{4.242}$$

The right-hand side is simply $|C|^2$ since Y_{lm+1}'s are also normalized. Thus we find from (4.240) that

$$C = \sqrt{(l+m)\,(l-m+1)}\hbar. \tag{4.243}$$

Hence,

$$L_-Y_{lm} = \sqrt{(l+m)\,(l-m+1)}\hbar Y_{lm-1}. \tag{4.244}$$

Similarly,

$$L_+Y_{lm} = \sqrt{(l-m)\,(l+m+1)}\hbar Y_{lm-1}. \tag{4.245}$$

Hence, L_+ acts as a "raising" operator. It is because of these properties that L_+ and L_- are called "ladder" operators. The role of the ladder operators will become clearer when we discuss the general properties of angular momentum in Chapter 26.

Finally, we note that we can express the state vectors corresponding to the angular wavefunctions simply as $|lm\rangle$, so that

$$\mathbf{L}^2 |lm\rangle = l(l+1)\hbar^2 |lm\rangle , \tag{4.246}$$

$$L_z |lm\rangle = m\hbar |lm\rangle , \tag{4.247}$$

$$L_- |lm\rangle = \sqrt{(l+m)(l-m+1)}\hbar |lm-1\rangle , \tag{4.248}$$

$$L_+ |lm\rangle = \sqrt{(l-m)(l+m+1)}\hbar |lm+1\rangle , \tag{4.249}$$

and the spherical harmonics $Y_{lm}(\theta,\phi)$ as the wavefunction

$$\langle \theta, \phi |lm\rangle = Y_{lm}(\theta,\phi). \tag{4.250}$$

4.14 Problems

1. Write down the free-particle Schrödinger equation for two dimensions in (i) Cartesian and (ii) polar coordinates. Obtain the corresponding wavefunction.
2. Use the orthogonality property (4.159) of $P_l(\cos\theta)$ and confirm that at least the first two terms on the right-hand side of (4.171) are correct.
3. Obtain the commutation relations $[L_i, L_j]$ by calculating the vector $\mathbf{L} \times \mathbf{L}$ using the definition $\mathbf{L} = \mathbf{r} \times \mathbf{p}$ directly instead of introducing a differential operator.
4. Consider a finite set of operators B_i. Let H be a Hamiltonian which commutes with the B_i's, i.e., $[B_i, H] = 0$. If $|a_n\rangle$'s are eigenstates of H satisfying

   $$H |a_n\rangle = a_n |a_n\rangle ,$$

   then show that $B_i |a_n\rangle$ are proportional to $|a_n\rangle$. That is, one can write

   $$B_i |a_n\rangle = b_i |a_n\rangle .$$

   Show that this result implies $[B_i, B_j] = 0$. How does this reconcile with the fact that $[L_i, H] = 0$ but $[L_i, L_j] \neq 0$ for the angular momentum operators L_i?
5. A free particle is moving along a path of radius R. Express the Hamiltonian in terms of the derivatives involving the polar angle of the particle and write down the Schrödinger equation. Determine the wavefunction and the energy eigenvalues of the particle.
6. Determine $[L_i, r]$ and $[L_i, \mathbf{r}]$.
7. Show that

   $$e^{-i\pi L_x/\hbar} |l, m\rangle = |l, -m\rangle .$$

5 Particles with spin ½

We previously considered angular momentum, which has a classical counterpart. We will continue with the free particle system but this time confine ourselves to a uniquely quantum-mechanical concept of particles of spin ½. A large number of particles in nature have spin ½, such as electrons, protons, neutrons, leptons, and quarks. To understand its mathematical origin one must go to relativistic quantum mechanics as outlined by Dirac's theory. This is discussed in Chapter 33 on the Dirac equation but, in the meantime, in this chapter, we consider the nonrelativistic, phenomenological aspect of this concept. We will also consider the Pauli exclusion principle for spin ½ particles and conclude with a description of Fermi levels which one observes in condensed matter systems.

5.1 Spin ½ system

Let us consider a collection of operators S_x, S_y, S_z which satisfy the same commutation relations as the ones satisfied by L_x, L_y, and L_z,

$$\left[S_x, S_y\right] = i\hbar S_z, \tag{5.1}$$

$$\left[S_y, S_z\right] = i\hbar S_x, \tag{5.2}$$

$$[S_z, S_x] = i\hbar S_y \tag{5.3}$$

with

$$\mathbf{S}^2 = s(s+1)\hbar^2\mathbf{1} \tag{5.4}$$

where

$$\mathbf{S}^2 = S_x^2 + S_y^2 + S_z^2 \tag{5.5}$$

and

$$S_i^\dagger = S_i. \tag{5.6}$$

Following the arguments that were used in the case of the orbital angular momentum operators L_i, one can show that $\mathbf{S}^2$ commutes with each of the components of $\mathbf{S}$, and that one can write the following compact relations,

$$\left[S_i, S_j\right] = i\hbar \sum_k \epsilon_{ijk} S_k \quad \text{and} \quad \left[\mathbf{S}^2, S_i\right] = 0 \tag{5.7}$$

where i, j, k $(= 1, 2, 3)$ stand for x-, y-, and z-components. We call S_i, the "spin angular momentum" or simply "spin."

Unlike the quantum number, l, for the orbital angular momentum, which has values 0, $1, 2, \ldots$ with z-components $m = -l, \ldots, +l$, the quantum number s of the spin is assumed to have a unique value $s = 1/2$ with z-components $m_s = -1/2, 1/2$. The eigenstates are referred to as "spin ½" states. They are characterized by the eigenvalues of two commuting operators, $\mathbf{S}^2$ and any one of the three $S_i's$, and designated by the kets $|1/2, m_s\rangle$.

A general eigenstate in this system is taken to be a column matrix of the type

$$\begin{pmatrix} \checkmark \\ \checkmark \end{pmatrix}. \tag{5.8}$$

The operators S_i will then be 2×2 matrices,

$$\begin{pmatrix} \checkmark & \checkmark \\ \checkmark & \checkmark \end{pmatrix}. \tag{5.9}$$

5.2 Pauli matrices

We define

$$S_i = \frac{\hbar}{2} \boldsymbol{\sigma}_i \tag{5.10}$$

where $\boldsymbol{\sigma}_i$ are called the Pauli matrices $(i = 1, 2, 3)$. As we will see in the following, our formalism will be simplified considerably by the introduction of these matrices. Using (5.3), (5.7), and (5.10) the commutation relations satisfied by the σ_i's can be expressed as

$$\left[\sigma_i, \sigma_j\right] = 2i \sum_k \epsilon_{ijk} \sigma_k \quad \text{and} \quad \left[\boldsymbol{\sigma}^2, \sigma_i\right] = 0 \tag{5.11}$$

where $\sigma_i^\dagger = \sigma_i$ and

$$\boldsymbol{\sigma}^2 = \sigma_x^2 + \sigma_y^2 + \sigma_z^2 = 3\begin{bmatrix} 1 & 0 \\ 0 & 1 \end{bmatrix}, \tag{5.12}$$

which is a multiple of a 2×2 unit operator.

5.3 The spin ½ eigenstates

We will obtain the common eigenstates of the two commuting operators, $\boldsymbol{\sigma}^2$ and σ_z. The eigenvalues of σ_z are $\pm$ 1, and, therefore, the matrix representing σ_z will be a diagonal matrix with $+1$ and -1 along the diagonal.

$$\sigma_z = \begin{bmatrix} 1 & 0 \\ 0 & -1 \end{bmatrix}. \tag{5.13}$$

The matrix representation of $\boldsymbol{\sigma}^2$ is given above.

The eigenstates of the operator σ_z given by (5.13) can be expressed as

$$\begin{bmatrix} 1 \\ 0 \end{bmatrix} \quad \text{and} \quad \begin{bmatrix} 0 \\ 1 \end{bmatrix}. \tag{5.14}$$

They have the properties

$$\sigma_z \begin{bmatrix} 1 \\ 0 \end{bmatrix} = (+1) \begin{bmatrix} 1 \\ 0 \end{bmatrix}, \tag{5.15}$$

$$\sigma_z \begin{bmatrix} 0 \\ 1 \end{bmatrix} = (-1) \begin{bmatrix} 0 \\ 1 \end{bmatrix}, \tag{5.16}$$

corresponding to eigenvalues $+1$ and -1 respectively. The two column matrices are variously designated as

$$\left| \frac{1}{2}, \frac{1}{2} \right\rangle = \ \uparrow \ = |z+\rangle = \begin{bmatrix} 1 \\ 0 \end{bmatrix} = \text{“spin-up”} \tag{5.17}$$

$$\left| \frac{1}{2}, -\frac{1}{2} \right\rangle = \ \downarrow \ = |z-\rangle = \begin{bmatrix} 0 \\ 1 \end{bmatrix} = \text{“spin-down.”} \tag{5.18}$$

We note that these eigenstates are orthonormal:

$$\langle z + |z+\rangle = \langle z - |z-\rangle = 1 \ \text{ and } \ \langle z + |z-\rangle = 0. \tag{5.19}$$

They also satisfy the completeness relation

$$|z+\rangle \langle z+| + |z-\rangle \langle z-| = \begin{bmatrix} 1 \\ 0 \end{bmatrix} \begin{bmatrix} 1 & 0 \end{bmatrix} + \begin{bmatrix} 0 \\ 1 \end{bmatrix} \begin{bmatrix} 0 & 1 \end{bmatrix} = \begin{bmatrix} 1 & 0 \\ 0 & 1 \end{bmatrix}. \tag{5.20}$$

5.4 Matrix representation of σ_x and σ_y

The matrix representation of σ_x and σ_y for the case where $\boldsymbol{\sigma}^2$ and σ_z are diagonal are derived below. Taking the trace of the first of the two relations in (5.11) we obtain

$$\mathrm{Tr}\left[\sigma_i, \sigma_j\right] = 2i \sum_k \epsilon_{ijk} \mathrm{Tr}(\sigma_k). \tag{5.21}$$

However, since the trace of a product of matrices is invariant under cyclic permutations, e.g., $\mathrm{Tr}(AB) = \mathrm{Tr}(BA)$,

$$\mathrm{Tr}\left[\sigma_i, \sigma_j\right] = \mathrm{Tr}(\sigma_i\sigma_j) - \mathrm{Tr}(\sigma_j\sigma_i) = \mathrm{Tr}(\sigma_i\sigma_j) - \mathrm{Tr}(\sigma_i\sigma_j) = 0. \tag{5.22}$$

Thus the right-hand side of (5.21) must also be zero. Since the matrices σ_i in the sum in (5.21) are independent of each other, each term on the right-hand side of (5.21) will vanish. Hence

$$\mathrm{Tr}(\sigma_k) = 0 \tag{5.23}$$

for all values of $k (= 1, 2, 3)$. The matrix, σ_z, of course, satisfies this relation.

Let us define the following operators:

$$\sigma_+ = \sigma_x + i\sigma_y, \tag{5.24}$$

$$\sigma_- = \sigma_x - i\sigma_y. \tag{5.25}$$

These are the counterparts of the ladder operators $L_\pm$ we considered earlier. We note that $\sigma_- = \sigma_+^\dagger$. One can derive the following commutation relations from (5.11):

$$[\sigma_z, \sigma_+] = 2\sigma_+, \tag{5.26}$$

$$[\sigma_z, \sigma_-] = -2\sigma_-, \tag{5.27}$$

$$[\sigma_+, \sigma_-] = 4\sigma_z. \tag{5.28}$$

Let us write σ_+ in the most general form,

$$\sigma_+ = \begin{bmatrix} a & b \\ c & d \end{bmatrix}. \tag{5.29}$$

We will now try to determine $a, b, c,$ and d from the above commutation relations. We note that

$$[\sigma_z, \sigma_+] = \sigma_z\sigma_+ - \sigma_+\sigma_z = \begin{bmatrix} 0 & 2b \\ -2c & 0 \end{bmatrix}. \tag{5.30}$$

From (5.26) and (5.29) we then conclude that

$$a = 0 = c = d. \tag{5.31}$$

Hence

$$\sigma_+ = \begin{pmatrix} 0 & b \\ 0 & 0 \end{pmatrix}. \tag{5.32}$$

where we take b to be real and positive. We find

$$\sigma_- = \sigma_+^\dagger = \begin{pmatrix} 0 & 0 \\ b & 0 \end{pmatrix}. \tag{5.33}$$

The left-hand side of the commutation relation (5.28) is now found to be

$$\sigma_+\sigma_- - \sigma_-\sigma_+ = \begin{pmatrix} b^2 & 0 \\ 0 & -b^2 \end{pmatrix}. \tag{5.34}$$

From (5.13), and the right-hand side of (5.28) we must have $b^2 = 4$. Therefore, $b = 2$. Hence

$$\sigma_+ = 2\begin{pmatrix} 0 & 1 \\ 0 & 0 \end{pmatrix} \quad \text{and} \quad \sigma_- = 2\begin{pmatrix} 0 & 0 \\ 1 & 0 \end{pmatrix}. \tag{5.35}$$

We can now determine σ_x and σ_y from σ_+ and σ_- as follows:

$$\sigma_x = \frac{\sigma_+ + \sigma_-}{2} = \begin{pmatrix} 0 & 1 \\ 1 & 0 \end{pmatrix}, \tag{5.36}$$

$$\sigma_y = \frac{\sigma_+ - \sigma_-}{2i} = \begin{pmatrix} 0 & -i \\ i & 0 \end{pmatrix}. \tag{5.37}$$

We note that these operators are Hermitian since they satisfy $\sigma_i^\dagger = \sigma_i$, where $i = x, y, z$. Thus we have the representation of all the Pauli matrices for the case when σ_z is diagonal. We note from (5.13), (5.36), and (5.37) that

$$\sigma_x^2 = \sigma_y^2 = \sigma_z^2 = \begin{pmatrix} 1 & 0 \\ 0 & 1 \end{pmatrix} \tag{5.38}$$

and hence $\boldsymbol{\sigma}^2$ is given by

$$\boldsymbol{\sigma}^2 = \sigma_x^2 + \sigma_y^2 + \sigma_z^2 = 3\begin{pmatrix} 1 & 0 \\ 0 & 1 \end{pmatrix}, \tag{5.39}$$

which reproduces (5.12).

5.5 Eigenstates of σ_x and σ_y

As for the matrices σ_+ and σ_-, defined in (5.32) and (5.33), we find

$$\sigma_+ \begin{pmatrix} 1 \\ 0 \end{pmatrix} = 0 \quad \text{and} \quad \sigma_+ \begin{pmatrix} 0 \\ 1 \end{pmatrix} = 2 \begin{pmatrix} 1 \\ 0 \end{pmatrix}. \tag{5.40}$$

Similarly

$$\sigma_- \begin{pmatrix} 1 \\ 0 \end{pmatrix} = 2 \begin{pmatrix} 0 \\ 1 \end{pmatrix} \quad \text{and} \quad \sigma_- \begin{pmatrix} 0 \\ 1 \end{pmatrix} = 0. \tag{5.41}$$

Thus σ_+ acts like a "raising" operator changing the spin-down state to spin-up state, while σ_- acts in exactly the opposite direction, as a "lowering" operator. The operators $\sigma_\pm$, like $L_\pm$, act as ladder operators.

We have thus far considered eigenstates of σ_z. Let us obtain, as a simple exercise, the eigenstates of the other Pauli matrices. If we designate $|x\pm\rangle$ as the spin-up and spin-down states in the x-direction, then

$$\sigma_x |x+\rangle = |x+\rangle \quad \text{and} \quad \sigma_x |x-\rangle = -|x-\rangle. \tag{5.42}$$

Let us obtain $|x+\rangle$ by writing it as a superposition of the eigenstates of σ_z,

$$|x+\rangle = c_1 \begin{pmatrix} 1 \\ 0 \end{pmatrix} + c_2 \begin{pmatrix} 0 \\ 1 \end{pmatrix} \tag{5.43}$$

with

$$|c_1|^2 + |c_2|^2 = 1. \tag{5.44}$$

The matrix σ_x is given by (5.36) and, from (5.42), the eigenstate $|x+\rangle$ satisfies the relation

$$\begin{pmatrix} 0 & 1 \\ 1 & 0 \end{pmatrix} \left[c_1 \begin{pmatrix} 1 \\ 0 \end{pmatrix} + c_2 \begin{pmatrix} 0 \\ 1 \end{pmatrix} \right] = \left[c_1 \begin{pmatrix} 1 \\ 0 \end{pmatrix} + c_2 \begin{pmatrix} 0 \\ 1 \end{pmatrix} \right], \tag{5.45}$$

which gives, together with (5.44), the result

$$c_1 = c_2 = \frac{1}{\sqrt{2}} \tag{5.46}$$

where we have taken the positive square roots. Thus,

$$|x+\rangle = \frac{1}{\sqrt{2}} \left[\begin{pmatrix} 1 \\ 0 \end{pmatrix} + \begin{pmatrix} 0 \\ 1 \end{pmatrix} \right] = \frac{1}{\sqrt{2}} \left[|z+\rangle + |z-\rangle \right] = \frac{1}{\sqrt{2}} \begin{pmatrix} 1 \\ 1 \end{pmatrix}. \tag{5.47}$$

Similarly one can obtain the eigenstates of σ_y.

5.6 Eigenstates of spin in an arbitrary direction

Let us now obtain eigenstates of $\boldsymbol{\sigma}$ pointing in an arbitrary direction. We designate a state corresponding to spin pointing in the direction of a unit vector $\mathbf{n}$ as

$$\left|\chi_{n+}\right\rangle. \tag{5.48}$$

This state is then an eigenstate of the projection of the spin operator $\boldsymbol{\sigma}$ in the $\mathbf{n}$ direction with eigenvalue $+1$, that is,

$$\boldsymbol{\sigma}\cdot\mathbf{n}\left|\chi_{\mathbf{n}+}\right\rangle = \left|\chi_{n+}\right\rangle. \tag{5.49}$$

If the polar angle of $\mathbf{n}$ is α and the azimuthal angle is β, then in terms of its coordinates in the Cartesian system one can write

$$\mathbf{n} = (\sin\alpha\cos\beta, \sin\alpha\sin\beta, \cos\alpha)\,; \tag{5.50}$$

then

$$\boldsymbol{\sigma}\cdot\mathbf{n} = \boldsymbol{\sigma}_x \sin\alpha\cos\beta + \boldsymbol{\sigma}_y \sin\alpha\sin\beta + \boldsymbol{\sigma}_z\cos\alpha \tag{5.51}$$

$$= \begin{bmatrix} 0 & 1 \\ 1 & 0 \end{bmatrix}\sin\alpha\cos\beta + \begin{bmatrix} 0 & -i \\ i & 0 \end{bmatrix}\sin\alpha\sin\beta + \begin{bmatrix} 1 & 0 \\ 0 & -1 \end{bmatrix}\cos\alpha \tag{5.52}$$

$$= \begin{bmatrix} \cos\alpha & \sin\alpha e^{-i\beta} \\ \sin\alpha e^{i\beta} & -\cos\alpha \end{bmatrix}. \tag{5.53}$$

Let

$$\left|\chi_{n+}\right\rangle = \begin{bmatrix} a \\ b \end{bmatrix} \tag{5.54}$$

with $|a|^2 + |b|^2 = 1$, then according to (5.49) we have

$$\begin{bmatrix} \cos\alpha & \sin\alpha e^{-i\beta} \\ \sin\alpha e^{i\beta} & -\cos\alpha \end{bmatrix}\begin{bmatrix} a \\ b \end{bmatrix} = \begin{bmatrix} a \\ b \end{bmatrix}, \tag{5.55}$$

which gives rise to the following two equations

$$a\cos\alpha + b\sin\alpha e^{-i\beta} = a, \tag{5.56}$$

$$a\sin\alpha e^{i\beta} - b\cos\alpha = b \tag{5.57}$$

The first equation gives

$$a(1-\cos\alpha) = b\sin\alpha e^{-i\beta}. \tag{5.58}$$

Since $(1 - \cos\alpha) = 2\sin^2(\alpha/2)$ and $\sin\alpha = 2\sin(\alpha/2)\cos(\alpha/2)$, we find a normalized solution as

$$a = \cos\left(\frac{\alpha}{2}\right) \quad \text{and} \quad b = \sin\left(\frac{\alpha}{2}\right)e^{i\beta}. \tag{5.59}$$

Hence

$$|\chi_{n+}\rangle = \cos\left(\frac{\alpha}{2}\right)\begin{bmatrix}1\\0\end{bmatrix} + \sin\left(\frac{\alpha}{2}\right)e^{i\beta}\begin{bmatrix}0\\1\end{bmatrix} = \begin{bmatrix}\cos\left(\frac{\alpha}{2}\right)\\ \sin\left(\frac{\alpha}{2}\right)e^{i\beta}\end{bmatrix}. \tag{5.60}$$

By taking appropriate values of α and β we can obtain $|x\pm\rangle$ and $|y\pm\rangle$. For example, if we take $\alpha = \pi/2$ and $\beta = -\pi/2$, we obtain $|y-\rangle$:

$$|y-\rangle = \frac{1}{\sqrt{2}}\left[\begin{pmatrix}1\\0\end{pmatrix} - i\begin{pmatrix}0\\1\end{pmatrix}\right] = \frac{1}{\sqrt{2}}\left[|z+\rangle - i\,|z-\rangle\right] = \frac{1}{\sqrt{2}}\begin{pmatrix}1\\-i\end{pmatrix}. \tag{5.61}$$

5.7 Some important relations for σ_i

The properties of the Pauli matrices that we have derived above can all be described by the following very useful and compact relation:

$$\sigma_i\sigma_j = \delta_{ij} + i\sum_k \epsilon_{ijk}\sigma_k. \tag{5.62}$$

In particular, we note that

$$\sigma_i\sigma_j + \sigma_j\sigma_i = 2\delta_{ij} \tag{5.63}$$

which states that the Pauli matrices anticommute for $i \neq j$.

A term one often comes across in the calculations involving Pauli matrices is the product $\boldsymbol{\sigma}\cdot\mathbf{A}\boldsymbol{\sigma}\cdot\mathbf{B}$. We can simplify this by writing it in the component form as follows:

$$\boldsymbol{\sigma}\cdot\mathbf{A}\boldsymbol{\sigma}\cdot\mathbf{B} = \sum_i\sum_j (\sigma_i A_i)(\sigma_j B_j) = \sum_i\sum_j (\sigma_i\sigma_j) A_i B_j. \tag{5.64}$$

From (5.62) we write

$$\begin{aligned}\sum_i\sum_j (\sigma_i\sigma_j) A_i B_j &= \sum_i\sum_j (\delta_{ij} + i\sum_k \epsilon_{ijk}\sigma_k) A_i B_j \\ &= \sum_i\sum_j \delta_{ij} A_i B_j + i\sum_k \sigma_k \sum_i\sum_j \epsilon_{kij} A_i B_j.\end{aligned} \tag{5.65}$$

In the second term we have made an even number (two) of permutations to go from ϵ_{ijk} to ϵ_{kij}, which involves no sign change. Furthermore,

$$\sum_i\sum_j \delta_{ij} A_i B_j = \mathbf{A}.\mathbf{B} \tag{5.66}$$

and

$$\sum_i \sum_j \epsilon_{kij} A_i B_j = (\mathbf{A} \times \mathbf{B})_k \, . \tag{5.67}$$

Hence

$$\boldsymbol{\sigma} \cdot \mathbf{A} \boldsymbol{\sigma} \cdot \mathbf{B} = \mathbf{A} \cdot \mathbf{B} + \mathbf{i}\boldsymbol{\sigma} \cdot (\mathbf{A} \times \mathbf{B}) \, . \tag{5.68}$$

In particular,

$$\boldsymbol{\sigma} \cdot \mathbf{A} \boldsymbol{\sigma} \cdot \mathbf{A} = \mathbf{A}^2. \tag{5.69}$$

5.8 Arbitrary 2×2 matrices in terms of Pauli matrices

We show below that any 2×2 matrix can be expressed in terms of four matrices: the unit matrix and the three Pauli matrices σ_i.

A general 2×2 matrix has four complex matrix elements. Let M be such a matrix. We write it as

$$M = a\mathbf{1} + \mathbf{b}.\boldsymbol{\sigma} \tag{5.70}$$

where $\mathbf{1}$ and $\boldsymbol{\sigma}$ are 2×2 matrices and a and b_j $(j = 1, \ldots, 3)$ are complex numbers. We will show below that these four numbers can be uniquely determined in terms of the matrix elements of M.

Since

$$\mathrm{Tr}(\sigma_i) = 0 \tag{5.71}$$

and $\mathrm{Tr}(\mathbf{1}) = 2$, taking the trace of M we find

$$\mathrm{Tr}(M) = 2a. \tag{5.72}$$

Similarly, since

$$\mathrm{Tr}(\sigma_i \sigma_j) = 2\delta_{ij}, \tag{5.73}$$

we have,

$$\mathrm{Tr}(\sigma_j M) = b_j \mathrm{Tr}(\sigma_j^2) = 2b_j. \tag{5.74}$$

Thus

$$a = \frac{1}{2}\mathrm{Tr}(M) \quad \text{and} \quad b_j = \frac{1}{2}\mathrm{Tr}(\sigma_j M). \tag{5.75}$$

From the above relations one can determine the four (complex) coefficients on the right-hand side of (5.70) in terms of the four (complex) matrix elements of M. If M happens to be Hermitian then, since $\mathbf{1}$ and σ_i are Hermitian, we must have

$$a^* = a \quad \text{and} \quad b_j^* = b_j. \tag{5.76}$$

5.9 Projection operator for spin ½ systems

Let us consider the properties of the projection operator and the density matrix that we introduced in Chapter 1 as they relate to the spin ½ systems. We start with an arbitrary ket vector in the spin-space as

$$|\chi\rangle = \begin{pmatrix} c_1 \\ c_2 \end{pmatrix} \tag{5.77}$$

with $|c_1|^2 + |c_2|^2 = 1$. The bra vector, $\langle\chi|$, is a row matrix given by

$$\langle\chi| = \left(c_1^*, c_2^*\right). \tag{5.78}$$

The projection operator P_χ that we defined in Chapter 1 is given, for this system, by

$$P_\chi = |\chi\rangle\langle\chi| = \begin{pmatrix} |c_1|^2 & c_1 c_2^* \\ c_2 c_1^* & |c_2|^2 \end{pmatrix}. \tag{5.79}$$

It is easy to show that

$$P_\chi^2 = P_\chi, \, P_\chi^\dagger = P_\chi \quad \text{and} \quad \text{Tr}(P_\chi) = 1, \tag{5.80}$$

which confirms our derivation in Chapter 1.

Let us consider some examples of the projection operator. We consider the following cases.

(i) The spin-up state $|z+\rangle$. The projection operator for this state is

$$P_{z+} = |z+\rangle\langle z+| = (1, 0)\begin{pmatrix} 1 \\ 0 \end{pmatrix} = \begin{pmatrix} 1 & 0 \\ 0 & 0 \end{pmatrix}. \tag{5.81}$$

(ii) The spin-down state $|z-\rangle$. The projection operator for this state is

$$P_{z-} = |z-\rangle\langle z-| = (0, 1)\begin{pmatrix} 0 \\ 1 \end{pmatrix} = \begin{pmatrix} 0 & 0 \\ 0 & 1 \end{pmatrix}. \tag{5.82}$$

We now determine some further properties of the projection operator. From the above results we find that P_χ is a 2×2 Hermitian matrix. We showed in the previous section that any 2×2 Hermitian matrix can be expressed as a linear combination of the unit matrix and the three Pauli matrices σ_i with $i = 1, 2, 3$. Thus P_χ can be represented as

$$P_\chi = a_0 \mathbf{1} + \mathbf{a} \cdot \boldsymbol{\sigma}. \tag{5.83}$$

Since $\text{Tr}(\sigma_i) = 0$ and $\text{Tr}(\mathbf{1}) = 2$, $\text{Tr}(P_\chi) = 1$, we have

$$a_0 = \frac{1}{2}. \tag{5.84}$$

Hence,

$$P_\chi = \frac{1}{2}\mathbf{1} + \mathbf{a} \cdot \boldsymbol{\sigma} \tag{5.85}$$

and

$$P_\chi^2 = \frac{1}{4}\mathbf{1} + \mathbf{a} \cdot \boldsymbol{\sigma} + (\mathbf{a} \cdot \sigma)^2. \tag{5.86}$$

From the relation (5.38) for the Pauli matrices we obtain

$$(\mathbf{a} \cdot \boldsymbol{\sigma})^2 = \mathbf{a}^2. \tag{5.87}$$

Hence,

$$P_\chi^2 = \frac{1}{4}\mathbf{1} + \mathbf{a} \cdot \boldsymbol{\sigma} + \mathbf{a}^2 \tag{5.88}$$

where we note that

$$\mathbf{a}^2 = \left(a_x^2 + a_y^2 + a_z^2\right)\begin{pmatrix} 1 & 0 \\ 0 & 1 \end{pmatrix} = a^2\mathbf{1}. \tag{5.89}$$

Taking the trace of both sides of (5.88) we find, since $\text{Tr}(\sigma_i) = 0$,

$$\text{Tr}(P_\chi^2) = \frac{1}{4}\text{Tr}(\mathbf{1}) + a^2\,\text{Tr}(\mathbf{1}). \tag{5.90}$$

Using $\text{Tr}(\mathbf{1}) = 2$, and $\text{Tr}(P_\chi^2) = 1$, we find

$$a^2 = \frac{1}{4}. \tag{5.91}$$

For the purposes of simplification let us write

$$\mathbf{a} = \frac{1}{2}\mathbf{p} \tag{5.92}$$

where the vector $\mathbf{p}$ (which should not be confused as the momentum vector) has components p_x, p_y, p_z and

$$\mathbf{p}^2 = \mathbf{1}. \tag{5.93}$$

Thus,

$$P_\chi = \frac{1}{2}\left(\mathbf{1} + \mathbf{p}\cdot\boldsymbol{\sigma}\right). \tag{5.94}$$

We write P_χ in terms of the components of $\boldsymbol{\sigma}$ and $\mathbf{p}$ as

$$P_\chi = \frac{1}{2}\begin{pmatrix} 1+p_z & p_x - ip_y \\ p_x + ip_y & 1-p_z \end{pmatrix} \tag{5.95}$$

where we have substituted the matrix representations of σ_x, σ_y, and σ_z obtained in (5.13), (5.36), and (5.37).

Let us discuss the significance of the vector $\mathbf{p}$. Based on our results in Chapter 1 we note that the expectation value of an operator A is related to the projection operator as follows:

$$\langle\chi|A|\chi\rangle = \mathrm{Tr}\left(AP_\chi\right). \tag{5.96}$$

The expectation value of the Pauli matrices with respect to the state $|\chi\rangle$ is given by

$$\langle\chi|\boldsymbol{\sigma}|\chi\rangle = \mathrm{Tr}\left(\boldsymbol{\sigma}P_\chi\right) = \frac{1}{2}\mathrm{Tr}\left(\boldsymbol{\sigma} + \boldsymbol{\sigma}\left(\mathbf{p}\cdot\boldsymbol{\sigma}\right)\right) = \frac{1}{2}\mathrm{Tr}\left(\boldsymbol{\sigma}\left(\mathbf{p}\cdot\boldsymbol{\sigma}\right)\right). \tag{5.97}$$

To determine the right-hand side in (5.97), let us first consider just the contribution of the x-component $\boldsymbol{\sigma}$:

$$\mathrm{Tr}\left(\sigma_x\left(\mathbf{p}\cdot\boldsymbol{\sigma}\right)\right) = \mathrm{Tr}\left(\sigma_x\left(p_x\sigma_x + p_y\sigma_y + p_z\sigma_z\right)\right) = p_x\mathrm{Tr}\left(\sigma_x^2\right) = 2p_x. \tag{5.98}$$

Therefore, in general,

$$\mathrm{Tr}\left(\boldsymbol{\sigma}\left(\mathbf{p}\cdot\boldsymbol{\sigma}\right)\right) = 2\mathbf{p} \tag{5.99}$$

and, hence,

$$\langle\chi|\boldsymbol{\sigma}|\chi\rangle = \mathbf{p}. \tag{5.100}$$

The vector $\mathbf{p}$ is thus the expectation value of the spin operator $\boldsymbol{\sigma}$. For this reason it is often called the polarization vector.

5.10 Density matrix for spin ½ states and the ensemble average

Following our discussion in Chapter 1, the density operator for spin ½ states in a mixed ensemble is written as

$$\rho = w_\alpha P_\alpha + w_\beta P_\beta \tag{5.101}$$

where w_α and w_β are the probabilities that the states $|\alpha\rangle$ and $|\beta\rangle$ are present in the mixed ensemble. The projection operators P_α and P_β are given by

$$P_\alpha = |\alpha\rangle\langle\alpha|, \quad P_\beta = |\beta\rangle\langle\beta|. \tag{5.102}$$

If the states $|\alpha\rangle$ and $|\beta\rangle$ are orthonormal and complete, then they will satisfy the completeness relation,

$$P_\alpha + P_\beta = \mathbf{1}. \tag{5.103}$$

Consider, for example, the case where the mixed ensemble contains the spin-up and spin-down states

$$|\alpha\rangle = \begin{pmatrix} 1 \\ 0 \end{pmatrix}, \quad |\beta\rangle = \begin{pmatrix} 0 \\ 1 \end{pmatrix}. \tag{5.104}$$

Then the projection operators are found to be

$$P_\alpha = \begin{pmatrix} 1 & 0 \\ 0 & 0 \end{pmatrix}, \quad P_\beta = \begin{pmatrix} 0 & 0 \\ 0 & 1 \end{pmatrix}, \tag{5.105}$$

which satisfy the completeness relation (5.103).

In terms of the polarization vector defined this time as $\mathbf{P} = \langle\alpha|\boldsymbol{\sigma}|\boldsymbol{\alpha}\rangle$, we find using (5.104) and (5.105) that

$$P_\alpha = \frac{1}{2}\left[\mathbf{1} + \mathbf{P}\cdot\boldsymbol{\sigma}\right], \tag{5.106}$$

$$P_\beta = \mathbf{1} - P_\alpha = \frac{1}{2}\left[\mathbf{1} - \mathbf{P}\cdot\boldsymbol{\sigma}\right]. \tag{5.107}$$

Hence the density operator is given by

$$\rho = w_\alpha P_a + w_\beta P_\beta \tag{5.108}$$

$$= \frac{1}{2}\left[\mathbf{1} + \left(w_\alpha - w_\beta\right)\mathbf{P}\cdot\boldsymbol{\sigma}\right] \tag{5.109}$$

where we have taken into account the fact that the sum of the probabilities must be 1, i.e., $w_\alpha + w_\beta = 1$.

The ensemble average of the Pauli matrix can now be obtained as

$$\langle\boldsymbol{\sigma}\rangle_{\text{av}} = \text{Tr}\left(\boldsymbol{\sigma}\rho\right) \tag{5.110}$$

$$= \text{Tr}\left[\boldsymbol{\sigma}\frac{1}{2}\left(\mathbf{1} + \left(w_\alpha - w_\beta\right)\mathbf{P}\cdot\boldsymbol{\sigma}\right)\right]. \tag{5.111}$$

Hence,

$$\langle\boldsymbol{\sigma}\rangle_{\text{av}} = \frac{1}{2}\text{Tr}\left(\boldsymbol{\sigma}\right) + \frac{1}{2}\left(w_\alpha - w_\beta\right)\text{Tr}\left[\boldsymbol{\sigma}\left(\mathbf{P}\cdot\boldsymbol{\sigma}\right)\right]. \tag{5.112}$$

Since $\text{Tr}(\boldsymbol{\sigma}) = 0$, we obtain, using (5.99), the following result:

$$\langle\boldsymbol{\sigma}\rangle_{\text{av}} = \left(w_\alpha - w_\beta\right)\mathbf{P}. \tag{5.113}$$

5.11 Complete wavefunction

Finally, in our discussions above we have been concerned only with the spin part of the wavefunction. The total wavefunction for a free particle with spin ½, described by a plane wave, can be written as the product wavefunction

$$u(\mathbf{r}) = \frac{e^{i\mathbf{k}.\mathbf{r}}}{\left(\sqrt{2\pi}\right)^3}|\chi\rangle \tag{5.114}$$

where $|\chi\rangle$ is the column matrix

$$|\chi\rangle = c_1\begin{pmatrix}1\\0\end{pmatrix} + c_2\begin{pmatrix}0\\1\end{pmatrix} = \begin{pmatrix}c_1\\c_2\end{pmatrix}, \tag{5.115}$$

which signifies that the particle has a probability $|c_1|^2$ that it is in a spin-up state, i.e., it has its spin pointing in the positive z-direction, and the probability $|c_2|^2$ that it is in a spin-down state with spin pointing in the opposite direction.

5.12 Pauli exclusion principle and Fermi energy

Let us consider the ground state consisting of N noninteracting electrons confined in one dimension of length L. Each electron will be described by a free wavefunction

$$u(x) = \frac{1}{\sqrt{L}}e^{ikx}\chi_\lambda \tag{5.116}$$

where the χ_λ's designate the spin-up and spin-down states

$$\chi_+ = \begin{bmatrix}1\\0\end{bmatrix}, \quad \chi_- = \begin{bmatrix}0\\1\end{bmatrix}, \tag{5.117}$$

which are normalized according to

$$\chi^\dagger_{\lambda_1}\chi_{\lambda_2} = \delta_{\lambda_1\lambda_2}. \tag{5.118}$$

We assume, as we have done before, that the wavefunction satisfies the periodic boundary condition

$$u(x+L) = u(x), \tag{5.119}$$

which implies that the momentum vector can take only discrete values given by

$$k_nL = 2n\pi. \tag{5.120}$$

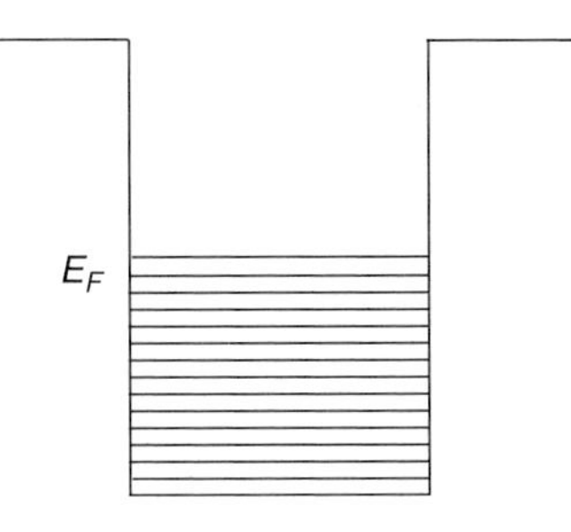

Fig. 5.1

The energy eigenvalues will then be

$$E_n = \frac{\hbar^2 k_n^2}{2m} = \frac{2n^2\pi^2\hbar^2}{mL^2}. \tag{5.121}$$

Since the electrons are noninteracting, the ground state of the N-electron system can be built up by putting electrons into different levels. Because of the exclusion principle, however, at most one electron can be placed in each level with a given value of k_n. Since an electron has spin that can take on two values, spin up and spin down, we are allowed to put no more than two electrons in a given level as long as their spins are in opposite directions.

Thus N electrons will fill up levels with $n = 1, 2, \ldots, N/2$ (see Fig. 5.1). The last level will consist of either one or two electrons depending on whether N is even or odd. The energy of the highest level can then be obtained by substituting $n = N/2$ in (5.121), which gives

$$E_F = \frac{N^2\pi^2\hbar^2}{2mL^2}. \tag{5.122}$$

This is the so called "Fermi energy" for the one-dimensional case. Fermi energy is a very important concept in condensed matter systems.

The total energy for the N-electron system is then

$$E_{\text{tot}} = 2\sum_{n=1}^{N/2} \frac{2n^2\pi^2\hbar^2}{mL^2} \tag{5.123}$$

where the factor 2 corresponds to the two spin states. Since N is assumed to be very large, the above sum can be converted to an integral:

$$\sum_{n=1}^{N/2} n^2 \approx \int_1^{N/2} dn\, n^2 \simeq \frac{N^3}{24}. \tag{5.124}$$

Thus the total energy of the electrons is

$$E_{\text{tot}} = \frac{N^3\pi^2\hbar^2}{6mL^2}. \tag{5.125}$$

The energy per electron is

$$E_e = \frac{E_{\text{tot}}}{N} = \frac{N^2\pi^2\hbar^2}{6mL^2}, \tag{5.126}$$

which implies that the energy of an individual electron increases as N^2.

Let us compare this result with a system consisting of N bosons, which are particles with integral spin. Since there is no exclusion principle preventing the bosons from occupying the same state, their ground state will consist of all N particles occupying the same state, $n = 1$. Thus the total energy of this system will be N times the ground-state energy. Hence,

$$E_{\text{tot}} = N\frac{2\pi^2\hbar^2}{mL^2} \tag{5.127}$$

and the energy of a single boson is

$$E_b = \frac{E_{\text{tot}}}{N} = \frac{\pi^2\hbar^2}{2mL^2} \tag{5.128}$$

which remains a constant. This result is in sharp contrast to the case of electrons.

We have confined the above to just the one-dimensional case. We will discuss the three-dimensional case in Chapter 38.

5.13 Problems

1. Determine the matrices S_x, S_y, S_z for spin 1 in the representation in which S_z and $\mathbf{S}^2$ are diagonal.
2. Obtain the eigenvalues and eigenstates of the operator $A = a\sigma_y + b\sigma_z$. Call the two eigenstates $|1\rangle$ and $|2\rangle$ and determine the probabilities that they will correspond to $\sigma_x = +1$.
3. Obtain the expectation values of S_x, S_y, and S_z for the case of a spin ½ particle with the spin pointed in the direction of a vector with azimuthal angle β and polar angle α.
4. Take the azimuthal angle, $\beta = 0$, so that the spin is in the $x - z$ plane at an angle α with respect to the z-axis, and the unit vector is $\mathbf{n} = (\sin\alpha, 0, \cos\alpha)$. Write

   $$|\chi_{n+}\rangle = |+\alpha\rangle$$

   for this case. Show that the probability that it is in the spin-up state in the direction θ with respect to the z-axis is

   $$|< +\theta\,|+\alpha\rangle|^2 = \cos^2\frac{(\alpha - \theta)}{2}.$$

   Also obtain the expectation value of $\boldsymbol{\sigma}\cdot\mathbf{n}$ with respect to the state $|+\theta\rangle$.
5. Consider an arbitrary density matrix, ρ, for a spin ½ system. Express each matrix element in terms of the ensemble averages $\langle S_i\rangle_{av}$ where $i = x, y, z$.

6. If a Hamiltonian is given by $\boldsymbol{\sigma}.\mathbf{n}$ where $\mathbf{n} = (\sin\alpha\cos\beta, \sin\alpha\sin\beta, \cos\alpha)$, determine the time evolution operator as a 2×2 matrix. If a state at $t = 0$ is given by

$$|\phi(0)\rangle = \begin{bmatrix} a \\ b \end{bmatrix},$$

then obtain $|\phi(t)\rangle$.
7. Consider a system of spin ½ particles in a mixed ensemble containing a mixture of 25% with spin given by $|z+\rangle$ and 75% with spin given by $|x-\rangle$. Determine the density matrix ρ and ensemble averages $\langle\sigma_i\rangle$ for $i = x, y, z$.
8. Show that the quantity, $\boldsymbol{\sigma}\cdot\mathbf{p}V(r)\,\boldsymbol{\sigma}\cdot\mathbf{p}$, when simplified, has a term proportional to $\mathbf{L}\cdot\sigma$.
9. Consider the following Hamiltonian for two spin ½ particles:

$$H = \lambda\boldsymbol{\sigma}_1\cdot\boldsymbol{\sigma}_2.$$

Designating the states as the product $|1/2, m_1\rangle\,|1/2, m_2\rangle$, and expressing H in terms of the raising and lowering operators for the individual spins, show that the eigenstates of H must have $m_1 = m_2 = 1/2$ or $m_1 = m_2 = -1/2$. Writing H as a matrix with respect to these two states, obtain the eigenvalues of the system.

6 Gauge invariance, angular momentum, and spin

Gauge invariance is a very important concept for electromagnetic interactions. It gives rise to some far-reaching conclusions when applied to quantum-mechanical problems of systems with angular momentum, particularly spin ½ systems. It also plays a fundamental role in the standard model of particle physics. Even though in doing this we will be departing from the subject of free particles, this is an important enough concept to be treated at this stage since it does not involve complicated details about the interactions.

6.1 Gauge invariance

The electric field, $\mathbf{E}$, and magnetic field, $\mathbf{B}$, are described in terms of the scalar potential, ϕ, and the vector potential, $\mathbf{A}$, respectively as

$$\mathbf{E} = -\boldsymbol{\nabla}\boldsymbol{\phi} - \frac{1}{c}\frac{\partial A}{\partial t} \tag{6.1}$$

and

$$\mathbf{B} = \nabla \times \mathbf{A}. \tag{6.2}$$

We note that there is a certain arbitrariness in these definitions since the expressions for $\mathbf{E}$ and $\mathbf{B}$ remain unchanged under the transformations

$$\mathbf{A} \to \mathbf{A} + \nabla\Lambda \tag{6.3}$$

and

$$\phi \to \phi - \frac{1}{c}\frac{\partial \Lambda}{\partial t}. \tag{6.4}$$

These are called gauge transformations. The electromagnetic field is then said to be invariant under gauge transformation. Gauge invariance is one of the most important subjects that we will discuss in greater detail in the later chapters. We will be concerned here only with a few of the special aspects of it.

6.2 Quantum mechanics

We now discuss the presence of electromagnetic interactions in quantum-mechanical problems and the consequences of imposing gauge invariance. We will consider only the time-independent case. The gauge transformation then corresponds to

$$\mathbf{A} \rightarrow \mathbf{A} + \nabla \Lambda \tag{6.5}$$

and

$$\phi \rightarrow \phi. \tag{6.6}$$

In classical mechanics the electromagnetic interactions are taken into account by the following replacements in the momentum $\mathbf{p}$ and energy E,

$$\mathbf{p} \rightarrow \mathbf{p} - \frac{e}{c}\mathbf{A}, \tag{6.7}$$

$$E \rightarrow E - e\phi. \tag{6.8}$$

which imply the following replacement in the Hamiltonian, H,

$$H = \frac{\mathbf{p}^2}{2m} \rightarrow H = \frac{\left(\mathbf{p} - \frac{e}{c}\mathbf{A}\right)^2}{2m} - e\phi \tag{6.9}$$

The Hamiltonian in quantum mechanics is also assumed to be

$$H = \frac{\left(\mathbf{p} - \frac{e}{c}\mathbf{A}\right)^2}{2m} \tag{6.10}$$

where we have taken $\phi = 0$, confining just to the magnetic fields. We note that $\mathbf{p}$ and $\mathbf{r}$ are operators, and $\mathbf{A}$ is a functions of $\mathbf{r}$ only. We show below that the Hamiltonian as written in (6.10) is gauge invariant.

Let $\mathbf{A}_G$ represent $\mathbf{A}$ under gauge transformation

$$\mathbf{A}_G = \mathbf{A} + \nabla \Lambda \tag{6.11}$$

where $\Lambda = \Lambda(\mathbf{r})$ is an arbitrary quantity which is a function of $\mathbf{r}$ only. We take $|\alpha\rangle$ to be a state vector, which, under gauge transformations, is given by

$$|\alpha\rangle \rightarrow |\alpha_G\rangle . \tag{6.12}$$

We assume that under gauge transformations the norm of the state vectors and the expectation values of the operators remain unchanged:

$$\langle \alpha_G | \alpha_G \rangle = \langle \alpha | \alpha \rangle , \tag{6.13}$$

$$\langle \alpha_G | \mathbf{r} | \alpha_G \rangle = \langle \alpha | \mathbf{r} | \alpha \rangle , \tag{6.14}$$

$$\langle \alpha_G | \left(\mathbf{p} - \frac{e}{c} \mathbf{A}_G \right) | \alpha_G \rangle = \langle \alpha | \left(\mathbf{p} - \frac{e}{c} \mathbf{A} \right) | \alpha \rangle . \tag{6.15}$$

To satisfy relation (6.13) we write

$$|\alpha_G\rangle = e^{i\Phi} |\alpha\rangle . \tag{6.15a}$$

If Φ is assumed to be a (real) constant then the relations (6.13), (6.14), and (6.15a) are trivially satisfied. This type of invariance is called "global" gauge invariance.

Let us, however, consider Φ to be a (real) function of $\mathbf{r}$, in which case the invariance is called "local" gauge invariance. The relations (6.13) and (6.14) are then automatically satisfied. However, we immediately note that

$$\langle \alpha_G | \mathbf{p} | \alpha_G \rangle \neq \langle \alpha | \mathbf{p} | \alpha \rangle , \tag{6.16}$$

$$\langle \alpha_G | \frac{e}{c} \mathbf{A}_G | \alpha_G \rangle \neq \langle \alpha | \frac{e}{c} \mathbf{A} | \alpha \rangle \tag{6.17}$$

because $\mathbf{p}$ $(= -i\hbar\boldsymbol{\nabla})$ while the relation (6.15a) between $|\alpha_G\rangle$ and $|\alpha\rangle$ involves a function of $\mathbf{r}$; and $\mathbf{A}_G$ given by (6.11) is different from $\mathbf{A}$. The combination $(\mathbf{p} - e/c\mathbf{A}_G)$, on the other hand, can be gauge invariant, as we will demonstrate below.

Writing the operator relation $\mathbf{p} = -i\hbar\boldsymbol{\nabla}$, we find

$$e^{-i\Phi} \mathbf{p} e^{i\Phi} |\alpha\rangle = e^{-i\Phi} (-i\hbar) \boldsymbol{\nabla} \left(e^{i\Phi} |\alpha\rangle \right) = [\hbar (\boldsymbol{\nabla} \Phi) + \mathbf{p}] |\alpha\rangle . \tag{6.18}$$

Therefore,

$$\langle \alpha_G | \left(\mathbf{p} - \frac{e}{c} \mathbf{A}_G \right) | \alpha_G \rangle = \langle \alpha | \left[\hbar \boldsymbol{\nabla} \Phi + \mathbf{p} - \frac{e}{c} \mathbf{A} - \frac{e}{c} \boldsymbol{\nabla} \Lambda \right] | \alpha \rangle . \tag{6.19}$$

To satisfy the above relation we must then have

$$\hbar \boldsymbol{\nabla} \Phi = \frac{e}{c} \boldsymbol{\nabla} \Lambda. \tag{6.20}$$

Hence, apart from an arbitrary constant, which we will neglect, our solution for Φ will be

$$\Phi = \frac{e\Lambda}{\hbar c}. \tag{6.21}$$

The relation between the two state vectors is then given by

$$|\alpha_G\rangle = e^{i\frac{e\Lambda}{\hbar c}} |\alpha\rangle . \tag{6.21a}$$

Having shown that

$$e^{-i\frac{e\Lambda}{\hbar c}}\left(\mathbf{p}-\frac{e}{c}\mathbf{A}_G\right)e^{i\frac{e\Lambda}{\hbar c}}=\mathbf{p}-\frac{e}{c}\mathbf{A}, \tag{6.22}$$

one can also show that

$$e^{-i\frac{e\Lambda}{\hbar c}}\left(\mathbf{p}-\frac{e}{c}\mathbf{A}_G\right)^2 e^{i\frac{e\Lambda}{\hbar c}}=\left(\mathbf{p}-\frac{e}{c}\mathbf{A}\right)^2. \tag{6.23}$$

From this relation one finds that

$$\langle\alpha_G|\,H_G\,|\alpha_G\rangle=\langle\alpha|\,H\,|\alpha\rangle\,. \tag{6.24}$$

Thus the Hamiltonian (6.10) is gauge invariant.

The physics will, therefore, be gauge invariant, so that the observable quantities such as energy levels will not depend on any specific gauge. However, for the purposes of calculations one may want to use the most convenient gauge available. The most commonly employed gauge is the transverse gauge,

$$\nabla\cdot\mathbf{A}=0. \tag{6.25}$$

Within this transverse gauge one also uses the so called Landau gauge and symmetric gauge. We will consider both types when we discuss the topic of Landau levels.

6.3 Canonical and kinematic momenta

If we denote the ith components of $\mathbf{r}$ and $\mathbf{p}$ by x_i and p_i, respectively, then the following Heisenberg relation can be written:

$$i\hbar\frac{dx_i}{dt}=[x_i,H] \tag{6.26}$$

From the fundamental commutator $\left[x_i,p_j\right]=i\hbar\delta_{ij}$, one can easily show that

$$\left[x_i,\mathbf{p}^2\right]=2i\hbar p_i. \tag{6.27}$$

Since $\left[x_i,A_j\right]=0$ for arbitrary i and j, from the Hamiltonian (6.10) we obtain

$$\frac{dx_i}{dt}=\frac{1}{m}\left[p_i-\frac{e}{c}A_i\right]. \tag{6.28}$$

We define

$$m\frac{dx_i}{dt}=\text{mechanical momentum }=P_i=p_i-\frac{e}{c}A_i, \tag{6.29}$$

$$p_i=\text{canonical momentum}, \tag{6.30}$$

and note that for the canonical momentum we have

$$\left[x_i, p_j\right] = i\hbar\delta_{ij} \quad \text{and} \quad \left[p_i, p_j\right] = 0. \tag{6.31}$$

But for P_i, while $\left[x_i, P_j\right] = i\hbar\delta_{ij}$, we have,

$$\left[P_i, P_j\right] = \left[\left(p_i - \frac{e}{c}A_i\right), \left(p_j - \frac{e}{c}A_j\right)\right] \tag{6.32}$$

$$= \left(-\frac{e}{c}\right)\left\{\left[p_i, A_j\right] + \left[A_i, p_j\right]\right\} \tag{6.33}$$

where

$$\left[p_i, A_j\right] = p_i A_j - A_j p_i. \tag{6.34}$$

We need to keep in mind that the terms on each side of the above equation operate on a wavefunction on the right-hand side; therefore, we will need to use the operator relation $\mathbf{p} = -i\hbar\nabla$. Thus, the product $p_i A_j$ gives the following result with ϕ $(= \langle x | \alpha \rangle)$ as the wavefunction:

$$\left[p_i, A_j\right]\phi = \left(p_i A_j - A_j p_i\right)\phi = -i\hbar\frac{\partial}{\partial x_i}\left(A_j\phi\right) + i\hbar A_j\left(\frac{\partial \phi}{\partial x_i}\right) = -i\hbar\left(\frac{\partial A_j}{\partial x_i}\right)\phi. \tag{6.35}$$

Hence, after removing the arbitrary ϕ, we have

$$\left[P_i, P_j\right] = \left(\frac{ie\hbar}{c}\right)\left(\frac{\partial A_j}{\partial x_i} - \frac{\partial A_i}{\partial x_j}\right) = \left(\frac{ie\hbar}{c}\right)\epsilon_{ijk}\left(\nabla \times \mathbf{A}\right)_k = \left(\frac{ie\hbar}{c}\right)\epsilon_{ijk}B_k \tag{6.36}$$

where we have used the relation (6.2). Thus, we obtain the commutator of the canonical momenta

$$\left[P_i, P_j\right] = \left(\frac{ie\hbar}{c}\right)\epsilon_{ijk}B_k. \tag{6.37}$$

6.4 Probability conservation

We consider probability conservation in the presence of the magnetic field. Let us take $\mathbf{A}$ to be real and assume the transverse gauge $\nabla \cdot \mathbf{A} = 0$. The time-dependent Schrödinger equation

$$i\hbar\frac{\partial \psi}{\partial t} = H\psi \tag{6.38}$$

can be written down from (6.10), after substituting $\mathbf{p} = -i\hbar\nabla$, as

$$-\frac{\hbar^2}{2m}\nabla^2\psi + \frac{ie\hbar}{mc}\mathbf{A}\cdot\nabla\psi + \frac{e^2}{2mc}\mathbf{A}^2\psi = i\hbar\frac{\partial \psi}{\partial t}. \tag{6.39}$$

Let us now take the complex conjugate of this equation,

$$-\frac{\hbar^2}{2m}\nabla^2\psi^* - \frac{ie\hbar}{mc}\mathbf{A}\cdot\nabla\boldsymbol{\psi}^* + \frac{e^2}{2mc}\mathbf{A}^2\psi^* = -i\hbar\frac{\partial\psi^*}{\partial t}. \tag{6.40}$$

We now multiply both sides of (6.39) and (6.40), respectively, by $\boldsymbol{\psi}^*$ and ψ and make a subtraction. We obtain

$$-\frac{\hbar^2}{2m}\left[\psi^*\nabla^2\psi - \psi\nabla^2\psi^*\right] + \frac{ie\hbar}{mc}\left[\boldsymbol{\psi}^*\mathbf{A}\cdot\nabla\psi + \psi\mathbf{A}\cdot\nabla\boldsymbol{\psi}^*\right] = i\hbar\left[\boldsymbol{\psi}^*\frac{\partial\psi}{\partial t} + \psi\frac{\partial\psi^*}{\partial t}\right]. \tag{6.41}$$

We can write this relation as the conservation relation

$$\nabla\cdot\mathbf{j} + \frac{\partial\rho}{\partial t} = 0 \tag{6.42}$$

where the probability density is the same as previously defined:

$$\rho = |\psi|^2. \tag{6.43}$$

However, the probability current density is then

$$\mathbf{j} = \frac{\hbar}{2im}\left[\psi^*\nabla\psi - \psi\nabla\psi^*\right] - \frac{e}{mc}\mathbf{A}\,|\psi|^2. \tag{6.44}$$

In comparing this with the expression we derived in Chapter 3, we note the additional term on the right-hand side. We will discuss this further when we come to the Meissner effect in Chapter 39.

6.5 Interaction with the orbital angular momentum

Let us express the Hamiltonian in the presence of a magnetic field as

$$H = \frac{\left(\mathbf{p} - \frac{e}{c}\mathbf{A}\right)^2}{2m} = \frac{\mathbf{p}^2}{2m} - \frac{e}{2mc}\left(\mathbf{p}\cdot\mathbf{A} + \mathbf{A}\cdot\mathbf{p}\right) + \frac{e^2}{2mc^2}\mathbf{A}^2. \tag{6.45}$$

The second term on the right-hand side gives, after substituting $\mathbf{p} = -i\hbar\nabla$ and introducing the wavefunction ϕ,

$$\begin{aligned}(\mathbf{p}\cdot\mathbf{A} + \mathbf{A}\cdot\mathbf{p})\,\phi &= -i\hbar\left[\nabla\cdot(\mathbf{A}\phi) + \mathbf{A}\cdot(\nabla\phi)\right]\\ &= -i\hbar\left[(\nabla\cdot\mathbf{A})\,\phi + 2\mathbf{A}\cdot(\nabla\phi)\right].\end{aligned} \tag{6.46}$$

We choose the transverse gauge so that the condition

$$\nabla\cdot\mathbf{A} = 0 \tag{6.47}$$

is satisfied. We then obtain

$$(\mathbf{p} \cdot \mathbf{A} + \mathbf{A} \cdot \mathbf{p}) = 2\mathbf{A} \cdot \mathbf{p}. \tag{6.48}$$

Let us now take

$$\mathbf{A} = \frac{1}{2}(\mathbf{B} \times \mathbf{r}) \tag{6.49}$$

where $\mathbf{r}$ is the radius vector. This expression satisfies the above gauge condition for $\mathbf{A}$ and reproduces $\nabla \times \mathbf{A} = \mathbf{B}$, where $\mathbf{B}$ is the magnetic field. Hence

$$(\mathbf{p} \cdot \mathbf{A} + \mathbf{A} \cdot \mathbf{p}) = (\mathbf{B} \times \mathbf{r}) \cdot \mathbf{p} \tag{6.50}$$

Using the vector identity $(\mathbf{A} \times \mathbf{B}) \cdot \mathbf{C} = \mathbf{A} \cdot (\mathbf{B} \times \mathbf{C})$ we obtain

$$(\mathbf{B} \times \mathbf{r}) \cdot \mathbf{p} = \mathbf{B} \cdot (\mathbf{r} \times \mathbf{p}) = \mathbf{B}.\mathbf{L} \tag{6.51}$$

where $\mathbf{L} = \mathbf{r} \times \mathbf{p}$ is the orbital angular momentum. Thus we have

$$\frac{\left(\mathbf{p} - \frac{e}{c}\mathbf{A}\right)^2}{2m} = \frac{\mathbf{p}^2}{2m} - \frac{e}{2mc}\mathbf{B}.\mathbf{L} + \frac{e^2}{2mc^2}\mathbf{A}^2. \tag{6.52}$$

Therefore, inherent in the Hamiltonian is a term that corresponds to the interaction of the magnetic field with the orbital angular momentum. The second term signifies an interaction of the type $-\boldsymbol{\mu} \cdot \mathbf{B}$, which corresponds to the interaction of the magnetic field with the magnetic moment due to the current generated through the particle's orbital motion, given by

$$\boldsymbol{\mu} = \frac{e}{2mc}\mathbf{L}. \tag{6.53}$$

6.6 Interaction with spin: intrinsic magnetic moment

In the Dirac theory of electrons for a free Dirac particle, the kinetic energy term in the Hamiltonian in the nonrelativistic approximation is found to be of the form

$$H = \frac{\boldsymbol{\sigma} \cdot \mathbf{p}\boldsymbol{\sigma} \cdot \mathbf{p}}{2m}. \tag{6.54}$$

If one uses the relation

$$\sigma_i\sigma_j = \delta_{ij} + i\sum_k \epsilon_{ijk}\sigma_k \tag{6.55}$$

then the Hamiltonian reduces to

$$H = \frac{\mathbf{p}^2}{2m}, \tag{6.56}$$

recovering the result from nonrelativistic mechanics.

The innocuous looking expression (6.54), however, has profound implications for the case when the electron, which is a spin ½ particle, is subjected to a magnetic field, $\mathbf{B} = \nabla \times \mathbf{A}$, where $\mathbf{A}$ is the vector potential. As we discussed earlier, one can incorporate the magnetic interactions by the replacement

$$\mathbf{p} \rightarrow \mathbf{p} - \frac{e}{c}\mathbf{A}, \tag{6.57}$$

which corresponds to the replacement

$$\frac{\mathbf{p}^2}{2m} \rightarrow \frac{\left(\mathbf{p} - \frac{e}{c}\mathbf{A}\right)^2}{2m}. \tag{6.58}$$

However, if before going to (6.58) from (6.56) we incorporate the transformation (6.57) in the nonrelativistic Dirac term (6.54), we will have

$$\frac{\boldsymbol{\sigma}.\left(\mathbf{p} - \frac{\mathbf{e}}{\mathbf{c}}\mathbf{A}\right)\boldsymbol{\sigma}.\left(\mathbf{p} - \frac{e}{c}\mathbf{A}\right)}{2m}. \tag{6.59}$$

This does not reduce simply to (6.59).

There are actually additional terms which we now discuss:

$$\begin{aligned} \sigma \cdot \left(\mathbf{p} - \frac{e}{c}\mathbf{A}\right) \boldsymbol{\sigma} \cdot \left(\mathbf{p} - \frac{e}{c}\mathbf{A}\right) &= \left(\mathbf{p} - \frac{e}{c}\mathbf{A}\right)^2 + i\sigma \cdot \left(\mathbf{p} - \frac{e}{c}\mathbf{A}\right) \times \left(\mathbf{p} - \frac{e}{c}\mathbf{A}\right) \\ &= \left(\mathbf{p} - \frac{e}{c}\mathbf{A}\right)^2 - i\frac{e}{c}\boldsymbol{\sigma} \cdot (\mathbf{p} \times \mathbf{A} + \mathbf{A} \times \mathbf{p}) \end{aligned} \tag{6.60}$$

The second term on the right-hand side can be simplified as follows:

$$\begin{aligned} (\mathbf{p} \times \mathbf{A} + \mathbf{A} \times \mathbf{p})\,\psi &= -i\hbar\left[\nabla \times (\mathbf{A}\psi) + (\mathbf{A} \times \nabla\psi)\right] \\ &= -i\hbar\left[(\nabla \times \mathbf{A})\,\psi + (\nabla\psi \times \mathbf{A}) + (\mathbf{A} \times \nabla\psi)\right] \\ &= -i\hbar\,(\nabla \times \mathbf{A})\,\psi. \end{aligned} \tag{6.61}$$

Since $\nabla \times \mathbf{A} = \mathbf{B}$, where $\mathbf{B}$ is the magnetic field, we obtain

$$\sigma \cdot \left(\mathbf{p} - \frac{e}{c}\mathbf{A}\right) \sigma \cdot \left(\mathbf{p} - \frac{e}{c}\mathbf{A}\right) = \left(\mathbf{p} - \frac{e}{c}\mathbf{A}\right)^2 - \frac{e\hbar}{c}\boldsymbol{\sigma} \cdot \mathbf{B}. \tag{6.62}$$

Therefore, substituting $\mathbf{S} = (\hbar/2)\,\boldsymbol{\sigma}$, we find for the Hamiltonian

$$H = \frac{\boldsymbol{\sigma}.\left(\mathbf{p} - \frac{\mathbf{e}}{\mathbf{c}}\mathbf{A}\right)\boldsymbol{\sigma}.\left(\mathbf{p} - \frac{e}{c}\mathbf{A}\right)}{2m} = \frac{1}{2m}\left(\mathbf{p} - \frac{e}{c}\mathbf{A}\right)^2 - \frac{e}{mc}\mathbf{S} \cdot \mathbf{B} \tag{6.63}$$

The first term was expected, but we now have a term of the form $\boldsymbol{\mu}_e \cdot \mathbf{B}$ with $\boldsymbol{\mu}_e$ as the magnetic moment,

$$\boldsymbol{\mu}_e = \frac{e}{mc}\mathbf{S} \tag{6.64}$$

This is called the intrinsic magnetic moment of the electron.

In the case of orbital angular momentum discussed in the previous section, a similar term, **B**.**L**, was present, which has a simple physical interpretation: the orbital motion of an electron creates a current loop which is responsible for generating a magnetic moment. The spin of the electron, on the other hand, is an abstract quantity as presented here since the electron is assumed to be a point particle. The fact that it generates a magnetic moment that can have physical effects, as we will see in the next chapter on Stern–Gerlach experiments, is a profound result and a confirmation of Dirac's theory. A fully relativistic Dirac theory, of which (6.56) is an approximation, will be considered in Chapter 33.

Combining the orbital and intrinsic terms, we can write

$$H = \frac{\mathbf{p}^2}{2m} - \frac{e}{2mc}\boldsymbol{\mu}\cdot\mathbf{B} + \frac{e^2}{2mc^2}\mathbf{A}^2 \tag{6.65}$$

where the magnetic moment can be expressed as a sum

$$\boldsymbol{\mu} = g_L\mathbf{L} + g_S\mathbf{S}. \tag{6.66}$$

The g_i's are called gyromagnetic ratios with

$$g_L = 1 \quad \text{and} \quad g_S = 2. \tag{6.67}$$

6.7 Spin-orbit interaction

In previous sections we took the electrostatic potential $\phi = 0$. Let us now include its contribution to the Hamiltonian. The presence of an electron spin creates a new term, which is called the spin–orbit term. Taking ϕ to be a central potential, e.g., Coulomb potential, between electron and proton in the hydrogen atom, we write

$$\phi = V_c(r). \tag{6.68}$$

The electric field due to V_c is

$$\mathbf{E} = -\nabla V_c = -\frac{\mathbf{r}}{r}\frac{dV_c}{dr}. \tag{6.69}$$

Since in the hydrogen atom the electron moves in an orbit around the proton, it will feel a magnetic field given by

$$\mathbf{B} = -\frac{\mathbf{v}}{c} \times \mathbf{E} \tag{6.70}$$

where $\mathbf{v}$ is the velocity of the electron. This magnetic field will interact with the electron's intrinsic magnetic moment. The interaction will be given by

$$H' = -\boldsymbol{\mu}\cdot\mathbf{B}. \tag{6.71}$$

Writing this in terms of $\mathbf{S}$ as given by (6.64) we obtain

$$H' = \frac{e}{mc}\mathbf{S}\cdot\left[\frac{\mathbf{v}}{c}\times\mathbf{E}\right]. \tag{6.72}$$

Inserting the nonrelativistic expression $\mathbf{v} = \mathbf{p}/m$, we find

$$H' = -\frac{e}{mc}\mathbf{S}.\left[\frac{\mathbf{p}}{mc}\times\left(\frac{\mathbf{r}}{r}\frac{dV_c}{dr}\right)\right] \tag{6.73}$$

$$= \frac{e}{m^2c^2}\frac{1}{r}\frac{dV_c}{dr}\left(\mathbf{L}\cdot\mathbf{S}\right) \tag{6.74}$$

where $\mathbf{L}$ is the angular momentum, $\mathbf{L} = \mathbf{r}\times\mathbf{p}$. This is the spin–orbit term, which must be included in the interaction Hamiltonian. We will formally derive this term when we consider the Dirac equation.

Thus the presence of both electric and magnetic fields will give rise to the following total Hamiltonian:

$$H = \frac{\mathbf{p}^2}{2m} - \frac{e}{2mc}\boldsymbol{\mu}\cdot\mathbf{B} + \frac{e}{m^2c^2}\frac{1}{r}\frac{dV_c}{dr}\left(\mathbf{L}\cdot\mathbf{S}\right) + \frac{e^2}{2mc^2}\mathbf{A}^2 \tag{6.75}$$

6.8 Aharonov-Bohm effect

The Hamiltonian in the presence of a vector potential $\mathbf{A}$ is given by

$$H = \frac{\left(\mathbf{p} - \frac{e}{c}\mathbf{A}\right)^2}{2m}. \tag{6.76}$$

We assume $\mathbf{A}$ to be time independent, so that $\mathbf{A} = \mathbf{A}(\mathbf{r})$. The time-dependent Schrödinger equation is then

$$i\hbar\frac{\partial\psi(\mathbf{r},t)}{\partial t} = -\frac{\hbar^2}{2m}\left(\boldsymbol{\nabla} - \frac{ie}{c\hbar}\mathbf{A}(\mathbf{r})\right)^2\psi(\mathbf{r},t). \tag{6.77}$$

Consider this equation in a region where the magnetic field vanishes. That is, where

$$\boldsymbol{\nabla}\times\mathbf{A} = 0. \tag{6.78}$$

One can write the solution of the equation (6.77) as

$$\psi(\mathbf{r},t) = \chi(\mathbf{r},t)e^{i\gamma(\mathbf{r})} \tag{6.79}$$

where

$$\gamma(\mathbf{r}) = \int_0 \frac{e}{\hbar}\mathbf{A}(\mathbf{r}')\cdot\mathbf{dr}', \tag{6.80}$$

which is a function of $\mathbf{r}$ only.

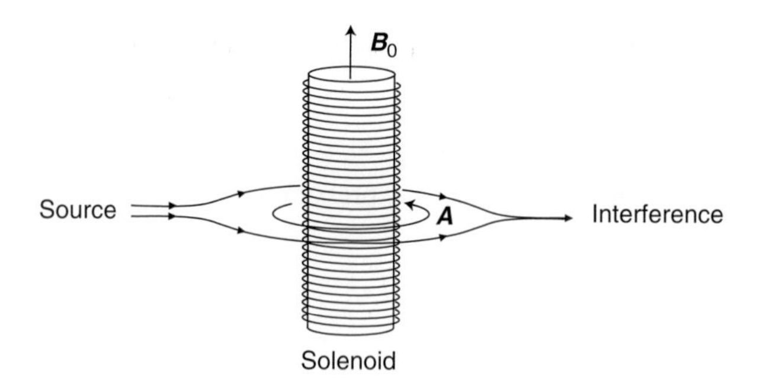

Fig. 6.1

To obtain the equation for $\chi(\mathbf{r},t)$ we note from (6.79) and (6.80) that

$$\nabla\psi = (\nabla\chi)\,e^{i\gamma} + i\chi\left(\frac{e}{\hbar}\mathbf{A}\right)e^{i\gamma}. \tag{6.81}$$

Hence,

$$\left[\nabla - \frac{ie}{c\hbar}\mathbf{A}(\mathbf{r})\right]\psi = (\nabla\chi)\,e^{i\gamma}. \tag{6.82}$$

Substituting this in equation (6.77) we obtain

$$i\hbar\frac{\partial\chi}{\partial t} = -\frac{\hbar^2}{2m}\nabla^2\chi. \tag{6.83}$$

Thus we have nicely separated out the function χ, which satisfies the Schrödinger equation without the presence of any interaction.

Consider a situation in which a solenoid carries a magnetic field $\mathbf{B}_0$ along its axis. The magnetic field is assumed to be uniform inside the solenoid but it vanishes outside. If the solenoid has a radius R, then the magnetic flux through it is given by

$$\Phi = \pi R^2 B_0 \tag{6.84}$$

and the vector potential at a point $r > R$ from the axis of the solenoid is given by

$$\mathbf{A} = \frac{\Phi}{2\pi r}\boldsymbol{\phi} \quad (r > R) \tag{6.85}$$

where $\boldsymbol{\phi}$ is a unit vector in the direction of the azimuthal angle ϕ. Note that the magnetic field vanishes in this region.

If a charge passes by the solenoid (see Fig. 6.1), then from the relation (6.80) the phase is given by

$$\gamma = \frac{e\Phi}{2\pi\hbar}\int\frac{\boldsymbol{\phi}}{r}\cdot r\,d\boldsymbol{\phi} = \pm\frac{e\Phi}{2\pi\hbar}\phi. \tag{6.86}$$

The $\pm$ sign depends on whether the electrons are traveling in the same direction as the current in the solenoid (which is the same direction as **A**) or the opposite.

Suppose a beam of electrons directed at the solenoid splits in two and goes around either side of the solenoid. The phase difference between the two parts of the beam when they recombine on the other side causes interference between them. The azimuthal angle ϕ covered by the beam in the direction of the current (which is in the same direction as **A**) will be π, and for the beam in the opposite direction will be $-\pi$. The difference between these two values is 2π. Taking the positive sign for the difference we obtain

$$\text{total phase difference} = \frac{e\Phi}{\hbar}. \tag{6.87}$$

This is the essence of the Aharonov–Bohm effect.

6.9 Problems

1. From the Hamiltonian given in terms of the mechanical momentum **P** , the commutation relations satisfied by the components P_i, and the relation

$$m\frac{d\mathbf{r}}{dt} = \mathbf{P},$$

show that one recovers the Lorentz force relation given by

$$\frac{d\mathbf{P}}{dt} = e\left[E + \frac{1}{2c}\left(\frac{d\mathbf{r}}{dt} \times \mathbf{B} - \mathbf{B} \times \frac{d\mathbf{r}}{dt}\right)\right].$$

Assume **P** to be a Heisenberg operator.

2. Starting with the Bohr–Sommerfeld condition for the mechanical momentum

$$\oint \mathbf{P}.\mathbf{dr} = \left(n + \frac{1}{2}\right) h,$$

for an electron subjected to a constant magnetic field **B** show that the magnetic flux $\phi = \oint \vec{\mathbf{B}} \cdot \mathbf{d}\vec{\mathbf{S}}$ satisfies the relation

$$\phi = \left(n + \frac{1}{2}\right) \frac{hc}{e}$$

where you may use the classical equation of motion satisfied by the charged particle in the presence of a magnetic field.

3. An electron of intrinsic magnetic moment $e\hbar/2mc$ pointing in an arbitrary direction is subjected to constant magnetic field which is in the x-direction. Use the Heisenberg equation of motion to determine $d\boldsymbol{\mu}/dt$.

4. For a charged particle moving along a ring of radius R, what are the energy eigenvalues and eigenfunctions if a long solenoid passes through the center of the ring with the magnetic flux given by Φ?
5. A particle is constrained to move along a circle of radius R. A constant magnetic field $\mathbf{B}$ perpendicular to the orbit is applied. Determine the energy eigenvalues and eigenfunctions.
6. Obtain the commutator $[v_i, L_z]$.

7 Stern–Gerlach experiments

This short chapter is devoted to the basic results of a very important experiment that firmly established the existence of spin one-half for the electron. It also provides an interesting insight into the concept of measurement in quantum mechanics.

7.1 Experimental set-up and electron's magnetic moment

Consider a beam of electrons traveling in the x-direction subjected to an inhomogeneous magnetic field, B_z, in the z-direction, and received on a detector on the opposite side. If the electron has a magnetic moment, $\boldsymbol{\mu}$, then its interaction energy will be $-\boldsymbol{\mu} \cdot \mathbf{B}$. The force on the electron due to the magnetic field will be given by

$$F_z = \frac{\partial}{\partial z}\left(\boldsymbol{\mu} \cdot \mathbf{B}\right) = \mu_z \frac{\partial B_z}{\partial z} \tag{7.1}$$

where μ_z is the projection of $\boldsymbol{\mu}$ in the z-direction. We assume the maximum value of μ_z to be μ_0.

The electron will then be deflected from its original path. If the electrons are randomly oriented as they enter the apparatus, and are treated as classical particles, then on the receiving screen one would expect to see a continuous band since, classically, μ_z is expected to take on continuous values, from $+\mu_0$ (upward deflection) to $-\mu_0$ (downward deflection) and all the values in between. Instead what one finds are only two distinct spots at $+\mu_0$ and $-\mu_0$.

This is one of the basic observations of the Stern–Gerlach (SG) experiment. In this experiment, carried out in 1921 by Stern and Gerlach, silver atoms heated in an oven were allowed to escape through a hole. The atomic beam went through a collimator and was then subjected to an inhomogeneous magnetic field. A silver atom consists of a nucleus

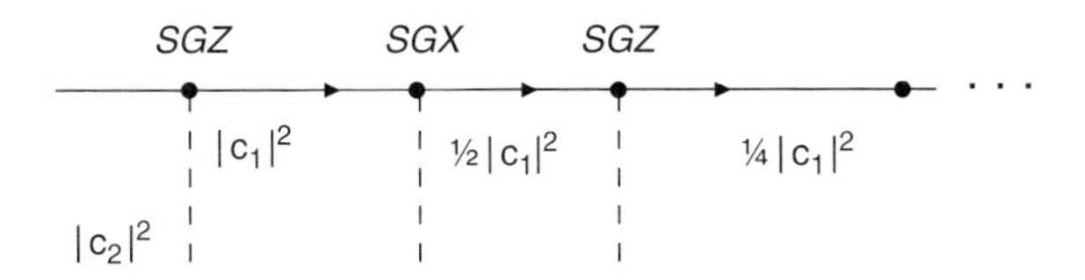

Fig. 7.1

and 47 electrons, 46 of which form a spherically symmetric system with no net angular momentum. Therefore, the spin of the leftover electron represents the angular momentum of the whole atom (ignoring the nuclear spin). Thus the magnetic moment of the atom is proportional to that of the electron.

The SG discovery was quite extraordinary in the early days of quantum mechanics. It showed that the electron has a magnetic moment and that its magnetic moment is quantized. The results were found not to depend on whether the SG apparatus was oriented in the z direction or any other direction, the quantization properties depended strictly on the direction of the magnetic field.

7.2 Discussion of the results

Quantum mechanically we have found that for an electron

$$\boldsymbol{\mu} = \mu_B \mathbf{S} \tag{7.2}$$

where $\mu_B = e/mc$ and $\mathbf{S}$ is the spin vector whose z-components are $S_{z+} = \hbar/2$ and $S_{z-} = -\hbar/2$. The SG discovery thus confirmed that the spin is quantized.

More remarkably, what was found through a series of experiments was the following. In the experiment described above which we call the SGZ experiment, corresponding to the magnetic field oriented in the z-direction, the outgoing electrons with $-z$ (negative z) components of the spin were blocked and only $+z$ (positive z) were allowed. When the leftover particles passed again through an SGZ, one observed only electrons in the $+z$ direction. However, when they passed through SGX, that is, were subjected to an inhomogeneous magnetic field in the x-direction, one found that 50% of the electrons were deflected along $+x$-direction and 50% along $-x$-direction. Naively, one would think that the electrons in the $\pm x$ directions would both have a memory of S_{z+}, and each would deflect accordingly. But that was found not to be so.

Now, to make things more interesting, suppose one puts a third apparatus, this time an SGZ again, in front of the electrons deflected in the $+x$-direction while blocking those that went in the $-x$-direction. One would, once again, naively, think that all of the electrons having the memory of their previous S_{z+} values will be deflected upward. However, that is not what is observed. One finds that the beam splits equally between $+z$ and $-z$-directions. The $-z$ component which was thought to have been blocked off, permanently, reappears. It is as if things started totally fresh, with all previous memory wiped out.

Let us discuss how the phenomenon is explained in quantum-mechanical terms. First let us designate the eigenstates with spin in $\pm z$ and $\pm x$ as $|z\pm\rangle$ and $|x\pm\rangle$ respectively. As previously discussed, they are written with respect to the z-axis as the quantization axis,

$$|z+\rangle = \begin{bmatrix} 1 \\ 0 \end{bmatrix}, \quad |z-\rangle = \begin{bmatrix} 0 \\ 1 \end{bmatrix}, \tag{7.3}$$

$$|x+\rangle = \frac{1}{\sqrt{2}} \begin{bmatrix} 1 \\ 1 \end{bmatrix}, \quad |x-\rangle = \frac{1}{\sqrt{2}} \begin{bmatrix} 1 \\ -1 \end{bmatrix}. \tag{7.4}$$

We also note that any one of the states $|z\pm\rangle$ can be represented in terms of $|x\pm\rangle$ and vice versa. For example,

$$|z+\rangle = \frac{1}{\sqrt{2}}\left[|x+\rangle + |x-\rangle\right] \tag{7.5}$$

and

$$|x+\rangle = \frac{1}{\sqrt{2}}\left[|z+\rangle + |z-\rangle\right]. \tag{7.6}$$

In the SG experiment, one electron per atom participates in the interaction. The beam consists of randomly distributed magnetic moments as it enters the SGZ apparatus. The magnetic field in the z-direction provides the quantization axis, and those electrons with spin up are deflected upward, while those with spin down are deflected downward. Let us assume that $|c_1|^2$ corresponds to the fraction that goes upward, represented by the state $|z+\rangle$, and $|c_2|^2$ to those that go downward, represented by the state $|z-\rangle$, with

$$|c_1|^2 + |c_2|^2 = 1. \tag{7.7}$$

As they enter the apparatus, the individual wavefunctions, including spatial dependence, are of the type

$$\left[c_1\,|z+\rangle + c_2\,|z-\rangle\right]\phi(\mathbf{r}). \tag{7.8}$$

After their interaction with the magnetic field they emerge as two separate wavefunctions

$$|z+\rangle\,\phi_+(\mathbf{r}) \quad \text{and} \quad |z-\rangle\,\phi_-(\mathbf{r}). \tag{7.9}$$

Suppose that after the beam comes out of SGZ one blocks off the downward component allowing only the fraction $|c_1|^2$ to proceed. If one puts in another SGZ apparatus, then all the leftover electrons are once again deflected upward, and there is no splitting. If, however, an SGX instead of SGZ is put in place, then the x-axis becomes the quantization axis, and one must express $|z+\rangle$ in terms of $|x\pm\rangle$ through the relation (7.5). Hence the state splits equally between $+x$ and $-x$ directions. The situation with each fraction is as follows:

$$\frac{1}{2}\,|c_1|^2 \text{ goes upward in the } +x\text{-direction,} \tag{7.10}$$

$$\frac{1}{2}\,|c_1|^2 \text{ goes downward in the } -x\text{-direction.} \tag{7.11}$$

Now suppose that the downward electrons are blocked and the leftover electrons, now described by $|x+\rangle$ alone are subjected to an SGZ apparatus. The quantization axis is now given by the z-axis and we must express $|x+\rangle$ in terms of $|z\pm\rangle$ given by (7.6). One then finds that the beam splits equally into $\pm z$-directions. That is,

$$\frac{1}{2}\cdot\frac{1}{2}\,|c_1|^2 = \frac{1}{4}\,|c_1|^2 \text{ goes upward in } +z \text{ direction,} \tag{7.12}$$

$$\frac{1}{2}\cdot\frac{1}{2}\,|c_1|^2 = \frac{1}{4}\,|c_1|^2 \text{ goes downward in } -z \text{ direction.} \tag{7.13}$$

The state $|z-\rangle$ has now reappeared. This phenomenon is described in Fig. 7.1.

We reach the following conclusions:

(i) The act of measurement changes the original system in a drastic way. When a measurement of an observable is made, the system is forced to be in an eigenstate of that observable.

(ii) In the SG experiment we found that the measurement of the observable S_x destroys the information of the previously made observation of the observable S_z. These two measurements are, therefore, incompatible. In mathematical terms, one says that S_x does not commute with S_z. This fact is reflected in quantum mechanical terms by the fact that S_x does not commute with S_z, i.e., $[S_x, S_z] \neq 0$.

In 1925 Uhlenbeck and Goudsmit postulated the existence of the electron spin to explain the results coming out of the SG experiments as well as other experiments. Stern–Gerlach experiments occupy a central role in quantum phenomena since their results signify pure quantum-mechanical effects that are completely different from classical predictions. Since it involves only two states, it is one of the simplest types of experiments to carry out and with which to discuss, conceptually, the fundamentals of quantum mechanics.

7.3 Problems

1. In the Stern–Gerlach setup assume the silver atoms to be traveling in the x-direction and subjected to a nonuniform magnetic field directed in the z-direction. If the y–z plane is assumed to be the symmetry plane, then show that both classical and quantum (using Ehrenfest's theorem) calculations give the same result for the separation between the spin-up and spin-down states at any arbitrary time t.
2. A beam of atoms with spin ½ passes through an SG apparatus with the magnet oriented in the direction **n** at an angle β with the z-axis. After emerging, the beam enters another apparatus, where this time the magnets are oriented in the z-direction. Determine the fraction of the atoms that come out with spin in the $+z$-direction and in the $-z$-direction.
3. In the Stern–Gerlach experiment where the particles pass between the magnets once, assume the magnetic field to be linear, $B = B_0 + zB_1$. Express the Hamiltonian as a 2×2 matrix and obtain the equations for the wavefunctions of the emerging particles as they travel with spin orientations in $\pm z$ directions.

8 Some exactly solvable bound-state problems

Having considered free systems, we now go to the more complicated problems involving interaction potentials. Specifically, we consider bound states, which are defined as states with energy eigenvalues less than the potential so that the particle's wavefunction vanishes at infinity. Those cases where the energy is greater than the potential come under the category of scattering, which will be considered in later chapters. We confine ourselves entirely to those cases that are exactly solvable. We will first examine some simple one-dimensional systems leading up to periodic potentials that are central to lattice problems and condensed matter physics. These topics provide excellent illustrations of boundary value problems, as well as of matrix methods, that prepare us to tackle potentials that are much more complicated. Finally, we will consider three-dimensional problems in both Cartesian and spherical coordinate systems. In the Cartesian system for potentials that are separable, one basically repeats the results from one dimension. The spherical system with spherically symmetric potentials brings in interesting and original problems, particularly the hydrogen-like atoms. These are solved in terms of the angular momentum eigenstates and radial wavefunctions.

8.1 Simple one-dimensional systems

8.1.1 Infinite potential barrier

We consider the problem where a free particle is confined between two infinite barriers that are separated by a finite distance. This situation is illustrated in Fig. 8.1 and is described by the following potential,

$$V(x) = 0, \quad -a < x < a, \tag{8.1}$$

$$V(x) = \infty, \quad x = \pm a. \tag{8.2}$$

Thus, within the barriers at $x = \pm a$, the wavefunction, $u(x)$, for the particle with mass m and energy E, satisfies the equation

$$-\frac{\hbar^2}{2m}\frac{d^2u}{dx^2} = Eu, \qquad |x| \leq a. \tag{8.3}$$

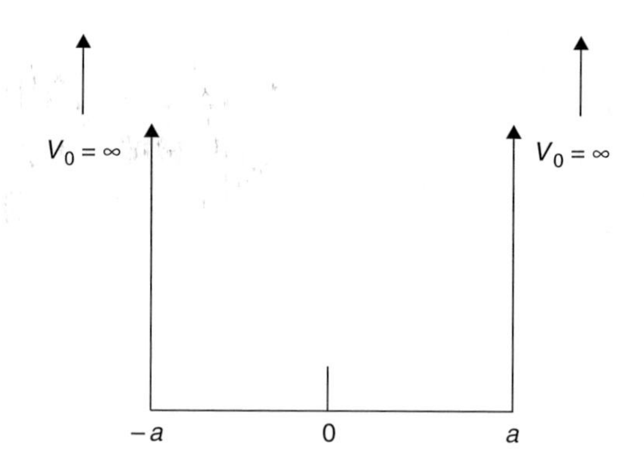

Fig. 8.1

Outside the barriers, $u(x)$ must vanish because, otherwise, the term $V(x)\,u(x)$ in the Schrödinger equation will become infinite. Hence we have

$$u \equiv 0, \quad |x| \geq a. \tag{8.4}$$

The differential equation is

$$\frac{d^2u}{dx^2} = -\alpha^2 u; \quad \alpha = \sqrt{\frac{2mE}{\hbar^2}}, \tag{8.5}$$

with boundary conditions

$$u(-a) = 0 = u(a). \tag{8.6}$$

The general solution is given by

$$u(x) = A\sin\alpha x + B\cos\alpha x. \tag{8.7}$$

The boundary conditions give

$$A\sin\alpha a + B\cos\alpha a = 0, \tag{8.8}$$

$$-A\sin\alpha a + B\cos\alpha a = 0. \tag{8.9}$$

By adding and subtracting the above two equations we obtain the following simplified relations:

$$A\sin\alpha a = 0, \tag{8.10}$$

$$B\cos\alpha a = 0. \tag{8.11}$$

We can divide these solutions into two categories, which lead to $u(x)$ being an even or odd function of x.

Type I

$$A = 0; \quad \cos \alpha a = 0, \tag{8.12}$$

with solutions

$$\alpha a = \frac{n\pi}{2}, \quad n = \text{odd number}. \tag{8.13}$$

The solution can then be written as

$$u_n(x) = B_n \cos \frac{n\pi x}{2a}, \quad n = 1, 3, \ldots. \tag{8.14}$$

The constant B_n is obtained through the normalization condition

$$\int_{-a}^{a} dx\, u_n^2(x) = 1, \tag{8.15}$$

which is found to be

$$B_n = \frac{1}{\sqrt{a}}. \tag{8.16}$$

Type II

$$B = 0; \quad \sin \alpha a = 0, \tag{8.17}$$

with solutions

$$\alpha a = \frac{n\pi}{2}, \quad n = \text{even number}. \tag{8.18}$$

The solution can be written as

$$u_n(x) = A_n \sin \frac{n\pi x}{2a}, \quad n = 2, 4, \ldots. \tag{8.19}$$

The normalization constant is then

$$A_n = \frac{1}{\sqrt{a}}. \tag{8.20}$$

It is easy to check, after inserting the normalization constants, that the wavefunctions satisfy the orthonormality condition as expected:

$$\int_{-a}^{a} dx\, u_m(x) u_n(x) = \delta_{mn}. \tag{8.21}$$

For both types of solutions the energy eigenvalues are given by

$$E_n = \frac{n^2 \pi^2 \hbar^2}{8ma^2}. \tag{8.22}$$

The fact that we have purely even and purely odd functions of x as our solutions is related to the symmetric (even) character of the potential, $V(x)$. We will discuss this in one of the later chapters on symmetries.

Matrix methods

We can also solve this problem through simple matrix methods if we express the two conditions in (8.8) and (8.9) in the following matrix form:

$$\begin{pmatrix} -\sin\alpha a & \cos\alpha a \\ \sin\alpha a & \cos\alpha a \end{pmatrix} \begin{pmatrix} A \\ B \end{pmatrix} = 0. \tag{8.23}$$

The determinant of the matrix must then vanish:

$$\det\begin{pmatrix} -\sin\alpha a & \cos\alpha a \\ \sin\alpha a & \cos\alpha a \end{pmatrix} = 0, \tag{8.24}$$

which gives $\sin 2\alpha a = 0$.

Therefore, we must have

$$2\alpha a = n\pi \quad \text{or} \quad \alpha = \frac{n\pi}{2a}. \tag{8.25}$$

Considering, separately, the $n =$ odd and $n =$ even cases, we find that for $n =$ odd, we have $\cos(n\pi/2) = 0$ and thus from (8.10) we deduce that $A = 0$. For $n =$ even, we have $\sin(n\pi/2) = 0$ and from (8.11) we deduce $B = 0$. We then obtain the same results that were derived earlier.

8.1.2 Finite potential barrier

We now consider the case where the height of the barrier is brought down to a finite value, with the potential given by

$$V(x) = 0, \qquad -a < x < a, \tag{8.26}$$

$$V(x) = V_0, \quad x < -a, x > a \tag{8.27}$$

where $V_0 > 0$. This potential is illustrated in Fig. 8.2. Since we are considering bound-state problems, the energy eigenvalues satisfy $E < V_0$. The Schrödinger equations in the two regions are

$$-\frac{\hbar^2}{2m}\frac{d^2u}{dx^2} = Eu, \qquad |x| \le a \tag{8.28}$$

$$-\frac{\hbar^2}{2m}\frac{d^2u}{dx^2} + V_0 u = Eu, \qquad |x| \ge a. \tag{8.29}$$

We solve the problem in the two regions separately and connect the solutions at the boundaries using the continuity conditions for the wavefunction and its derivative.

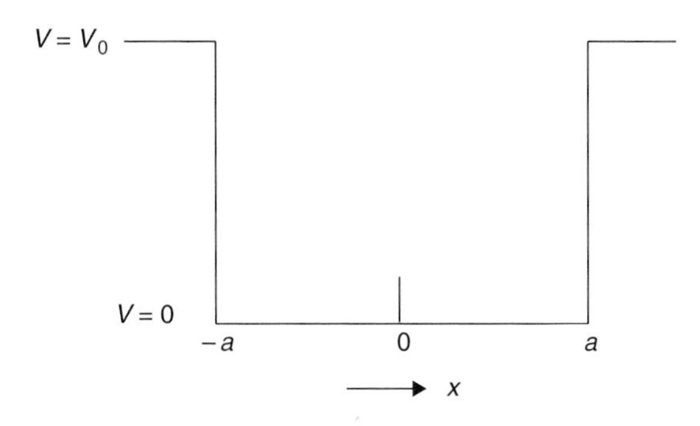

Fig. 8.2

(i) For $|x| \le a$, we have, once again, the equation

$$\frac{d^2u}{dx^2} = -\alpha^2 u \tag{8.30}$$

with

$$\alpha = \sqrt{\frac{2mE}{\hbar^2}}, \tag{8.31}$$

which has the solution

$$u(x) = A \sin \alpha x + B \cos \alpha x. \tag{8.32}$$

(ii) For $|x| \ge a$, we write the equation as

$$\frac{d^2u}{dx^2} = \beta^2 u. \tag{8.33}$$

where

$$\beta = \sqrt{\frac{2m(V_0 - E)}{\hbar^2}}. \tag{8.34}$$

Writing the solutions in the two regions separately, we have

$$u(x) = C_2 e^{-\beta x} + D_2 e^{\beta x} \quad \text{for } x \le -a \text{ (region 2)}, \tag{8.35}$$

$$u(x) = C_3 e^{-\beta x} + D_3 e^{\beta x} \quad \text{for } x \ge a \text{ (region 3)}. \tag{8.36}$$

We note that in order for the wavefunction to have a finite normalization as $x \to \pm\infty$, the term with $e^{-\beta x}$ should be absent in the region $x < -a$, and similarly $e^{\beta x}$ should be absent for $x > a$. Hence we have

$$u(x) = D_2 e^{\beta x} \quad \text{for } x \le -a \text{ (region 2)}, \tag{8.37}$$

$$u(x) = C_3 e^{-\beta x} \quad \text{for } x \ge a \text{ (region 3)}. \tag{8.38}$$

The boundary conditions corresponding to the continuity conditions for u and du/dx at $x = -a$, and at $x = a$, give:

At $x = -a$,

$$-A \sin \alpha a + B \cos \alpha a = D_2 e^{-\beta a}, \tag{8.39}$$

$$A\alpha \cos \alpha a + B\alpha \sin \alpha a = D_2 \beta e^{-\beta a}. \tag{8.40}$$

At $x = a$,

$$A \sin \alpha a + B \cos \alpha a = C_3 e^{-\beta a}, \tag{8.41}$$

$$A\alpha \cos \alpha a - B\alpha \sin \alpha a = -C_3 \beta e^{-\beta a}. \tag{8.42}$$

Adding (8.39) and (8.41), and subtracting (8.42) from (8.40), one obtains, respectively,

$$2B \cos \alpha a = (C_3 + D_2)\, e^{-\beta a}, \tag{8.43}$$

$$2B\alpha \sin \alpha a = (C_3 + D_2)\, \beta e^{-\beta a}. \tag{8.44}$$

Dividing (8.44) by (8.43), the following result is obtained:

$$\alpha \tan \alpha a = \beta \quad \textbf{Type I}. \tag{8.45}$$

Similarly, one obtains

$$2A \sin \alpha a = (C_3 - D_2)\, e^{-\beta a}, \tag{8.46}$$

$$2A\alpha \cos \alpha a = -\,(C_3 - D_2)\, \beta e^{-\beta a}. \tag{8.47}$$

Dividing (8.47) by (8.46), the following result is obtained:

$$\alpha \cot \alpha a = -\beta \quad \textbf{Type II}. \tag{8.48}$$

Equations (8.45) and (8.48) cannot both be satisfied simultaneously. One can demonstrate this by multiplying the left-hand sides of (8.45) and (8.48) and equating the product to the product of the right-hand sides of these equations. We obtain $\alpha^2 = -\beta^2$, i.e., $V_0 = 0$ which is not what we have assumed for the potential. By dividing, we obtain $\tan^2 \alpha a = -1$ (which is untenable). Hence only one or the other type of solution is allowed, and, therefore, as in the case of the infinite barriers, we have two types of solutions that separate the even and odd wavefunctions. Again, this is due to the even character of the potential. We summarize the results for the wavefunctions as follows.

Type I

$$C_3 = D_2, \quad A = 0, \tag{8.49}$$

$$B = C_3 \frac{\beta e^{-\beta a}}{\alpha \sin \alpha a}, \tag{8.50}$$

and even wavefunctions

$$u(x) = C_3 e^{\beta x}, \quad x < -a, \tag{8.51}$$

$$u(x) = B \cos \alpha x, \quad -a < x < a, \tag{8.52}$$

$$u(x) = C_3 e^{-\beta x}, \quad x > a. \tag{8.53}$$

Type II

$$C_3 = -D_2, \quad B = 0, \tag{8.54}$$

$$A = -C_3 \frac{\beta e^{-\beta a}}{\alpha \cos \alpha a}, \tag{8.55}$$

and odd wavefunctions

$$u(x) = -C_3 e^{\beta x}, \quad x < -a, \tag{8.56}$$

$$u(x) = A \sin \alpha x, \quad -a < x < a, \tag{8.57}$$

$$u(x) = C_3 e^{\beta x}, \quad x > a. \tag{8.58}$$

Analytical solutions for α and β are almost impossible to achieve, but one can obtain the energy eigenvalues by graphical methods. To that end we define

$$\xi = \alpha a, \quad \eta = \beta a \tag{8.59}$$

and take note of the fact that

$$\left(\alpha^2 + \beta^2\right) a^2 = \frac{2mV_0 a^2}{\hbar^2}. \tag{8.60}$$

Thus, we have

$$\xi \tan \xi = \eta \quad \textbf{Type I} \tag{8.61}$$

and

$$\xi \cot \xi = -\eta \quad \textbf{Type II}. \tag{8.62}$$

Both satisfy

$$\xi^2 + \eta^2 = \frac{2mV_0 a^2}{\hbar^2}. \tag{8.63}$$

We note from the graphs in Fig. 8.3a and Fig. 8.3b that there will be a finite number of intersections between the graph for (8.63) and either of the graphs for the other two, (8.61) and (8.62). Furthermore, we make note of the following:

(i) There will always be a solution for **Type I**, no matter how small the value of V_0 is, as long as it is positive.

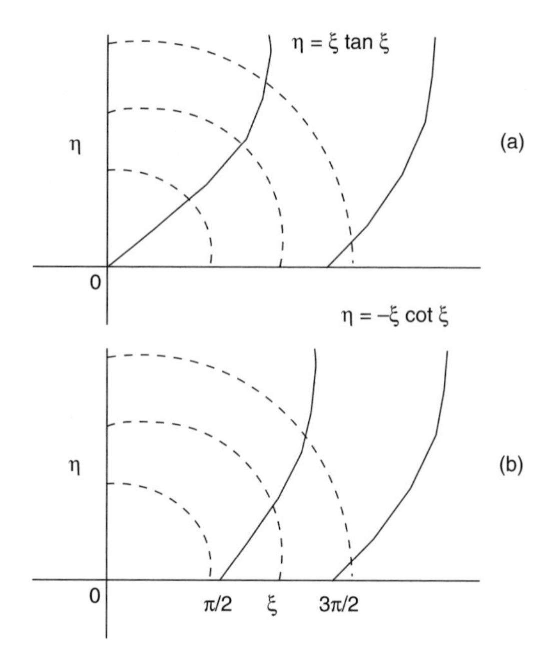

Fig. 8.3

(ii) The solution of **Type II** appears as soon as $\xi = \pi/2$ and $\eta = 0$.
(iii) In general we will have the following conditions on the potentials:

$$0 < V_0 a^2 < \frac{\pi^2 \hbar^2}{8m} \quad \text{one solution of } \textbf{Type I}, \tag{8.64}$$

$$\frac{\pi^2 \hbar^2}{8m} < V_0 a^2 < \frac{\pi^2 \hbar^2}{2m} \quad \text{one solution each of } \textbf{Type I} \text{ and } \textbf{Type II}. \tag{8.65}$$

and so on.

Matrix methods

This problem can be solved also by matrix methods. The conditions (8.39) and (8.40) at $x = -a$, after inverting them, give

$$\begin{bmatrix} A \\ B \end{bmatrix} = M \begin{bmatrix} 0 \\ D_2 \end{bmatrix}, \tag{8.66}$$

where

$$M = \begin{bmatrix} 0 & \left(-\sin \alpha a + \dfrac{\beta}{\alpha} \cos \alpha a \right) e^{-\beta a} \\ 0 & \left(\cos \alpha a + \dfrac{\beta}{\alpha} \sin \alpha a \right) e^{-\beta a} \end{bmatrix}. \tag{8.67}$$

The boundary conditions (8.41) and (8.42) at $x = a$, after inverting them, give

$$\begin{bmatrix} A \\ B \end{bmatrix} = N \begin{bmatrix} C_3 \\ 0 \end{bmatrix} \tag{8.68}$$

where

$$N = \begin{bmatrix} \left(\sin \alpha a + \frac{\beta}{\alpha} \cos \alpha a\right) e^{\beta a} & 0 \\ \left(\cos \alpha a - \frac{\beta}{\alpha} \sin \alpha a\right) e^{\beta a} & 0 \end{bmatrix}. \tag{8.69}$$

Thus, by equating the right-hand sides of (8.66) and (8.68), we find

$$M \begin{bmatrix} 0 \\ D_2 \end{bmatrix} = N \begin{bmatrix} C_3 \\ 0 \end{bmatrix}. \tag{8.70}$$

This gives us two simultaneous equations

$$\left(-\sin \alpha a + \frac{\beta}{\alpha} \cos \alpha a\right) e^{-\beta a} D_2 = \left(\sin \alpha a - \frac{\beta}{\alpha} \cos \alpha a\right) e^{-\beta a} C_3, \tag{8.71}$$

$$\left(\cos \alpha a + \frac{\beta}{\alpha} \sin \alpha a\right) e^{-\beta a} D_2 = \left(\cos \alpha a + \frac{\beta}{\alpha} \sin \alpha a\right) e^{-\beta a} C_3. \tag{8.72}$$

We then have two possible solutions:

$$\text{I. } D_2 = C_3, \quad -\sin \alpha a + \frac{\beta}{\alpha} \cos \alpha a = 0, \text{ i.e., } \alpha \tan \alpha a = \beta, \tag{8.73}$$

$$\text{II. } D_2 = -C_3, \quad \sin \alpha a + \frac{\beta}{\alpha} \cos \alpha a = 0, \quad \text{i.e., } \alpha \cot \alpha a = -\beta. \tag{8.74}$$

These are precisely the Type I and Type II solutions.

8.2 Delta-function potential

Let us consider an idealized but very instructive potential described by an attractive δ-function (Fig. 8.4 gives a sketch of how this unconventional mathematical function would look)

$$V(x) = -g\delta(x). \tag{8.75}$$

The corresponding Schrödinger equation is given by

$$-\frac{\hbar^2}{2m} \frac{d^2u}{dx^2} - g\delta(x)u = Eu. \tag{8.76}$$

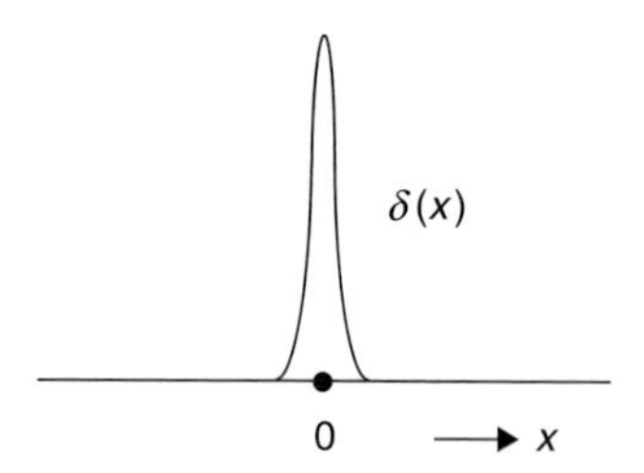

Fig. 8.4

We will consider the problem of bound states. Therefore,

$$E = -E_B, \quad \beta = \sqrt{\frac{2mE_B}{\hbar^2}}. \tag{8.77}$$

The equation (8.76) now reads

$$\frac{d^2u}{dx^2} + \frac{2mg}{\hbar^2}\delta(x)u = \beta^2 u. \tag{8.78}$$

For $x < 0$ and $x > 0$ where $\delta(x)$ vanishes, the solution of (8.78) is given by

$$u(x) = Ae^{-\beta x}, \quad x > 0 \tag{8.79}$$

$$= Ae^{\beta x}, \quad x < 0 \tag{8.80}$$

where we have imposed the condition that the wavefunction is continuous at $x = 0$. Next we integrate the differential equation over the interval $(-\epsilon, \epsilon)$ to include the contribution of the δ-function,

$$\int_{-\epsilon}^{\epsilon} dx \frac{d^2u}{dx^2} + \int_{-\epsilon}^{\epsilon} dx \frac{2mg}{\hbar^2}\delta(x)u = \beta^2 \int_{-\epsilon}^{\epsilon} dx\, u, \tag{8.81}$$

which gives

$$\left[u'(\epsilon) - u'(-\epsilon)\right] + \frac{2mg}{\hbar^2}u(0) = 0. \tag{8.82}$$

Substituting (8.79) and (8.80) we obtain

$$-2A\beta + \frac{2mg}{\hbar^2}A = 0. \tag{8.83}$$

Hence,

$$\beta = \frac{mg}{\hbar^2}. \tag{8.84}$$

Since the wavefunction is normalized we obtain

$$A^2 \int_{-\infty}^{\infty} dx\, e^{-2\beta|x|} = 1. \tag{8.85}$$

Therefore,

$$A = \sqrt{\beta}. \tag{8.86}$$

Hence,

$$u(x) = \sqrt{\beta} e^{-\beta|x|}, \qquad -\infty < x < \infty \tag{8.87}$$

and the binding energy from (8.77) and (8.84) is given by

$$E_B = \frac{mg^2}{\hbar^2}. \tag{8.88}$$

8.3 Properties of a symmetric potential

In the previous sections we have considered several cases that are consequences of the symmetry

$$V(x) = V(-x). \tag{8.89}$$

Let us now elaborate on it. Consider the Schrödinger equation

$$-\frac{\hbar^2}{2m}\frac{\partial^2 \psi(x,t)}{\partial x^2} + V(x)\psi(x,t) = i\hbar\frac{\partial \psi(x,t)}{\partial t}. \tag{8.90}$$

Taking $x \to -x$ and using the relation (8.89), we can write another equation,

$$-\frac{\hbar^2}{2m}\frac{\partial^2 \psi(-x,t)}{\partial x^2} + V(x)\psi(-x,t) = i\hbar\frac{\partial \psi(-x,t)}{\partial t}. \tag{8.91}$$

Let us write the equations obtained by adding and subtracting the relations (8.90) and (8.91). We obtain

$$-\frac{\hbar^2}{2m}\frac{\partial^2 \psi_e(x,t)}{\partial x^2} + V(x)\psi_e(x,t) = i\hbar\frac{\partial \psi_e(x,t)}{\partial t}, \tag{8.92}$$

$$-\frac{\hbar^2}{2m}\frac{\partial^2 \psi_0(x,t)}{\partial x^2} + V(x)\psi_0(x,t) = i\hbar\frac{\partial \psi_0(x,t)}{\partial t} \tag{8.93}$$

where

$$\psi_e(x,t) = \psi(x,t) + \psi(-x,t), \tag{8.94}$$

$$\psi_0(x,t) = \psi(x,t) - \psi(-x,t). \tag{8.95}$$

Each combination has a definite "parity," that is, they are either even or odd under reflection $x \to -x$, and each satisfies the Schrödinger equation.

Thus if the potential is invariant under parity transformation then the original Schrödinger equation gives rise to solutions that are either odd or even under parity. This is exactly what we discovered in the previous cases. We will return to the subject of symmetry in greater detail in Chapter 26.

8.4 The ammonia molecule

The ammonia molecule, NH_3, consists of three hydrogen atoms that are found to be in a plane forming an equilateral triangle. The nitrogen atom, N, is located at a point away from the plane. If we consider the three hydrogen atoms in the plane as a single entity of mass $3m_H$ and designate the distance of N from the plane as x, then to a good approximation the potential looks like Fig. 8.5.

The potential $V(x)$ is symmetric. It has two parts to it. At the outside edges it becomes infinite. In the inside, it vanishes except in a small region. We assume that the height of the potential in this region is V_0. If we designate the wavefunction as $u(x)$ then it must vanish at the outside edges. We consider energies $E < V_0$. We then have the even and odd solutions as follows, where the three distances a, b, and Δ are related by

$$\Delta = a - 2b \tag{8.96}$$

Even solution

$$u(x) = A \sin k\left(b + \frac{a}{2} + x\right), \quad -b - \frac{a}{2} < x < -\frac{\Delta}{2} \tag{8.97}$$

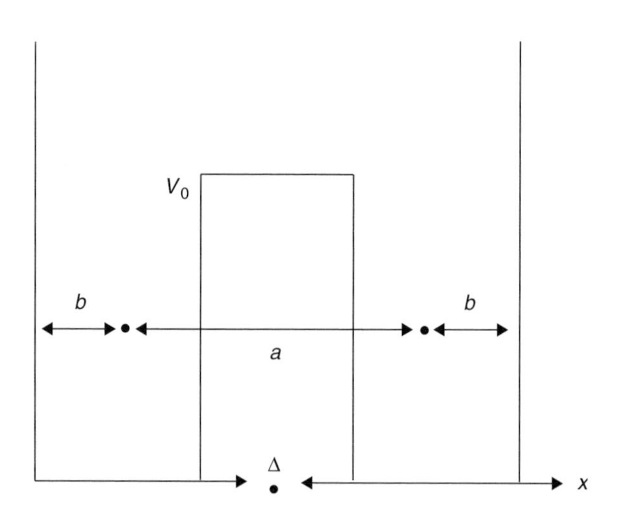

Fig. 8.5

$$= B \cosh Kx, \qquad -\frac{\Delta}{2} < x < \frac{\Delta}{2} \tag{8.98}$$

$$= A \sin k\left(b + \frac{a}{2} - x\right), \qquad \frac{\Delta}{2} < x < b + \frac{a}{2}. \tag{8.99}$$

Odd solution

$$u(x) = -A' \sin k\left(b + \frac{a}{2} + x\right), \quad -b - \frac{a}{2} < x < -\frac{\Delta}{2} \tag{8.100}$$

$$= B' \sinh Kx, \qquad -\frac{\Delta}{2} < x < \frac{\Delta}{2} \tag{8.101}$$

$$= A' \sin k\left(b + \frac{a}{2} - x\right), \qquad \frac{\Delta}{2} < x < b + \frac{a}{2} \tag{8.102}$$

where

$$k = \sqrt{\frac{2mE}{\hbar^2}}, \quad K = \sqrt{\frac{2m(V_0 - E)}{\hbar^2}}. \tag{8.103}$$

First we consider the even solutions. The continuity of $u(x)$ and its derivative at $x = \Delta/2$ give the following conditions

$$A \sin ka = B \cosh \frac{K\Delta}{2}, \tag{8.104}$$

$$Ak \cos ka = -BK \sinh \frac{K\Delta}{2} \tag{8.105}$$

where we have used the relation (8.96). Dividing the two relations, we obtain

$$\tan ka = -\frac{k}{K} \coth \frac{K\Delta}{2}. \tag{8.106}$$

Similarly, the odd solutions will give

$$\tan ka = -\frac{k}{K} \tanh \frac{K\Delta}{2}. \tag{8.107}$$

Let us solve the equation (8.106) under the following simplifying assumptions,

$$E \ll V_0, \tag{8.108}$$

and therefore,

$$K \approx \sqrt{\frac{2mV_0}{\hbar^2}} \gg k. \tag{8.109}$$

We also assume that

$$K\Delta \gg 1. \tag{8.110}$$

We then find that for the even solutions,

$$\coth\frac{K\Delta}{2}=\frac{e^{\frac{K\Delta}{2}}+e^{-\frac{K\Delta}{2}}}{e^{\frac{K\Delta}{2}}-e^{-\frac{K\Delta}{2}}}\simeq\left(1+2e^{-K\Delta}\right); \tag{8.111}$$

while for the odd solutions,

$$\tanh\frac{K\Delta}{2}\simeq\left(1-2e^{-K\Delta}\right). \tag{8.112}$$

Hence the two equations (8.106) and (8.107) are given by

$$\tan ka=-\frac{k}{K}\left(1\pm 2e^{-K\Delta}\right). \tag{8.113}$$

In view of the assumptions (8.109) and (8.110) we find that $\tan ka$ must be very small and negative. Hence,

$$ka\sim\pi. \tag{8.114}$$

For a better determination of k we write

$$ka=\pi-\pi\epsilon \tag{8.115}$$

where $\epsilon\ll 1$. Taking $\tan\pi\epsilon=\pi\epsilon$, we obtain from (8.113)

$$\epsilon=\frac{1}{Ka}\left(1\pm 2e^{-K\Delta}\right). \tag{8.116}$$

From now on, instead of calling the solutions even and odd we will use the conventional terms "symmetric" and "antisymmetric" respectively, i.e., ϵ_S and ϵ_A. Thus,

$$\epsilon_S=\frac{1}{Ka}\left(1+2e^{-K\Delta}\right), \tag{8.117}$$

$$\epsilon_A=\frac{1}{Ka}\left(1-2e^{-K\Delta}\right). \tag{8.118}$$

We note that

$$\epsilon_A<\epsilon_S. \tag{8.119}$$

The two bound state energies are given by

$$E_S=\frac{\hbar^2k_S^2}{2m}=\frac{\pi^2\hbar^2}{2ma^2}(1-2\epsilon_S), \tag{8.120}$$

$$E_A=\frac{\hbar^2k_A^2}{2m}=\frac{\pi^2\hbar^2}{2ma^2}(1-2\epsilon_A) \tag{8.121}$$

where we have used (8.115). Thus we find from (8.117) and (8.118) that $E_S<E_A$. Their difference is given by

$$E_A-E_S=\frac{4\pi^2\hbar^2}{ma^2}\left(\frac{e^{-K\Delta}}{Ka}\right)\ll 1. \tag{8.122}$$

8.5 Periodic potentials

The problem of periodic potentials is of great practical importance in condensed matter physics as it relates to the conduction of electrons. The periodic potential is created by the lattice that is formed by the atoms in a crystal. As a useful approximation to this potential one can imagine a succession of square barriers of the type we considered earlier and placed in a periodic manner, continuing indefinitely in both directions.

We will consider the following periodic potential:

$$V(x+d) = V(x), \qquad -\infty < x < \infty. \tag{8.123}$$

Here the states will be designated by the ket vectors $|x\rangle$ or $|\phi\rangle$ depending on the problem at hand. In order to discuss the consequences of the periodic symmetry, let us consider an operator $F(d)$ such that

$$F(d)\,|x\rangle = |x+d\rangle\,. \tag{8.124}$$

This operator is not constructed out of infinitesimal displacements, but it corresponds to a single step in which there is translation by a discrete amount d. It is, however, unitary, since the norms of the states $|x\rangle$ and $|x+d\rangle$ are assumed to be the same. Therefore,

$$F^{\dagger}(d)\,F(d) = \mathbf{1}. \tag{8.125}$$

Multiplying (8.124) on the left by $F^{\dagger}(d)$, one can then deduce that

$$F^{\dagger}(d)\,|x+d\rangle = |x\rangle\,. \tag{8.126}$$

Let us now consider the product $F^{\dagger}(d)\,XF(d)$, where $X\,|x\rangle = x\,|x\rangle$. This product when operating on $|x\rangle$ gives rise to the following result:

$$F^{\dagger}(d)XF(d)\,|x\rangle = F^{\dagger}(d)X\,|x+d\rangle = (x+d)F^{\dagger}(d)\,|x+d\rangle = (x+d)\,|x\rangle\,. \tag{8.127}$$

We can write the last relation as $(X + d\mathbf{1})\,|x\rangle$. Hence we have the following operator relation:

$$F^{\dagger}(d)XF(d) = X + d\mathbf{1}. \tag{8.128}$$

The periodic property of $V(x)$ given by (8.123) can then be described as follows

$$F^{\dagger}(d)V(X)F(d) = V\,(X + d\mathbf{1}) = V(X). \tag{8.129}$$

Thus

$$[F(d)\,, V(X)] = 0. \tag{8.130}$$

Since $F(d)$ corresponds to displacement by a constant amount, d, it does not depend on X, and hence it will commute with the momentum operator P, as well as the Hamiltonian, $H = P^2/2m + V(X)$:

$$[F(d), H] = 0. \tag{8.131}$$

Thus the eigenstates of H will also be the eigenstates of $F(d)$.

If $|\phi\rangle$ is an eigenstate of H and $F(d)$ then we can write

$$F(d)\,|\phi\rangle = \lambda\,|\phi\rangle\,. \tag{8.132}$$

Since $F(d)$ is not necessarily Hermitian, we need not have λ real. Multiplying (8.132) with its Hermitian conjugate and using (8.125) we obtain

$$|\lambda|^2 = 1. \tag{8.133}$$

Since λ has unit magnitude, we write it in the exponential form

$$\lambda = e^{-iKd} \tag{8.134}$$

where K is a real constant. Thus

$$\langle x|\,F(d)\,|\phi\rangle = e^{-iKd}\,\langle x\,|\phi\rangle\,, \tag{8.135}$$

and, therefore, from (8.124),

$$\phi(x-d) = e^{-iKd}\phi(x) \tag{8.136}$$

where $\langle x\,|\phi\rangle = \phi\,(x)$ is the wavefunction of the particle. Let us now define a function

$$u_K(x) = e^{-iKx}\phi(x). \tag{8.137}$$

We note that

$$u_K(x-d) = e^{-iK(x-d)}\phi(x-d) = e^{-iKx}e^{iKd}\phi(x-d) = e^{-iKx}\phi(x) = u_K(x) \tag{8.138}$$

where we have used the relation (8.136). Thus $u_K\,(x)$ is a periodic function, known as the Bloch wavefunction.

Consider now a specific type of periodic potential, of period d, defined as follows (see Fig. 8.6):

$$V(x) = 0, \quad 0 < x < a, \tag{8.139}$$

$$V(x) = V_0, \quad a < x < d, \tag{8.140}$$

$$V(x+d) = V(x). \tag{8.141}$$

The treatment for the region $0 < x < d$ is similar to the finite barrier problem we considered earlier with $E > 0$. Since we are considering bound states, we must also have $E < V_0$. We then obtain the following in the different regions:

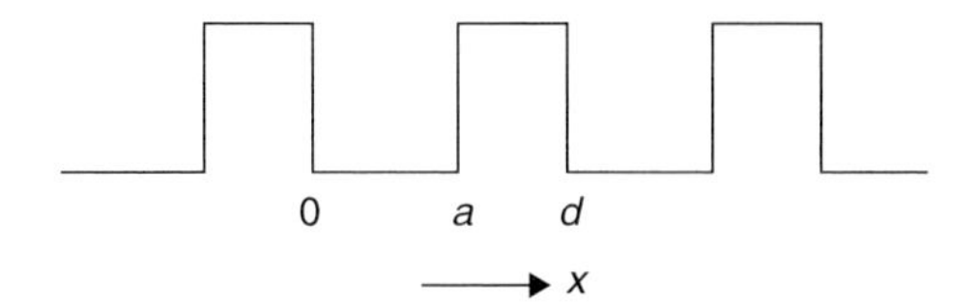

Fig. 8.6

$0 < x < a$

Here we have

$$\phi(x) = A\cos\alpha x + B\sin\alpha x, \tag{8.142}$$

$$\frac{d\phi(x)}{dx} = -A\alpha\sin\alpha x + B\alpha\cos\alpha x \tag{8.143}$$

where $\alpha = \sqrt{2mE/\hbar^2}$. From these equations we can write down the following results for the Bloch wavefunctions in $0 < x < a$ defined in (8.137):

$$u_K(x) = (A\cos\alpha x + B\sin\alpha x)\,e^{-iKx}, \tag{8.144}$$

$$\frac{du_K(x)}{dx} = (-A\alpha\sin\alpha x + B\alpha\cos\alpha x)\,e^{-iKx} - iKu_K(x). \tag{8.145}$$

$a < x < d$

Here we find $E < V_0$ and, therefore, the wavefunction will have an exponential dependence,

$$\phi(x) = Ce^{-\beta x} + De^{\beta x}, \tag{8.146}$$

$$\frac{d\phi(x)}{dx} = -C\beta e^{-\beta x} + D\beta e^{\beta x} \tag{8.147}$$

where $\beta = \sqrt{2m(V_0 - E)/\hbar^2}$. The corresponding Bloch functions satisfy

$$u_K(x) = \left(Ce^{-\beta x} + De^{\beta x}\right)e^{-iKx}, \tag{8.148}$$

$$\frac{du_K(x)}{dx} = \left(-C\beta e^{-\beta x} + D\beta e^{\beta x}\right)e^{-iKx} - iKu_K(x). \tag{8.149}$$

We impose the periodic property of the Bloch wavefunctions at the end points of the period at $x = 0$ and $x = d$, and the continuity relation satisfied by the wavefunction $\phi(x)$ at the finite barrier at $x = a$.

Boundary conditions

(1) From the periodic property of u_K, $u_K(0) = u_K(d)$ at the points $x = 0$ and $x = d$,

$$A = \left(Ce^{-\beta d} + De^{\beta d}\right)e^{-iKd}, \tag{8.150}$$

$$B = \frac{1}{\alpha}\left(-C\beta e^{-\beta d} + D\beta e^{\beta d}\right)e^{-iKd}. \tag{8.151}$$

(2) From the continuity property of $\phi(x)$ and $\partial\phi/\partial x$ at $x = a$,

$$A\cos\alpha a + B\sin\alpha a = Ce^{-\beta a} + De^{\beta a}, \tag{8.152}$$

$$-A\alpha\sin\alpha a + B\alpha\cos\alpha a = -C\beta e^{-\beta a} + D\beta e^{\beta a}. \tag{8.153}$$

Matrix relations
Rewriting the above relations in the matrix form we obtain, with appropriate inversions,

(1)
$$\begin{pmatrix} A \\ B \end{pmatrix} = \begin{pmatrix} e^{-\beta d} & e^{\beta d} \\ -\frac{\beta}{\alpha}e^{-\beta d} & \frac{\beta}{\alpha}e^{\beta d} \end{pmatrix} e^{-iKd} \begin{pmatrix} C \\ D \end{pmatrix}, \tag{8.154}$$

which we write as

$$\begin{pmatrix} A \\ B \end{pmatrix} = M \begin{pmatrix} C \\ D \end{pmatrix}, \tag{8.155}$$

$$M = \begin{pmatrix} e^{-\beta d}e^{-iKd} & e^{-\beta d}e^{-iKd} \\ -\frac{1}{\alpha}\beta e^{-\beta d}e^{-iKd} & \frac{1}{\alpha}\beta e^{-\beta d}e^{-iKd} \end{pmatrix}. \tag{8.156}$$

(2)
$$\begin{pmatrix} \cos\alpha a & \sin\alpha a \\ -\sin\alpha a & \cos\alpha a \end{pmatrix} \begin{pmatrix} A \\ B \end{pmatrix} = \begin{pmatrix} e^{-\beta a} & e^{\beta a} \\ \frac{\beta}{\alpha}e^{-\beta a} & \frac{\beta}{\alpha}e^{\beta a} \end{pmatrix} \begin{pmatrix} C \\ D \end{pmatrix}. \tag{8.157}$$

Therefore, inverting the matrix on the left-hand side, we obtain

$$\begin{pmatrix} A \\ B \end{pmatrix} = \begin{pmatrix} \cos\alpha a & -\sin\alpha a \\ \sin\alpha a & \cos\alpha a \end{pmatrix} \begin{pmatrix} e^{-\beta a} & e^{\beta a} \\ \frac{\beta}{\alpha}e^{-\beta a} & \frac{\beta}{\alpha}e^{\beta a} \end{pmatrix} \begin{pmatrix} C \\ D \end{pmatrix}. \tag{8.158}$$

We write this relation as

$$\begin{pmatrix} A \\ B \end{pmatrix} = N \begin{pmatrix} C \\ D \end{pmatrix} \tag{8.159}$$

where

$$N = \begin{pmatrix} (\cos a\alpha)\, e^{-a\beta} - \frac{1}{\alpha}\beta\,(\sin a\alpha)\, e^{-a\beta} & (\cos a\alpha)\, e^{a\beta} - \frac{1}{\alpha}\beta\,(\sin a\alpha)\, e^{a\beta} \\ (\sin a\alpha)\, e^{-a\beta} + \frac{1}{\alpha}\beta\,(\cos a\alpha)\, e^{-a\beta} & (\sin a\alpha)\, e^{a\beta} + \frac{1}{\alpha}\beta\,(\cos a\alpha)\, e^{a\beta} \end{pmatrix}. \tag{8.160}$$

Thus subtracting (8.159) from (8.155) we obtain

$$(M - N)\begin{pmatrix} C \\ D \end{pmatrix} = 0. \tag{8.161}$$

This implies that

$$\det(M - N) = 0. \tag{8.162}$$

In order to calculate this determinant we make note of the following relation:

$$\det(M - N) = \det M + \det N - M_{11}N_{22} + M_{12}N_{21} + M_{21}N_{12} - M_{22}N_{11} \tag{8.163}$$

where M_{ij} and N_{ij} are the matrix elements of M and N.

It is quite straightforward to obtain the two determinants, $\det M$ and $\det N$. Their sum is found to be

$$\det M + \det N = \frac{2}{\alpha}\beta\,(2\cos Kd)\,e^{-iKd}. \tag{8.164}$$

After a lengthy calculation, and some cancellations and rearrangement of terms, we find

$$- M_{11}N_{22} + M_{12}N_{21} + M_{21}N_{12} - M_{22}N_{11} \tag{8.165}$$

$$= \left[\left(1 - \frac{\beta^2}{\alpha^2}\right)(2\sinh\beta b)(\sin\alpha a) - \left(\frac{2\beta}{\alpha}\right)(2\cosh\beta b)(\cos\alpha a)\right] e^{-iKd} \tag{8.166}$$

where $b = d - a$.

According to (8.162) and (8.163), the sum of the two terms given by (8.164) and (8.166) must vanish. Hence

$$\cos Kd = (\cosh\beta b)(\cos\alpha a) + \left(\frac{\beta^2 - \alpha^2}{2\alpha\beta}\right)(\sinh\beta b)(\sin\alpha a). \tag{8.167}$$

Because $-1 \le \cos Kd \le 1$, the left-hand side is constrained to remain within -1 and $+1$. The right-hand side, however, has no such constraints. In fact, it can go outside these limits as $\cosh\beta b \ge 1$ and $\sinh\beta b \ge 0$. Since β and α are related to the energy, E, the magnitude of the right-hand side depends crucially on the magnitude of E. Hence, only for those values of E for which the right-hand side stays within the limits can there be bound states. This band structure is shown in Fig. 8.7.

The energy levels can then be determined in terms of the parameters of the model, e.g., V_0, a, and d. Since this involves a range of continuous values of E, they form what are known as the "allowed bands" in the energy spectrum. For those continuous values of E for which the right-hand side goes outside the limits, there will be no bound states. For example, bound states will be absent when $\alpha a = N\pi$ and $\cosh\beta b > 1$. One expects that there will be a narrow but continuous range of E in the proximity of these discrete points where there will also be no bound states. These are called the "forbidden bands."

This band structure with alternately arranged allowed and forbidden zones is a striking characteristic of periodic potentials. It allows one to understand the basic properties of materials, e.g., electric conduction in metals. The band structure was based on the motion of a single particle (an electron), but in a more realistic case one must consider a many-electron system in which the Pauli exclusion principle plays an essential role. When supplemented with the exclusion principle, the results we have obtained allow one to predict whether the material would be a conductor, a semiconductor, or an insulator, as well as to understand the presence of any defects in the periodic structure. The agreement between experiment and theory that one finds here constitutes a great triumph of quantum theory.

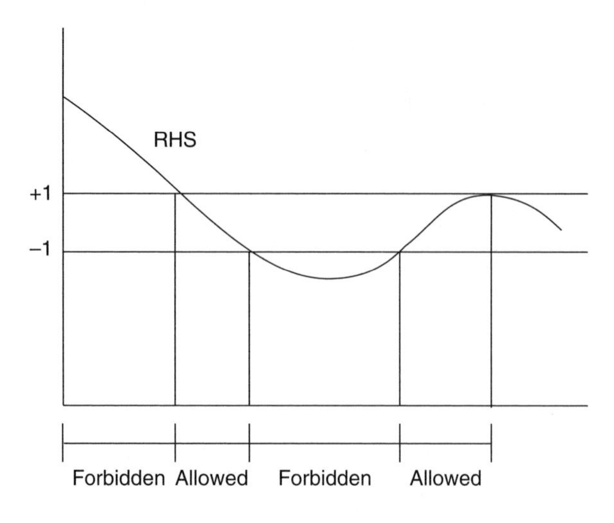

Fig. 8.7

8.6 Problems in three dimensions

8.6.1 Separable potentials in Cartesian coordinates

The Schrödinger equation for the wavefunction $u(\mathbf{r})$ with the eigenvalue E is given by

$$-\frac{\hbar^2}{2m}\nabla^2 u(\mathbf{r}) + V(\mathbf{r})\,u(\mathbf{r}) = Eu(\mathbf{r}), \tag{8.168}$$

which in Cartesian coordinates is written as

$$-\frac{\hbar^2}{2m}\left(\frac{\partial^2}{\partial x^2}+\frac{\partial^2}{\partial y^2}+\frac{\partial^2}{\partial z^2}\right)u(x,y,z) + V(x,y,z)u(x,y,z) = Eu(x,y,z). \tag{8.169}$$

This equation can be solved for the so-called separable potentials, which have the property that they can be written as a sum of three potentials, with each potential being a function of only one variable, i.e., one can write

$$V(x,y,z) = V_1(x) + V_2(y) + V_3(z). \tag{8.170}$$

We can now solve this in the same manner as the free particle equation in Chapter 2 by writing

$$u(x,y,z) = X(x)Y(y)Z(z), \tag{8.171}$$

so that we have

$$-\frac{\hbar^2}{2m}\left[YZ\frac{d^2X}{dx^2} + ZX\frac{d^2Y}{dy^2} + XY\frac{d^2Z}{dz^2}\right] + [V_1(x) + V_2(y) + V_3(z)]XYZ = EXYZ. \tag{8.172}$$

This equation can be reduced further by dividing it by XYZ to obtain

$$-\frac{\hbar^2}{2m}\left[\frac{1}{X}\frac{d^2X}{dx^2}+\frac{1}{Y}\frac{d^2Y}{dy^2}+\frac{1}{Z}\frac{d^2Z}{dz^2}\right]+[V_1(x)+V_2(y)+V_3(z)]=E. \tag{8.173}$$

Now combining the terms with individual variables we find

$$\left[-\frac{\hbar^2}{2m}\frac{1}{X}\frac{d^2X}{dx^2}+V_1(x)\right]+\left[-\frac{\hbar^2}{2m}\frac{1}{Y}\frac{d^2Y}{dy^2}+V_2(y)\right]+\left[-\frac{\hbar^2}{2m}\frac{1}{Z}\frac{d^2Z}{dz^2}+V_3(z)\right]=E. \tag{8.174}$$

This can be written as three one-dimensional equations,

$$-\frac{\hbar^2}{2m}\frac{d^2X}{dx^2}+V_1(x)X=E_1X, \tag{8.175}$$

$$-\frac{\hbar^2}{2m}\frac{d^2Y}{dy^2}+V_2(y)Y=E_2Y, \tag{8.176}$$

$$-\frac{\hbar^2}{2m}\frac{d^2Z}{dz^2}+V_3(z)Z=E_3Z, \tag{8.177}$$

with

$$E_1+E_2+E_3=E. \tag{8.178}$$

Since we now have three one-dimensional equations, we can solve each of them separately following the procedure outlined for the free particle case in Chapter 2. If $X_n(x)$, $Y_m(y)$ and $Z_p(z)$ are the solutions with eigenvalues E_{1n}, E_{2m}, and E_{3p}, respectively then a solution for the three-dimensional system is

$$u_{nmp}(x,y,z)=X_n(x)Y_m(y)Z_p(z) \tag{8.179}$$

with energy eigenvalues,

$$E_{nmp}=E_{1n}+E_{2m}+E_{3p}. \tag{8.180}$$

A number of interesting problems exist, namely, a three-dimensional cubic well or three-dimensional harmonic oscillators. We will not consider these since they are mostly extensions of the one-dimensional problems that we have already discussed.

8.6.2 Potentials depending on relative distance between particles

We have thus far been considering problems involving the motion of a single particle in an external potential. Consider now the case where we have two particles of masses m_1 and m_2

located respectively at distances $\mathbf{r}_1(=x_1, y_1, z_1)$ and $\mathbf{r}_2(=x_2, y_2, z_2)$ from the origin with a potential that depends on their relative separation.

The time-dependent Schrödinger equation is now

$$i\hbar\frac{\partial}{\partial t}\psi(\mathbf{r}_1, \mathbf{r}_2; t) = \left[-\left(\frac{\hbar^2}{2m_1}\nabla_1^2 + \frac{\hbar^2}{2m_2}\nabla_2^2\right) + V(\mathbf{r}_1 - \mathbf{r}_2)\right]\psi(\mathbf{r}_1, \mathbf{r}_2; t) \tag{8.181}$$

where

$$\nabla_i^2 = \left(\frac{\partial^2}{\partial x_i^2} + \frac{\partial^2}{\partial y_i^2} + \frac{\partial^2}{\partial z_i^2}\right) \quad \text{for } i = 1, 2. \tag{8.182}$$

We now make the following changes of variables,

$$\mathbf{r} = \mathbf{r}_1 - \mathbf{r}_2 \tag{8.183}$$

and

$$\mathbf{R} = \frac{m_1\mathbf{r}_1 + m_2\mathbf{r}_2}{M} \tag{8.184}$$

where

$$M = m_1 + m_2 = \text{total mass.} \tag{8.185}$$

We identify

$$\mathbf{R} = (X, Y, Z) = \text{center of mass coordinates,} \tag{8.186}$$

$$\mathbf{r} = (x, y, z) = \text{relative coordinates.} \tag{8.187}$$

We also introduce the quantity

$$m = \frac{m_1 m_2}{m_1 + m_2} = \text{reduced mass.} \tag{8.188}$$

It is now straightforward to show that

$$\frac{\hbar^2}{2m_1}\nabla_1^2 + \frac{\hbar^2}{2m_2}\nabla_2^2 = \frac{\hbar^2}{2M}\nabla_R^2 + \frac{\hbar^2}{2m}\nabla^2 \tag{8.189}$$

where ∇_R^2 and ∇^2 have the same expression as ∇^2 except that in the former case we replace (x, y, z) by (X, Y, Z) while in the latter case (x, y, z) refer to the relative coordinates (8.187).

The equation now reads

$$i\hbar\frac{\partial}{\partial t}\psi(\mathbf{r}, \mathbf{R}; t) = \left[-\left(\frac{\hbar^2}{2M}\nabla_R^2 + \frac{\hbar^2}{2m}\nabla^2\right) + V(\mathbf{r})\right]\psi(\mathbf{r}, \mathbf{R}; t). \tag{8.190}$$

To solve the energy eigenvalue problem we use the separation of variables method and write, replacing the energy term by $(E_M + E)$,

$$\psi(\mathbf{r}, \mathbf{R}; t) = u(\mathbf{r})U(\mathbf{R})e^{-i(E_M+E)t/\hbar} \tag{8.191}$$

where the individual wavefunctions satisfy

$$-\frac{\hbar^2}{2M}\nabla_R^2 U(\mathbf{R}) = E_M U(\mathbf{R}). \tag{8.192}$$

Thus, the center of mass wavefunction $U(\mathbf{R})$ satisfies a free particle equation. We will ignore the motion of the center of mass in any future problems since it is irrelevant as far as the effects of the potential are concerned and concentrate only on the wavefunction $u(\mathbf{r})$, which now satisfies the equation

$$\left[-\frac{\hbar^2}{2m}\nabla^2 + V(\mathbf{r})\right] u(\mathbf{r}) = Eu(\mathbf{r}), \tag{8.193}$$

which is of the usual form except that the coordinates refer to the relative distances between the interacting particles and m is the reduced mass.

8.6.3 Formalism for spherically symmetric potentials

In Chapter 2 we considered the radial differential equation for a free particle through the separation of variables technique in which we expressed the wavefunction $u(\mathbf{r})$ as

$$u(\mathbf{r}) = u(r, \theta, \phi) = R(r)Y(\theta, \phi) \tag{8.194}$$

and found that the individual functions had the forms

$$R(r) = R_l(r) \tag{8.195}$$

and

$$Y(\theta, \phi) = Y_{lm}(\theta, \phi) \tag{8.196}$$

where Y_{lm} are the spherical harmonics. The quantum numbers l and m correspond to the angular momentum operator and take on values $l = 0, 1, 2, \ldots$ with $m = -l, \ldots, l$. In the case of the Schrödinger equation with a potential, $V(r)$, which we will be considering below, since $V(r)$ is spherically symmetric, i.e., since it depends only on the radial coordinate, r, the dependence on θ and ϕ of the wavefunction $u(\mathbf{r})$ will still be given by $Y_{lm}(\theta, \phi)$. Hence we will be considering only the radial equation.

The radial wavefunction, R_l, for a spherically symmetric potential, $V(r)$, is given by

$$-\frac{\hbar^2}{2m}\frac{1}{r}\frac{d^2(rR_l)}{dr^2} + \left[V(r) + \frac{\hbar^2 l(l+1)}{2mr^2}\right] R_l = ER_l. \tag{8.197}$$

Since the radial equation for the problems to be discussed below can be cast into a form similar to the free particle radial equation that we considered in Chapter 2, we summarize below the essential properties of the free particle solutions. As discussed in Chapter 2, this equation is of the form

$$\frac{d^2R_l}{d\rho^2} + \frac{2}{\rho}\frac{dR_l}{d\rho} - \frac{l(l+1)R_l}{\rho^2} + R_l = 0 \tag{8.198}$$

where

$$\rho = kr, \quad \text{and} \quad k = \sqrt{2mE/\hbar}. \tag{8.199}$$

There are two solutions to equation (8.198): the spherical Bessel function, $\mathrm{j}_l(\rho)$, and spherical Neumann function, $\mathrm{n}_l(\rho)$. One can also construct two linear combinations of these, the spherical Hankel functions of the first and second kinds, $\mathrm{h}_l^{(1)}(\rho)$ and $\mathrm{h}_l^{(2)}(\rho)$ respectively. The properties of these special functions have already been discussed in Chapter 2.

The free particle wavefunction that we discussed in Chapter 2 can only involve $\mathrm{j}_l(\rho)$, which is well behaved at $\rho = 0$, while $\mathrm{n}_l(\rho)$ is excluded because it becomes infinite at $\rho = 0$. Hence the free particle radial wavefunction is

$$R_l = A_l \mathrm{j}_l(\rho). \tag{8.200}$$

8.7 Simple systems

8.7.1 Spherical wall

Consider a potential which is very much like the infinite barrier in one-dimension which we discussed earlier in Section 8.3, except this time it involves the r-variable. This configuration is often called a spherical wall inside which the particle is trapped, and it is given by

$$V(r) = 0, \quad r < a \tag{8.201}$$

$$= \infty, \quad r \geq a, \tag{8.202}$$

as illustrated in Fig. 8.8. Thus the wavefunction R_l vanishes at $r = a$.

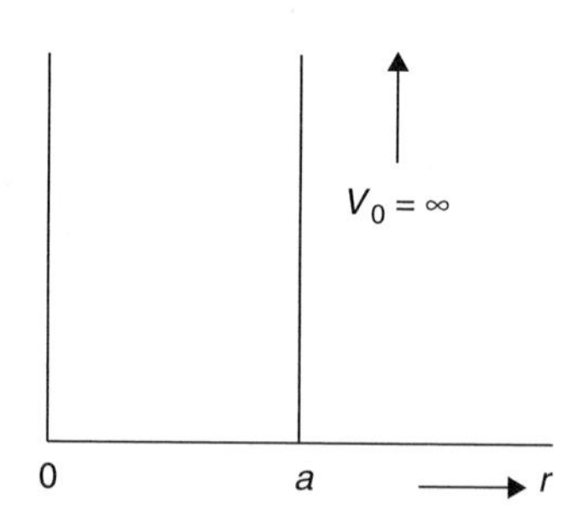

Fig. 8.8

The radial equation for $0 < r < a$ is

$$-\frac{\hbar^2}{2m}\frac{1}{r}\frac{d^2(rR_l)}{dr^2} + \frac{\hbar^2 l(l+1)}{2mr^2}R_l = ER_l. \tag{8.203}$$

As in the one-dimensional case, we take

$$\rho = \alpha r \quad \text{and} \quad \alpha = \sqrt{\frac{2mE}{\hbar^2}} \tag{8.204}$$

and express the equation in terms of ρ, and write

$$\frac{d^2R_l}{d\rho^2} + \frac{2}{\rho}\frac{dR_l}{d\rho} - \frac{l(l+1)R_l}{\rho^2} + R_l = 0. \tag{8.205}$$

The solutions to this equation, as we mentioned earlier, are $\mathrm{j}_l(\rho)$ and $\mathrm{n}_l(\rho)$. However, since $\mathrm{n}_l(\rho)$ diverges as $\rho \to 0$, the only possible solutions we can have are given by

$$R_l = A_l \mathrm{j}_l(\rho) = A_l \mathrm{j}_l(\alpha r), \quad r < a, \tag{8.206}$$

$$R_l = 0, \qquad r > a \tag{8.207}$$

where the constant A_l can be obtained if we normalize the wavefunction inside the wall. Therefore,

$$A_l = \frac{1}{\int_0^a dr\, r^2 \mathrm{j}_l^2(\alpha r)}. \tag{8.208}$$

The boundary condition at $r = a$ implies that j_l must vanish there. As we already know, j_l has an infinite number of zeros. The energy eigenvalues are then given by

$$\mathrm{j}_l(\alpha_{nl} a) = 0 \quad \text{for } n = 1, 2, \ldots . \tag{8.209}$$

Therefore, for each l there are an infinite number of discrete eigenvalues given by the zeros of j_l. For $l = 0$, these are given quite simply by the zeros of the sine function, $\sin(\alpha_{nl} a)$, i.e., at $\alpha_{nl} a = \pi, 2\pi, \ldots$; for $l = 1$, the zeros, numerically, are at $\alpha_{nl} a = 0, 4.5, 7.7, \ldots$; and so on for other l-values.

8.7.2 Finite barrier

For the case of a finite barrier we have (see Fig. 8.9)

$$V(r) = 0, \quad r < a \tag{8.210}$$

$$= V_0, \quad r > a. \tag{8.211}$$

The radial equation for $r < a$ will be the same as that of a free particle in the same region,

$$\frac{d^2R_l}{d\rho^2} + \frac{2}{\rho}\frac{dR_l}{d\rho} - \frac{l(l+1)R_l}{\rho^2} + R_l = 0, \tag{8.212}$$

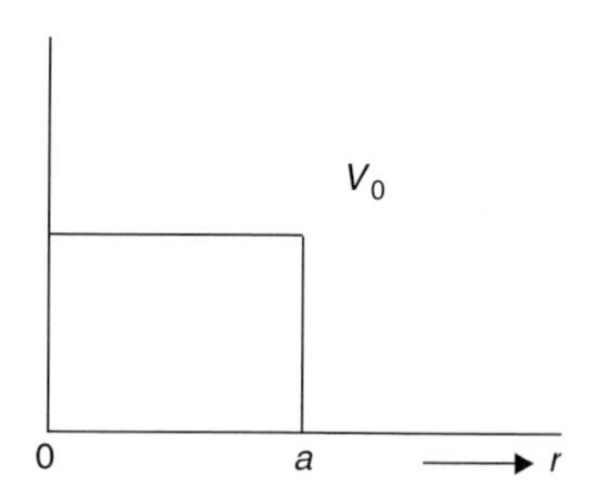

Fig. 8.9

where

$$\rho = \alpha r \quad \text{and} \quad \alpha = \sqrt{\frac{2mE}{\hbar^2}}, \tag{8.213}$$

and will have the same solution,

$$R_l = A_l \mathrm{j}_l(\alpha r) \; r < a. \tag{8.214}$$

However, for $r > a$, the radial equation is

$$-\frac{\hbar^2}{2m}\frac{1}{r}\frac{d^2(rR_l)}{dr^2} + \left[V_0 + \frac{\hbar^2 l(l+1)}{2mr^2}\right] R_l = ER_l. \tag{8.215}$$

Since we are considering bound states, we must have $E < V_0$. Let us define the following quantities:

$$\beta = \sqrt{\frac{2m(V_0 - E)}{\hbar^2}} \quad \text{and} \quad \rho = i\beta r. \tag{8.216}$$

The above equation is then of the form

$$\frac{d^2 R_l}{d\rho^2} + \frac{2}{\rho}\frac{dR_l}{d\rho} - \frac{l(l+1)R_l}{\rho^2} + R_l = 0, \quad r > a. \tag{8.217}$$

The solution of this equation is

$$R_l(r) = B_l \mathrm{h}_l^{(1)}(i\beta r). \tag{8.218}$$

Since $h_l^{(1)}(i\beta r) \to \frac{1}{\beta r} e^{-\beta r}$ as $r \to \infty$ this solution gives a convergent solution that vanishes at infinity.

The boundary conditions at $r = a$ give

$$A_l \mathrm{j}_l(\alpha a) = B_l \mathrm{h}_l^{(1)}(i\beta a), \tag{8.219}$$

$$A_l \frac{d\mathrm{j}_l(\alpha a)}{dr} = B_l \frac{d\mathrm{h}_l^{(1)}(i\beta a)}{dr} \tag{8.220}$$

where it is understood that one puts $r = a$, after the derivative is taken. The solutions of these equations can be obtained graphically, as we did in the one-dimensional case, though now, except for $l = 0$, it is quite complicated. Since it does not add any interesting information, we do not pursue it any further.

8.7.3 Square-well potential

Here what we mean by a square well is actually a spherical well since we are considering the problem in radial coordinates. The potential is given by

$$V(r) = -V_0, \quad r < a \tag{8.221}$$

$$= 0, \quad r > a. \tag{8.222}$$

Since we are considering bound states we take $E = -E_B$, with $E_B > 0$. For $r < a$, the radial equation is

$$-\frac{\hbar^2}{2m}\frac{1}{r}\frac{d^2(rR_l)}{dr^2} + \left[-V_0 + \frac{\hbar^2 l(l+1)}{2mr^2}\right]R_l + E_B R_l = 0, \tag{8.223}$$

which can be reduced to the known form

$$\frac{d^2R_l}{d\rho^2} + \frac{2}{\rho}\frac{dR_l}{d\rho} - \frac{l(l+1)R_l}{\rho^2} + R_l = 0 \tag{8.224}$$

by taking

$$\rho = \alpha r \quad \text{where} \quad \alpha = \sqrt{\frac{2m(V_0 - E_B)}{\hbar^2}}. \tag{8.225}$$

The solution, as in previous cases, is

$$R_l(r) = A_l \mathrm{j}_l(\alpha r), \quad r < a. \tag{8.226}$$

The radial equation for $r > a$ is

$$-\frac{\hbar^2}{2m}\frac{1}{r}\frac{d^2(rR_l)}{dr^2} + \frac{\hbar^2 l(l+1)}{2mr^2}R_l + E_B R_l = 0, \tag{8.227}$$

which can be cast into the familiar form

$$\frac{d^2R_l}{d\rho^2} + \frac{2}{\rho}\frac{dR_l}{d\rho} - \frac{l(l+1)R_l}{\rho^2} + R_l = 0 \tag{8.228}$$

by taking

$$\rho = i\beta r \quad \text{where} \quad \beta = \sqrt{\frac{2mE_B}{\hbar^2}}. \tag{8.229}$$

The solution here will once again be the Hankel function $h_l^{(1)}(i\beta r)$ which vanishes at infinity,

$$R_l(r) = B_l h_l^{(1)}(i\beta r). \tag{8.230}$$

The boundary conditions at $r = a$ are

$$A_l j_l(\alpha r) = B_l h_l^{(1)}(i\beta r) \tag{8.231}$$

$$A_l \frac{dj_l(\alpha r)}{dr} = B_l \frac{dh_l^{(1)}(i\beta r)}{dr}. \tag{8.232}$$

These conditions are similar to the finite-barrier case, except that the definitions of α and β are different. The equations can be solved by the same type of graphical methods.

8.8 Hydrogen-like atom

Let us now consider hydrogen-like atoms where a bound state is formed because of the attractive Coulomb potential between the nucleus of charge Ze, consisting of Z protons of charge $+e$, and an electron of charge $-e$. The corresponding potential is given by

$$V(r) = -\frac{Ze^2}{r}, \tag{8.233}$$

which is represented in Fig. 8.10. Since the proton mass is much greater than the mass of the electron, the reduced mass that enters the Schrödinger equation will be approximately the same as that of the electron, which we take to be m. The radial Schrödinger equation will be given by

$$-\frac{\hbar^2}{2m}\frac{1}{r}\frac{d^2(rR_l)}{dr^2} + \left[-\frac{Ze^2}{r} + \frac{\hbar^2 l(l+1)}{2mr^2}\right] R_l + E_B R_l = 0, \tag{8.234}$$

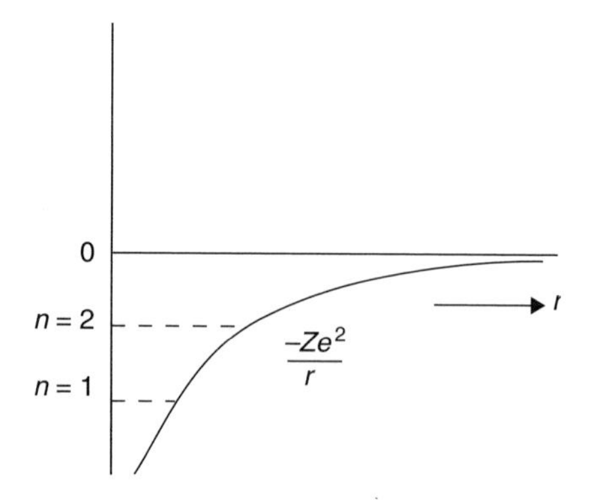

Fig. 8.10

where R_l is the radial wavefunction. For the energy eigenvalue we have written $E = -E_B$, where E_B, binding energy, is positive since we are considering bound states.

To simplify the equation we write $\rho = \alpha r$, where α is chosen such that the equation can be written in the form

$$\frac{d^2 R_l}{d\rho^2} + \frac{2}{\rho}\frac{dR_l}{d\rho} + \left[\frac{\lambda}{\rho} - \frac{l(l+1)}{\rho^2} - \frac{1}{4}\right] R_l = 0 \tag{8.235}$$

where we define the constants λ and α by

$$\lambda = \frac{2mZe^2}{\hbar^2 \alpha} \quad \text{and} \quad \frac{2mE_B}{\hbar^2 \alpha^2} = \frac{1}{4}. \tag{8.236}$$

We will determine these constants so that we have a bound state with binding energy E_B.

Let us examine the behavior of (8.235) as $\rho \to \infty$. Keeping only the leading terms, the radial equation (8.235) reduces to

$$\frac{d^2 R_l}{d\rho^2} - \frac{1}{4} R_l = 0, \tag{8.237}$$

whose solution is $R_l = e^{\pm \frac{1}{2}\rho}$. The negative sign in the exponent is the correct choice as it allows the probability to be finite. We note that

$$R_l \to \rho^n e^{-\frac{1}{2}\rho} \tag{8.238}$$

as $\rho \to \infty$ is also a solution, where n is a finite integer, since the behavior of the exponential dominates over that of any power of ρ. This implies that a finite sum of powers of ρ, i.e., a polynomial in ρ, can be multiplied to the above exponential function and R_l will still remain a solution in the limit $\rho \to \infty$. Let $F(\rho)$ be such a polynomial so that we can write

$$R_l = F(\rho) e^{-\frac{1}{2}\rho}. \tag{8.239}$$

If ρ^s is the leading power in $F(\rho)$ then we take it out as a common factor to simplify the problem and write

$$F(\rho) = \rho^s L(\rho) \tag{8.240}$$

where $L(\rho)$ will be a polynomial of the type

$$L(\rho) = a_0 + a_1 \rho + \cdots + a_\nu \rho^\nu + \cdots \tag{8.241}$$

and thus R_l can be expressed as

$$R_l = \rho^s L(\rho) e^{-\frac{1}{2}\rho}. \tag{8.242}$$

In order for R_l to be finite at $\rho = 0$ (so that the probability remains finite) we must have s as a positive quantity. We substitute the above expression into the radial equation (8.235) for R_l and consider the limit $\rho \to 0$. In this limit the term ρ^s is the leading contributor since it is the smallest power. It is easy to check that the terms $\left(d^2R_l/d\rho^2\right)$ and $\left[l(l+1)R_l/\rho^2\right]$ in (8.235) give the leading terms. Their combined contribution is

$$[s(s+1) - l(l+1)]\rho^{s-2}. \tag{8.243}$$

Since the radial equation is valid for continuous values of ρ, the coefficient of ρ^{s-2} must vanish. Therefore,

$$s(s+1) - l(l+1) = 0, \tag{8.244}$$

leading to the solutions

$$s = l \quad \text{or} \quad -(l+1). \tag{8.245}$$

Clearly the first solution for positive $s\ (= l)$ is the correct one for the probability to remain finite. Hence we write

$$F(\rho) = \rho^l L(\rho). \tag{8.246}$$

and, therefore, we can write

$$R_l = \rho^l L(\rho) e^{-\frac{1}{2}\rho}. \tag{8.247}$$

Substituting R_l from (8.247) into equation (8.235), we obtain the following equation for $L(\rho)$:

$$\rho L'' + [2(l+1) - \rho]L' + (\lambda - l - 1)L = 0. \tag{8.248}$$

After substituting the series (8.241) for $L(\rho)$ into (8.248) we combine the coefficients of equal powers of ρ. This series will look like

$$A_0 + A_1\rho + \cdots\cdots + A_\nu\rho^\nu + \cdots = 0 \tag{8.249}$$

Since the equation is valid for continuous values of ρ, each of the coefficients must vanish, i.e.,

$$A_0 = A_1 = \cdots A_\nu = \cdots = 0. \tag{8.250}$$

The vanishing of A_ν leads to

$$a_{\nu+1}[(\nu+1)(\nu+2l+2)] - a_\nu[\nu - \lambda + l + 1] = 0. \tag{8.251}$$

Taking the ratio of the consecutive terms, and examining the behavior for large ν, we find

$$\frac{a_{\nu+1}}{a_\nu} = \frac{[\nu - \lambda + l + 1]}{[(\nu+1)(\nu+2l+2)]} \to \frac{1}{\nu} \quad \text{as} \quad \nu \to \infty. \tag{8.252}$$

This ratio turns out to be the same as of the ratio of the consecutive terms for the exponential function

$$e^{\rho} = 1 + \rho + \frac{\rho^2}{2!} + \cdots + \frac{\rho^{\nu}}{\upsilon!} + \frac{\rho^{\nu+1}}{(\upsilon + 1)!} + \cdots \tag{8.253}$$

which is

$$\frac{\nu!}{(\nu + 1)!} = \frac{1}{\nu + 1} \to \frac{1}{\nu} \quad \text{as } \nu \to \infty. \tag{8.254}$$

Thus $L(\rho)$ will mimic the behavior of e^{ρ} as ρ becomes large, because when ρ is large the higher-order terms with large values ν become significant. For $L(\rho) \to e^{\rho}$, we find

$$R_l \to \rho^l e^{\rho} e^{-\frac{1}{2}\rho} = \rho^l e^{+\frac{1}{2}\rho}, \tag{8.255}$$

which is untenable as it makes R_l diverge as $\rho \to \infty$. Therefore, the only alternative is that the series (8.241) for $L(\rho)$ terminates at some value of ν.

If the series (8.241) terminates at the power ρ^{ν}, it implies that $a_{\nu+1}$ must vanish. From the recursion relation (8.251) we note that the terms $a_{\nu+2}, a_{\nu+3} \ldots$ that follow will then also vanish and we will be left with a finite polynomial, which is what we require. If, for example, the series terminates at $\nu = n'$(=integer), then, $a_{n'}$ is nonzero while $a_{n+1'}$ must vanish, which, from the second term in the recursion relation (8.251) or from the numerator on the right-hand side of (8.252), implies that

$$n' - \lambda + l + 1 = 0. \tag{8.256}$$

Therefore, λ must be an integer ($= n$),

$$\lambda = n' + l + 1 = n. \tag{8.257}$$

The function L will then be a finite polynomial of the form

$$L = a_0 + a_1\rho + \cdots + a_{n'}\rho^{n'} \tag{8.258}$$

where one calls

$$n' = \text{radial quantum number} = 0, 1, 2, \cdots. \tag{8.259}$$

Substituting $\lambda = n$ in the relation (8.236) between λ and α, we obtain

$$n = \frac{2mZe^2}{\hbar^2\alpha}, \tag{8.260}$$

which allows us to obtain α and, therefore, from the relation (8.236) between α and E_B, we obtain the binding energy

$$E_B = \frac{mZ^2e^4}{2\hbar^2 n^2}. \tag{8.261}$$

To make this relation more transparent we introduce a quantity that has the dimensions of radius, called the Bohr radius, a_0, given by

$$a_0 = \frac{\hbar^2}{me^2}. \tag{8.262}$$

We can write down the energy eigenvalues in terms of the Bohr radius as

$$E_n = -\frac{Z^2e^2}{2a_0}\frac{1}{n^2} \tag{8.263}$$

where we call

$$n = \text{principal quantum number}. \tag{8.264}$$

Since n' takes on values $0, 1, 2, \ldots$, the values of n and l will be given by

$$n = 1, 2, 3, \ldots, \tag{8.265}$$

$$l \leq n - 1. \tag{8.266}$$

Therefore, for each n, we will have

$$l = 0, 1, \ldots, n - 1. \tag{8.267}$$

And, for each l we will have

$$m = -l, -l + 1, \cdots, l - 1, l \quad \text{i.e., } (2l + 1) \text{ different states}. \tag{8.268}$$

Since E_n depends only on n, we have a degeneracy in the energy eigenstates given by

$$\sum_{l=0}^{n-1} (2l + 1) = n^2. \tag{8.269}$$

The polynomial $L(\rho)$ is a well-known mathematical function called the Laguerre polynomial, written as

$$L(\rho) = L_{n+l}^{2l+1}(\rho), \tag{8.270}$$

which satisfies the normalization condition

$$\int_0^\infty d\rho\ \rho^{2l+2} e^{-\rho} \left[L_{n+l}^{2l+1}(\rho)\right]^2 = \frac{2n\left[(n+l)!\right]^3}{(n-l-1)!}. \tag{8.271}$$

Consequently, the radial wavefunction, which now depends on n as well as l, is written as

$$R_{nl}(r) = \left[\left(\frac{2Z}{na_0}\right)^3 \frac{(n-l-1)!}{2n\left[(n+l)!\right]^3}\right] \rho^l L_{n+l}^{2l+1}(\rho) e^{-\frac{1}{2}\rho} \tag{8.272}$$

where

$$\rho = \alpha r \text{ with } \alpha = \frac{2Z}{na_0} \tag{8.273}$$

with normalization

$$\int_0^\infty dr\, r^2 R_{nl}^2(r) = 1. \tag{8.274}$$

.

The complete three-dimensional wavefunction $u_{nlm}(\mathbf{r})$ can now be written down as

$$u_{nlm}(\mathbf{r}) = R_{nl}(r) Y_{lm}(\theta, \phi). \tag{8.275}$$

Below we summarize our major results for a small set of wavefunctions and energy eigenvalues, where a state is designated by n and l, with the spectroscopic terminology of calling a state with $l = 0$ an S-state; $l = 1$ a P-state; $l = 2$ a D-state, and so on. Thus, writing in a tabulated form, we have for the lowest two values, $n = 1$ and $n = 2$, the following.

(1) The energy eigenvalues, E_n

$$\begin{bmatrix} \underline{n} & \underline{E_n} \\ 1 & -\dfrac{Z^2 e^2}{2a_0} \\ 2 & -\dfrac{Z^2 e^2}{2a_0} \cdot \dfrac{1}{4} \\ . & . \end{bmatrix} \tag{8.276}$$

(2) Radial wavefunctions, $R_{nl}(r)$

$$\begin{bmatrix} \underline{l} & \underline{n=1} & \underline{n=2} \\ 0 & \left(\dfrac{Z}{a_0}\right)^{3/2} 2e^{-Zr/a_0} \quad (1S \text{ state}) & \left(\dfrac{Z}{2a_0}\right)^{3/2} \left(2 - \dfrac{Zr}{a_0}\right) e^{-Zr/2a_0} \quad (2S \text{ state}) \\ 1 & . & \left(\dfrac{Z}{2a_0}\right)^{3/2} \dfrac{Zr}{a_0\sqrt{3}} e^{-Zr/2a_0} \quad (2P \text{ state}) \\ . & . & . \end{bmatrix} \tag{8.277}$$

(3) Spherical harmonics, $Y_{lm}(\theta, \phi)$

$$\begin{bmatrix} \underline{m} & \underline{l=0} \ (S \text{ state}) & \underline{l=1} \ (P \text{ state}) \\ 0 & \frac{1}{\sqrt{4\pi}} & \sqrt{\dfrac{3}{4\pi}} \cos\theta \\ 1 & . & -\sqrt{\dfrac{3}{8\pi}} \sin\theta e^{i\phi} \\ -1 & . & \sqrt{\dfrac{3}{8\pi}} \sin\theta e^{-i\phi} \\ . & . & . \end{bmatrix} \tag{8.278}$$

8.9 Problems

1. Consider the energy eigenvalue equation for an $l = 0$ radial wavefunction $\chi_0(r)$ with a spherically symmetric potential $V(r)$. Multiplying this equation by $d\chi_0/dr$ and rearranging the terms, and then integrating the resulting equation from 0 to ∞, show that

$$|R_0(0)|^2 = \frac{2m}{\hbar^2}\left\langle\frac{dV}{dr}\right\rangle$$

where $R_l = \chi_l/r$. Generalize this result for the case when $l \neq 0$.

2. For one-dimensional problems, show that the spectrum for bound-state energies is always nondegenerate.
3. For one-dimensional bound-state problems, show that if the energy eigenvalues are placed in the order $E_1 < E_2 < \cdots$ then the corresponding eigenfunctions will have increasing number of zeros so that the nth eigenfunction will have $n-1$ zeros.
4. Consider a series of δ-functions given by

$$V(x) = \frac{\lambda\hbar^2}{2m}\sum_{n=-\infty}^{n=+\infty}\delta(x-nd)$$

as an approximation to the periodic potential considered in this chapter. Determine the forbidden energy bands at low energies for this configuration.

5. Consider a rigid rotator consisting of two spherical balls attached by a rigid rod whose center is at the origin in a three-dimensional space. Express the classical Hamiltonian in terms of the moment of inertia, I, of the system. Write the quantum-mechanical Hamiltonian by expressing I in terms of the angular momentum l of the system. Obtain the energy eigenvalues and eigenfunction for this problem.
6. Consider the radial Schrödinger equation for $l = 0$, with a potential which, as $r \to \infty$, has the form

$$V(r) = \frac{\lambda}{r^s}$$

with $s \geq 1$. Show that for bound states the asymptotic expression for the radial wavefunction, R_l, can be written as

$$R_l \sim e^{-\beta r}F(r)$$

where $\beta = \sqrt{2mE_B/\hbar^2}$. Determine the functional form of $F(r)$ for arbitrary values of s. Show that, for the Coulomb potential, R_l does not have a purely exponential form.

7. A particle is trapped between two infinite walls at $x = -a$ and $x = a$ respectively. Determine the bound-state energies and wavefunctions for the particle if an additional potential $\lambda\delta(x)$ is present.

8. Determine the bound-state energies and wavefunctions for a particle subjected to the potential

$$V(r) = -\lambda\delta(r-a).$$

Obtain the minimum value of λ for which there is a bound state.

9. For a potential given by a symmetric pair of δ-functions,

$$V(x) = -g\left[\delta(x-a) + \delta(x+a)\right],$$

determine the energy eigenvalues. It may be easier to solve this problem by treating even and odd parity states separately.

10. Consider the problem of periodic potentials relevant to lattice sites of dimension d where the potentials correspond to the infinite wall configuration. If there are n sites and $|0_n\rangle$ corresponds to the ground states for individual sites with energy E_0, then the translation operator satisfies $F(d)\,|0_n\rangle = |0_{n+1}\rangle$. Consider now the state

$$|\alpha\rangle = \sum_{-\infty}^{\infty} e^{in\alpha}\,|0_n\rangle$$

where α is a real number. Show that $|\alpha\rangle$ is an eigenstate of H and

$$F(d)\,|\alpha\rangle = e^{-i\alpha}\,|\alpha\rangle\,.$$

Also show that

$$\langle x\,|F(d)|\,\alpha\rangle = \langle x-d\,|\alpha\rangle = e^{-i\alpha}\langle x\,|\alpha\rangle\,.$$

11. In problem 10 assume that the walls are high but not infinite, so that tunneling from the state $|0_n\rangle$ to its nearest neighbors $|0_{n\pm1}\rangle$ is possible. If $\langle 0_{n\pm1}\,|H|\,0_n\rangle = -\Delta$, then show that

$$H\,|\alpha\rangle = (E_0 - 2\Delta\cos\alpha)\,|\alpha\rangle\,.$$

12. A particle moving in one dimension is subjected to a potential given by

$$\begin{aligned} V(x) &= -\frac{\lambda}{x}, \quad x > 0 \\ &= \infty, \quad x \le 0. \end{aligned}$$

Show that the Schrödinger equation for this system is mathematically equivalent to that of a hydrogen atom in an $l = 0$ state. Using this result, obtain its energy eigenvalue and the corresponding eigenfunction.

13. Consider the orbital motion of an electron classically with the power, P, emitted by the electron, given by $P = (2/3)e^2/c^3a^2$, where a is the acceleration. Use the classical

formula $ma = e^2/a_0^2$ where a_0 is the Bohr radius. How much time will it take for the electron to lose the energy given by the difference between the $n = 2$ and $n = 1$ levels?

14. Consider the following infinite barrier problem:

$$\begin{aligned} V(x) &= 0, \quad 0 < x < a \\ &= \infty, \quad x < 0 \\ &= \infty, \quad x > a. \end{aligned}$$

Obtain the energy eigenvalues and eigenfunctions. Also obtain the expectation value of the momentum operator for the ground-state wavefunction.

15. For the following square-well potential,

$$\begin{aligned} V(x) &= -V_0, \quad |x| \leq a \\ V(x) &= 0, \quad |x| \geq a \end{aligned}$$

show that a bound state will exist for arbitrarily small V_0. Obtain an analytical solution that relates the binding energy, E_B, and V_0 for the case when $2mV_0a^2/\hbar^2 = \epsilon \ll 1$.

16. There are two degenerate states with wavefunctions $u_1(\mathbf{r})$ and $u_2(\mathbf{r})$, with a common eigenvalue E_0, given by

$$u_1(\mathbf{r}) = f_1(r), \quad \text{and} \quad u_2(\mathbf{r}) = f_2(r)\cos\theta.$$

A perturbation H' is applied of the form

$$H' = V(r)\left(1 + \cos\theta + \cos^2\theta\right).$$

Obtain the perturbed energies to first order. Assume $\int_0^\infty dr\, r^2 f_i^*(r)V(r)f_j(r) = g_{ij}$.

17. A wavefunction in a one-dimensional bound state problem with energy variable λ is given by

$$u(x) = \left(a_0 + a_1x + \cdots + a_\upsilon x^\upsilon + \cdots\right)e^{-x} \qquad 0 < x < \infty$$

where the coefficients satisfy the recursion relation

$$\frac{a_{\upsilon+1}}{a_\upsilon} = \frac{2\upsilon - \lambda}{(\upsilon+1)(\upsilon+2)}.$$

Determine the behavior of this series as $\upsilon \to \infty$. From this information obtain the eigenvalues of λ.

18. Consider the radial equation

$$\frac{d^2R}{dr^2} + \frac{2}{r}\frac{dR}{dr} + \left[\frac{\lambda}{r} - 1\right]R = 0,$$

which corresponds to a bound-state problem. Determine $f(r)$, the asymptotic form of $R(r)$. Select the correct $f(r)$ that is appropriate to this problem and write $R(r) = f(r)h(r)$. Obtain the equation for $h(r)$. Write

$$h(r) = r^s(a_0 + a_1 r + \cdots + a_\nu r^\nu + \cdots)$$

and obtain the value of s. Also obtain the recursion relation for the coefficients above and, from it, the eigenvalues of λ.

9 Harmonic oscillator

The simple harmonic oscillator plays a key role in classical as well as quantum systems. It is found that complicated physical systems can often be approximated in ways that allows one to cast them as harmonic oscillators. The framework we develop here, involving raising and lowering operators in the Heisenberg picture, is of fundamental importance in many branches of physics including condensed matter physics and quantum field theory.

9.1 Harmonic oscillator in one dimension

9.1.1 Formalism in the Heisenberg representation

The harmonic oscillator potential is described (see Fig. 9.1) by

$$V(x) = \frac{1}{2}Kx^2, \quad -\infty < x < \infty. \tag{9.1}$$

The corresponding Hamiltonian is given by

$$H = \frac{P^2}{2m} + \frac{1}{2}KX^2 = \frac{P^2}{2m} + \frac{1}{2}m\omega^2X^2 \tag{9.2}$$

where we have taken $K = m\omega^2$, where ω is called the classical frequency, sometimes designated as ω_c. The X and P operators will satisfy the fundamental commutation relation

$$[X, P] = i\hbar\mathbf{1}. \tag{9.3}$$

We will work within the framework of the Heisenberg representation in which case these operators will then depend on time.

In order to simplify the structure of the Hamiltonian we define the following operators:

$$a = \sqrt{\frac{m\omega}{2\hbar}}\left(X + i\frac{P}{m\omega}\right), \quad a^\dagger = \sqrt{\frac{m\omega}{2\hbar}}\left(X - i\frac{P}{m\omega}\right) \tag{9.4}$$

where we have used the fact that X and P are Hermitian. Inverting these relations we find

$$X = \sqrt{\frac{\hbar}{2m\omega}}\left(a + a^\dagger\right), \quad P = -i\sqrt{\frac{m\omega\hbar}{2}}\left(a - a^\dagger\right). \tag{9.5}$$

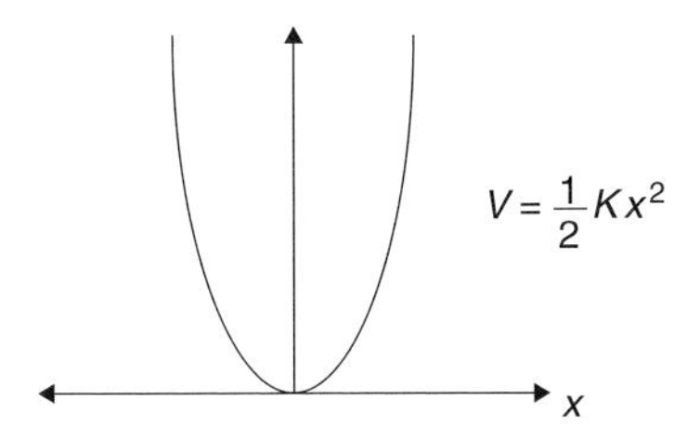

Fig. 9.1

Taking the products between a and $a^\dagger$, we obtain

$$aa^\dagger = \frac{m\omega}{2\hbar}\left(X^2 + \frac{P^2}{m^2\omega^2} + \frac{\hbar}{m\omega}\mathbf{1}\right), \tag{9.6}$$

$$a^\dagger a = \frac{m\omega}{2\hbar}\left(X^2 + \frac{P^2}{m^2\omega^2} - \frac{\hbar}{m\omega}\mathbf{1}\right) \tag{9.7}$$

where we have made use of relation (9.3).

Subtracting and adding the above two equations we obtain, respectively,

$$aa^\dagger - a^\dagger a = \mathbf{1} \tag{9.8}$$

and

$$aa^\dagger + a^\dagger a = \frac{2H}{\hbar\omega}. \tag{9.9}$$

These two expressions can be rewritten as

$$\left[a, a^\dagger\right] = \mathbf{1} \tag{9.10}$$

and

$$H = \left(a^\dagger a + \frac{1}{2}\mathbf{1}\right)\hbar\omega. \tag{9.11}$$

The harmonic oscillator problem has thus been simplified by introducing the operators a and $a^\dagger$.

$a^\dagger$ and a as raising and lowering operators

Let us consider the eigenstates of H, designated by an orthonormal set of states $|n\rangle$, with energy eigenvalues E_n. The relevant equation is then

$$H\,|n\rangle = E_n\,|n\rangle\,. \tag{9.12}$$

In view of the expression (9.11) relating H to $a^\dagger a$, we can simplify (9.12) by writing

$$a^\dagger a\,|n\rangle = \lambda_n\,|n\rangle \tag{9.13}$$

where λ_n will be related to E_n by the relation

$$E_n = \left(\lambda_n + \frac{1}{2}\right)\hbar\omega. \tag{9.14}$$

We now obtain the properties of the state vector, $a^\dagger |n\rangle$. For that purpose we note the following:

$$\begin{aligned} a^\dagger a\left(a^\dagger |n\rangle\right) &= a^\dagger\left(aa^\dagger |n\rangle\right) = a^\dagger\left(\mathbf{1} + a^\dagger a\right)|n\rangle \\ &= a^\dagger (1 + \lambda_n)|n\rangle = (\lambda_n + 1)\left(a^\dagger |n\rangle\right) \end{aligned} \tag{9.15}$$

where we have used the commutation relation $\left[a, a^\dagger\right] = \mathbf{1}$. The relations (9.13) and (9.15) tell us that when $a^\dagger a$ operates on $|n\rangle$ its eigenvalue is λ_n but when it operates on the state $a^\dagger |n\rangle$ its eigenvalue is $(\lambda_n + 1)$. The operator $a^\dagger$, therefore, has the property that when it operates on the state $|n\rangle$ it increases its energy eigenvalue from λ_n to $(\lambda_n + 1)$. For this reason, one calls $a^\dagger$ the "raising" operator. Similarly, we obtain

$$a^\dagger a\,(a\,|n\rangle) = (\lambda_n - 1)\,(a\,|n\rangle)\,. \tag{9.16}$$

Hence, a acts as a "lowering" operator.

Below we summarize the connection between the various eigenstates and eigenvalues:

$$|n\rangle \to \lambda_n, \tag{9.17}$$

$$a^\dagger |n\rangle \to \lambda_n + 1, \tag{9.18}$$

$$a\,|n\rangle \to \lambda_n - 1. \tag{9.19}$$

Determination of the energy eigenvalue, E_n

The above simplifications will now enable us to obtain E_n. We expect the lowest energy eigenvalue to be finite since the potential is bounded from below. In other words, since the minimum of the harmonic potential, $V(x) = (1/2)\,Kx^2$, is $V = 0$, one cannot have an energy level below this value. On the other hand, there is no restriction on how high the energy levels can go since $V(x) \to \infty$ as $x \to \infty$.

Thus there must be a lowest state in this problem. If we designate this state by $|0\rangle$ then applying the lowering operator, a, to this state we will have

$$a\,|0\rangle = 0, \tag{9.20}$$

which implies, by operating on the left by $a^\dagger$, that

$$a^\dagger a\,|0\rangle = 0. \tag{9.21}$$

Hence, from (9.13),

$$\lambda_0 = 0. \tag{9.22}$$

Since $a^\dagger|n\rangle$ corresponds to eigenvalue λ_n+1 of the operator $a^\dagger a$, then for the next step up from $|0\rangle$, i.e., for $a^\dagger|0\rangle$, the eigenvalue will be

$$\lambda_1 = 1. \tag{9.23}$$

The corresponding eigenstate of $a^\dagger a$ can be designated as $|1\rangle$ with an appropriate normalization constant. Continuing this process, we conclude that for $(a^\dagger)^n|0\rangle$ we will have

$$\lambda_n = n, \tag{9.24}$$

corresponding to an eigenstate proportional to $|n\rangle$. Hence, from (9.14) the energy eigenvalues are simply given by

$$E_n = \left(n+\frac{1}{2}\right)\hbar\omega. \tag{9.25}$$

Thus, the harmonic oscillator will have an infinite number of discrete levels given by $n = 0, 1, 2 \ldots$. We note, in particular, that for $n = 0$, we have $E_0 = \frac{1}{2}\hbar\omega$. This is the so-called "zero point energy."

From (9.13) and (9.24) we conclude that

$$a^\dagger a|n\rangle = \lambda_n|n\rangle = n|n\rangle. \tag{9.26}$$

Hence, $a^\dagger a$ is called the "number" operator:

$$N = a^\dagger a. \tag{9.27}$$

We can write the Hamiltonian in terms of it as

$$H = \left(N+\frac{1}{2}\right)\hbar\omega \tag{9.28}$$

with eigenstates $|n\rangle$ which satisfy the orthonormality condition $\langle m|n\rangle = \delta_{mn}$.

a and $a^\dagger$, again

From the properties of a and $a^\dagger$ that we have already discussed, one can write

$$a|n\rangle = C_n|n-1\rangle. \tag{9.29}$$

Multiplying the above expression on both sides with the respective complex conjugates, and using the orthonormality properties for the eigenstates, we find

$$\langle n|a^\dagger a|n\rangle = |C_n|^2. \tag{9.30}$$

However, since the eigenvalue of $a^\dagger a$ is n we obtain, assuming C_n to be real,

$$C_n = \sqrt{n}, \tag{9.31}$$

so that

$$a\,|n\rangle = \sqrt{n}\,|n-1\rangle\,. \tag{9.32}$$

Similarly,

$$a^\dagger\,|n\rangle = \sqrt{n+1}\,|n+1\rangle\,. \tag{9.33}$$

From (9.33), a general state $|n\rangle$ can be constructed from $|0\rangle$ by repeated applications of $a^\dagger$ as follows:

$$|n\rangle = \frac{a^\dagger}{\sqrt{n}}\,|n-1\rangle = \left(\frac{a^\dagger}{\sqrt{n}}\right)\left(\frac{a^\dagger}{\sqrt{n-1}}\right)|n-2\rangle = \cdots = \frac{\left(a^\dagger\right)^n}{\sqrt{(n)\,.\,(n-1)\ldots 2.1}}\,|0\rangle\,. \tag{9.34}$$

Thus,

$$|n\rangle = \frac{\left(a^\dagger\right)^n}{\sqrt{n!}}\,|0\rangle\,, \tag{9.35}$$

which implies that one can generate an arbitrary state $|n\rangle$ by an appropriate number of applications of the operator $a^\dagger$ on the lowest energy state, $|0\rangle$.

Matrix representation

Since n runs from 0 to ∞, the matrix representation of $|n\rangle$ will be an infinite column matrix and a, $a^\dagger$ and N will be square matrices of infinite dimensions:

$$|n\rangle = \begin{bmatrix} |0\rangle \\ |1\rangle \\ |2\rangle \\ \cdot \\ \cdot \end{bmatrix}, \tag{9.36}$$

$$a = \begin{bmatrix} 0 & \sqrt{1} & 0 & 0 & \cdot \\ 0 & 0 & \sqrt{2} & 0 & \cdot \\ 0 & 0 & 0 & \cdot & \cdot \\ 0 & 0 & 0 & 0 & \cdot \\ \cdot & \cdot & \cdot & \cdot & \cdot \end{bmatrix}, \quad a^\dagger = \begin{bmatrix} 0 & 0 & 0 & 0 & \cdot \\ \sqrt{1} & 0 & 0 & 0 & \cdot \\ 0 & \sqrt{2} & 0 & 0 & \cdot \\ 0 & 0 & \cdot & 0 & \cdot \\ \cdot & \cdot & \cdot & \cdot & \cdot \end{bmatrix}, \quad N = \begin{bmatrix} 0 & 0 & 0 & 0 & \cdot \\ 0 & 1 & 0 & 0 & \cdot \\ 0 & 0 & 2 & 0 & \cdot \\ 0 & 0 & 0 & \cdot & \cdot \\ \cdot & \cdot & \cdot & \cdot & \cdot \end{bmatrix}. \tag{9.37}$$

Time dependence

The time dependence of the operators is given by the Heisenberg relation

$$i\hbar\frac{da}{dt} = [a, H]\,. \tag{9.38}$$

Inserting expression (9.11) for H and using the commutation relation (9.10) between a and $a^\dagger$ we obtain

$$i\hbar\frac{da}{dt} = \hbar\omega a, \tag{9.39}$$

which then gives as a solution

$$a(t) = a(0)e^{-i\omega t} \tag{9.40}$$

and similarly,

$$a^\dagger(t) = a^\dagger(0)e^{i\omega t}. \tag{9.41}$$

The operators N and H remain independent of time since they involve the product $a^\dagger(t)a(t)$ which, from (9.40) and (9.41), is the same as $a^\dagger(0)a(0)$.

The wavefunctions

The energy eigenfunction, $u_n(x)$, for a particle undergoing harmonic oscillations is given by

$$u_n(x) = \langle x| n\rangle. \tag{9.42}$$

We first calculate the ground-state wavefunction, $u_0(x)$, by noting from (9.20) that

$$a\,|0\rangle = 0, \tag{9.43}$$

which, from relation (9.20) for a, gives

$$\left(X + i\frac{P}{m\omega}\right)|0\rangle = 0. \tag{9.44}$$

Expressing P in the form of the derivative operator in the x-space, and multiplying the above relation by $\langle x|$, we have the following equation:

$$\left(x + \frac{\hbar}{m\omega}\frac{d}{dx}\right)\langle x\,|0\rangle = 0, \tag{9.45}$$

which leads to the differential equation

$$\frac{du_0(x)}{dx} = -\frac{m\omega x}{\hbar}u_0(x). \tag{9.46}$$

The solution is easily obtained as

$$u_0(x) = A_0 \exp\left(-\frac{m\omega x^2}{2\hbar}\right). \tag{9.47}$$

The constant A_0 can be obtained from the normalization condition

$$\int_{-\infty}^{\infty} dx\, u_0^2(x) = 1. \tag{9.48}$$

Therefore,

$$A_0^2 \int_{-\infty}^{\infty} dx \exp\left(-\frac{m\omega x^2}{\hbar}\right) = 1. \tag{9.49}$$

The integration over the Gaussian can be easily carried out. We find

$$A_0 = \left(\frac{m\omega}{\pi\hbar}\right)^{\frac{1}{4}}. \tag{9.50}$$

Therefore,

$$u_0(x) = \left(\frac{m\omega}{\pi\hbar}\right)^{\frac{1}{4}} \exp\left(-\frac{m\omega x^2}{2\hbar}\right). \tag{9.51}$$

From the expression for $u_0(x)$ one can obtain a general form for $u_n(x)$ by first obtaining $u_1(x)$, where

$$\begin{aligned} u_1(x) = \langle x|1\rangle = \langle x| a^\dagger |0\rangle &= \sqrt{\frac{m\omega}{\hbar}} \langle x| \left(X - i\frac{P}{m\omega}\right)|0\rangle \\ &= \sqrt{\frac{m\omega}{2\hbar}} \left(x - \frac{\hbar}{m\omega}\frac{d}{dx}\right) u_0(x). \end{aligned} \tag{9.52}$$

We can simplify this to

$$u_1(x) = -\sqrt{\frac{\hbar}{2m\omega}} \exp\left(\frac{m\omega x^2}{2\hbar}\right) \frac{d}{dx}\left[\exp\left(-\frac{m\omega x^2}{2\hbar}\right) u_0(x)\right]. \tag{9.53}$$

Substituting the functional form of $u_0(x)$, we obtain

$$u_1(x) = \sqrt{\frac{2m\omega}{\hbar}} x u_0(x). \tag{9.54}$$

We rewrite the above relations in terms of the dimensionless variable

$$\xi = \alpha x \tag{9.55}$$

where

$$\alpha = \sqrt{\frac{m\omega}{\hbar}}. \tag{9.56}$$

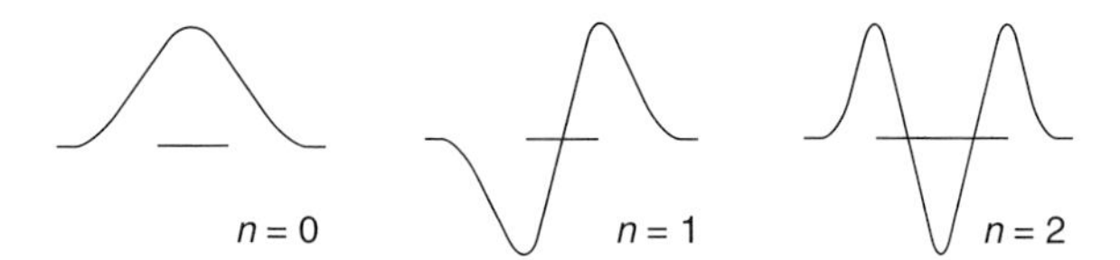

Fig. 9.2

The expression (9.53) after substituting the expression for $u_0(x)$, is then of the form

$$u_1(x) = (-1)\left(\frac{\alpha}{\sqrt{\pi}}\right)^{\frac{1}{2}} \exp\left(\frac{\xi^2}{2}\right)\frac{d}{d\xi}\left[\exp\left(-\xi^2\right)\right]. \tag{9.57}$$

Wavefunctions for $n = 0$, $n = 1$, and $n = 2$ are illustrated in Fig. 9.2.

By recursion, the expression for $u_n(x)$ can be generalized to

$$u_n(x) = (-1)^n \left(\frac{\alpha}{\sqrt{\pi}2^n n!}\right)^{\frac{1}{2}} \exp\left(\frac{\xi^2}{2}\right)\frac{d^n}{d\xi^n}\left[\exp(-\xi^2)\right]. \tag{9.58}$$

9.1.2 Formalism in the Schrödinger picture

The Schrödinger equation with a harmonic potential is given by

$$-\frac{\hbar^2}{2m}\frac{d^2u}{dx^2} + \frac{1}{2}m\omega^2x^2u = Eu, \quad -\infty < x < \infty. \tag{9.59}$$

This equation can be simplified by making the substitution

$$\xi = \alpha x \tag{9.60}$$

where α is given by (9.56), so that the new equation reads

$$\frac{d^2u}{d\xi^2} + \left(\lambda - \xi^2\right)u = 0, \quad -\infty < \xi < \infty \tag{9.61}$$

with

$$\lambda = \frac{2E}{\hbar\omega}. \tag{9.62}$$

To obtain the complete solution of (9.61), we first determine the behavior of u at infinity. We note that the equation (9.61) and the equation

$$\left(\frac{d}{d\xi} - \xi\right)\left(\frac{d}{d\xi} + \xi\right)u = 0 \tag{9.63}$$

have the same form as $\xi \to \infty$, namely,

$$\frac{d^2u}{d\xi^2} - \xi^2 u = 0 \tag{9.64}$$

and, therefore, will have the same solution at infinity. Hence the equation

$$\left(\frac{d}{d\xi}+\xi\right)u=0, \tag{9.65}$$

solves (9.63) and will also describe the solution of (9.61) at infinity. The above equation yields

$$u=C\exp\left(-\frac{1}{2}\xi^2\right) \quad \text{as } \xi\to\infty. \tag{9.66}$$

Indeed, if $H(\xi)$ is a polynomial, then

$$u=NH(\xi)\exp\left(-\frac{1}{2}\xi^2\right) \tag{9.67}$$

is also a solution as $\xi\to\infty$, where N is the overall normalization. Substituting the above in (9.61) we obtain

$$\frac{d^2H(\xi)}{d\xi^2}-2\xi\frac{dH(\xi)}{d\xi}+(\lambda-1)H(\xi)=0. \tag{9.68}$$

Our task now is to determine H in order that we obtain the complete solution of (9.61). We first note that if $H(\xi)$ satisfies the above equation then so does $H(-\xi)$ .We can thus conveniently divide the solutions into two types: even and odd given by $(H(\xi)+H(-\xi))$, and $(H(\xi)-H(-\xi))$, respectively.

From now on, H in equation (9.68) will signify an even or odd combination. Since it is a polynomial, we can obtain H by expanding it in the neighborhood of $\xi=0$. The series will be given as

$$H=\xi^s\left(a_0+a_2\xi^2+\cdots+a_\nu\xi^\nu+\cdots\right) \quad \text{with } a_0\neq 0 \tag{9.69}$$

where ν is an even integer.

Substituting (9.69) into (9.68) and collecting terms with the same powers in ξ, we obtain

$$s(s-1)\xi^{s-2}a_0+\cdots+[(s+\nu+2)(s+\nu+1)a_{\nu+2}$$
$$-(2s+2\nu-\lambda+1)a_\nu)]\,\xi^{s+\nu}+\cdots=0. \tag{9.70}$$

Since ξ is a continuous variable the only way in which the above equation can be valid is if the coefficient of each power of ξ vanishes. From the first term, since $a_0\neq 0$, we obtain two possible values for s

$$s=0 \quad \text{or} \quad s=1 \tag{9.71}$$

which implies, as expected, that we have either an even series or odd series for H in (9.69). From the vanishing of the coefficient for $\xi^{s+\nu}$ we find

$$\frac{a_{\nu+2}}{a_\nu}=\frac{(2s+2\nu-\lambda+1)}{(s+\nu+2)(s+\nu+1)}. \tag{9.72}$$

This ratio will allow us to determine whether the series for H converges or not. To determine the convergence we note that

$$\frac{a_{\nu+2}}{a_\nu} \to \frac{2}{\nu} \quad \text{as} \quad \nu \to \infty. \tag{9.73}$$

We can compare this ratio to the corresponding ratio in the well-known series for $\exp\left(\xi^2\right)$, which we write as

$$\exp\left(\xi^2\right) = b_0 + b_2\xi^2 + \cdots + b_\nu\xi^\nu + b_{\nu+2}\xi^{\nu+2} + \cdots. \tag{9.74}$$

Since the terms in this series are known, the ratio of the coefficients of the consecutive terms are also known and are given by

$$\frac{b_{\nu+2}}{b_\nu} = \frac{(\nu/2)!}{((\nu/2)+1)!} \to \frac{2}{\nu} \quad \text{as} \quad \nu \to \infty. \tag{9.75}$$

This ratio is identical to the ratio in (9.73) as ν becomes large.

Thus the series for $H(\xi)$ given in (9.69) will behave as $\exp\left(\xi^2\right)$, which means that the wavefunction $u(\xi)$ in (9.67) will behave like $\exp\left(\frac{1}{2}\xi^2\right)$. This expression, however, diverges as $\xi \to \infty$ and, is therefore, untenable since $u(\xi)$ is expected to be finite in that limit. The only remedy then is to demand that the series for $H(\xi)$ terminates at some point so that it remains a finite polynomial. For this to happen, one of the coefficients in the series must vanish, as then all the subsequent terms will vanish also through (9.72).

If the series for $H(\xi)$ in (9.69) terminates at the term with ξ^ν (i.e., $a_{\nu+2} = 0$ with $a_\nu \neq 0$), then one must have

$$(2s + 2\nu - \lambda + 1) = 0 \tag{9.76}$$

which then determines the value of λ,

$$\lambda_n = 2(s+\nu) + 1 = 2n + 1. \tag{9.77}$$

Hence the energy eigenvalues are given by

$$E_n = \left(n + \frac{1}{2}\right)\hbar\omega. \tag{9.78}$$

To determine the wavefunction, we note that since ν is an even integer, n takes the values

$$n = \text{ even integer, for } s = 0, \tag{9.79}$$

$$n = \text{ odd integer, for } s = 1. \tag{9.80}$$

Substituting the value of λ_n in the equation for H in (9.68) we obtain

$$\frac{d^2H_n(\xi)}{d\xi^2} - 2\xi\frac{dH_n(\xi)}{d\xi} + 2nH_n(\xi) = 0. \tag{9.81}$$

The solution $H_n(\xi)$ is then a finite polynomial,

$$H_n(\xi) = \xi^s \left(a_0 + a_2\xi^2 + \cdots + a_n\xi^n\right) \tag{9.82}$$

where all the coefficients are determined in terms of a_0 through the relation (9.72). The wavefunction is then given by

$$u_n = N_n H_n(\xi) \exp\left(-\frac{1}{2}\xi^2\right) \tag{9.83}$$

where $H_n(\xi)$ is called the Hankel function.

Some typical values of $H_n(\xi)$ are given by

$$H_0(\xi) = 1, \tag{9.84}$$

$$H_1(\xi) = 2\xi, \tag{9.85}$$

$$H_2(\xi) = 4\xi^2 - 2. \tag{9.86}$$

The general form for $H_n(\xi)$ is given by

$$H_n(\xi) = (-1)^n e^{\xi^2} \frac{d^n}{d\xi^n}(e^{-\xi^2}). \tag{9.87}$$

The normalization constant N_n can be found from the relation

$$\int_{-\infty}^{\infty} d\xi\, H_n^2(\xi)\, e^{-\xi^2} = \sqrt{\pi} 2^n n!. \tag{9.88}$$

Hence

$$N_n = \left(\frac{\alpha}{\sqrt{\pi} 2^n n!}\right)^{\frac{1}{2}}. \tag{9.89}$$

The wavefunction u_n is then

$$u_n(x) = (-1)^n \left(\frac{\alpha}{\sqrt{\pi} 2^n n!}\right)^{\frac{1}{2}} e^{\frac{1}{2}\xi^2} \frac{d^n}{d\xi^n}(e^{-\xi^2}). \tag{9.90}$$

Thus we obtain the same results as we did previously through the Heisenberg picture. The wavefunctions $u_n(x)$ are plotted in Fig. 9.2 for $n = 0, 1, 2$.

9.2 Problems

1. Assume $x(t)$ and $p(t)$ to be Heisenberg operators with $x(0) = x_0$ and $p(0) = p_0$. For a Hamiltonian corresponding to the harmonic oscillator show that

$$x(t) = x_0 \cos\omega t + \frac{p_0}{m\omega} \sin\omega t$$

and

$$p_0(t) = p_0 \cos \omega t - m\omega x_0 \sin \omega t.$$

2. On the basis of the results already derived for the harmonic oscillator, determine the energy eigenvalues and the ground-state wavefunction for the truncated oscillator

$$\begin{aligned} V(x) &= \frac{1}{2}Kx^2, \quad 0 \le x \le \infty \\ &= \infty, \qquad x < 0. \end{aligned}$$

3. Show that for a harmonic oscillator in the state $|n\rangle$, the following uncertainty product holds:

$$\Delta x \Delta p = \left(n + \frac{1}{2}\right)\hbar.$$

4. Consider the following two-dimensional harmonic oscillator problem:

$$-\frac{\hbar^2}{2m}\frac{\partial^2 u}{\partial x^2} - \frac{\hbar^2}{2m}\frac{\partial^2 u}{\partial y^2} + \frac{1}{2}K_1 x^2 u + \frac{1}{2}K_2 y^2 u = Eu$$

where (x, y) are the coordinates of the particle. Use the separation of variables technique to obtain the energy eigenvalues. Discuss the degeneracy in the eigenvalues if $K_1 = K_2$.

5. Consider now a variation on Problem 4 in which we have a coupled oscillator with the potential given by

$$V(x, y) = \frac{1}{2}K\left(x^2 + y^2 + 2\lambda xy\right).$$

Obtain the energy eigenvalues by changing the variables (x, y) to (x', y') such that the new potential is quadratic in (x', y'), without the coupling term.

6. Solve the above problem by matrix methods by writing

$$V(x, y) = \tilde{X}MX$$

and obtaining the eigenvalues, where

$$X = \begin{bmatrix} x \\ y \end{bmatrix}, \quad M = \text{symmetric } 2 \times 2 \text{ matrix}, \quad \tilde{X} = \begin{bmatrix} x & y \end{bmatrix}.$$

7. Consider two coupled harmonic oscillators in one dimension of natural length a and spring constant K connecting three particles located at x_1, x_2, and x_3. The corresponding Schrödinger equation is given as

$$-\frac{\hbar^2}{2m}\frac{\partial^2 u}{\partial x_1^2} - \frac{\hbar^2}{2m}\frac{\partial^2 u}{\partial x_2^2} - \frac{\hbar^2}{2m}\frac{\partial^2 u}{\partial x_3^2} + \frac{K}{2}\left[(x_2 - x_1 - a)^2 + (x_3 - x_2 - a)^2\right]u = Eu.$$

Obtain the energy eigenvalues using the matrix method.

8. As a variation on Problem 7 assume that the middle particle at x_2 has a different mass, M. Reduce this problem to the form of Problem 7 by a scale change in x_2 and then use the matrix method to obtain the energy eigenvalues.
9. By expanding the exponential $\exp(ikx)$ in powers of x determine the ground-state expectation value $\langle 0 |\exp(ikx)| 0\rangle$, and the transition probability amplitude $\langle n |\exp(ikx)| 0\rangle$.
10. Determine $a(t)$ for a charged particle harmonic oscillator subjected to an electric field $E(t)$ given by

$$\begin{aligned} E(t) &= 0, & t < 0 \\ &= E_0 e^{-t/\tau}, & t > 0. \end{aligned}$$

Use the Green's function technique and impose causality.
11. Show that

$$\left[a, F(a^\dagger)\right] = \frac{\partial F}{\partial a^\dagger}, \quad \left[a^\dagger, F(a)\right] = -\frac{\partial F}{\partial a}$$

by first proving it for $F(b) = b^n$, where $b = a, a^\dagger$, and then generalizing.
12. A pendulum formed by a particle of mass m attached to a rod is executing small oscillations. The rod is massless of length l. Let θ be the angle, assumed to be small. Considering this to be a quantum-mechanical system, write down the canonical variables corresponding to x and p and their commutator. Express the Hamiltonian in terms of these two variables. Comparing this expression with that in the harmonic oscillator problem, obtain the energy eigenvalues and the ground-state wavefunction.

10 Coherent states

A coherent state is a quantum state of a harmonic oscillator whose properties closely resemble those of a classical harmonic oscillator. It is an eigenstate of the destruction operator in a harmonic oscillator system with the remarkable property that it behaves as a semiclassical state. We explain this in detail below. Coherent states arise in many physical systems. In particular, they play an important role in the quantum theory of light; for example, a laser wave can be represented as a coherent state of the electromagnetic field.

10.1 Eigenstates of the lowering operator

The eigenstates of the lowering operator, a, have some remarkable properties as we will discover below. The basic properties of this operator for the harmonic oscillator Hamiltonian, H, are given by

$$a\left|n\right\rangle = \sqrt{n}\left|n-1\right\rangle, \quad \left[a, a^{+}\right] = 1, \quad \text{with } H = \left(a^{+}a + \frac{1}{2}\right)\hbar\omega. \tag{10.1}$$

We designate an eigenstate of a as $\left|\alpha\right\rangle$,

$$a\left|\alpha\right\rangle = \alpha\left|\alpha\right\rangle. \tag{10.2}$$

In Chapter 1 we showed that when an operator is Hermitian its eigenvalues will be real and the corresponding eigenstates will be orthogonal. However, since $a^{\dagger} \neq a$, a is not Hermitian. Hence the eigenvalues α will not necessarily be real; they can be complex, and the eigenstates $\left|\alpha\right\rangle$ need not be orthogonal.

In order to relate the states $\left|\alpha\right\rangle$ to the eigenstates $\left|n\right\rangle$ of the Hamiltonian, we use the completeness theorem, and expand $\left|\alpha\right\rangle$ in terms of $\left|n\right\rangle$ as follows:

$$\left|\alpha\right\rangle = \sum_{n=0}^{\infty} c_n \left|n\right\rangle. \tag{10.3}$$

We operate the state $\left|\alpha\right\rangle$ by a, and write the following sequence of results:

$$a\left|\alpha\right\rangle = \sum_{n=0}^{\infty} c_n a\left|n\right\rangle = \sum_{n=1}^{\infty} c_n \sqrt{n}\left|n-1\right\rangle = \sum_{n=0}^{\infty} c_{n+1}\sqrt{n+1}\left|n\right\rangle \tag{10.4}$$

where in the first equality we have used (10.1). The second equality starts with $n = 1$, since, in the previous sum, $a|0\rangle = 0$. In the third equality we shift the lower limit back to $n = 0$ with the necessary changes in the other terms. From (10.2) and (10.3) we write

$$a|\alpha\rangle = \alpha|\alpha\rangle = \alpha \sum_{n=0}^{\infty} c_n |n\rangle . \tag{10.5}$$

Equating the right-hand sides of (10.4) and (10.5), we obtain

$$c_{n+1}\sqrt{n+1} = \alpha c_n \quad \text{or} \quad c_{n+1} = \frac{\alpha}{\sqrt{n+1}} c_n . \tag{10.6}$$

Writing c_{n+1} in a recursive series of products, we find

$$c_{n+1} = \left(\frac{\alpha}{\sqrt{n+1}}\right)\left(\frac{\alpha}{\sqrt{n}}\right)\cdots\left(\frac{\alpha}{\sqrt{1}}\right) c_0 = \frac{\alpha^{n+1}}{\sqrt{(n+1)!}} c_0 . \tag{10.7}$$

Substituting this result in (10.3) we obtain

$$|\alpha\rangle = c_0 \sum_n \frac{\alpha^n}{\sqrt{n!}} |n\rangle . \tag{10.8}$$

If $|\alpha\rangle$'s are normalized, i.e., if $\langle\alpha|\alpha\rangle = 1$, then, since $|n\rangle$'s are orthonormal, we obtain

$$c_0^2 \sum_{n=0}^{\infty} \frac{|\alpha|^{2n}}{n!} = 1 \tag{10.9}$$

where c_0 is assumed to be real. The above infinite sum is a well-known series for the exponential $e^{|\alpha|^2}$. Therefore,

$$c_0^2 e^{|\alpha|^2} = 1 \quad \text{or} \quad c_0 = e^{-\frac{1}{2}|\alpha|^2} . \tag{10.10}$$

Hence, from (10.7), we obtain

$$c_n = \frac{\alpha^n}{\sqrt{n!}} e^{-\frac{1}{2}|\alpha|^2} . \tag{10.11}$$

Consequently, (10.8) can be written as

$$|\alpha\rangle = e^{-\frac{1}{2}|\alpha|^2} \sum_n \frac{\alpha^n}{\sqrt{n!}} |n\rangle . \tag{10.12}$$

We note from our earlier result for the harmonic oscillator that

$$|n\rangle = \frac{\left(a^\dagger\right)^n}{\sqrt{n!}} |0\rangle . \tag{10.13}$$

Substituting this expression in (10.12), we obtain

$$|\alpha\rangle = e^{-\frac{1}{2}|\alpha|^2}\left[\sum_n \frac{\left(\alpha a^\dagger\right)^n}{n!}\right]|0\rangle. \tag{10.14}$$

The infinite series once again sums to an exponential, this time to $\exp\left(\alpha a^\dagger\right)$, so that

$$|\alpha\rangle = e^{-\frac{1}{2}|\alpha|^2} e^{\alpha a^\dagger}\,|0\rangle\,. \tag{10.15}$$

Thus, the operator $\exp\left(-\frac{1}{2}|\alpha|^2 + \alpha a^\dagger\right)$ transforms the ground state $|0\rangle$ to the state $|\alpha\rangle$. However, this operator is not unitary, since the product

$$(e^{-\frac{1}{2}|\alpha|^2}e^{\alpha a^\dagger})(e^{-\frac{1}{2}|\alpha|^2}e^{\alpha a^\dagger})^\dagger = (e^{-\frac{1}{2}|\alpha|^2}e^{\alpha a^\dagger})(e^{\alpha^* a}e^{-\frac{1}{2}|\alpha|^2}) = e^{-|\alpha|^2}e^{\alpha a^\dagger}e^{\alpha^* a} \tag{10.16}$$

is not the unit operator, where we have noted that $|\alpha|^2$ is a number.

In order to preserve the norm of the states under the transformation (10.15) we need to construct an operator that is unitary but, at the same time, we need to ensure that this new operator leaves the relation (10.15) intact. We can accomplish this by multiplying the operator on the right by a factor like $e^{\beta a}$ and taking advantage of the fact that, since $a\,|0\rangle = 0$,

$$e^{\beta a}\,|0\rangle = \left(1 + \beta a + \frac{(\beta a)^2}{2!} + \cdots\right)|0\rangle = |0\rangle\,. \tag{10.17}$$

Designating this unitary operator by $D(\alpha)$, we write

$$D(\alpha) = e^{-\frac{1}{2}|\alpha|^2}e^{\alpha a^\dagger}e^{\beta a}. \tag{10.18}$$

To be unitary $D(\alpha)$ must satisfy

$$D(\alpha)D^\dagger(\alpha) = D^\dagger(\alpha)D(\alpha) = \mathbf{1}. \tag{10.19}$$

This relation will allow us to determine β. We take α and β to be infinitesimal, and expand the exponential in (10.18), keeping only the linear terms. The relation (10.19) then yields

$$\left(\mathbf{1} + \alpha a^\dagger + \beta a + \alpha^* a + \beta^* a^\dagger\right) = \mathbf{1}, \tag{10.20}$$

which implies that $\beta = -\alpha^*$. Including higher orders in the expansion will not change this result. Hence, we obtain

$$D(\alpha) = e^{-\frac{1}{2}|\alpha|^2}e^{\alpha a^\dagger}e^{-\alpha^* a}. \tag{10.21}$$

From the identity

$$e^{A+B} = e^A e^B e^{-\frac{1}{2}[A,B]} \tag{10.22}$$

we find, by taking $A = \alpha a^{\dagger}$, $B = -\alpha^{*} a$, and using the relation $\left[a, a^{\dagger}\right] = 1$, that $[A, B] = \frac{1}{2}|\alpha|^2$. Hence,

$$D(\alpha) = e^{\alpha a^{\dagger} - \alpha^{*} a}. \tag{10.23}$$

The operator $D(\alpha)$ is now unitary and does the requisite task of converting the ground state, $|0\rangle$, to $|\alpha\rangle$ .The relation (10.15) can be replaced by

$$|\alpha\rangle = D(\alpha)\,|0\rangle\,. \tag{10.24}$$

Thus, in some sense the eigenstate of the operator a is related to the ground state of the harmonic oscillator. We explore this connection below.

Let us now obtain the transformation of the operators a and $a^{\dagger}$ due to the operator $D(\alpha)$ as it will be required in the discussions to follow. The product $D^{\dagger}(\alpha) a D(\alpha)$ can be written for infinitesimal α (keeping terms up to α) as

$$\left(1 + \alpha^{*} a - \alpha a^{\dagger}\right) a \left(1 + \alpha a^{\dagger} - \alpha^{*} a\right) = a + \alpha\left[a, a^{\dagger}\right] = a + \alpha. \tag{10.25}$$

From this result we deduce that

$$D^{\dagger}(\alpha) a D(\alpha) = a + \alpha, \tag{10.26}$$

as the higher order terms in α will not change the result in (10.25). Taking the Hermitian conjugate of (10.26) we get

$$D^{\dagger}(\alpha) a^{\dagger} D(\alpha) = a^{\dagger} + \alpha^{*}. \tag{10.27}$$

Furthermore, from (10.26) and (10.27), we obtain

$$D^{\dagger}(\alpha) a^{\dagger} a D(\alpha) = \left[D^{\dagger}(\alpha) a^{\dagger} D(\alpha)\right]\left[D^{\dagger}(\alpha) a D(\alpha)\right] = a^{\dagger} a + \alpha^{*} a + \alpha a^{\dagger} + |\alpha|^2 \tag{10.28}$$

where we have used the fact that D is unitary.

Let us now evaluate the transformation of the X and P operators under $D(\alpha)$. As we know from the discussions of the harmonic oscillator, these operators can be expressed in terms of a and $a^{\dagger}$ through the relations

$$X = \sqrt{\frac{\hbar}{2m\omega}}\left(a + a^{\dagger}\right) \tag{10.29}$$

and

$$P = -i\sqrt{\frac{\hbar m\omega}{2}}\left(a - a^{\dagger}\right). \tag{10.30}$$

We then find from the transformations of a and $a^\dagger$ in (10.26) and (10.27) respectively that

$$D^\dagger(\alpha)XD(\alpha) = X + \sqrt{\frac{\hbar}{2m\omega}}\left(\alpha + \alpha^*\right), \tag{10.31}$$

$$D^\dagger(\alpha)PD(\alpha) = P - i\sqrt{\frac{\hbar m\omega}{2}}\left(\alpha - \alpha^*\right). \tag{10.32}$$

Thus, $D(\alpha)$ acts as a displacement operator for both space and momentum operators. For this reason one often calls $D(\alpha)$ a phase space displacement operator. It is then natural to define the constants α and α^* in terms of the corresponding displacements. We can accomplish this by writing

$$\sqrt{\frac{\hbar}{2m\omega}}\left(\alpha + \alpha^*\right) = x_0, \tag{10.33}$$

$$-i\sqrt{\frac{\hbar m\omega}{2}}\left(\alpha - \alpha^*\right) = p_0. \tag{10.34}$$

The exponent $(\alpha a^\dagger - \alpha^* a)$ that appears in (10.22) for the operator $D(\alpha)$ can be simplified accordingly,

$$\alpha a^\dagger - \alpha^* a = \alpha\sqrt{\frac{m\omega}{2\hbar}}\left(X - i\frac{P}{m\omega}\right) - \alpha^*\sqrt{\frac{m\omega}{2\hbar}}\left(X + i\frac{P}{m\omega}\right) = \frac{i}{\hbar}\left(p_0 X - x_0 P\right), \tag{10.35}$$

which leads to

$$D(\alpha) = e^{\frac{i}{\hbar}(p_0 X - x_0 P)}. \tag{10.36}$$

This result confirms the results we derived earlier for the transformations of X and P operators, namely,

$$D^\dagger(\alpha)XD(\alpha) = X + x_0 \tag{10.37}$$

$$D^\dagger(\alpha)PD(\alpha) = P + p_0. \tag{10.38}$$

Generalizing the above results, if $f(X,P)$ is a function of the operators X and P, then

$$D^\dagger(\alpha)f(X,P)D(\alpha) = f(X + x_0, P + p_0). \tag{10.39}$$

Taking the expectation value of each of the above terms with respect to the ground state $|0\rangle$, we find, since $\langle 0|X|0\rangle = 0 = \langle 0|P|0\rangle$,

$$\langle 0|D^\dagger(\alpha)XD(\alpha)|0\rangle = \langle\alpha|X|\alpha\rangle = x_0, \tag{10.40}$$

$$\langle 0|D^\dagger(\alpha)PD(\alpha)|0\rangle = \langle\alpha|P|\alpha\rangle = p_0. \tag{10.41}$$

On the left hand side of the equality sign we have used the relation $D(\alpha)|0\rangle = |\alpha\rangle$. The quantities $\langle\alpha|X|\alpha\rangle$ and $\langle\alpha|P|\alpha\rangle$ correspond to the expectation values of the space and momentum operators. The above relations show that $|\alpha\rangle$ corresponds to a state whose space and momentum eigenvalues are displaced to x_0 and p_0 compared with the ground-state values, which are both zero.

We note that while the states are normalized, i.e., $\langle\alpha|\alpha\rangle = 1$, any two different states $|\alpha\rangle$ and $|\alpha'\rangle$ are not necessarily orthogonal. This is not surprising, as we stated earlier, since the orthogonality results derived in Chapter 1 apply only to eigenstates of a Hermitian operator, which the operator a is not. We can determine the product $\langle\alpha|\alpha'\rangle$ from the state vectors determined in (10.12), where

$$|\alpha\rangle = e^{-\frac{1}{2}|\alpha|^2}\sum_n \frac{\alpha^n}{\sqrt{n!}}|n\rangle \quad \text{and} \quad |\alpha'\rangle = e^{-\frac{1}{2}|\alpha'|^2}\sum_n \frac{\alpha'^n}{\sqrt{n!}}|n\rangle \tag{10.42}$$

which gives, from the orthonormality of $|n\rangle$,

$$\langle\alpha|\alpha'\rangle = e^{-\frac{1}{2}|\alpha|^2}e^{-\frac{1}{2}|\alpha'|^2}\sum_n \frac{(\alpha\alpha^{*\prime})^n}{n!} = e^{-\frac{1}{2}|\alpha|^2}e^{-\frac{1}{2}|\alpha'|^2}e^{\alpha\alpha^{*\prime}} \tag{10.43}$$

The last equality follows from the fact that the infinite series sums to an exponential $e^{\alpha\alpha^{*\prime}}$. Therefore,

$$|\langle\alpha|\alpha'\rangle|^2 = e^{-|\alpha|^2}e^{-|\alpha'|^2}e^{2\alpha\alpha^{*\prime}} \tag{10.44}$$

where we note that α and α' are numbers, not operators. The exponent can, therefore, be simplified by using the relation

$$-|\alpha|^2 - |\alpha'|^2 + 2\alpha\alpha^{*\prime} = -|\alpha - \alpha'|^2 \tag{10.45}$$

to give

$$|\langle\alpha|\alpha'\rangle|^2 = e^{-|\alpha-\alpha'|^2}. \tag{10.46}$$

This result shows that the states $|\alpha\rangle$ and $|\alpha'\rangle$ are not orthogonal although their inner-product does becomes exponentially small as the eigenvalue α' moves farther away from α.

10.2 Coherent states and semiclassical description

From expansion (10.12) for $|\alpha\rangle$ one can write the time-dependent state $|\alpha(t)\rangle$ quite simply as

$$|\alpha(t)\rangle = e^{-\frac{1}{2}|\alpha|^2}\sum_n \frac{\alpha^n}{\sqrt{n!}}|n\rangle\, e^{-\frac{iE_nt}{\hbar}} \tag{10.47}$$

where E_n is the eigenvalue of $|n\rangle$, which for the harmonic oscillator is given by $E_n = \left(n + \frac{1}{2}\right)\hbar\omega$. Substituting this in (10.47) we find

$$|\alpha(t)\rangle = e^{-\frac{1}{2}|\alpha|^2} e^{-\frac{i\omega t}{2}} \sum_n \frac{\alpha^n}{\sqrt{n!}} |n\rangle\, e^{-in\omega t} = e^{-\frac{1}{2}|\alpha|^2} e^{-\frac{i\omega t}{2}} \sum \frac{\left(\alpha e^{-i\omega t}\right)^n}{\sqrt{n!}} |n\rangle\,. \qquad (10.48)$$

Hence, apart from a constant multiplicative term, $|\alpha(t)\rangle$ is obtained from $|\alpha\rangle$ by replacing α by $\alpha\exp(-i\omega t)$, i.e.,

$$|\alpha(t)\rangle = e^{-\frac{i\omega t}{2}} \left|\alpha e^{-i\omega t}\right\rangle. \qquad (10.49)$$

Let us express this relation by invoking the product $D\left(\alpha e^{-i\omega t}\right)|0\rangle$. Instead of using the relation (10.23) for D, it is simpler to use the equivalent expression (10.21) since it involves the factor $e^{-\alpha^* a}$ which does not affect $|0\rangle$. Thus, replacing α by $\alpha e^{-i\omega t}$ in (10.21) we write (10.49) as

$$|\alpha(t)\rangle = e^{-\frac{i\omega t}{2}} D\left(\alpha e^{-i\omega t}\right)|0\rangle = e^{-\frac{i\omega t}{2}} e^{-\frac{1}{2}|\alpha|^2} \exp\left[\alpha a^\dagger e^{-i\omega t}\right]|0\rangle\,. \qquad (10.50)$$

Let us return to expression (10.36) for $D(\alpha)$ in terms of X and P and consider the wavefunction $\psi_\alpha(x,t)$ at $t = 0$, corresponding to the state $|\alpha\rangle$. If $\psi_0(x,t)$ is the ground-state wavefunction, then

$$\psi_\alpha(x,0) = \langle x|\alpha\rangle = \langle x|D(\alpha)|0\rangle = \langle x|\, e^{\frac{i}{\hbar}(p_0 X - x_0 P)}\,|0\rangle = e^{\frac{i}{\hbar}p_0 x}\psi_0(x - x_0, 0) \qquad (10.51)$$

where we have made use of the fact that the operator $e^{-\frac{i}{\hbar}x_0 P}$ displaces the x-coordinate by x_0, and correspondingly, $\psi_0(x - x_0, 0)$ is the ground-state wavefunction of the harmonic oscillator centered at $x = x_0$. The time dependence of this wavefunction is then given by

$$\psi_\alpha(x,t) = \langle x|\alpha(t)\rangle = \left\langle x\middle|\alpha e^{-i\omega t}\right\rangle. \qquad (10.52)$$

As $\alpha \to \alpha e^{-i\omega t}$, we can determine $x_0(t)$ and $p_0(t)$ from our earlier relations as follows:

$$\begin{aligned} x_0 &= \sqrt{\frac{\hbar}{2m\omega}}\left(\alpha + \alpha^*\right) \to x_0(t) = \sqrt{\frac{\hbar}{2m\omega}}\left(\alpha e^{-i\omega t} + \alpha^* e^{i\omega t}\right) \\ &= x_0\cos\omega t + \left(\frac{p_0}{m\omega}\right)\sin\omega t \end{aligned} \qquad (10.53)$$

and

$$\begin{aligned} p_0 &= -i\sqrt{\frac{\hbar m\omega}{2}}\left(\alpha - \alpha^*\right) \to p_0(t) = -i\sqrt{\frac{\hbar m\omega}{2}}\left(\alpha e^{-i\omega t} - \alpha^* e^{i\omega t}\right) \\ &= p_0\cos\omega t - m\omega x_0\sin\omega t \end{aligned} \qquad (10.54)$$

where we have used the relations (10.33) and (10.34) connecting x_0, p_0 to α, α^*.

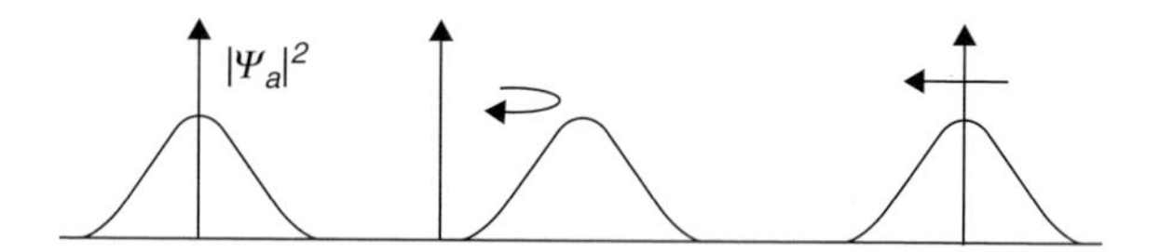

Fig. 10.1

With the above values of $x_0(t)$ and $p_0(t)$ we have

$$\langle x| e^{\frac{i}{\hbar}(p_0(t)X - x_0(t)P)} |0\rangle = e^{\frac{i}{\hbar}p_0(t)x}\psi_0\left[x - \left(x_0 \cos\omega t + \left(\frac{p_0}{m\omega}\right)\sin\omega t\right), 0\right]. \tag{10.55}$$

Thus,

$$\psi_\alpha(x,t) = e^{\frac{i}{\hbar}p_0(t)x}\psi_0\left[x - \left(x_0 \cos\omega t + \left(\frac{p_0}{m\omega}\right)\sin\omega t\right), 0\right]. \tag{10.56}$$

The ground-state wavefunction is now centered at the point where, classically, the particle undergoing harmonic oscillations will find itself at time, t, as we show below.

The classical equation of motion for a harmonic oscillator with spring constant $K = m\omega^2$ is

$$\ddot{x} = -\omega^2 x, \tag{10.57}$$

whose solution is

$$x = A\cos\omega t + B\sin\omega t. \tag{10.58}$$

If the values of the position ($= x$) and momentum ($= m\,dx/dt$) of the particle at $t = 0$ are x_0, and p_0, respectively, then the constants A and B can be determined, and we obtain

$$x_0(t) = x_0 \cos\omega t + \left(\frac{p_0}{m\omega}\right)\sin\omega t, \tag{10.59}$$

$$p_0(t) = p_0\cos\omega t - m\omega x_0 \sin\omega t. \tag{10.60}$$

Thus the time dependence of the eigenstate $\psi_\alpha(x,t)$ is given by that of an undistorted ground-state wavefunction centered at $x_0(t)$ moving as if it described a classical particle, as depicted in Fig. 10.1. This is quite a remarkable result. The state $|\alpha\rangle$ is often called a coherent state or a semiclassical state.

10.3 Interaction of a harmonic oscillator with an electric field

We will now consider the problem of the interaction of an electric field with a charged particle undergoing simple harmonic motion. This problem will highlight the crucial role of the operator $D(\alpha)$ that we discussed in the previous section.

Let H_0 be the (unperturbed) Hamiltonian corresponding to the harmonic oscillator in one dimension,

$$H_0 = \frac{P^2}{2m} + \frac{1}{2}m\omega_0^2 X^2 = \left(a^\dagger a + \frac{1}{2}\right)\hbar\omega_0. \tag{10.61}$$

The eigenstates of H_0 will be designated by $|n\rangle$, so that $H_0 |n\rangle = E_n |n\rangle$ where $E_n = (n + \frac{1}{2})\hbar\omega_0$. We will be concerned only with the ground state $|0\rangle$ with $E_0 = \frac{1}{2}\hbar\omega_0$.

First let us consider the simple example of a time-independent electric field, $\widetilde{E}$. Its contribution to the Hamiltonian will be given by

$$-e\widetilde{E}X \tag{10.62}$$

where X is the operator for the space coordinate of the particle and e is its charge. We write X in terms of a and $a^\dagger$:

$$X = \sqrt{\frac{\hbar}{2m\omega_0}}\left(a + a^\dagger\right). \tag{10.63}$$

The total Hamiltonian including the perturbation due to the interaction of the electric field can be written as

$$H = H_0 - e\widetilde{E}X = H_0 + \lambda\left(a + a^\dagger\right)\hbar\omega_0 \tag{10.64}$$

where

$$\lambda = -\frac{e\widetilde{E}}{\hbar\omega_0}\sqrt{\frac{\hbar}{2m\omega_0}}. \tag{10.65}$$

From equation (10.28) we know that

$$D^\dagger(\alpha)a^\dagger a D(\alpha) = a^\dagger a + \alpha^* a + \alpha a^\dagger + |\alpha|^2 \tag{10.66}$$

where $D(\alpha)$ is the operator we discussed in the previous section, given by

$$D(\alpha) = e^{\alpha a^\dagger - \alpha^* a}. \tag{10.67}$$

From (10.66) we obtain

$$D^\dagger(\alpha)H_0 D(\alpha) = H_0 + \left[\alpha^* a + \alpha a^\dagger + |\alpha|^2\right]\hbar\omega_0. \tag{10.68}$$

In order to compare this Hamiltonian with the Hamiltonian H given in (10.64) we take

$$\alpha = \lambda \tag{10.69}$$

which is assumed to be real. Thus, from (10.67),

$$D(\alpha) = e^{\lambda(a^\dagger - a)}. \tag{10.70}$$

From (10.64), (10.68), and (10.69) we obtain

$$D^{\dagger}(\alpha)\left[H_0 - \lambda^2\hbar\omega_0\right]D(\alpha) = H_0 + \lambda\left(a + a^{\dagger}\right)\hbar\omega_0 = H. \tag{10.71}$$

Thus, effectively, $D(\alpha)$ transforms H_0 to H. We take $|0\rangle'$ as the eigenstate of H with eigenvalue E_0', and, as stated earlier, $|0\rangle$ as the eigenstate of H_0 with eigenvalue E_0. From our earlier discussions it is clear that $D(\alpha)$ will satisfy the relation

$$|0\rangle' = D^{\dagger}(\alpha)\,|0\rangle\,. \tag{10.72}$$

Sandwiching both sides of the equation (10.71) between the eigenstates $|0\rangle'$ then gives

$$E_0' = E_0 - \lambda^2\hbar\omega_0 = \frac{1}{2}\hbar\omega_0 - \lambda^2\hbar\omega_0 = \frac{1}{2}\hbar\omega_0 - \frac{e^2\widetilde{E}^2}{2m\omega_0^2}. \tag{10.73}$$

Substituting (10.70) for $D(\alpha)$,we write,

$$|0\rangle' = e^{-\lambda(a^{\dagger}-a)}\,|0\rangle\,. \tag{10.74}$$

From our earlier results for the ground state we can write down the wavefunction

$$\psi_0'(x,0) = \langle x\,|0\rangle' = \langle x|\,e^{-\lambda(a^{\dagger}-a)}\,|0\rangle = \psi_0(x - x_0, 0). \tag{10.75}$$

Substituting for λ, given by (10.65), we have

$$x_0 = \sqrt{\frac{\hbar}{2m\omega_0}}2\lambda = -\frac{e\widetilde{E}}{m\omega_0^2}. \tag{10.76}$$

We point out the same results as above were obtained in Chapter 9 simply by changing variables and writing the term $(\frac{1}{2})m\omega_0^2x^2 - e\widetilde{E}x$, in the Hamiltonian, H, in terms of $(\frac{1}{2})m\omega_0^2x'^2$.

Let us consider the more complicated case of an electric field, $\widetilde{E}(t)$, which is time-dependent. Here the advantage of the above approach involving the operator $D(\alpha)$ will become evident since the change of variables technique will no longer work. We assume that initially, at $t = 0$, the oscillator is in its ground state when the electric field is applied, and we assume further that the field is applied only for a short duration.

The total Hamiltonian is then

$$H = H_0 - e\widetilde{E}(t)X = H_0 + \lambda(t)\left(a + a^{\dagger}\right)\hbar\omega_0 \tag{10.77}$$

with

$$\lambda(t) = -\frac{e\widetilde{E}(t)}{\hbar\omega_0}\sqrt{\frac{\hbar}{2m\omega_0}}. \tag{10.78}$$

In the Heisenberg representation, as discussed in Chapter 3, the time dependence of a, in the absence of perturbation, is given by

$$i\hbar\frac{da}{dt} = [a, H_0] = a\hbar\omega_0, \tag{10.79}$$

which gives rise to the solution

$$a(t) = a(0)e^{-i\omega_0 t}. \tag{10.80}$$

In the presence of perturbation with the Hamiltonian given by (10.64), the equation becomes

$$i\hbar\frac{da}{dt} = [a, H] = a\hbar\omega_0 + \lambda(t)\hbar\omega_0. \tag{10.81}$$

The equation of motion for a is then

$$\frac{da}{dt} + i\omega_0 a = -i\omega_0\lambda(t). \tag{10.82}$$

This equation can be solved using the Green's function technique (see Appendix 20.13 for details) to give

$$a(t) = a(0)e^{-i\omega_0 t} - i\omega_0\int_{-\infty}^{\infty} dt' g(t-t')\lambda(t'), \tag{10.83}$$

where the first term is the homogeneous solution, i.e., the solution in the absence of perturbation, and $g(t-t')$ is the Green's function that satisfies the equation

$$\frac{dg(t-t')}{dt} + i\omega_0 g(t-t') = \delta\left(t-t'\right). \tag{10.84}$$

Even though in the integral in (10.83) we have kept the lower limit at $t' = -\infty$, and upper limit at $t' = +\infty$, it is understood that the electric field included in the factor $\lambda(t)$ appears for only a short duration.

We write $g(t-t')$ and the Dirac δ-function, $\delta\left(t-t'\right)$, in the form of a Fourier integral

$$g(t-t') = \frac{1}{\sqrt{2\pi}}\int_{-\infty}^{\infty} d\omega\, e^{-i\omega(t-t')}g(\omega) \tag{10.85}$$

and, as we have discussed in Chapter 1, the δ-function is given by

$$\delta(t-t') = \frac{1}{2\pi}\int_{-\infty}^{\infty} d\omega\, e^{-i\omega(t-t')}. \tag{10.86}$$

Substituting (10.85) and (10.86) in equation (10.84) we obtain

$$g(\omega) = \frac{i}{\sqrt{2\pi}} \left(\frac{1}{\omega - \omega_0} \right). \tag{10.87}$$

Hence

$$g(t - t') = \frac{i}{2\pi} \int_{-\infty}^{\infty} d\omega \, \frac{e^{-i\omega(t-t')}}{\omega - \omega_0}. \tag{10.88}$$

To calculate this integral we impose causality. That is, we impose the condition that the cause as signified by the presence of the electric field, $\widetilde{E}(t')$, at time t' (therefore, $\lambda(t')$ in (10.83) at t') will have its effect only at a time $t > t'$. Thus we demand that $g(t-t') = 0$ for $t < t'$. The above integral must then vanish for $t < t'$. This implies that the pole at $\omega = \omega_0$ be shifted down to $\omega_0 - i\epsilon$, i.e.,

$$\frac{1}{\omega - \omega_0} \rightarrow \frac{1}{\omega - \omega_0 + i\epsilon}. \tag{10.89}$$

One can then use Cauchy's residue theorem and carry out the integration in the complex plane to obtain

$$g(t - t') = e^{-i\omega_0(t-t')}, \qquad \text{for } t > t' \tag{10.90}$$

$$= 0, \qquad \text{for } t < t'. \tag{10.91}$$

The solution for $a(t)$ now reads

$$a(t) = a(0)e^{-i\omega_0 t} - i\omega_0 \int_{-\infty}^{t} dt' \, e^{-i\omega_0(t-t')}\lambda(t'). \tag{10.92}$$

As we stated earlier, we assume that the electric field $\widetilde{E}(t)$ and, therefore $\lambda(t)$, appears only for a short duration, in which case we can safely extend the upper limit along a region in which $\lambda(t')$ is zero, all the way to $t' = \infty$. Thus, we write

$$a(t) = a(0)e^{-i\omega_0 t} - i\omega_0 e^{-i\omega_0 t} \int_{-\infty}^{\infty} dt' \, e^{i\omega_0 t'}\lambda(t'). \tag{10.93}$$

However, the integral on the right is simply the (inverse) Fourier transform of $\lambda(t)$ which we write as $\lambda(\omega_0)$. Thus,

$$a(t) = \left[a(0) - i\sqrt{2\pi}\,\omega_0\lambda(\omega_0) \right] e^{-i\omega_0 t}. \tag{10.94}$$

Comparing this with the time dependence of the unperturbed $a(t)$ given in (10.80) we find that there is a shift in the unperturbed value $a(0)$ to $\left[a(0) - i\sqrt{2\pi}\,\omega_0\lambda(\omega_0) \right]$ due to the interaction with the electric field.

This shift can be accomplished through the operator $D(\alpha)$ since it satisfies the property

$$D^{\dagger}(\alpha)a(0)D(\alpha) = a(0) + \alpha \tag{10.95}$$

if we choose

$$\alpha = -i\sqrt{2\pi}\,\omega_0\lambda(\omega_0). \tag{10.96}$$

Thus, an oscillator initially in a ground state $|0\rangle$ when subjected to a time-dependent electric field over a finite duration is found, at large values of time t after the interaction is turned off, to be in the state

$$|\alpha(t)\rangle = e^{-\frac{i\omega_0 t}{2}} D\left(\alpha e^{-i\omega_0 t}\right)|0\rangle = e^{-\frac{i\omega_0 t}{2}} e^{-\frac{1}{2}|\alpha|^2} \exp\left[\alpha a^{\dagger} e^{-i\omega_0 t}\right]|0\rangle, \tag{10.97}$$

which is a coherent state.

10.4 Appendix to Chapter 10

10.4.1 Some important identities involving exponentials of operators

First we note that the exponential of an operator is given by the infinite series

$$e^A = 1 + \frac{A}{1!} + \frac{A^2}{2!} + \cdots. \tag{10.98}$$

Let us now prove the following identities

$$1.\ e^A B e^{-A} = B + [A,B] + \frac{1}{2!}[A,[A,B]] + \frac{1}{3!}[A,[A\,[A,B]]] + \cdots. \tag{10.99}$$

$$2.\ e^A e^B = e^{A+B+(1/2)[A,B]} \tag{10.100}$$

where in (2) we assume that both A and B commute with the commutator $[A,B]$.

The proofs are as follows.

1. Let us define a function $F(x)$ in terms of the operators A and B given by

$$F(x) = e^{xA} B e^{-xA} \tag{10.101}$$

where x is a variable, not an operator. We expand it in a Taylor series

$$F(x) = F(0) + \frac{x}{1!}F'(0) + \frac{x^2}{2!}F''(0) + \cdots. \tag{10.102}$$

We note from (10.101) that

$$F(0) = B \tag{10.103}$$

and

$$F'(x) = A\left[e^{xA}Be^{-xA}\right] - \left[e^{xA}BAe^{-xA}\right] = [A, F(x)] \tag{10.104}$$

where we have used the fact that A and $\exp(-xA)$ commute. Hence,

$$F'(0) = [A, B]. \tag{10.105}$$

From (10.104) we find

$$F''(x) = \left[A, F'(x)\right]. \tag{10.106}$$

Hence,

$$F''(0) = \left[A, F'(0)\right] = [A, [A, B]]. \tag{10.107}$$

Similarly, continuing with the higher derivatives we obtain the general result

$$F(x) = B + \frac{x}{1!}[A, B] + \frac{x^2}{2!}[A, [A, B]] + \cdots. \tag{10.108}$$

Putting $x = 1$ we obtain the result (10.99)

2. Consider the function

$$G(x) = e^{xA}e^{xB}. \tag{10.109}$$

Then

$$G'(x) = Ae^{xA}e^{xB} + Ae^{xA}Be^{xB} = \left[A + e^{xA}Be^{-xA}\right]G(x). \tag{10.110}$$

From the expansion (10.99) for the exponential and the assumption that both A and B commute with $[A, B]$ we can simplify the second term above and write

$$G'(x) = [A + B + x[A, B]]\, G(x). \tag{10.111}$$

This differential equation can be solved. We obtain, since $G(0) = 1$,

$$G(x) = e^{[A+B]x}e^{(1/2)[A,B]x^2}. \tag{10.112}$$

Putting $x = 1$ we obtain (10.100).

10.5 Problems

1. Show that for coherent states

$$\Delta x \Delta p = \frac{\hbar}{2}$$

where the expectation values are taken with respect to the coherent state $|\alpha\rangle$.

2. Prove the following result:

$$D(\alpha_1)D(\alpha_2) = D(\alpha_1 + \alpha_2)\, e^{i\,\mathrm{Im}(\alpha_1\alpha_2^*)}.$$

3. Consider an arbitrary state $|\phi\rangle$ as a linear superposition of the harmonic oscillator states $|n\rangle$,

$$|\phi\rangle = \sum_n c_n\, |n\rangle\,.$$

Express the superposition for $|\phi\rangle$ in terms of the coherent states $|\alpha\rangle$. Since α is continuous, write

$$|\phi\rangle = \int d\alpha\, f(\alpha)\, |\alpha\rangle\,.$$

Obtain $f(\alpha)$.

4. For a coherent state $|\alpha\rangle$ show that if $n = \langle\alpha\, |N|\, \alpha\rangle$ where $N = a^\dagger a$ then

$$n = |\alpha|^2\,.$$

Also obtain Δn defined by

$$\Delta n = \sqrt{\langle N^2\rangle - \langle N\rangle^2}$$

where the expectation values are taken with respect to the state $|\alpha\rangle$.

5. If θ is the phase of the complex quantity α,

$$\alpha = |\alpha|\, e^{i\theta}$$

then show that

$$-i\frac{\partial}{\partial\theta}\, |\alpha\rangle = N\, |\alpha\rangle\,.$$

6. If operators A, B, and C are related as

$$[A, B] = iC \quad \text{and} \quad [C, A] = iB$$

then show that

$$e^{iAt}Be^{-iAt} = B\cos t - C\sin t.$$

7. For a coherent state show that the probability of finding a particle in the state $|n\rangle$ is given by the Poisson distribution

$$P(n) = e^{-\langle n\rangle}\frac{\langle n\rangle^n}{n!}.$$

8. Show that the following completeness relation is satisfied by the coherent states

$$\int (d\,\mathrm{Re}\,\alpha)\,(d\,\mathrm{Im}\,\alpha)\,|\alpha\rangle\,\langle\alpha| = \frac{1}{\pi}.$$

(Hint: substitute the expression for $|\alpha\rangle$ in terms of $|n\rangle$; use polar coordinates for integration and completeness condition for $|n\rangle$.)

9. Obtain the eigenfunction $u_\alpha(x) = \langle x\,|\alpha\rangle$ from the equation

$$a\,|\alpha\rangle = \alpha\,|\alpha\rangle$$

by writing a in terms of x and p and solving the corresponding first-order differential equation. Compare your result with the result obtained in the text through the relation $\langle x\,|\alpha\rangle = \langle x\,|D(\alpha)|\,0\rangle$.

10. Show through the superposition relation in terms of the states $|n\rangle$ that one cannot construct an eigenstate of the raising operator $a^\dagger$.
11. Determine the propagator function between two coherent states given by $\left\langle\alpha\,\left|e^{iH_0t/\hbar}\right|\,\alpha'\right\rangle$.
12. Take $\alpha = |\alpha|\,e^{i\theta}$ and obtain $u_\alpha(x,t) = \langle x|\,\alpha e^{-i\omega t}\rangle$. Show that it is a Gaussian centered at $\sqrt{2\hbar}\,|\alpha|\,/m\omega\cos\left(\theta - \omega t\right)$.
13. Show that

$$\left[e^{ix_0p/\hbar}, a\right] = x_0\sqrt{m\omega/2\hbar}e^{ix_0p/\hbar}.$$

From this obtain $\left\langle n\,\left|e^{ix_0p/\hbar}\right|\,0\right\rangle$.

14. Let

$$a = be^{i\phi}$$

where b and ϕ are Hermitian operators. Show that (i) $\left[e^{i\phi}, b^2\right] = e^{i\phi}$ and (ii) $b^2 = N+1$.

15. A harmonic oscillator in its ground state acquires a momentum p_0. What is the probability that it will remain in the ground state?
16. Show that $[a, D(\alpha)] = \alpha D(\alpha)$.

11 Two-dimensional isotropic harmonic oscillator

This is a simple extension of the one-dimensional harmonic oscillator problem. It will play an essential role when we consider the famous Landau levels.

11.1 The two-dimensional Hamiltonian

Let us consider a harmonic oscillator in a two-dimensional x–y plane where the spring constant $K\ (= m\omega^2)$ is the same in both directions. The Hamiltonian is then given by

$$H = \frac{p_x^2}{2m} + \frac{p_y^2}{2m} + \frac{1}{2}m\omega^2\left(x^2 + y^2\right). \tag{11.1}$$

The operators satisfy the commutation relations

$$[x, p_x] = i\hbar\mathbf{1}, \quad [y, p_y] = i\hbar\mathbf{1}, \quad \text{all other commutators} = 0. \tag{11.2}$$

Using the separation of variables technique we can reduce this problem to that of two one-dimensional harmonic oscillators. As with the one-dimensional problems, we introduce the lowering and raising operators a_x, a_y and $a_x^\dagger, a_y^\dagger$ respectively. Specifically, we write

$$a_x = \sqrt{\frac{m\omega}{2\hbar}}\left(x + i\frac{p_x}{m\omega}\right), \tag{11.3}$$

$$a_y = \sqrt{\frac{m\omega}{2\hbar}}\left(y + i\frac{p_y}{m\omega}\right). \tag{11.4}$$

11.1.1 Eigenstates $|n_x, n_y\rangle$ of H

From the above commutation relations for x, y, p_x, and p_y we obtain

$$\left[a_x, a_x^\dagger\right] = \mathbf{1} = \left[a_y, a_y^\dagger\right], \quad \text{all other commutators} = 0. \tag{11.5}$$

Using the same techniques that we employed in the one-dimensional case, we write the following sequence of relations and definitions:

$$H = \left(a_x^\dagger a_x + \frac{1}{2}\right)\hbar\omega + \left(a_y^\dagger a_y + \frac{1}{2}\right)\hbar\omega, \tag{11.6}$$

$$N_x = a_x^\dagger a_x, \quad N_y = a_y^\dagger a_y, \tag{11.7}$$

$$H = \left(N_x + \frac{1}{2}\right)\hbar\omega + \left(N_y + \frac{1}{2}\right)\hbar\omega. \tag{11.8}$$

If we designate the eigenstates of H as $|n_x, n_y\rangle$ then, based on the discussions on the one-dimensional systems, we have the following results:

$$a_x^\dagger |n_x, n_y\rangle = \sqrt{(n_x+1)}\,|n_x+1, n_y\rangle, \quad a_y^\dagger |n_x, n_y\rangle = \sqrt{(n_y+1)}\,|n_x, n_y+1\rangle, \tag{11.9}$$

$$N_x |n_x, n_y\rangle = n_x |n_x, n_y\rangle, \quad N_y |n_x, n_y\rangle = n_y |n_x, n_y\rangle, \tag{11.10}$$

$$H |n_x, n_y\rangle = (n_x + n_y + 1)\,\hbar\omega\,|n_x, n_y\rangle = (n+1)\,\hbar\omega\,|n_x, n_y\rangle, \tag{11.11}$$

where we have written, $n_x + n_y = n$. The last relation implies that the energy eigenvalues are given by

$$E_n = (n+1)\,\hbar\omega. \tag{11.12}$$

We note that different values of n_x and n_y may add up to the same value of n. That is, the state $|1, 0\rangle$, for example, corresponding to $n_x = 1$ and $n_y = 0$, will have the same value $(n = 1)$, as $|0, 1\rangle$, which has $n_x = 0$ and $n_y = 1$. Thus we have a 2-fold degeneracy where two different states will have the same energy eigenvalue.

The fact that $|n_x, n_y\rangle$ correspond to two unrelated eigenvalues n_x and n_y leads one to suspect that there may be a second operator that commutes with the Hamiltonian. We will go to the next section to address this question.

11.1.2 Eigenstates $|n_+, n_-\rangle$ of H and L_z

Let us consider the operator L_z corresponding to the angular momentum around the z-axis:

$$L_z = xp_y - yp_x. \tag{11.13}$$

Substituting the relations

$$x = \sqrt{\frac{\hbar}{2m\omega}}\left(a_x + a_x^\dagger\right), \quad p_x = -i\sqrt{\frac{m\hbar\omega}{2}}\left(a_x - a_x^\dagger\right), \tag{11.14}$$

$$y = \sqrt{\frac{\hbar}{2m\omega}}\left(a_y + a_y^\dagger\right), \quad p_y = -i\sqrt{\frac{m\hbar\omega}{2}}\left(a_y - a_y^\dagger\right), \tag{11.15}$$

and after taking into account the commutation relations between the operators a_x, a_y, $a_x^\dagger$, and $a_y^\dagger$, we obtain

$$L_z = i\left(a_x a_y^\dagger - a_x^\dagger a_y\right)\hbar. \tag{11.16}$$

It is easy to check that L_z is Hermitian, i.e., $L_z^\dagger = L_z$.

To determine the commutation relations between L_z and H, we find after some lengthy calculations that

$$\left[\left(a_x a_y^\dagger - a_x^\dagger a_y\right), a_x^\dagger a_x\right] = a_x a_y^\dagger + a_x^\dagger a_y \tag{11.17}$$

and

$$\left[\left(a_x a_y^\dagger - a_x^\dagger a_y\right), a_y^\dagger a_y\right] = -\left(a_x a_y^\dagger + a_x^\dagger a_y\right). \tag{11.18}$$

Adding the above two relations we find from (11.6) and (11.16), that L_z and H commute,

$$[L_z, H] = 0. \tag{11.19}$$

Hence L_z and H will have common eigenstates. It can be shown that $\left|n_x, n_y\right\rangle$ are not the eigenstates of L_z.

To obtain the relevant eigenstates we define a new set of raising and lowering operators:

$$a_+ = \frac{1}{\sqrt{2}}\left(a_x + ia_y\right) = \frac{1}{2}\sqrt{\frac{m\omega}{\hbar}}\left[(x+iy) + i\left(\frac{p_x + ip_y}{m\omega}\right)\right], \tag{11.20}$$

$$a_- = \frac{1}{\sqrt{2}}\left(a_x - ia_y\right) = \frac{1}{2}\sqrt{\frac{m\omega}{\hbar}}\left[(x-iy) + i\left(\frac{p_x - ip_y}{m\omega}\right)\right]. \tag{11.21}$$

From the commutation relations of a_x and a_y one can verify that

$$\left[a_+, a_+^\dagger\right] = \mathbf{1} = \left[a_-, a_-^\dagger\right], \quad \text{all other commutators} = 0. \tag{11.22}$$

To obtain H, we note that

$$a_x^\dagger a_x + a_y^\dagger a_y = \frac{1}{2}\left(a_+^\dagger + a_-^\dagger\right)(a_+ + a_-) + \frac{1}{2}\left(a_+^\dagger - a_-^\dagger\right)(a_+ - a_-) \tag{11.23}$$

$$= a_+^\dagger a_+ + a_-^\dagger a_-. \tag{11.24}$$

Substituting this into (11.6), we obtain

$$H = \left(a_+^\dagger a_+ + a_-^\dagger a_- + 1\right)\hbar\omega = (N_+ + N_- + 1)\,\hbar\omega \tag{11.25}$$

where

$$N_+ = a_+^\dagger a_+ \quad \text{and} \quad N_- = a_-^\dagger a_-. \tag{11.26}$$

Similarly, to obtain L_z, we write

$$a_x a_y^\dagger - a_x^\dagger a_y = a_y^\dagger a_x - a_x^\dagger a_y \tag{11.27}$$

since a_x, $a_y^\dagger$ commute. Thus

$$a_y^\dagger a_x - a_x^\dagger a_y = \frac{i}{2}\left(a_+^\dagger - a_-^\dagger\right)(a_+ + a_-) + \frac{i}{2}\left(a_+^\dagger + a_-^\dagger\right)(a_+ - a_-) \tag{11.28}$$

$$= i\left(a_+^\dagger a_+ - a_-^\dagger a_-\right). \tag{11.29}$$

Putting this relation into (11.16) gives

$$L_z = -\left(a_+^\dagger a_+ - a_-^\dagger a_-\right)\hbar = (N_- - N_+)\,\hbar. \tag{11.30}$$

Let $|n_+, n_-\rangle$ be eigenstates of the mutually commuting operators N_+ and N_-:

$$N_+ |n_+, n_-\rangle = n_+ |n_+, n_-\rangle \quad \text{and} \quad N_- |n_+, n_-\rangle = n_- |n_+, n_-\rangle. \tag{11.31}$$

Based on the arguments for the one-dimensional case, one can easily show that

$$a_+^\dagger |n_+, n_-\rangle = \sqrt{(n_+ + 1)}\, |n_+ + 1, n_-\rangle, \quad a_-^\dagger |n_+, n_-\rangle = \sqrt{(n_- + 1)}\, |n_+, n_- + 1\rangle. \tag{11.32}$$

One also finds

$$H |n_+, n_-\rangle = (n_+ + n_- + 1)\,\hbar\omega\, |n_+, n_-\rangle \tag{11.33}$$

and

$$L_z |n_+, n_-\rangle = (n_- - n_+)\,\hbar\, |n_+, n_-\rangle. \tag{11.34}$$

If, instead of n_+, and n_- we select the quantum numbers

$$n = n_+ + n_- \quad \text{and} \quad m = n_- - n_+ \tag{11.35}$$

that correspond to the eigenvalues for H and L_z, respectively, then we can designate a new eigenstate as $|n, m\rangle$, where

$$H |n, m\rangle = (n + 1)\,\hbar\omega\, |n, m\rangle, \tag{11.36}$$

$$L_z |n, m\rangle = m\hbar\, |n, m\rangle. \tag{11.37}$$

Since n_+ and n_- are positive integers and $n_+ + n_- = n$, when n takes on values

$$n = 0, 1, 2, \ldots \tag{11.38}$$

then n_+ and n_- go over the values

$$n_+ = 0, 1, \ldots \quad \text{and} \quad n_- = n, n-1, \ldots \tag{11.39}$$

Hence for each n, the quantum number m will vary as given by (11.35), which gives rise to the same energy eigenvalues. Hence the degeneracy in the energy levels is $n+1$. Thus, the degeneracy for the two-dimensional oscillator can be understood in terms of the eigenvalues of L_z.

We will not proceed further to determine the eigenfunctions. However, we will revisit this problem when we discuss the so-called Landau levels in Chapter 12.

Finally, two-dimensional oscillators are of great interest in the case where conduction electrons in a solid are confined within nanometric dimensions. These configurations are referred to as quantum dots.

11.2 Problems

1. Show that

$$[ab, cd] = a\,[b,c]\,d + ac\,[b,d] + [a,c]\,db + c\,[a,d]\,b.$$

2. Consider two independent harmonic oscillators for which the raising and lowering operators are $a_+^\dagger$ and a_+ and $a_-^\dagger$ and a_- respectively. Let

$$A_+ = a_+^\dagger a_-, \quad A_- = a_-^\dagger a_+, \quad A_z = \left(a_+^\dagger a_+ - a_-^\dagger a_-\right).$$

Using the results from the previous problem show that

$$[A_\pm, A_z] = \pm A_\pm.$$

If one defines A_x and A_y in the same manner as in the case of the angular momentum operators, then show that

$$\left[\mathbf{A}^2, A_z\right] = 0$$

where $\mathbf{A}^2 = A_x^2 + A_y^2 + A_z^2$ (which can be written in terms of $A_\pm$ and A_z). If $N_+ = a_+^\dagger a_+$ and $N_- = a_-^\dagger a_-$ and $N = N_+ + N_-$ then show that

$$\mathbf{A}^2 = \frac{N}{2}\left(\frac{N}{2} + 1\right)\mathbf{1}.$$

12 Landau levels and quantum Hall effect

In this chapter we consider the motion of a charged particle in a magnetic field. This is intrinsically a three-dimensional problem but for a charged particle in a uniform magnetic field it can be recast as a two-dimensional problem since the particle acts as a free particle in the direction of the magnetic field, while in the plane perpendicular to the magnetic field the particle's motion can be described in terms of an equivalent two-dimensional harmonic oscillator. It is found that when charged particles are subjected to a magnetic field their orbital motion is quantized. Thus the particles can only occupy orbits with discrete energy eigenvalues called Landau levels named after the physicist Lev Landau. A characteristic property of these levels is that they are highly degenerate. We will also briefly discuss the fascinating subject of the quantum Hall effect.

12.1 Landau levels in symmetric gauge

12.1.1 Basic equations

The Hamiltonian for a free particle is given by

$$H = \frac{\mathbf{p}^2}{2m}. \tag{12.1}$$

As we discussed in Chapter 6, to include the magnetic interaction we will replace $\mathbf{p}$ by $(\mathbf{p}-e/c\mathbf{A})$ in the above Hamiltonian, where $\mathbf{A}$ is the vector potential, related to the magnetic field, $\mathbf{B}$, by

$$\nabla \times \mathbf{A} = \mathbf{B}. \tag{12.2}$$

Thus the Hamiltonian for a charged particle in the presence of a magnetic field is given by

$$H = \frac{1}{2m}\left(\mathbf{p} - \frac{e}{c}\mathbf{A}\right)^2. \tag{12.3}$$

We will discuss our results in terms of the velocity, $\mathbf{v}$, given by

$$\mathbf{v} = \frac{1}{m}\left(\mathbf{p}-\frac{e}{c}\mathbf{A}\right) \tag{12.4}$$

and consider a uniform magnetic field $\mathbf{B}$ in the z-direction,

$$\mathbf{B} = B_0\mathbf{k}. \tag{12.5}$$

At the moment we will not commit to a specific gauge for $\mathbf{A}$.

We note that the motion in the z-direction (i.e., parallel to the magnetic field) is unaffected by the magnetic field and hence the corresponding Hamiltonian $H_{\parallel}$ is simply that of a free particle. In the following discussion we will ignore this motion and consider only $H_{\perp}$.

Thus, the motion in a plane perpendicular to $\mathbf{B}$, i.e., in the x–y plane is described by

$$H_{\perp} = \frac{\left(p_x - \frac{e}{c}A_x\right)^2}{2m} + \frac{\left(p_y - \frac{e}{c}A_y\right)^2}{2m} = \frac{m}{2}\left(v_x^2 + v_y^2\right). \tag{12.6}$$

In Chapter 6 we considered this Hamiltonian and showed that

$$\left[v_x, v_y\right] = i\frac{e\hbar}{m^2c}B_0 = i\frac{\hbar\omega_c}{m} \tag{12.7}$$

where ω_c is the cyclotron frequency

$$\omega_c = \frac{eB_0}{mc}. \tag{12.8}$$

Since the right-hand side in (12.7) acts like a unit operator times a number, we find that v_x and v_y satisfy quantum conditions of the type

$$[x, p] = i\hbar. \tag{12.9}$$

Expression (12.6) is reminiscent of the harmonic oscillator Hamiltonian. Let us, therefore, define the following operators similar to the raising and lowering operators for a harmonic oscillator:

$$b = \sqrt{\frac{m}{2\hbar\omega_c}}\left(v_x + iv_y\right), \quad b^{\dagger} = \sqrt{\frac{m}{2\hbar\omega_c}}\left(v_x - iv_y\right). \tag{12.10}$$

We then obtain

$$b^{\dagger}b = \left(\frac{m}{2\hbar\omega_c}\right)\left[v_x^2 + v_y^2 + i\left(v_xv_y - v_yv_x\right)\right] \tag{12.11}$$

$$= \left(\frac{m}{2\hbar\omega_c}\right)\left(v_x^2 + v_y^2\right) - \frac{1}{2} \tag{12.12}$$

where we have used (12.7). Hence the Hamiltonian, $H_{\perp}$, is simply given by

$$H_{\perp} = \left(b^{\dagger}b + \frac{1}{2}\right)\hbar\omega_c. \tag{12.13}$$

This is now a familiar expression from the harmonic oscillator calculations and the consequences of using $b^{\dagger}$ and b as raising and lowering operators follow immediately.

The eigenvalues of $H_\perp$ are then given by

$$E_n = \left(n + \frac{1}{2}\right)\hbar\omega_c, \quad n = 0, 1, 2 \ldots. \tag{12.14}$$

These are the familiar Landau levels. However, since this is a two-dimensional system, we need to examine this problem in the context of a two-dimensional configuration. As expected, the energy eigenstates are highly degenerate as we will elaborate below.

12.1.2 Symmetric gauge

Consider the following gauge, called the symmetric gauge,

$$\mathbf{A} = \frac{1}{2}(\mathbf{B} \times \mathbf{r}). \tag{12.15}$$

For $\mathbf{B} = B_0\mathbf{k}$, we obtain

$$A_x = -\frac{1}{2}B_0 y, \quad A_y = \frac{1}{2}B_0 x, \quad A_z = 0. \tag{12.16}$$

The velocities are then given by

$$v_x = \frac{1}{m}\left(p_x - \frac{e}{c}A_x\right) = \frac{p_x}{m} + \frac{\omega_c}{2}y, \tag{12.17}$$

$$v_y = \frac{1}{m}\left(p_y - \frac{e}{c}A_y\right) = \frac{p_y}{m} - \frac{\omega_c}{2}x. \tag{12.18}$$

The corresponding Hamiltonian (12.6) now reads

$$H = \frac{p_x^2}{2m} + \frac{p_y^2}{2m} + \frac{1}{8}m\omega_c^2(x^2 + y^2) - \frac{\omega_c}{2}\left(xp_y - yp_x\right) \tag{12.19}$$

where the last term is proportional to the angular momentum, which we write as

$$(L_z)_2 = \left(xp_y - yp_x\right). \tag{12.20}$$

We compare H with the two-dimensional isotropic Hamiltonian, which we designate as H_2, considered in the previous chapter. This will allow us to discuss the degeneracies observed in Landau levels more clearly.

12.1.3 Comparison with the two-dimensional oscillator and the degeneracy question

The Hamiltonian, H_2, is given by (replacing ω by $\omega_c/2$),

$$H_2 = \frac{p_x^2}{2m} + \frac{p_y^2}{2m} + \frac{1}{8}m\omega_c^2(x^2 + y^2).$$

Hence,

$$H = H_2 - \frac{\omega_c}{2} L_z. \tag{12.21}$$

Let us express the operators H, H_2, and $(L_z)_2$ in terms of the number operators N_+ and N_- defined in the previous chapter. From the previous chapter we have

$$H_2 = (N_+ + N_- + 1)\frac{\hbar\omega_c}{2} \tag{12.22}$$

and

$$(L_z)_2 = (N_- - N_+)\hbar. \tag{12.23}$$

To obtain H, which is given in (12.3) in terms of b and $b^\dagger$, we note that

$$\mathbf{v} = \frac{1}{m}\left(\mathbf{p} - \frac{e}{c}\mathbf{A}\right). \tag{12.24}$$

Therefore, from (12.17) and (12.18) we have the following relation:

$$b = \sqrt{\frac{m}{2\hbar\omega_c}}\left(v_x + iv_y\right) = -\frac{i}{2}\sqrt{\frac{m\omega_c}{2\hbar}}\left[(x+iy) + \frac{2i}{m\omega_c}\left(p_x + ip_y\right)\right]. \tag{12.25}$$

Comparing this with a_+ and a_- of the two-dimensional harmonic oscillator considered in Chapter 11, we identify

$$b = -ia_+ \tag{12.26}$$

where we have replaced ω of Chapter 11 by $\omega_c/2$. If we define

$$N_+ = a_+^\dagger a_+, \tag{12.27}$$

then our H is of the form

$$H = \left(N_+ + \frac{1}{2}\right)\hbar\omega_c. \tag{12.28}$$

If n_+ and n_- are the quantum numbers corresponding to the operators N_+ and N_-, respectively, then for the two-dimensional oscillator the quantum numbers corresponding to H_2, and $(L_z)_2$ are

$$n = n_+ + n_- \quad \text{and} \quad m = n_- - n_+, \tag{12.29}$$

respectively, where n takes the values

$$n = 0, 1, 2, \ldots. \tag{12.30}$$

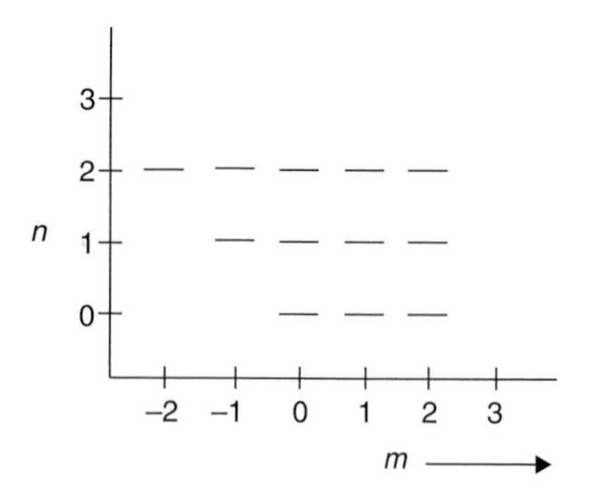

Fig. 12.1

For a given n, the quantum numbers n_+ and n_- will vary as

$$0 \le n_+ \le n \quad \text{and} \quad 0 \le n_- \le n \quad \text{with} \quad n_+ + n_- = n. \tag{12.31}$$

Therefore, m will vary within the limits given by

$$m = -n, \dots, n. \tag{12.32}$$

In the Hamiltonian H which involves only the operator N_+, for a given eigenvalue $n_+ = n$, the missing quantum number m is now given by

$$m = n_- - n. \tag{12.33}$$

There are no restrictions now on n_- as it does not contribute to H, except that it be an integer, i.e, $n_- = 0, 1, \dots$. Therefore,

$$m = -n, -n+1, \dots \tag{12.34}$$

with no upper limit, indicating an infinite degeneracy. This is the most striking feature of Landau levels. For example, for the ground state, $n = 0$

$$m = 0, 1, \dots$$

indefinitely. These ground state levels are called the lowest Landau levels (LLLs). Fig. 12.1 describes the Landau levels.

12.2 Wavefunctions for the LLL

We continue with the symmetric gauge, which enables one to write the wavefunction in a more elegant and transparent way. The lowering operator b defined in (12.10) can be written as

$$b = \sqrt{\frac{1}{2\hbar m\omega_c}}\left[\left(p_x + ip_y\right) - i\frac{m\omega_c}{2}(x+iy)\right]. \tag{12.35}$$

The ground state for the Landau levels $|0\rangle$ satisfies the relation

$$b\,|0\rangle = 0. \tag{12.36}$$

In the two-dimensional x–y space the ground-state wavefunction $u_0(x,y)$ is given by

$$\langle x,y\,|0\rangle = u_0(x,y), \tag{12.37}$$

and b is given by

$$b = \sqrt{\frac{\hbar}{2m\omega_c}}\left[\left(\frac{\partial}{\partial x} + i\frac{\partial}{\partial y}\right) + \frac{m\omega_c}{2}(x+iy)\right]. \tag{12.38}$$

The equation satisfied by $u_0(x,y)$ is then

$$\sqrt{\frac{\hbar}{2m\omega_c}}\left[\left(\frac{\partial}{\partial x} + i\frac{\partial}{\partial y}\right) + \frac{m\omega_c}{2}(x+iy)\right]u_0(x,y) = 0. \tag{12.39}$$

There is a very elegant way of obtaining u_0 if we change variables as follows:

$$z = x + iy, \tag{12.40}$$

$$z^* = x - iy. \tag{12.41}$$

Then it is straightforward to show that

$$\frac{\partial}{\partial x} + i\frac{\partial}{\partial y} = \frac{\partial}{\partial z^*}, \tag{12.42}$$

$$b = \sqrt{\frac{\hbar}{2m\omega_c}}\left[\frac{\partial}{\partial z^*} + \frac{m\omega_c}{2}z\right], \tag{12.43}$$

and

$$u_0(x,y) \to u_0(z,z^*). \tag{12.44}$$

Hence the equation for u_0, now written as $u_0(z,z^*)$, becomes

$$\left[\frac{\partial}{\partial z^*} + \frac{m\omega_c}{2}z\right]u_0(z,z^*) = 0. \tag{12.45}$$

The solution for $u_0(z,z^*)$ is more easily obtained if we write it as

$$u_0(z,z^*) = f_0\left(z,z^*\right)\exp\left(-\frac{m\omega_c}{2\hbar}zz^*\right). \tag{12.46}$$

Substituting this in (12.45) gives

$$\frac{\partial}{\partial z^*} f_0\left(z, z^*\right) = 0, \tag{12.47}$$

which implies that f_0 must be a constant as a function of z^*. If we assume that f_0 is analytic as a function of z, then any polynomial in z will satisfy the above equation. Typically, therefore,

$$f_0\left(z, z^*\right) \sim z^q \tag{12.48}$$

where q is an integer. Thus,

$$u_0(z, z^*) = Cz^q \exp\left(-\frac{m\omega_c}{2\hbar} |z|^2\right) \tag{12.49}$$

describes the degenerate lowest LLLs for $q = 0, 1, 2, \ldots$.

12.3 Landau levels in Landau gauge

We once again consider a uniform magnetic field **B** applied in the z-direction:

$$\mathbf{B} = B_0 \mathbf{k}. \tag{12.50}$$

For simplicity we take the Landau gauge for **A** given by

$$\mathbf{A} = (0, B_0 x, 0). \tag{12.51}$$

It is easily verified that the relation, $\mathbf{B} = \nabla \times \mathbf{A}$, is satisfied. Writing the Hamiltonian in the component form, we have

$$H = \frac{p_x^2}{2m} + \frac{\left(p_y - \frac{e}{c} B_0 x\right)^2}{2m} + \frac{p_z^2}{2m}. \tag{12.52}$$

The eigenvalue equation

$$H\psi = E\psi \tag{12.53}$$

can be solved by writing the wavefunction in the separation of variables form discussed in Chapter 2 as $\phi(x, y)Z(z)$. It can be easily shown that $Z(z)$ will correspond to a free particle wavefunction

$$Z(z) = N_1 e^{ik_z z} \tag{12.54}$$

with energy eigenvalue $\hbar^2 k_z^2 / 2m$, where N_1 is the normalization constant. Thus, the motion of the particle in the z-direction is decoupled from the influence of the magnetic field. We will, therefore, focus entirely on the two-dimensional x–y plane, and confine the particle within a rectangle of lengths L_x and L_y respectively.

The Schrödinger equation for ϕ is given by

$$H\phi = \left[-\frac{\hbar^2}{2m}\frac{\partial^2 \phi}{\partial x^2} + \frac{1}{2m}\left(-i\hbar\frac{\partial}{\partial y} - \frac{eB_0}{c}x \right)^2 \phi \right] = E\phi. \tag{12.55}$$

Since this is a differential equation in the x-variable, the solution in the y-variable can be taken to be that of a free particle. We write

$$\phi(x, y) = u(x)e^{ik_y y}. \tag{12.56}$$

Hence $u(x)$ satisfies the equation

$$\left[-\frac{\hbar^2}{2m}\frac{d^2 u}{dx^2} + \frac{\left(\hbar k_y - \frac{eB_0}{c}x \right)^2 u}{2m} \right] = Eu. \tag{12.57}$$

We take

$$\omega_c = \frac{eB_0}{mc} \quad \text{and} \quad x_0 = \frac{c\hbar k_y}{eB_0} \tag{12.58}$$

where ω_c is the cyclotron frequency. Thus, the equation is further simplified to

$$-\frac{\hbar^2}{2m}\frac{d^2 u}{dx^2} + \frac{1}{2}m\omega_c^2 (x - x_0)^2 u = Eu. \tag{12.59}$$

This is the familiar one-dimensional equation for the harmonic oscillator centered at $x = x_0$, for which the energy eigenvalues are given by

$$E_n = \left(n + \frac{1}{2} \right) \hbar\omega_c, \quad n = 0, 1, 2, \ldots. \tag{12.60}$$

These are the Landau levels. The lowest Landau level corresponding to $n = 0$ is generally designated as LLL. The wavefunction for the LLL is of the form

$$u_0(x) = C_0 e^{-\alpha(x - x_0)^2} \tag{12.61}$$

where C_0 and α are parameters that have already been obtained in Chapter 9 but whose specific values do not concern us.

We note that the energy eigenvalues are independent of x_0; therefore, they will not depend on k_y. There is, therefore, an enormous degeneracy in the energy eigenvalues due to the spread in the allowed values of k_y. This is a striking characteristic of Landau levels as we already know. This degeneracy can be determined as follows.

Let us now consider the practical consequences of our results by taking the finite extension of our system into account. We assume our two-dimensional system to be confined in a rectangle of sides L_x and L_y. We will, therefore, have

$$0 < x_0 < L_x. \tag{12.62}$$

Thus,

$$(x_0)_{\max} = L_x. \tag{12.63}$$

Hence,

$$\frac{c\hbar(k_y)_{\max}}{eB_0} = L_x. \tag{12.64}$$

The plane waves in the y-direction satisfy the periodic boundary conditions

$$k_y L_y = 2n_y\pi. \tag{12.65}$$

Let

$$N = (n_y)_{\max}. \tag{12.66}$$

Then N will determine the amount of degeneracy since it corresponds to the number of allowed states for which the energy E_n is the same, given by (12.37). From (12.65) we find

$$N = \frac{(k_y)_{\max} L_y}{2\pi} = \left(\frac{eB_0 L_x}{c\hbar}\right)\frac{L_y}{2\pi} = \left(\frac{eB_0}{2\pi c\hbar}\right) L_x L_y. \tag{12.67}$$

If we write

$$L_x L_y = A \tag{12.68}$$

then A is the area within which the particle is confined. Thus, the maximum allowed number of degenerate states is, writing $2\pi\hbar = h$,

$$N = \frac{A}{\left(\dfrac{ch}{eB_0}\right)}. \tag{12.69}$$

12.4 Quantum Hall effect

It is found in some materials described by a system of electrons in two dimensions, e.g., in the x–y plane, that if an electric field is applied along the x-direction and a magnetic field is applied along the z-direction then a current appears in the y-direction. This is the essence of the Hall effect, which was named after Edwin Hall who discovered it in 1879. The effect is described in terms of electrical conductivity which, as we will discuss below, is a tensor σ_{ij} represented by an off-diagonal 2×2 matrix.

In the integer quantum Hall effect the conductivity is found to be quantized. Its value is determined entirely by fundamental constants: the Planck constant h and the elementary charge e. It is found, remarkably, not to depend on the properties of the substance containing the electrons, e.g., the dielectric constant, the magnetic permeability, or even the size of the system.

To discuss this problem let us now return to the Landau gauge, which we considered earlier.

$$\mathbf{A} = (0, B_0 x, 0). \tag{12.70}$$

Since $\mathbf{A}$ is entirely in the y-direction, the velocity v_y is given by

$$v_y = \left(p_y - \frac{e}{c}A_y\right). \tag{12.71}$$

and the Hamiltonian is

$$H = \frac{p_x^2}{2m} + \frac{\left(p_y - \frac{e}{c}B_0 x\right)^2}{2m}. \tag{12.72}$$

We found earlier that the energy eigenvalues are given by

$$E_n = \left(n + \frac{1}{2}\right)\hbar\omega_c \tag{12.73}$$

with the ground state ($n = 0$) wavefunction, corresponding to the lowest Landau level (LLL), given by

$$u_0(x, y) = u_0(x)\frac{e^{ik_y y}}{\sqrt{L_y}} \tag{12.74}$$

where we have once again assumed the system to be confined within a rectangle of dimensions L_x, L_y. The wavefunction $u_0(x)$ corresponding to the ground state is given by

$$u_0(x) = C_0 e^{-\alpha(x - x_0)^2} \tag{12.75}$$

with

$$\omega_c = \frac{eB_0}{mc}, \quad p_y = \hbar k_y, \quad \text{and} \quad x_0 = \frac{c\hbar k_y}{eB_0} \tag{12.76}$$

and $0 < x_0 < L_x$.

The electromagnetic current of the particle is of the form

$$j_y \sim ev_y = \frac{e\hbar}{m}\left(k_y - \frac{eB_0}{c\hbar}x\right). \tag{12.77}$$

It is instructive to derive this form of the current directly from the time-dependent Schrödinger equation given by

$$-\frac{\hbar^2}{2m}\frac{\partial^2\psi}{\partial x^2} + \frac{1}{2m}\left[-i\hbar\frac{\partial}{\partial y} - \frac{eB_0 x}{c}\right]^2\psi = i\hbar\frac{\partial\psi}{\partial t}, \tag{12.78}$$

i.e.,

$$-\frac{\hbar^2}{2m}\frac{\partial^2\psi}{\partial x^2}-\frac{\hbar^2}{2m}\frac{\partial^2\psi}{\partial y^2}+i\frac{e\hbar}{mc}B_0x\frac{\partial\psi}{\partial y}+\frac{e^2B_0^2x^2}{2mc^2}\psi=i\hbar\frac{\partial\psi}{\partial t}. \tag{12.79}$$

The complex conjugate of (12.79) is

$$-\frac{\hbar^2}{2m}\frac{\partial^2\psi^*}{\partial x^2}-\frac{\hbar^2}{2m}\frac{\partial^2\psi^*}{\partial y^2}-i\frac{e\hbar}{mc}B_0x\frac{\partial\psi^*}{\partial y}+\frac{e^2B_0^2x^2}{2mc^2}\psi^*=-i\hbar\frac{\partial\psi^*}{\partial t}. \tag{12.80}$$

We multiply (12.79) by ψ^* and subtract from it (12.80) multiplied by ψ,

$$\begin{aligned}&-\frac{\hbar^2}{2m}\frac{\partial}{\partial x}\left[\psi^*\frac{\partial\psi}{\partial x}-\psi\frac{\partial\psi^*}{\partial x}\right]-\frac{\hbar^2}{2m}\frac{\partial}{\partial y}\left[\psi^*\frac{\partial\psi}{\partial y}-\psi\frac{\partial\psi^*}{\partial y}\right]\\&+\frac{ie\hbar}{mc}B_0x\left[\psi^*\frac{\partial\psi}{\partial y}+\psi\frac{\partial\psi^*}{\partial y}\right]=i\hbar\frac{\partial}{\partial t}\left(\psi^*\psi\right).\end{aligned} \tag{12.81}$$

If ψ is an eigenstate of energy for LLL (i.e., $n=0$) then

$$\psi=u_0(x,y)e^{-iEt/\hbar} \tag{12.82}$$

where $u_0(x,y)$ has been given in (12.74) and (12.75),

$$u_0\left(x,y\right)=\frac{C_0}{\sqrt{L_y}}e^{-\alpha(x-x_0)^2}e^{ik_yy}, \tag{12.83}$$

and the right-hand side of (12.81) vanishes. We write equation (12.81) in the form

$$\nabla\cdot\mathbf{j}=0 \tag{12.84}$$

where $\mathbf{j}$ is the current density vector. In terms of the components of $\mathbf{j}$ it can be written as

$$\frac{\partial j_x}{\partial x}+\frac{\partial j_y}{\partial y}=0 \tag{12.85}$$

where

$$j_x=\frac{\hbar}{2im}\left[u_0^*\frac{\partial u_0}{\partial x}-u_0\frac{\partial u_0^*}{\partial x}\right]. \tag{12.86}$$

Since u_0 is a real function (it is a bound state wavefunction), the right-hand side is zero. Thus the current density along the x-direction vanishes. To obtain j_y we can write the second and third terms in (12.81) as

$$\frac{\partial j_y}{\partial y}=-\frac{\hbar^2}{2m}\frac{\partial}{\partial y}\left[\left(u_0^*\frac{\partial u_0}{\partial y}-u_0\frac{\partial u_0^*}{\partial y}\right)-\frac{i2eB_0x}{\hbar c}u_0^*u_0\right]. \tag{12.87}$$

Therefore, j_y is given by

$$j_y=\frac{e\hbar}{2im}\left\{\left[u_0^*\frac{\partial u_0}{\partial y}-u_0\frac{\partial u_0^*}{\partial y}\right]-\frac{i2eB_0x}{\hbar c}u_0^*u_0\right\}, \tag{12.88}$$

which from (12.83) gives

$$j_y = \frac{e\hbar}{2im}\left[2ik_y - \frac{i2eB_0x}{c\hbar}\right]u_0^*u_0 \tag{12.89}$$

$$= \frac{e\hbar}{m}\left[k_y - \frac{eB_0x}{c\hbar}\right]\frac{|C_0|^2}{L_y}e^{-2\alpha(x-x_0)^2}. \tag{12.90}$$

We can then write j_y as

$$j_y = \frac{e^2B_0}{mc}\left[x_0 - x\right]\frac{|C_0|^2}{L_y}e^{-2\alpha(x-x_0)^2}, \tag{12.91}$$

where x_0 has been defined in (12.76).

The total current in the x-direction is an integral of j_y along the x-direction. However, since j_y is an odd function of x centered at $x = x_0$, the integral will vanish.

Let us now subject the charged particle to an electric field $\mathbf{E}_0$ in the x-direction. The new Hamiltonian is

$$H = \frac{p_x^2}{2m} + \frac{1}{2m}\left[p_y - \frac{eB_0x}{c}\right]^2 - eE_0x. \tag{12.92}$$

We change variables so that the new equation is

$$-\frac{\hbar^2}{2m}\frac{d^2u}{dx^2} + \frac{1}{2}m\omega_c^2\left[x - \bar{x}_0\right]^2 u = E'u \tag{12.93}$$

where

$$\bar{x}_0 = x_0 + \frac{eE_0}{m\omega_c^2} \tag{12.94}$$

$$= \frac{c\hbar k_y}{eB_0} + \frac{mc^2E_0}{eB_0^2}. \tag{12.95}$$

The current density is then of the form

$$j_y = \frac{e^2B_0}{mc}\left[x_0 - x\right]e^{-2\alpha(x-\bar{x}_0)^2}\frac{|C|^2}{L_y}. \tag{12.96}$$

The integration along the x-direction can now be carried out to give the current

$$I_x = \frac{eE_0}{B_0L_y}. \tag{12.97}$$

Since, as we found in (12.67), the number of states in the LLL is

$$N = \left(\frac{eB_0}{2\pi c\hbar}\right)L_xL_y, \tag{12.98}$$

the corresponding total current is given by $(I_x)_{\text{LLL}} = NI_x$. Hence,

$$(I_x)_{\text{LLL}} = \frac{e^2 E_0}{2\pi c\hbar} L_x = \frac{e^2}{ch} V_H \tag{12.99}$$

where $h = 2\pi\hbar$, and where V_H ($= E_0 L_x$) is the potential difference across the surface.

The above result is independent of the Landau level quantum number, n. This is so because the wavefunction u_n for the nth level is normalized and remains a function of $(x - x_0)$, hence the corresponding term $(x - x_0)u_n^* u_n$ in the expression for j_y is an odd function that vanishes upon integration over x.

Hence, for the case where ν Landau levels below the Fermi energy are filled we have

$$I_{\text{tot}} = \frac{\nu e^2}{ch} V_H. \tag{12.100}$$

Thus I_{tot} is quantized. This is the essence of the quantum Hall effect.

We note that while the electric field is in the x-direction, the current is in the y-direction, the relation between the current and electric field is, therefore, tensorial in character,

$$j_y = \sigma_{yx} E_x \tag{12.101}$$

where σ_{yx} is a tensor corresponding to electrical conductivity. Expression (12.100) is simply the integral of (12.96) along the x-axis since the integral of j_y along the x-axis gives I_{tot} and the integral of E_0 also along the x-axis gives V_H. Hence we conclude that

$$\sigma_{yx} = \frac{\nu e^2}{ch}, \quad \nu = 1, 2, \ldots. \tag{12.102}$$

This is the off-diagonal element of the Hall conductivity, which is quantized. The diagonal element σ_{xx} vanishes since there is no current in the direction of the applied field. The experimental determination of this conductivity, and the evidence of quantization, has been astonishingly precise, almost one part in 10^9. This effect also provides an incredibly precise measurement of the fine structure constant, $e^2/\hbar c$. There exists also a fractional quantum Hall effect in which one finds $\nu = 2/7, 1/3, 2/5 \ldots$.

12.5 Wavefunction for filled LLLs in a Fermi system

Suppose now that all the energy levels corresponding to LLLs are filled up to an energy that is less than or equal to the Fermi energy. The wavefunction is then the product of each of the wavefunctions. Since, according to the Pauli principle, no two states can occupy the same level, there can be only one electron per level. There is then only one way to construct a product of wavefunctions: the wavefunction for each level must pick a different value of q. Hence the wavefunction u_{LLL} for N filled levels will be

$$\left[z_1^0 z_2^1 \ldots z_N^{N-1}\right] \exp\left(-\frac{m\omega_c}{2\hbar} \sum_{i=1}^{N-1} z_i z_i^*\right) \tag{12.103}$$

where z_i's are the z-values of the individual levels. Since the total wavefunction must be an antisymmetric function, we need to obtain the Slater determinant

$$\det \begin{vmatrix} 1 & z_1 & . & . & z_1^{N-1} \\ 1 & z_2 & . & . & z_2^{N-1} \\ . & . & . & . & . \\ . & . & . & . & . \\ 1 & z_N & . & . & z_N^{N-1} \end{vmatrix} . \tag{12.104}$$

Since the above determinant vanishes for $z_i = z_j$, it must involve products of the form $(z_i - z_j)$. Thus, the determinant will be the following for different values of N:

$$N = 2: \ (z_2 - z_1), \tag{12.105}$$

$$N = 3: \ (z_2 - z_1)(z_3 - z_2)(z_1 - z_3), \tag{12.106}$$

and so on. One finds then that u_{LLL} will be given by

$$u_{\mathrm{LLL}} = \prod_{i<j}^{N} \left(z_i - z_j\right) \exp\left(-\frac{m\omega_c}{2\hbar} \sum_{k=1}^{N} |z_k|^2 \right). \tag{12.107}$$

The phenomenon described above corresponds to what is called the integer quantum Hall effect in which we have considered the case where all N levels are filled.

However, one may have a situation where not all levels are filled in which case the wavefunction cannot be represented by a unique single product but will be represented as a superposition of different states. An interesting phenomenon occurs if only

$$\frac{N}{2p+1} \tag{12.108}$$

states are occupied, where $p = 1, 2, \ldots$ ($p = 0$ corresponds to fully filled levels). This gives rise to the so called fractional quantum Hall effect where $1/\left(2p+1\right)$ is called the filling factor. The wavefunction for this system is conjectured by Laughlin to be

$$\prod_{i<j}^{N} \left(z_i - z_j\right)^{(2p+1)}, \tag{12.109}$$

which is found in many cases to be a good approximation to the actual ground-state wavefunction.

12.6 Problems

1. From expression (6.37) for the commutator $\left[P_i, P_j\right]$ converted to $\left[v_i, v_j\right]$ through the relation $\mathbf{v} = \mathbf{P}/m$, and the Hamiltonian given by $H = (1/2)\, m\mathbf{v}^2$, use the Heisenberg

equation of motion to show that if the magnetic field **B** is a constant then

$$\frac{d\mathbf{v}}{dt} = \frac{e}{mc}\left(\mathbf{v} \times \mathbf{B}\right).$$

Take $\mathbf{B} = B_0\mathbf{k}$ and derive the equations for v_x, v_y, v_z. Show that v_x and v_y satisfy the harmonic oscillator equations, while v_z satisfies the free particle equation.

2. Show that if **B** is not a constant then

$$\left[B_i, v_j\right] = \left(\frac{i\hbar}{m}\right)\frac{\partial B_i}{\partial x_j}.$$

From the Heisenberg equation of motion and the above relation show that

$$\frac{d\mathbf{v}}{dt} = \frac{e}{mc}\left(\mathbf{v} \times \mathbf{B}\right) + i\frac{e\hbar}{2m^2c}\nabla \times \mathbf{B}.$$

3. If $\mathbf{B} = B_0\mathbf{k}$ then for the gauge $\mathbf{A} = (-B_0 y, 0, 0)$ derive the formula for the Landau levels.

13 Two-level problems

Having obtained exact solutions for the one-level problems, we turn now to exact solutions in two-channel problems. Many problems, particularly those involving particles with spin ½, which we will discuss in the next chapter are, indeed, two-level systems. In fact, as we will see in this chapter and in Chapters 14 and 15 to follow, that two-channel problems combine to form perhaps the richest manifestation of the success of quantum mechanics in providing explanations of some far-reaching and puzzling problems.

13.1 Time-independent problems

13.1.1 Basic formalism

We assume that in a system of two levels, described in Fig. 13.1, the Hamiltonian, H_0, has eigenstates $\left|\psi_i^0\right\rangle$ with eigenvalues E_i^0,

$$H_0 \left|\psi_i^0\right\rangle = E_i^0 \left|\psi_i^0\right\rangle, \qquad \text{with } i = 1, 2 \tag{13.1}$$

with $\langle\psi_i^0|\psi_j^0\rangle = \delta_{ij}$, where the scalar product includes possible integrations over the space variables. Let us assume that the eigenstates and eigenvalues of H_0, the "unperturbed" Hamiltonian, are known. From the knowledge of these facts we need to determine the eigenvalues and eigenvectors of a more complicated Hamiltonian that has an additional, "perturbative," term, H', that does not depend on time:

$$H = H_0 + H' \tag{13.2}$$

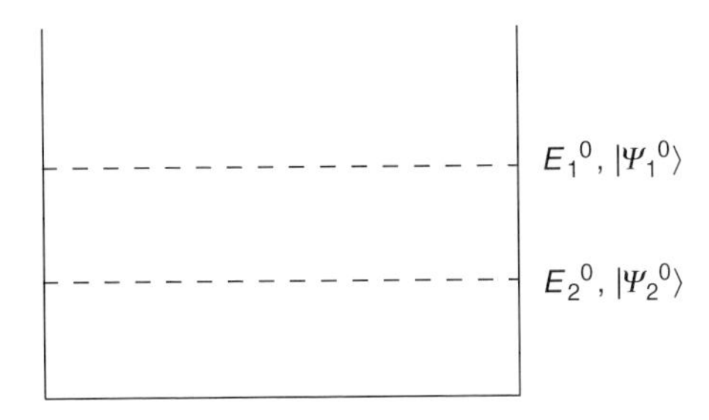

Fig. 13.1

where H_0 and H' are Hermitian. If $|\psi\rangle$ is an eigenstate of H then it will satisfy

$$H|\psi\rangle = E|\psi\rangle. \tag{13.3}$$

Our aim is to determine $|\psi\rangle$ and E. From our discussions in Chapter 1 it is clear that we will need to make a unitary transformation to change basis from $|\psi_i^0\rangle$ to $|\psi_i\rangle$. First we consider the problem more simply by using the superposition principle.

Since $|\psi_i^0\rangle$, with $i = 1, 2$, form a complete orthonormal set then, from the superposition principle, one can write

$$|\psi\rangle = c_1 |\psi_1^0\rangle + c_2 |\psi_2^0\rangle \tag{13.4}$$

where, the probability amplitudes, c_1 and c_2 are constants satisfying the relation

$$|c_1|^2 + |c_2|^2 = 1. \tag{13.4a}$$

To obtain $|\psi\rangle$ we need to obtain c_1 and c_2.

Substituting (13.4) and (13.2) in (13.3) we find

$$(H_0 + H')\left(c_1|\psi_1^0\rangle + c_2|\psi_2^0\rangle\right) = E\left(c_1|\psi_1^0\rangle + c_2|\psi_2^0\rangle\right). \tag{13.5}$$

We define the matrix elements of H' as

$$\langle\psi_i^0|H'|\psi_j^0\rangle = H'_{ij} \tag{13.6}$$

and note from the Hermiticity condition of H' (i.e., $H'^{\dagger} = H'$) that

$$H'_{ji} = \langle\psi_j^0|H'|\psi_i^0\rangle = \langle\psi_j^0|H'^{\dagger}|\psi_i^0\rangle = \langle\psi_i^0|H'|\psi_j^0\rangle^* = H'^*_{ij}. \tag{13.7}$$

Since H_0, H', and $|\psi_i^0\rangle$ are presumed to be known, these matrix elements are calculable.

After multiplying both sides of (13.5) by $\langle\psi_1^0|$ and by $\langle\psi_2^0|$ and using the orthonormality of the two states, we obtain two simultaneous equations for the coefficients c_1 and c_2:

$$c_1(E_1^0 - E + H'_{11}) + c_2 H'_{12} = 0, \tag{13.8}$$

$$c_1 H'_{21} + c_2(E_2^0 - E + H'_{22}) = 0. \tag{13.9}$$

The determinant formed by the coefficients of c_1 and c_2 must, therefore, vanish,

$$\left\| \begin{bmatrix} (E_1^0 - E + H'_{11}) & H'_{12} \\ H'_{21} & (E_2^0 - E + H'_{22}) \end{bmatrix} \right\| = 0 \tag{13.10}$$

which leads to the following quadratic equation in E:

$$\left(E_1^0 - E + H'_{11}\right)\left(E_2^0 - E + H'_{22}\right) - |H'_{12}|^2 = 0 \tag{13.11}$$

where we have used the property (13.7). In the following, for the purposes of simplification, we consider H'_{12} to be real, and designate

$$A = E_1^0 + H'_{11}, \quad B = E_2^0 + H'_{22}, \quad D = H'_{12} \quad \text{and} \quad \Delta = A - B. \tag{13.12}$$

Equation (13.11) is then found to be

$$E^2 - (A + B)E + AB - D^2 = 0. \tag{13.13}$$

The solutions of this equation are given by

$$E_+ = \frac{A + B + \sqrt{\Delta^2 + 4D^2}}{2}, \tag{13.14}$$

$$E_- = \frac{A + B - \sqrt{\Delta^2 + 4D^2}}{2}. \tag{13.15}$$

Since, by definition, the square-roots are positive, i.e., $\sqrt{\alpha^2} = +|\alpha|$, we have

$$E_+ > E_-. \tag{13.16}$$

For each solution of E we obtain the corresponding values of c_1 and c_2 from either of the equations (13.8) or (13.9) supplemented by the normalization condition (13.4a). We will then have two orthonormal solutions, $|\psi_+\rangle$ and $|\psi_-\rangle$ given by

$$H|\psi_+\rangle = E_+|\psi_+\rangle, \tag{13.17}$$

$$H|\psi_-\rangle = E_-|\psi_-\rangle. \tag{13.18}$$

13.1.2 Simple example

For the purposes of illustration we consider a particularly simple example where

$$H'_{11} = 0 = H'_{22} \quad \text{and} \quad E_1^0 = 0 = E_2^0, \tag{13.19}$$

which imply $A = 0$, $B = 0$, and $\Delta = 0$. The equations (13.8) and (13.9) now become

$$-c_1E + c_2H'_{12} = 0, \tag{13.20}$$

$$c_1H'_{21} - c_2E = 0. \tag{13.21}$$

And the quadratic equation (13.11) is found to be

$$E^2 - |H'_{12}|^2 = 0. \tag{13.22}$$

This gives two simple eigenvalues:

$$E_+ = +|H'_{12}| \quad \text{and} \quad E_- = -|H'_{12}|. \tag{13.23}$$

From (13.20) we note that for the eigenvalue E_+ we have $c_1/c_2 = 1$, while for E_- we have $c_1/c_2 = -1$ (note that because of the determinant condition (13.22), the equation (13.21) will give the same answers). Incorporating the normalization condition (13.4a) we find $c_1 = c_2 = 1/\sqrt{2}$ and

$$|\psi_+\rangle = \frac{|\psi_1^0\rangle + |\psi_2^0\rangle}{\sqrt{2}} \quad \text{with eigenvalue } E_+ = +\left|H'_{12}\right|. \tag{13.24}$$

For the second solution, we take, $c_1 = -c_2 = -1/\sqrt{2}$ and

$$|\psi_-\rangle = \frac{-|\psi_1^0\rangle + |\psi_2^0\rangle}{\sqrt{2}} \quad \text{with eigenvalue } E_- = -\left|H'_{12}\right|. \tag{13.25}$$

As expected, $|\psi_+\rangle$ and $|\psi_-\rangle$ satisfy the orthonormality condition.

13.1.3 Formalism in terms of mixing angles

Returning to a general two-level problem, we could simplify the equations by expressing c_1 and c_2 in a particularly convenient form, in terms of cosine and sine functions, which makes the orthonormality of the state vectors manifest. We express (13.4) for the two eigenstates as

$$|\psi_1\rangle = \cos\theta \left|\psi_1^0\right\rangle + \sin\theta \left|\psi_2^0\right\rangle \quad \text{and} \quad |\psi_2\rangle = -\sin\theta \left|\psi_1^0\right\rangle + \cos\theta \left|\psi_2^0\right\rangle \tag{13.26}$$

where we have assumed that for $\theta = 0$ there is no interaction and $|\psi_i\rangle$ coincide with $|\psi_i^0\rangle$. This equation can be written in a compact matrix form as

$$\begin{bmatrix} |\psi_1\rangle \\ |\psi_2\rangle \end{bmatrix} = \begin{bmatrix} \cos\theta & \sin\theta \\ -\sin\theta & \cos\theta \end{bmatrix} \begin{bmatrix} |\psi_1^0\rangle \\ |\psi_2^0\rangle \end{bmatrix}. \tag{13.27}$$

The above matrix equation can be expressed as

$$|\psi\rangle = R(\theta)\left|\psi^0\right\rangle \tag{13.28}$$

where $|\psi\rangle$ and $|\psi^0\rangle$ are column matrices appearing in (13.27) and the rotation matrix $R(\theta)$ is given by

$$R(\theta) = \begin{bmatrix} \cos\theta & \sin\theta \\ -\sin\theta & \cos\theta \end{bmatrix}. \tag{13.29}$$

This implies that we are rotating the basis from the unperturbed states to the perturbed states in order to solve our eigenvalue problem. Since $R^\dagger R = \mathbf{1}$, the rotation preserves the norms of the state vectors, $\langle\psi|\psi\rangle = \langle\psi^0|\psi^0\rangle$. This is an example of the unitary transformation that was discussed in Chapter 1.

The eigenvalue equation (13.3) can be written as $H\left|\psi_j\right\rangle = E_j\left|\psi_j\right\rangle$ and, therefore, we can write the following relation for the matrix elements of H using orthonormality of the eigenstates,

$$\left\langle\psi_i\right| H \left|\psi_j\right\rangle = E_j\delta_{ij}. \tag{13.30}$$

This implies that H is diagonal in the $\left|\psi_i\right\rangle$ basis and its diagonal values are E_i. Since the matrix elements $\left\langle\psi_i^0\right| H \left|\psi_j^0\right\rangle$ are calculable we need to express the left-hand side of the above equation in terms of it. This can be accomplished using completeness.

We first introduce a complete set of states in (13.30),

$$\sum_{l,m}\left\langle\psi_i\middle|\psi_l^0\right\rangle\left\langle\psi_l^0\right| H \left|\psi_m^0\right\rangle\left\langle\psi_m^0\middle|\psi_j\right\rangle = E_i\delta_{ij} \tag{13.31}$$

where $\left\langle\psi_l^0\right| H \left|\psi_m^0\right\rangle$ corresponds to the matrix elements of H with respect to $\left|\psi_i^0\right\rangle$ as the basis. We can write this matrix as

$$\left\{\left\langle\psi_l^0\right| H \left|\psi_m^0\right\rangle\right\} = \begin{bmatrix} E_1^0 + H'_{11} & H'_{12} \\ H'_{21} & E_2^0 + H'_{22} \end{bmatrix}. \tag{13.32}$$

Using the notations A, B, and D defined in (13.12), and also assuming D (i.e., H'_{12}) to be real, the above matrix can be expressed as

$$\left\{\left\langle\psi_l^0\right| H \left|\psi_m^0\right\rangle\right\} = \begin{bmatrix} A & D \\ D & B \end{bmatrix}. \tag{13.33}$$

The matrix relation (13.31) can then be written as

$$\begin{bmatrix} \cos\theta & \sin\theta \\ -\sin\theta & \cos\theta \end{bmatrix}\begin{bmatrix} A & D \\ D & B \end{bmatrix}\begin{bmatrix} \cos\theta & -\sin\theta \\ \sin\theta & \cos\theta \end{bmatrix} = \begin{bmatrix} E_1 & 0 \\ 0 & E_2 \end{bmatrix}. \tag{13.34}$$

The product of the matrices on the left-hand side above can be carried out to obtain

$$\begin{bmatrix} A\cos^2\theta + 2D\sin\theta\cos\theta + B\sin^2\theta & -\left(A-B\right)\sin\theta\cos\theta + D(\cos^2\theta - \sin^2\theta)\right) \\ -\left(A-B\right)\sin\theta\cos\theta + D(\cos^2\theta - \sin^2\theta)\right) & A\sin^2\theta - 2D\sin\theta\cos\theta + B\cos^2\theta \end{bmatrix}. \tag{13.35}$$

Since the right-hand side of (13.35) is diagonal, on the left-hand side we must have

$$-(A-B)\sin\theta\cos\theta + D(\cos^2\theta - \sin^2\theta) = 0, \tag{13.36}$$

which will give us a relation for the rotation angle, θ. This relation, in terms of $\tan 2\theta$, is given by

$$\tan 2\theta = \frac{2D}{A-B} = \frac{2D}{\Delta} \tag{13.37}$$

where $\Delta = A - B$, as defined in (13.12). One can also determine $\tan\theta$ from (13.37),

$$\tan\theta = \frac{-\Delta + \sqrt{\Delta^2 + 4D^2}}{2D} \tag{13.38}$$

where the "+" sign is chosen so that for $\Delta > 0$, and $D > 0$ we will have $\theta = 0$ in the limit when $D \to 0$. This can be shown by expanding $\sqrt{\Delta^2 + 4D^2}$, which for small values of D gives

$$\tan\theta \to \frac{D}{\Delta} \to 0. \tag{13.39}$$

Once the mixing angle θ is calculated, one can express $|\psi_1\rangle$ and $|\psi_2\rangle$ in terms of $|\psi_1^0\rangle$ and $|\psi_2^0\rangle$, given by

$$|\psi_1\rangle = \cos\theta\left|\psi_1^0\right\rangle + \sin\theta\left|\psi_2^0\right\rangle, \tag{13.40}$$

$$|\psi_2\rangle = -\sin\theta\left|\psi_1^0\right\rangle + \cos\theta\left|\psi_2^0\right\rangle. \tag{13.41}$$

We can also now obtain the energy eigenvalues,

$$E_1 = A\cos^2\theta + 2D\sin\theta\cos\theta + B\sin^2\theta, \tag{13.42}$$

$$E_2 = A\sin^2\theta - 2D\sin\theta\cos\theta + B\cos^2\theta$$

and determine E_+ and E_-.

We note that if $E_1 > E_2$ then $(E_+, E_-) = (E_1, E_2)$ and $\left(|\psi_+\rangle, |\psi_-\rangle\right) = \left(|\psi_1\rangle, |\psi_2\rangle\right)$, while if $E_1 < E_2$ then relations are changed accordingly. This connection will be discussed further in Section 13.1.5.

For the example considered earlier, $H'_{11} = 0 = H'_{22}$, and $E_1^0 = 0 = E_2^0$, assuming H'_{12} to be real, we have

$$A = 0 = B = \Delta \quad \text{and} \quad D = \left|H'_{12}\right|. \tag{13.43}$$

Thus, from (13.37) and (13.38) we obtain

$$\tan 2\theta = +\infty \quad \text{and} \quad \tan\theta = +1.$$

Hence the rotation angle is found to be

$$\theta = \frac{\pi}{4}. \tag{13.44}$$

The eigenvalues and eigenfunctions are then given by

$$E_+ = E_1 = \left|H'_{12}\right|, \tag{13.45}$$

$$E_- = E_2 = -\left|H'_{12}\right| \tag{13.46}$$

and

$$|\psi_+\rangle = \frac{|\psi_1^0\rangle + |\psi_2^0\rangle}{\sqrt{2}}, \tag{13.47}$$

$$|\psi_-\rangle = \frac{-|\psi_1^0\rangle + |\psi_2^0\rangle}{\sqrt{2}}. \tag{13.48}$$

Below we consider some further examples using (13.37) and (13.38).

13.1.4 More examples

Example 1

$$\left\{\left\langle \psi_i^0 \middle| H \middle| \psi_j^0 \right\rangle\right\} = \begin{bmatrix} A & D \\ D & A \end{bmatrix}. \tag{13.49}$$

This is the degenerate case with $A = B$ and, therefore, $\Delta = 0$. The eigenvalues are given by

$$E_+ = A + |D|, \tag{13.50}$$

$$E_- = A - |D| \tag{13.51}$$

where we have written $\sqrt{D^2} = |D|$. The mixing angle is obtained as

$$\tan 2\theta = \infty, \tag{13.52}$$

$$\tan\theta = \frac{|D|}{D}. \tag{13.53}$$

(i) For $D > 0$

$$\tan\theta = +1 \quad \text{and} \quad \theta = \frac{\pi}{4}, \tag{13.54}$$

$$|\psi_+\rangle = \frac{1}{\sqrt{2}}\left[\left|\psi_1^0\right\rangle + \left|\psi_2^0\right\rangle\right] \quad \text{and} \quad |\psi_-\rangle = \frac{1}{\sqrt{2}}\left[-\left|\psi_1^0\right\rangle + \left|\psi_2^0\right\rangle\right]. \tag{13.55}$$

(ii) For $D < 0$, i.e., $D = -|D|$

$$\tan\theta = -1 \quad \text{and} \quad \theta = \frac{3\pi}{4}, \tag{13.56}$$

$$|\psi_+\rangle = \frac{1}{\sqrt{2}}\left[-\left|\psi_1^0\right\rangle + \left|\psi_2^0\right\rangle\right] \quad \text{and} \quad |\psi_-\rangle = -\frac{1}{\sqrt{2}}\left[\left|\psi_1^0\right\rangle + \left|\psi_2^0\right\rangle\right]. \tag{13.57}$$

But the energy eigenvalues remain unchanged:

$$E_+ = A + |D|, \tag{13.58}$$

$$E_- = A - |D|. \tag{13.59}$$

Example 2

$$\left\{\left\langle\psi_i^0\right|H\left|\psi_j^0\right\rangle\right\} = \begin{bmatrix} 0 & D \\ D & B \end{bmatrix} \tag{13.60}$$

with $D \ll B$. We note that $\Delta = -B$ and

$$E_+ = \frac{B + \sqrt{B^2 + 4D^2}}{2} \approx \frac{B + B + \frac{2D^2}{B}}{2} \approx B, \tag{13.61}$$

$$E_- = -\frac{D^2}{B} \tag{13.62}$$

where we have expanded the square root in (13.61) since $D \ll B$. The mixing angle is given by

$$\tan 2\theta = -\frac{2D}{B}, \tag{13.63}$$

$$\tan\theta = \frac{B + \sqrt{B^2 + 4D^2}}{2D} \approx \frac{B}{D} \gg 1. \tag{13.64}$$

Here we have a case where $\tan 2\theta$ is small and negative while $\tan\theta$ is large and positive. Thus θ must be close to but slightly less than $\pi/2$. The state vectors are then

$$\left|\psi_+\right\rangle \simeq \left|\psi_2^0\right\rangle \quad \text{and} \quad \left|\psi_-\right\rangle \simeq -\left|\psi_1^0\right\rangle. \tag{13.65}$$

An interesting outcome of the condition $D \ll B$ is that the mixing angle can be simply related to the eigenvalues; for example, if we write

$$\theta = \pi/2 - \epsilon \tag{13.66}$$

then the relation for $\tan 2\theta$ yields $\tan 2\epsilon = 2D/B$. However, ϵ is very small since $D \ll B$, so that one can write $\tan 2\epsilon \approx 2\epsilon$. We then obtain

$$\epsilon = \frac{D}{B} = \sqrt{\frac{|E_-|}{E_+}}. \tag{13.67}$$

This result has some interesting implications in quark physics and for the neutrinos. In the relativistic context the energy E of a particle is proportional to mc^2, where m is the mass of the particle. If we relate the smaller of the two energy values, E_-, to the particle with a smaller mass, m_1, and the larger, E_+, to m_2, then

$$\epsilon = \sqrt{\frac{m_1}{m_2}} \tag{13.68}$$

This is a very well-known result in neutrino physics, where it goes under the name "see-saw mechanism."

Further examples are considered in the Problems in Section 13.3.

13.1.5 Level crossing and switching of eigenstates

Let us consider the energy levels $E_\pm$ and their trajectories for arbitrary values of A and B in the limit $\Delta \to \pm\infty$, as illustrated in Fig. 13.2. The expressions for $E_\pm$ are given by

$$E_+ = \frac{A + B + \sqrt{\Delta^2 + 4D^2}}{2}, \tag{13.69}$$

$$E_- = \frac{A + B - \sqrt{\Delta^2 + 4D^2}}{2} \tag{13.70}$$

where $\Delta = A - B$. Consider first the case when $D = 0$, and

$$E_+ = \frac{A + B + |\Delta|}{2}, \tag{13.71}$$

$$E_- = \frac{A + B - |\Delta|}{2}. \tag{13.72}$$

In this case there is no coupling between the two levels since the matrix element $D \, (= H'_{12})$ that connects the two channels is zero. The trajectories for $E_\pm$ as $\Delta \to \pm\infty$ are given by the dotted lines in Fig. 13.2. These are straight lines which intersect at a point where $\Delta = 0$ (i.e., $A = B$), where $E_+ = E_-$

However when $D \neq 0$ these levels "repel" each other. We find that at the point of closest separation, when $\Delta = 0$, we have $E_+ - E_- = 2\,|D|$.

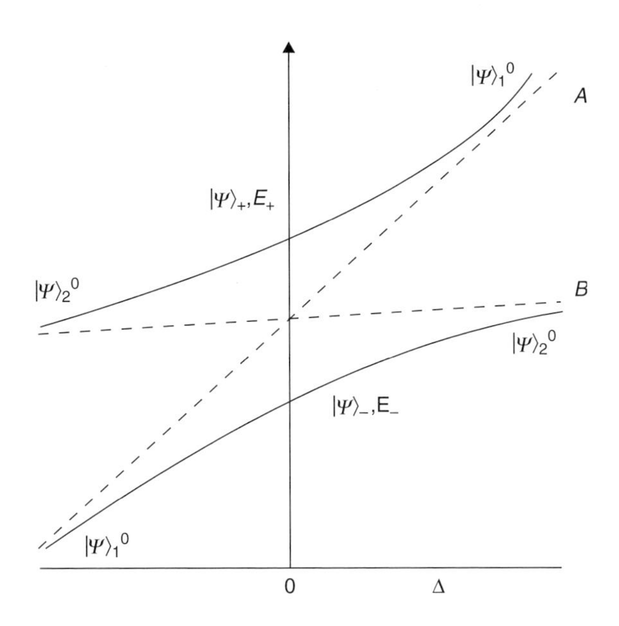

Fig. 13.2

For a fixed value of D, we now examine the behavior of $E_\pm$ and $|\psi_\pm\rangle$ as $\Delta \to +\infty$, and as $\Delta \to -\infty$. The trajectories are given by the solid lines in Fig.13.2. We note the following:

(i) In the limit $\Delta \to +\infty$, since $\Delta = A - B$, we must have $A > B$ and, since D is finite and much less than Δ, $|\Delta| = A - B$, and

$$E_+ = \frac{A + B + |\Delta|}{2} = A, \tag{13.73}$$

$$E_- = \frac{A + B - |\Delta|}{2} = B. \tag{13.74}$$

The mixing angle then has the behavior

$$\tan\theta = \frac{-|\Delta| + \sqrt{\Delta^2 + 4D^2}}{2D} \approx \frac{|D|}{|\Delta|} \to 0 \tag{13.75}$$

since $|D| \ll |\Delta|$. Therefore, in this limit, $\theta \to 0$ and the wavefunctions have the behavior

$$|\psi_+\rangle \to |\psi_1^0\rangle \quad \text{and} \quad |\psi_-\rangle \to |\psi_2^0\rangle. \tag{13.76}$$

(ii) In the limit $\Delta \to -\infty$, we must have $A < B$ and thus $|\Delta| = B - A$, and

$$E_+ = \frac{A + B + |\Delta|}{2} = B, \tag{13.77}$$

$$E_- = \frac{A + B - |\Delta|}{2} = A. \tag{13.78}$$

The mixing angle then has the behavior

$$\tan\theta = \frac{|\Delta| + \sqrt{\Delta^2 + 4D^2}}{2D} \approx \frac{|\Delta|}{D} \to \infty. \tag{13.79}$$

Therefore, $\theta \to \pi/2$, with the wavefunctions given by

$$|\psi_+\rangle \to |\psi_2^0\rangle \quad \text{and} \quad |\psi_-\rangle \to -|\psi_1^0\rangle. \tag{13.80}$$

Thus, we have a rather remarkable situation that as Δ varies from $+\infty$ to $-\infty$, the eigenstates "switch" sides:

$$|\psi_+\rangle \to |\psi_1^0\rangle \quad \text{as } \Delta \to +\infty \tag{13.81}$$

$$\to |\psi_2^0\rangle \quad \text{as } \Delta \to -\infty \tag{13.82}$$

and

$$|\psi_-\rangle \to |\psi_2^0\rangle \quad \text{as } \Delta \to +\infty \tag{13.83}$$

$$\to -|\psi_1^0\rangle \quad \text{as } \Delta \to -\infty. \tag{13.84}$$

These results have some very interesting implications in important physical phenomena such as the solar neutrino puzzle, which will be discussed in Chapter 15.

13.1.6 Relation between energy eigenstates and eigenvalues in two different frameworks

As we mentioned earlier, for $E_+ > E_-$ one must identify E_+ with the larger of the two eigenvalues E_1, E_2, and E_- with the smaller. We will make this connection a bit more formal in the following way.

We start with the following product involving the energy eigenstates $|\psi_i\rangle$ of the total Hamiltonian H, with the eigenvalues E_1 and E_2,

$$\left[\langle\psi_1| \quad \langle\psi_2|\right]\begin{bmatrix} E_1 & 0 \\ 0 & E_2 \end{bmatrix}\begin{bmatrix} |\psi_1\rangle \\ |\psi_2\rangle \end{bmatrix}. \tag{13.85}$$

Let us now express $|\psi_i\rangle$'s in terms of $|\psi_i^0\rangle$'s using (13.27). The above product is then given by

$$\left[\langle\psi_1^0| \quad \langle\psi_2^0|\right]\begin{bmatrix} \cos\theta & -\sin\theta \\ \sin\theta & \cos\theta \end{bmatrix}\begin{bmatrix} E_1 & 0 \\ 0 & E_2 \end{bmatrix}\begin{bmatrix} \cos\theta & \sin\theta \\ -\sin\theta & \cos\theta \end{bmatrix}\begin{bmatrix} |\psi_1^0\rangle \\ |\psi_2^0\rangle \end{bmatrix}. \tag{13.86}$$

The product of the middle three matrices can be expressed as

$$\frac{1}{2}\begin{bmatrix} E_1 + E_2 - \delta\cos 2\theta & -\delta\sin 2\theta \\ -\delta\sin 2\theta & E_1 + E_2 + \delta\cos 2\theta \end{bmatrix} \tag{13.87}$$

where $\delta = E_2 - E_1$.

The eigenvalues of the above are what we have defined as E_+ and E_-. Even though all this looks like a roundabout way of getting the same results that we started with, it provides an important comparison between $E_\pm$ and $E_{1,2}$. The eigenvalues then are

$$E_\pm = \frac{1}{2}\left(E_1 + E_2 \pm \sqrt{\delta^2\cos^2 2\theta + \delta^2\sin^2 2\theta}\right) = \frac{1}{2}\left(E_1 + E_2 \pm |\delta|\right) \tag{13.88}$$

where we have written $\sqrt{\delta^2\cos^2 2\theta + \delta^2\sin^2 2\theta} = |\delta|$.

Let us now consider two cases.

(i) $E_1 > E_2$, in which case $|\delta| = E_1 - E_2$. We then have

$$E_+ = E_1, \; E_- = E_2, \tag{13.89}$$

$$|\psi_+\rangle = |\psi_1\rangle, \; |\psi_-\rangle = |\psi_2\rangle, \tag{13.90}$$

and

$$\left[\langle\psi_1| \quad \langle\psi_2|\right]\begin{bmatrix} E_1 & 0 \\ 0 & E_2 \end{bmatrix}\begin{bmatrix} |\psi_1\rangle \\ |\psi_2\rangle \end{bmatrix} = \left[\langle\psi_+| \quad \langle\psi_-|\right]\begin{bmatrix} E_+ & 0 \\ 0 & E_- \end{bmatrix}\begin{bmatrix} |\psi_+\rangle \\ |\psi_-\rangle \end{bmatrix}. \tag{13.91}$$

(ii) $E_2 > E_1$, in which case $|\delta| = E_2 - E_1$. We then have

$$E_+ = E_2,\ E_- = E_1, \tag{13.92}$$

$$|\psi_+\rangle = |\psi_2\rangle,\ |\psi_-\rangle = |\psi_1\rangle, \tag{13.93}$$

and

$$\begin{bmatrix}\langle\psi_1| & \langle\psi_2|\end{bmatrix}\begin{bmatrix}E_1 & 0\\ 0 & E_2\end{bmatrix}\begin{bmatrix}|\psi_1\rangle\\ |\psi_2\rangle\end{bmatrix} = \begin{bmatrix}\langle\psi_-| & \langle\psi_+|\end{bmatrix}\begin{bmatrix}E_- & 0\\ 0 & E_+\end{bmatrix}\begin{bmatrix}|\psi_-\rangle\\ |\psi_+\rangle\end{bmatrix}. \tag{13.94}$$

These results provide important information on the relation between the two basis sets and their eigenvalues. They will be useful when we discuss the solar neutrino puzzle in Chapter 15.

13.2 Time-dependent problems

13.2.1 Basic formalism

Once again, we start with an unperturbed Hamiltonian, H_0, for which the eigenstates and eigenvalues are known with a perturbing Hamiltonian, $H'(t)$, which is now time-dependent. The total Hamiltonian is then given by

$$H = H_0 + H'(t) \tag{13.95}$$

with the state vectors $|\phi(t)\rangle$, which are time-dependent. The time-evolution equation of $|\phi(t)\rangle$ is given by

$$i\hbar\frac{\partial}{\partial t}|\phi(t)\rangle = H|\phi(t)\rangle. \tag{13.96}$$

From the superposition principle we know that if $|\phi_1(t)\rangle$ and $|\phi_2(t)\rangle$ are two orthonormal state vectors in the two-level Hilbert space, then another state vector $|\phi(t)\rangle$ can be expressed in terms of these as

$$|\phi(t)\rangle = c_1|\phi_1(t)\rangle + c_2|\phi_2(t)\rangle. \tag{13.97}$$

The coefficients, c_1 and c_2, will be functions of time depending on the time-evolution equations that the three state vectors satisfy. If all three satisfy the same equation, then it can easily be shown that c_1 and c_2 will be independent of time. On the other hand, if, for example, $|\phi_1(t)\rangle$ and $|\phi_2(t)\rangle$ satisfy the time-evolution equation with H_0 as the Hamiltonian, while $|\phi(t)\rangle$ satisfies the equation with H as the Hamiltonian, then c_1 and c_2 will depend on time.

We will elaborate on the above comments by considering the following problem.

Let $\left|\psi_i^0(t)\right\rangle$ with $i = 1, 2$ be the eigenstates with eigenvalues E_i^0. They will satisfy the time-dependent relation

$$i\hbar\frac{\partial}{\partial t}\left|\psi_i^0\right\rangle = H_0\left|\psi_i^0\right\rangle = E_i^0\left|\psi_i^0\right\rangle. \tag{13.98}$$

The solution is given by

$$\left|\psi_i^0(t)\right\rangle = \left|\psi_i^0(0)\right\rangle e^{-\frac{iE_i^0 t}{\hbar}}. \tag{13.99}$$

We note that $\left|\psi_i^0(t)\right\rangle$ are orthonormal, if, as we will assume in the following, $\left|\psi_i^0(0)\right\rangle$ are also orthonormal.

The equation satisfied by a general state vector $|\psi(t)\rangle$ is then

$$i\hbar\frac{\partial}{\partial t}|\psi(t)\rangle = H|\psi(t)\rangle = \left(H_0 + H'(t)\right)|\psi(t)\rangle. \tag{13.100}$$

The matrix representation of the Hamiltonian with $\left|\psi_1^0(t)\right\rangle$ and $\left|\psi_2^0(t)\right\rangle$ as the basis will be

$$\left\langle\psi_i^0(t)\right|H\left|\psi_j^0(t)\right\rangle = \begin{bmatrix} E_1^0 + H'_{11}(t) & H'_{12}(t) \\ H'_{21}(t) & E_2^0 + H'_{22}(t) \end{bmatrix}. \tag{13.101}$$

We express $|\psi(t)\rangle$ as a superposition of $\left|\psi_i^0(t)\right\rangle$,

$$|\psi(t)\rangle = c_1(t)\left|\psi_1^0(t)\right\rangle + c_2(t)\left|\psi_2^0(t)\right\rangle \tag{13.102}$$

where

$$|c_1(t)|^2 + |c_2(t)|^2 = 1. \tag{13.103}$$

We point out that $\left|\psi_1^0(t)\right\rangle$ and $\left|\psi_2^0(t)\right\rangle$ satisfy the equation (13.98), which is different from the equation (13.100) for $|\psi(t)\rangle$, and thus c_1 and c_2 will be functions of t. Substituting this expression into (13.100) we obtain

$$i\hbar\frac{\partial}{\partial t}\left(c_1(t)\left|\psi_1^0(t)\right\rangle + c_2(t)\left|\psi_2^0(t)\right\rangle\right) = \left(H_0 + H'(t)\right)\left(c_1(t)\left|\psi_1^0(t)\right\rangle + c_2(t)\left|\psi_2^0(t)\right\rangle\right). \tag{13.104}$$

Using (13.98) for $\left|\psi_i^0(t)\right\rangle$ on the left-hand side we note that the derivative terms $\left(\partial\left|\psi_i^0(t)\right\rangle/\partial t\right)$ cancel the contribution of H_0 on the right-hand side. Equating the leftover terms we obtain

$$i\hbar\left(\dot{c}_1\left|\psi_1^0(t)\right\rangle + \dot{c}_2\left|\psi_2^0(t)\right\rangle\right) = c_1(t)H'(t)\left|\psi_1^0(t)\right\rangle + c_2(t)H'(t)\left|\psi_2^0(t)\right\rangle \tag{13.105}$$

where

$$\dot{c}_i = \frac{dc_i}{dt}. \tag{13.106}$$

Multiplying both sides of (13.105), consecutively by $\langle \psi_1^0(t)|$ and $\langle \psi_2^0(t)|$, respectively, we obtain two equations,

$$i\hbar\dot{c}_1 = c_1 H'_{11}(t) + c_2 H'_{12}(t)\exp(i\omega_{12}t), \tag{13.107}$$

$$i\hbar\dot{c}_2 = c_1 H'_{21}(t)\exp(-i\omega_{12}t) + c_2 H'_{22}(t), \tag{13.108}$$

where we have defined

$$H'_{ij}(t) = \left\langle \psi_i^0(0)\middle| H'(t) \middle|\psi_j^0(0)\right\rangle \quad \text{and} \quad \omega_{12} = \frac{\left(E_1^0 - E_2^0\right)}{\hbar}. \tag{13.109}$$

In the following sections we will discuss the solutions of (13.107) and (13.108) for different types of $H'_{ij}(t)$.

13.2.2 Constant perturbation

Let us consider the above equations for the case where the perturbation $H'(t)$ is applied at $t = 0$ and is a constant for $t \geq 0$, as in Fig. 13.3. In particular, we take

$$H'_{12}(t) = H'_{21}(t) = \text{constant} = \hbar\gamma \qquad \text{for } t \geq 0, \tag{13.110}$$

$$H'_{12}(t) = H'_{21}(t) = 0 \qquad \text{for } t < 0. \tag{13.111}$$

We first consider an example where

$$H'_{11} = H'_{22} = 0, \;\; E_2^0 = E_1^0 = E_0. \tag{13.112}$$

The matrix representation of H will be

$$\left\langle \psi_i^0(t)\middle| H \middle|\psi_j^0(t)\right\rangle = \begin{bmatrix} E_0 & \hbar\gamma \\ \hbar\gamma & E_0 \end{bmatrix}. \tag{13.113}$$

Here we find that $\omega_{12} = 0$. From (13.107) and (13.108) we obtain the following coupled equations:

$$i\dot{c}_1 = \gamma c_2, \tag{13.114}$$

$$i\dot{c}_2 = \gamma c_1. \tag{13.115}$$

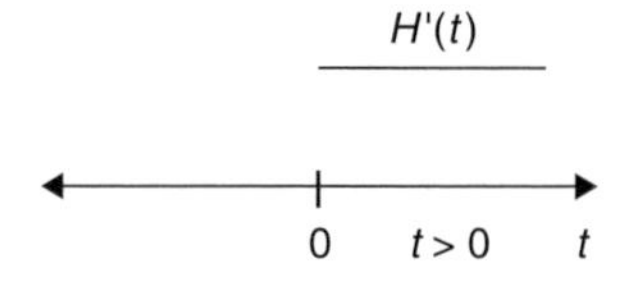

Fig. 13.3

Taking derivatives on both sides of (13.114) we obtain, using (13.115),

$$\ddot{c}_1 = -\gamma^2 c_1, \tag{13.116}$$

whose solution is

$$c_1 = A\cos\gamma t + B\sin\gamma t. \tag{13.117}$$

Similarly, we obtain

$$\ddot{c}_2 = -\gamma^2 c_2 \tag{13.118}$$

with solution

$$c_2 = C\cos\gamma t + D\sin\gamma t. \tag{13.119}$$

We assume the following conditions at $t = 0$,

$$c_1(0) = 1 \quad \text{and} \quad c_2(0) = 0. \tag{13.120}$$

With this initial condition, it is easy to show that we must have

$$c_1 = \cos\gamma t \quad \text{and} \quad c_2 = -i\sin\gamma t. \tag{13.121}$$

We obtain

$$|\psi(t)\rangle = \cos\gamma t\left|\psi_1^0(t)\right\rangle - i\sin\gamma t\left|\psi_2^0(t)\right\rangle. \tag{13.122}$$

We thus see that $|\psi(t)\rangle$, which at $t = 0$ is given entirely by $\left|\psi_1^0(t)\right\rangle$, will oscillate between the two states for $t > 0$ and, in fact, for times t such that γt is an odd integral multiple of $\pi/2$ it will be given entirely by $\left|\psi_2^0(t)\right\rangle$.

The probability of finding the state $|\psi(t)\rangle$ to be in $\left|\psi_i^0(0)\right\rangle$ is then given by $\left|\langle\psi_i^0(0)|\psi(t)\rangle\right|^2 = |c_i|^2$. From the relation (13.121) we find

$$\left|\langle\psi_1^0(0)|\psi(t)\rangle\right|^2 = \cos^2\gamma t, \quad \left|\langle\psi_2^0(0)|\psi(t)\rangle\right|^2 = \sin^2\gamma t. \tag{13.123}$$

As expected, the probability that $|\psi(t)\rangle$ will be either in $\left|\psi_1^0(0)\right\rangle$ or in $\left|\psi_2^0(0)\right\rangle$, given by the sum of the above two terms, is unity.

13.2.3 Mixing angles and Rabi's formula

We continue to consider a perturbation H' that is a constant for $t \geq 0$ while vanishing for $t < 0$. This time we will involve the mixing angles that we defined earlier.

We concentrate specifically on the state vectors at $t = 0$. As in the time-independent case, the matrix elements of the total Hamiltonian with respect to the unperturbed states $\{|\psi_1^0(0)\rangle, |\psi_2^0(0)\rangle\}$ as the basis can be written as

$$\left\{\left\langle\psi_i^0\right|H\left|\psi_j^0\right\rangle\right\} = \begin{bmatrix} A & D \\ D & B \end{bmatrix} \tag{13.124}$$

where $A = E_1^0 + H'_{11}$, $B = E_2^0 + H'_{22}$, $\Delta = A - B$, and $D = H'_{12}$ (assumed real), with the assumption

$$H'_{ij}(t) = \text{constant} \quad \text{for } t \geq 0, \tag{13.125}$$

$$H'_{ij}(t) = 0 \quad \text{for } t < 0. \tag{13.126}$$

Since the perturbation H' is a constant for $t \geq 0$ we can use the time-independent methods considered earlier and obtain the eigenstates $|\psi_\pm(t)\rangle$ of H for $t \geq 0$ with the eigenvalues $E_\pm$ given by

$$E_+ = \frac{A + B + \sqrt{\Delta^2 + 4D^2}}{2}, \tag{13.127}$$

$$E_- = \frac{A + B - \sqrt{\Delta^2 + 4D^2}}{2}. \tag{13.128}$$

The time-evolution equations for these states are

$$i\hbar\frac{\partial}{\partial t}|\psi_+(t)\rangle = H|\psi_+(t)\rangle = E_+|\psi_+(t)\rangle \quad \text{and}$$

$$i\hbar\frac{\partial}{\partial t}|\psi_-(t)\rangle = H|\psi_-(t)\rangle = E_-|\psi_-(t)\rangle. \tag{13.129}$$

Their time dependence is given by

$$|\psi_+(t)\rangle = |\psi_+(0)\rangle e^{-\frac{iE_+t}{\hbar}}, \tag{13.130}$$

$$|\psi_-(t)\rangle = |\psi_-(0)\rangle e^{-\frac{iE_-t}{\hbar}}. \tag{13.131}$$

Let $|\psi(t)\rangle$ be a general solution of the time-dependent equation given by

$$i\hbar\frac{\partial}{\partial t}|\psi(t)\rangle = \left(H_0 + H'(t)\right)|\psi(t)\rangle. \tag{13.132}$$

Expressing $|\psi(t)\rangle$ as a superposition of $|\psi_+(t)\rangle$ and $|\psi_-(t)\rangle$, we have

$$|\psi(t)\rangle = c_+|\psi_+(t)\rangle + c_-|\psi_-(t)\rangle. \tag{13.133}$$

Since all three state vectors in the above relation satisfy the same time-evolution equation, the coefficients c_+ and c_- will be independent of time.

Let us now consider the unperturbed states $\left|\psi_i^0(0)\right\rangle$. Since they are presumed to be known, we can, as we did for the time-independent case, obtain $\left|\psi_\pm(0)\right\rangle$ in terms of $\left|\psi_i^0(0)\right\rangle$ through the mixing angles, e.g.,

$$\left|\psi_+(0)\right\rangle = \cos\theta\left|\psi_1^0(0)\right\rangle + \sin\theta\left|\psi_2^0(0)\right\rangle, \tag{13.134}$$

$$\left|\psi_-(0)\right\rangle = -\sin\theta\left|\psi_1^0(0)\right\rangle + \cos\theta\left|\psi_2^0(0)\right\rangle. \tag{13.135}$$

We invert this relation and write

$$\left|\psi_1^0(0)\right\rangle = \cos\theta\left|\psi_+(0)\right\rangle - \sin\theta\left|\psi_-(0)\right\rangle, \tag{13.136}$$

$$\left|\psi_2^0(0)\right\rangle = \sin\theta\left|\psi_+(0)\right\rangle + \cos\theta\left|\psi_-(0)\right\rangle. \tag{13.137}$$

We assume that at $t = 0$, $|\psi(t)\rangle$, the solution of the equation (13.132), is given by

$$|\psi(0)\rangle = \left|\psi_1^0(0)\right\rangle. \tag{13.138}$$

From this initial condition and the relations (13.136) we have

$$|\psi(0)\rangle = \left|\psi_1^0(0)\right\rangle = \cos\theta\left|\psi_+(0)\right\rangle - \sin\theta\left|\psi_-(0)\right\rangle. \tag{13.139}$$

Since $\left|\psi_\pm\right\rangle$ are eigenstates of H with eigenvalues $E_\pm$, the t-dependence of $|\psi\rangle$ can be written down in terms of the t-dependence of $\left|\psi_\pm\right\rangle$ as

$$|\psi(t)\rangle = \cos\theta\left|\psi_+(t)\right\rangle - \sin\theta\left|\psi_-(t)\right\rangle \tag{13.140}$$

$$= \cos\theta\left|\psi_+(0)\right\rangle e^{-\frac{iE_+t}{\hbar}} - \sin\theta\left|\psi_-(0)\right\rangle e^{-\frac{iE_-t}{\hbar}} \tag{13.141}$$

where we have used (13.130) and (13.131). This implies that $|\psi(t)\rangle$ will oscillate as a function of time. In this formalism the oscillation will be between the eigenstates $|\psi_\pm\rangle$ of the total Hamiltonian, H, and not the eigenstates $|\psi_{1,2}^0\rangle$ of the unperturbed Hamiltonian, H_0. This is possible only because the perturbing Hamiltonian is constant in time.

The probability amplitude that the state $|\psi(t)\rangle$ at some time $t > 0$ will once again be in the state $\left|\psi_1^0(0)\right\rangle$ is then

$$\langle\psi_1^0(0)\,|\psi(t)\rangle = \cos\theta\langle\psi_1^0(0)\left|\psi_+(0)\right\rangle e^{-\frac{iE_+t}{\hbar}} - \sin\theta\langle\psi_1^0(0)\left|\psi_-(0)\right\rangle e^{-\frac{iE_-t}{\hbar}}. \tag{13.142}$$

From the relation (13.134) we obtain

$$\langle\psi_1^0(0)\left|\psi_+(0)\right\rangle = \cos\theta, \tag{13.143}$$

$$\langle\psi_1^0(0)\left|\psi_-(0)\right\rangle = -\sin\theta. \tag{13.144}$$

Thus the above probability amplitude is

$$\langle \psi_1^0(0) | \psi(t) \rangle = \cos^2\theta e^{-\frac{iE_+ t}{\hbar}} + \sin^2\theta e^{-\frac{iE_- t}{\hbar}}. \tag{13.145}$$

By taking the absolute square of this result we obtain the probability

$$\left| \langle \psi_1^0(0) | \psi(t) \rangle \right|^2 = \cos^4\theta + \sin^4\theta + 2\sin^2\theta\cos^2\theta\cos\left[\frac{(E_+ - E_-)\,t}{\hbar}\right], \tag{13.146}$$

which can be simplified to

$$\left| \langle \psi_1^0(0) | \psi(t) \rangle \right|^2 = 1 - \sin^2 2\theta \sin^2\left[\frac{(E_+ - E_-)\,t}{2\hbar}\right]. \tag{13.147}$$

The probability amplitude $\langle \psi_2^0(0) | \psi(t) \rangle$ can also be calculated and is found to be

$$\langle \psi_2^0(0) | \psi(t) \rangle = \sin\theta\cos\theta\left[e^{-\frac{iE_+ t}{\hbar}} - e^{-\frac{iE_- t}{\hbar}}\right]. \tag{13.148}$$

After some simplifications we find

$$\left| \langle \psi_2^0(0) | \psi(t) \rangle \right|^2 = \sin^2 2\theta \sin^2\left[\frac{(E_+ - E_-)\,t}{2\hbar}\right]. \tag{13.149}$$

As expected, the two probabilities (13.147) and (13.149) add to 1.

We can simplify things further by utilizing the formula for the mixing angle:

$$\tan 2\theta = \frac{2D}{\Delta}. \tag{13.150}$$

From this we obtain

$$\sin^2 2\theta = \frac{4D^2}{\Delta^2 + 4D^2}. \tag{13.151}$$

Furthermore, from the expressions for E_+ and E_- we have

$$(E_+ - E_-) = \sqrt{\Delta^2 + 4D^2}. \tag{13.152}$$

Thus,

$$\left| \langle \psi_2^0(0) | \psi(t) \rangle \right|^2 = \frac{4D^2}{\Delta^2 + 4D^2}\sin^2\left[\frac{\sqrt{\Delta^2 + 4D^2}}{2\hbar}t\right]. \tag{13.153}$$

Notice that we recover our results as (13.123) by taking $\Delta = 0$ and $D = \hbar\gamma$. The above expression is called the Rabi formula, which we will discuss further in the sections below when we consider harmonic time dependence for H'.

13.2.4 Harmonic time dependence

Let us now consider a somewhat more general case where $H'(t)$ is no longer a constant as a function of time but has a harmonic dependence for $t > 0$ while vanishing for $t < 0$. Specifically, we assume the following behavior for the matrix elements for $t \geq 0$:

$$H'_{12} = H'^{*}_{21} = \hbar\gamma e^{i\omega t}, \tag{13.154}$$

$$H'_{11} = H'_{22} = 0. \tag{13.155}$$

The Hamiltonian now is

$$\left\{\left\langle \psi_i^0 \middle| H \middle| \psi_j^0 \right\rangle\right\} = \begin{bmatrix} E_1^0 & \hbar\gamma e^{i\omega t} \\ \hbar\gamma e^{-i\omega t} & E_2^0 \end{bmatrix}. \tag{13.156}$$

Because of this explicit time dependence in the perturbing Hamiltonian H' we cannot use the results from the previous sections involving the mixing angles. So we go back to the time-dependent equations for c_1 and c_2 derived earlier. One finds the following:

$$i\dot{c}_1 = c_2\gamma e^{i\delta t}, \tag{13.157}$$

$$i\dot{c}_2 = c_1\gamma e^{-i\delta t} \tag{13.158}$$

where $\delta = \omega - \omega_{21}$ and $\omega_{21} = \left(E_2^0 - E_1^0\right)/\hbar$. We wish to obtain $c_1(t)$ and $c_2(t)$ for $t > 0$. For the initial condition we take

$$c_1(0) = 1, \quad c_2(0) = 0. \tag{13.159}$$

From these two coupled equations we first derive the equation for c_2 through the following steps:

$$i\ddot{c}_2 = \dot{c}_1\gamma e^{-i\delta t} + c_1\gamma\,(-i\delta)\,e^{-i\delta t}. \tag{13.160}$$

Substituting for c_1 and $\dot{c}_1$ from the above equations we obtain

$$i\ddot{c}_2 = \left(\frac{c_2\gamma e^{i\delta t}}{i}\right)\gamma e^{-i\delta t} + \left(\frac{i\dot{c}_2 e^{i\delta t}}{\gamma}\right)\gamma\,(-i\delta)\,e^{-i\delta t}. \tag{13.161}$$

Therefore,

$$\ddot{c}_2 + i\dot{c}_2\delta + c_2\gamma^2 = 0. \tag{13.162}$$

To solve this equation we substitute $c_2 = e^{i\alpha t}$, and obtain the following quadratic equation for the exponent:

$$\alpha^2 + \alpha\delta - \gamma^2 = 0 \tag{13.163}$$

with solutions

$$\alpha = \frac{-\delta \pm \sqrt{\delta^2 + 4\gamma^2}}{2}, \tag{13.164}$$

which gives

$$c_2(t) = Ae^{\frac{i\left[-\delta+\sqrt{\delta^2+4\gamma^2}\right]t}{2}} + Be^{\frac{i\left[-\delta-\sqrt{\delta^2+4\gamma^2}\right]t}{2}}. \tag{13.165}$$

The condition $c_2(0) = 0$ implies $B = -A$ in the above expression and, hence,

$$c_2(t) = Ae^{-\frac{i\delta t}{2}}\left[e^{\frac{i\sqrt{\delta^2+4\gamma^2}t}{2}} - e^{\frac{-i\sqrt{\delta^2+4\gamma^2}t}{2}}\right], \tag{13.166}$$

which simplifies to

$$c_2(t) = 2iAe^{-\frac{i\delta t}{2}}\sin\left(\frac{\sqrt{\delta^2+4\gamma^2}}{2}t\right). \tag{13.167}$$

To determine A, we consider the equations for c_1 and c_2 at $t = 0$ to give

$$i\dot{c}_2(0) = \gamma \tag{13.168}$$

where we have used the relation $c_1(0) = 1$. From the equation for $c_2(t)$ above we have

$$\dot{c}_2(0) = 2iA\frac{\sqrt{\delta^2+4\gamma^2}}{2}. \tag{13.169}$$

Equating the two expressions for $\dot{c}_2(0)$ we find

$$A = -\frac{\gamma}{\sqrt{\delta^2+4\gamma^2}}. \tag{13.170}$$

If we take

$$\Omega = \frac{\sqrt{\delta^2+4\gamma^2}}{2} \tag{13.171}$$

then $c_2(t)$ and $c_1(t)$ will be given respectively by

$$c_2(t) = -\frac{i\gamma}{\Omega}\sin(\Omega t)\, e^{-\frac{i\delta t}{2}}, \tag{13.172}$$

$$c_1(t) = \left[\frac{-i\delta}{2\Omega}\sin(\Omega t) + \cos(\Omega t)\right] e^{\frac{i\delta t}{2}}. \tag{13.173}$$

Writing it more explicitly, the probability $|c_2(t)|^2$ is given by

$$|c_2(t)|^2 = \frac{4\gamma^2}{(\omega-\omega_{21})^2+4\gamma^2}\sin^2\left(\frac{\sqrt{(\omega-\omega_{21})^2+4\gamma^2}}{2}t\right). \tag{13.174}$$

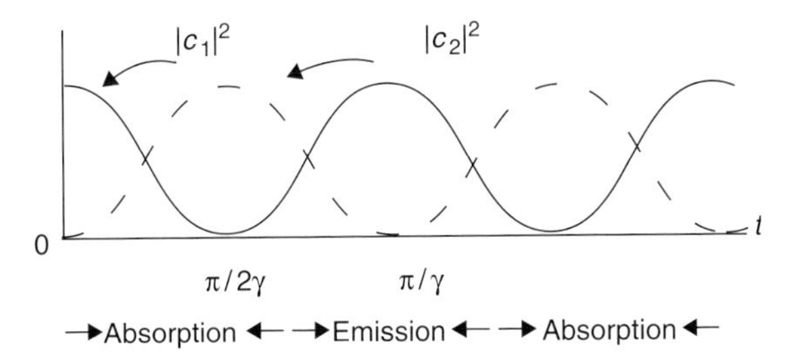

Fig. 13.4

The expression for $|c_2(t)|^2$ is in the form of the Rabi formula. The quantity

$$\frac{4\gamma^2}{(\omega-\omega_{21})^2+4\gamma^2} \tag{13.175}$$

is the amplitude for the oscillation of $|c_2(t)|^2$. For ω that satisfies the relation

$$\omega=\omega_{21}=\frac{E_2^0-E_1^0}{\hbar} \tag{13.176}$$

there is a peak in the amplitude where it reaches the maximum allowed value of 1. At this point one says that there is a "resonance" (see Fig. 13.4). At a resonance then, we have

$$|c_2(t)|^2=\sin^2(\gamma t) \tag{13.177}$$

$$|c_1(t)|^2=\cos^2(\gamma t). \tag{13.178}$$

The oscillatory behaviors of $|c_1(t)|^2$ and $|c_2(t)|^2$ at a resonance are depicted in Fig. 13.4. Starting with $c_1=1$ and $c_2=0$ at $t=0$, we find the following.

(i) During the interval between $t=0$ and $t=\pi/2\gamma$, $c_1(t)$ becomes smaller and the state $|\psi_1^0\rangle$ gets depleted as the system absorbs energy from the external interaction. The system achieves full absorption at $t=\pi/2\gamma$, as the higher energy level $|\psi_2^0\rangle$ gets fully populated and $c_1(t)=0$, $c_2(t)=1$. This interval corresponds to the so called "absorption cycle."

(ii) From $t=\pi/2\gamma$ to $t=\pi/\gamma$, the cycle reverses, $c_2(t)$ becomes smaller as the system gives up excess energy from the upper level to the external potential while it descends down to the lower level $|\psi_1^0\rangle$. At $t=\pi/\gamma$ we have, once again, $c_1(t)=1$, $c_2(t)=0$. This is called the "emission cycle."

(iii) This absorption–emission cycle continues indefinitely.

(iv) The maximum value (= 1) of the above amplitude is achieved at $\omega=\omega_{21}$, while at $\omega=\omega_{21}\pm 2\gamma$ the amplitude reaches half the maximum value. The quantity 4γ is called the full width at half-maximum (see Fig. 13.5). The absorption–emission cycle exists away from the resonance but one never achieves full absorption ($c_2(t)=1$) or full emission ($c_1(t)=1$).

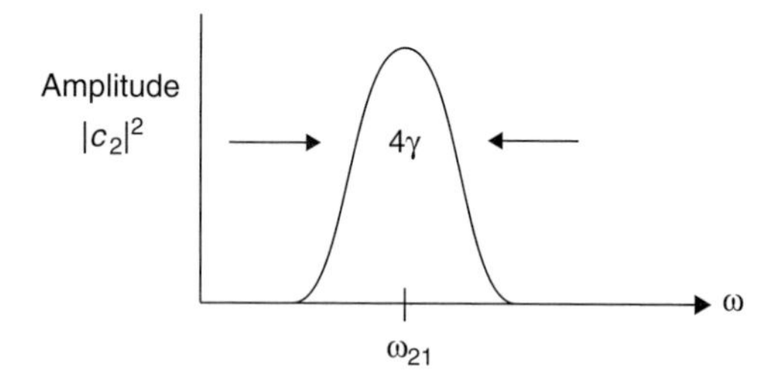

Fig. 13.5

13.2.5 Ammonia maser

In Chapter 8 we considered the energy eigenstates of the ammonia molecule NH_3 treated as a one-dimensional bound-state problem with a symmetric potential, $V(x)$. We found that there are two lowest eigenstates, a symmetric state $|\psi_S\rangle$ with eigenvalue E_S and an antisymmetric state $|\psi_A\rangle$, with eigenvalue E_A such that $E_A > E_S$, with their difference given by

$$E_A - E_S = 2\delta \tag{13.179}$$

where δ is a very small quantity for normal configurations of the potentials. We will write

$$E_A = E_0 + \delta, \tag{13.180}$$

$$E_S = E_0 - \delta \tag{13.181}$$

where E_0 is the average value of E_A and E_S.

If we subject the molecule to an electric field $\mathbf{E}$ then a dipole is generated as the electrons and the nucleus are stretched apart under the influence of the field. The interaction between the dipole and the electric field is given

$$H' = -\mathbf{d}.\mathbf{E} \tag{13.182}$$

where $\mathbf{d}$ is the dipole moment of the molecule. In this one-dimensional approximation, $\mathbf{d}$ will be in the x-direction and hence

$$H' \sim x. \tag{13.183}$$

Therefore, from symmetry considerations we have

$$\langle\psi_S| H' |\psi_S\rangle = 0 = \langle\psi_A| H' |\psi_A\rangle, \tag{13.184}$$

while

$$\langle\psi_S| H' |\psi_A\rangle \neq 0. \tag{13.185}$$

Thus only the off-diagonal element of H' will be nonzero.

We assume the electric field to have a harmonic time dependence,

$$\mathbf{E} = \mathbf{E}_0 e^{i\omega t}, \tag{13.186}$$

and take

$$\eta_E = \mathbf{d}.\mathbf{E}_0. \tag{13.187}$$

Then the matrix representation of $H (= H_0 + H')$ with respect to $|\psi_S\rangle$ and $|\psi_A\rangle$ is given by

$$H = \begin{bmatrix} E_0 - \delta & -\eta_E e^{i\omega t} \\ -\eta_E e^{i\omega t} & E_0 + \delta \end{bmatrix}. \tag{13.188}$$

This is similar to the problem we have encountered in the previous section.

If we take the initial condition for the probability amplitudes as $c_A(0) = 1, c_S(0) = 0$, then the transition probability for $A \to S$ can be written down from the previous section as

$$P_{A\to S} = |c_S(t)|^2 = \frac{4\eta_E^2}{(\omega - \omega_{21})^2 + 4\eta_E^2} \sin^2\left(\frac{\sqrt{(\omega - \omega_{21})^2 + 4\eta_E^2}}{2} t\right) \tag{13.189}$$

where

$$\omega_{21} = E_A - E_S = 2\delta. \tag{13.190}$$

Thus at

$$\omega = \omega_{21} = 2\delta \tag{13.191}$$

we have a resonance and the accompanying emission and absorption cycles described previously.

Consider the case when the ammonia molecules in the state $|\psi_A\rangle$ enter an apparatus subjected to a harmonically varying electric field during the emission cycle. If the frequency ω of the oscillating electric field is tuned to the level difference, 2δ, of the molecule then $|\psi_A\rangle$ will give up energy to the radiation field and convert to $|\psi_S\rangle$. We note that the transition $|\psi_A\rangle \to |\psi_S\rangle$ will also happen naturally through the tunneling of the middle barrier; that is, a spontaneous emission that has a rate which is much smaller than the "stimulated" emission we are considering here.

In practice the ammonia beam will contain an equal mixture of $|\psi_A\rangle$ and $|\psi_S\rangle$. But prior to entering the apparatus and being subjected to the oscillating electric field, it is made to pass through a nonhomogeneous time-independent electric field in order that $|\psi_A\rangle$ and $|\psi_S\rangle$ are separated. This separation comes about as follows.

Since the interaction is time-independent in this region, we can calculate the energy eigenvalues of the total Hamiltonian,

$$H = \begin{bmatrix} E_0 - \delta & -\eta_E \\ -\eta_E & E_0 + \delta \end{bmatrix}, \tag{13.192}$$

which are

$$E_+ = E_0 + \sqrt{\delta^2 + \eta_E^2}, \tag{13.193}$$

$$E_- = E_0 - \sqrt{\delta^2 + \eta_E^2}. \tag{13.194}$$

If the electric field is very weak such that η_E, which is proportional to $|\mathbf{E}|$, is $\ll \delta$, then

$$E_+ = E_0 + \delta + \frac{\eta_E^2}{2\delta}, \tag{13.195}$$

$$E_- = E_0 - \delta - \frac{\eta_E^2}{2\delta}. \tag{13.196}$$

We note by taking the limit $\eta_E \to 0$ that E_+ should be identified as the energy of $|\psi_A\rangle$ and E_- that of $|\psi_S\rangle$. The additional term $\eta_E^2/2\delta$ can be identified as the potential energy contributed by the electric field. If the electric field is nonhomogeneous then the molecules will be subjected to a force

$$\pm\nabla(\eta_E^2/2\delta). \tag{13.197}$$

Since the sign of the force is different between the states $|\psi_A\rangle$ and $|\psi_S\rangle$, these two types of particles will be deflected differently, much the same way as in the Stern–Gerlach experiment for spin-up and spin-down particles. This is then the basic mechanism of separating the two states.

In practice, then, after the separation has been achieved, the pure $|\psi_A\rangle$ beam enters a microwave cavity that has the dimensions adjusted so that the beam spends exactly the same time as the emission cycle ($t = \pi\hbar/2\eta_E$). The microwave is tuned to the energy difference, $E_A - E_S$, in order that the entering state $|\psi_A\rangle$ gives out all the energy to the radiation energy, which then gains in strength.

This mechanism is the essence of the maser, which is the acronym for microwave amplification by simulated emission of radiation.

13.3 Problems

1. In the problem of the ammonia molecule let $|\psi_L\rangle$ and $|\psi_R\rangle$ be the state vectors that indicate that the N atom is predominantly on the left side and right side, respectively. Show from parity invariance that the expectation values of the Hamiltonian satisfy

$$\langle\psi_L|H|\psi_L\rangle = \langle\psi_R|H|\psi_R\rangle.$$

Let E_0 be the common expectation value and assume that there is tunneling between the two sides such that

$$\langle\psi_L|H|\psi_R\rangle = \langle\psi_R|H|\psi_L\rangle = A.$$

Show that this interaction will induce the switch $|\psi_L\rangle \rightleftarrows |\psi_R\rangle$. Determine the eigenvalues of H and show that the two eigenstates can be written as their symmetric and antisymmetric combinations $|\psi_S\rangle$ and $|\psi_A\rangle$, respectively. If at $t = 0$, the state vector $|\psi(t)\rangle$ of N is given by

$$|\psi(0)\rangle = |\psi_L\rangle$$

then determine $|\psi(t)\rangle$ for arbitrary times (i) in terms of $|\psi_{S,A}\rangle$ and (ii) in terms of $|\psi_{L,R}\rangle$.

2. In the above problem, express all your answers for the state vectors in terms of column matrices by taking

$$|\psi_L\rangle = \begin{bmatrix} 1 \\ 0 \end{bmatrix}, \quad |\psi_R\rangle = \begin{bmatrix} 0 \\ 1 \end{bmatrix}.$$

3. Consider the $2S$ and $2P$ levels (with $m = 0$) for the hydrogen atom as a two-level system. At $t = 0$ the atom is subjected to a constant electric field, $\mathbf{E}_0$, in the z-direction that lasts only for a finite time t_0. If the atom starts out in the $2S$ state, determine the probabilities that for $t > 0$ it is in the $2S$ and $2P$ states.
4. Do the same problem as problem 3 except that now the electric field has a time dependence given by $E_0 \sin \omega t$.
5. Consider the harmonic oscillator as a three-level system made up of $n = 0, 1, 2$ states. If a perturbing potential

$$V'(x) = g_1 x + g_2 x^2$$

is applied, obtain the energy eigenvalues.

6. For a two-level time-dependent problem the total Hamiltonian is given by

$$\begin{aligned} H &= \hbar\omega_0 \sigma_z, && \text{for } t \geq 0 \\ &= 0, && \text{for } t < 0. \end{aligned}$$

If a state $|\psi(t)\rangle$ at $t = 0$ is in the "spin down" state in the x-direction, $|\psi_{x-}(0)\rangle$, then determine the probablitity that at times $t. > 0$ it is in a state $\psi_{x+}(0)$.

7. For a two-level time-dependent problem assume that the total Hamiltonian is given by

$$\begin{aligned} H &= \hbar\omega_0 \sigma_x, && \text{for } t \geq 0 \\ &= 0, && \text{for } t < 0. \end{aligned}$$

If a state $|\psi(t)\rangle$ at $t = 0$ has spin in the positive z-direction, determine the probability that at times $t > 0$ it is in the direction of a unit vector $\mathbf{n}$ whose x, y, z components are $(\sin\alpha \cos\beta, \sin\alpha \sin\beta, \cos\alpha)$.

8. Consider the following Hamiltonian

$$
\begin{aligned}
H &= \hbar\omega_0\sigma_x, && \text{for} \quad 0 < t < t_0 \\
&= 0, && \text{for} \quad t_0 < t < 2t_0 \\
&= \hbar\omega_0\sigma_x, && \text{for} \quad 2t_0 < t < 3t_0
\end{aligned}
$$

with $H = 0$ for $t > 3t_0$. If a state $|\psi(t)\rangle$ at $t = 0$ is in the "spin up" state in the z-direction, $|\psi_{z+}(0)\rangle$, determine the state vector for $t > 3t_0$.

9. Consider the Hamiltonian matrix

$$
\left\{\left\langle\psi_i^0\right| H \left|\psi_j^0\right\rangle\right\} = \begin{bmatrix} A & D \\ D & 0 \end{bmatrix}
$$

where A and D are > 0. Obtain $\tan 2\theta$ and $\tan\theta$ in the two limits (i) $A \ll D$, (ii) $D \ll A$ (wherever necessary be sure to expand the square roots that appear in your solution).

10. For a two-level time-dependent problem the total Hamiltonian is given by

$$
\begin{aligned}
H &= \hbar\omega_0\sigma_z, && \text{for} \quad t \geq 0 \\
&= 0, && \text{for} \quad t < 0.
\end{aligned}
$$

If a state $|\psi(t)\rangle$ at $t = 0$ is in the "spin down" state in the x-direction, $|\psi_{x-}(0)\rangle$, determine the probability that at times $t > 0$ it is in a state $|\psi_{y+}(0)\rangle$.

11. For the Hamiltonian

$$
H = \begin{bmatrix} 0 & i\delta \\ -i\delta & 0 \end{bmatrix}
$$

where δ is real, obtain the energy eigenvalues. If

$$
i\hbar\frac{d}{dt}\begin{bmatrix} c_1 \\ c_2 \end{bmatrix} = H\begin{bmatrix} c_1 \\ c_2 \end{bmatrix},
$$

determine $c_1(t)$ and $c_2(t)$ given that $c_1(0) = 1$ and $|c_1(t)|^2 + |c_2(t)|^2 = 1$.

12. The Hamiltonian

$$
H = \begin{bmatrix} 0 & \delta \\ -\delta & 0 \end{bmatrix}
$$

where δ is real is not Hermitian. Obtain the eigenvalues. If

$$
i\hbar\frac{d}{dt}\begin{bmatrix} c_1 \\ c_2 \end{bmatrix} = H\begin{bmatrix} c_1 \\ c_2 \end{bmatrix},
$$

determine $c_1(t)$ and $c_2(t)$ given that $c_1(t), c_2(t) \to 0$ as $t \to \infty$.

13. Consider the problem of two infinite barriers, located respectively at $x = -a$ and $x = a$, as a two-level problem in which only the ground state and the first excited states given

by $u_1(x)$ and $u_2(x)$, with energies E_1^0 and E_2^0, respectively, are considered and the rest are ignored. The following perturbative potential is applied,

$$\begin{aligned} V'(x) &= c_1 + c_2 x, \quad \text{for } |x| \le a \\ &= 0, \quad \text{for } |x| \ge a. \end{aligned}$$

If

$$H = \begin{bmatrix} A & D \\ D & B \end{bmatrix},$$

obtain A, B, D for this problem. Obtain the energy eigenvalues $E_\pm$ and the corresponding eigenfunctions $u_\pm(x)$. Express these eigenfunctions in terms of $u_{1,2}(x)$.

14. Consider now the time dependence for the above problem and replace the u's by ψ's. What is the energy dependence of $\psi_\pm(x,t)$? Express an arbitrary state of the system, $\psi(x,t)$, as a linear combination of $\psi_\pm(x,t)$. If at $t = 0$, $\psi(x,0) = u_1(x)$, obtain $\psi(x,t)$ for arbitrary t.
15. For the problem of a two-level system with regard to a particle confined inside two infinite walls (at $\pm a$) (see the above problem), consider instead the following two interactions separately ($b < a$):

$$\text{(i) } V'(x) = g\left[\delta(x+b) - \delta(x-b)\right],$$

$$\text{(ii) } V'(x) = g\delta(x).$$

For each case obtain (i) A, B, D, (ii) the mixing angles, (iii) the energy eigenvalues $E_\pm$, and (iv) the corresponding eigenfunctions $u_\pm(x)$. Express these eigenfunctions in terms of $u_{1,2}(x)$. If the perturbation is turned on at $t = 0$, what is the energy dependence of $\psi_\pm(x,t)$? Express an arbitrary state of the system, $\psi(x,t)$, as a linear combination of $\psi_\pm(x,t)$. If at $t = 0$, $\psi(x,0) = u_1(x)$, obtain $\psi(x,t)$ for arbitrary t.
16. Consider a spin ½ system, with

$$H_0 = g_3 \hbar \sigma_3$$

and

$$H' = g_1 \hbar \sigma_1 + g_2 \hbar \sigma_2$$

where the g_i's are constants and σ_i's are the Pauli matrices. (The above expressions allow us to write the total Hamiltonian as $H = \hbar \mathbf{g} \cdot \boldsymbol{\sigma}$.) Take the unperturbed states as the spin-up and spin-down states. Obtain (i) the energy eigenvalues and (ii) the eigenfunctions in terms of the unperturbed states. Show that the coefficients are complex (i.e., they have a phase) (take $g = \sqrt{g_1^2 + g_2^2 + g_3^3}$). Also show that the mixing angles, which are real by definition, cannot be defined as in the text, even though H is Hermitian (see below).

17. Consider the Hamiltonian matrix

$$\begin{bmatrix} A & D \\ D^* & B \end{bmatrix}$$

where A and B are real but D is complex. Show that this matrix is Hermitian. Also show that the above matrix cannot be diagonalized by

$$\begin{bmatrix} \cos\theta & \sin\theta \\ -\sin\theta & \cos\theta \end{bmatrix}.$$

Determine whether both of the following are unitary and whether they will diagonalize the above matrix:

$$\text{(a)} \quad \begin{bmatrix} e^{i\delta}\cos\theta & \sin\theta \\ -\sin\theta & e^{-i\delta}\cos\theta \end{bmatrix}$$

or

$$\text{(b)} \begin{bmatrix} \cos\theta & e^{i\delta}\sin\theta \\ -e^{-i\delta}\sin\theta & \cos\theta \end{bmatrix}$$

where

$$D = |D|\, e^{i\delta}.$$

For the case where diagonalization is possible, obtain the formula for $\tan 2\theta$. Which of the two rotation matrices is physically relevant?

14 Spin ½ systems in the presence of magnetic fields

Spin ½ systems are intrinsically two-level systems. They provide a very fertile field in which we can obtain exact solutions to some very interesting problems involving the interactions with a magnetic field.

14.1 Constant magnetic field

As we mentioned earlier, the spin ½ systems lend themselves easily to be a part of the two-level problems since the system has, in the spin-space, just two states: spin-up and spin-down.

Consider now the Hamiltonian, $H = H_0 + H'(t)$, corresponding to a charged, spin ½, particle at rest, subjected to a constant magnetic field $\mathbf{B}$ for $t \geq 0$. Since the particle is at rest we have

$$H_0 = 0. \tag{14.1}$$

Because of its spin properties, a charged particle with spin ½, as we discussed in Chapter 5, will possess a magnetic moment, μ, that is related to the spin $\mathbf{S}$ by

$$\boldsymbol{\mu} = \frac{e}{mc}\mathbf{S} \tag{14.2}$$

where e is the charge and m the mass of the particle. This magnetic moment will interact with the magnetic field $\mathbf{B}$. If this interaction is turned on at $t = 0$, then the perturbed Hamiltonian will be given by

$$H'(t) = -\boldsymbol{\mu} \cdot \mathbf{B}, \quad \text{for } t > 0 \tag{14.3}$$

$$= 0, \quad \text{for } t \leq 0. \tag{14.4}$$

Let us consider the case where the magnetic field is directed along the z-direction, $\mathbf{B} = B_0\mathbf{k}$, and where B_0 is a constant. We then have

$$\boldsymbol{\mu} \cdot \mathbf{B} = \frac{e}{mc}\mathbf{S} \cdot \mathbf{B} = \frac{eB_0}{mc}S_z. \tag{14.5}$$

From Chapter 2 we note that $\mathbf{S}$ is related to the Pauli matrices, $\boldsymbol{\sigma}$, by

$$\mathbf{S} = \frac{\hbar}{2}\boldsymbol{\sigma}. \tag{14.6}$$

Writing in terms of $\boldsymbol{\sigma}$ we note that, since $H_0 = 0$, the total Hamiltonian is given by

$$H(t) = H'(t) = \frac{1}{2}\hbar\omega_0\sigma_z, \quad \text{for } t \geq 0 \tag{14.7}$$

$$= 0, \quad \text{for } t < 0 \tag{14.8}$$

where from (14.3),

$$\omega_0 = -\frac{eB_0}{mc} \tag{14.9}$$

and σ_z is given by

$$\sigma_z = \begin{pmatrix} 1 & 0 \\ 0 & -1 \end{pmatrix}. \tag{14.10}$$

The Hamiltonian for $t \geq 0$ is then given by

$$H = \begin{pmatrix} \dfrac{\hbar\omega_0}{2} & 0 \\ 0 & -\dfrac{\hbar\omega_0}{2} \end{pmatrix}. \tag{14.11}$$

Below we will obtain the eigenstates of H for various simple examples.

14.1.1 Initial spin in the z-direction

The eigenstates at $t = 0$ corresponding to "spin-up" and "spin-down" in the z-direction, respectively, are given by

$$|z+(0)\rangle = \begin{pmatrix} 1 \\ 0 \end{pmatrix} = |+\rangle, \tag{14.12}$$

$$|z-(0)\rangle = \begin{pmatrix} 0 \\ 1 \end{pmatrix} = |-\rangle \tag{14.13}$$

where

$$\sigma_z |+\rangle = |+\rangle, \tag{14.14}$$

$$\sigma_z |-\rangle = -|-\rangle. \tag{14.15}$$

The time evolution equations for $t > 0$ for $|z+(t)\rangle$ and $|z-(t)\rangle$ are, respectively,

$$i\hbar\frac{\partial}{\partial t}|z+(t)\rangle = H(t)|z+(t)\rangle = \frac{1}{2}\hbar\omega_0 |z+(t)\rangle, \tag{14.16}$$

$$i\hbar\frac{\partial}{\partial t}|z-(t)\rangle = H(t)|z-(t)\rangle = -\frac{1}{2}\hbar\omega_0 |z-(t)\rangle, \tag{14.17}$$

with the solutions

$$|z+(t)\rangle = e^{-i\frac{1}{2}\omega_0 t}\,|+\rangle\,, \tag{14.18}$$

$$|z-(t)\rangle = e^{i\frac{1}{2}\omega_0 t}\,|-\rangle\,. \tag{14.19}$$

Therefore, if the particle at $t = 0$ has spin-up (spin-down) in the z-direction, then under the interaction (14.7) it continues to have spin-up (spin-down) for $t > 0$. This is as expected.

14.1.2 Initial spin in the x-direction

Consider now a problem where at $t = 0$, the particle has "spin-up" in the x-direction. We wish to consider its time-development under the influence of the interaction given by (14.7).

We first construct the spin-up and spin-down eigenstates along the x-axis, $|x_\pm(0)\rangle$ in terms of $|z_\pm(0)\rangle$, where

$$\sigma_x = \begin{pmatrix} 0 & 1 \\ 1 & 0 \end{pmatrix} \tag{14.20}$$

and

$$\sigma_x\,|x+(0)\rangle = |x+(0)\rangle\,, \tag{14.21}$$

$$\sigma_x\,|x-(0)\rangle = -\,|x-(0)\rangle\,. \tag{14.22}$$

Writing $|x+(0)\rangle$ and $|x-(0)\rangle$ as a superposition of the states $|+\rangle$ and $|-\rangle$, we obtain

$$|x+(0)\rangle = \frac{1}{\sqrt{2}}\,|+\rangle + \frac{1}{\sqrt{2}}\,|-\rangle\,, \tag{14.23}$$

$$|x-(0)\rangle = \frac{1}{\sqrt{2}}\,|+\rangle - \frac{1}{\sqrt{2}}\,|-\rangle\,. \tag{14.24}$$

The state $|x+(t)\rangle$ is then given by

$$|x+(t)\rangle = \frac{1}{\sqrt{2}}\,|z+(t)\rangle + \frac{1}{\sqrt{2}}\,|z-(t)\rangle\,. \tag{14.25}$$

Since the perturbation is independent of time and both sides satisfy the same equation, we find, from (14.18) and (14.19),

$$|x+(t)\rangle = \frac{e^{-i\frac{1}{2}\omega_0 t}}{\sqrt{2}}\,|+\rangle + \frac{e^{\frac{1}{2}i\omega_0 t}}{\sqrt{2}}\,|-\rangle\,.$$

The probability amplitude that the state $|x_\pm(t)\rangle$ at arbitrary time t is in the spin-up state $|x+(0)\rangle$ is given by

$$|\langle x+(0)\,|x+(t)\rangle|^2 = \left|\left[\frac{1}{\sqrt{2}}\,\langle +| + \frac{1}{\sqrt{2}}\,\langle -|\right]\left[\frac{e^{-i\frac{1}{2}\omega_0 t}}{\sqrt{2}}\,|+\rangle + \frac{e^{i\frac{1}{2}\omega_0 t}}{\sqrt{2}}\,|-\rangle\right]\right|^2 = \cos^2\frac{1}{2}\omega_0 t. \tag{14.26}$$

Similarly, the probability for the state $|x+(t)\rangle$ to be the spin-down $|\psi_{z-}(0)\rangle$ state at time t is given by

$$|\langle x-(0)|x+(t)\rangle|^2 = \sin^2\frac{1}{2}\omega_0 t. \tag{14.27}$$

We see that at $t=0$ the state has "spin-up" in the x-direction, as it should be since that is the initial condition. At a later time it can change to "spin-down" due to the presence of the magnetic field in the z-direction. Since both the positive and negative x-directions are perpendicular to the z-direction, the state $|x+(t)\rangle$ is said to "precess" in the x–y plane as a function of time with a period $4\pi/\omega_0$ about the direction of the magnetic field, which in our case is the z-direction. This is called the Larmor precession and ω_0 is called the Larmor frequency. Note that the frequency dependence is given by $(\omega_0/2)$, which is, of course, directly related to the relation $\mathbf{S}=(1/2)\boldsymbol{\sigma}$.

14.1.3 Initial spin with polar angle α

Let us consider the case in which, at $t=0$, the state vector $|\psi(t)\rangle$ corresponds to spin pointing in a direction with polar angle α and azimuthal angle $=0$. We have already constructed such a state in Chapter 5. It is given by

$$|\psi(0)\rangle = \cos\left(\frac{\alpha}{2}\right)|+\rangle + \sin\left(\frac{\alpha}{2}\right)|-\rangle \tag{14.28}$$

where $|\psi(0)\rangle$ is the state vector $|\psi(t)\rangle$ at $t=0$. The angle α then corresponds to the direction of the spin with respect to the positive z-axis, which is the direction of $|+\rangle$ (angle $\pi-\alpha$ then corresponds to the angle with respect to the negative z-direction given by $|-\rangle$).

For $t\geq 0$, from our previous arguments, this state can be expressed as

$$|\psi(t)\rangle = \cos\left(\frac{\alpha}{2}\right)e^{-i\frac{1}{2}\omega_0 t}|+\rangle + \sin\left(\frac{\alpha}{2}\right)e^{i\frac{1}{2}\omega_0 t}|-\rangle. \tag{14.29}$$

This implies that $|\psi(t)\rangle$ continues to subtend the angle α along the positive z-axis but, because of the oscillating time-dependence, it precesses around the z-axis with period $4\pi/\omega_0$.

14.2 Spin precession

We now discuss the time dependence of the expectation values of σ_x, σ_y, and σ_z for the problem we have discussed above. The expectation value of σ_z for an arbitrary state $|\psi(t)\rangle$,

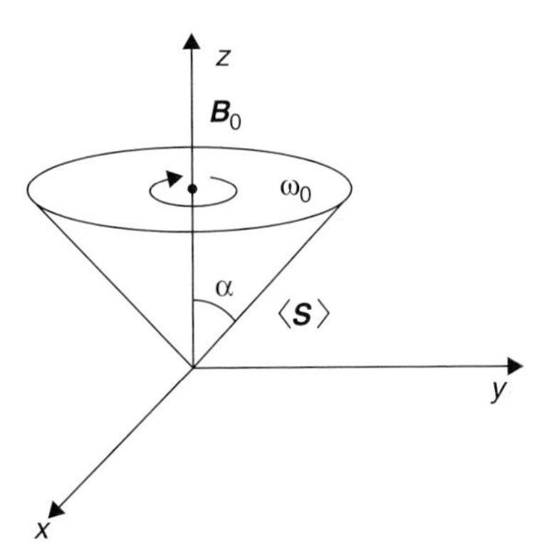

Fig. 14.1

as expressed above, is found to be

$$\langle \psi(t) | \sigma_z | \psi(t) \rangle = \left[\cos\left(\frac{\alpha}{2}\right) e^{+i\frac{1}{2}\omega_0 t} \langle + | + \sin\left(\frac{\alpha}{2}\right) e^{-i\frac{1}{2}\omega_0 t} \langle - | \right] \left[\cos\left(\frac{\alpha}{2}\right) e^{-i\frac{1}{2}\omega_0 t} | + \rangle - \sin\left(\frac{\alpha}{2}\right) e^{i\frac{1}{2}\omega_0 t} | - \rangle \right] \tag{14.30}$$

$$= \cos^2\left(\frac{\alpha}{2}\right) - \sin^2\left(\frac{\alpha}{2}\right) \tag{14.31}$$

$$= \cos\alpha, \tag{14.32}$$

which is independent of time, as expected.

One can similarly obtain the expectation values of σ_x and σ_y for arbitrary t, which are found to be

$$\langle \psi(t) | \sigma_x | \psi(t) \rangle = \sin\alpha \cos\omega_0 t \tag{14.33}$$

and

$$\langle \psi(t) | \sigma_y | \psi(t) \rangle = \sin\alpha \sin\omega_0 t. \tag{14.34}$$

Thus the spin itself precesses about the z-axis, but with a period $2\pi/\omega_0$ that is half the period with which the state $|\psi(t)\rangle$ precesses (see Fig. 14.1).

14.3 Time-dependent magnetic field: spin magnetic resonance

A problem of considerable importance is that of a charged particle with spin ½ which is subjected to a time-dependent rotating magnetic field. This problem differs from the earlier ones in which the magnetic field was assumed to be independent of time.

14.3.1 Initial spin in the z-direction

Assuming that the particle is at rest (i.e., $H_0 = 0$), the total Hamiltonian for this process is given by

$$H = -\boldsymbol{\mu} \cdot \mathbf{B}, \quad \text{for } t \geq 0 \tag{14.35}$$

$$= 0, \quad \text{for } t < 0 \tag{14.36}$$

where $\mathbf{B}$ is the magnetic field, and $\boldsymbol{\mu}$ is the magnetic moment. As we have derived before,

$$\boldsymbol{\mu} = \frac{e}{mc}\mathbf{S} = \frac{e\hbar}{2mc}\boldsymbol{\sigma}. \tag{14.37}$$

We consider a magnetic field whose components in the x- and y-directions are functions of time that correspond to a rotating field in the x–y plane with frequency ω, while along the z-direction the field is a constant. Thus we write

$$B = B_0\mathbf{k} + B_1\left(\mathbf{i}\cos\omega t + \mathbf{j}\sin\omega t\right) \tag{14.38}$$

where, $\mathbf{i}$, $\mathbf{j}$, and $\mathbf{k}$ are unit vectors in the x-, y- and z-directions, respectively, and B_0 and B_1 are constants. The Hamiltonian for $t \geq 0$ can then be written as

$$H = \frac{1}{2}\hbar\omega_0\sigma_z + \frac{1}{2}\hbar\omega_1\left(\sigma_x\cos\omega t + \sigma_y\sin\omega t\right) \tag{14.39}$$

where

$$\omega_0 = -\frac{eB_0}{mc}, \quad \omega_1 = -\frac{eB_1}{mc}. \tag{14.40}$$

We write the Hamiltonian as

$$H = \begin{pmatrix} \hbar\overline{\omega}_0 & \hbar\overline{\omega}_1 e^{-i\omega t} \\ \hbar\overline{\omega}_1 e^{i\omega t} & -\hbar\overline{\omega}_0 \end{pmatrix} \tag{14.41}$$

where

$$\overline{\omega}_0 = \frac{1}{2}\omega_0 \quad \text{and} \quad \overline{\omega}_1 = \frac{1}{2}\omega_1. \tag{14.42}$$

We can directly obtain the corresponding probability amplitudes c_1 and c_2, with $c_1(0) = 1$, and $c_2(0) = 0$, from Chapter 13 by making the following substitutions in (13.172):

$$\omega \to -\omega, \tag{14.43}$$

$$\delta \to 2\overline{\omega}_0 - \omega = \omega_0 - \omega, \tag{14.44}$$

$$\gamma \to \overline{\omega}_1 = \frac{1}{2}\omega_1, \tag{14.45}$$

$$\Omega \to \frac{\sqrt{4\overline{\omega}_1^2 + \delta^2}}{2} = \frac{\sqrt{\omega_1^2 + (\omega_0 - \omega)^2}}{2}. \tag{14.46}$$

The probability $|c_2(t)|^2$ is then

$$|c_2(t)|^2 = \frac{\omega_1^2}{(\omega - \omega_0)^2 + \omega_1^2} \sin^2\left(\frac{\sqrt{(\omega - \omega_0)^2 + \omega_1^2}}{2} t\right). \tag{14.47}$$

We have then a resonance at $\omega = \omega_0$.

The results we have derived are of fundamental importance in molecular beam experiments where, by varying the magnetic field, one can make an accurate determination of the magnetic moment. We note that ω_0, which is proportional to the magnetic moment, is just the precession frequency of the rotating field. Thus when the applied frequency equals the precession frequency, a resonance occurs since, in a sense, the spin now sees a constant magnetic field and thus the system oscillates between the up and down states with a frequency $(1/2)\,\omega_1$.

14.3.2 Spin aligned with the magnetic field

Let us consider the case where the spin is oriented in the same direction as the magnetic field. To facilitate the calculation we will make the following changes in the Hamiltonian given by (14.41):

$$\hbar\overline{\omega}_0 \to \hbar\overline{\omega}_0 \cos\theta, \quad \hbar\overline{\omega}_1 \to \hbar\overline{\omega}_1 \sin\theta. \tag{14.48}$$

This establishes the magnetic field as pointing along a direction with polar angle θ and azimuthal angle ωt. The Hamiltonian is then

$$H = \hbar\overline{\omega}_1 \begin{pmatrix} \cos\theta & \sin\theta e^{-i\omega t} \\ \sin\theta e^{i\omega t} & -\cos\theta \end{pmatrix}. \tag{14.49}$$

We note that the energy eigenvalues at $t = 0$ are given by

$$E_\pm = \pm\hbar\overline{\omega}_1 = \pm\hbar\frac{\omega_1}{2}. \tag{14.50}$$

We will take the spin to be in the same direction as the magnetic field; thus the corresponding spin-up and spin-down wavefunctions are

$$\chi_+(t) = \begin{bmatrix} \cos(\theta/2) \\ e^{i\omega t}\sin(\theta/2) \end{bmatrix} \tag{14.51}$$

and

$$\chi_-(t) = \begin{bmatrix} \sin(\theta/2) \\ -e^{i\omega t}\cos(\theta/2) \end{bmatrix}. \tag{14.52}$$

We write a general solution as

$$\chi(t) = c_+(t)\chi_+(t) + c_-(t)\chi_-(t). \tag{14.53}$$

If $c_+(0) = 1$ and $c_-(0) = 0$ then one can show that for $t > 0$

$$c_+(t) = \left[\cos(\Omega t) + i\left(\frac{\omega_1 + \omega\cos\theta}{2\Omega}\right)\sin(\Omega t)\right]e^{-i\omega t/2}, \tag{14.54}$$

$$c_-(t) = \left[\frac{i\omega}{2\Omega}\sin\theta\sin(\Omega t)\right]e^{-i\omega t/2}, \tag{14.55}$$

where

$$\Omega = \frac{\sqrt{\omega^2 + \omega_1^2 + 2\omega\omega_1\cos\theta}}{2}. \tag{14.56}$$

From these results one can obtain the respective probability amplitudes $|c_+(t)|^2$ and $|c_-(t)|^2$.

14.4 Problems

1. An electron has its spin directed along a vector with azimuthal angle α and polar angle β. It is subjected to a constant, time-independent magnetic field $\mathbf{B}_0$ in the positive z-direction. Obtain the probability, as a function of time, that it is a state with $s_y = +\hbar/2$. Also obtain the expectation values of the operators $\mathbf{S}_x$, $\mathbf{S}_y$, and $\mathbf{S}_z$.
2. A spin ½ particle at rest is subjected to a constant magnetic field $\mathbf{B}_0$ in the z-direction. Determine the time dependence of the expectation value of the magnetic moment $\boldsymbol{\mu}$ assuming it to be an operator in the Heisenberg representation. Show that the same result is obtained by directly calculating the expectation value.
3. Consider an electron traveling with velocity $\mathbf{v} = (v_x, v_y, 0)$. A constant magnetic field is applied in the z-direction. Determine the time dependence of the expectation values $\langle \boldsymbol{\sigma}.\mathbf{v}\rangle$ and $\langle (\boldsymbol{\sigma} \times \mathbf{v})_z \rangle$.

4. The Hamiltonian for a spin ½ system subjected to a magnetic field $\mathbf{B}$ is given by

$$H = \mu \boldsymbol{\sigma} \cdot \mathbf{B}.$$

Show that

$$\frac{d\boldsymbol{\sigma}(t)}{dt} = \mathbf{B} \times \boldsymbol{\sigma}(t).$$

For a uniform magnetic field $\mathbf{B} = \mathbf{B}_0$, solve for $\boldsymbol{\sigma}(t)$ by expressing it as a linear combination of three orthogonal vectors: $\mathbf{B}_0$, $\mathbf{B}_0 \times \boldsymbol{\sigma}(0)$, $\mathbf{B}_0 \times (\mathbf{B}_0 \times \boldsymbol{\sigma}(0))$.

15 Oscillation and regeneration in neutrinos and neutral K-mesons as two-level systems

Here we consider three fascinating problems in particle physics that can be approximated as two-level systems with a somewhat phenomenological Hamiltonian. Two of them involve neutrinos and one involves neutral K-mesons. Together they provide a remarkable success story of the applications of simple quantum-mechanical principles.

15.1 Neutrinos

Neutrinos are spin ½ particles with no charge and a minuscule mass. They interact with other elementary particles only through the so-called weak interaction. Neutrinos come in three different species (called flavors): electron-neutrino (ν_e), muon-neutrino (ν_μ), and tau-neutrino (ν_τ) and form a part of the so-called lepton family. They are neutral accompaniments to the charged leptons: electrons (e), muons (μ), and tau leptons (τ). In the following two sections we will ignore ν_τ and discuss the solutions of some rather fundamental problems in neutrino physics within the framework of two-level systems.

15.2 The solar neutrino puzzle

To understand the solar neutrino puzzle we need first to note that the energy that is radiated from the solar surface comes from intense nuclear reactions that produce fusion of different nuclei in the interior of the sun. Among the by-products of these reactions are photons, electrons, and neutrinos. In the interior it is mostly electron-neutrinos, ν_e's, that are produced. The shell of the sun is extraordinarily dense, so that the electrons and photons are absorbed. However, because neutrinos undergo only weak interactions they are able to escape from the solar surface and reach the earth. The rate at which the ν_e's are expected to reach earth can be calculated from the standard solar model. However, only one-third of what is expected from this calculation is actually observed. This is the essence of the solar neutrino puzzle.

A possible solution to this problem was proposed by Mikheyev, Smirnov, and Wolfenstein (called the MSW effect). We discuss below the outline of this solution, elaborated by Bethe, who considered this as a two-level problem involving the interaction between ν_e and ν_μ, in the presence of solar matter, in their passage from the interior of the sun to the solar surface.

In this problem the counterparts of $|\psi_1^0\rangle$ and $|\psi_2^0\rangle$ discussed in Chapter 13 are $|\nu_e\rangle$ and $|\nu_\mu\rangle$ respectively. These are the so-called "flavor eigenstates."

The neutrinos are found to have extremely small mass, as stated earlier. Since in the relativistic case the energy is related to mass, one speaks of "mass eigenstates" instead of energy eigenstates. The counterparts of the eigenstates of $E_1$and E_2 in the neutrino interaction are the mass eigenstates $|\nu_1\rangle$ and $|\nu_2\rangle$ with eigenvalues m_1 and m_2. Thus, the interaction between flavor eigenstates generates the mass eigenstates.

The connection between the two types of states can be written following our earlier discussions:

$$|\nu_1\rangle = \cos\theta\,|\nu_e\rangle + \sin\theta\,|\nu_\mu\rangle \quad \text{and} \quad |\nu_2\rangle = -\sin\theta\,|\nu_e\rangle + \cos\theta\,|\nu_\mu\rangle. \tag{15.1}$$

(This convention, which is generally used in quantum mechanics texts, is opposite to the convention used in particle physics obtained by changing $\theta \to -\theta$.)

In the interior of the sun, the electrons and electron-neutrinos are produced in great abundance. The ν_e's interact with each other through the so-called current–current interaction, which is a part of the basic weak interaction, whose value can be calculated. The matrix element for the interaction Hamiltonian H' is given by

$$\langle\nu_e|\,H'\,|\nu_e\rangle = \delta_e \tag{15.2}$$

where δ_e is positive, of dimension $(\text{mass})^2$, and approximately a constant, proportional to the density of the electrons and the weak-interaction coupling constant (Fermi constant). It becomes smaller and ultimately vanishes as one goes from the interior of the sun to the solar surface where the electron density goes to zero. All other matrix elements of H' are negligibly small.

The matrix representation of the total Hamiltonian for the neutrinos is given by

$$\begin{bmatrix}\langle\nu_1| & \langle\nu_2|\end{bmatrix}\begin{bmatrix} m_1^2 & 0 \\ 0 & m_2^2 \end{bmatrix}\begin{bmatrix} |\nu_1\rangle \\ |\nu_2\rangle \end{bmatrix} + \langle\nu_e|\,H'\,|\nu_e\rangle \tag{15.3}$$

where the first term is expressed in terms of the mass eigenstates, while the second, corresponding to the interaction Hamiltonian, is expressed in terms of the flavor eigenstates. We note that the second term above appears as a consequence of the solar matter with which ν_e interacts. In the absence of any matter, the first term alone will describe the total Hamiltonian. We take $m_2 = m_{\nu\mu}$, the mass of ν_μ and $m_1 = m_{\nu e}$, the mass of ν_e, and note that $m_2 \gg m_1$. Our purpose here is not to determine the masses, which remains a fundamental unexplained question, but to discuss the eigenstates and their behavior in the presence of solar matter.

We combine both the matrices above in terms of the same basis formed by $|\nu_e\rangle$ and $|\nu_\mu\rangle$:

$$\begin{bmatrix}\langle\nu_e| & \langle\nu_\mu|\end{bmatrix}\left\{R^\dagger(\theta)\begin{bmatrix} m_1^2 & 0 \\ 0 & m_2^2 \end{bmatrix}R(\theta) + \begin{bmatrix} \delta_e & 0 \\ 0 & 0 \end{bmatrix}\right\}\begin{bmatrix} |\nu_e\rangle \\ |\nu_\mu\rangle \end{bmatrix} \tag{15.4}$$

where $R(\theta)$ is the rotation matrix connecting the two bases as discussed in the preceding chapter. As in the previous discussion, we express the sum of the matrix elements as

$$\frac{1}{2}\begin{bmatrix} \left(m_1^2+m_2^2+\delta_e\right)+(\delta_e-\delta\cos 2\theta) & -\delta\sin 2\theta \\ -\delta\sin 2\theta & \left(m_1^2+m_2^2+\delta_e\right)-(\delta_e-\delta\cos 2\theta) \end{bmatrix} \tag{15.5}$$

where $\delta = m_2^2 - m_1^2 > 0$. Note that in the 2, 2 matrix element above even though δ_e actually cancels out, it is left there for reasons which will be evident below.

The new mass eigenvalues are

$$m_{\nu\pm}^2 = \frac{1}{2}\left[\left(m_1^2+m_2^2+\delta_e\right) \pm \sqrt{(\delta_e-\delta\cos 2\theta)^2+\delta^2\sin^2 2\theta}\right] \tag{15.6}$$

with the corresponding eigenstates designated as $|\nu\pm\rangle$. We consider the variation of these eigenvalues, adiabatically, as δ_e varies very slowly from a very large value at the interior, where the electron density is extremely high, to zero at the surface of the sun. For simplicity we also take the mixing angle θ to be small.

In the absence of solar matter, when $\delta_e = 0$, the second term in (15.4) is absent, and $m_{\nu+}^2 = m_2^2$ and $m_{\nu-}^2 = m_1^2$.

In the interior of the sun, it is found that $\delta_e \gg \delta$, because of the large electron density; therefore,

$$m_{\nu+}^2 \simeq \frac{1}{2}\left[\left(m_1^2+m_2^2+\delta_e\right)+(\delta_e-\delta)\right] = m_1^2+\delta_e, \tag{15.7}$$

$$m_{\nu-}^2 \simeq \frac{1}{2}\left[\left(m_1^2+m_2^2+\delta_e\right)-(\delta_e-\delta)\right] = m_2^2, \tag{15.8}$$

with $\delta_e > 0$. Thus $m_{\nu+}^2$ and, therefore, $|\nu+\rangle$ represent the electron-neutrino whose mass is sufficiently enhanced by the attractive interaction in the interior of the sun so that $m_{\nu+}^2 > m_{\nu-}^2$ where $m_{\nu-}^2$ and therefore $|\nu-\rangle$ represent the muon neutrino.

At the surface of the sun, however, $\delta_e = 0$ and we obtain

$$m_{\nu+}^2 = \frac{1}{2}\left[\left(m_1^2+m_2^2\right)+|\delta|\right] = m_2^2, \tag{15.9}$$

$$m_{\nu-}^2 = \frac{1}{2}\left[\left(m_1^2+m_2^2\right)-|\delta|\right] = m_1^2. \tag{15.10}$$

Thus, remarkably, the electron-neutrinos produced in the sun's interior and identified as $|\nu+\rangle$ appear at the solar surface as muon-neutrinos, while the muon-neutrinos, which are much less abundant in sun's interior, and identified as $|\nu-\rangle$, come out at the solar surface as electron-neutrinos.

The point is simply this: if there were no solar matter to influence the events, the electron would have taken the lower curve and muon the upper curve with each curve as straight lines. See Fig. 13.2 for comparison. In the presence of a large δ_e, however, the electron is boosted to the upper curve and the muon is demoted to the lower one. Once there, both are stuck on their respective curves, which are not allowed to cross. Since on earth the particle identification is made based on the mass values, the electron-neutrino coming out with higher mass and bigger abundance is taken to be a muon-neutrino and the muon-neutrino coming out with lower mass and lower abundance is classified as an electron-neutrino.

This represents the basis of a possible solution, though not necessarily a complete one, to the solar neutrino puzzle. The solution as presented here is clearly oversimplified since, among other things, the mixing angle is not negligible. What is interesting, however, is that it involves simple two-channel concepts.

15.3 Neutrino oscillations

The above explanation of the solar neutrino puzzle involves switching of the eigenstates during the passage of ν_e from the sun's interior to the surface. This switching is sometimes called "matter oscillation." Oscillations will also occur in vacuum between $|\nu_e\rangle$ and $|\nu_\mu\rangle$ as ν_e's are streaming down through the atmosphere from the sun. These are the "vacuum" oscillations. This could also be the solution or part of the solution of the solar neutrino puzzle. We note that oscillations will also occur between neutrinos that are by-products of weak interaction decays.

We develop the formalism for these oscillations in the mixing angle framework by writing the following superposition at $t = 0$:

$$|\nu_1(0)\rangle = \cos\theta\,|\nu_e(0)\rangle + \sin\theta\,|\nu_\mu(0)\rangle \tag{15.11}$$

$$|\nu_2(0)\rangle = -\sin\theta\,|\nu_e(0)\rangle + \cos\theta\,|\nu_\mu(0)\rangle, \tag{15.12}$$

which can be inverted to give

$$|\nu_e(0)\rangle = \cos\theta\,|\nu_1(0)\rangle - \sin\theta\,|\nu_2(0)\rangle, \tag{15.13}$$

$$|\nu_\mu(0)\rangle = \sin\theta\,|\nu_2(0)\rangle + \cos\theta\,|\nu_1(0)\rangle, \tag{15.14}$$

where $|\nu_1(0)\rangle$ and $|\nu_2(0)\rangle$ are mass eigenstates at $t = 0$.

Consider the situation where at $t = 0$ the state $|\nu(t)\rangle$ representing a neutrino coincides with $|\nu_e(0)\rangle$. The time evolution of $|\nu(t)\rangle$ is then written as

$$|\nu(t)\rangle = \cos\theta\,|\nu_1(0)\rangle\, e^{-\frac{iE_1 t}{\hbar}} - \sin\theta\,|\nu_2(0)\rangle\, e^{-\frac{iE_2 t}{\hbar}}. \tag{15.15}$$

From the earlier discussions, we can write the probabilities that $|\nu(t)\rangle$ will reappear as $|\nu_e(0)\rangle$ or that it will appear as $|\nu_\mu(0)\rangle$ as, respectively,

$$P_{\nu_e\to\nu_e} = |\langle\nu_e(0)\,|\nu(t)\rangle|^2 = 1 - \sin^2 2\theta \sin^2\left[\frac{(E_1 - E_2)\,t}{2\hbar}\right], \tag{15.16}$$

$$P_{\nu_e\to\nu_\mu} = |\langle\nu_\mu(0)\,|\nu(t)\rangle|^2 = \sin^2 2\theta \sin^2\left[\frac{(E_1 - E_2)\,t}{2\hbar}\right]. \tag{15.17}$$

Since neutrinos are relativistic particles they satisfy the following energy–momentum relation

$$E = \sqrt{c^2p^2 + m^2c^4} \tag{15.18}$$

where p and m are the momentum and mass of the particle. Since the masses are extremely small compared with the momenta in the situations we are considering, we can make an expansion as follows

$$E \simeq cp\left[1 + \frac{m^2c^2}{2p^2}\right] = cp + \frac{m^2c^3}{2p}. \tag{15.19}$$

Hence,

$$E_1 - E_2 = \frac{\left(m_1^2 - m_2^2\right)c^3}{2p} \tag{15.20}$$

where we have assumed the momenta of the two neutrinos to be the same as they belong to the same beam with a fixed momentum.

Since it is easier and more instructive to measure the oscillation in terms of the distance traveled by the neutrino from the point of origin, rather than the time elapsed, one introduces an "oscillation length," l_{12}, given by

$$l_{12} = \frac{2\pi\hbar c}{|E_1 - E_2|}. \tag{15.21}$$

Because the sine functions in (15.16) appear quadratically, we can replace $(E_1 - E_2)$ by $|E_1 - E_2|$. Thus,

$$\frac{|E_1 - E_2|\,t}{2\hbar} = \left(\frac{\pi}{l_{12}}\right)\frac{c}{t}. \tag{15.22}$$

Since the speed of a neutrino will be essentially the same as the speed of light, c, the ratio (c/t) is just the distance the neutrino will travel from the point of origin. We denote this distance by x.

Therefore,

$$P_{\nu_e\to\nu_e} = 1 - \sin^2 2\theta\sin^2\left(\frac{\pi x}{l_{12}}\right), \tag{15.23}$$

$$P_{\nu_e\to\nu_\mu} = \sin^2 2\theta\sin^2\left(\frac{\pi x}{l_{12}}\right). \tag{15.24}$$

If neutrinos have traveled a long distance then it is more instructive to consider the average value of the probabilities. For that purpose we write

$$\sin^2\left(\frac{\pi x}{l_{12}}\right) = \frac{1}{2}\left[1 - \cos\left(\frac{2\pi x}{l_{12}}\right)\right]. \tag{15.25}$$

Since the oscillating cosine function will average to zero, we have

$$\left\langle \sin^2\left(\frac{\pi x}{l_{12}}\right)\right\rangle = \frac{1}{2}. \tag{15.26}$$

Hence,

$$\left\langle P_{\nu_e\to\nu_e}\right\rangle = 1 - \frac{1}{2}\sin^2 2\theta, \tag{15.27}$$

$$\left\langle P_{\nu_e\to\nu_\mu}\right\rangle = \frac{1}{2}\sin^2 2\theta. \tag{15.28}$$

Thus, the probability of finding the different flavor eigenstates is given by the mixing angles.

As a final note, we emphasize that the neutrinos undergo oscillations only if they have mass. It is the observation of such oscillations that led to the discovery that neutrinos that were thought to be massless in fact have mass, although quite minuscule. As we stated earlier, we have ignored the tau-neutrino. In a complete calculation they must, of course, be included. Not all the answers are in yet with regard to the masses and mixing angles of the neutrinos. This continues to remain a rich and exciting subject.

15.4 Decay and regeneration

Decays occur often in the regime of particle physics. Neutrons decay into protons, π^0-mesons decay into photons and, as we will consider below, K-mesons decay in a variety of different forms. And that is just a small set of examples. A particle left alone, or better still in the presence of other particles that can provide extra energy, can disappear. Even the proton may not be absolutely stable; its lifetime as currently measured is of the order of at least 10^{25} years, which is a huge number but still finite. Let us then pursue a very small portion of this subject further.

15.4.1 Basic formalism

If a particle at time t is described by an energy eigenstate $|\psi(t)\rangle = |\psi(0)\rangle \exp(-iE_n t/\hbar)$, corresponding to an energy E_n, then the probability of finding it, $|\langle\psi(t)\,|\psi(t)\rangle|^2$, is the same for all values of time. Consider now a situation where

$$|\langle\psi(t)\,|\psi(t)\rangle|^2 = 1 \quad \text{at} \quad t = 0 \quad \text{but} \quad |\langle\psi(t)\,|\psi(t)\rangle|^2 < 1 \;\; \text{for } t > 0. \tag{15.29}$$

We then say that the particle described by the state $|\psi(t)\rangle$ undergoes "decay" for $t > 0$. Such decays as we mentioned above are a common phenomenon in processes such as α-decay, β-decay, photon emission in atomic transitions, as well as in decays of π- and K-mesons.

In a decay process $A \to B+C$, one has a situation in which the particle A disappears and in its place particles B and C are created. One then requires that in any probability conservation relation we include $|\psi_A\rangle$, $|\psi_B\rangle$, and $|\psi_C\rangle$ as well as the quantum of the field that triggers the decay. We therefore need a formalism that allows for the creation and destruction of particles, which happens to be the subject of quantum field theory that includes strong and weak interactions. Except for basic quantum electrodynamics discussed in Chapters 41–43, this is beyond the scope of this book. For our purposes, we will, therefore, follow a more phenomenological path.

If we write

$$|\psi(t)\rangle = |\psi(0)\rangle\, e^{-i\alpha t/\hbar} \tag{15.30}$$

with the normalization

$$|\langle\psi(0)\,|\psi(0)\rangle|^2 = 1, \tag{15.31}$$

then to satisfy the condition that the particle decays for $t > 0$, we must have α complex. If we write

$$\alpha = a + ib \tag{15.32}$$

where a and b are real, then

$$|\psi(t)\rangle = |\psi(0)\rangle\, e^{-iat/\hbar} e^{-bt/\hbar} \tag{15.33}$$

and

$$|\langle\psi(t)\,|\psi(t)\rangle|^2 = e^{2bt/\hbar}. \tag{15.34}$$

In order for the decay to occur we must have $b < 0$. If we relate a to the energy eigenvalue, E, then we can write

$$\alpha = E - i\frac{\Gamma}{2} \quad \text{with} \quad \Gamma > 0. \tag{15.35}$$

We note that Γ like E has the dimensions of energy. The state vector is then given by

$$|\psi(t)\rangle = |\psi(0)\rangle \exp\left(-i\frac{Et}{\hbar} - \frac{\Gamma t}{2\hbar}\right), \tag{15.36}$$

with the probability expressed as

$$|\langle\psi(t)\,|\psi(t)\rangle|^2 = \exp\left(-\frac{\Gamma t}{\hbar}\right). \tag{15.37}$$

One calls $\hbar/\Gamma$ the "lifetime" of the particle and Γ the "decay width."

Let the total Hamiltonian responsible for the decay process be written as $H = H_0 + H'$; then the Schrödinger equation has the form

$$i\hbar\frac{\partial}{\partial t}|\psi(t)\rangle = \left(H_0 + H'\right)|\psi(t)\rangle. \tag{15.38}$$

Taking the derivative of $|\psi(t)\rangle$ given in (15.36) we find

$$i\hbar\frac{\partial}{\partial t}|\psi(t)\rangle = \left(E - i\frac{\Gamma}{2}\right)|\psi(t)\rangle . \tag{15.39}$$

While H_0 is Hermitian, we will, in this phenomenological approach, express H' in terms of real and imaginary parts and write $H' = H'_R + iH'_I$. If, furthermore, H'_I is a constant as a function of time, then it can be identified as simply $(-i\Gamma/2)$, so H is given by

$$H = H_0 + H'_R - i\frac{\Gamma}{2} \tag{15.40}$$

with

$$\left(H_0 + H'_R\right)|\psi(t)\rangle = E\,|\psi(t)\rangle . \tag{15.41}$$

By making the choice (15.40) for the Hamiltonian we have given up the requirement that H be Hermitian. But as we explained before, we are following a phenomenological approach to tackle this problem.

15.4.2 Current conservation with complex potentials

Let us go back to the beginning when we considered the Schrödinger equation with a potential $V(r)$ and obtained the following current conservation relation:

$$\nabla\cdot\mathbf{j} = -\frac{\partial\rho}{\partial t}. \tag{15.42}$$

As we mentioned at the time, the left-hand side is related to the flow of current crossing a closed surface around a volume and the right-hand side with a negative sign indicates a decrease in charge. In other words, the equation tells us that the current leaving a surface amounts to a decrease in the overall charge. Basically, in terms of number of particles, if there are 5 particles at rest inside a volume then, since they have zero velocity, there will be no current crossing the surface surrounding the volume. Thus the left-hand side will be zero. Then, according to the equation, the right-hand side will also be zero, confirming that there is no change in the number of particles from the initial number 5. However if, say, 3 of 5 particles inside the volume decay into products that are also essentially stationary, the left-hand is again zero. The right-hand side, on the other hand, will continue to indicate that the number of particles has not changed from the initial count of 5. Thus the relation (15.42) cannot be correct if the particles undergo decay. If we examine the above relation closely, we note that the assumption that the potential, $V(r)$, is real played a crucial part in the derivation of this equation.

Let us, therefore, take another look at our derivation of current conservation. The Schrödinger equation is given by

$$-\frac{\hbar^2}{2m}\nabla^2\psi + V(r)\,\psi = i\hbar\frac{\partial\psi}{\partial t}. \tag{15.43}$$

Taking the complex conjugate, we find

$$-\frac{\hbar^2}{2m}\nabla^2\psi^* + V^*(r)\,\psi^* = -i\hbar\frac{\partial\psi^*}{\partial t}. \tag{15.44}$$

Multiplying (15.43) by ψ^* and (15.44) by ψ and subtracting, we obtain

$$-\frac{\hbar^2}{2m}\left(\psi^*\nabla^2\psi - \psi\nabla^2\psi^*\right) + (V - V^*)\,\psi\psi^* = i\hbar\left(\psi^*\frac{\partial\psi}{\partial t} + \psi\frac{\partial\psi^*}{\partial t}\right). \tag{15.45}$$

From our discussion in Chapter 1, the probability density and probability current density are given by

$$\rho = \psi^*\psi, \tag{15.46}$$

$$\mathbf{j} = \frac{\hbar}{2im}\left(\psi^*\nabla\psi - \psi\nabla\psi^*\right), \tag{15.47}$$

respectively. If we take the potential to be complex,

$$V(r) = V_R + iV_I, \tag{15.48}$$

then the differential equation (15.45) can be expressed as

$$\nabla\cdot\mathbf{j} + \frac{\partial\rho}{\partial t} = \frac{2V_I}{\hbar}\rho. \tag{15.49}$$

For a situation where the particle is at rest, i.e., in the static case, we have $\mathbf{j} \approx 0$; then equation (15.49) becomes

$$\frac{\partial\rho}{\partial t} = \frac{2V_I}{\hbar}\rho, \tag{15.50}$$

which gives rise to the solution

$$\rho = \rho_0 \exp\left(\frac{2V_I}{\hbar}t\right) \tag{15.51}$$

where we assume V_I to be a constant independent of $\mathbf{r}$ and t. To prevent ρ from increasing indefinitely with time, we must have $V_I < 0$, i.e., $V_I = -|V_I|$. The probability density is then

$$\rho = \rho_0 \exp\left(-\frac{2|V_I|}{\hbar}t\right). \tag{15.52}$$

This result then corresponds to the "decay" of the system. If, as in the previous case, we identify $|V_I| = \Gamma/2$, then we can write

$$\rho = \rho_0 \exp\left(-\frac{\Gamma}{\hbar}t\right). \tag{15.53}$$

The wavefunction is now of the form

$$\psi(\mathbf{r}) = \psi_0(\mathbf{r}) \exp\left(-i\frac{Et}{\hbar} - \frac{\Gamma t}{2\hbar}\right), \tag{15.54}$$

with the potential given by

$$V(r) = V_R(r) - i\frac{\Gamma}{2}. \tag{15.55}$$

15.5 Oscillation and regeneration of stable and unstable systems

In the previous section, we described how a decay process can be incorporated into the formalism phenomenologically. In the following we begin with the problem at $t = 0$ so that we can apply the results that were obtained from our earlier time-independent considerations.

We will frame this as a two-level problem. We will assume, as we did in our previous problems, that the Hamiltonian is a 2×2 matrix, which now has complex matrix elements and will not necessarily be Hermitian.

Consider the states $|\psi(t)\rangle$ and $|\overline{\psi}(t)\rangle$ as representing a two-level system and, furthermore, assume that at $t = 0$ these states are related to each other by a discrete, unitary, transformation U, with $U^2 = 1$:

$$U|\psi(0)\rangle = |\overline{\psi}(0)\rangle, \tag{15.56}$$

$$U|\overline{\psi}(0)\rangle = |\psi(0)\rangle. \tag{15.57}$$

If U commutes with the Hamiltonian, i.e., $[U, H] = UH - HU = 0$, then we write the following relations involving the diagonal and off-diagonal elements of the matrix, H, expressed as a complex quantity of the type $E - i1/2\Gamma$. That is, we take

$$\langle\psi(0)|H|\psi(0)\rangle = E_0 - i\frac{1}{2}\Gamma_0, \tag{15.58}$$

$$\langle\psi(0)|H|\overline{\psi}(0)\rangle = E_0' - i\frac{1}{2}\Gamma_0'. \tag{15.59}$$

Using the property $UH - HU = 0$, we obtain

$$\langle\overline{\psi}(0)|H|\overline{\psi}(0)\rangle = \langle\overline{\psi}(0)|U^{-1}HU|\overline{\psi}(0)\rangle = \langle\psi(0)|H|\psi(0)\rangle = E_0 - i\frac{1}{2}\Gamma_0, \tag{15.60}$$

$$\langle\overline{\psi}(0)|H|\psi(0)\rangle = \langle\overline{\psi}(0)|U^{-1}HU|\psi(0)\rangle = \langle\psi(0)|H|\overline{\psi}(0)\rangle = E_0' - i\frac{1}{2}\Gamma_0'. \tag{15.61}$$

The relation (15.60) is equivalent to $H_{11} = H_{22}$ for a description in terms of a 2×2 matrix and the relation (15.61) is equivalent to $H_{12} = H_{21}$ (but not H_{21}^*).

To obtain the eigenvalues of H in this two-level system consisting of $|\psi(0)\rangle$, and $|\overline{\psi}(0)\rangle$, we need to diagonalize the following Hamiltonian matrix, H, in the $\{|\psi(0)\rangle, |\overline{\psi}(0)\rangle\}$ basis, written in the familiar form

$$\begin{bmatrix} A & D \\ D & A \end{bmatrix} \tag{15.62}$$

where A and D are the complex quantities

$$A = E_0 - i\frac{1}{2}\Gamma_0, \tag{15.63}$$

$$D = E_0' - i\frac{1}{2}\Gamma_0'. \tag{15.64}$$

We will assume that the diagonalization process we pursued previously is valid even though the Hamiltonian is complex and non-Hermitian.

The eigenvalues of the matrix (15.62) are

$$E_+ = A + D = E_0 + E_0' - i\frac{1}{2}(\Gamma_0 + \Gamma_0'), \tag{15.65}$$

$$E_- = A - D = E_0 - E_0' - i\frac{1}{2}(\Gamma_0 - \Gamma_0'). \tag{15.66}$$

We assume the imaginary parts to be small. The mixing angle from our previous (time-independent) considerations is found to be $\theta = \pi/2$, neglecting the contribution of the imaginary parts. Thus, the eigenstates of H are

$$|\psi_+(0)\rangle = \frac{|\psi(0)\rangle + |\overline{\psi}(0)\rangle}{\sqrt{2}}, \tag{15.67}$$

$$|\psi_-(0)\rangle = \frac{|\psi(0)\rangle - |\overline{\psi}(0)\rangle}{\sqrt{2}}. \tag{15.68}$$

We note from (15.56) and (15.57) that

$$U|\psi_+(0)\rangle = |\psi_+(0)\rangle, \tag{15.69}$$

$$U|\psi_-(0)\rangle = -|\psi_-(0)\rangle. \tag{15.70}$$

We then say that $|\psi_+(0)\rangle$ has a positive U-parity and $|\psi_-(0)\rangle$ has a negative U-parity.

Consider the following specific example where

$$\Gamma_0' \lesssim \Gamma_0, \tag{15.71}$$

which corresponds to the well-known problem of $K^0 - \overline{K}^0$ mixing in particle physics. Thus, $|\psi_+(0)\rangle$ will correspond to a state with a large decay width $(\Gamma_0 + \Gamma_0')$, which we will designate as $|\psi_S(0)\rangle$ and Γ_S, respectively, signifying a "short-lived" state, and write the real part of the energy eigenvalue, $(E_0 + E_0')$, as E_S. Similarly $|\psi_-(0)\rangle$ will be a "long-lived" state, designated by $|\psi_L(0)\rangle$, with real and imaginary parts of the eigenvalues, $(E_0 - E_0')$ and $(\Gamma_0 - \Gamma_0')$, written as E_L and Γ_L, respectively. To summarize then, we have

$$\Gamma_L < \Gamma_S \tag{15.72}$$

and

$$|\psi_S(0)\rangle = \frac{|\psi(0)\rangle + |\overline{\psi}(0)\rangle}{\sqrt{2}}, \tag{15.73}$$

$$|\psi_L(0)\rangle = \frac{|\psi(0)\rangle - |\overline{\psi}(0)\rangle}{\sqrt{2}}. \tag{15.74}$$

The time dependence of these state vectors is given by

$$|\psi_S(t)\rangle = |\psi_S(0)\rangle e^{-\frac{i}{\hbar}(E_S - i\frac{1}{2}\Gamma_S)t}, \tag{15.75}$$

$$|\psi_L(t)\rangle = |\psi_L(0)\rangle e^{-\frac{i}{\hbar}(E_L - i\frac{1}{2}\Gamma_L)t}. \tag{15.76}$$

Conversely,

$$|\psi(0)\rangle = \frac{|\psi_S(0)\rangle + |\psi_L(0)\rangle}{\sqrt{2}}, \tag{15.77}$$

$$|\overline{\psi}(0)\rangle = \frac{|\psi_S(0)\rangle - |\psi_L(0)\rangle}{\sqrt{2}}, \tag{15.78}$$

with their time dependence given by

$$|\psi(t)\rangle = \frac{|\psi_S(t)\rangle + |\psi_L(t)\rangle}{\sqrt{2}}, \tag{15.79}$$

$$|\overline{\psi}(t)\rangle = \frac{|\psi_S(t)\rangle - |\psi_L(t)\rangle}{\sqrt{2}}, \tag{15.80}$$

where the time dependences of $|\psi_S(t)\rangle$ and $|\psi_L(t)\rangle$ are already given in (15.75) and (15.76).

(i) *Stable case*

We first consider the case where $|\psi_S(t)\rangle$ and $|\psi_L(t)\rangle$ are stable, that is, $\Gamma_S = \Gamma_L = 0$:

$$|\psi_S(t)\rangle = |\psi_S(0)\rangle e^{-\frac{i}{\hbar}E_S t}, \tag{15.81}$$

$$|\psi_L(t)\rangle = |\psi_L(0)\rangle e^{-\frac{i}{\hbar}E_L t}.$$

We will then have

$$|\psi(t)\rangle = \frac{|\psi_S(0)\rangle e^{-\frac{i}{\hbar}E_S t} + |\psi_L(0)\rangle e^{-\frac{i}{\hbar}E_L t}}{\sqrt{2}}. \tag{15.82}$$

To understand further the time development of $|\psi(t)\rangle$ we write

$$|\psi(t)\rangle = e^{-\frac{i}{\hbar}E_S t}\frac{\left[|\psi_S(0)\rangle + |\psi_L(0)\rangle e^{-\frac{i}{\hbar}(E_L-E_S)t}\right]}{\sqrt{2}}, \tag{15.83}$$

where we assume $E_L > E_S$. Note that at $t = \pi\hbar/(E_L - E_S)$ we have

$$|\psi(t)\rangle \sim \frac{[|\psi_S(0)\rangle - |\psi_L(0)\rangle]}{\sqrt{2}} = |\overline{\psi}(0)\rangle. \tag{15.84}$$

Thus we have a conversion

$$|\psi\rangle \rightarrow |\overline{\psi}\rangle. \tag{15.85}$$

Consequently, there will be an oscillation of $|\psi\rangle \rightleftarrows |\overline{\psi}\rangle$ as time progresses, a process we are already familiar with from a number of examples considered earlier, including neutrino oscillations. This particular case corresponds to the well-known $|K^0\rangle \rightleftarrows |\overline{K}^0\rangle$ oscillation.

(ii) *Unstable case*

If $|\psi_S\rangle$ and $|\psi_L\rangle$ are unstable then for $|\psi(t)\rangle$ we obtain

$$|\psi(t)\rangle = \frac{|\psi_S(0)\rangle e^{-iE_S t/\hbar}e^{-\Gamma_S t/2\hbar} + |\psi_L(0)\rangle e^{-iE_L t/\hbar}e^{-\Gamma_L t/2\hbar}}{\sqrt{2}}. \tag{15.86}$$

The probability of finding a particle in a state $|\psi(t)\rangle$ is given by

$$N = |\langle\psi(t)|\psi(t)\rangle|^2 \tag{15.87}$$

and the corresponding probability for $|\overline{\psi}(t)\rangle$ is

$$\overline{N} = |\langle\overline{\psi}(t)|\overline{\psi}(t)\rangle|^2. \tag{15.88}$$

Since $|\psi(t)\rangle$ and $|\overline{\psi}(t)\rangle$ correspond to unstable states, neither N nor $\overline{N}$ is expected to be 1.

We obtain the following:

$$N = \frac{1}{4}\left\{e^{-\frac{\Gamma_S}{\hbar}t} + e^{-\frac{\Gamma_L}{\hbar}t} + 2e^{-\frac{(\Gamma_S+\Gamma_L)}{2\hbar}t}\cos\left[\frac{(E_S-E_L)}{\hbar}t\right]\right\}, \tag{15.89}$$

$$\overline{N} = \frac{1}{4}\left\{e^{-\frac{\Gamma_S}{\hbar}t} + e^{-\frac{\Gamma_L}{\hbar}t} - 2e^{-\frac{(\Gamma_S+\Gamma_L)}{2\hbar}t}\cos\left[\frac{(E_S-E_L)}{\hbar}t\right]\right\}. \tag{15.90}$$

We thus see that there is an oscillation between N and $\overline{N}$ while, at the same time, each decreases in time.

We now consider the time-development of $|\psi(t)\rangle$. Since $\Gamma_L < \Gamma_S$, then for $\hbar/\Gamma_S < t < \hbar/\Gamma_L$, we have

$$|\psi(t)\rangle \approx |\psi_L(0)\rangle e^{-\frac{i}{\hbar}(E_L - i\frac{1}{2}\Gamma_L)t}. \tag{15.91}$$

Thus after a sufficient lapse of time we have only the $|\psi_L\rangle$ state left since it is relatively a long-lived state. We express it in terms of $|\psi(0)\rangle$ and $|\overline{\psi}(0)\rangle$ as

$$|\psi_L(0)\rangle = \frac{|\psi(0)\rangle - |\overline{\psi}(0)\rangle}{\sqrt{2}}. \tag{15.92}$$

Let us assume that since the particle represented by $|\psi_L\rangle$ is traveling through matter, $|\psi\rangle$ and $|\overline{\psi}\rangle$ undergo "scattering," that is, the interaction in matter forces a change in their motion that depends on whether it is $|\psi(0)\rangle$ or $|\overline{\psi}(0)\rangle$. Hence, this interaction will lead to $|\psi(0)\rangle \to f\,|\psi(0)\rangle$ and $|\overline{\psi}(0)\rangle \to \overline{f}\,|\overline{\psi}(0)\rangle$. We then have

$$|\psi_L(0)\rangle \to \frac{f\,|\psi(0)\rangle - \overline{f}\,|\overline{\psi}(0)\rangle}{\sqrt{2}}, \tag{15.93}$$

where f and $\overline{f}$ are the so-called scattering amplitudes.

Writing $|\psi\rangle$ and $|\overline{\psi}\rangle$ back in terms of $|\psi_S\rangle$ and $|\psi_L\rangle$, we find that

$$|\psi_L(0)\rangle \to \frac{1}{2}\left[\left(f - \overline{f}\right)|\psi_S(0)\rangle + \left(f + \overline{f}\right)|\psi_L(0)\rangle\right]. \tag{15.94}$$

If $f \neq \overline{f}$ then $|\psi_S\rangle$, which had decayed and was assumed long gone, is now "regenerated." The quantity

$$f - \overline{f} = f_{21} \tag{15.95}$$

is called the regeneration amplitude. Clearly this is possible only if particles represented by $|\psi\rangle$ and $|\overline{\psi}\rangle$ scatter differently.

The results we obtained in this section are confirmed in the $|K^0\rangle - |\overline{K}^0\rangle$ system in particle physics. We elaborate on this below.

15.6 Neutral K-mesons

All particles have corresponding antiparticles with the same mass, lifetime, spin, and a few other characteristics but with one or two other quantum numbers that have opposite signs. Thus, electrons and protons have positrons and antiprotons that are identical in every way except for electric charge in both cases and, in the case of protons, opposite so-called baryon number. Neutrons have no charge, but the antineutrons have opposite baryon number. The truly neutral particles such as photons and π^0-mesons are identical to their antiparticles. The neutral K-mesons belong to a special class of elementary particles.

15.6.1 The K^0, $\overline{K}^0$, K_S, and K_L mesons

$K^0$and $\overline{K}^0$ are spin 0 particles that form a particle–antiparticle pair that is degenerate in mass and have many other properties that are identical, except for the quantum number called "strangeness." K^0 has positive strangeness while $\overline{K}^0$ has negative strangeness. These mesons undergo both strong and weak interactions. While strong interactions, which include nuclear interactions, conserve strangeness, weak interactions do not. As a consequence, while strong interactions allow K^0 and $\overline{K}^0$ to maintain their individual identity, the presence of weak interactions can trigger transitions between K^0 and $\overline{K}^0$. It is found that K^0 and $\overline{K}^0$ can be related to each other by a transformation that is a product of the transformations called charge conjugation, (C) and parity (P):

$$CP\left|K^0\right\rangle = \left|\overline{K}^0\right\rangle. \tag{15.96}$$

Thus CP acts as the U operator discussed earlier. The weak interaction is assumed to preserve the CP quantum number, but not strangeness. The strong interaction, on the other hand, preserves C and P (therefore, also the product CP), as well as strangeness. One therefore forms linear combinations according to a specific CP quantum number (also called CP "parity"). Hence, one forms

$$|K_S\rangle = \frac{\left|K^0\right\rangle + \left|\overline{K}^0\right\rangle}{\sqrt{2}} \quad \text{with} \quad CP\,|K_S\rangle = (+)\,|K_S\rangle, \tag{15.97}$$

which has a positive CP parity, and

$$|K_L\rangle = \frac{\left|K^0\right\rangle - \left|\overline{K}^0\right\rangle}{\sqrt{2}} \quad \text{with} \quad CP\,|K_L\rangle = (-)\,|K_L\rangle, \tag{15.98}$$

which has a negative CP parity. Thus $|K_S\rangle$ and $|K_L\rangle$ eigenstates are relevant in a Hamiltonian that contains terms corresponding to weak interactions.

Comparing the K-meson formalism with the formalism in the previous section, we make the following identifications:

$$|\psi\rangle \to \left|K^0\right\rangle, \quad \left|\overline{\psi}\right\rangle \to \left|\overline{K}^0\right\rangle, \quad |\psi_L\rangle \to |K_L\rangle, \quad |\psi_S\rangle \to |K_S\rangle \tag{15.99}$$

and

$$U = CP. \tag{15.100}$$

The state $|K_S\rangle$ is short-lived with a predominant decay, $K_S \to 2\pi$, and a lifetime of $\tau_S = .89 \times 10^{-10}$ s, while the long-lived, $|K_L\rangle$ decays as $K_L \to 3\pi$ with a lifetime $\tau_L = 5.17 \times 10^{-8}$ s. Their mass difference is found to be $\Delta m = 3.49 \times 10^{-12}$ MeV.

If, again, we consider the particles at rest, and take the relativistic result $E = mc^2$ then the expectation value of the Hamiltonian H gives the masses of K_S and K_L :

$$m_S = \langle K_S \, |H| \, K_S \rangle = \frac{1}{2} \langle (K + \overline{K}) \, |H| \, (K + \overline{K}) \rangle \tag{15.101}$$

and

$$m_L = \langle K_L \, |H| \, K_L \rangle = \frac{1}{2} \langle (K - \overline{K}) \, |H| \, (K - \overline{K}) \rangle . \tag{15.102}$$

Hence the mass difference is given by

$$\Delta m = m_S - m_L = \langle K \, |H| \, \overline{K} \rangle + \langle \overline{K} \, |H| \, K \rangle . \tag{15.103}$$

Thus, the mass difference is given by the matrix element responsible for the $K^0 \rightleftarrows \overline{K}^0$ oscillations.

If we consider the particles at rest, we can replace E_S, E_L by the masses m_S, m_L respectively. The decay probabilities are then given by

$$N(K) = \frac{1}{4} \left\{ e^{-\frac{\Gamma_S}{\hbar} t} + e^{-\frac{\Gamma_L}{\hbar} t} + 2e^{-\frac{(\Gamma_S + \Gamma_L)}{2\hbar} t} \cos \Delta m t \right\} , \tag{15.104}$$

$$\overline{N}(K) = \frac{1}{4} \left\{ e^{-\frac{\Gamma_S}{\hbar} t} + e^{-\frac{\Gamma_L}{\hbar} t} - 2e^{-\frac{(\Gamma_S + \Gamma_L)}{2\hbar} t} \cos \Delta m t \right\} . \tag{15.105}$$

15.6.2 Regeneration of K-mesons

In a typical example of the regeneration process, one starts with a beam of K^0-mesons, which is an equal mixture of K_S and K_L. After 10^{-8}s but before 10^{-10}s, all K_S disappear leaving behind essentially a pure beam of K_L, which is an equal mixture of $K^0$and $\overline{K}^0$. If now a nuclear material is interposed then, because of the very different interactions of K^0 and $\overline{K}^0$ with matter, it turns out that $\overline{K}^0$ is almost entirely absorbed and one is left with a K^0, and contained in it the K_S, providing a startling case of regeneration! (See Fig. 15.1.)

This "schizophrenic" property of neutral K-mesons is a quintessentially quantum-mechanical phenomenon. The K^0- or $\overline{K}^0$-meson is a superposition of K_L and K_S, which in turn, are superpositions of K^0 and $\overline{K}^0$ and so on. What gets measured depends on the type of the measuring device and what it is supposed to measure. If it is a nuclear (strong)

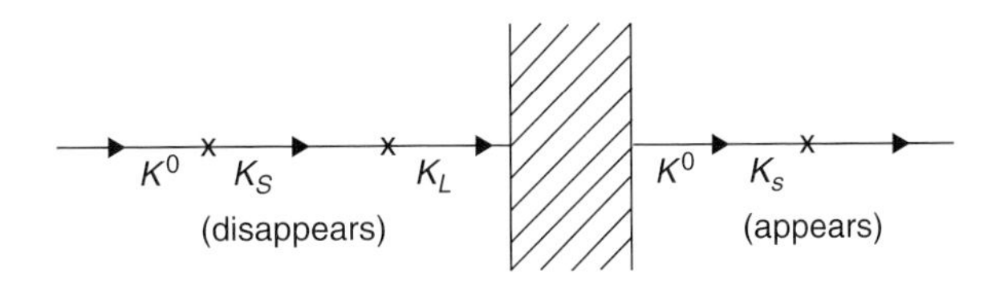

Fig. 15.1

interaction or an interaction that distinguishes particles from antiparticles and conserves strangeness, then it involves K^0 or $\overline{K}^0$. If the interaction allows decays into channels with definite CP parity, but not necessarily strangeness, like decays into 2π's or 3π's, then it involves K_S and K_L.

15.7 Problems

1. Consider the following Hamiltonian for a system of three neutrinos $(\nu_e, \nu_\mu, \nu_\tau)$. Assume the particles to be at rest, with the Hamiltonian (also called the mass matrix) given by the 3×3 matrix

$$\hbar\omega_0 \begin{bmatrix} 1 & \lambda & \lambda \\ \lambda & 1 & \lambda \\ \lambda & \lambda & 1 \end{bmatrix}$$

and the unperturbed state vectors as

$$\begin{bmatrix} |\nu_e\rangle \\ |\nu_\mu\rangle \\ |\nu_\tau\rangle \end{bmatrix}.$$

Determine the eigenvalues and eigenfunctions for this system.

2. Assume in Problem 1 that at $t = 0$ only $|\nu_e(t)\rangle$ is present. Determine the time dependence of $|\nu_e(t)\rangle$, $|\nu_\mu(t)\rangle$, $|\nu_\tau(t)\rangle$ and the probability that one will find ν_μ and ν_τ for $t > 0$.

16 Time-independent perturbation for bound states

Here we tackle problems that cannot be solved exactly, which means that one must resort to approximations. Among the methods available for this purpose, perturbation methods are the most commonly used provided certain conditions are met. One starts with the unperturbed system that is solvable and closest to the problem at hand. The interaction Hamiltonian is expressed as a sum of the unperturbed, solvable, part and the new interaction term. The problem is solved essentially as a power series in the strength of this interaction term. In this chapter we consider those cases that do not depend on time.

16.1 Basic formalism

In Chapter 8 we considered bound-state problems that were solvable exactly. With the Hamiltonian that was provided we were able to obtain exact results for the eigenvalues and eigenstates. In many practical situations, however, this is not feasible because the Hamiltonian is often too complicated to render an exact solution. The perturbation method we will outline below allows one to obtain reliable approximate solutions provided the new Hamiltonian, which we call the "perturbed" Hamiltonian, is not too different from the original Hamiltonian, which we call the "unperturbed" Hamiltonian. Here again we will confine ourselves to bound states.

Let H_0 be the unperturbed and H the perturbed Hamiltonians such that

$$H = H_0 + \lambda H' \tag{16.1}$$

where H' is the "perturbation" which, in some sense, is assumed to be small. H_0 and H satisfy the eigenvalue equations

$$H_0 \left| \psi_s^{(0)} \right\rangle = E_s^{(0)} \left| \psi_s^{(0)} \right\rangle, \tag{16.2}$$

$$H \left| \psi_s \right\rangle = E_s \left| \psi_s \right\rangle, \tag{16.3}$$

with $\left|\psi_s^0\right\rangle$ and $\left|\psi_s\right\rangle$ as the eigenstates and E_s^0 and E_s as the eigenvalues of H_0 and H, respectively. We have put a multiplicative factor λ in (16.1) primarily for bookkeeping purposes, so that

$$\left|\psi_s\right\rangle \to \left|\psi_s^{(0)}\right\rangle \quad \text{and} \quad E_s \to E_s^{(0)} \quad \text{as } \lambda \to 0 \tag{16.4}$$

when the contribution of H' vanishes. Using perturbation theory, therefore, makes good sense if the problem at hand is sufficiently similar to the original problem whose solutions are known exactly.

The perturbation method relies on obtaining the solutions as a power series in λ. The central assumption is that such a power series converges and that the solutions are smooth and continuous functions of λ, so that one can obtain the required solution at $\lambda = 1$. The energy eigenvalue E_s is therefore expressed as an infinite series (a "perturbative" series) in λ,

$$E_s = E_s^{(0)} + \lambda E_s^{(1)} + \lambda^2 E_s^{(2)} + \cdots . \tag{16.5}$$

Consequently, since $|\psi_n^0\rangle$ form a complete set, the perturbed states $|\psi_n\rangle$ can be expressed as follows

$$|\psi_s\rangle = \sum_n c_{ns} \left|\psi_n^{(0)}\right\rangle \tag{16.6}$$

where the coefficients c_{ns} are also expanded in a perturbative series,

$$c_{ns} = c_{ns}^{(0)} + \lambda c_{ns}^{(1)} + \lambda^2 c_{ns}^{(2)} + \cdots . \tag{16.7}$$

From (16.4) we obtain

$$c_{ns}^{(0)} = \delta_{ns}. \tag{16.8}$$

We then have

$$|\psi_s\rangle = \left|\psi_s^{(0)}\right\rangle + \lambda \sum_n c_{ns}^{(1)} \left|\psi_n^{(0)}\right\rangle + \lambda^2 \sum_n c_{ns}^{(2)} \left|\psi_n^{(0)}\right\rangle + \cdots . \tag{16.9}$$

Pulling out the term with $|\psi_s^0\rangle$ from each of the above sums we have

$$|\psi_s\rangle = \left(1 + \lambda c_{ss}^{(1)} + \cdots\right) \left|\psi_s^{(0)}\right\rangle + \lambda \sum_{n \neq s} c_{ns}^{(1)} \left|\psi_n^{(0)}\right\rangle + \cdots . \tag{16.10}$$

The first term contains an infinite series which we will normalize to unity. Thus one can write

$$1 + \lambda c_{ss}^{(1)} + \cdots = e^{i\lambda\xi}. \tag{16.11}$$

The first term will then acquire a phase. In the discussion to follow we will take $\xi = 0$. Hence we write

$$|\psi_s\rangle = \left|\psi_s^{(0)}\right\rangle + \lambda \sum_{n \neq s} c_{ns}^{(1)} \left|\psi_n^{(0)}\right\rangle + \cdots . \tag{16.12}$$

Since $\left|\psi_s^{(0)}\right\rangle$ are assumed to be normalized to unity, it is clear that $|\psi_s\rangle$ cannot be so normalized. Let us write

$$|\psi_s\rangle = Z_s^{-1/2} |\psi_s\rangle_R \tag{16.13}$$

where we take $|\psi_s\rangle_R$ to be normalized to unity. Thus from (16.12) using the orthonormality property of $\left|\psi_n^{(0)}\right\rangle$ we obtain

$$Z_s^{-1} = 1 + \lambda^2 \sum_{n \neq s} \left|c_{ns}^{(1)}\right|^2 + \cdots . \tag{16.14}$$

Z_s is called the renormalization constant for the state vector $|\psi_s\rangle$, and $|\psi_s\rangle_R$ is called the renormalized state.

Inverting this relation (assuming λ to be small) we find

$$Z_s = 1 - \lambda^2 \sum_{n \neq s} \left|c_{ns}^{(1)}\right|^2 \cdots . \tag{16.15}$$

Thus Z_s is less than 1. The amount by which it is reduced from 1 is related to the probability of transitions from the initial unperturbed state $\left|\psi_s^{(0)}\right\rangle$ to other unperturbed states.

Also, from (16.13), since $\langle\psi_s^{(0)}|\psi_s\rangle = 1$, therefore,

$$Z_s^{1/2} = \langle\psi_s^{(0)}|\psi_s\rangle_R . \tag{16.16}$$

Thus one can interpret Z_s as the probability of finding the (renormalized) perturbed state in the original unperturbed state.

In the following we will write the series (16.12) as

$$|\psi_s\rangle = \left|\psi_s^{(0)}\right\rangle + \lambda\left|\psi_s^{(1)}\right\rangle + \lambda^2\left|\psi_s^{(2)}\right\rangle + \cdots \tag{16.17}$$

where

$$\left|\psi_s^{(1)}\right\rangle = \sum_{n \neq s} c_{ns}^{(1)}\left|\psi_n^{(0)}\right\rangle, \quad \left|\psi_s^{(2)}\right\rangle = \sum_{n \neq s} c_{ns}^{(2)}\left|\psi_n^{(0)}\right\rangle, \ldots \tag{16.18}$$

Equation (16.3) can be written in a fully expanded series as follows:

$$\begin{aligned}&\left(H_0 + \lambda H'\right)\left\{\left|\psi_s^{(0)}\right\rangle + \lambda\left|\psi_s^{(1)}\right\rangle + \lambda^2\left|\psi_s^{(2)}\right\rangle + \cdots\right\}\\ &\quad = \left(E_s^{(0)} + \lambda E_s^{(1)} + \lambda^2 E_s^{(2)} + \cdots\right)\left\{\left|\psi_s^{(0)}\right\rangle + \lambda\left|\psi_s^{(1)}\right\rangle + \lambda^2\left|\psi_s^{(2)}\right\rangle + \cdots\right\}.\end{aligned} \tag{16.19}$$

The above series, after we move all the terms to the left, will be of the form

$$A + B\lambda + C\lambda^2 + \cdots = 0. \tag{16.20}$$

Our basic assumption is that the above equation is valid for continuous values of λ in which case the only possible solution is

$$A = 0 = B = C = \cdots . \tag{16.21}$$

This result implies that we equate the coefficients of equal powers of λ on the two sides of the equation (16.19). Hence,

$$\left(H_0 - E_s^{(0)}\right)\left|\psi_s^{(0)}\right\rangle = 0, \tag{16.22}$$

$$\left(H_0 - E_s^{(0)}\right)\left|\psi_s^{(1)}\right\rangle = \left(E_s^{(1)} - H'\right)\left|\psi_s^{(0)}\right\rangle, \tag{16.23}$$

$$\left(H_0 - E_s^{(0)}\right)\left|\psi_s^{(2)}\right\rangle = \left(E_s^{(1)} - H'\right)\left|\psi_s^{(1)}\right\rangle + E_s^{(2)}\left|\psi_s^{(0)}\right\rangle, \tag{16.24}$$

and so on.

Equation (16.22) is trivial since it reproduces the known result (16.2). In the remaining equations, we use the orthogonality property

$$\left\langle \psi_s^{(0)} | \psi_s^{(j)} \right\rangle = 0 \ \text{ for } j \neq 0 \tag{16.25}$$

to determine the constants, $c_{ms}^{(i)}$ $(i = 1, 2, \ldots)$. Multiplying both sides of (16.23) by $\left\langle \psi_s^{(0)} \right|$ we obtain

$$E_s^{(1)} = \left\langle \psi_s^{(0)} \right| H' \left| \psi_s^{(0)} \right\rangle = H'_{ss}. \tag{16.26}$$

Thus the first-order correction to the energy eigenvalues is simply given by the expectation value of H'.

Multiplying equation (16.23) by $\left\langle \psi_m^{(0)} \right|$ with $m \neq s$, we obtain the coefficient

$$c_{ms}^{(1)} = \frac{H'_{ms}}{\left(E_s^{(0)} - E_m^{(0)}\right)}. \tag{16.27}$$

Similarly, by multiplying (16.24) by $\left\langle \psi_s^{(0)} \right|$ we get

$$E_s^{(2)} = \sum_{n \neq s} \frac{\left|H'_{ns}\right|^2}{\left(E_s^{(0)} - E_n^{(0)}\right)} \tag{16.28}$$

where we have used the fact that H' is Hermitian, $H'^{\dagger} = H'$.

We notice that in the second order term, $E_s^{(2)}$, if the levels E_n^0 and E_s^0 are adjacent then the lower of the two becomes even lower and the higher of the two becomes higher. There is, thus, no crossing of the levels due to perturbation.

One can now continue in a systematic way to obtain the eigenstates and eigenvalues to all orders in λ.

Summarizing, then, we have the following results for E_s to second order in λ, and $|\psi_s\rangle$ to first order:

$$E_s = E_s^{(0)} + \lambda H'_{ss} + \lambda^2 \sum_{n \neq s} \frac{|H'_{ns}|^2}{\left(E_s^{(0)} - E_n^{(0)}\right)} + \cdots, \tag{16.29}$$

$$|\psi_s\rangle = \left|\psi_s^{(0)}\right\rangle + \lambda \sum_{n \neq s} \frac{H'_{ns}}{\left(E_s^{(0)} - E_n^{(0)}\right)} \left|\psi_n^{(0)}\right\rangle + \cdots. \tag{16.30}$$

We can put $\lambda = 1$ assuming that the infinite series converges.

Let us consider some examples.

16.2 Harmonic oscillator: perturbative vs. exact results

16.2.1 Energy levels

The Schrödinger equation for the harmonic oscillator is

$$-\frac{\hbar^2}{2m}\frac{d^2u}{dx^2} + \frac{1}{2}m\omega^2 x^2 u = Eu. \tag{16.31}$$

Consider a linear perturbation

$$H' = x.$$

Here x is to be considered as an operator even though we have not written it in capital letters. This problem can be solved quite simply in terms of the raising and lowering operators $a^\dagger$ and a, respectively, which we have already discussed in Chapter 9. We write x in terms of them:

$$x = \sqrt{\frac{\hbar}{2m\omega}}\left(a + a^\dagger\right). \tag{16.32}$$

A typical matrix element of H' is then

$$H'_{ns} = \langle n| H' |s\rangle = \sqrt{\frac{\hbar}{2m\omega}}[\langle n| a |s\rangle + \langle n| a^\dagger |s\rangle] \tag{16.33}$$

$$= \sqrt{\frac{\hbar}{2m\omega}}\left[\sqrt{s}\delta_{n,s-1} + \sqrt{s+1}\delta_{n,s+1}\right] \tag{16.34}$$

where we have used the properties of $a^\dagger$ and a.

We now wish to calculate the energy eigenvalues and the ground-state energy eigenfunction in perturbative expansion. As it turns out, this problem can also be solved exactly. This will enable us to compare the exact results when expanded in powers of λ with the results obtained from perturbative expansion in terms of λ.

First we consider the perturbative terms. The unperturbed energy eigenvalue is

$$E_s^{(0)} = \left(s + \frac{1}{2}\right)\hbar\omega. \tag{16.35}$$

From (16.26) and (16.34) it is clear that the first-order correction vanishes:

$$E_s^{(1)} = H'_{ss} = 0. \tag{16.36}$$

The second-order term is given by

$$E_s^{(2)} = \sum_{n \neq s} \frac{\left|H'_{ns}\right|^2}{\left(E_s^{(0)} - E_n^{(0)}\right)}. \tag{16.37}$$

From (16.34) we can write

$$E_s^{(2)} = \frac{\left|H'_{s+1,s}\right|^2}{\left(E_s^{(0)} - E_{s+1}^{(0)}\right)} + \frac{\left|H'_{s-1,s}\right|^2}{\left(E_s^{(0)} - E_{s-1}^{(0)}\right)}, \tag{16.38}$$

which gives

$$E_s^{(2)} = -\frac{1}{2m\omega^2}. \tag{16.39}$$

Thus, to second order in λ,

$$E_s = \left(s + \frac{1}{2}\right)\hbar\omega - \frac{\lambda^2}{2m\omega^2}. \tag{16.40}$$

Let us treat this problem exactly. The Schrödinger equation with H' is given by

$$-\frac{\hbar^2}{2m}\frac{d^2u}{dx^2} + \frac{1}{2}m\omega^2x^2u + \lambda xu = Eu. \tag{16.41}$$

We can write this equation as follows:

$$-\frac{\hbar^2}{2m}\frac{d^2u}{dx^2} + \frac{1}{2}m\omega^2\left(x + \frac{\lambda}{m\omega^2}\right)^2 u - \frac{\lambda^2}{2m\omega^2}u = Eu. \tag{16.42}$$

We now change variables and write

$$y = x + \frac{\lambda}{m\omega^2}. \tag{16.43}$$

The new Schrödinger equation is

$$-\frac{\hbar^2}{2m}\frac{d^2u}{dy^2} + \frac{1}{2}m\omega^2y^2u = E'u \tag{16.44}$$

where

$$E' = E + \frac{\lambda^2}{2m\omega^2}. \tag{16.45}$$

Equation (16.44) is of the same form as the Schrödinger equation without any perturbative terms. The energy eigenvalues for this problem are given by

$$E'_s = \left(s + \frac{1}{2}\right)\hbar\omega. \tag{16.46}$$

Therefore,

$$E_s = \left(s + \frac{1}{2}\right)\hbar\omega - \frac{\lambda^2}{2m\omega^2}. \tag{16.47}$$

Interestingly, the energy eigenvalues to second order are the same as the exact result. The reason is simply that all higher-order corrections vanish. This can be understood by noting, for example, that for the third order we need to obtain a matrix element of the type

$$H'_{sn}H'_{nm}H'_{ms} \tag{16.48}$$

with $m \neq n \neq s$. From (16.34) it is evident that to obtain nonvanishing matrix elements the level differences must be 1. Thus $|n - s| = 1 = |m - n|$, which implies that m and s must differ by either 0 or 2; in either case the corresponding matrix element will vanish. This argument can be repeated for all higher-order corrections. Thus the perturbative results to second order provide the complete answer.

16.2.2 Wavefunctions

From perturbation theory results we obtain the first-order correction to the wavefunction as

$$|\psi_s\rangle = \left|\psi_s^{(0)}\right\rangle + \lambda \sum_{n \neq s} \frac{H'_{ns}}{\left(E_s^{(0)} - E_n^{(0)}\right)} \left|\psi_n^{(0)}\right\rangle. \tag{16.49}$$

Inserting the results for H'_{ns} for the ground state ($s = 0$) and rewriting the above in terms of the wavefunctions $u_n^{(0)}(x)$, we obtain for the ground state wavefunction to first order in λ

$$u_0(x) = u_0^{(0)}(x) - \frac{\lambda}{\hbar\omega}\sqrt{\frac{\hbar}{2m\omega}}u_1^{(0)}(x). \tag{16.50}$$

Substituting $u_1^{(0)}(x)$ in terms of $u_0^{(0)}(x)$ obtained in Chapter 9,

$$u_1^{(0)}(x) = \sqrt{\frac{2m\omega}{\hbar}}xu_0^{(0)}(x), \tag{16.51}$$

we obtain

$$u_0(x) = \left(1 - \frac{\lambda x}{\hbar\omega}\right) u_0^{(0)}(x). \tag{16.52}$$

The exact ground state eigenfunction in terms of the variable y is given by

$$u_0(y) = \left(\frac{m\omega}{\pi\hbar}\right)^{\frac{1}{4}} \exp\left(-\frac{m\omega y^2}{2\hbar}\right). \tag{16.53}$$

Expressing y in terms of x we find

$$u_0(y) = \left(\frac{m\omega}{\pi\hbar}\right)^{\frac{1}{4}} \exp\left[-\frac{m\omega}{2\hbar}\left(x + \frac{\lambda}{m\omega^2}\right)^2\right]. \tag{16.54}$$

Expanding this to order λ we find:

$$u_0(x) = \left(1 - \frac{\lambda x}{\hbar\omega}\right) u_0^{(0)}(x). \tag{16.55}$$

We find that the perturbative result is the same as the exact result to order λ. Here one finds that the perturbative series will not terminate at a specific power of λ. This is consistent with the exact result, which involves an infinite expansion of an exponential in terms of λ. One finds that the two series, perturbative and exact, agree term by term.

This result, therefore, gives credence to the basic assumption of perturbation theory in relying on perturbation expansions.

16.3 Second-order Stark effect

Let us consider another example, this time of second-order perturbation that arises when an hydrogen atom is subjected to an electric field. The total Hamiltonian for this case is

$$H = H_0 + H' \tag{16.56}$$

where H_0 is the unperturbed Hamiltonian responsible for the formation of the hydrogen atom, and H' is given by

$$H' = -e\mathbf{E}_0 \cdot \mathbf{r} \tag{16.57}$$

where $\mathbf{E}_0$ is the electric field, and $\mathbf{r}$ corresponds to coordinates of the electron in the hydrogen atom. This is the so-called Stark effect. We do not put the multiplicative factor λ in (16.56) because the charge, e, in H' serves as an expansion parameter intrinsic to the interaction.

Taking the direction of $\mathbf{E}_0$ to be in the z-direction, and, therefore, $H' = -eE_0z$, the first-order shift in the energy is given by

$$E_s^{(1)} = H'_{ss} = -e\,|\mathbf{E}_0|\left\langle \psi_s^{(0)}\,|z|\,\psi_s^{(0)}\right\rangle = -e\,|\mathbf{E}_0|\,z_{ss} \tag{16.58}$$

where $\left|\psi_s^{(0)}\right\rangle$ corresponds to the hydrogen atom wavefunction and

$$z_{ss} = \int d^3r\, z \left|\psi_s^0(\mathbf{r})\right|^2. \tag{16.59}$$

This integral vanishes because it is not invariant under (parity) transformation, $\mathbf{r} \to -\mathbf{r}$, as we show below.

The hydrogen atom wavefunction is of the form, as derived in Chapter 8,

$$u_{lmn} = R_l(r) Y_{lm}(\theta, \phi) \tag{16.60}$$

where $R_l(r)$ is the radial wavefunction and $Y_{lm}(\theta, \phi)$ is the spherical harmonic function. Under parity transformation, the radial coordinates transform as $r \to r$, $\theta \to \pi - \theta$ and $\phi \to \pi + \phi$. Thus $R_l(r)$ remains the same but

$$Y_{lm} \to (-1)^m Y_{lm} \tag{16.61}$$

where l is an integer. Hence

$$\left|\psi_s^{(0)}(\mathbf{r})\right|^2 \to (-1)^{2l} \left|\psi_s^{(0)}(\mathbf{r})\right|^2 = \left|\psi_s^{(0)}(\mathbf{r})\right|^2. \tag{16.62}$$

The integral (16.59) under $\mathbf{r} \to -\mathbf{r}$ gives

$$z_{ss} \to -z_{ss}. \tag{16.63}$$

Hence $z_{ss} = 0$. Thus there is no first-order Stark effect. The situation is different if the states are degenerate, as we will discuss in the next section.

Let us then consider the second-order correction,

$$E_s^{(2)} = \sum_{n \neq s} \frac{\left|H'_{ns}\right|^2}{\left(E_s^{(0)} - E_n^{(0)}\right)} = e^2 \left|\mathbf{E}_0\right|^2 \sum_{n \neq s} \frac{\left|z_{ns}\right|^2}{\left(E_s^{(0)} - E_n^{(0)}\right)}. \tag{16.64}$$

This infinite sum can be evaluated by an ingenious method due to Dalgarno and Lewis (D-L) in which the energy to second order, given by

$$E_0^{(2)} = \sum_{n \neq 0} \frac{\left|H'_{n0}\right|^2}{\left(E_0^0 - E_n^0\right)}, \tag{16.65}$$

is expressed in terms of an operator F that satisfies the relation

$$H' = [F, H_0]. \tag{16.66}$$

Thus,

$$H'_{n0} = \left\langle n \left| H' \right| 0 \right\rangle = \langle n \left| [F, H_0] \right| 0 \rangle = \left(E_0^0 - E_n^0\right) \langle n \left| F \right| 0 \rangle \tag{16.67}$$

and

$$E_0^{(2)} = \sum_{n \neq 0} \langle 0 |H'| n \rangle \langle n |F| 0 \rangle . \tag{16.68}$$

Assuming further that

$$\langle 0 |F| 0 \rangle = 0, \tag{16.69}$$

one can write

$$E_0^{(2)} = \sum_{n=0} \langle 0 |H'| n \rangle \langle n |F| 0 \rangle = \langle 0 |H'F| 0 \rangle \tag{16.70}$$

where the term with $n = 0$ is included in the summation. The last equality follows from the completeness relation. Thus the D-L method reduces an infinite sum to a single expectation value.

For our problem, we have

$$E_0^{(2)} = e^2 |\mathbf{E}_0|^2 \langle 0 |zF| 0 \rangle . \tag{16.71}$$

An F that satisfies (16.66) and (16.69) is found by D-L to be

$$F = -\frac{ma_0}{\hbar^2} \left(\frac{r}{2} + a_0 \right) z. \tag{16.72}$$

Hence,

$$\langle 0 |zF| 0 \rangle = -\frac{ma_0}{\hbar^2} \left\langle 0 \left| \left(\frac{r}{2} + a_0 \right) z^2 \right| 0 \right\rangle . \tag{16.73}$$

The ground state of the hydrogen atom is a $1S$ state with principal quantum number $n = 1$, $l = 0$, and $m = 0$. The wavefunction is, therefore, symmetric and, as a result, one finds the following simplifications:

$$\langle 0|z^2|0\rangle = \frac{1}{3}\langle 0|r^2|0\rangle , \tag{16.74}$$

while

$$\langle 0|rz^2|0\rangle = \frac{1}{3}\langle 0|r^3|0\rangle . \tag{16.75}$$

These integrals are the expectation values with respect to the ground states, which can be evaluated leading to the result

$$E_0^{(2)} = -\frac{9}{4} a_0^3 |\mathbf{E}_0|^2 . \tag{16.76}$$

This is then the second-order correction to the energy eigenvalue of the ground state.

The first-order correction to the wavefunction can also be evaluated quite simply using the D-L method. The wavefunction is given by

$$|\psi_s^{(1)}\rangle = \sum_{n\neq s} \frac{H'_{ns}}{\left(E_s^0 - E_n^0\right)} |\psi_n^0\rangle. \tag{16.77}$$

For the ground state we substitute $s = 0$ in (16.77) and obtain

$$\sum_{n\neq 0} \frac{\langle n\,|H'|\,0\rangle}{E_0 - E_n} |n\rangle = \sum_{n\neq 0} \langle n\,|F|\,0\rangle |n\rangle \tag{16.78}$$

where we have written $|\psi_n^0\rangle = |n\rangle$. Since $\langle 0\,|F|\,0\rangle = 0$ we can write the right-hand side of (16.78) as

$$\sum_{n} \langle n\,|F|\,0\rangle |n\rangle = \sum_{n} |n\rangle \langle n\,|F|\,0\rangle = F|0\rangle \tag{16.79}$$

where we have used the completeness relation. From the F given in (16.22) we find

$$|\psi_0^{(1)}\rangle = -\left(\frac{ma_0}{\hbar^2}\right)\left(\frac{r}{2} + a_0\right) |\psi_0^{(0)}\rangle. \tag{16.80}$$

This is then the first-order correction to the ground-state wavefunction obtained in a closed form. We finally note that this problem can also be solved using parabolic coordinates.

16.4 Degenerate states

If the unperturbed states are degenerate, we must proceed somewhat differently. The classic case of degenerate states is provided by the hydrogen atom levels with principal quantum number $n \geq 2$.

Suppose we have two unperturbed states $\left|\psi_{s1}^{(0)}\right\rangle$ and $\left|\psi_{s2}^{(0)}\right\rangle$ with the same energy eigenvalue (two-fold degeneracy), with $\left\langle \psi_{s1}^{(0)} \middle| \psi_{s2}^{(0)}\right\rangle = 0$, then

$$H_0 \left|\psi_{s1}^{(0)}\right\rangle = E_s^0 \left|\psi_{s1}^{(0)}\right\rangle, \tag{16.81}$$

$$H_0 \left|\psi_{s2}^{(0)}\right\rangle = E_s^0 \left|\psi_{s2}^{(0)}\right\rangle. \tag{16.82}$$

As before, we have a perturbation due to H' so that

$$H = H_0 + \lambda H' \tag{16.83}$$

with

$$H\left|\psi_s\right\rangle = E_s \left|\psi_s\right\rangle \tag{16.84}$$

where $|\psi_s\rangle$ and E_s are the eigenstate and eigenvalue of the perturbed Hamiltonian H. We expand E_s as we did before,

$$E_s = E_s^{(0)} + \lambda E_s^{(1)} + \lambda^2 E_s^{(2)} + \cdots . \tag{16.85}$$

With regard to $|\psi_s\rangle$, we have to take into account the fact that, in the limit $\lambda \to 0$, one may recover either of the two unperturbed states. Since the unperturbed states are indistinguishable as far as their energy eigenvalues are concerned, there is no way of knowing *a priori* where the perturbed states will end up in the limit $\lambda \to 0$. Thus, instead of a single unperturbed state as a leading term in the perturbation expansion we take a linear combination of the two:

$$|\psi_s\rangle = a_{s1}\left|\psi_{s1}^{(0)}\right\rangle + a_{s2}\left|\psi_{s2}^{(0)}\right\rangle + \lambda\left|\psi_s^{(1)}\right\rangle + \lambda^2\left|\psi_s^{(2)}\right\rangle + \cdots . \tag{16.86}$$

We have assumed that the degeneracy is removed to first order in λ. We then have from (16.84), (16.85), and (16.86),

$$\begin{aligned}(H_0 + \lambda H')&\left\{a_{s1}\left|\psi_{s1}^{(0)}\right\rangle + a_{s2}\left|\psi_{s2}^{(0)}\right\rangle + \lambda\left|\psi_s^{(1)}\right\rangle + \lambda^2\left|\psi_s^{(2)}\right\rangle + \cdots\right\}\\ &= \left(E_s^{(0)} + \lambda E_s^{(1)} + \lambda^2 E_s^{(2)} + \cdots\right)\\ &\quad \times \left\{a_{s1}\left|\psi_{s1}^{(0)}\right\rangle + a_{s2}\left|\psi_{s2}^{(0)}\right\rangle + \lambda\left|\psi_s^{(1)}\right\rangle + \lambda^2\left|\psi_s^{(2)}\right\rangle + \cdots\right\}.\end{aligned} \tag{16.87}$$

Multiplying the above by $\left\langle\psi_{s1}^{(0)}\right|$ and $\left\langle\psi_{s2}^{(0)}\right|$ in succession and using the orthonormality relations, we obtain the following, after equating the coefficients of λ on both sides of (16.87),

$$a_{s1}\left(H'_{11} - E_s^{(1)}\right) + a_{s2}H'_{12} = 0, \tag{16.88}$$

$$a_{s1}H'_{21} + a_{s2}\left(H'_{22} - E_s^{(1)}\right) = 0. \tag{16.89}$$

We have a situation similar to the two-channel problems discussed in Chapter 13. Evaluating the following determinant will then give the energy eigenvalues $E_s^{(1)}$:

$$\left\|\begin{bmatrix}\left(H'_{11} - E_s^{(1)}\right) & H'_{12}\\ H'_{21} & \left(H'_{22} - E_s^{(1)}\right)\end{bmatrix}\right\| = 0. \tag{16.90}$$

For each value of $E_s^{(1)}$ one obtains the coefficients a_{s1} and a_{s2} through (16.88) and (16.89) and, hence, the linear combination of unperturbed states we need to zeroth order.

If there is an n-fold degeneracy then we have the following $n \times n$ determinant to solve:

$$\left\| \begin{bmatrix} \left(H'_{11} - E_s^{(1)}\right) & H'_{12} & H'_{13} & . & H'_{1n} \\ H'_{21} & \left(H'_{22} - E_s^{(1)}\right) & . & . & . \\ H'_{31} & . & \left(H'_{33} - E_s^{(1)}\right) & . & . \\ . & . & . & . & . \\ H'_{n1} & . & . & . & \left(H'_{nn} - E_s^{(1)}\right) \end{bmatrix} \right\| = 0. \quad (16.91)$$

In degenerate perturbation theory, generally, one is not interested in the higher-order terms in the wavefunctions.

16.5 Linear Stark effect

We return to the subject of a hydrogen atom subjected to a uniform electric field $\mathbf{E}_0$. The interaction Hamiltonian is given by

$$H' = -e\mathbf{E}_0 \cdot \mathbf{r} \quad (16.92)$$

$$= -eE_0 r \cos\theta \quad (16.93)$$

where we take the z-axis to be along the direction of $\mathbf{E}_0$, and θ the angle between $\mathbf{r}$ and $\mathbf{E}_0$. As we discussed earlier, the first-order correction vanishes:

$$E_s^{(1)} = H'_{ss} = -e\left|\mathbf{E}_0\right|\left\langle\psi_s^{(0)}\left|z\right|\psi_s^{(0)}\right\rangle = -e\left|\mathbf{E}_0\right| z_{ss} = 0. \quad (16.94)$$

In the Stark effect problem for the hydrogen atom with $n \geq 2$, the situation is different as we will discuss below, since we will need matrix elements between two states that are degenerate but which are not necessarily identical. In this case we apply the degenerate perturbation theory discussed in the previous section.

The wavefunctions for the hydrogen atom have already been obtained in Chapter 8. They are of the form

$$u_{nlm}(\mathbf{r}) = R_{nl}(r)Y_{lm}(\theta, \phi) \quad (16.95)$$

where n is the principal quantum number, l the angular momentum, and m the projection of the angular momentum in the z-direction. The only nonzero matrix elements of H' are those where the difference between the m-values of the two states vanishes and the difference between the l-values is ± 1. That is,

$$\Delta m = 0, \quad \Delta l = \pm 1. \quad (16.96)$$

For $n = 1$ there is only one state, the $1S$ state. There are four degenerate states for $n = 2$, two of which satisfy the above criteria: $2S$ and the $2P$ state with $m = 0$. As explained above, the diagonal matrix elements of H' with respect to these states vanish. The off-diagonal matrix element is given by

$$\langle 2S\,|(-eE_0 r\cos\theta)|\,2P(m=0)\rangle = -eE_0\int d^3r\, u_{210}(\mathbf{r})\,(r\cos\theta)\,u_{100}(\mathbf{r}) \tag{16.97}$$

$$= -3eE_0 a_0 \tag{16.98}$$

where a_0 is the Bohr radius and the two wavefunctions are given in Chapter 8.

The matrix $\{H'_{ij}\}$ between the two states is then

$$\begin{bmatrix} & 2S & 2P(m=0) & 2P(m=1) & 2P(m=-1) \\ 2S & 0 & -3eE_0a_0 & 0 & 0 \\ 2P(m=0) & -3eE_0a_0 & 0 & 0 & 0 \\ 2P(m=1) & 0 & 0 & 0 & 0 \\ 2P(m=-1) & 0 & 0 & 0 & 0 \end{bmatrix} \tag{16.99}$$

where we have substituted

$$H'_{11} = 0 = H'_{22};\ H'_{12} = H'_{12} = -3eE_0a_0;\ \text{remaining } H'_{ij} = 0. \tag{16.100}$$

Hence the linear energy correction $E_s^{(1)}$ is found from the condition

$$\det\begin{bmatrix} E_s^{(1)} & 3eE_0a_0 \\ 3eE_0a_0 & E_s^{(1)} \end{bmatrix} = 0. \tag{16.101}$$

This gives

$$E_{s\pm}^{(1)} = \pm 3eE_0a_0. \tag{16.102}$$

The remaining two roots are zero and, therefore, the states $2P(m=1)$ and $2P(m=-1)$ remain degenerate.

The wavefunctions corresponding to each energy eigenvalue can be easily determined. We find

$$u_+(\mathbf{r}) = \frac{1}{\sqrt{2}}(u_{210}(\mathbf{r}) + u_{100}(\mathbf{r})) \quad \text{for } E_{s+}^{(1)} = 3eE_0a_0, \tag{16.103}$$

$$u_-(\mathbf{r}) = \frac{1}{\sqrt{2}}(u_{210}(\mathbf{r}) - u_{100}(\mathbf{r})) \quad \text{for } E_{s-}^{(1)} = -3eE_0a_0. \tag{16.104}$$

16.6 Problems

1. Consider a particle trapped within two infinite barriers at $x = -a$ and $x = a$ respectively. A perturbing potential of the type

$$V'(x) = x$$

is applied. Obtain the energy eigenvalues to first and second orders and the eigenfunction to first order. Do the same if x is replaced by $|x|$. Explain the difference between the two results based on the symmetry of the potential.

2. Designate the state vector for the ground state in a given problem as $|0\rangle$. Show that the second-order correction term for the energy satisfies the relation $E_0^{(2)} < 0$. If, furthermore, the expectation value $\langle 0| H' |0\rangle = 0$ then show that

$$E_0^{(2)} > \frac{\langle 0| H'^2 |0\rangle}{E_0^{(0)} - E_1^{(0)}}$$

where $E_1^{(0)}$ is the energy of the first excited state.

3. Two atoms, which are hydrogen-like, are placed with their protons a distance R apart. Assume the unperturbed eigenfunction for the system as the product of ground-state functions for the two atoms. Approximating each atom as an electric dipole, show that the interaction energy between the two atoms to first order vanishes. Without explicitly calculating the matrix elements, show that in second order the interaction between the two atoms is attractive and goes as $(1/R^6)$ as $R \to \infty$.

4. Consider a hydrogen atom subjected to a magnetic field in the z-direction, $\mathbf{B} = B_0 z$. We have shown that it interacts with the magnetic moment generated by the orbital motion of the electron with an interaction Hamiltonian

$$H' = -\frac{e}{mc}\mathbf{L} \cdot \mathbf{B}$$

where $\mathbf{L}$ is the orbital angular momentum operator. Use degenerate perturbation theory to determine the energy eigenvalues of the $n = 2$ state.

5. Consider the Hamiltonian given as a 3×3 matrix,

$$H = \begin{bmatrix} 0 & 0 & a \\ 0 & 0 & a \\ a & a & b \end{bmatrix}.$$

where a represents perturbation. Obtain the energy eigenvalues and eigenfunctions (1) in the lowest order perturbation (with $a \ll b$) and (2) exactly. Compare these two results for $a \ll b$.

6. A charged particle trapped between two infinite walls located at $x = \pm a$ in a ground state is subjected to an electric field E_0 in the positive x-direction. Determine the probability that it will be in the first excited state.

7. A particle between two infinite walls located at $x = \pm a$ is subjected to a perturbation in the form of a finite repulsive barrier $V = V_0$ between $-b$ and $+b$, where $b < a$. What is the first-order correction to the ground-state energy?

8. A harmonic oscillator with spring constant K is subjected to a perturbation of the form $\left(\lambda_1 x + \lambda_2 x^2 + \lambda_3 x^3\right)$. Determine the perturbed energies $E_0^{(1)}$ and $E_0^{(2)}$ for the ground state.

9. A free particle satisfying periodic boundary conditions with period L is subjected to a perturbation

$$V' = \lambda \cos\left(\frac{2\pi x N}{L}\right)$$

where N is an integer assumed to be large. Expressing the even and odd parity free particle states as $u_+(x)$ and $u_-(x)$ respectively, obtain the energy levels of each state to first- and second-order perturbation.

10. A hydrogen atom is subjected to the potential

$$V(\mathbf{r}) = V_0 \delta^{(3)}(\mathbf{r}).$$

Determine the energy corrections to first order for $1S$ and $2P$ states (this potential is typical of the interaction that gives rise to the so-called Lamb shift).

11. Consider an electron in a hydrogen atom where the Coulomb potential is given by

$$\begin{aligned} V(r) &= -\frac{Ze^2}{r}, \quad R < r < \infty \\ &= -\frac{Ze^2}{R}, \quad 0 < r < R \end{aligned}$$

where R is the radius of the nucleus. Determine the energy eigenvalues of $1S$, $2S$ and $2P$ states.

12. In problem 11 consider the nucleus to be a uniform sphere of charge Ze and radius R. Using Gauss's theorem determine the corresponding electrostatic field and from it the potential due to the finite size of the nucleus. Obtain the correction to the energy eigenvalue of the $1S$ state and the shift in the energy levels between $2S$ and $2P$ states.

13. A hydrogen atom is subjected to a uniform magnetic field. Use first-order perturbation theory to obtain the correction to the ground-state energy. Take the vector potential to be given by $\mathbf{A} = (1/2)\mathbf{B} \times \mathbf{r}$ and keep the $\mathbf{A}^2$ term in the Hamiltonian.

14. A particle is trapped between two infinite walls located at $x = 0$ and $x = a$, respectively. While the particle is in the ground state at $t = 0$, the walls are suddenly moved so the spacing between them is doubled. What is the probability that the particle is now in the ground state of the new system? Consider both cases where the uncertainty in time when the expansion occurs satisfies $\Delta E \Delta t \gg \hbar$ and $\Delta E \Delta t \ll \hbar$. Determine the probability distribution (in terms of the momentum) if the walls disappear altogether.

15. Solve a two-level problem in perturbation theory by writing $H = H_0 + \lambda H'$, so that one writes

$$H = \begin{bmatrix} A & D \\ D & B \end{bmatrix}$$

where $A = E_1^0 + \lambda H'_{11}$, $B = E_2^0 + \lambda H'_{22}$, $D = \lambda H'_{12}$, and $\Delta = A - B$. Write down the expressions for the energy eigenstates $E_\pm$ derived in Chapter 13. Expand these expressions in powers of λ up to λ^2. Compare these results to the perturbative results to order λ^2.

17 Time-dependent perturbation

In this chapter we consider the more complicated time-dependent perturbation case. We are now going to look at the problems that are closer to the practical level, concentrating on sudden perturbation and harmonic time dependence. We will consider, in particular, the effect of the onset of interactions, scattering, and decays of composite systems

17.1 Basic formalism

As in the case of the time-independent problems, we will obtain the solutions through perturbation methods, where the perturbing potential $H'(t)$ is time-dependent. Once again we write

$$H = H_0 + \lambda H'(t) \tag{17.1}$$

where H_0 is the unperturbed Hamiltonian and λ the parameter in terms of which we will carry out the perturbation expansion.

We assume that at $t = -\infty$ the system is given by the unperturbed Hamiltonian. The perturbation may be applied gently (adiabatically) when, as we will see below, the situation is close to time-independent perturbation. Or it could be turned on suddenly, a subject we will discuss at length. Our basic interest is then to determine the behavior of the wavefunction for finite times and as $t \to \infty$.

The perturbed state $\left|\psi_s(t)\right\rangle$ will satisfy the time evolution equation

$$i\hbar\frac{\partial}{\partial t}\left|\psi_s(t)\right\rangle = \left(H_0 + \lambda H'(t)\right)\left|\psi_s(t)\right\rangle \tag{17.2}$$

with

$$H'(t) \to 0 \quad \text{as } t \to -\infty \tag{17.3}$$

so that at $t = -\infty$ we recover the unperturbed solution.

Therefore,

$$\left|\psi_s(t)\right\rangle \to \left|\psi_s^{(0)}(t)\right\rangle \quad \text{as } t \to -\infty \tag{17.4}$$

where $\left|\psi_s^{(0)}(t)\right\rangle$ is an eigenstate of H_0 with eigenvalue $E_s^{(0)}$ and satisfies the equation

$$i\hbar\frac{\partial\left|\psi_s^{(0)}(t)\right\rangle}{\partial t} = H_0\left|\psi_s^{(0)}(t)\right\rangle = E_s^{(0)}\left|\psi_s^{(0)}(t)\right\rangle \tag{17.5}$$

with the orthogonality condition

$$\langle\psi_m^{(0)}(t)\left|\psi_s^{(0)}(t)\right\rangle = \delta_{ms}. \tag{17.6}$$

The solution of the equation (17.5) is given by

$$\left|\psi_s^{(0)}(t)\right\rangle = \left|\psi_s^{(0)}(0)\right\rangle e^{-\frac{iE_s^{(0)}}{\hbar}t}. \tag{17.7}$$

We also assume the following condition in the limit when the perturbation is turned off,

$$\left|\psi_s(t)\right\rangle \to \left|\psi_s^{(0)}(t)\right\rangle \quad \text{as } \lambda \to 0 \tag{17.8}$$

which means that in the absence of perturbation one returns to the unperturbed state.

Thus the system is a function of both t and λ with each playing an essential role. We write $\left|\psi_s(t)\right\rangle$ in terms of a complete set of eigenstates $\left|\psi_n^0(t)\right\rangle$ as follows:

$$\left|\psi_s(t)\right\rangle = \sum_n c_{ns}(t)\left|\psi_n^{(0)}(t)\right\rangle \tag{17.9}$$

where $c_{ns}(t)$ will depend on time. Substituting this in (17.2) we obtain

$$i\hbar\frac{\partial}{\partial t}\left(\sum_n c_{ns}(t)\left|\psi_n^{(0)}(t)\right\rangle\right) = \left(H_0 + \lambda H'(t)\right)\left(\sum_n c_{ns}(t)\left|\psi_n^{(0)}(t)\right\rangle\right). \tag{17.10}$$

We note that the following two terms,

$$i\hbar\frac{\partial\left|\psi_n^{(0)}(t)\right\rangle}{\partial t} \quad \text{and} \quad H_0\left|\psi_n^{(0)}(t)\right\rangle \tag{17.11}$$

one on each side of (17.10) cancel each other since both give $E_n^{(0)}$.

Inserting the result (17.7) for $\left|\psi_n^{(0)}(t)\right\rangle$ in (17.10) and multiplying both sides of the above equation by $\left\langle\psi_m^0(t)\right|$ and using orthonormality, we obtain

$$i\hbar\dot{c}_{ms} = \lambda\sum_n c_{ns}H'_{mn}(t)\exp(i\omega_{mn}t) \tag{17.12}$$

where $\omega_{mn} = \frac{E_m^{(0)}-E_n^{(0)}}{\hbar}$, and $\dot{c}_{ms}$ is the time derivitive of c_{ms}.

We will now expand the coefficient c_{ps} in a perturbation series in λ,

$$c_{ps}(t) = c_{ps}^{(0)}(t) + \lambda c_{ps}^{(1)}(t) + \lambda^2 c_{ps}^{(2)}(t) + \cdots. \tag{17.13}$$

From the condition (17.4) for the limit $t \to -\infty$, it is clear that

$$c_{ms}^{(0)}(-\infty) = \delta_{ms}, \tag{17.14}$$

$$c_{ms}^{(1)}(-\infty) = c_{ms}^{(2)}(-\infty) = \cdots = 0. \tag{17.15}$$

Substituting the series (17.13) in (17.12) and equating the coefficients of the equal powers of λ on both sides of the equation, we obtain, for $c_{ms}^{(0)}, c_{ms}^{(1)}, c_{ms}^{(2)}, \ldots,$

$$\dot{c}_{ms}^{(0)}(t) = 0, \tag{17.16}$$

$$i\hbar \dot{c}_{ms}^{(1)}(t) = \sum_n c_{ns}^{(0)}(t) H'_{mn}(t) \exp(i\omega_{mn} t), \tag{17.17}$$

$$i\hbar \dot{c}_{ms}^{(2)}(t) = \sum_n c_{ns}^{(1)}(t) H'_{mn}(t) \exp(i\omega_{mn} t), \tag{17.18}$$

and so on. From (17.16) we conclude that $c_{ms}^{(0)}(t) =$ constant for all values of t and (17.14) tells us that this constant must be δ_{ms}. Thus

$$c_{ms}^{(0)}(t) = \delta_{ms} \quad \text{for all values of } t. \tag{17.19}$$

Hence

$$i\hbar \dot{c}_{ms}^{(1)} = H'_{ms}(t) \exp(i\omega_{ms} t). \tag{17.20}$$

Therefore,

$$c_{ms}^{(1)}(t) = -\frac{i}{\hbar} \int_{-\infty}^{t} dt' H'_{ms}(t') \exp(i\omega_{ms} t'). \tag{17.21}$$

This is one of the basic relations of time-dependent perturbation theory. Since one is often interested in first-order transitions only, the relation (17.21) allows one to calculate the transition probability amplitude in terms of H'_{ms}. We call $c_{ms}^{(1)}(t)$ the transition probability amplitude to first order for transitions from the initial state s to an arbitrary state m.

We then have the following result for the state vector to first order in λ:

$$|\psi_s(t)\rangle = \left|\psi_s^0(t)\right\rangle + \lambda \sum_n c_{ns}^{(1)}(t) \left|\psi_n^0(t)\right\rangle \tag{17.22}$$

where $c_{ns}^{(1)}(t)$ is determined through the integral (17.21), and

$$\left|c_{ns}^{(1)}(t)\right|^2 = \text{probability of transitions } s \to n \text{ to first order}. \tag{17.23}$$

Let us try to clarify the difference between the approaches involved in time-dependent and time-independent formalisms. In the latter case the problem was basically mathematical in nature. There the perturbing Hamiltonian H' was supposed to have existed at all times rather than being switched on and off at different times. The mathematical form of the total Hamiltonian H, however, was such that it was not possible to obtain exact solutions, and so we had to resort to approximate solutions through perturbation methods.

For a time-dependent H' there are definite times when it is turned on and off. Can one then relate the two approaches in some limiting sense? The answer is, yes, provided the time-dependent perturbation starts at $t = -\infty$, and is switched on sufficiently "gently" or, using the proper mathematical term, "adiabatically." We discuss in the Appendix how this can be done.

The case where H' is turned on at $t = 0$, remaining constant thereafter, is one of the examples of a sudden onset of perturbation; it is given by

$$H'_{ms}(t) = H'_{ms}(0)\theta(t). \tag{17.24}$$

Taking the derivative of the above expression we obtain

$$\frac{dH'}{dt'} \sim \delta(t). \tag{17.25}$$

The difference between a δ-function above and a more gentle behavior, e.g., a Gaussian, for the derivative of H' reflects, in essence, the difference between the formalisms for time-dependent and time-independent perturbations.

17.2 Harmonic perturbation and Fermi's golden rule

Let us consider a specific type of perturbing potential called "harmonic perturbation" that is turned on at $t = 0$ and oscillates with a definite frequency thereafter. One expresses it as follows, with the time dependence factored out,

$$H'_{ms}(t) = \left[Ae^{-i\omega t} + A^{\dagger}e^{i\omega t}\right]\theta(t) \tag{17.26}$$

where A is independent of time. The Hamiltonian as written is Hermitian. We will consider a more specific example,

$$H'_{ms}(t) = 2H'_{ms}(0)\sin\omega t \tag{17.27}$$

$$= iH'_{ms}(0)\left[e^{-i\omega t} - e^{i\omega t}\right], \quad t > 0, \tag{17.28}$$

$$H'_{ms}(t) = 0, \quad t \leq 0. \tag{17.29}$$

We assume $H'_{ms}(0)$ is Hermitian, and, therefore, from (17.27) and (17.28) so is $H'_{ms}(t)$. We note that $H'_{ms}(0)$ is independent of time, but it still involves integration over space variables that enter into the calculation of the matrix element $\langle\psi_m^{(0)}\left|H'(0)\right|\psi_s^{(0)}\rangle$.

Substituting (17.26) in (17.21) yields

$$c_{ms}^{(1)}(t) = \frac{H'_{ms}(0)}{\hbar} \int_0^t dt' \left[\exp(i(\omega_{ms} - \omega)t') - \exp(i(\omega_{ms} + \omega)t')\right] \tag{17.30}$$

$$= \frac{H'_{ms}(0)}{\hbar} \left\{ \begin{array}{c} \dfrac{[\exp i(\omega_{ms} - \omega)t - 1]}{i(\omega_{ms} - \omega)} \\ -\dfrac{[\exp i(\omega_{ms} + \omega)t - 1]}{i(\omega_{ms} + \omega)}. \end{array} \right\} \tag{17.31}$$

At this stage it is conventional to take the index s to signify the "initial" state and m to signify the "final" state. Thus, taking $m = f$ and $s = i$, we write expression (17.31) as

$$c_{fi}^{(1)}(t) = A_{fi}(\omega, t) - A_{fi}(-\omega, t) \tag{17.32}$$

where

$$A_{fi}(\omega, t) = \frac{H'_{fi}(0)}{\hbar} \frac{\left[\exp i\left(\omega_{fi} - \omega\right)t - 1\right]}{i(\omega_{fi} - \omega)}. \tag{17.33}$$

The probability of transition from i to f is then given by

$$\left|c_{fi}^{(1)}(t)\right|^2 = \left|A_{fi}(\omega, t)\right|^2 + \left|A_{fi}(-\omega, t)\right|^2 - 2\,\mathrm{Re}\left[A_{fi}^*(\omega, t)A_{fi}(-\omega, t)\right]. \tag{17.34}$$

One can simplify the expression for $A_{fi}(\omega, t)$ as

$$A_{fi}(\omega, t) = \frac{2H'_{fi}(0)}{\hbar} \frac{\sin\left[\left(\omega_{fi} - \omega\right)\dfrac{t}{2}\right]}{(\omega_{fi} - \omega)} \left[\exp i\left(\omega_{fi} - \omega\right)\frac{t}{2}\right] \tag{17.35}$$

and obtain

$$\left|A_{fi}(\omega, t)\right|^2 = \frac{4\left|H'_{fi}(0)\right|^2}{\hbar^2} \frac{\sin^2\left[\left(\omega_{fi} - \omega\right)t/2\right]}{(\omega_{fi} - \omega)^2}. \tag{17.36}$$

Often a more interesting quantity is transition "rate," i.e., transitions per unit time. Assuming the system to be continuous in time, the rate is simply given by the time derivative. That is, by

$$\frac{d}{dt}\left|c_{fi}^{(1)}(t)\right|^2 = \frac{d}{dt}\left|A_{fi}(\omega, t)\right|^2 + \frac{d}{dt}\left|A_{fi}(-\omega, t)\right|^2 - 2\frac{d}{dt}\,\mathrm{Re}\left[A_{fi}^*(\omega, t)A_{fi}(-\omega, t)\right]. \tag{17.37}$$

Consider the time derivative of the first term above. We obtain from (17.36)

$$\frac{d}{dt}\left|A_{fi}(\omega, t)\right|^2 = \frac{2\left|H'_{fi}(0)\right|^2}{\hbar^2} \frac{\sin\left[\left(\omega_{fi} - \omega\right)t\right]}{(\omega_{fi} - \omega)} \tag{17.38}$$

where we have used the identity $2\sin\alpha\cos\alpha = \sin 2\alpha$. A quantity of great physical interest is the rate when t is very large, when the system is supposed to have settled down. Denoting this quantity as λ_{fi}, we have

$$\lambda_{fi} = \lim_{t\to\infty}\frac{d}{dt}\left|c_{fi}^{(1)}(t)\right|^2. \tag{17.39}$$

Considering first the contribution coming from $A_{fi}(\omega,t)$ we note that, in this limit, the sine-function that appears on the right-hand side of (17.38) can be expressed as a δ-function, from the well-known relation

$$\lim_{L\to\infty}\frac{1}{\pi}\frac{\sin Lx}{x} = \delta(x), \tag{17.40}$$

which we discuss in the Appendix.

We thus obtain the result

$$\lim_{t\to\infty}\frac{d}{dt}\left|A_{fi}(\omega,t)\right|^2 = \frac{2\pi\left|H'_{fi}(0)\right|^2}{\hbar^2}\delta(\omega_{fi}-\omega). \tag{17.41}$$

This can be further simplified by using the property $\delta(ax) = (1/|a|)\,\delta(x)$. We then obtain

$$\lim_{t\to\infty}\frac{d}{dt}\left|A_{fi}(\omega,t)\right|^2 = \frac{2\pi}{\hbar}\left|H'_{fi}(0)\right|^2\delta\left[E_f^0-\left(E_i^0+\hbar\omega\right)\right]. \tag{17.42}$$

Similarly,

$$\lim_{t\to\infty}\frac{d}{dt}\left|A_{fi}(-\omega,t)\right|^2 = \frac{2\pi}{\hbar}\left|H'_{fi}(0)\right|^2\delta\left[E_f^0-\left(E_i^0-\hbar\omega\right)\right]. \tag{17.43}$$

It is easy to show that

$$\frac{d}{dt}\mathrm{Re}\left[A_{fi}^*(\omega,t)A_{fi}(-\omega,t)\right] = \frac{\left|H'_{fi}(0)\right|^2}{\hbar^2}\left[\frac{2\omega\sin\omega t}{(\omega_{fi}^2-\omega^2)}+\frac{\sin\left(\omega-\omega_{fi}\right)t}{(\omega_{fi}+\omega)}+\frac{\sin\left(\omega+\omega_{fi}\right)t}{(\omega_{fi}-\omega)}\right]. \tag{17.44}$$

In taking the limit $t\to\infty$ we come across the following type of behavior in each term on the right-hand side:

$$\lim_{L\to\infty}\sin Lx. \tag{17.45}$$

However, this term vanishes because the consecutive zeros of $\sin Lx$ separated by a distance π/L get crammed in closer and closer as $L\to\infty$. When integrating over $(-L,L)$, the function $\sin Lx$ will not make any contribution. Another way to understand this result is to write

$$\lim_{L\to\infty}\sin Lx = \pi x\lim_{L\to\infty}\frac{1}{\pi}\frac{\sin Lx}{x} = \pi x\delta(x) = 0, \tag{17.46}$$

which follows from the properties of $\delta(x)$. The difference between the behavior of $\sin Lx$ and $\sin Lx/x$ in the limit $L \to \infty$ is principally due to the fact that $\sin Lx/x$ at $x = 0$ is proportional to L, and becomes very large as $L \to \infty$, resulting in a δ-function behavior, the function $\sin Lx$, on the other hand vanishes at that point for any fixed value of L.

Hence (17.44) vanishes in the limit $t \to \infty$, leaving us, after combining (17.42) and (17.43), with the following result:

$$\lambda_{fi} = \frac{2\pi \left|H'_{fi}(0)\right|^2}{\hbar} \left\{\delta\left[E_f^0 - \left(E_i^0 + \hbar\omega\right)\right] + \delta\left[E_f^0 - \left(E_i^0 - \hbar\omega\right)\right]\right\}. \tag{17.47}$$

This is then the general expression for the transition rate for harmonic potentials in the limit $t \to \infty$.

The first δ-function in (17.47) corresponds to the transition from an initial state to an excited state by absorption of a quantum of energy $\hbar\omega$ (for example, a single photon). This process is called absorption. Here the external interaction supplies the energy. The second δ-function in (17.47) corresponds to the transition from an excited state down to a lower state by emitting a quantum of energy $\hbar\omega$ (again, for example, a single photon). This is called stimulated emission. Here the excited state gives up the energy. The probability for the two is the same, which reflects the symmetry between absorption and emission called "detailed balancing." There is a third process in which an excited state makes a transition to a lower state without the influence of an external perturbation as in the stimulated case. This is called spontaneous emission and is peculiar to quantum field theory, in which a particle is constantly emitting and reabsorbing quanta that are associated with the so-called "zero point" energy.

17.3 Transitions into a group of states and scattering cross-section

17.3.1 Harmonic perturbation

We continue with the harmonic perturbation and consider the case where $\hbar\omega \approx E_f^{(0)} - E_i^{(0)}$. Thus only the first term in (17.47) will contribute. That is, we will be considering only the interaction of the type

$$H'_{fi}(t) = H'_{fi}(0)e^{-i\omega t} \tag{17.48}$$

where $H'_{fi}(0)$ is Hermitian.

In this process the perturbation raises the initial state to a higher energy level by supplying an energy $\hbar\omega$. The transition rate λ_{fi} is then given by

$$\lambda_{fi} = \frac{2\pi}{\hbar} \left|H'_{fi}(0)\right|^2 \delta\left[E_f - (E_i + \hbar\omega)\right] \tag{17.49}$$

where we have removed the superscript "0" in E_f and E_i.

We specifically discuss the case where the final state corresponds to a free particle. We now introduce the quantity w_{fi}, which determines the transition rate into a group of final free particle states in the limit $t \to \infty$. It is given by

$$w_{fi} = \frac{2\pi}{\hbar} \sum_{f \neq i} \left| H'_{fi}(0) \right|^2 \delta \left[E_f - (E_i + \hbar\omega) \right] \tag{17.50}$$

where the summation is carried out over all final states, except the initial state, consistent with the constraints due to the δ-function. We express the summation over the final states as

$$w_{fi} = \int dN_f \, \lambda_{fi} \tag{17.51}$$

where dN_f designates the number of free particle states.

For a box normalization, the free particle wavefunction is given by

$$u(\mathbf{r}) = \frac{1}{\sqrt{V}} e^{i\mathbf{k}.\mathbf{r}}. \tag{17.52}$$

In terms of a cube of length, L, we can write this wavefunction as

$$u(x, y, z) = \frac{1}{\sqrt{L^3}} e^{i(k_x x + k_y y + k_z z)} \tag{17.53}$$

where the periodic boundary condition

$$k_i L = 2\pi n_i, \quad i = x, y, z \tag{17.54}$$

inherent in the box normalization is implied. The total number of states is then given by

$$dN = dn_x \, dn_y \, dn_z = \frac{d^3 k L^3}{(2\pi)^3} = \frac{d^3 k V}{(2\pi)^3} \tag{17.55}$$

where $d^3 k = dk_x \, dk_y \, dk_z$. If Ω is the solid angle into which the particles scatter, then one can write in spherical coordinates, $d^3 k = k^2 dk \, d\Omega$, where k is the magnitude of the center of mass momentum $|\mathbf{k}|$.

We define the density of states $\rho(E)$ as

$$\rho(E) = \frac{dN}{dE}. \tag{17.56}$$

The transition probability into a group of final states can be expressed from (17.51) as

$$w_{fi} = \int \lambda_{fi} dN_f = \int \lambda_{fi} \rho(E_f) \, dE_f \tag{17.57}$$

Substituting λ_{fi} from (17.49) and using the property of δ-function we obtain

$$w_{fi} = \frac{2\pi}{\hbar} \left| H'_{fi}(0) \right|^2 \rho(E_f) \quad \text{with} \quad E_f = E_i + \hbar\omega. \tag{17.58}$$

The above expression, called "Fermi's Golden Rule," gives the transition rate, as $t \to \infty$, from an initial state $|i\rangle$ into a group of free-particle final states, $|f\rangle$, under the influence of the interaction given by (17.28) and (17.29).

For the case of free particles the density of states obtained from (17.55) and (17.56) is

$$\rho(E) = \frac{d^3k}{dE}\frac{V}{(2\pi)^3} = \frac{mkV}{\hbar^2 (2\pi)^3} d\Omega \tag{17.59}$$

where we have taken the free particle value for the energy E to be

$$E = \frac{p^2}{2m} = \frac{\hbar^2 k^2}{2m}. \tag{17.60}$$

In principle one should write $d\rho(E)$ for the left-hand side in (17.59) since the right-hand side involves the differential element $d\Omega$. However, we will follow convention and write it simply as $\rho(E)$.

The cross-section $d\sigma$, for scattering into a solid angle $d\Omega$, for this process is then defined as

$$d\sigma = \left(\frac{w_{fi}}{J_i}\right) \tag{17.61}$$

where J_i is the incident flux determined by the states that are incident per unit area per unit time. From (17.58), (17.59), and (17.61) we obtain

$$\frac{d\sigma_{fi}}{d\Omega} = \frac{1}{J_i}\frac{mk_f V}{4\pi^2\hbar^3}\left|H'_{fi}(0)\right|^2. \tag{17.61a}$$

We will determine the flux J_i for each scattering problem separately. The case where the initial state corresponds to free particles is discussed below.

17.3.2 Free particle scattering

We now consider the case where we have scattering of a free incoming particle into a free outgoing particle in the presence of a radial potential. There is no harmonic time dependence or any type of explicit time dependence in this case as the potential is only a function of the distance between the particles. The time dependence enters implicitly. The interaction is turned on, or is felt, when the particle is close to the center of the potential. This is then simple elastic scattering with

$$E_f^0 = E_i^0 \tag{17.62}$$

couched in the language of time dependent perturbation.

We therefore take $\omega = 0$ in (17.26) and write

$$H'_{fi}(t) = H'_{fi}(0) \quad \text{for } t > 0 \tag{17.63}$$

$$= 0, \quad \text{for } t < 0. \tag{17.64}$$

If the particles interact via a potential $V(r)$, then

$$H'(0) = V(r). \tag{17.65}$$

The total Hamiltonian is then given by

$$H = H_0 + V(r) \tag{17.66}$$

where, as before, H_0 is the unperturbed Hamiltonian:

$$H_0 = \frac{p^2}{2m} = \frac{\hbar^2 k^2}{2m}. \tag{17.67}$$

The free particle wavefunctions for the initial and final states are given respectively by

$$u_i(\mathbf{r}) = \frac{1}{\sqrt{V}} e^{i\mathbf{k}_i.\mathbf{r}} \quad \text{and} \quad u_f(\mathbf{r}) = \frac{1}{\sqrt{V}} e^{i\mathbf{k}_f.\mathbf{r}} \tag{17.68}$$

where V is the volume in which the wavefunction is normalized, which should not be confused with the potential $V(r)$.

The matrix element $H'_{fi}(0)$ is then given by

$$H'_{fi}(0) = \int d^3r \left(\frac{1}{\sqrt{V}} e^{-i\mathbf{k}_f.\mathbf{r}}\right) V(\mathbf{r}) \left(\frac{1}{\sqrt{V}} e^{i\mathbf{k}_i.\mathbf{r}}\right) = \frac{1}{V} \int d^3r e^{-i\mathbf{q}.\mathbf{r}} V(\mathbf{r}) \tag{17.69}$$

where $\mathbf{q} \left(= \left(\mathbf{k}_f - \mathbf{k}_i\right)\right)$ is the momentum-transfer vector.

Since we have the process, free particle $\to$ free particle, in the presence of a (heavy) scattering center, e.g., a heavy atom, which acts as the source of the potential, the energies will be conserved, $E_f = E_i$, where we have removed the superscript (0). In this process only the direction of the momentum will be changed in going from, $\mathbf{k}_i$ to $\mathbf{k}_f$ (with $|\mathbf{k}_i| = |\mathbf{k}_f|$).

To obtain the cross-section we must evaluate the incident flux J_i given by

$$J_i = \text{no. of particles incident per unit area per unit time.}$$

We assume that the incident particles are traveling in the z-direction, covering a distance Δz in time Δt, and crossing an area A. The corresponding flux J_i can be written as

$$J_i = \left(\frac{A\Delta z}{V}\right)\left(\frac{1}{A}\right)\left(\frac{1}{\Delta t}\right) = \frac{\Delta z}{\Delta t}\frac{1}{V} = \frac{v}{V}. \tag{17.70}$$

We now explain each of the above factors. First, using (17.58) we have assumed the free particle wavefunctions to be normalized to one particle in volume V. Since the term $A\Delta z$ corresponds to the volume covered by the incident particle, the first factor in (17.70) corresponds to the fraction of the incident particles. The other two factors correspond to the fraction being evaluated per unit area and per unit time. Finally, the factor $\Delta z/\Delta t$ corresponds to the velocity, v, with which the particle is traveling. The product of the middle three factors, therefore, reproduces the right-hand side.

Substituting the above expression in (17.61a), and inserting w_{fi} from (17.50) (with $\omega = 0$), we obtain

$$\frac{d\sigma}{d\Omega} = \frac{m^2}{4\pi^2\hbar^4}\left|\int d^3r\, e^{-i\mathbf{q}.\,\mathbf{r}}V(\mathbf{r})\right|^2. \tag{17.71}$$

The above expression is derived in the first-order approximation, that is, in the approximation in which we consider only the first-order term, $c_{ms}^{(1)}(t)$. This is commonly called the "Born approximation." The differential cross-section, given in (17.71), is generally expressed in terms of the scattering amplitude $f(\theta)$, as

$$\frac{d\sigma}{d\Omega} = |f(\theta)|^2 \tag{17.72}$$

and

$$f_B(\theta) = -\frac{m}{2\pi\hbar^2}\int d^3r\, e^{-i\mathbf{q}.\,\mathbf{r}}V(\mathbf{r}) \tag{17.73}$$

where, $f^B(\theta)$ is the scattering amplitude in the Born approximation. The negative sign will be discussed in Chapter 21 when it will be shown that an attractive potential (which has an overall negative sign) will give a positive sign for the scattering amplitude. The scattering cross-section is then written as

$$\sigma_B = \int d\Omega\left|f^B(\theta)\right|^2 \tag{17.74}$$

where $d\Omega = d\cos\theta\, d\phi$. The angle θ is the angle between $\mathbf{k}_f$ and $\mathbf{k}_i$, and it enters in $f^B(\theta)$ through the momentum transfer $\mathbf{q}$ in (17.69). Typically, we have

$$|\mathbf{q}| = \left|\left(\mathbf{k}_f - \mathbf{k}_i\right)\right| = \sqrt{\left|\mathbf{k}_f\right|^2 + |\mathbf{k}_i|^2 - 2\left|\mathbf{k}_f\right||\mathbf{k}_i|\cos\theta} = \sqrt{2k^2(1-\cos\theta)} \tag{17.75}$$

where, as stated earlier, because of energy conservation, we take $\left|\mathbf{k}_f\right| = |\mathbf{k}_i| = k$.

17.4 Resonance and decay

17.4.1 Basic formalism

In the previous sections we considered transitions from the initial state, $|i\rangle$, to a final state $|f\rangle \neq |i\rangle$. Now let us find out what happens to the initial state itself. Clearly, as transitions out of the initial state occur, the initial state will begin to show signs of depletion.

We revisit Section 17.1 on the basic formalism for time-dependent perturbation, but this time with an approximation that does not resort to an expansion in λ. Let us consider the state vector $\left|\psi_s(t)\right\rangle$, which satisfies the equation

$$i\hbar\frac{\partial\left|\psi_s(t)\right\rangle}{\partial t} = \left(H_0 + H'\right)\left|\psi_s(t)\right\rangle \tag{17.76}$$

where H_0 is the unperturbed Hamiltonian and H' is the time-dependent perturbation applied at $t = 0$. We express $\left|\psi_s(t)\right\rangle$ in terms of the eigenstates $\left|\psi_n^0(t)\right\rangle$ of H_0 as we did before,

$$\left|\psi_s(t)\right\rangle = \sum_n c_{ns}(t)\left|\psi_n^0(t)\right\rangle \tag{17.77}$$

where $\left|\psi_n^0(t)\right\rangle$ are normalized as

$$\left|\psi_n^0(t)\right\rangle = \left|\psi_n^0(0)\right\rangle e^{-i\frac{E_n^0}{\hbar}t}. \tag{17.78}$$

Substituting (17.77) in (17.76) we obtain

$$i\hbar\frac{\partial}{\partial t}\left(\sum_n c_{ns}(t)\left|\psi_n^0(t)\right\rangle\right) = \left(H_0 + H'\right)\left(\sum_n c_{ns}(t)\left|\psi_n^0(t)\right\rangle\right). \tag{17.79}$$

Multiplying both sides of the above equation by $\left\langle\psi_m^0(t)\right|$ we obtain

$$i\hbar\dot{c}_{ms} = \sum_n c_{ns}H'_{mn}(t)\exp(i\omega_{mn}t) \tag{17.80}$$

where $\omega_{mn} = \left(E_m^0 - E_n^0\right)/\hbar$. Thus far we have obtained the same equation as in (17.10) but without the expansion parameter λ.

We consider the case where

$$H'_{mn}(t) = H'_{mn}(0) = \text{constant}, \quad t \geq 0 \tag{17.81}$$

$$= 0, \quad t < 0. \tag{17.82}$$

Thus we will be studying the problem of sudden perturbation.

We treat this problem effectively as a two-level problem with two coupled equations. The first equation will correspond to taking $m \neq s$ on the left-hand side of (17.80), in which we include only the contribution of $n = s$ on the right-hand side. In other words, we consider the "feedback" coming only from the initial state, which will be the dominant state at least for small values of t. We then write

$$i\hbar\dot{c}_{ms} = c_{ss}(t)\,H'_{ms}e^{i\omega_{ms}t} \quad \text{for} \quad m \neq s. \tag{17.83}$$

For the second equation we take $m = s$ in (17.80) and consider the feedback from both $n = s$ and $n \neq s$:

$$i\hbar\dot{c}_{ss} = c_{ss}H'_{ss} + \sum_{n\neq s} c_{ns}H'_{sn}e^{i\omega_{sn}t}. \tag{17.84}$$

We assume the following initial conditions:

$$c_{ss}(0) = 1, \tag{17.85}$$

$$c_{ms}(0) = 0, \quad \text{for} \quad m \neq s \tag{17.86}$$

$$c_{ms}(t) = 0, \quad \text{for} \quad t < 0 \quad \text{for all} \quad m. \tag{17.87}$$

These conditions are consistent with a sudden onset of perturbation, which we have assumed here, but with only $c_{ss}(0) = 1$. Integrating (17.83) we obtain

$$c_{ms} = -\frac{i}{\hbar}\int_0^t dt' c_{ss}\left(t'\right) H'_{ms} e^{i\omega_{ms}t'}, \quad m \neq s. \tag{17.88}$$

Substituting this expression, after replacing m by n, in the right-hand side of (17.84) we obtain the relation

$$i\hbar \dot{c}_{ss} = c_{ss}H'_{ss} + \sum_{n\neq s}\left[-\frac{i}{\hbar}\int_0^t dt' c_{ss}\left(t'\right) H'_{ns} e^{i\omega_{ns}t'}\right] H'_{sn} e^{i\omega_{sn}t}. \tag{17.89}$$

Therefore,

$$i\hbar \dot{c}_{ss} = c_{ss}H'_{ss} - \frac{i}{\hbar}\sum_{n\neq s}\left|H'_{ns}\right|^2 \int_0^t dt' c_{ss}\left(t'\right) e^{i\omega_{ns}\left(t'-t\right)}. \tag{17.90}$$

Thus, we have reduced the problem to solving a single-channel equation for c_{ss}.

To simplify notations we write

$$c_{ss} = c_i, \tag{17.91}$$

$$c_{ns} = c_f, \quad \text{for } n \neq s \tag{17.92}$$

and use the index f instead of n in equation (17.90) with an explicit connotation that

$$i = \text{initial}, \quad \text{and} \quad f = \text{final}. \tag{17.93}$$

Equation (17.90) now reads

$$i\hbar \dot{c}_i = c_i H'_{ii} - \frac{i}{\hbar}\sum_{f\neq i}\left|H'_{fi}\right|^2 \int_0^t dt' c_i\left(t'\right) e^{i\omega_{fi}\left(t'-t\right)}. \tag{17.94}$$

We once again note that

$$c_i(0) = 1 \quad \text{and} \quad c_i(t) = 0 \ \text{ for } t < 0. \tag{17.95}$$

To obtain the solution for c_i we write it in the form of a Fourier transform,

$$c_i(t) = \frac{1}{\sqrt{2\pi}}\int_{-\infty}^{\infty} d\omega\, c_i(\omega)\, e^{i\omega t} \tag{17.96}$$

and evaluate $c_i(\omega)$.

In view of condition (17.95), the inverse Fourier transform will involve only the limits 0 to ∞ for the integration over t:

$$c_i(\omega) = \frac{1}{\sqrt{2\pi}} \int_0^\infty dt\, c_i(t)\, e^{-i\omega t}. \tag{17.97}$$

We next write $\dot{c}_i$ in the form of a Fourier transform. Because of the vanishing of c_i for $t < 0$ as given by (17.95), we cannot simply take the derivative of (17.96) to determine $\dot{c}_i$. Instead we write it in a new Fourier transform relation:

$$\dot{c}_i = \frac{1}{\sqrt{2\pi}} \int_{-\infty}^{\infty} d\omega\, D_i(\omega)\, e^{i\omega t}. \tag{17.98}$$

We now evaluate $D_i(\omega)$. We note that $\dot{c}_i$ will vanish for $t < 0$ since c_i vanishes identically in this region. The inverse Fourier transform is then given by

$$D_i(\omega) = \frac{1}{\sqrt{2\pi}} \int_0^\infty dt\, \dot{c}_i(t) e^{-i\omega t}. \tag{17.99}$$

Let us integrate the above by parts. We will assume that c_i vanishes as $t \to \infty$; therefore, we have

$$D_i(\omega) = \frac{1}{\sqrt{2\pi}} \left[-c_i(0) + i\omega \int_0^\infty dt\, c_i(t)\, e^{-i\omega t} \right] = -\frac{1}{\sqrt{2\pi}} \left[1 - i\omega\sqrt{2\pi}\, c_i(\omega) \right] \tag{17.100}$$

and hence substituting $D_i(\omega)$ in (17.98) we obtain

$$\dot{c}_i = -\frac{1}{2\pi} \int_{-\infty}^{\infty} d\omega \left[1 - i\omega\sqrt{2\pi}\, c_i(\omega) \right] e^{i\omega t}. \tag{17.101}$$

We now compare the Fourier transforms of both sides of (17.94). The right-hand side of (17.94) is given by

$$H'_{ii} \frac{1}{\sqrt{2\pi}} \int_{-\infty}^{\infty} d\omega\, c_i(\omega)\, e^{i\omega t} - \frac{i}{\hbar} \sum_{f \neq i} \left| H'_{fi} \right|^2 \left[\int_0^t dt' \frac{1}{\sqrt{2\pi}} \int_{-\infty}^{\infty} d\omega\, c_i(\omega)\, e^{i\omega t'} e^{i\omega_{fi}(t'-t)} \right]. \tag{17.102}$$

After the integration over t' is carried out in the second term, we obtain

$$H'_{ii} \frac{1}{\sqrt{2\pi}} \int_{-\infty}^{\infty} d\omega\, c_i(\omega)\, e^{i\omega t} - \frac{i}{\hbar} \sum_{f \neq i} \left| H'_{fi} \right|^2 \frac{1}{\sqrt{2\pi}} \int_{-\infty}^{\infty} d\omega\, c_i(\omega) \left[\frac{e^{i\omega t} - e^{-i\omega_{fi} t}}{i(\omega + \omega_{fi})} \right]. \tag{17.103}$$

Comparing this term to the right-hand side of (17.101) by equating the coefficients of $e^{i\omega t}$ on both sides of the equations, we obtain

$$-i\hbar\frac{1}{2\pi}\left[1-i\sqrt{2\pi}\omega c_i(\omega)\right]=\frac{1}{\sqrt{2\pi}}\left[H'_{ii}-\sum_{f\neq i}\frac{\left|H'_{fi}\right|^2}{\hbar\left(\omega+\omega_{fi}\right)}\right]c_i(\omega), \tag{17.104}$$

which gives

$$-\frac{i}{2\pi}=\frac{1}{\sqrt{2\pi}}\left[\omega+\frac{H'_{ii}}{\hbar}-\frac{1}{\hbar^2}\sum_{f\neq i}\frac{\left|H'_{fi}\right|^2}{\omega+\omega_{fi}}\right]ic_i(\omega). \tag{17.105}$$

Thus,

$$\sqrt{2\pi}c_i(\omega)=\frac{-i}{\omega+\frac{H'_{ii}}{\hbar}-\frac{1}{\hbar^2}\sum_{f\neq i}\frac{\left|H'_{fi}\right|^2}{\omega+\omega_{fi}}}. \tag{17.106}$$

From (17.96) we then obtain

$$c_i(t)=\frac{i}{2\pi}\int_{-\infty}^{\infty}d\omega\frac{e^{i\omega t}}{\omega+\frac{H'_{ii}}{\hbar}-\frac{1}{\hbar^2}\sum_{f\neq i}\frac{\left|H'_{fi}\right|^2}{\omega+\omega_{fi}}}. \tag{17.107}$$

Since $c_i(t)=0$ for $t<0$, the presence of $e^{i\omega t}$ in the integrand implies that we must move the pole in ω, which is on the real axis, to the upper half-plane to obtain a convergent result. To accomplish this, we change the position of the pole, $\omega\to\omega-i\varepsilon$. We also make the change

$$\frac{1}{\omega+\omega_{fi}}\to\frac{1}{\omega-i\varepsilon+\omega_{fi}}. \tag{17.108}$$

The roots of the denominator above in the complex ω-plane are very complicated, but since we are interested in $c_i(t)$ as $t\to\infty$, only small values of ω will contribute to the integral. Hence we write

$$\frac{1}{\omega+\omega_{fi}}\to\frac{1}{\omega_{fi}-i\varepsilon}=P\left(\frac{1}{\omega_{fi}}\right)+i\pi\delta\left(\omega_{fi}\right). \tag{17.109}$$

Thus,

$$c_i(t)=\frac{i}{2\pi}\int_{-\infty}^{\infty}d\omega\frac{e^{i\omega t}}{\omega-i\epsilon+\frac{H'_{ii}}{\hbar}-\frac{1}{\hbar^2}\left[\sum_{f\neq i}\frac{\left|H'_{fi}\right|^2}{\omega_{fi}}+i\pi\sum_{f\neq i}\left|H'_{fi}\right|^2\delta\left(\omega_{fi}\right)\right]} \tag{17.110}$$

where we have replaced $P(1/\omega_{fi})$ by $(1/\omega_{fi})$ since the summation does not include $f = i$. The denominator above can be written (reverting back to writing the energies E_i and E_f as $E_i^{(0)}$ and $E_f^{(0)}$ respectively) as

$$\omega - i\epsilon + \frac{H'_{ii}}{\hbar} + \frac{1}{\hbar}\sum_{f\neq i}\frac{\left|H'_{fi}\right|^2}{E_i^{(0)} - E_f^{(0)}} - \frac{i\pi}{\hbar}\sum_{f\neq i}\left|H'_{fi}\right|^2 \delta\left(E_i^{(0)} - E_f^{(0)}\right). \tag{17.111}$$

We note from time-independent perturbation theory that the perturbed energies to first and second order are given respectively by

$$E_i^{(1)} = H'_{ii} \quad \text{and} \quad E_i^{(2)} = \sum_{f\neq i}\frac{\left|H'_{fi}\right|^2}{E_i^{(0)} - E_f^{(0)}}. \tag{17.112}$$

We write the combined perturbed energy as E'_i,

$$E'_i = H'_{ii} + \sum_{f\neq i}\frac{\left|H'_{fi}\right|^2}{E_i^{(0)} - E_f^{(0)}}. \tag{17.113}$$

If $E_i^{(0)}$ is the unperturbed energy, then the total energy, E_i, is given by

$$E_i = E_i^{(0)} + H'_{ii} + \sum_{f\neq i}\frac{\left|H'_{fi}\right|^2}{E_i^{(0)} - E_f^{(0)}}. \tag{17.114}$$

To determine the last term in (17.111) we note that the probability for the transition $i \to f$ per unit time as $t \to \infty$ is given by

$$\lambda_{fi} = \frac{2\pi}{\hbar}\left|H'_{fi}\right|^2 \delta\left(E_i^{(0)} - E_f^{(0)}\right). \tag{17.115}$$

For transitions to a group of states we sum over the final states, and define the result as

$$\Gamma = \frac{2\pi}{\hbar}\sum_{f\neq i}\left|H'_{fi}\right|^2 \delta\left(E_i^{(0)} - E_f^{(0)}\right). \tag{17.116}$$

For the case when the final states are free particles we have designated the above quantity as w_{fi}.

Since Γ_{fi} equals $\sum\left(d\left|c_f\right|^2/dt\right)$ for $f \neq i$, one obtains an approximate relation

$$\sum_{f\neq i}\left|c_f\right|^2 \approx \Gamma t. \tag{17.117}$$

Inserting the above results in the denominator of (17.110) we can write $c_i(t)$ as follows, ignoring ϵ compared to Γ,

$$c_i(t) = \frac{-i}{2\pi}\int_{-\infty}^{\infty} d\omega \frac{e^{i\omega t}}{\omega + \frac{E_i'}{\hbar} - i\frac{\Gamma}{2}} \tag{17.118}$$

After carrying out the Cauchy integration in the complex ω-plane we obtain

$$c_i(t) = e^{-i\frac{E_i' t}{\hbar}} e^{-\frac{\Gamma}{2}t} \quad \text{for } t \geq 0. \tag{17.119}$$

We need to point out here that the wavefunction is given by the product $c_i(t)\left|\psi_i^{(0)}(t)\right\rangle$, which can be written as

$$c_i(t)\left|\psi_i^{(0)}(t)\right\rangle = e^{-i\frac{E_i' t}{\hbar}} e^{-\frac{\Gamma}{2}t} e^{-i\frac{E_i^{(0)} t}{\hbar}}\left|\psi_i^{(0)}(0)\right\rangle$$
$$= e^{-i\frac{E_i t}{\hbar}} e^{-\frac{\Gamma}{2}t}\left|\psi_i^{(0)}(0)\right\rangle$$

where E_i is the total energy defined in (17.114).

The probability is found to be

$$|c_i(t)|^2 = e^{-\Gamma t}. \tag{17.120}$$

This is the famous "exponential law" that one refers to in radioactive and other decay processes. Hence the initial state is depleted with a lifetime, τ_i, given by

$$\tau_i = \frac{1}{\Gamma}. \tag{17.121}$$

We note that for small values of t,

$$|c_i(t)|^2 = 1 - \Gamma t. \tag{17.122}$$

From (17.117), we conclude that

$$|c_i|^2 + \sum_{f\neq i}|c_f|^2 = 1, \tag{17.123}$$

as one expects.

In summary, the eigenstate of the unperturbed Hamiltonian, H_0, under the influence of perturbation, H', applied at $t = 0$, is depleted as the particle makes transitions to other states in the system. Thus the initial state undergoes "decay." The rate of decay depends on the rate of transitions to the other states.

We can also obtain $c_f(t)$ by substituting $c_i(t)$ in the equation

$$c_f(t) = -\frac{i}{\hbar}\int_0^t dt' c_i(t') H_{fi}' e^{i\omega_{fi}t'} \tag{17.124}$$

where

$$\omega_{fi} = \frac{E_f^{(0)} - E_i^{(0)}}{\hbar}. \tag{17.125}$$

We will replace $E_f^{(0)}$ by E_f but keep $E_i^{(0)}$. We find, after taking the limit $t \to \infty$ and doing the integration, that

$$c_f(\infty) = \frac{H'_{fi}(0)}{E_f - E_i - i\Gamma/2} \tag{17.126}$$

where E_i is now the total energy given by (17.114). The transition probability is then

$$|c_f(\infty)|^2 = \frac{\left|H'_{fi}(0)\right|^2}{(E_f - E_i)^2 + \Gamma^2/4}. \tag{17.127}$$

The above results were derived for constant perturbations; however, they can be generalized to harmonic perturbations of the type

$$H'_{fi}(t) = H'_{fi}(0)e^{-i\omega t}, \quad t \geq 0 \tag{17.128}$$

$$= 0, \quad t < 0. \tag{17.129}$$

The corresponding transition probability is given by

$$|c_f(\infty)|^2 = \frac{\left|H'_{fi}(0)\right|^2}{(E_f - E_i - \hbar\omega)^2 + \Gamma^2/4}. \tag{17.130}$$

Clearly then, if the energy $\hbar\omega$ supplied by the perturbation is equal to the energy difference between the initial state and one of the higher states in the system, the transition probability is very large and we have a classic example of what is called a "resonance." We will be discussing this phenomenon further in the later chapters.

17.5 Appendix to Chapter 17

17.5.1 Adiabatic perturbation

Let us assume that the matrix elements of H' have the behavior

$$\begin{aligned} H'_{ms}(t) &\to 0 \text{ as } t \to -\infty, \\ H'_{ms}(t) &\to \text{const. as } t \to +\infty. \end{aligned} \tag{17.131}$$

In other words, H'_{ms} starts at zero and gradually reaches a constant. We further assume that it varies very slowly with time once it achieves its maximum level, i.e.,

$$\left(1/H'_{ms}\right) dH'_{ms}(t)/dt \approx 0 \tag{17.132}$$

at large t. The first-order transition probability amplitude for $m = s$ is found to be

$$c_{ss}^{(1)} = \left(\frac{-i}{\hbar}\right) \int_{-\infty}^{t} dt' \, H'_{ss}(t'). \tag{17.133}$$

Integrating by parts, we obtain

$$c_{ss}^{(1)} = \left(\frac{-i}{\hbar}\right) \left[H'_{ss} t - \int_{-\infty}^{t} dt' \, t' H'_{ss}(t') \right]. \tag{17.134}$$

For $m \neq s$ the probability amplitude is given by

$$c_{ms}^{(1)} = \left(\frac{-i}{\hbar}\right) \int_{-\infty}^{t} dt' \, H'_{ms}(t') \, e^{i\omega_{ms} t'}, \tag{17.135}$$

which is found, after integrating by parts, to be

$$c_{ms}^{(1)} = \frac{H'_{ms}(t) \, e^{i\omega_{ms} t'}}{E_m^{(0)} - E_s^{(0)}} + \frac{1}{E_m^{(0)} - E_s^{(0)}} \int_{-\infty}^{t} dt' \frac{d}{dt'} H'_{ms}(t') \, e^{i\omega_{ms} t'}. \tag{17.136}$$

Because of the relation (17.131) for large values of t, we can extend the upper integration limit from t to ∞ for both types of probability amplitudes. We consider for illustration purposes the following Gaussian behavior,

$$\frac{dH'}{dt'} = \frac{dH'(0)}{dt'} e^{-\frac{t^2}{2\Delta t^2}}, \tag{17.137}$$

which is consistent with the criteria (17.131) and (17.132). We can identify Δt as the time period over which H' achieves its maximum value. The integral on the right-hand side of the expression for $c_{ss}^{(1)}$ is then

$$\int_{-\infty}^{\infty} dt' \, t' H'_{ss}(t') = \frac{dH'(0)}{dt'} \int_{-\infty}^{\infty} dt' \, t' e^{-\frac{t^2}{2\Delta t^2}} = 0, \tag{17.138}$$

since the integrand is a product of odd and even functions.

Hence we can write

$$c_{ss}^{(1)} = \left(\frac{-i}{\hbar}\right) H'_{ss} t. \tag{17.139}$$

For $m \neq s$ we find

$$\int_{-\infty}^{\infty} dt' \frac{d}{dt'} H'_{ms}(t')\, e^{i\omega_{ms}t'} = \text{Fourier transform of } \frac{dH'}{dt'}, \tag{17.140}$$

$$\int_{-\infty}^{\infty} dt' \frac{dH'_{ms}(t')}{dt'}\, e^{i\omega_{ms}t'} = \frac{dH'_{ms}(0)}{dt}\, e^{-\frac{\Delta t^2 \omega_{ms}^2}{2}}. \tag{17.141}$$

We note that if $\Delta t \gg 1/\omega_{ms}$ then the above integral vanishes. This implies that if the perturbation is turned on very slowly, with Δt much greater than the period of oscillation between the states $(1/\omega_{ms})$, then for $m \neq s$,

$$c_{ms}^{(1)} = -\frac{H'_{ms}(t)\, e^{i\omega_{ms}t}}{E_m^{(0)} - E_s^{(0)}} \tag{17.142}$$

and

$$|\psi_s\rangle = \left(1 - \frac{\lambda H'_{ss}}{\hbar}\right) \left|\psi_s^0\right\rangle e^{-iE_s^0 t/\hbar} - \lambda \sum_{n\neq s} \frac{H'_{ns}(t)\, e^{i\omega_{ns}t}}{E_n^{(0)} - E_s^{(0)}} \left|\psi_n^0\right\rangle e^{-iE_n^0 t/\hbar}. \tag{17.143}$$

Since we are interested in the results only up to the first power in λ, the above relation is the same as the following to order λ,

$$|\psi_s\rangle = \left[\left|\psi_s^0\right\rangle + \lambda \sum_{n\neq s} \frac{H'_{ns}(t)}{E_n^{(0)} - E_s^{(0)}} \left|\psi_n^0\right\rangle\right] e^{-i(E_s^0 + \lambda H'_{ss})t/\hbar}. \tag{17.144}$$

This is the same result as in the time-independent case, as it reproduces the perturbed eigenstate and eigenvalue to first order. Thus if the perturbation takes effect gently and over a sufficiently long period of time, we will have the same result as for the case of time-independent perturbation. In other words, specifically, an energy eigenstate remains essentially an energy eigenstate if the perturbation varies with time in an adiabatic manner. If the perturbation is small, then the energy eigenvalue remains essentially unchanged.

17.5.2 Berry's phase

Let us begin with the time-dependent Schrödinger equation in which the state $|\psi_n\rangle$ is an eigenstate of the Hamiltonian with eigenvalue E_n,

$$i\hbar \frac{\partial |\psi_n\rangle}{\partial t} = H |\psi_n\rangle = E_n |\psi_n\rangle. \tag{17.145}$$

Here H and E_n are independent of time. The solution is simply given by

$$|\psi_n(t)\rangle = |\psi_n(0)\rangle e^{-iE_n t/\hbar}. \tag{17.146}$$

Now consider the case where the Hamiltonian varies slowly with time. If this variation is adiabatic then we expect E_n to also vary with time but still remain an energy eigenstate. Hence we write

$$i\hbar\frac{\partial\left|\psi_n(t)\right\rangle}{\partial t}=H(t)\left|\psi_n(t)\right\rangle=E_n(t)\left|\psi_n(t)\right\rangle. \tag{17.147}$$

The solution of this equation can be written in the most general form as

$$\left|\psi_n(t)\right\rangle=\left|\phi_n(t)\right\rangle e^{-i\alpha_n(t)}e^{i\gamma_n(t)} \tag{17.148}$$

where we have introduced an additional phase $\gamma_n(t)$ to allow for adiabatic time variation, since such a possibility cannot be ruled out.

The two phases have a distinct origin and are designated as follows:

$$\alpha_n(t)=\int_0^t dt' E_n(t')=\text{dynamical phase} \tag{17.149}$$

and

$$\gamma_n(t)=\text{geometrical phase.} \tag{17.150}$$

If we substitute (17.148) in equation (17.147) we obtain the equation

$$\frac{\partial\left|\phi_n(t)\right\rangle}{\partial t}+i\left|\phi_n(t)\right\rangle\frac{d\gamma_n(t)}{dt}=0. \tag{17.151}$$

The solution of this can readily be obtained in terms of the integral

$$\gamma_n(t)=\int_0^t dt' i\left\langle\phi_n\left|\frac{\partial\left|\phi_n\right\rangle}{\partial t}\right.\right\rangle. \tag{17.152}$$

Often the variation in time is dictated by a parameter or a set of parameters. If λ is such a parameter, then

$$\frac{\partial\phi_n}{\partial t}=\frac{\partial\phi_n}{\partial\lambda}\frac{d\lambda}{dt}. \tag{17.153}$$

Hence one can write

$$\gamma_n(t)=\int_{\lambda_i}^{\lambda_f}\left\langle\phi_n\left|\frac{\partial\left|\phi_n\right\rangle}{\partial\lambda}\right.\right\rangle d\lambda. \tag{17.154}$$

Example: Spin aligned with the magnetic field

A simple illustration of Berry's phase is provided by the example we considered in Chapter 14 of a particle whose spin is aligned in the direction of the applied magnetic field which precesses about the z-axis with frequency ω. The magnetic field is taken to be

$$\mathbf{B} = \frac{\hbar\omega_1}{2}\left[\mathbf{i}\sin\theta\cos\omega t + \mathbf{j}\sin\theta\sin\omega t + \mathbf{k}\cos\theta\right] \tag{17.155}$$

where θ denotes the polar angle for the magnetic field. The Hamiltonian for this process is then

$$H = \frac{\hbar\omega_1}{2}\begin{pmatrix}\cos\theta & \sin\theta e^{-i\omega t}\\ \sin\theta e^{i\omega t} & -\cos\theta\end{pmatrix}. \tag{17.156}$$

The energy eigenvalues at any fixed t are

$$E_\pm = \pm\frac{\hbar\omega_1}{2}. \tag{17.157}$$

The spin wavefunctions are given by

$$\chi_+(t) = \begin{bmatrix}\cos(\theta/2)\\ e^{i\omega t}\sin(\theta/2)\end{bmatrix} \quad \text{and} \quad \chi_-(t) = \begin{bmatrix}\sin(\theta/2)\\ -e^{i\omega t}\cos(\theta/2)\end{bmatrix}. \tag{17.158}$$

If the particle starts out at $t = 0$ in the spin-up position (in the direction of the magnetic field) then the probability amplitude of finding it in the spin-up and -down positions at $t > 0$ is given, respectively, by

$$c_+(t) = \left[\cos(\Omega t) + i\left(\frac{\omega_1 + \omega\cos\theta}{2\Omega}\right)\sin(\Omega t)\right]e^{-i\omega t/2} \tag{17.159}$$

and

$$c_-(t) = \left[\frac{i\omega}{2\Omega}\sin\theta\sin(\Omega t)\right]e^{-i\omega t/2} \tag{17.160}$$

where

$$\Omega = \frac{\sqrt{\omega^2 + \omega_1^2 + 2\omega\omega_1\cos\theta}}{2}. \tag{17.161}$$

The adiabatic condition we outlined above is met if we take

$$\omega \approx 0, \tag{17.162}$$

in which case c_- can be ignored and the spin maintains its direction aligned with the magnetic field. In this limit,

$$\Omega \approx \frac{\omega_1 + \omega\cos\theta}{2} \tag{17.163}$$

and

$$c_+(t) \approx e^{\Omega t} e^{-i\omega t/2} = e^{\frac{i\omega_1 t}{2}} e^{i\omega(\cos\theta - 1)t/2}. \tag{17.164}$$

The first factor corresponds to the dynamical phase since, from (17.157), $\hbar\omega_1/2$ is the energy of the particle. The second term corresponds to the geometric phase

$$\gamma(t) = \omega(\cos\theta - 1)\frac{t}{2}. \tag{17.165}$$

For a complete cycle, $t = 2\pi/\omega$, Berry's phase will then be

$$\gamma(2\pi/\omega) = \pi(\cos\theta - 1). \tag{17.166}$$

17.6 Problems

1. Consider a one-dimensional harmonic oscillator that is in its ground state ($n = 0$). A time-dependent perturbation

$$H' = \lambda x e^{-\alpha t}$$

is applied at $t = 0$. Obtain the probability and λ_{fi} for the cases that it makes transitions to $n = 1$ and $n = 2$ states. What is λ_{fi} if the exponential factor is replaced by the harmonic function $e^{-i\omega t}$? For the harmonic case determine λ_{fi} if the final state is a free particle state given by $\left(1/\sqrt{2\pi}\right) \exp ikx$.

2. A spin ½ system is subjected to a perturbation at $t = 0$,

$$H'(t) = \begin{bmatrix} 0 & ge^{-i\omega t} \\ ge^{i\omega t} & 0 \end{bmatrix}.$$

If the unperturbed energies are E_1^0 and E_2^0 and the initial state corresponds to "spin-up", then use time-dependent perturbation theory to obtain the probability that for $t > 0$ the system is in "spin-down" state. Complete this problem exactly.

3. A particle of charge e is carrying out linear harmonic oscillation. It is subjected to a uniform time-dependent electric field, $\mathbf{E}$, with a Gaussian time-dependence of the form

$$\mathbf{E}(t) = \mathbf{E}_0 e^{-\alpha t^2}, \quad -\infty < t < \infty.$$

If the field is applied at $t = -\infty$, determine the probability that it makes a transition from the ground state to the first excited state at $t = \infty$.

4. Consider the same time-dependent electric field as in Problem 3 applied to an electron in the hydrogen atom. Determine the transition probability that it is goes from $1S$ state at $t = -\infty$ to $2P$ state at $t = \infty$.

5. A particle of charge e carrying out linear harmonic oscillation is subjected to a constant electric field $\mathbf{E}_0$ at $t = 0$. Solve the problem exactly. If the particle is in the ground state at $t = 0$, determine the probability that it is in the first excited state for $t > 0$.
6. The charge of the nucleus in the hydrogen atom is changed from Z at $t = 0$ to $(Z + 1)$ for $t > 0$. Determine the probability that the electron makes a transition from the $1S$ to the $2S$ state.
7. A particle between two infinite walls located at $x = \pm a$ is subjected to a perturbation at $t = 0$ in the form of a finite repulsive barrier $V = V_0$ between $-b$ and $+b$, which is removed at $t = t_0$. If the particle is in the ground state, $n = 1$, at $t \leq t_0$ what is the probability that it will be in the first excited state $n = 2$ for $t > t_0$?
8. A hydrogen atom is subjected to a constant electric field $\mathbf{E}_0$ that lasts for a time $0 < t < \tau$. If at $t = 0$ the atom is in the $2S$ state, determine the time dependence of the system in the interval $0 < t < \tau$. What is the probability that it will be in the $2P$ state for $t > 0$.
9. For the above problem assume that the time dependence of $\mathbf{E}_0$ is given by

$$\mathbf{E} = \mathbf{E}_0 e^{-\Gamma t} \quad \text{for } t \geq 0$$
$$= 0 \quad \text{for } t < 0.$$

Determine the probability in the limit $t \to \infty$ that the atom will make a transition from a $1S$ state to a $2P$ state. Determine λ_{fi}.

10. A harmonic oscillator is subjected to an electromagnetic field (laser) such that the interaction Hamiltonian is given by

$$H'(t) = \frac{eE_0}{2}\left[\frac{p}{m\omega}\sin\omega t - x\cos\omega t\right]$$

for $0 < t < \infty$. Determine the probability at time t that the oscillator will make a transition from the ground state, $|0\rangle$, to the state $|1\rangle$. Also obtain λ_{fi}.

11. Based on first-order time-dependent perturbation theory, show that the probability of transition from a state $|i\rangle$ to a state $|j\rangle$ is the same as the probability of transition from $|j\rangle$ to $|i\rangle$ (this is the first-order version of the principle of detailed balancing).
12. An electron passing by a hydrogen atom in the ground state excites it to the level $2S$. Determine the differential cross-section for this process.
13. For a time-dependent problem we have

$$H'(t) = 0, \quad -\infty < t < T_0$$
$$= V(r) \quad T_0 < t < T_1$$
$$= 0, \quad T_1 < t < \infty.$$

Obtain $c_{fi}(t)$ and λ_{fi}. Assume $\langle f|\, V(r)\, |i\rangle = \mu$.

14. A particle confined within two infinite walls experiences a time-dependent perturbing potential of the form

$$V'(x) = 0, \qquad t < 0$$
$$= \lambda (1 + x) \exp(-i\omega_0 t), \quad t \geq 0.$$

Obtain the transition probability and λ_{fi} for transitions from the ground state to the first excited state.

15. For a harmonic oscillator defined in terms of the operators a and $a^\dagger$,

$$H = \left(a^\dagger a + \frac{1}{2}\right)\hbar\omega,$$

use time-dependent perturbation theory to obtain the transition probability amplitude for transitions from the ground state to a state $|n\rangle$, under the perturbing potential given (with $b > 0$) by

$$H'(t) = 0 \qquad \text{for } t < 0$$
$$= gx^2 \exp(-(b + i\alpha)t) \quad \text{for } t > 0.$$

18 Interaction of charged particles and radiation in perturbation theory

We consider here some specific and well-known cases involving the interactions of charged particles and electromagnetic field which are solved using time-dependent perturbation theory. These include Coulomb excitation of atoms, photoelectric effect, ionization of atoms due to electromagnetic interaction, plus, in a first use of second-order time-dependent perturbation, obtaining the cross-sections for Thomson scattering and the related Rayleigh and Raman scattering.

18.1 Electron in an electromagnetic field: the absorption cross-section

As discussed in Chapter 6, the Hamiltonian for an electron of charge e in an atom interacting with an external electromagnetic field is given by

$$H = \frac{1}{2m}\left(\mathbf{p} - \frac{e}{c}\mathbf{A}\right)^2 + e\phi(\mathbf{r}) + V_0(r) \tag{18.1}$$

where $\mathbf{A}$ and ϕ are the vector and scalar potentials, respectively, for the electromagnetic field. The potential representing the binding of the electron to the rest of the atom is represented by $V_0(r)$. Keeping in mind that $\mathbf{p}$ and $\mathbf{r}$ are operators, we obtain

$$H = \frac{1}{2m}\left[\mathbf{p}^2 - \frac{e}{c}\mathbf{p}\cdot\mathbf{A} - \frac{e}{c}\mathbf{A}\cdot\mathbf{p} + \frac{e^2}{c^2}\mathbf{A}^2\right] + e\phi(\mathbf{r}) + V_0(r). \tag{18.2}$$

We note that $\mathbf{p} = -i\hbar\nabla$ and H operate on a wavefunction on the right. If $f(\mathbf{r})$ is the wavefunction then the second term in the square bracket in (18.2) corresponds to

$$\mathbf{p}\cdot\mathbf{A}f = -i\hbar\nabla\cdot(\mathbf{A}f) = -i\hbar(\nabla\cdot\mathbf{A})f + \mathbf{A}\cdot\mathbf{p}f. \tag{18.3}$$

Therefore,

$$H = \frac{1}{2m}\left[\mathbf{p}^2 + \frac{i\hbar e}{c}\nabla\cdot\mathbf{A} - \frac{2e}{c}\mathbf{A}\cdot\mathbf{p} + \frac{e^2}{c^2}\mathbf{A}^2\right] + e\phi(\mathbf{r}) + V_0(r). \tag{18.4}$$

We take $\phi = 0$ and choose the transverse gauge, $\nabla\cdot\mathbf{A} = 0$, for the vector potential, $\mathbf{A}$. To apply perturbation theory we take e, the charge of the particle, as a parameter that will take the place of λ in the perturbation expansion. It is numerically a small quantity and, therefore, perturbation expansion is justified. To first order in e we will then neglect the contribution of the $\mathbf{A}^2$ term and write

$$H = \frac{p^2}{2m} - \frac{e}{mc}\mathbf{A} \cdot \mathbf{p} + V_0(r). \tag{18.5}$$

This equation has the form $H = H_0 + H'$. The unperturbed Hamiltonian is given by

$$H_0 = \frac{p^2}{2m} + V_0(r), \tag{18.6}$$

which includes the kinetic energy term as well as $V_0(r)$, which is responsible for atomic binding. The perturbation due to the interaction of the electron with the electromagnetic field is identified as

$$H'(t) = -\frac{e}{mc}\mathbf{A} \cdot \mathbf{p}, \tag{18.7}$$

which is time-dependent since we take the vector potential to be time-dependent given by

$$\mathbf{A}\,(\mathbf{r}, t) = \mathbf{A}_0[e^{i(\mathbf{k}\cdot\mathbf{r}-\,\omega\, t)} + e^{-i(\mathbf{k}\cdot\mathbf{r}-\,\omega\, t)}] \tag{18.8}$$

where $\mathbf{A}_0$ is real. We also assume that this interaction is turned on at $t = 0$.

The first term in $\mathbf{A}\,(\mathbf{r}, t)$ given by (18.8) with $e^{-i\omega\, t}$ corresponds to a process in which the electron absorbs the energy $\hbar\omega$. It contributes a factor $\delta[E_n - (E_i + \hbar\omega)]$ to the transition probability as we discussed in the previous chapter. It is this term in which we are interested. We classify this as a "radiation absorption" process. The second term proportional to $\exp(i\omega t)$ is responsible for the so-called "stimulated emission" in which the atom goes down to a lower level by emission of radiation. The wavevector $\mathbf{k}$ for electromagnetic radiation is related to the frequency ω by $k = \omega/c$, where c is the velocity of light. If $\mathbf{n}$ is a unit vector in the direction of $\mathbf{k}$, then $\mathbf{k} = (\omega/c)\mathbf{n}$. The matrix element of $H'(t)$ at $t = 0$ connecting the initial state $|i\rangle$ with energy E_i of the atomic electron to a higher atomic level $|n\rangle$ with energy E_n is given by

$$H'_{ni}(0) = -\frac{e}{mc}\,\langle n|\, e^{i(\frac{\omega}{c}\mathbf{n}\cdot\mathbf{r})}\, \mathbf{A}_0 \cdot \mathbf{p}\,|i\rangle\,. \tag{18.9}$$

The transition probability λ_{ni} for transitions $i \to n$ per unit time as $t \to \infty$ is then given by

$$\lambda_{ni} = \frac{2\pi}{\hbar}\left|H'_{ni}(0)\right|^2 \delta(E_n - E_i - \hbar\omega). \tag{18.10}$$

For transitions to occur, $\hbar\omega$ must be of the order of $E_n - E_i$, which is of the order of the atomic levels. In the hydrogen atom the levels are proportional to e^2/a_0, where a_0 is the Bohr radius. For an atom we replace a_0 by the corresponding atomic radius, which we denote R. Hence one expects transitions to occur for

$$\hbar\omega \sim \frac{Ze^2}{R}. \tag{18.11}$$

This implies that the wavevector will be of the order

$$k = \frac{\omega}{c} \sim \frac{Ze^2}{\hbar c R} = \left(\frac{Z}{137}\right)\frac{1}{R} \tag{18.12}$$

where in the last step we have substituted the numerical value $1/137$ for the fine-structure constant $\left(e^2/\hbar c\right)$.

The exponent in (18.9) is then estimated to be

$$\left|\frac{\omega}{c}\mathbf{n}\cdot\mathbf{r}\right| \sim \left(\frac{Z}{137}\right)\frac{r}{R}. \tag{18.13}$$

Since r involved in the integrations in (18.9) is less than or of the order of R, $|(\omega/c)\widehat{n}\cdot\mathbf{r}| \ll 1$ for light atoms with small values of Z. Hence, we can approximate

$$e^{i\frac{\omega}{c}\mathbf{n}\cdot\mathbf{r}} \simeq 1. \tag{18.14}$$

We take $\mathbf{A}_0$ to be in the x-direction, and write

$$\langle n \mid \mathbf{A}\cdot\mathbf{p} \mid i\rangle = |\mathbf{A}_0|\,\langle n\, |p_x|\, \mathrm{i}\rangle.$$

To evaluate the above matrix element we first calculate the commutator $[x, H_0]$ and note that x will commute with V_0 since it is a function of the coordinates. To obtain $\langle n\, |p_x|\, \mathrm{i}\rangle$ we use the result

$$[x, H_0] = \frac{1}{2m}\left[x, p^2\right] = i\hbar\frac{p_x}{m}. \tag{18.15}$$

The last step follows from the commutation relations $[x, p_x] = i\hbar$ and $\left[x, p_y\right] = 0 = [x, p_z]$. Thus we obtain

$$\langle n \mid p_x \mid \mathrm{i}\rangle = \frac{m}{i\hbar}\langle n \mid [x, H_0] \mid i\rangle = \frac{m}{i\hbar}\langle n|\,(xH_0 - H_0x)\mid \mathrm{i}\rangle = \mathrm{i}m\,\omega_{ni}\langle n \mid x \mid \mathrm{i}\rangle \tag{18.16}$$

where

$$\omega_{ni} = \frac{E_n - E_i}{\hbar} \tag{18.17}$$

We have made use of the fact that $|i\rangle$ and $|n\rangle$ are eigenstates of H_0 with eigenvalues E_i and E_n respectively. Substituting (18.16) into (18.9) gives

$$H'_{ni}(0) = -\frac{\mathrm{i}e}{c}\,|\mathbf{A}_0|\,\omega_{ni}\langle n \mid x \mid i\rangle.$$

Since $\langle n \mid x \mid i\rangle$ is proportional to the dipole moment of the atom, our approximation is called the "electric dipole approximation." The transition probability per unit time, λ_{ni}, corresponding to the absorption of radiation is then given by

$$\lambda_{ni} = \frac{2\pi e^2\,|\mathbf{A}_0|^2}{\hbar c^2}\omega_{ni}^2\,|\langle n \mid x \mid i\rangle|^2\,\delta(E_n - E_i - \hbar\omega). \tag{18.18}$$

Let us now calculate the absorption cross-section, σ_{abs}, for the above process, which is given by

$$\sigma_{\text{abs}} = \frac{\lambda_{ni}}{J_i} \tag{18.19}$$

where J_i is the incident photon flux. We will use the language of second quantization in which the radiation field is described in terms of discrete number of particles called photons carrying energy $\hbar\omega$. The energy absorbed per unit time is then equal to $\lambda_{ni}\hbar\omega$. If we define the incident energy flux as the radiation energy incident per unit area per unit time, then this flux is simply cU where c is the velocity of light, which corresponds to the velocity of the incident radiation, and U is the energy density of incident radiation. Since the energy of a single photon is $\hbar\omega$, the corresponding photon flux J_i is the ratio, $(cU/\hbar\omega)$. Therefore, the absorption cross-section is given by

$$\sigma_{\text{abs}} = \frac{\hbar\omega\lambda_{ni}}{cU}. \tag{18.20}$$

The electromagnetic energy density from classical electrodynamics is found to be

$$U = \frac{1}{2\pi}\frac{\omega^2}{c^2}|A_0|^2. \tag{18.21}$$

Thus

$$\sigma_{\text{abs}} = \lambda_{ni}\frac{2\pi\hbar c}{\omega_{ni}|A_0|^2} \tag{18.22}$$

where, because of the δ-function in (18.18), ω must satisfy

$$\omega = \omega_{ni}. \tag{18.23}$$

Substituting the expression for λ_{ni} in (18.22) we obtain, with α, the fine-structure constant, being $e^2/\hbar c$,

$$\sigma_{abs} = 4\pi^2\hbar\omega_{ni}\alpha\,|\langle n|x|i\rangle|^2\,\delta(E_n - E_i - \hbar\omega) \tag{18.24}$$

The total absorption cross-section summed over all final states and integrated over ω is given by

$$\sum_f \int \sigma_{abs}d\omega = \frac{4\pi^2\alpha}{\hbar}\sum_f (E_f - E_i)\,|\langle f|x|i\rangle|^2 \tag{18.25}$$

We can perform the above summation through the following series of steps.

First, we consider the double commutator $[[x, H_0], x]$, which is given by

$$[[x, H_0], x] = 2xH_0x - H_0x^2 - x^2H_0. \tag{18.26}$$

Taking the diagonal elements of the above commutator we get

$$\sum_f \int \sigma_{abs} d\omega = \frac{2\pi^2 e^2}{mc} \tag{18.27}$$

where we have used the fact that x is Hermitian and $H_0 |i\rangle = E_i |i\rangle$. To determine the first term on the right-hand side of (18.27), we insert a complete set of intermediate states $|n\rangle$,

$$\langle i |xH_0 x| i\rangle = \sum_n \langle i |x| n\rangle \langle n |H_0 x| i\rangle \tag{18.28}$$

$$= \sum_n |\langle n |x| i\rangle|^2 E_n \tag{18.29}$$

where we have used the relation $\langle n| H_0 = E_n \langle n|$. To determine the second term on the right-hand side of (18.27), we note that

$$\left\langle i \left| x^2 \right| i \right\rangle = \sum_n \langle i |x| n\rangle \langle n |x| i\rangle \tag{18.30}$$

$$= \sum_n |\langle n |x| i\rangle|^2 . \tag{18.31}$$

Hence from (18.29) and (18.31) we obtain

$$\langle i |[[x, H_0], x]| \mathrm{i}\rangle = 2 \sum_n (E_n - E_i) |\langle n |x| i\rangle|^2 . \tag{18.32}$$

From (18.15) we know that

$$[x, H_0] = i\hbar \frac{p_x}{m} \tag{18.33}$$

and therefore

$$\langle \mathrm{i} |[[x, H_0], x]| i\rangle = \frac{i\hbar}{m} \langle i |[p_x, x]| i\rangle = \frac{\hbar^2}{m} \tag{18.34}$$

where we have used the fundamental commutation relation

$$[p_x, x] = -i\hbar. \tag{18.35}$$

Thus from (18.32) we have

$$\sum_n (E_n - E_i) |\langle n |x| i\rangle|^2 = \frac{\hbar^2}{2m}. \tag{18.36}$$

This relation is called the Thomas–Reiche–Khun sum rule. Substituting it in (18.25) we obtain the integrated absorption cross-section,

$$\int d\omega \sigma_{abs} = \frac{2\pi^2 e^2}{mc}, \tag{18.37}$$

where we have substituted $\alpha = e^2/\hbar c$. This result is remarkable in that it depends neither on $\hbar$ nor on the details of the Hamiltonian. Actually, the same result is obtained in classical electrodynamics.

18.2 Photoelectric effect

The situation here is similar to the radiation absorption phenomena we discussed in the previous section except that this time the incident radiation is assumed to have sufficiently high energy that the absorption leads to the ejection of the electron into a free particle state.

The form of the matrix element for the perturbing potential is, once again,

$$H'_{fi}(0) = -\frac{e}{mc}\left\langle f \left| e^{i\frac{\omega}{c}\mathbf{n}\cdot\mathbf{r}}\mathbf{A}_0 \cdot \mathbf{p} \right| i \right\rangle \tag{18.38}$$

where $|i\rangle$ is the initial atomic state while $|f\rangle$ is a free particle state, not one of the levels of the atomic system. Thus the transitions in this case will be to a continuum of states rather than discrete states. We can utilize the formulas developed previously for the cross-section given by

$$\frac{d\sigma_{fi}}{d\Omega} = \frac{1}{J_i}\left|H'_{fi}(0)\right|^2 \frac{mk_f V}{4\pi^2\hbar^3}. \tag{18.39}$$

Here J_i is the incident particle (photon) flux, which can be shown to be

$$J_i = \frac{\omega\left|A_0\right|^2}{2\pi\hbar c} \tag{18.40}$$

and V is the volume of normalization. Writing $\mathbf{A}_0 = A_0\boldsymbol{\epsilon}$ in (18.38), we obtain

$$\frac{d\sigma_{fi}}{d\Omega} = \frac{Vk_f\alpha}{2\pi m\hbar\omega}\left|\left\langle f \left| e^{i\frac{\omega}{c}\mathbf{n}\cdot\mathbf{r}}\boldsymbol{\epsilon} \cdot \mathbf{p} \right| i \right\rangle\right|^2. \tag{18.41}$$

We consider a hydrogen atom in its ground state as the initial state and assume the ejected electron to be a free particle described by a plane wave (see Fig. 18.1). Thus the two wavefunctions are

$$u_i(\mathbf{r}) = \frac{1}{\sqrt{\pi}}\left(\frac{Z}{a_0}\right)^{3/2} e^{-\frac{Zr}{a_0}}, \quad u_f(\mathbf{r}) = \frac{1}{\sqrt{V}}e^{i\mathbf{k}_f\cdot\mathbf{r}}. \tag{18.42}$$

The matrix element is then

$$\left\langle f \left| e^{i\frac{\omega}{c}\mathbf{n}\cdot\mathbf{r}}\boldsymbol{\epsilon} \cdot \mathbf{p} \right| i \right\rangle = \frac{1}{\sqrt{V}}\int d^3r\, e^{-i\mathbf{k}_f\cdot\mathbf{r}}e^{i\frac{\omega}{c}\mathbf{n}\cdot\mathbf{r}}\left[-i\hbar\boldsymbol{\epsilon}\cdot\nabla u_i(\mathbf{r})\right] \tag{18.43}$$

where we have substituted $\mathbf{p} = -i\hbar\nabla$. The above integral can be obtained through integration by parts, utilizing the fact that $u_i \to 0$ as $r \to \infty$. The integral then becomes, after

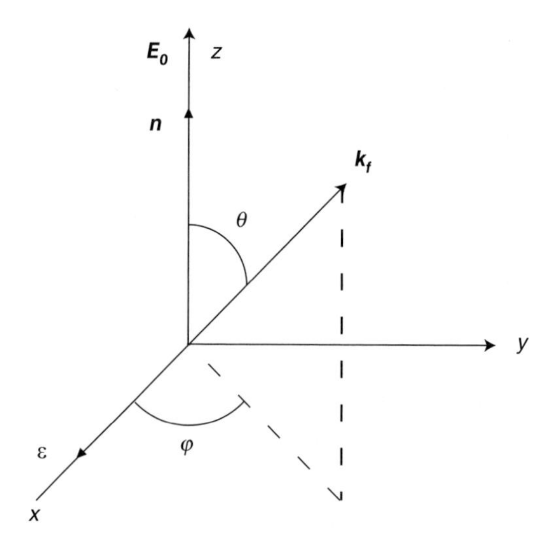

Fig. 18.1

substituting the expression for u_i,

$$\frac{i\hbar}{\sqrt{V}}\int d^3r\boldsymbol{\epsilon}\cdot\left[\boldsymbol{\nabla}\left(e^{-i\mathbf{q}\cdot\mathbf{r}}\right)\right]\frac{1}{\sqrt{\pi}}\left(\frac{Z}{a_o}\right)^{\frac{3}{2}}e^{-\frac{Zr}{a_0}} \tag{18.44}$$

where $\mathbf{q}=\mathbf{k}_f-(\omega/c)\mathbf{n}$. Furthermore,

$$\boldsymbol{\epsilon}\cdot\boldsymbol{\nabla}\left(e^{-i\mathbf{q}\cdot\mathbf{r}}\right)=-i\boldsymbol{\epsilon}\cdot\mathbf{q}e^{-i\mathbf{q}\cdot\mathbf{r}}. \tag{18.45}$$

We assume the vector potential $\mathbf{A}$ to be transverse, $\boldsymbol{\nabla}\cdot\mathbf{A}=0$, which implies that $\boldsymbol{\epsilon}\cdot\mathbf{n}=0$ and hence $\boldsymbol{\epsilon}\cdot\mathbf{q}=\boldsymbol{\epsilon}\cdot\mathbf{k}_f$.

Expression (18.44) is then simplified to

$$\frac{1}{\sqrt{\pi}}\left(\frac{Z}{a_0}\right)^{\frac{3}{2}}\frac{\hbar}{\sqrt{V}}\boldsymbol{\epsilon}\cdot\mathbf{k}_f\int d^3r'\,e^{-iqr'\cos\theta'}e^{-\frac{Zr'}{a_0}} \tag{18.46}$$

where $d^3r'=r'^2dr'd\cos\theta'd\phi'$. The integral over ϕ' gives 2π. We find after integrating over $\cos\theta'$ that we have an integral of the following form, which is calculated through partial integration,

$$\int_0^\infty dr\,re^{-\alpha r}=\frac{1}{\alpha^2}. \tag{18.47}$$

Substituting these results in the expression for the differential cross-section given by (18.41) we obtain

$$\frac{d\sigma_{fi}}{d\Omega}=\frac{32e^2k_f\left(\boldsymbol{\epsilon}\cdot\mathbf{k}_f\right)^2}{mc\omega}\left(\frac{Z}{a_0}\right)^5\left(\frac{Z^2}{a^2}+q^2\right)^{-4}. \tag{18.48}$$

If we take $\mathbf{n}$ in the z-direction, $\boldsymbol{\epsilon}$ in the x-direction and $\mathbf{k}_f$ as a vector with polar angle θ and azimuthal angle ϕ, then in the differential cross-section one may substitute the following expressions:

$$\boldsymbol{\epsilon} \cdot \mathbf{k}_f = k_f \sin\theta \cos\phi, \tag{18.49}$$

$$q^2 = k_f^2 - 2\mathbf{k}_f \cdot \mathbf{n}\frac{\omega}{c} + \frac{\omega^2}{c^2} \tag{18.50}$$

$$= k_f^2 - 2k_f \frac{\omega}{c} \cos\theta + \frac{\omega^2}{c^2}, \tag{18.51}$$

with $d\Omega = d\cos\theta\, d\phi$.

18.3 Coulomb excitations of an atom

Consider a particle of charge e passing by a stationary atom as shown in Fig. 18.2. The interaction between the charged particle and an electron in the atom will be characterized by the Coulomb force, which will be time-dependent because of the motion of the particle. Instead of radiation as we have considered previously, this time we have a Coulomb field that will cause excitations in the atom. As it imparts energy to the atom, the charged particle will lose energy. This is the dominant process that describes the passage of a charged particle through matter.

The motion of the charged particle, which we call the projectile, is treated classically. We also assume that the distance between the projectile and the target, the atom, is large compared with the dimensions of the atom.

We apply the techniques of time-dependent perturbation theory to determine the probability for the atom to make a transition from an initial state $|i\rangle$, which we assume to be the ground state, to one of the excited states, which we designate as $|f\rangle$. We are assuming here that the time of passage by the projectile, that is, the collision time, is shorter than the natural period of oscillation of the electron in the atom given by $1/\omega_{fi}$, where $\hbar\omega_{fi}$ is the

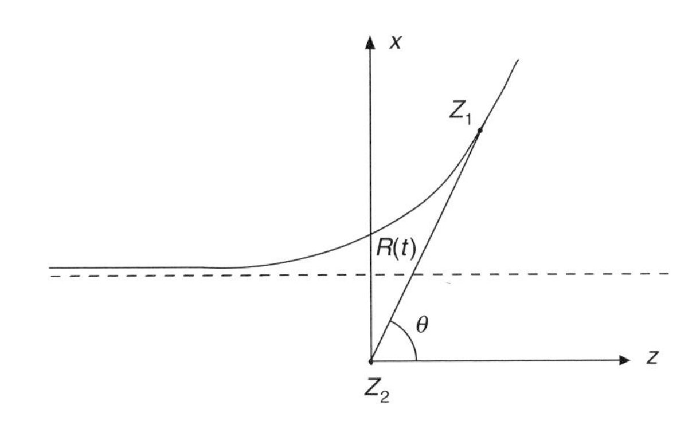

Fig. 18.2

energy difference between the states $|i\rangle$ and $|f\rangle$ of the electron. Thus, we isolate a single electron in an atom to study the effects of the projectile. In the process we also neglect, in the first approximation, the energy loss of the projectile due to the Coulomb interaction. Through energy conservation we then determine the energy loss in terms of the excitation energies.

Here the interaction Hamiltonian is given by the Coulomb potential,

$$H' = \frac{e^2}{|\mathbf{R}(t) - \mathbf{r}_i|}, \tag{18.52}$$

where $\mathbf{R}(t)$ is the distance of the charged particle from the origin, which depends on time, and $\mathbf{r}_i$ is the position of the electron. The ratio $|\mathbf{r}/\mathbf{R}| \ll 1$ since $\mathbf{r}$ is quite small, of atomic dimensions. We can, therefore, make the following expansion

$$\frac{1}{|\mathbf{R}(t) - \mathbf{r}_i|} = \frac{1}{R} + \frac{\mathbf{r}_i \cdot \mathbf{R}}{R^3} + \cdots. \tag{18.53}$$

We keep only the first two terms. Since R is independent of the atomic variables, the matrix element of the first term $\langle f \left| e^2/R \right| i\rangle$ vanishes unless $|f\rangle = |i\rangle$, which corresponds to elastic scattering. This is the familiar Rutherford scattering, which we have considered in Chapter 17. We are here concerned with "inelastic" scattering, where the final state can be different from the initial state, which the second term in (18.53) allows as we will show below.

The projectile starts on the left at infinity as shown in Fig. 18.2 and, because of a repulsive Coulomb interaction with electrons in the atom, the projectile curves upward as shown. The distance of shortest approach as indicated in the figure is called the impact parameter, which we designate as b. We take the direction along the impact parameter as the fixed z-axis, with the x-axis is as indicated. The plane of the orbit is taken as the x–z plane. With respect to these coordinate axes we write $\mathbf{R} = (R, \theta, \phi)$ and $\mathbf{r}_i = \left(r_i, \theta'', \phi''\right)$ in spherical coordinates, while we denote by θ' the angle between $\mathbf{R}$ and $\mathbf{r}_i$. The second term in (18.53) is now

$$\frac{\mathbf{r}_i \cdot \mathbf{R}}{R^3} = \frac{r_i}{R^2}\cos\theta' = \frac{r_i}{R}\left[\cos\theta\cos\theta'' + \sin\theta\sin\theta''\cos\left(\phi - \phi''\right)\right]. \tag{18.54}$$

This term corresponds to the interaction Hamiltonian H'. The matrix element of (18.54) is given by

$$\left\langle f \left| \frac{\mathbf{r}_i \cdot \mathbf{R}}{R^3} \right| i \right\rangle = \frac{\cos\theta}{R^2}\langle f \left| r_i \cos\theta'' \right| i\rangle + \frac{\sin\theta}{R^2}\langle f \left| r_i \sin\theta'' \cos\left(\phi - \phi''\right) \right| i\rangle \tag{18.55}$$

where we note that R is time-dependent. The individual matrix elements in (18.55) only involve the internal coordinates of the atom. The integral involving $\cos\left(\phi - \phi''\right)$ vanishes,

$$\int_0^{2\pi} d\phi'' \cos\left(\phi - \phi''\right) = 0. \tag{18.56}$$

Hence only the first term in (18.55) will contribute.

We write

$$r_i \cos\theta'' = z_i. \tag{18.57}$$

Therefore,

$$\left\langle f \left| \frac{\mathbf{r}_i \cdot \mathbf{R}}{R^3} \right| i \right\rangle = \frac{\cos\theta}{R^2} \langle f \left| z_i \right| i \rangle . \tag{18.58}$$

The matrix element of the interaction Hamiltonian is

$$H'_{fi} = \frac{e^2 \cos\theta}{R^2} \langle f \left| z_i \right| i \rangle . \tag{18.59}$$

The transition probability amplitude is then given by

$$c_{fi}(t) = -\frac{i}{\hbar} \int_{-\infty}^{t} dt' H'_{fi} e^{i\omega_{fi} t'} \approx -\frac{ie^2}{\hbar} \int_{-\infty}^{t} dt' \frac{\cos\theta\left(t'\right)}{R^2\left(t'\right)} \langle f \left| z_i \right| i \rangle \tag{18.60}$$

where we have made explicit the fact that θ depends on time and where, as mentioned earlier, we have assumed $t' \ll 1/\omega_{fi}$, and have, therefore, neglected the contribution of the exponential.

To obtain the integrals we first use classical angular momentum conservation, where the angular momentum is defined as $\mathbf{L} = \mathbf{r} \times \mathbf{p}$. The conservation relation implies

$$(b)(mv) = (R)\left(mR\dot{\theta}\right) \tag{18.61}$$

where v is the velocity with which the projectile is traveling. The left-hand side above corresponds to the angular momentum for the charged particle at infinity, as indicated by the figure, and the term on the right is the angular momentum at finite values of time. Therefore,

$$\frac{d\theta}{dt'} = \frac{vb}{R^2\left(t'\right)}. \tag{18.62}$$

That is,

$$\frac{dt'}{R^2\left(t'\right)} = \frac{d\theta}{vb}. \tag{18.63}$$

Since $\langle f \left| z_i \right| i \rangle$ involves the internal system of the atom, it does not depend on t'. Hence the transition probability amplitude at $t = \infty$ is given by

$$c_{fi}(t) = -\frac{ie^2}{\hbar vb} \langle f \left| z_i \right| i \rangle \int_{\theta(-\infty)}^{\theta(t)} d\theta \cos\theta\left(t'\right) \tag{18.64}$$

where θ depends on t'. We note that $\theta = -\pi/2$ at $t' = -\infty$; and $\theta = \pi/2$ at $t' = +\infty$ for the projectile's trajectory. Hence, at $t = \infty$,

$$c_{fi}(\infty) = -\frac{ie^2 \langle f |z_i| i\rangle}{\hbar v b} \int_{-\pi/2}^{\pi/2} d\theta \cos\theta = -\frac{2ie^2}{\hbar v b} \langle f |z_i| i\rangle . \tag{18.65}$$

Until now we have taken the projectile to have a unit charge. If Z_1 is the charge of the projectile then we simply multiply the Coulomb term given by (18.52) by a factor of Z_1, which gives

$$c_{fi}(\infty) = -\frac{2iZ_1 e^2}{\hbar v b} \langle f |z_i| i\rangle . \tag{18.66}$$

The projectile will lose energy, giving up part of it to kick the electron to a higher level. An interesting quantity to calculate is the energy loss of the projectile in its motion, which, through energy conservation, is equal to the energy imparted to the atom

$$T = \text{Energy loss} = \sum_f (E_f - E_i) \left|c_{fi}(\infty)\right|^2 = \frac{4Z_1^2 e^4}{\hbar^2 v^2 b^2} \sum_f (E_f - E_i) \left|\langle f |z_i| i\rangle\right|^2 . \tag{18.67}$$

The Thomas–Reiche–Khun sum rule obtained earlier, after replacing x by z, is given by

$$\frac{2m}{\hbar^2} \sum_f (E_f - E_i) \left|\langle f |z| i\rangle\right|^2 = 1. \tag{18.68}$$

If the atom has Z_2 electrons then the above sum must include all the electrons, in which case the right-hand side will be multiplied by Z_2 and we have

$$\frac{2m}{\hbar^2} \sum_f (E_f - E_i) \left|\langle f |z| i\rangle\right|^2 = Z_2. \tag{18.69}$$

The energy loss for this case is then given by

$$T = \frac{2Z_1^2 Z_2 e^4}{mv^2 b^2}. \tag{18.70}$$

This is then the energy that a fast-moving particle of charge $Z_1 e$ will lose in its collisions with an atom containing Z_2 electrons. This expression is essentially the same as the classical result.

18.4 Ionization

In the problems of absorption and photoelectric effect we investigated the interaction of radiation with the electron in an atom. There the field depended on both space and time through the vector potential $\mathbf{A}_0 e^{\pm i(\mathbf{k}\cdot\mathbf{r}-\omega t)}$. Let us consider now the case that does not involve radiation; instead an atom is subjected to an electric field oscillating in time, $\mathbf{E}_0 e^{\pm i\omega t}$, with $\mathbf{E}_0$ as a constant vector. This would be the case when the atom is placed between the plates of a capacitor to which alternating voltage is applied. The perturbed Hamiltonian is then given by

$$H'(t) = e\mathbf{E}_0 \cdot \mathbf{r} e^{-i\omega t} \tag{18.71}$$

where we keep only the harmonic time dependence $e^{-i\omega t}$ since we will be considering a situation in which the energy is absorbed by the atom. As we discussed earlier, this type of potential will induce transitions to states with $E_f = E_i + \hbar\omega$. If $\hbar\omega$ is sufficiently large then the electron will become free (with $E_f > 0$), i.e., the atom will ionize. We will neglect any residual Coulomb interaction between the ionized electron and the atom and take the wavefunction of the electron in the final state as

$$\frac{e^{i\mathbf{k}_f\cdot\mathbf{r}}}{\sqrt{V}}. \tag{18.72}$$

The matrix element for this process is then given by

$$\langle f |H'(0)| i\rangle = e \int d^3r \, \frac{e^{-i\mathbf{k}_f\cdot\mathbf{r}}}{\sqrt{V}} \mathbf{E}_0 \cdot \mathbf{r} u_i(\mathbf{r}) \tag{18.73}$$

where $u_i(\mathbf{r})$ is the wavefunction of the bound electron in the initial state. If the electron is in an $n = 1$ S-state, then $u_i(\mathbf{r})$ is given by

$$u_i(\mathbf{r}) = \frac{1}{\sqrt{\pi a_0^3}} e^{-r/a_0}. \tag{18.74}$$

The ionization process is described in Fig. 18.3.

The differential cross-section for ionization refers to the direction $\mathbf{k}_f$ of the ejected electron with respect to a fixed direction, which we take to be the direction of $\mathbf{E}_0$. We will take the z-axis in the same direction as $\mathbf{E}_0$. There are three different angles involved in this problem: θ' between $\mathbf{r}$ and $\mathbf{k}_f$; θ'' between $\mathbf{r}$ and $\mathbf{E}_0$; θ between $\mathbf{k}_f$ and $\mathbf{E}_0$. The integration (18.73), however, involves angles between $\mathbf{r}$ and $\mathbf{k}_f$, and $\mathbf{r}$ and $\mathbf{E}_0$. Thus we have

$$\langle f |H'(0)| i\rangle = \frac{eE_0}{\sqrt{\pi a_0^3 V}} \int d^3r \, e^{-ikr\cos\theta'} \left(r\cos\theta''\right) e^{-r/a_0}. \tag{18.75}$$

For integration purposes we take $\mathbf{k}_f$ to be the polar axis, so that

$$d^3r = r^2 dr \, d\cos\theta' d\phi'. \tag{18.76}$$

To determine θ'' that enters in (18.75), we note that if θ and ϕ are the polar and azimuthal angles of $\mathbf{k}_f$ (with respect to $\mathbf{E}_0$) then one can relate θ'' to angles θ' and θ by the following well-known formula:

$$\cos\theta'' = \cos\theta\cos\theta' - \sin\theta\sin\theta'\cos(\phi - \phi') \tag{18.77}$$

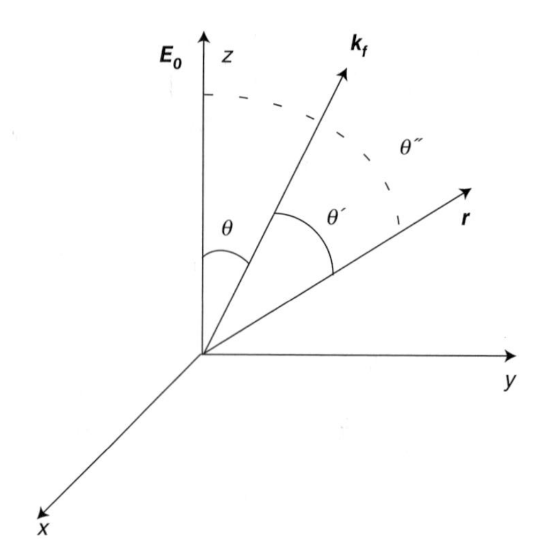

Fig. 18.3

where ϕ' is the azimuthal angle of $\mathbf{r}$ (note: if $\mathbf{k}_f$, $\mathbf{r}$ and $\mathbf{E}_0$ were in the same plane then $\theta'' = \theta + \theta'$). We substitute the expression for $\cos\theta''$ in (18.77).

The integration over ϕ' can be carried out trivially since

$$\int_0^{2\pi} d\phi' \cos(\phi - \phi') = 0. \tag{18.78}$$

Hence we obtain

$$\langle f | H'(0) | i \rangle = \frac{eE_0}{\sqrt{\pi a_0^3 V}} \cos\theta \int_0^\infty dr\, r^3 \int_{-1}^1 d\cos\theta' \cos\theta' e^{-ik_f r \cos\theta'} e^{-r/a_0}. \tag{18.79}$$

This complicated-looking integral can be calculated through the following steps. First we do the integral

$$I_1 = \int_0^\infty dr\, \mathrm{re}^{-r/a_0} \int_{-1}^1 d\cos\theta' e^{-ik_f r\cos\theta'} \tag{18.80}$$

$$= \frac{2}{\left(\frac{1}{a_0^2} + k_f^2\right)}. \tag{18.81}$$

We then take the derivative of I_1, with respect to $1/a_0$, to obtain

$$I_2 = -\frac{dI_1}{d(1/a_0)} = \int_0^\infty dr\, r^2 e^{-r/a_0} \int_{-1}^1 d\cos\theta'\, e^{-ik_f r\cos\theta'} \tag{18.82}$$

$$= \frac{4}{a_0} \frac{1}{\left(\frac{1}{a_0^2} + k_f^2\right)^2}. \tag{18.83}$$

We now take the derivative of I_2 with respect to k_f to find

$$I_3 = -\frac{dI_2}{dk_f} = i\int_0^\infty dr\, r^3 e^{-r/a_0}\int_{-1}^{1} d\cos\theta'\ \cos\theta' e^{-ik_f r\cos\theta'} \tag{18.84}$$

$$= \frac{16k_f}{a_0}\frac{1}{\left(\frac{1}{a_0^2}+k_f^2\right)^3} = \frac{16k_f a_0^5}{\left(1+k_f^2 a_0\right)^3}. \tag{18.85}$$

We note that (18.79) is related to (18.84). Hence,

$$\langle f\left|H'(0)\right| i\rangle = \frac{eE_0}{\sqrt{\pi a_0^3 V}}\cos\theta\frac{16k_f a_0^5}{\left(1+k_f^2 a_0^2\right)^3}. \tag{18.86}$$

The transition probability for ionization per unit solid angle is given by

$$\frac{dw_{fi}}{d\Omega} = \frac{2\pi}{\hbar}\left|\langle f\left|H'(0)\right| i\rangle\right|^2. \tag{18.87}$$

From (18.86) we obtain

$$\frac{dw_{fi}}{d\Omega} = \frac{256me^2k_f^3a_0^7E_0^2}{\pi\hbar^3\left(1+k_f^2a_0^2\right)^6}\cos^2\theta \tag{18.88}$$

where the solid angle $d\Omega = d\cos\theta d\phi$. Integrating out the angles θ and ϕ. From (18.86) we obtain a factor of $(4\pi/3)$ and, hence,

$$w_{fi} = \frac{1024mk_f^3a_0^7E_0^2}{3\hbar^3\left(1+k_f^2a_0^2\right)^6}. \tag{18.89}$$

This is the ionization probability.

18.5 Thomson, Rayleigh, and Raman scattering in second-order perturbation

In this section we will consider three important and closely related scattering phenomena involving the interaction of a radiation field with an atom. These are of the type

$$\gamma + A \to \gamma + B \tag{18.90}$$

where γ represents a photon (radiation field) and A and B are atomic states.

The interaction Hamiltonian is the same as we have considered before,

$$H' = -\frac{e}{2mc}\left[\mathbf{p}\cdot\mathbf{A}+\mathbf{A}\cdot\mathbf{p}\right]+\frac{e^2}{2mc^2}\mathbf{A}\cdot\mathbf{A}, \tag{18.91}$$

where we have left the two terms $\mathbf{p}\cdot\mathbf{A}$ and $\mathbf{A}\cdot\mathbf{p}$ in place without making further simplifications.

We will try to solve the problem of scattering using perturbation theory, in which e will once again play the role of the perturbation parameter. We have already considered the consequences of this interaction in a variety of problems, but what sets this apart is that the radiation field as described by the photon occurs twice, resulting in a scattering of the type described by (18.91). One can imagine the terms in the square bracket as providing a two-step process where first an absorption of a photon occurs, a phenomenon we have already treated in the earlier sections, which one can represent by $\gamma + A \to A'$ (excited state), followed by an emission of a photon. The quadratic term, which we are considering for the first time in this chapter provides a direct interaction.

Even though we have not yet discussed the formalism of second quantization, which will be the subject of Chapter 37, we will use some of the simplified elements of it. In this formalism a radiation field is described as a collection of a discrete number of photons. We write $\mathbf{A}$ in the form of a Fourier transform as follows,

$$\mathbf{A} = \int d^3k \frac{\boldsymbol{\epsilon}}{\sqrt{2\omega V}}\left[a_k e^{i(k\cdot r-\omega_k t)}+a_k^\dagger e^{-i(k\cdot r-\omega_k t)}\right] \tag{18.92}$$

where $\boldsymbol{\epsilon}$ is the polarization vector, which is assumed to be real, $\mathbf{k}$ is the momentum of the photon, and $\omega = c\,|\mathbf{k}|$. The energy of each photon is $\hbar\omega$. The field is normalized inside a volume V.

One calls the operators a_k the destruction operators and $a_k^\dagger$ the creation operators. They have properties very similar to that of the raising and lowering operators for the harmonic oscillator. For example, they satisfy the commutation relation

$$\left[a_k, a_{k'}^\dagger\right] = \delta_{kk'}, \quad \left[a_k, a_{k'}\right] = 0, \quad \left[a_k^\dagger, a_{k'}^\dagger\right] = 0. \tag{18.93}$$

The difference here is that these operators are defined in the so-called multiparticle Hilbert space. If we designate states with no photons and with one photon as $|0\rangle$ and $|1\rangle$, respectively, which are part of the Hilbert space that represents multiparticle systems, then, for example,

$$a_k^\dagger\,|0\rangle = |1\rangle \quad \text{and} \quad a_k\,|1\rangle = |0\rangle\,. \tag{18.94}$$

The above two relations are all we will need in order to calculate the matrix elements for H'.

We consider the approximation, called the long-wavelength approximation, in which the wavelength of the radiation is much larger than the typical dimensions of the atom, $kr \ll 1$.

Hence we write

$$\mathbf{A} = \int d^3k \frac{\boldsymbol{\epsilon}}{\sqrt{2\omega_k V}} \left[a_k e^{-i\omega_k t} + a_k^\dagger e^{i\omega_k t} \right], \tag{18.95}$$

$$\mathbf{p} \cdot \mathbf{A} = \int d^3k \frac{\mathbf{p} \cdot \boldsymbol{\epsilon}}{\sqrt{2\omega_k V}} \left[a_k e^{-i\omega_k t} + a_k^\dagger e^{i\omega_k t} \right], \tag{18.96}$$

and write the matrix element of H' as

$$H'_{fi} = -\frac{e}{2mc} \langle f | (\mathbf{p} \cdot \mathbf{A} + \mathbf{A} \cdot \mathbf{p}) | i \rangle + \frac{e^2}{2mc^2} \langle f | \mathbf{A} \cdot \mathbf{A} | i \rangle \tag{18.97}$$

where for the photons we take $\mathbf{k}_f$ and $\boldsymbol{\epsilon}^{(f)}$ as the final momentum and polarization vectors, respectively, with $\omega_f = ck_f$, while the corresponding parameters for the initial state are $\mathbf{k}_i$ and $\boldsymbol{\epsilon}^{(i)}$ with $\omega_i = ck_i$. The internal variables for the atom are absorbed in the designations i and f, respectively.

$$\left| f, \mathbf{k}_f, \boldsymbol{\epsilon}^{(f)} \right\rangle = | f \rangle \quad \text{and} \quad \left| i, \mathbf{k}_i, \boldsymbol{\epsilon}^{(i)} \right\rangle = | i \rangle . \tag{18.98}$$

At this stage it is much more convenient to look at the scattering process pictorially as given by the three diagrams in Fig. 18.4a, b, c. The time is supposed to flow from bottom up. These diagrams are all different but they describe the same process (18.90). The solid lines correspond to the atom and the wavy lines to the photon. In Fig. 18.4a, the initial state $|i\rangle$ indicated by incoming arrows consists of the atom A and the photon of momentum $\mathbf{k}_i$ and

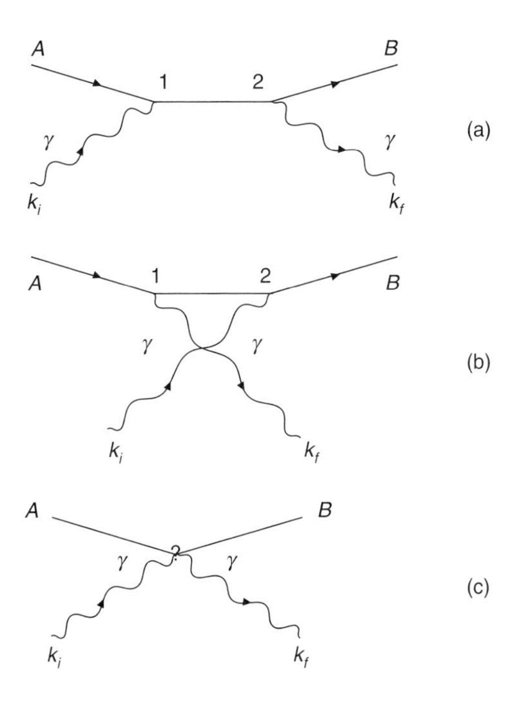

Fig. 18.4

polarization $\boldsymbol{\epsilon}^{(i)}$. Similarly, the final state $|f\rangle$ indicated by the outgoing arrows consists of the atom B and the photon with momentum $\mathbf{k}_f$ and polarization vector $\epsilon^{(f)}$. In this figure an incoming photon is destroyed at the vertex 1, corresponding to the term $a_{k_i}|1\rangle = |0\rangle$ (i.e., absorbed by the atom A, which converts to an excited atomic state indicated by the middle solid line). A photon is then created at the vertex 2 (i.e., the excited state emits a photon, leaving a residual atomic state B, corresponding to $a^{\dagger}_{k_f}|0\rangle$). Thus H'_{fi} for this process will be proportional to

$$H'_{fi} \sim \left\langle 1 \left| a^{\dagger}_{k_f} a_{k_i} \right| 1 \right\rangle. \tag{18.99}$$

In Fig. 18.4b, a photon is created at vertex 1 and destroyed at vertex 2 corresponding to

$$H'_{fi} \sim \left\langle 1 \left| a_{k_i} a^{\dagger}_{k_f} \right| 1 \right\rangle. \tag{18.100}$$

In Fig. 18.4c, these processes happen at the same vertex.

The transition probability must then include $\mathbf{A}$ twice. In the interaction Hamiltonian (18.97) the terms in the bracket, which are linear in $\mathbf{A}$ and in e, will contribute to diagrams 18.4a and 18.4b. Therefore, we will need to solve this part of the problem using second-order perturbation theory. On the other hand, the second term in H', which is quadratic in $\mathbf{A}$, will correspond to Fig. 18.4c. Since this term already contains e^2, we only solve the problem to first-order, not in e but in e^2.

The first- and second-order transition probability amplitudes are given as

$$c^{(1)}_{fi} = \left(\frac{-i}{\hbar}\right) \int_{-\infty}^{t} dt'\, H'_{fi}(t')\, e^{i\omega_{fi}t'}, \tag{18.101}$$

$$c^{(2)}_{fi}(t) = \left(-\frac{i}{\hbar}\right)^2 \sum_n \int_{-\infty}^{t} dt'' \left[\int_{-\infty}^{t''} dt' H'_{ni}(t') \exp(i\omega_{ni}t') \right] H'_{fn}(t'') \exp(i\omega_{fn}t''). \tag{18.102}$$

The matrix element of H' for the $\mathbf{A}.\mathbf{A}$ term is given by

$$\langle f | H' | i \rangle \tag{18.103}$$

$$= \left\langle f \left| \frac{e^2}{2mc^2} \mathbf{A} \cdot \mathbf{A} \right| i \right\rangle \tag{18.104}$$

$$= \int d^3k_i \int d^3k_f \left\langle f \left| \frac{e^2}{2mc^2} \left(a_{\mathbf{k}_i} a^{\dagger}_{\mathbf{k}_f} + a^{\dagger}_{\mathbf{k}_f} a_{\mathbf{k}_i} \right) \frac{1}{2V\sqrt{\omega_i \omega_f}} \boldsymbol{\epsilon}^{(i)} \cdot \boldsymbol{\epsilon}^{(f)} \right. \right. \tag{18.105}$$

$$\left. \times \exp\left[-i\left(\omega_i - \omega_f\right)t\right] \right| i \Big\rangle$$

$$= \frac{e^2}{2mc^2} \frac{1}{2V\sqrt{\omega_i \omega_f}} 2\boldsymbol{\epsilon}^{(i)} \cdot \boldsymbol{\epsilon}^{(f)} \exp\left[-i\left(\omega_i - \omega_f\right)t\right] \langle f | i \rangle. \tag{18.106}$$

This will enter into the first-order transition amplitude, which is then found to be

$$c_{fi}^{(1)}(t) = \left(\frac{-i}{\hbar}\right)\frac{e^2}{2mc^2}\frac{1}{2V\sqrt{\omega_i\omega_f}}2\delta_{fi}\boldsymbol{\epsilon}^{(i)}\cdot\boldsymbol{\epsilon}^{(f)}\int_0^t \exp\left[i\left(\omega_{fi}+\omega_f-\omega_i\right)t_1\right]dt_1 \tag{18.107}$$

where we have taken $\langle f|i\rangle = \delta_{fi}$ to account for the fact the states $|i\rangle$ and $|f\rangle$ may not be the same. Also, we have used the definition $\omega_{ba} = \omega_b - \omega_a$.

The second-order transition amplitude, after completing the operation of a and $a^\dagger$, is given by

$$c_{fi}^{(2)}(t) = \left(\frac{-i}{\hbar}\right)^2 \frac{1}{2V\sqrt{\omega_i\omega_f}}\left(-\frac{e}{mc}\right)^2\int_0^t dt_2\int_0^{t_2} dt_1 \tag{18.108}$$

$$\times\left\{\sum_n \left\langle f|\mathbf{p}\cdot\boldsymbol{\epsilon}^{(f)}|n\right\rangle \exp\left[i\left(\omega_{fn}+\omega_f\right)t_2\right]\times\left\langle n\left|\mathbf{p}\cdot\boldsymbol{\epsilon}^{(i)}\right|i\right\rangle \exp\left[i\left(\omega_{ni}-\omega_i\right)t_1\right]\right.$$

$$\left.+\sum_n \left\langle f\left|\mathbf{p}\cdot\boldsymbol{\epsilon}^{(i)}\right|n\right\rangle \exp i\left(\omega_{fn}-\omega_i\right)t_2\times\left\langle n\left|\mathbf{p}\cdot\boldsymbol{\epsilon}^{(f)}\right|i\right\rangle \exp\left[i\left(\omega_{ni}+\omega_f\right)t_1\right]\right\} \tag{18.109}$$

$$= -\frac{c^2}{i\hbar 2V\sqrt{\omega_i\omega_f}}\left(\frac{e}{mc}\right)^2$$

$$\times\sum_n\left(\frac{\left\langle f\left|\mathbf{p}\cdot\boldsymbol{\epsilon}^{(f)}\right|n\right\rangle\left\langle n\left|\mathbf{p}\cdot\boldsymbol{\epsilon}^{(i)}\right|i\right\rangle}{\omega_{ni}-\omega_i}+\frac{\left\langle f\left|\mathbf{p}\cdot\boldsymbol{\epsilon}^{(i)}\right|n\right\rangle\left\langle n\left|\mathbf{p}\cdot\boldsymbol{\epsilon}^{(f)}\right|i\right\rangle}{\omega_{ni}+\omega_f}\right) \tag{18.110}$$

$$\times\int_0^t dt_2 \exp\left[i\left(\omega_{fn}+\omega_{ni}+\omega_f-\omega_i\right)t_2\right] \tag{18.111}$$

where we have inserted a complete set of intermediate states $|n\rangle$.

The transition probability is then given by

$$\text{Transition probability} = \left|c_{fi}^{(1)}(t) + c_{fi}^{(2)}(t)\right|^2$$

and the transition rate is

$$\lambda_{fi} = \lim_{t\to\infty}\frac{d}{dt}\left|c_{fi}^{(1)}(t) + c_{fi}^{(2)}(t)\right|^2 \tag{18.112}$$

where terms of order e^2 will be relevant to our problem.

To calculate λ_{fi} we note that the exponentials appearing in $c_{fi}^{(1)}(t)$ and $c_{fi}^{(2)}(t)$ are the same since the two exponents satisfy

$$\omega_{fn} + \omega_{ni} = \omega_f - \omega_i. \tag{18.113}$$

Thus the δ-function that normally results after the t-integrations in the transition amplitude will be

$$\delta\left(\omega_{fi} + \omega_f - \omega_i\right). \tag{18.114}$$

The formula for the Fermi rule gives

$$w_{fi} = \frac{2\pi}{\hbar}\left(\frac{1}{2V\sqrt{\omega_i\omega_f}}\right)^2\left(\frac{e^2}{mc^2}\right)^2\frac{V}{(2\pi)^3}\frac{\omega_f^2}{\hbar c^3}d\Omega \tag{18.115}$$

$$\times\left|\delta_{fi}\boldsymbol{\epsilon}^{(i)}\cdot\boldsymbol{\epsilon}^{(f)} - \frac{1}{m\hbar}\sum_n\left(\frac{\langle f\left|\mathbf{p}\cdot\boldsymbol{\epsilon}^{(f)}\right|n\rangle\langle n\left|\mathbf{p}\cdot\boldsymbol{\epsilon}^{(i)}\right|i\rangle}{\omega_{ni}-\omega_i} + \frac{\langle f\left|\mathbf{p}\cdot\boldsymbol{\epsilon}^{(i)}\right|n\rangle\langle n\left|\mathbf{p}\cdot\boldsymbol{\epsilon}^{(f)}\right|i\rangle}{\omega_{ni}+\omega_f}\right)\right| \tag{18.116}$$

The differential cross-section for an incident radiation flux corresponding to the incoming photons has been calculated before. We obtain

$$\frac{d\sigma}{d\Omega} = r_0^2\left(\frac{\omega_f}{\omega_i}\right)\left|\delta_{fi}\boldsymbol{\epsilon}^{(i)}\cdot\boldsymbol{\epsilon}^{(f)} - \frac{1}{m\hbar}\sum_n\left(\frac{\langle f\left|\mathbf{p}\cdot\boldsymbol{\epsilon}^{(f)}\right|n\rangle\langle n\left|\mathbf{p}\cdot\boldsymbol{\epsilon}^{(i)}\right|i\rangle}{\omega_{ni}-\omega_i}\right.\right.$$
$$\left.\left. + \frac{\langle f\left|\mathbf{p}\cdot\boldsymbol{\epsilon}^{(i)}\right|n\rangle\langle n\left|\mathbf{p}\cdot\boldsymbol{\epsilon}^{(f)}\right|i\rangle}{\omega_{ni}+\omega_f}\right)\right|^2 \tag{18.117}$$

where

$$r_0 = \frac{e^2}{4\pi mc^2} \tag{18.118}$$

is called the classical radius of the electron. The above result is called the Kramers–Heisenberg formula.

18.5.1 Thomson scattering

Consider the case of elastic scattering where

$$E_i = E_f, \quad \omega_f = \omega_i = \omega, \quad \text{and } B = A. \tag{18.119}$$

One finds

$$\left(c_{fi}^{(1)} + c_{fi}^{(2)}\right) \to \left[\boldsymbol{\epsilon}^{(i)}\cdot\boldsymbol{\epsilon}^{(f)} - \frac{1}{m\hbar}\sum_n\left(\frac{\langle i\left|\mathbf{p}\cdot\boldsymbol{\epsilon}^{(f)}\right|n\rangle\langle n\left|\mathbf{p}\cdot\boldsymbol{\epsilon}^{(i)}\right|i\rangle}{\omega_{ni}-\omega}\right.\right.$$
$$\left.\left. + \frac{\langle i\left|\mathbf{p}\cdot\boldsymbol{\epsilon}^{(i)}\right|n\rangle\langle n\left|\mathbf{p}\cdot\boldsymbol{\epsilon}^{(f)}\right|i\rangle}{\omega_{ni}+\omega}\right)\right]. \tag{18.120}$$

In the limit

$$\omega \gg \omega_{ni}, \tag{18.121}$$

that is, in the limit in which the energy of the incoming radiation is much larger than ω_{ni} $(= \omega_n - \omega_i)$, which is effectively the binding energy of the excited state, n, the terms inside the summation above go to zero compared with the first term. Therefore,

$$\left(c_{fi}^{(1)} + c_{fi}^{(2)}\right) \to \boldsymbol{\epsilon}^{(i)} \cdot \boldsymbol{\epsilon}^{(f)}. \tag{18.122}$$

The cross-section is then given by

$$\frac{d\sigma}{d\Omega} = r_0^2 \left|\boldsymbol{\epsilon}^{(i)} \cdot \boldsymbol{\epsilon}^{(f)}\right|^2. \tag{18.123}$$

This is called Thomson scattering where r_0 has already been defined.

18.5.2 Rayleigh scattering

In the opposite limit,

$$\omega \ll \omega_{ni}, \tag{18.124}$$

which corresponds to very low-frequency radiation, we expand the denominators in (18.120) in powers of ω. Thus the second term in (18.120) is

$$\begin{aligned}
&\sum_n \frac{1}{m\hbar}\left(\frac{\langle i\,|\mathbf{p}\cdot\boldsymbol{\epsilon}^{(f)}|\,n\rangle\langle n\,|\mathbf{p}\cdot\boldsymbol{\epsilon}^{(i)}|\,i\rangle}{\omega_{ni}-\omega} + \frac{\langle i\,|\mathbf{p}\cdot\boldsymbol{\epsilon}^{(i)}|\,n\rangle\langle n\,|\mathbf{p}\cdot\boldsymbol{\epsilon}^{(f)}|\,i\rangle}{\omega_{ni}+\omega}\right)\\
&= \sum_n \frac{1}{m\hbar}\left\{\frac{1}{\omega_{ni}}\left[1+\frac{\omega}{\omega_{ni}}+\left(\frac{\omega}{\omega_{ni}}\right)^2\right]\left\langle i\left|\mathbf{p}\cdot\boldsymbol{\epsilon}^{(f)}\right|n\right\rangle\left\langle n\left|\mathbf{p}\cdot\boldsymbol{\epsilon}^{(i)}\right|i\right\rangle\right.\\
&\qquad\left.+\frac{1}{\omega_{ni}}\left[1-\frac{\omega}{\omega_{ni}}+\left(\frac{\omega}{\omega_{ni}}\right)^2\right]\left\langle i\left|\mathbf{p}\cdot\boldsymbol{\epsilon}^{(i)}\right|n\right\rangle\left\langle n\left|\mathbf{p}\cdot\boldsymbol{\epsilon}^{(f)}\right|i\right\rangle\right\}.
\end{aligned} \tag{18.125}$$

We will now consider the coefficients of different powers of ω. The following relation was derived in Section 18.1:

$$\langle a\,|p_i|\,b\rangle = -\frac{im\,(E_b - E_a)}{\hbar}\langle a\,|x_i|\,b\rangle = -im\omega_{ba}\,\langle a\,|\mathbf{r}|\,b\rangle\,. \tag{18.126}$$

Therefore, we will have

$$\left\langle i\left|\mathbf{p}\cdot\boldsymbol{\epsilon}^{(f)}\right|n\right\rangle = -im\omega_{ni}\left\langle i\left|\mathbf{r}\cdot\boldsymbol{\epsilon}^{(f)}\right|n\right\rangle \text{ and } \left\langle n\left|\mathbf{p}\cdot\boldsymbol{\epsilon}^{(f)}\right|i\right\rangle = im\omega_{ni}\left\langle n\left|\mathbf{r}\cdot\boldsymbol{\epsilon}^{(f)}\right|i\right\rangle. \tag{18.127}$$

Let us first consider the term independent of ω,

$$\sum_n \left\{ \frac{1}{\omega_{ni}} \left\langle i \left| \mathbf{p}\cdot\boldsymbol{\epsilon}^{(f)} \right| n \right\rangle \left\langle n \left| \mathbf{p}\cdot\boldsymbol{\epsilon}^{(i)} \right| i \right\rangle + \frac{1}{\omega_{ni}} \left\langle i \left| \mathbf{p}\cdot\boldsymbol{\epsilon}^{(i)} \right| n \right\rangle \left\langle n \left| \mathbf{p}\cdot\boldsymbol{\epsilon}^{(f)} \right| i \right\rangle \right\}. \tag{18.128}$$

From (18.126) it can be written as

$$\sum_n -im \left\{ \left\langle i \left| \mathbf{r}\cdot\boldsymbol{\epsilon}^{(f)} \right| n \right\rangle \left\langle n \left| \mathbf{p}\cdot\boldsymbol{\epsilon}^{(i)} \right| i \right\rangle + \left\langle i \left| \mathbf{p}\cdot\boldsymbol{\epsilon}^{(i)} \right| n \right\rangle \left\langle n \left| \mathbf{r}\cdot\boldsymbol{\epsilon}^{(f)} \right| i \right\rangle \right\} \tag{18.129}$$

$$= -im \left\{ \langle i | \, \mathbf{r}\cdot\boldsymbol{\epsilon}^{(f)}\mathbf{p}\cdot\boldsymbol{\epsilon}^{(i)} - \mathbf{p}\cdot\boldsymbol{\epsilon}^{(i)}\mathbf{r}\cdot\boldsymbol{\epsilon}^{(f)} \right\} |i\rangle \tag{18.130}$$

where we have carried out the sum over intermediate states. Writing in component form we find the above expression in braces to be

$$\sum_{i,j} -im(x_i p_j - p_j x_i)\epsilon_i^{(f)}\epsilon_j^{(i)}. \tag{18.131}$$

If we use the fundamental commutator

$$\left[x_i, p_j\right] = i\hbar\delta_{ij}, \tag{18.132}$$

then for the second term in (18.120) we obtain

$$\boldsymbol{\epsilon}^{(f)}\cdot\boldsymbol{\epsilon}^{(i)}, \tag{18.133}$$

which cancels the first term in (18.120) exactly. Therefore, the term $\boldsymbol{\epsilon}^{(f)}\cdot\boldsymbol{\epsilon}^{(i)}$ will no longer be present in the expression for $(c_{fi}^{(1)} + c_{fi}^{(2)})$.

Let us now consider the linear term in ω. We obtain

$$\sum_n \left\{ \frac{\omega}{\mathbf{p}\cdot\boldsymbol{\epsilon}^{(i)}} \left\langle i \left| \mathbf{p}\cdot\boldsymbol{\epsilon}^{(f)} \right| n \right\rangle \left\langle n \left| \mathbf{p}\cdot\boldsymbol{\epsilon}^{(i)} \right| i \right\rangle - \frac{\omega}{\omega_{ni}^2} \left\langle i \left| \mathbf{p}\cdot\boldsymbol{\epsilon}^{(i)} \right| n \right\rangle \left\langle n \left| \mathbf{p}\cdot\boldsymbol{\epsilon}^{(f)} \right| i \right\rangle \right\} \tag{18.134}$$

$$= \omega \sum_n \frac{1}{\omega_{ni}^2} \left[\left\langle i \left| \mathbf{p}\cdot\boldsymbol{\epsilon}^{(f)} \right| n \right\rangle \left\langle n \left| \mathbf{p}\cdot\boldsymbol{\epsilon}^{(i)} \right| i \right\rangle - \left\langle i \left| \mathbf{p}\cdot\boldsymbol{\epsilon}^{(i)} \right| n \right\rangle \left\langle n \left| \mathbf{p}\cdot\boldsymbol{\epsilon}^{(f)} \right| i \right\rangle \right]. \tag{18.135}$$

Using the relation (18.126) we can replace each of the above factors by factors involving $\mathbf{r}\cdot\boldsymbol{\epsilon}^{(f)}$ and $\mathbf{r}\cdot\boldsymbol{\epsilon}^{(i)}$ but without the term ω_{ni}^2 in the denominator. After using the completeness theorem and summing over the intermediate states, one finds that the two terms cancel each other. So the linear term gives no contribution.

Let us now turn to the last term, quadratic in ω, which is given by

$$\sum_n \left\{ \frac{\omega^2}{\omega_{ni}^3} \left\langle i \left| \mathbf{p}\cdot\boldsymbol{\epsilon}^{(f)} \right| n \right\rangle \left\langle n \left| \mathbf{p}\cdot\boldsymbol{\epsilon}^{(i)} \right| i \right\rangle + \frac{\omega^2}{\omega_{ni}^3} \left\langle i \left| \mathbf{p}\cdot\boldsymbol{\epsilon}^{(i)} \right| n \right\rangle \left\langle n \left| \mathbf{p}\cdot\boldsymbol{\epsilon}^{(f)} \right| i \right\rangle \right\}. \tag{18.136}$$

After converting the above matrix elements to the matrix elements involving $\mathbf{r}\cdot\boldsymbol{\epsilon}^{(f)}$ and $\mathbf{r}\cdot\boldsymbol{\epsilon}^{(i)}$ through (18.126), we rewrite this term as

$$m^2\omega^2 \sum_n \frac{1}{\omega_{ni}} \left\{ \left\langle i \left| \mathbf{r}\cdot\boldsymbol{\epsilon}^{(f)} \right| n \right\rangle \left\langle n \left| \mathbf{r}\cdot\boldsymbol{\epsilon}^{(i)} \right| i \right\rangle + \left\langle i \left| \mathbf{r}\cdot\boldsymbol{\epsilon}^{(i)} \right| n \right\rangle \left\langle n \left| \mathbf{r}\cdot\boldsymbol{\epsilon}^{(f)} \right| i \right\rangle \right\}. \tag{18.137}$$

Hence the differential cross-section for $\omega \ll \omega_{ni}$ is found to be

$$\frac{d\sigma}{d\Omega} = r_0^2 \left(\frac{m\omega^2}{\hbar}\right)^2 \left|\sum_n \left(\frac{1}{\omega_{ni}}\right) \left[\left\langle i \left| \mathbf{r} \cdot \boldsymbol{\epsilon}^{(f)} \right| n \right\rangle \left\langle n \left| \mathbf{r} \cdot \boldsymbol{\epsilon}^{(i)} \right| i \right\rangle + \left\langle i \left| \mathbf{r} \cdot \boldsymbol{\epsilon}^{(i)} \right| n \right\rangle \left\langle n \left| \mathbf{r} \cdot \boldsymbol{\epsilon}^{(f)} \right| i \right\rangle \right]\right|^2 . \tag{18.138}$$

This is the differential cross-section for the well-known Rayleigh scattering. As we note, it varies as ω^4.

18.5.3 Raman scattering

The above results were derived for the case of elastic scattering with $\omega_f = \omega_i$. For $\omega_f \neq \omega_i$ the Kramers–Heisenberg formula was first verified in an atomic physics experiment by Raman. This process, therefore, goes by the name of Raman scattering, or the Raman effect.

Our discussions thus far have been largely concentrated on two extreme limits: $\omega \gg \omega_{ni}$ or $\omega \ll \omega_{ni}$. However, for the case where the energy of the incident photon is of the order of one of the binding energies, $\omega \approx \omega_{ni}$, then the terms in (18.120) become very large, providing a classic case of what are called resonances. We mentioned this phenomenon in Section 17.4. We will discuss it further in Chapter 22 and particularly in Chapter 23 when we consider the scattering on composite objects.

Finally, if we allow the energies of the photons to reach extremely high limits then we have to treat the entire problem relativistically and involve the complete machinery of second quantization. The special case where photons scatter off the electrons is called Compton scattering, which we will discuss in Chapter 43.

18.6 Problems

1. A hydrogen atom in its ground state is subjected to a time-dependent perturbation of the type

$$H' = \lambda f(z)\, e^{-i\omega t}$$

   where $f(z)$ is a polynomial in z. Determine the transition probability for ionization in terms of an appropriate integral over $f(z)$.
2. In the problem of the Coulomb excitation of an atom, write $\mathbf{R} = \mathbf{i}vt + \mathbf{k}b$, where $\mathbf{i}$ and $\mathbf{k}$ are unit vectors in the x-direction and z-direction, respectively. Use the relation

$$K_0(z) = \int_0^\infty dt \frac{\cos(zt)}{\sqrt{1+t^2}}$$

and the limit $K_0(z) \to -z \ln z$ for $z \ll 1$ to show that the same relation for the energy transfer is obtained as in the text.

3. Show that a charged harmonic oscillator can only absorb (or emit) radiation if the radiation frequency is the same as the natural frequency of the oscillator.
4. To take into account spontaneous emission in radiative transitions in atoms one can resort to quantum electrodynamics (QED) and start with designating the combined radiation and atomic states as $|n_k, \alpha\rangle$, where n_k corresponds to the number of photons and the subscript k stands for the quantum numbers (momentum, polarization, etc.) of the photon, while α indicates the state of the atom. Let the vector potential be written as

$$\mathbf{A}(\mathbf{r}, t) = \mathbf{A}_0 \sum_k \left[a_k e^{i(\mathbf{k}\cdot\mathbf{r} - \omega t)} + a_k^\dagger e^{-i(\mathbf{k}\cdot\mathbf{r} - \omega t)} \right]$$

where a_k's are operators on the states $|n_k, \alpha\rangle$ and satisfy the same commutation relations as for the harmonic oscillator case such that $a_k^\dagger a_k = n_k$ and $a_k |n_k, \alpha\rangle = \sqrt{n_k}\, |n_k - 1, \alpha\rangle$. Insert the $\mathbf{A}(\mathbf{r}, t)$ in the expression for the matrix element $\langle f \mid \mathbf{A} \cdot \mathbf{p} \mid i\rangle$. Show that the absorption term is proportional to n_k while the emission term is proportional to $(n_k + 1)$. Identify the part of the emission term that corresponds to stimulated emission and the part that corresponds to spontaneous emission.
5. Consider a system consisting of radiation, in thermal equilibrium, and atoms in a cavity. If N_e is the number of atoms undergoing emission and N_a is the number undergoing absorption, then show that, in equilibrium conditions, the number of photons in the system can be written as

$$n_k = \frac{1}{\dfrac{N_a}{N_e} - 1}.$$

If one assumes the Boltzmann distribution $\left(\sim e^{-E/kT}\right)$ for the atoms then, since $E_e - E_a = \hbar\omega$,

$$n_k = \frac{1}{e^{\hbar\omega/kT} - 1}.$$

After converting n_k to the density of radiation, show that one arrives at the famous Planck black-body radiation formula.
6. A neutron, which is a spin-½ particle, is subjected to a constant magnetic field $\mathbf{B}_0$ in the z-direction. It travels along the x-direction with velocity v starting at the origin. A time-dependent perturbation is applied represented by the magnetic field $\mathbf{B}_1$ in the x–y plane given by

$$B_{1x} = |\mathbf{B}_1| \exp(-x/a) \cos \omega t,$$

$$B_{1y} = |\mathbf{B}_1| \exp(-x/a) \sin \omega t.$$

The interaction Hamiltonian, as usual, is given by

$$H' = -\boldsymbol{\mu}.\mathbf{B}$$

with $\boldsymbol{\mu} = \gamma(1/2)\boldsymbol{\sigma}$, $\gamma\,|\mathbf{B}_0| = \hbar\omega_0$ and $\gamma\,|\mathbf{B}_1| = \hbar\omega_1$. Represent the neutron state as a superposition

$$|\psi(t)\rangle = c_1(t)\,|+\rangle + c_2(t)\,|-\rangle$$

where $|+\rangle$ and $|-\rangle$ are the time-independent "spin-up" and "spin-down" states, respectively, along the z-direction. Use first order time-dependent perturbation to obtain $c_1(t)$ assuming that at $t = 0$ the neutron is in the $|-\rangle$ state. Assume $c_1(0) = 0$. You may take the unperturbed value $c_2(t) = 1$ to calculate the integral for $c_1(t)$. Obtain $c_1(\infty)$and show that there is a resonance in ω. Determine its position.

7. Use Fermi's golden rule to obtain, w_{fi}, the transition probability per unit time as $t \to \infty$, for transitions to a group of free particle states when an electron in the hydrogen atom in the ground state ($n = 1, l = 0$) is ejected as a free particle by the potential

$$\begin{aligned} H'(t) &= 0, && t < 0 \\ &= -g\frac{1}{r}e^{-\mu r}e^{-i\omega t} && t > 0. \end{aligned}$$

Take the energy of the ejected electron as E_0. What is the minimum value for ω such that this process takes place?

19 Scattering in one dimension

We now go into the full details of scattering problems, starting in this chapter with some typical one-dimensional systems. We describe the scattering problem in terms of the so-called reflection and transmission coefficients. We consider a number of interesting problems ending with the typically quantum-mechanical question of tunneling through a barrier.

19.1 Reflection and transmission coefficients

Let us discuss the effect of a potential of arbitrary shape but of finite range, a, as depicted in Fig. 19.1:

$$V(x) \neq 0, \quad 0 < x < a \tag{19.1}$$

$$V(x) = 0, \quad x < 0 \text{ and } x > a. \tag{19.2}$$

We consider now a particle that is traveling from left to right along the x-axis. The wavefunctions in the regions where the potential vanishes are given by

$$u(x) = Ae^{ikx} + Be^{-ikx}, \quad x < 0 \tag{19.3}$$

$$u(x) = Ce^{ikx}, \qquad x > a \tag{19.4}$$

where

$$k = \sqrt{\frac{2mE}{\hbar^2}}. \tag{19.5}$$

We are at the moment not concerned with the region where the potential is nonzero. The results we derive will not depend on whether the energy is less than or greater than the maximum height of $V(x)$.

It should be noted that for $x < 0$, the second term in (19.3) is included to take account of possible reflections at the boundary. For $x > a$, on the other hand, no reflections will occur. One often calls the wavefunction as written an "outgoing wave."

The current is given by

$$j(x) = \frac{\hbar}{2im}\left(u^* \frac{du}{dx} - u\frac{du^*}{dx}\right). \tag{19.6}$$

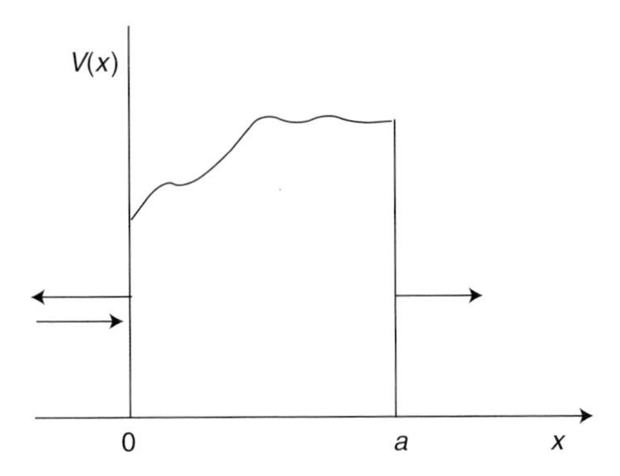

Fig. 19.1

Substituting the above wavefunctions in the two regions we find

$$j(x) = v\left(|A|^2 - |B|^2\right), \quad x < 0 \tag{19.7}$$

$$j(x) = v\,|C|^2, \quad x > a, \tag{19.8}$$

where v is the velocity given by

$$v = \frac{\hbar k}{m}. \tag{19.9}$$

Since the current is continuous, we must have

$$|A|^2 - |B|^2 = |C|^2. \tag{19.10}$$

That is,

$$|B|^2 + |C|^2 = |A|^2. \tag{19.11}$$

Let us define the following two quantities.
Reflection coefficient

$$R = \frac{|B|^2}{|A|^2}. \tag{19.12}$$

Transmission coefficient

$$T = \frac{|C|^2}{|A|^2}. \tag{19.13}$$

From (19.11) we have

$$R + T = 1. \tag{19.14}$$

This is a mathematical statement of an obvious fact that everything is accounted for: either particles that are coming in from the left are turned back (reflected) or they pass through (transmitted). Nothing disappears into thin air. The disappearance of the particles, called absorption, does happen in practical situations. The disappearing particle then appears in the form of energy or as another particle or a collection of particles. This is a much more complicated mechanism, which is often approximated by a complex potential. We will, however, continue to concern ourselves only with real potentials.

We can also write the currents in terms of R and T as reflected current, $j_r(x)$, and transmitted current, $j_t(x)$:

$$j_r(x) = v\,|A|^2\,(1-R)\,, \quad x<0 \tag{19.15}$$

$$j_t(x) = v\,|A|^2\,T, \quad x>a. \tag{19.16}$$

If we normalize the incident wave to have unit probability, then we can write

$$u(x) = e^{ikx} + S_{11}e^{-ikx}, \quad x<0 \tag{19.17}$$

$$u(x) = S_{12}e^{ikx}, \quad x>a \tag{19.18}$$

where the letter S corresponds to the quantity called the S-matrix, the subscript 1 refers to the incident channel, and as subscript 2 to the transmitted channel. The relation (19.14) is then given by

$$|S_{11}|^2 + |S_{12}|^2 = 1. \tag{19.19}$$

This relation reflects the unitary property of the S-matrix, a subject we will discuss at length in the later sections. For the present we will express our results in terms of R and T.

Below we consider a series of examples.

19.2 Infinite barrier

Here the potential is given by (see Fig. 19.2)

$$V(x) = 0 \quad x<0 \tag{19.20}$$

$$= \infty \quad x>0. \tag{19.21}$$

The corresponding wavefunctions are then

$$u(x) = Ae^{ikx} + Be^{-ikx}, \quad x<0 \tag{19.22}$$

$$u(x) = 0, \quad x>0. \tag{19.23}$$

The boundary condition at $x=0$ gives

$$A + B = 0, \tag{19.24}$$

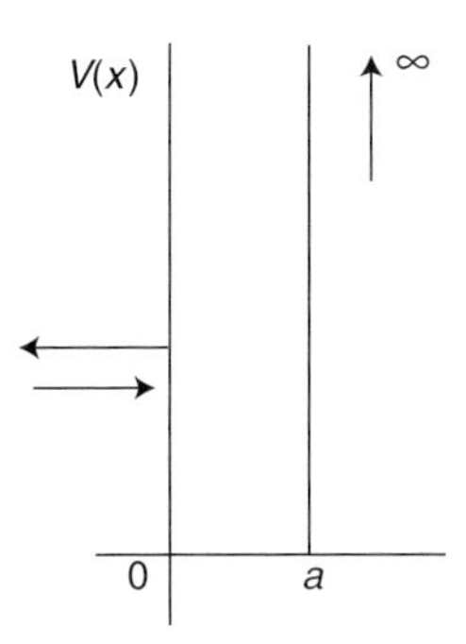

Fig. 19.2

and hence

$$R = 1, \tag{19.25}$$

$$T = 0, \tag{19.26}$$

which indicates that the entire incident wave is reflected. The wavefunctions are then given by

$$u(x) = A(e^{ikx} - e^{-ikx}), \quad x < 0 \tag{19.27}$$

$$u(x) = 0, \qquad x > 0. \tag{19.28}$$

19.3 Finite barrier with infinite range

There are two cases we need to consider for the incident energy since the barrier is finite: $E < V_0$ and $E > V_0$.

For $E < V_0$ (see Fig. 19.3) the wavefunctions are

$$u(x) = Ae^{ikx} + Be^{-ikx}, \quad x < 0 \tag{19.29}$$

$$u(x) = Fe^{-\beta x}, \qquad x > 0 \tag{19.30}$$

where

$$k = \sqrt{\frac{2mE}{\hbar^2}}, \quad \beta = \sqrt{\frac{2m\left(V_0 - E\right)}{\hbar^2}}. \tag{19.31}$$

Boundary conditions at $x = 0$ give

$$A + B = F, \tag{19.32}$$

$$(A - B)ik = -\beta F. \tag{19.33}$$

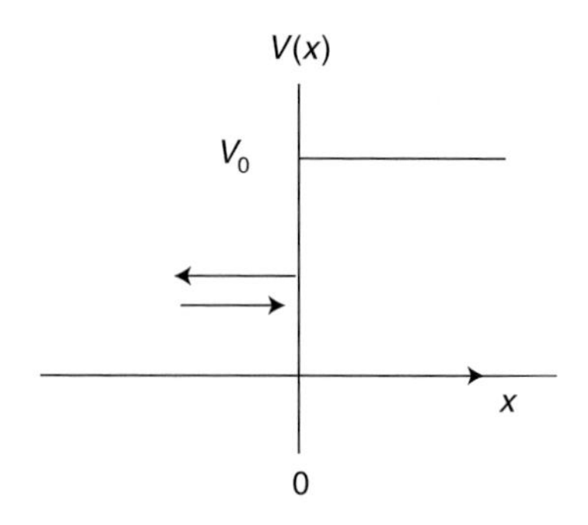

Fig. 19.3

Thus,

$$A = \frac{1}{2}\left(1 + \frac{i\beta}{k}\right)F, \qquad B = \frac{1}{2}\left(1 - \frac{i\beta}{k}\right)F. \tag{19.34}$$

Hence,

$$|A|^2 = |B|^2 = \frac{1}{4}\left(1 + \frac{\beta^2}{k^2}\right)|F|^2. \tag{19.35}$$

Therefore, the magnitudes of A and B are the same though there may be a phase difference between them. Hence we write

$$B = Ae^{2i\delta} \tag{19.36}$$

where

$$e^{2i\delta} = \frac{\left(1 - \frac{i\beta}{k}\right)}{\left(1 + \frac{i\beta}{k}\right)}. \tag{19.37}$$

The wavefunctions are

$$u(x) = A\left[e^{ikx} + e^{2i\delta}\, e^{-ikx}\right], \quad x < 0 \tag{19.38}$$

$$u(x) = Fe^{-\beta x}, \qquad x > 0. \tag{19.39}$$

We note that the reflected wave acquires a phase but the amplitude still remains the same as that of the incident wave,

$$R = \frac{\left|Ae^{2i\delta}\right|^2}{|A|^2} = 1 \tag{19.40}$$

The reflected current is, therefore,

$$j_r(x) = v\,|A|^2\,(1 - R) = 0, \quad x < 0. \tag{19.41}$$

Since the barrier is of infinite range, the particle will not appear as a free, transmitted, particle in any region in $x > 0$. This is reflected in the fact that $u(x)$ is real in this region and, hence, $j_t(x) = 0$.

For the case $E > V_0$:

$$u(x) = Ae^{ikx} + Be^{-ikx}, \quad x < 0 \tag{19.42}$$

$$u(x) = Fe^{i\alpha x}, \quad x > 0 \tag{19.43}$$

where

$$k = \sqrt{\frac{2mE}{\hbar^2}}, \quad \alpha = \sqrt{\frac{2m\left(E - V_0\right)}{\hbar^2}}. \tag{19.44}$$

The boundary conditions at $x = 0$ give

$$A + B = F, \tag{19.45}$$

$$(A - B)ik = i\alpha F. \tag{19.46}$$

Thus

$$A = \frac{1}{2}\left(1 + \frac{\alpha}{k}\right)F, \quad B = \frac{1}{2}\left(1 - \frac{\alpha}{k}\right)F. \tag{19.47}$$

Here we have both the reflected and transmitted waves with nonzero currents in both regions, and

$$R = \left|\frac{1 - \frac{\alpha}{k}}{1 + \frac{\alpha}{k}}\right|^2, \tag{19.48}$$

$$T = \left|\frac{2}{1 + \frac{\alpha}{k}}\right|^2. \tag{19.49}$$

The wavefunctions in the two regions can be written as

$$u(x) = A\left[e^{ikx} + \left[\frac{1 - \frac{\alpha}{k}}{1 + \frac{\alpha}{k}}\right], e^{-ikx}\right], \quad x < 0 \tag{19.50}$$

$$u(x) = \left(\frac{2}{1 + \frac{\alpha}{k}}\right)e^{i\alpha x}, \quad x > 0. \tag{19.51}$$

19.4 Rigid wall preceded by a potential well

The potential is described as follows (see Fig. 19.4):

$$V(x) = 0 \qquad \text{for } x < -a \tag{19.52}$$

$$= -V_0 \qquad \text{for } -a < x < 0 \tag{19.53}$$

$$= \infty \qquad \text{for } x > 0. \tag{19.54}$$

The corresponding wavefunctions are

$$u(x) = Ae^{ikx} + Be^{-ikx}, \qquad x < -a \tag{19.55}$$

$$u(x) = Fe^{i\alpha x} + Ge^{-i\alpha x}, \qquad -a < x < 0 \tag{19.56}$$

$$u(x) = 0, \qquad x > 0. \tag{19.57}$$

Since it leads to considerable simplifications, we apply boundary conditions at $x = 0$ first. We find

$$F + G = 0. \tag{19.58}$$

Therefore, for $-a < x < 0$,

$$u(x) = F\left(e^{i\alpha x} - e^{-i\alpha x}\right) = 2iF\sin(\alpha x) = F'\sin(\alpha x) \tag{19.59}$$

where $F' = 2iF$.

At $x = -a$ the boundary conditions give

$$Ae^{-ika} + Be^{ika} = -F'\sin(\alpha a), \tag{19.60}$$

$$(Ae^{-ika} - Be^{ika})ik = F'\alpha\cos(\alpha a). \tag{19.61}$$

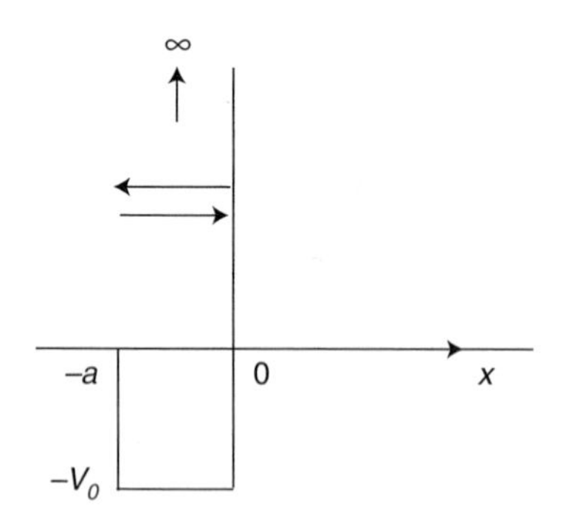

Fig. 19.4

Let

$$Ae^{-ika} = A', \tag{19.62}$$

$$Be^{ika} = B'. \tag{19.63}$$

Then

$$A' + B' = -F' \sin(\alpha a), \tag{19.64}$$

$$A' - B' = -i\frac{\alpha F'}{k} \cos(\alpha a). \tag{19.65}$$

Hence,

$$A' = -\left(\sin(\alpha a) + i\frac{\alpha}{k} \cos(\alpha a)\right)\frac{F'}{2} \tag{19.66}$$

$$B' = -\left(\sin(\alpha a) - i\frac{\alpha}{k} \cos(\alpha a)\right)\frac{F'}{2} \tag{19.67}$$

which leads to

$$B' = \frac{\sin(\alpha a) - i\frac{\alpha}{k}\cos(\alpha a)}{\sin(\alpha a) + i\frac{\alpha}{k}\cos(\alpha a)} A'. \tag{19.68}$$

We note that

$$\left|\frac{\sin(\alpha a) - i\frac{\alpha}{k}\cos(\alpha a)}{\sin(\alpha a) + i\frac{\alpha}{k}\cos(\alpha a)}\right| = 1 \tag{19.69}$$

and will therefore write

$$-\frac{\sin(\alpha a) - i\frac{\alpha}{k}\cos(\alpha a)}{\sin(\alpha a) + i\frac{\alpha}{k}\cos(\alpha a)} = e^{2i\delta'}. \tag{19.70}$$

The reasons behind the negative sign will be clear later. The coefficients A' and B' have the same magnitude but differ in phase. Hence from (19.68),

$$B' = -e^{2i\delta'} A'. \tag{19.71}$$

The above equality can be written as

$$\frac{\cos(\alpha a) + i\frac{k}{\alpha}\sin(\alpha a)}{\cos(\alpha a) - i\frac{k}{\alpha}\sin(\alpha a)} = \frac{\cos\delta' + i\sin\delta'}{\cos\delta' - i\sin\delta'}. \tag{19.72}$$

Hence

$$\tan(\delta') = \frac{k}{\alpha}\tan(\alpha a) \tag{19.73}$$

We also find

$$B = -e^{2i\delta'}e^{-2ika}A = -e^{2i\delta}A, \qquad \text{where} \quad \delta = \delta' - ka. \tag{19.74}$$

Thus,

$$R = \frac{|B|^2}{|A|^2} = 1, \tag{19.75}$$

as expected.

In the limit that $V_0 \to 0$, $\alpha \to k$,

$$e^{2i\,\delta'} \to -\frac{\sin(ka) - i\cos(ka)}{\sin(ka) + i\cos(ka)} = e^{2ika}. \tag{19.76}$$

The negative sign has been compensated and we obtain the expected result:

$$V_0 \to 0, \quad \delta' \to ka, \quad \delta \to 0. \tag{19.77}$$

Let us define the scattering coefficient,

$$\left|\frac{e^{2i\delta} - 1}{2ik}\right|^2 = \text{scattering coefficient}, \tag{19.78}$$

which is the same as the probability of reflection due to V_0. We note that for $V_0 = 0$ the phase vanishes and so does the scattering coefficient.

In general, one can write

$$\left|\frac{e^{2i\delta} - 1}{2ik}\right|^2 = \frac{\sin^2\delta}{k^2} = \frac{1}{k^2(1 + \cot^2\delta)}. \tag{19.79}$$

The scattering coefficient will then have peaks when

$$\delta = \frac{n\pi}{2}, \quad \text{with } n = 1, 3, \ldots. \tag{19.80}$$

These are the counterparts of the resonances we discussed earlier.

The phase δ' is then found to be

$$\delta' - ka = \frac{n\pi}{2}. \tag{19.81}$$

For $n = 1$, since

$$\tan(\delta') = -\cot(ka), \tag{19.82}$$

we find

$$\alpha \cot(\alpha a) = -k \tan(ka) \tag{19.83}$$

We can write this relation as

$$\alpha \cot(\alpha a) = -k \cot(ka). \tag{19.84}$$

This transcendental equation provides a relation between k and V_0. The solution represents the energies at which the resonances occur.

19.5 Square-well potential and resonances

Consider a square-well potential of finite range (see Fig. 19.5):

$$V = -V_0, \quad \text{for } 0 \leq x \leq a \tag{19.85}$$

$$V = 0, \quad \text{for } x < 0, x > a. \tag{19.86}$$

The incident wave is given by

$$u(x) = Ae^{ikx} + Be^{-ikx}, \quad \text{for } x < 0. \tag{19.87}$$

In the region of the potential well the wavefunction has the form

$$u(x) = Fe^{i\alpha x} + Ge^{-i\alpha x}, \quad \text{for } 0 \leq x \leq a \tag{19.88}$$

where

$$k = \sqrt{\frac{2mE}{\hbar^2}} \quad \text{and} \quad \alpha = \sqrt{\frac{2m(E+V_0)}{\hbar^2}}, \tag{19.89}$$

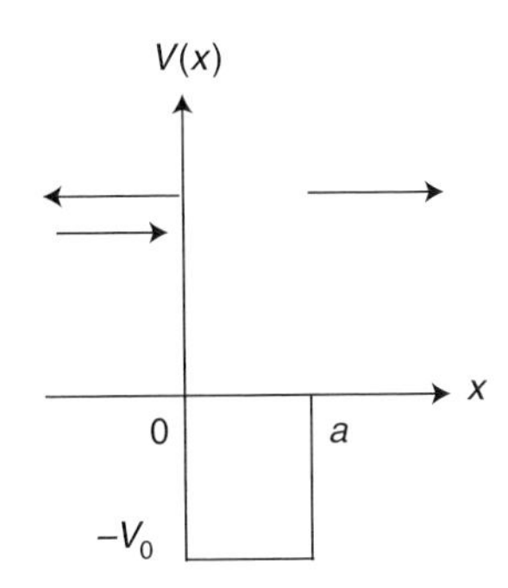

Fig. 19.5

while the outgoing wave is given by

$$u_k(x) = Ce^{ikx}, \quad \text{for } x > a. \tag{19.90}$$

The boundary conditions at $x = 0$ lead to

$$A + B = F + G, \tag{19.91}$$

$$(A - B)k = \alpha\,(F - G)\,. \tag{19.92}$$

Writing this in the matrix form we have

$$\begin{bmatrix} 1 & 1 \\ k & -k \end{bmatrix} \begin{bmatrix} A \\ B \end{bmatrix} = \begin{bmatrix} 1 & 1 \\ \alpha & -\alpha \end{bmatrix} \begin{bmatrix} F \\ G \end{bmatrix}. \tag{19.93}$$

The boundary conditions at $x = a$ give, in the matrix form,

$$\begin{bmatrix} e^{i\alpha a} & e^{-i\alpha a} \\ i\alpha e^{i\alpha a} & -i\alpha e^{-i\alpha a} \end{bmatrix} \begin{bmatrix} F \\ G \end{bmatrix} = \begin{bmatrix} 1 \\ ik \end{bmatrix} e^{ika} C. \tag{19.94}$$

From this relation we can obtain F and G, by inverting the matrix on the left-hand side; substitution in (19.93) gives

$$\frac{A}{C} = \frac{e^{i(k+\alpha)a}\left[(k+\alpha)^2 e^{-2i\alpha a} - (k-\alpha)^2\right]}{4k\alpha}, \tag{19.95}$$

$$\frac{B}{C} = \frac{e^{i(k-\alpha)a}\left[(k^2-\alpha^2)(1-e^{2i\alpha a})\right]}{4k\alpha}. \tag{19.96}$$

Hence the reflection and transmission coefficients are given by

$$R = \left|\frac{B}{A}\right|^2 = \left|\frac{\left[(k^2-\alpha^2)(1-e^{2i\alpha a})\right]}{\left[(k+\alpha)^2 - (k-\alpha)^2 e^{2i\alpha a}\right]}\right|^2 \tag{19.97}$$

and

$$T = \left|\frac{C}{A}\right|^2 = \left|\frac{4k\alpha}{\left[(k+\alpha)^2 - (k-\alpha)^2 e^{2i\alpha a}\right]}\right|^2. \tag{19.98}$$

Let us now consider the transmission coefficient in some detail. We can rewrite (19.98) as

$$T = \frac{1}{1 + \dfrac{V_0^2 \sin^2(\alpha a)}{4E(E+V_0)}}. \tag{19.99}$$

We note that

$$T = 1 \quad \text{for} \quad \alpha a = n\pi \quad \text{where} \quad n = 1, 2, \ldots. \tag{19.100}$$

In other words, there is, remarkably, perfect transmission for specific values of the energy. The reflection coefficient, which can be written as

$$R = \frac{1}{1 + \dfrac{4E(E+V_0)}{V_0^2 \sin^2(\alpha a)}}, \tag{19.101}$$

vanishes, as expected, at the same values of $\alpha a (= n\pi)$. This type of effect is seen in the transmission of light through refracting layers and is related to the so called Ramseur–Townsend effect for the scattering of electrons through noble gases. We will return to this subject when we discuss three-dimensional scattering. Physically one understands it as the interference of waves reflected at $x = 0$ and $x = a$. This phenomena is also called transmission resonance.

Let us obtain the expression for T for energies in close proximity to $\alpha a = n\pi$. These give us what we will call the resonance energies. At one of these energies, $E = E_0$, we will make an expansion of αa, using the Taylor expansion technique,

$$f(E) = f(E_0) + (E - E_0)\frac{df}{dE}. \tag{19.102}$$

In the low energy limit, $E \ll V_0$ when the potential is strong, i.e.,

$$g = \sqrt{\frac{2mV_0 a^2}{\hbar^2}} \gg 1, \tag{19.103}$$

we find

$$\sin(\alpha a) = (-1)^n \sin\left(g\frac{E - E_0}{V_0}\right) = (-1)^n \frac{g}{2V_0}(E - E_0). \tag{19.104}$$

The transmission coefficient is given by

$$T = \frac{1}{1 + \dfrac{g^2(E - E_0)^2}{16E_0V_0}}. \tag{19.105}$$

We write this as

$$T = \frac{\frac{1}{4}\Gamma^2}{(E - E_0)^2 + \frac{1}{4}\Gamma^2}, \tag{19.106}$$

which is in the so-called resonance form with E_0 the resonance energy as already stated, where $T = 1$, and

$$\Gamma = \frac{8\sqrt{E_0V_0}}{g} \tag{19.107}$$

as the width.

19.6 Tunneling

We will continue with the same problem as the previous one except that instead of an attractive potential, we will now consider a potential barrier (see Fig. 19.6). This implies that we change the sign $V_0 \to -V_0$. At the same time we will consider energies that are smaller than the potential, $E < V_0$. This will lead us to the problem of "tunneling," which is a purely quantum-mechanical phenomenon. Effectively then, the above changes imply that we make the change

$$\alpha \to i\beta \tag{19.108}$$

in the above formulas, where

$$\beta = \sqrt{\frac{2m(V_0 - E)}{\hbar^2}}. \tag{19.109}$$

Hence the transmission coefficient will be given by

$$T = \frac{1}{1 + \dfrac{V_0^2 \sinh^2(\beta a)}{4E(V_0 - E)}} \tag{19.110}$$

which is nonzero in contrast to the classical case where there will be no transmission for $E < V_0$.

Let us now consider the limit

$$V_0 \gg E, \quad \text{i.e.,} \quad \beta a \gg 1. \tag{19.111}$$

We find

$$T \simeq \frac{16E(V_0 - E)}{V_0^2} e^{-2\beta a}. \tag{19.112}$$

Thus, the transmission through a barrier decreases exponentially. The behavior of T as a function of E/V_0 is given in Fig. 19.7.

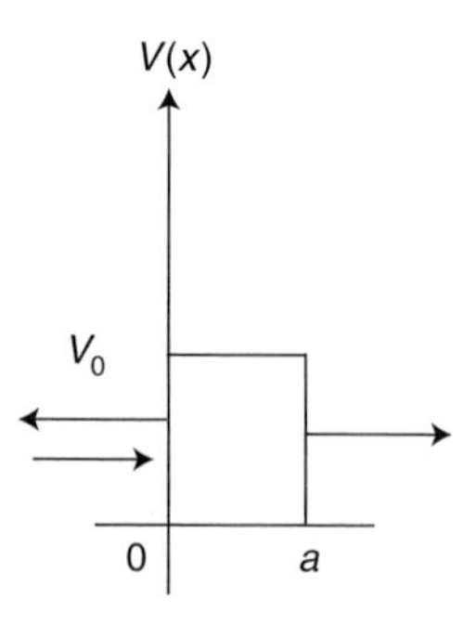

Fig. 19.6

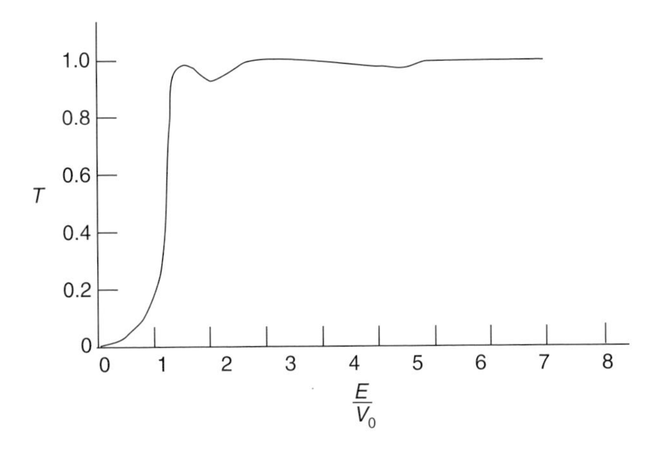

Fig. 19.7

The above result for T is of the form that can be extended to the case where the potential is of arbitrary shape described by $V(x)$, which is assumed to be a smoothly varying function of x. This is possible if $V(x)$ can be described in terms of a series of potentials each of a rectangular shape with infinitesimal range, where by rectangular shape we mean that the potential is of the form of a finite barrier which we have considered earlier.

First we take the logarithm of T,

$$\log T \simeq \log\left[\frac{16E(V_0 - E)}{V_0^2}\right] - 2\beta a. \tag{19.113}$$

The second term will dominate unless $E = V_0$, which is called the classical turning point. For $E \ll V_0$ then

$$\log T \approx -2\beta a. \tag{19.114}$$

Since the potential can be described in terms of rectangular potentials each of infinitesimal range, the transmission coefficient for a particle tunneling through such a barrier is effectively a product of the transmission coefficients through each barrier. We can then write

$$T \approx T_1 T_2 \cdots . \tag{19.115}$$

Taking the logarithms of both sides we obtain

$$\log T \approx \log T_1 + \log T_2 + \cdots \tag{19.116}$$

where T_i corresponds to the transmission coefficient through the ith barrier. Substituting expression (19.114) for each rectangular barrier, we obtain

$$\log T \approx -2\beta_1 \Delta x - 2\beta_2 \Delta x \cdots = -2\int_{x_1}^{x_2} \sqrt{\frac{2m(V(x) - E)}{\hbar^2}}\, dx \tag{19.117}$$

where β_i's are the values of β given in (19.109) for each interval and Δx is the barrier width, while x_1 and x_2 are the classical turning points at the two ends. We have replaced the infinite sum by an integral. Thus the transmission coefficient for tunneling through a barrier is given by

$$T \approx \exp\left[-2\int_{x_1}^{x_2}\sqrt{\frac{2m(V(x)-E)}{\hbar^2}}dx\right]. \tag{19.118}$$

We will return to this formula when we consider the WKB approximation, where proper modifications are made to the above result to take account of the problem with regard to the turning points.

19.7 Problems

1. Consider a one-dimensional problem with the following potential:

$$\begin{aligned} V(x) &= -V_0 \quad \text{for } x < 0 \\ &= 0 \quad \text{for } x > 0. \end{aligned}$$

   If the particle is moving from left to right, obtain the wavefunctions in the two regions as well as R and T for (i) $E > 0$, (ii) $-V_0 < E < 0$.

2. Determine the transmission coefficient for the potential

$$\begin{aligned} V(x) &= \lambda\left(a^2 - x^2\right), \quad \text{for } x < a \\ &= 0, \quad \text{for } x > a. \end{aligned}$$

3. Solve the Schrödinger equation in one dimension given by

$$\frac{d^2u}{dx^2} + k^2u(x) = \frac{2m}{\hbar^2}V(x)u(x)$$

   using the Green's function formalism by writing

$$u(x) = u_0(x) + \frac{2m}{\hbar^2}\int dx'\, G_0(x-x')V(x')u(x').$$

   Show that for the outgoing wave boundary condition G_0 is given by

$$\begin{aligned} G_0(x-x') &= \frac{i}{2k}e^{ik(x-x')}, \quad \text{for } x > x' \\ &= \frac{i}{2k}e^{-ik(x-x')}, \quad \text{for } x < x'. \end{aligned}$$

4. Using the Green's function obtained in problem 3, determine the wavefunction, $u(x)$, for $x > 0$ and $x < 0$, for an attractive delta function potential given by

$$\frac{2m}{\hbar^2} V(x) = -\lambda\delta(x).$$

Also obtain the reflection and transmission coefficients R and T, respectively.

5. Consider the double-delta potential

$$V(x) = -g\left[\delta(x-a) + \delta(x+a)\right]$$

and obtain the corresponding scattering solutions (as in the case of the bound states, it may be easier to treat the even and odd parities separately). Determine the reflection and transmission coefficients. Determine the location of resonances.

20 Scattering in three dimensions – a formal theory

In this chapter we go to three dimensions and formalize the scattering theory without going through time-dependent perturbation theory. We solve our problems through the function that plays a pivotal role in physics, the Green's function. We define the S-matrix, the T-matrix, and their connection to the scattering amplitude.

20.1 Formal solutions in terms of Green's function

We will present here a formal way, invoking the Green's function technique, to obtain the scattering solutions in terms of abstract state vectors and operators. This will also allow us to define two very important quantities in scattering theory, the T- and S-matrices.

We start with the energy eigenvalue equation for the abstract state vector $|\phi\rangle$

$$H|\phi\rangle = E|\phi\rangle \tag{20.1}$$

where H is the total Hamiltonian given by

$$H = H_0 + V \tag{20.2}$$

with H_0 the unperturbed Hamiltonian and V the potential.

Equation (20.1) is then of the form

$$(E - H_0)|\phi\rangle = V|\phi\rangle . \tag{20.3}$$

A formal solution of this can be written as

$$|\phi\rangle = |\phi_0\rangle + \frac{1}{(E - H_0)}V|\phi\rangle , \tag{20.4}$$

where $|\phi_0\rangle$ is the homogeneous solution that satisfies the equation

$$(E - H_0)|\phi_0\rangle = 0. \tag{20.5}$$

One can easily verify by multiplying (20.4) on both sides by $(E - H_0)$ that equation (20.1) is recovered.

Let us write

$$G_0 = \frac{1}{E - H_0}. \tag{20.6}$$

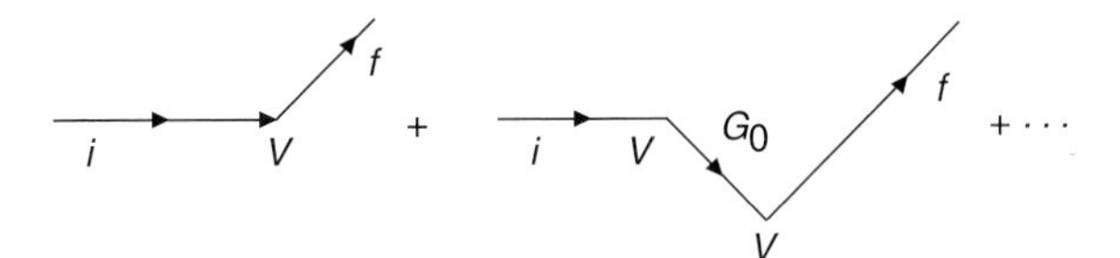

Fig. 20.1

This is the so called "free particle" Green's function in terms of which one can write equation (20.4) as

$$|\phi\rangle = |\phi_0\rangle + G_0 V |\phi\rangle . \tag{20.7}$$

To obtain $|\phi\rangle$ one can resort to perturbation expansion in terms of V and write the following series through iteration:

$$|\phi\rangle = |\phi_0\rangle + G_0 V |\phi_0\rangle + G_0 V G_0 V |\phi_0\rangle + \cdots . \tag{20.8}$$

This expansion is described pictorially in Fig. 20.1.

We now define the "total" Green's function

$$G = \frac{1}{E - H} \tag{20.9}$$

where H is given by (20.2). Hence

$$G = \frac{1}{E - H_0 - V}. \tag{20.10}$$

It can easily be shown that $|\phi\rangle$ satisfies the equation

$$|\phi\rangle = |\phi_0\rangle + GV |\phi_0\rangle . \tag{20.11}$$

To express G in terms of G_0 and V we can compare (20.11) and (20.8) and obtain

$$G = G_0 + G_0 V G_0 + \cdots . \tag{20.12}$$

This result can also be obtained by the expansion

$$G = \frac{1}{(E - H_0) - V} = \frac{1}{(E - H_0)} + \frac{1}{(E - H_0)} V \frac{1}{(E - H_0)} + \cdots \tag{20.13}$$

$$= G_0 + G_0 V G_0 + \cdots . \tag{20.14}$$

The infinite expansion (20.12) and (20.14) can be written compactly as

$$G = G_0 + G_0 V G. \tag{20.15}$$

Another possible way to describe $|\phi\rangle$ is to write it as

$$|\phi\rangle = |\phi_0\rangle + G_0 T |\phi_0\rangle \tag{20.16}$$

where T is called the "scattering matrix". Comparing (20.16) and (20.7), we obtain

$$T\left|\phi_0\right\rangle = V\left|\phi\right\rangle . \tag{20.17}$$

If we express $|\phi\rangle$ as an expansion given in (20.8) and insert it in (20.17) then one finds

$$T = V + VG_0V + \cdots . \tag{20.18}$$

The series on the right-hand side is described pictorially in Fig. 20.1.

Expression (20.18) can be written compactly in terms of G as

$$T = V + VGV. \tag{20.19}$$

We note that (20.7), (20.11), and (20.16) are all equivalent ways to write $|\phi\rangle$.

20.2 Lippmann–Schwinger equation

We are familiar with G_0 from our previous calculations. We will elaborate on its role through formal steps to obtain the scattering amplitude. We start with equation (20.4) and define an "outgoing" state $\left|\phi^{(+)}\right\rangle$ by adding an $i\epsilon$ term in the denominator of G_0 as follows:

$$\left|\phi^{(+)}\right\rangle = \left|\phi_0\right\rangle + \frac{1}{E - H_0 + i\epsilon} V \left|\phi^{(+)}\right\rangle \tag{20.20}$$

$$= \left|\phi_0\right\rangle + G_0(E + i\epsilon) V \left|\phi^{(+)}\right\rangle . \tag{20.21}$$

This is called the Lippmann–Schwinger equation. The presence of $i\epsilon$ will ensure that we have an outgoing solution.

We can write (20.20) in coordinate representation by multiplying it on the left by the bra vector $\langle \mathbf{r}|$ and inserting a complete set of states in appropriate places. Thus,

$$\langle \mathbf{r}\left|\phi^{(+)}\right\rangle = \langle \mathbf{r}\left|\phi_0\right\rangle + \int d^3r' \int d^3r'' \langle \mathbf{r}\left|G_0(E+i\epsilon)\right|\mathbf{r}'\rangle\langle \mathbf{r}'\left|V\right|\mathbf{r}''\rangle \left\langle \mathbf{r}''\right|\phi^{+}\rangle . \tag{20.22}$$

Since V is a local operator, as the interactions occur at a single unique point, we write

$$\langle \mathbf{r}'\left|V\right|\mathbf{r}''\rangle = V(\mathbf{r}'')\delta^{(3)}(\mathbf{r}' - \mathbf{r}''). \tag{20.23}$$

If, furthermore, we write

$$\langle \mathbf{r}\left|\phi^{(+)}\right\rangle = \phi^{(+)}(\mathbf{r}), \quad \langle \mathbf{r}\left|\phi_0\right\rangle = \phi_0(\mathbf{r}) \quad \text{and} \quad \left\langle r\left|G_0(E+i\epsilon)\right|r'\right\rangle = G_0^{(+)}(\mathbf{r}, \mathbf{r}') \tag{20.24}$$

then we have

$$\phi^{(+)}(\mathbf{r}) = \phi_0(\mathbf{r}) + \int d^3r'\, G_0^{(+)}(\mathbf{r}, \mathbf{r}')V(\mathbf{r}')\phi^{(+)}(\mathbf{r}') \tag{20.25}$$

which is the same relation as (20.7) in coordinate representation except that here we have introduced the outgoing boundary condition explicitly from the beginning.

Let us now evaluate the corresponding Green's function $G_0^{(+)}(\mathbf{r}, \mathbf{r}')$, given by

$$G_0^{(+)}(\mathbf{r}, \mathbf{r}') = <\mathbf{r}\left|\frac{1}{E - H_0 + i\epsilon}\right|\mathbf{r}'> = <\mathbf{r}\left|\frac{1}{\dfrac{\hbar^2 k^2}{2m} - H_0 + i\epsilon}\right|\mathbf{r}'> \tag{20.26}$$

where we have taken the energy, $E = \hbar^2 k^2/2m$. To make the dependence on the momentum vector $\mathbf{k}$ more explicit, we write

$$|\phi_0\rangle = |\phi_0(\mathbf{k})\rangle . \tag{20.27}$$

Therefore,

$$H_0 |\phi_0(\mathbf{k})\rangle = \frac{\hbar^2 k^2}{2m} |\phi_0(\mathbf{k})\rangle \tag{20.28}$$

and

$$\langle\mathbf{r} |\phi_0(\mathbf{k})\rangle = \frac{e^{i\mathbf{k}\cdot\mathbf{r}}}{\left(\sqrt{2\pi}\right)^3}. \tag{20.29}$$

We then obtain, after inserting complete sets of states,

$$G_0^{(+)}(\mathbf{r}, \mathbf{r}') = \int d^3k' \int d^3k'' \langle\mathbf{r} |\phi_0(\mathbf{k}')\rangle \langle\phi_0(\mathbf{k}')\left|\frac{1}{\frac{\hbar^2 k^2}{2m} - H_0 + i\epsilon}\right|\phi_0(\mathbf{k}'')\rangle \langle\phi_0(\mathbf{k}'' |\mathbf{r}'\rangle$$

$$= \int d^3k' \int d^3k'' \frac{e^{i\mathbf{k}''\cdot\mathbf{r}}}{(\sqrt{2\pi})^3} \left[\frac{\delta^{(3)}\left(\mathbf{k}' - \mathbf{k}''\right)}{\dfrac{\hbar^2 k^2}{2m} - \dfrac{\hbar^2 k''^2}{2m} + i\epsilon}\right] \frac{e^{-i\mathbf{k}''\cdot\mathbf{r}'}}{(\sqrt{2\pi})^3} \tag{20.30}$$

where we have used the relation $\langle\phi_0(\mathbf{k}') |\phi_0(\mathbf{k}'')\rangle = \delta^{(3)}\left(\mathbf{k}' - \mathbf{k}''\right)$.

Thus,

$$G_0^{(+)}(\mathbf{r}, \mathbf{r}') = \frac{1}{(2\pi)^3} \int d^3k' \frac{e^{i\mathbf{k}'\cdot(\mathbf{r}-\mathbf{r}')}}{\dfrac{\hbar^2}{2m}\left(k^2 - k'^2\right) + i\epsilon}$$

$$= -\frac{2m}{(2\pi)^3 \hbar^2} \int d^3k' \frac{e^{i\mathbf{k}'\cdot(\mathbf{r}-\mathbf{r}')}}{k'^2 - k^2 - i\epsilon'}. \tag{20.31}$$

Using the integration technique outlined in the Appendix for an outgoing wave we obtain

$$G_0^{(+)}(\mathbf{r}, \mathbf{r}') = -\frac{m}{2\pi\hbar^2} \frac{e^{ik'|\mathbf{r}-\mathbf{r}'|}}{|\mathbf{r} - \mathbf{r}'|}. \tag{20.32}$$

In our earlier discussions of the three-dimensional Schrödinger equation we have used the notation $u(\mathbf{r})$ for the wavefunction. If we replace $\phi^{(+)}(\mathbf{r})$ by $u(\mathbf{r})$ and $\phi_0(\mathbf{r})$ by $u_i(\mathbf{r})$, to indicate an "incoming wave", we obtain

$$u(\mathbf{r}) = u_i(\mathbf{r}) - \left(\frac{m}{2\pi\hbar^2}\right)\int d^3r' \frac{e^{ik'|\mathbf{r}-\mathbf{r}'|}}{|\mathbf{r}-\mathbf{r}'|} V(r')\, u(\mathbf{r}'). \tag{20.33}$$

Let us consider this result in the limit $r \to \infty$, which is relevant to scattering. We find in this limit that, since $r \gg r'$,

$$\left|\mathbf{r}-\mathbf{r}'\right| = \sqrt{r^2 + r'^2 - 2rr'\cos\theta'} = r\sqrt{1 + \left(\frac{r'}{r}\right)^2 - 2\left(\frac{r'}{r}\right)\cos\theta'} \to r - r'\cos\theta' \tag{20.34}$$

where we have kept only the first two leading terms. Notice that the angle between $\mathbf{r}$ and $\mathbf{r}'$ is the same as the angle between the scattered momentum $\mathbf{k}_f$ and $\mathbf{r}'$ with $\left|\mathbf{k}_f\right| = |\mathbf{k}| = k$. We can then write

$$k\left(r - r'\cos\theta'\right) = kr - kr'\cos\theta' = kr - \mathbf{k}_f.\mathbf{r}' \tag{20.35}$$

and

$$\frac{e^{ik|\mathbf{r}-\mathbf{r}'|}}{|\mathbf{r}-\mathbf{r}'|} \to \frac{e^{ikr}}{r} e^{-i\mathbf{k}_f.\mathbf{r}'}. \tag{20.36}$$

If we take the incident (incoming) wave to be traveling in the z-direction, we obtain the following asymptotic behavior:

$$u(\mathbf{r}) \to \frac{1}{(\sqrt{2\pi})^3}\left[e^{ikz} - \left(\frac{m\sqrt{2\pi}}{\hbar^2}\right)\frac{e^{ikr}}{r}\int d^3r' e^{-i\mathbf{k}_f.\mathbf{r}'} V(r')\, u(\mathbf{r}')\right]. \tag{20.37}$$

We define the scattering amplitude, $f(\theta)$, through the relation

$$u(\mathbf{r}) \to \frac{1}{(\sqrt{2\pi})^3}\left[e^{ikz} + f(\theta)\frac{e^{ikr}}{r}\right]. \tag{20.38}$$

From (20.37) we obtain the following expression for $f(\theta)$:

$$f(\theta) = -\left(\frac{m\sqrt{2\pi}}{\hbar^2}\right)\int d^3r' e^{-i\mathbf{k}_f.\mathbf{r}'} V(r')\, u(\mathbf{r}'). \tag{20.39}$$

Thus, $f(\theta)$ measures the effect of the potential in causing the particle to undergo scattering. We note that $f(\theta) = 0$ when $V = 0$. In other words, no scattering takes place in the absence of potential, as it should be.

We obtain the cross-section using the same derivation that we used in time-dependent perturbation. We now return to box normalization inside a volume V, since it is more convenient to define the cross-section in terms of it. Of course, the expression for the cross-section will not depend explicitly on V. We therefore write the asymptotic behavior for the wavefunction as

$$u(\mathbf{r}) \to \frac{1}{\sqrt{V}}\left[e^{ikz} + f(\theta)\frac{e^{ikr}}{r}\right]. \tag{20.40}$$

The scattering cross-section into an area $r^2\, d\Omega$ is given by

$$d\sigma = \frac{\text{scattered flux into area } r^2 d\Omega}{\text{incident flux}} = \frac{\left[\frac{1}{V}\left|f(\theta)\right|^2\frac{1}{r^2}\right] r^2 d\Omega}{1/V}. \tag{20.41}$$

Thus, the differential cross-section is given by

$$\frac{d\sigma}{d\Omega} = \left|f(\theta)\right|^2 \tag{20.42}$$

and the total cross-section by

$$\sigma = \int \left|f(\theta)\right|^2 d\Omega. \tag{20.43}$$

20.3 Born approximation

An approximate solution to the scattering problem can be achieved if the potential is weak. As we saw in Chapter 17 on time-dependent perturbation theory, one can make an expansion in terms of $V(r)$. The first-order term then corresponds to the Born approximation, which we discussed in that chapter. We can obtain the first-order term by replacing $u(\mathbf{r}')$ inside the integral on the right-hand side of (20.39) by $u_i(\mathbf{r}')$ $(= e^{ikz'})$. We write

$$\frac{1}{(\sqrt{2\pi})^3}e^{ikz'} = \frac{1}{(\sqrt{2\pi})^3}e^{i\mathbf{k}.\mathbf{r}'} \tag{20.44}$$

where $\mathbf{k}_i = k\mathbf{e}_3$, with $\mathbf{e}_3$ being a unit vector in the z-direction. Thus we obtain the following Born approximation result

$$f_B(\theta) = -\frac{m}{2\pi\hbar^2}\int d^3r' e^{-i\mathbf{q}.\mathbf{r}'} V\left(r'\right) \tag{20.45}$$

where $\mathbf{q} = \mathbf{k}_f - \mathbf{k}_i$ is the momentum transfer. This is the same result as the one derived previously from time-dependent perturbation theory.

Below we obtain the scattering amplitude in the Born approximation for two very important potentials that we have already considered previously: Yukawa and Coulomb potentials.

20.4 Scattering from a Yukawa potential

A Yukawa potential is given by

$$V(r) = -g\frac{e^{-\mu r}}{r}. \tag{20.46}$$

Inserting this in (20.45), we obtain

$$f_B(\theta) = \frac{mg}{2\pi\hbar^2}\int d^3r\frac{e^{-\mu r}}{r}e^{-i\mathbf{q}\cdot\mathbf{r}}. \tag{20.47}$$

Let us write

$$\mathbf{q}\cdot\mathbf{r} = |\mathbf{q}|\, r\cos\theta', \qquad d^3r = r^2 dr d\cos\theta' d\phi'. \tag{20.48}$$

We can carry out the integration over ϕ' and obtain

$$f_B(\theta) = \frac{mg}{\hbar^2}\int_0^\infty dr \quad re^{-\mu r}\int_{-1}^{1} d\cos\theta' e^{-iqr\cos\theta'} \tag{20.49}$$

where $q = |\mathbf{q}|$. The integration over $\cos\theta'$ gives

$$f_B(\theta) = \frac{img}{\hbar^2 q}\int_0^\infty dr\left[e^{-(\mu+iq)r} - e^{-(\mu-iq)r}\right] \tag{20.50}$$

$$= \frac{2mg}{\hbar^2}\left(\frac{1}{\mu^2+q^2}\right) \tag{20.51}$$

where

$$q^2 = \left|\mathbf{k}_f\right|^2 + |\mathbf{k}_i|^2 - 2\left|\mathbf{k}_f\right||\mathbf{k}_i|\cos\theta \tag{20.52}$$

For the elastic scattering that we are considering,

$$\left|\mathbf{k}_f\right| = |\mathbf{k}_i| = k. \tag{20.53}$$

Therefore,

$$q^2 = 2k^2(1-\cos\theta). \tag{20.54}$$

Hence,

$$f_B(\theta) = \frac{2mg}{\hbar^2}\frac{1}{\mu^2+2k^2(1-\cos\theta)}. \tag{20.55}$$

The differential cross-section is given by

$$\frac{d\sigma}{d\Omega} = |f_B(\theta)|^2 = \frac{4m^2g^2}{\hbar^4}\frac{1}{\left[\mu^2+2k^2(1-\cos\theta)\right]^2}. \tag{20.56}$$

The cross-section can be obtained by writing $d\Omega = d\cos\theta d\phi$. Hence, integrating out ϕ, we obtain

$$\sigma = \frac{8\pi m^2 g^2}{\hbar^4} \int_{-1}^{1} \frac{d\cos\theta}{\left[\mu^2 + 2k^2(1-\cos\theta)\right]^2}. \tag{20.57}$$

The integral over $\cos\theta$ can be carried out to give

$$\sigma = \frac{16\pi m^2 g^2}{\hbar^4} \frac{1}{\mu^2\left(\mu^2 + 4k^2\right)}. \tag{20.58}$$

20.5 Rutherford scattering

Rutherford scattering corresponds to the scattering of an electron off a nucleus of charge Ze, say. The potential, $V(r)$, then is the familiar Coulomb potential:

$$V(r) = -\frac{Ze^2}{r}. \tag{20.59}$$

We can obtain $f_B(\theta)$ and $d\sigma/d\Omega$ directly from the previous Yukawa result by taking

$$\mu = 0, \quad g = Ze^2. \tag{20.60}$$

From the previous section we find

$$f_B(\theta) = \frac{2mZe^2}{2\hbar^2 k^2} \frac{1}{1-\cos\theta}. \tag{20.61}$$

Writing $1 - \cos\theta = 2\sin^2\theta/2$ we obtain

$$f_B(\theta) = \frac{mZe^2}{2\hbar^2 k^2} \frac{1}{\sin^2(\theta/2)} \tag{20.62}$$

and

$$\frac{d\sigma}{d\Omega} = \left(\frac{mZe^2}{2\hbar^2 k^2}\right)^2 \frac{1}{\sin^4(\theta/2)}. \tag{20.63}$$

We have thus reproduced the formula derived in the chapter on time-dependent perturbation theory for the Rutherford scattering.

20.6 Charge distribution

If instead of a point charge, the Coulomb potential corresponds to a charge distribution, e.g.,

$$V(r) = -e^2 \int \frac{d^3r' \rho(\mathbf{r}')}{|\mathbf{r}-\mathbf{r}'|} \tag{20.64}$$

where $e\rho(\mathbf{r})$ is the charge density of the source with total charge Ze, then the scattering amplitude is given by

$$f_B(\theta) = \frac{me^2}{2\pi\hbar^2} \int d^3r \int \frac{d^3r' \rho(\mathbf{r}')}{|\mathbf{r}-\mathbf{r}'|} e^{-i\mathbf{q}\cdot\mathbf{r}}. \tag{20.65}$$

We can change variables by letting

$$\mathbf{r}'' = \mathbf{r} - \mathbf{r}' \tag{20.66}$$

while keeping $\mathbf{r}' = \mathbf{r}'$. Hence,

$$f_B(\theta) = \frac{me^2}{2\pi\hbar^2} \int \frac{d^3r''}{r''} e^{-i\mathbf{q}\cdot\mathbf{r}''} \int d^3r' \rho(\mathbf{r}') e^{-i\mathbf{q}\cdot\mathbf{r}'}. \tag{20.67}$$

We write

$$F\left(q^2\right) = \int d^3r' \rho(\mathbf{r}') e^{-i\mathbf{q}\cdot\mathbf{r}'}, \tag{20.68}$$

which is the Fourier transform of the charge density, called the charge form factor, normalized as

$$F(0) = \int d^3r' \rho(\mathbf{r}') = Z \tag{20.69}$$

if Z is the total charge. Therefore,

$$f_B(\theta) = \frac{me^2}{2\pi\hbar^2} F\left(q^2\right) \int \frac{d^3r'' e^{-i\mathbf{q}\cdot\mathbf{r}''}}{r''}. \tag{20.70}$$

The integral has already been evaluated earlier and we obtain

$$f_B(\theta) = \frac{me^2}{2\hbar^2 k^2} \frac{F(q^2)}{\sin^2(\theta/2)}. \tag{20.71}$$

20.7 Probability conservation and the optical theorem

The probability conservation relation gives

$$\frac{\partial \rho}{\partial t} + \nabla \cdot \mathbf{j} = 0 \tag{20.72}$$

where ρ is the probability density and $\mathbf{j}$ is the probability current density. Since the particles undergoing scattering are energy eigenstates, the probability density, ρ, is a constant and its time derivative vanishes. Hence,

$$\nabla \cdot \mathbf{j} = 0. \tag{20.73}$$

The integral of $\nabla \cdot \mathbf{j}$, therefore, also vanishes:

$$\int_V d^3r \, (\nabla \cdot \mathbf{j}) = 0 \tag{20.74}$$

where V is the volume over which the integration takes place. Using Gauss's theorem we can express the above relation as

$$\oint d\mathbf{S} \cdot \mathbf{j} = 0 \tag{20.75}$$

where the integration is carried out over the surface covering the volume V.

The current is given in terms of the wavefunction $u(\mathbf{r})$ as

$$\mathbf{j} = \frac{\hbar}{2im} \left(u^* \nabla u - u \nabla u^* \right) = \frac{\hbar}{m} \operatorname{Im} \left(u^* \nabla u \right) \tag{20.76}$$

where "Im" means imaginary part. Relation (20.75) then implies

$$\operatorname{Im} \oint d\mathbf{S} \cdot \left(u^* \nabla u \right) = 0. \tag{20.77}$$

We show below that this relation leads to the optical theorem.

If we consider V to be a sphere of radius r, then the relation (20.77) becomes

$$\operatorname{Im} \oint \left(u^* \frac{\partial u}{\partial r} \right) r^2 d\,\Omega = 0 \tag{20.78}$$

where $d\Omega$ is the solid angle element. We will now use the expression (20.38) for $u\,(\mathbf{r})$.

The free particle wavefunction corresponding to the incident wave traveling in the z-direction is

$$u_{fr}\,(\mathbf{r}) = \frac{1}{(\sqrt{2\pi})^3} e^{ikz} \tag{20.79}$$

and the outgoing spherical wave is

$$u_{out}(\mathbf{r}) = \frac{1}{(\sqrt{2\pi})^3} f(\theta) \frac{e^{ikr}}{r}. \tag{20.80}$$

The asymptotic expression (20.38) for u is then of the form

$$u = u_{fr} + u_{out}. \tag{20.81}$$

A wavefunction in the absence of potentials will also satisfy (20.78). Therefore, u_{fr} satisfies

$$\text{Im} \oint d\Omega \left(u_{fr}^* \frac{\partial u_{fr}}{\partial r} \right) r^2 = 0. \tag{20.82}$$

Subtracting (20.82) from (20.78), we obtain, using expression (20.81) for u,

$$\text{Im} \oint r^2 d\Omega \left(u_{fr}^* \frac{\partial u_{out}}{\partial r} + u_{out}^* \frac{\partial u_{fr}}{\partial r} + u_{out}^* \frac{\partial u_{out}}{\partial r} \right) = 0. \tag{20.83}$$

To leading order in $1/r$,

$$u_{out}^* \frac{\partial u_{out}}{\partial r} = ik \frac{|f(\theta)|^2}{r^2}. \tag{20.84}$$

The coefficient $(1/2\pi)^3$ is omitted since it can be factored out of the equation. Therefore,

$$\text{Im} \oint r^2 d\Omega \left(u_{out}^* \frac{\partial u_{out}}{\partial r} \right) = k \oint d\Omega\, |f(\theta)|^2. \tag{20.85}$$

The integral on the right-hand side, however, is just the cross-section, σ, hence

$$\text{Im} \oint r^2 d\Omega \left(u_{out}^* \frac{\partial u_{out}}{\partial r} \right) = k\sigma. \tag{20.86}$$

Thus (20.83) can be written as

$$\text{Im} \oint r^2 d\Omega \left(u_{fr}^* \frac{\partial u_{out}}{\partial r} + u_{out}^* \frac{\partial u_{fr}}{\partial r} \right) = -k\sigma. \tag{20.87}$$

Let us first consider the integral on the left-hand side without explicitly writing "Im". It can be written as

$$\oint r^2 d\Omega \left(u_{fr}^* \frac{\partial u_{out}}{\partial r} + u_{out}^* \frac{\partial u_{fr}}{\partial r} \right) = \oint d\mathbf{S} \cdot \left(u_{fr}^* \nabla u_{out} + u_{out}^* \nabla u_{fr} \right). \tag{20.88}$$

We now use the property Im $(A^*B) = -$Im (AB^*) for the second term on the right and then convert the surface integral to a volume integral through Gauss's theorem.

$$\begin{aligned} \oint d\mathbf{S} \cdot \left(u_{fr}^* \nabla u_{out} + u_{out}^* \nabla u_{fr} \right) &= \oint d\mathbf{S} \cdot \left(u_{fr}^* \nabla u_{out} - u_{out} \nabla u_{fr}^* \right) \\ &= \int_V d^3r \left(u_{fr}^* \nabla^2 u_{out} - u_{out} \nabla^2 u_{fr}^* \right). \end{aligned} \tag{20.89}$$

We now write from (20.81)

$$u_{out} = u - u_{fr}, \tag{20.90}$$

and use the result for the free particle wavefunction:

$$\nabla^2 u_{fr} = -k^2 u_{fr}. \tag{20.91}$$

The integral on the right-hand side of (20.89) then becomes

$$\int_V d^3 r u_{fr}^* \left[\nabla^2 u + k^2 u\right] = \frac{2m}{\hbar^2} \int_V d^3 r \, u_{fr}^* V(r) u \tag{20.92}$$

where we have used the fact that u satisfies the Schrödinger equation with $V(r)$ as the potential:

$$\nabla^2 u - \frac{2m}{\hbar^2} V(r) u + k^2 u = 0. \tag{20.93}$$

We note that the scattering amplitude is defined by

$$f(\theta) = -\frac{m}{2\pi\hbar^2} \int_V d^3 r \, e^{-i\mathbf{k}_f \cdot \mathbf{r}} V(r) u(r) \tag{20.94}$$

where θ is the angle between the initial and final momenta $\mathbf{k}_i$ and $\mathbf{k}_f$, while u_{fr} refers to the incident free particle wavefunction, which can be written as

$$u_{fr} = e^{i\mathbf{k}_i \cdot \mathbf{r}}. \tag{20.95}$$

Hence the right-hand side of (20.87) will be proportional to $f(\theta)$ with $\mathbf{k}_f$ replaced by $\mathbf{k}_i$, which will correspond to $\theta = 0$. Thus,

$$\frac{2m}{\hbar^2} \int_V d^3 r \, u_{fr}^* V(r) u = -4\pi f(0). \tag{20.96}$$

Relation (20.87) then gives

$$\operatorname{Im} f(0) = \frac{k}{4\pi} \sigma, \tag{20.97}$$

which is the optical theorem. It is a consequence of probability conservation.

Physically, one can interpret the above relation as being due to the destructive interference between the incident wave and the scattered wave in the probability calculation. This interference term, which is proportional to

$$e^{ikr} . e^{-ikz} = e^{ikr(1-\cos\theta)}, \tag{20.98}$$

oscillates very rapidly as $r \to \infty$ and does not contribute to the solid angle integration at infinity, except at $\theta = 0$. This is the "shadow" region, behind the target. What one finds after imposing probability conservation is that the scattered wave removes from the incident wave an amount proportional to σ, leading to (20.97).

20.8 Absorption

In our previous discussions we found that

$$\oint d\mathbf{S} \cdot \mathbf{j} = 0. \tag{20.99}$$

This corresponds to the fact that the total flux of the particle vanishes, where total flux is defined by

$$\text{flux} = \frac{1}{V} \oint d\mathbf{S} \cdot \mathbf{j}. \tag{20.100}$$

If we take the surface to infinity in the above integral then, physically, this relation corresponds to the observation that there is no net loss of particles. Indeed, that is what was expected since this relation was derived from the probability conservation relation. However, if one found that the above relation was not satisfied and, in fact, one found

$$\oint d\mathbf{S} \cdot \mathbf{j} = \text{negative}, \tag{20.101}$$

then it would imply that there was a net loss of particles. This means that the particles are taken away or absorbed from the incident beam. This is the phenomenon of absorption.

As an example, consider the case where we have two particles A and B and, say, B is very heavy, which we will call the target. A typical scattering process then corresponds to

$$A + B \to A + B. \tag{20.102}$$

If this is all that happens when A scatters off B, then we have a purely elastic scattering. On the other hand, if in addition to (20.102) we also have, for example, the process

$$A + B \to C + B \tag{20.103}$$

then the incident beam represented by A disappears, and in the final state another particle, C, appears. This is the classic case of absorption in a scattering experiment.

A negative flux is generated if the potential is complex,

$$V(r) = V_R(r) - iV_I(r), \tag{20.104}$$

with $V_I > 0$. The conservation equation changes to

$$\frac{\partial \rho}{\partial t} + \nabla \cdot \mathbf{j} = -\frac{2V_I}{\hbar}\rho. \tag{20.105}$$

Since ρ is a constant, we have

$$\nabla \cdot \mathbf{j} = -\frac{2V_I}{\hbar}\rho. \tag{20.106}$$

We then find

$$\oint d\mathbf{S}\cdot\mathbf{j}=-\frac{2}{\hbar}\int_V d^3r\, V_I\rho, \tag{20.107}$$

which is negative. If we take the volume V to be infinite, then the integral on the left will be over a surface at infinity. We thus see that there is a net loss of particles due to the presence of (positively signed) V_I.

The flux of the absorbed particles is

$$-\frac{1}{V}\oint d\mathbf{S}\cdot\mathbf{j} \tag{20.108}$$

where V is, once again, the volume. The incident flux is v/V, where v is the velocity of the incident particle. Therefore, the absorption cross-section can be defined as their ratio,

$$\sigma_{abs}=-\frac{\oint d\mathbf{S}\cdot\mathbf{j}/V}{v/V}=-\frac{1}{v}\oint d\mathbf{S}\cdot\mathbf{j}. \tag{20.109}$$

Substituting $\mathbf{j}$ given by

$$\mathbf{j}=\frac{\hbar}{m}\operatorname{Im}\left(u^*\nabla u\right) \tag{20.110}$$

in (20.109), we obtain

$$\frac{\hbar}{m}\operatorname{Im}\oint d\mathbf{S}\cdot\left(u^*\nabla u\right)=-\mathbf{v}\sigma_{abs}. \tag{20.111}$$

Or, since $mv=\hbar k$, we have

$$\operatorname{Im}\oint d\mathbf{S}\cdot\left(u^*\nabla u\right)=-k\sigma_{abs}. \tag{20.112}$$

From our previous discussion we then have

$$4\pi\operatorname{Im}f(0)-k\sigma=k\sigma_{abs}. \tag{20.113}$$

The cross-section σ should now be identified as the elastic cross-section, σ_{el}, and $f(\theta)$ as the elastic scattering amplitude. In the discussions above we have used the term absorption cross-section, but we could also use the more general term inelastic cross-section, $\sigma_{inel}=\sigma_{abs}$.

Therefore, if we define the total cross-section, σ_T, as

$$\sigma_T=\sigma_{el}+\sigma_{inel} \tag{20.114}$$

then we have

$$\operatorname{Im}f(0)=\frac{k}{4\pi}\sigma_T. \tag{20.115}$$

This is called the generalized optical theorem.

20.9 Relation between the T-matrix and the scattering amplitude

The T-operator was defined earlier as

$$T\left|\phi_0\right\rangle = V\left|\phi\right\rangle. \tag{20.116}$$

We once again substitute $\left|\phi_0\right\rangle = \left|\phi_0(\mathbf{k})\right\rangle$. Then $T\left|\phi_0(\mathbf{k})\right\rangle = V\left|\phi\right\rangle$, and

$$\begin{aligned}\left\langle\phi_0(\mathbf{k}')\right| T\left|\phi_0(\mathbf{k})\right\rangle &= \left\langle\phi_0(\mathbf{k}')\left|V\right|\phi\right\rangle = \int d^3r\, \left\langle\phi_0(\mathbf{k}')\,|\mathbf{r}\right\rangle V(r)\left\langle\mathbf{r}|\phi\right\rangle \\ &= \int d^3r \frac{e^{-i\mathbf{k}'\cdot\mathbf{r}}}{(2\pi)^{3/2}} V(r)\,\phi(\mathbf{r}).\end{aligned} \tag{20.117}$$

The above integral is related to the scattering amplitude, $f(\theta)$, given by (20.39). Therefore,

$$\left\langle\phi_0(\mathbf{k}')\right| T\left|\mathbf{k}\right\rangle = -\frac{1}{(2\pi)^3}\left(\frac{2\pi\hbar^2}{m}\right)f(\theta) \tag{20.118}$$

where θ is now the angle describing the direction of the vector $\mathbf{k}'$.

20.9.1 The optical theorem

We once again derive the optical theorem, this time using the T-matrix formalism without introducing a potential. We begin with the relation (20.21) for $\left|\phi^{(+)}\right\rangle$. We have, writing once again $\left|\phi_0\right\rangle = \left|\phi_0(\mathbf{k})\right\rangle$,

$$\left|\phi_0(\mathbf{k})\right\rangle = \left|\phi^{(+)}\right\rangle - G_0 V\left|\phi^{(+)}\right\rangle \tag{20.119}$$

where

$$G_0 = G_0(E + i\epsilon). \tag{20.120}$$

Taking the complex conjugate of (20.119) we have

$$\left\langle\phi_0(\mathbf{k})\right| = \left\langle\phi^{(+)}\right| - \left\langle\phi^{(+)}\right| VG_0^{\dagger}. \tag{20.120a}$$

Let us now consider scattering in the forward direction, for which $\mathbf{k}' = \mathbf{k}$. Since

$$T\left|\phi_0(\mathbf{k})\right\rangle = V\left|\phi^{(+)}\right\rangle, \tag{20.121}$$

the matrix element of T in the forward direction is

$$\begin{aligned}\left\langle\phi_0(\mathbf{k})\right| T\left|\phi_0(\mathbf{k})\right\rangle &= \left\langle\phi_0(\mathbf{k})\left|V\right|\phi^{(+)}\right\rangle \\ &= \left\langle\phi^{(+)}\left|V\right|\phi^{(+)}\right\rangle - \left\langle\phi^{(+)}\left|VG_0^{\dagger}V\right|\phi^{(+)}\right\rangle\end{aligned} \tag{20.122}$$

where we have used (20.120a). To obtain the optical theorem, we need to obtain the imaginary part of $\langle\phi_0(\mathbf{k})|\,T\,|\phi_0(\mathbf{k})\rangle$. We note that since V is Hermitian,

$$\left\langle \phi^{(+)}\,|V|\,\phi^{(+)}\right\rangle^{\dagger} = \left\langle \phi^{(+)}\,|V|\,\phi^{(+)}\right\rangle. \tag{20.123}$$

Hence $\langle \phi^{(+)}\,|V|\,\phi^{(+)}\rangle$ is real. The second term in (20.122), however, can be complex. We write

$$\left\langle \phi^{(+)}\left|VG_0^{\dagger}V\right|\phi^{(+)}\right\rangle = \left\langle \phi^{(+)}\left|V\left(\frac{1}{E-H_0-i\epsilon}\right)V\right|\phi^{(+)}\right\rangle. \tag{20.124}$$

But

$$\frac{1}{E-H_0-i\epsilon} = \left[P\left(\frac{1}{E-H_0}\right) + i\pi\,\delta\left(E-H_0\right)\right]. \tag{20.125}$$

We take the imaginary part of both sides of (20.122) and note that since the first term is the principal part and hence real, only the second term will contribute:

$$\begin{aligned}\operatorname{Im}\langle\phi_0(\mathbf{k})|\,T\,|\phi_0(\mathbf{k})\rangle &= -\pi\left\langle \phi^{(+)}\,|V\delta\left(H_0-E\right)V|\,\phi^{(+)}\right\rangle\\ &= -\pi\left\langle\phi_0(\mathbf{k})\right|T^{\dagger}\delta\left(H_0-E\right)T\left|\phi_0(\mathbf{k})\right\rangle.\end{aligned} \tag{20.126}$$

One can write, after inserting complete sets of sets,

$$\begin{aligned}&\langle\phi_0(\mathbf{k})|\,T^{\dagger}\delta\left(H_0-E\right)T\,|\phi_0(\mathbf{k})\rangle\\ &= \int d^3k'\int d^3k''\,\langle\phi_0(\mathbf{k})|\,T^{\dagger}\,|\phi_0(\mathbf{k}')\rangle\langle\phi_0(\mathbf{k}')|\,\delta\left(H_0-E\right)|\phi_0(\mathbf{k}'')\rangle\langle\phi_0(\mathbf{k}')|\,T\,|\phi_0(\mathbf{k})\rangle.\end{aligned} \tag{20.127}$$

Now,

$$\begin{aligned}\langle\phi_0(\mathbf{k}')|\,\delta\left(H_0-E\right)|\phi_0(\mathbf{k}'')\rangle &= \langle\phi_0(\mathbf{k}')\,|\phi_0(\mathbf{k}'')\rangle\,\delta\left(\frac{\hbar^2k'^2}{2m}-\frac{\hbar^2k^2}{2m}\right)\\ &= \delta^{(3)}\left(\mathbf{k}'-\mathbf{k}''\right)\left(\frac{2m}{\hbar^2}\right)\delta\left(k'^2-k^2\right).\end{aligned} \tag{20.128}$$

Therefore, from (20.126) the imaginary part is given by

$$\operatorname{Im}\langle\phi_0(\mathbf{k})|\,T\,|\phi_0(\mathbf{k})\rangle = -\frac{2\pi m}{\hbar^2}\int d^3k'\,\langle\phi_0(\mathbf{k})|\,T^{+}\,|\phi_0(\mathbf{k}')\rangle\langle\phi_0(\mathbf{k}')|\,T\,|\phi_0(\mathbf{k})\rangle\,\delta(k'^2-k^2). \tag{20.129}$$

We note that $\int d^3k' = \int_0^{\infty} k'^2dk'\int d\Omega'$ and

$$\delta\left(k'^2-k^2\right) = \frac{\delta\left(k'-k\right)+\delta\left(k'+k\right)}{2k} \tag{20.130}$$

where, by definition, k and k' are positive. Therefore, the second delta-function will not contribute to the integral. We then have

$$\text{Im}\left\langle \phi_0(\mathbf{k}) \right| T \left| \phi_0(\mathbf{k}) \right\rangle = -\left(\frac{\pi mk}{\hbar^2}\right) \int d\Omega' \left| \left\langle \phi_0(\mathbf{k}) \right| T \left| \phi_0(\mathbf{k}') \right\rangle \right|^2. \tag{20.131}$$

Using the relation between $f(\theta)$ and T, and the expression for the total cross-section σ,

$$\sigma = \int d\Omega \left| f(\theta) \right|^2, \tag{20.132}$$

we conclude that

$$\text{Im} f(0) = \frac{k}{4\pi}\sigma. \tag{20.133}$$

This is the famous optical theorem. This time it is derived by using the Green's function formalism.

20.10 The *S*-matrix

20.10.1 Basic Formalism

We now consider the time-dependent Schrödinger equation. Let $\psi(\mathbf{r}, t)$ be the wavefunction. It will satisfy the equation

$$i\hbar \frac{\partial \psi(\mathbf{r}, t)}{\partial t} = H\psi(\mathbf{r}, t) \tag{20.134}$$

where

$$H = H_0 + V \quad \text{and} \quad H_0 = \frac{\mathbf{p}^2}{2m} \tag{20.135}$$

where $\mathbf{p}$ is the momentum operator, H_0 is the free particle Hamiltonian, and V is the potential, which we assume to be a function of $r = |\mathbf{r}|$ only.

We write the solution for $\psi(\mathbf{r}, t)$ in the Green's function formalism,

$$\psi(\mathbf{r}, t) = \psi_0(\mathbf{r}, t) + \int d^3r' \int dt' G_0(\mathbf{r}, t; \mathbf{r}', t') V(r') \psi(\mathbf{r}', t'), \tag{20.136}$$

where G_0 is the Green's function, which now involves time, and ψ_0 is the homogeneous solution which satisfies the time-dependent equation

$$i\hbar \frac{\partial \psi_0(\mathbf{r}, t)}{\partial t} = H_0 \psi_0(\mathbf{r}, t). \tag{20.137}$$

We assume that $V(r)$ vanishes for large values of r. Inserting (20.136) into (20.134) and using (20.135) and (20.137), we find that G_0 must satisfy the equation

$$\left(i\hbar\frac{\partial}{\partial t} - H_0\right) G_0(\mathbf{r}, t; \mathbf{r}', t') = \delta^{(3)}(\mathbf{r} - \mathbf{r}')\delta(t - t'). \tag{20.138}$$

As in the previous problems, we note that G_0 will depend only on the differences of the coordinates,

$$G_0(\mathbf{r}, t; \mathbf{r}', t') = G_0(\mathbf{r} - \mathbf{r}', t - t'). \tag{20.139}$$

To simplify the equations we take $\mathbf{r}' = 0$ and $t' = 0$ and write

$$G_0 = G_0(\mathbf{r}, t). \tag{20.140}$$

At the end of the calculation we will replace $\mathbf{r}$ by $\mathbf{r} - \mathbf{r}'$ and t by $t - t'$. Thus (20.138) gives

$$\left(i\hbar\frac{\partial}{\partial t} - H_0\right) G_0(\mathbf{r}, t) = \delta^{(3)}(\mathbf{r})\delta(t). \tag{20.141}$$

Writing G_0 in the form of a Fourier transform, we have

$$G_0(\mathbf{r}, t) = \frac{1}{(\sqrt{2\pi})^4}\int d^3k \int d\omega\, e^{i(\mathbf{k}\cdot\mathbf{r} - \omega t)} g_0(\mathbf{k}, \omega). \tag{20.142}$$

We are already familiar with the expression for the δ-function in the form of a Fourier transform. The product of the δ-functions on the right-hand side of (20.141) is then

$$\begin{aligned}\delta^{(3)}(\mathbf{r})\delta(t) &= \left[\frac{1}{(2\pi)^3}\int d^3k\, e^{i\mathbf{k}\cdot\mathbf{r}}\right]\left[\frac{1}{2\pi}\int d\omega\, e^{-i\omega t}\right] \\ &= \frac{1}{(2\pi)^4}\int d^3k \int d\omega\, e^{i(\mathbf{k}\cdot\mathbf{r} - \omega t)}. \end{aligned} \tag{20.143}$$

Substituting the two Fourier transform equations (20.142) and (20.143) in (20.141), we find

$$\frac{1}{(\sqrt{2\pi})^4}\int d^3k \int d\omega \left[\hbar\omega - \frac{\hbar^2 k^2}{2m}\right] e^{i(\mathbf{k}\cdot\mathbf{r} - \omega t)} g_0(\mathbf{k}, \omega) = \frac{1}{(2\pi)^4}\int d^3k \int d\omega\, e^{i(\mathbf{k}\cdot\mathbf{r} - \omega t)}. \tag{20.144}$$

Thus,

$$g_0(k, \omega) = \frac{1}{(\sqrt{2\pi})^4}\left(\frac{1}{\hbar\omega - \dfrac{\hbar^2 k^2}{2m}}\right) \tag{20.145}$$

and $G_0(\mathbf{r}, t)$ is given by

$$G_0(\mathbf{r}, t) = \frac{1}{(2\pi)^4}\int d^3k\, e^{i\mathbf{k}\cdot\mathbf{r}} \int_{-\infty}^{\infty} d\omega \frac{e^{-i\omega t}}{\left(\hbar\omega - \dfrac{\hbar^2 k^2}{2m}\right)}. \tag{20.146}$$

Equation (20.136) implies that the Green's function $G_0(\mathbf{r}, t; \mathbf{r}', t')$ connects or transfers information from the space-time points $(\mathbf{r}', t')$ to the point $(\mathbf{r}, t)$. If we identify the events occurring at t' as "cause," then these must precede the "effect" occurring at t as described by the wavefunction $\psi(\mathbf{r}, t)$. Thus, we impose the condition that $G_0(\mathbf{r}, t; \mathbf{r}', t')$ must vanish unless $t < t'$. This is the statement of causality. For $G_0(\mathbf{r}, t)$ it implies that

$$G_0(\mathbf{r}, t) = 0 \quad \text{and} \quad t < 0. \tag{20.147}$$

The integral over ω in (20.146) for $G_0(\mathbf{r}, t)$ can be obtained through Cauchy's residue theorem. To implement the causality condition, which involves $t < 0$, we must choose the upper half of the complex ω-plane to do the integration since the integrand vanishes along the infinite semi-circle in the upper half-plane when $t < 0$. Since $G_0 = 0$ for $t < 0$, the pole in (20.146) which is on the real axis must be moved to the lower half-plane. It will then not contribute to the Cauchy's integral, which is along the closed contour formed by the real axis and the infinite semi-circle. To accomplish all this we write

$$G_0(\mathbf{r}, t) = \frac{1}{(2\pi)^4} \int d^3k\, e^{i\mathbf{k}\cdot\mathbf{r}} \int_{-\infty}^{\infty} d\omega \frac{e^{-i\omega t}}{(\hbar\omega - \hbar\omega_k + i\epsilon)} \tag{20.148}$$

where, to simplify the calculations, we have defined $\hbar\omega_k = \hbar^2 k^2/2m$. Cauchy's theorem then gives

$$\int_{-\infty}^{\infty} d\omega \frac{e^{-i\omega t}}{\hbar\omega - \hbar\omega_k + i\epsilon} = 0, \qquad \text{if } t < 0 \tag{20.149}$$

$$= \frac{-2\pi i}{\hbar} e^{-i\omega_k t}, \qquad \text{if } t > 0. \tag{20.150}$$

The expression for $G_0(\mathbf{r}, t)$ will then be

$$G_0(\mathbf{r}, t) = \frac{-i\theta(t)}{(2\pi)^3 \hbar} \int d^3k\, e^{i(\mathbf{k}\cdot\mathbf{r} - \omega_k t)} \tag{20.151}$$

where $\theta(t) = 1$ for $t > 0$, and $\theta(t) = 0$ for $t < 0$.

Returning to $G_0(\mathbf{r}, t; \mathbf{r}', t')$ by making the replacements $\mathbf{r} \to \mathbf{r} - \mathbf{r}'$ and $t \to t - t'$, we obtain

$$G_0(\mathbf{r}, t; \mathbf{r}', t') = \frac{-i\theta(t - t')}{(2\pi)^3 \hbar} \int d^3k\, e^{i\mathbf{k}\cdot(\mathbf{r} - \mathbf{r}')} e^{-i\omega_k (t - t')}. \tag{20.152}$$

The free particle wavefunction with momentum $\hbar\mathbf{k}$, and energy $\hbar\omega_k (= \hbar^2 k^2/2m)$ is given by

$$\psi_{0k}(\mathbf{r}, t) = \frac{1}{\left(\sqrt{2\pi}\right)^3 \hbar} e^{i(\mathbf{k}\cdot\mathbf{r} - \omega_k t)}. \tag{20.153}$$

Thus G_0 can be written as

$$G_0(\mathbf{r}, t; \mathbf{r}', t') = -i\frac{\theta(t - t')}{\hbar} \int d^3k \psi_{0k}^*(\mathbf{r}', t') \psi_{0k}(\mathbf{r}, t). \tag{20.154}$$

This G_0 is the same as the Feynman propagator defined in Chapter 1. For our purposes, it can be left in the integral form.

20.10.2 The S-matrix

Consider a situation in which the particle at $t \to -\infty$ is given by a free wavefunction of momentum k_i. Let us denote this wavefunction $\psi_{oi}(\mathbf{r}, t)$ and the complete wavefunction as $\psi_i(\mathbf{r}, t)$. Then

$$\psi_i(\mathbf{r}, t) \to \psi_{oi}(\mathbf{r}, t) = \frac{1}{(\sqrt{2\pi})^3} e^{i(\mathbf{k}_i \cdot \mathbf{r} - \omega_{ik} t)} \quad \text{as } t \to -\infty. \tag{20.155}$$

Let us rewrite the integral equation (20.136) by replacing $\psi(\mathbf{r}, t)$ by $\psi_i(\mathbf{r}, t)$ and $\psi_0(\mathbf{r}, t)$ by $\psi_{oi}(\mathbf{r}, t)$

$$\psi_i(\mathbf{r}, t) = \psi_{oi}(\mathbf{r}, t) + \int d^3r' \int dt' G_0(\mathbf{r}, t; \mathbf{r}', t') V(\mathbf{r}') \psi_i(\mathbf{r}', t). \tag{20.156}$$

The above equation with subscript i in place describes the following physical situation: an incident particle starts out at $t = -\infty$ as a free particle given by (20.155), since $G_0 = 0$ at that point. As it moves forward in time t, it begins to undergo interaction. Its wavefunction at every stage will be given by the above integral equation. After undergoing interaction, this particle, at $t \to +\infty$, will emerge, once again, as a free particle (note that we have assumed V to vanish at $r' = \infty$), which can be expressed as

$$\psi_{of}(\mathbf{r}, t) = \frac{1}{(\sqrt{2\pi})^3} e^{i(\mathbf{k}_f \cdot \mathbf{r} - \omega_f t)} \tag{20.157}$$

with the subscript f signifying a final free particle but with $\mathbf{k}_f$ not necessarily in the same direction as $\mathbf{k}_i$.

The probability that $\psi_i(\mathbf{r}, t)$ will emerge at $t \to +\infty$ as a free particle with wavefunction $\psi_{of}(\mathbf{r}, t)$ is given by the so-called S-matrix, defined as

$$S_{fi} = \lim_{t \to +\infty} \int d^3r \, \psi^*_{of}(\mathbf{r}, t) \psi_i(\mathbf{r}, t). \tag{20.158}$$

We note that as $t \to +\infty$, the θ-function in the expression for G_0 in (20.154) will be unity and hence in that limit

$$\psi_i(\mathbf{r}, t) = \psi_{oi}(\mathbf{r}, t) + \frac{1}{\hbar} \int d^3r' \int dt' \left[(-i) \int d^3k \psi^*_{ok}(\mathbf{r}', t') \psi_{ok}(\mathbf{r}, t) \right] V(\mathbf{r}') \psi_i(\mathbf{r}', t'). \tag{20.159}$$

The S-matrix then reads

$$S_{fi} = \int d^3r\, \psi^*_{of}(\mathbf{r}, t)\psi_{oi}(\mathbf{r}, t) \tag{20.160}$$

$$+ \frac{1}{\hbar}\int d^3r \int d^3r' \int dt' \int d^3k\, \psi^*_{of}(\mathbf{r}, t)\psi^*_{ok}(\mathbf{r}', t')\psi_{ok}(\mathbf{r}, t)V(\mathbf{r}')\psi_i(\mathbf{r}', t'). \tag{20.161}$$

We note that

$$\int d^3r\, \psi^*_{of}(\mathbf{r}, t)\psi_{oi}(\mathbf{r}, t) = \delta^{(3)}(\mathbf{k}_f - \mathbf{k}_i), \tag{20.162}$$

which simplifies the first term in S_{fi}. For the second term we use the relations

$$\int d^3r\, \psi^*_{of}(\mathbf{r}, t)\psi_{ok}(\mathbf{r}, t) = \delta^{(3)}(\mathbf{k}_f - \mathbf{k}), \tag{20.163}$$

$$\int d^3k\, \psi^*_{ok}(\mathbf{r}', t')\delta^{(3)}(\mathbf{k}_f - \mathbf{k}) = \psi^*_{of}(\mathbf{r}', t'). \tag{20.164}$$

Hence,

$$S_{fi} = \delta^{(3)}(\mathbf{k}_f - \mathbf{k}_i) - \frac{i}{\hbar}\int d^3r' \int dt'\, \psi^*_{of}(\mathbf{r}', t')V(r')\psi_i(\mathbf{r}', t'). \tag{20.165}$$

This is then the expression for the S-matrix for the scattering of a free particle under the influence of the potential $V(r')$.

20.11 Unitarity of the S-matrix and the relation between S and T

Once again we will make explicit the fact that we are concerned only with the outgoing waves by introducing the superscript (+). Thus,

$$\psi^{(+)}(\mathbf{r}, t) \to e^{i\mathbf{k}.\mathbf{r}}e^{-\mathbf{i}\omega\mathbf{t}} \tag{20.166}$$

as $r(= |\mathbf{r}|)$ and t both go to ∞. The S-matrix written in (20.165) can be defined in terms of this wavefunction by replacing $\psi_i(\mathbf{r}', t')$ by $\psi_i^{(+)}(\mathbf{r}', t')$. We will then have

$$S_{fi} = \delta^{(3)}(\mathbf{k}_f - \mathbf{k}_i) - \frac{i}{\hbar}\int d^3r' \int dt'\, \psi^*_{of}(\mathbf{r}', t')V(r')\psi_i^{(+)}(\mathbf{r}', t'). \tag{20.167}$$

It will be more convenient at this stage to invoke the abstract state vectors. Let us define, for the purposes of clarity, the initial and final free particle states respectively, as follows:

$$|i(t)\rangle = |i\rangle e^{-iE_i t/\hbar}, \quad |f(t)\rangle = |f\rangle e^{-iE_f t/\hbar} \tag{20.168}$$

where $|i\rangle$ and $|f\rangle$ represent the $t = 0$ values of the state vectors.

Therefore,

$$\psi_{oi}(\mathbf{r}, t) = \langle \mathbf{r}|i(t)\rangle \quad \text{and} \quad \psi_{of}(\mathbf{r}, t) = \langle \mathbf{r}|f(t)\rangle. \tag{20.169}$$

We will define the state vector corresponding to $\psi^{(+)}(\mathbf{r}, t)$ as $|\psi^{(+)}(t)\rangle$. Hence,

$$\psi_i^{(+)}(\mathbf{r}, t) = \langle \mathbf{r}|\psi_i^{(+)}(t)\rangle \quad \text{and} \quad \psi_f^{(+)}(\mathbf{r}, t) = \langle \mathbf{r}|\psi_f^{(+)}(t)\rangle. \tag{20.170}$$

In terms of the total Green's function $G_i^{(+)}$ we write

$$|\psi_i^{(+)}\rangle = |i\rangle + G_i^{(+)} V|i\rangle, \tag{20.171}$$

$$|\psi_f^{(+)}\rangle = |f\rangle + G_f^{(+)} V|f\rangle, \tag{20.172}$$

with

$$G^{(+)} = \frac{1}{E - H + i\epsilon} \tag{20.173}$$

where H is the total Hamiltonian $H = H_0 + V$. One can put the appropriate subscripts for $G^{(+)}$ and E corresponding to whether we have the initial (i) or final (f) states.

We write the S-matrix in the following abbreviated form

$$S_{fi} = \delta_{fi} - \frac{i}{\hbar} \int_{-\infty}^{\infty} dt' \langle f(t')\,|V|\,\psi_i^{(+)}(t')\rangle. \tag{20.174}$$

One can also write it in terms of the T-matrix, defined in (20.116), as

$$S_{fi} = \delta_{fi} - \frac{i}{\hbar} \int_{-\infty}^{\infty} dt' \langle f(t')\,|T|\,i(t')\rangle. \tag{20.175}$$

Using the relation (20.168) and the integral

$$\int_{-\infty}^{\infty} dt' e^{i(E_f - E_i)t'/\hbar} = 2\pi\hbar\delta(E_f - E_i), \tag{20.176}$$

the equation (20.175) for the S-matrix can be written as

$$S_{fi} = \delta_{fi} - i2\pi\delta(E_f - E_i)T_{fi} \tag{20.177}$$

where

$$T_{fi} = \langle f\,|T|\,i\rangle. \tag{20.178}$$

Let us write

$$S_{fi} = \langle f | S | i \rangle. \tag{20.179}$$

We show below that the S-matrix is unitary, i.e.,

$$SS^\dagger = \mathbf{1}. \tag{20.180}$$

This equation will also lead to some important relations involving the T-matrix.

Taking the matrix element of $SS^\dagger$ and expressing it in terms of a complete set of states, we write

$$\begin{aligned}\langle f \left| SS^\dagger \right| i \rangle &= \sum_n \langle f | S | n \rangle \langle n \left| S^\dagger \right| i \rangle \\ &= \sum_n S_{fn} S^*_{in} \\ &= \sum_n \left\{ \left[\delta_{fn} - 2\pi i \delta(E_f - E_n) T_{fn} \right] \right. & (20.181) \\ &\qquad \left. \times \left[\delta_{in} + 2\pi i \delta(E_i - E_n) T^*_{in} \right] \right\} & (20.182)\end{aligned}$$

where we have inserted the expression (20.177) for the S-matrix. We note that

$$\sum_n \delta_{fn} \delta_{in} = \delta_{fi}.$$

Hence we obtain

$$\begin{aligned}\langle f \left| SS^\dagger \right| i \rangle = \delta_{fi} - 2\pi i \delta(E_f - E_i) \left[T_{fi} - T^*_{if} \right] & \quad (20.183) \\ - \sum_n \left[2\pi i \delta(E_f - E_n) T_{fn} \right] \left[2\pi i \delta(E_i - E_n) T^*_{in} \right]. & \quad (20.184)\end{aligned}$$

The third term (including the minus sign) on the right-hand side of the above equation sums to

$$4\pi^2 \delta(E_f - E_i) \sum_n \delta(E_f - E_n) T_{fn} T^*_{in}. \tag{20.185}$$

Let us now calculate the second term in the square bracket in (20.183). We note that

$$T_{fi} = \langle f | V | \psi_i^{(+)} \rangle \quad \text{and} \quad T^*_{if} = \langle i | V | \psi_f^{(+)} \rangle^* = \langle \psi_f^{(+)} | V | i \rangle. \tag{20.186}$$

In the last relation we have made use of the fact that the operator corresponding to V is Hermitian, $V^\dagger = V$. Hence,

$$T_{fi} - T^*_{if} = \langle f | V | \psi_i^{(+)} \rangle - \langle \psi_f^{(+)} | V | i \rangle = \langle f \left| V G_i^{(+)} V \right| i \rangle - \langle f \left| V G_f^{(+)\dagger} V \right| i \rangle \tag{20.187}$$

where we have used the relations (20.171) and (20.172) for $\left|\psi_i^{(+)}\right\rangle$ and $\left\langle\psi_f^{(+)}\right|$. Thus we have, after inserting the expression (20.173) for G, the following:

$$T_{fi} - T_{if}^* = \left\langle f \left| V \frac{1}{E_i - H + i\epsilon} V \right| i \right\rangle - \left\langle f \left| V \frac{1}{E_f - H - i\epsilon} V \right| i \right\rangle$$
$$= \left\langle f \left| V \left[\frac{1}{E_i - H + i\epsilon} - \frac{1}{E_f - H - i\epsilon} \right] \right| i \right\rangle. \tag{20.188}$$

We note that there is a factor $\delta(E_f - E_i)$ multiplying $(T_{fi} - T_{if}^*)$ in the second term of the relation (20.183) for $\langle f \left| SS^\dagger \right| i\rangle$. Thus we must put $E_i = E_f$ in (20.188). We write

$$\frac{1}{E_i - H \pm i\epsilon} = P\left(\frac{1}{E_i - H}\right) \mp i\pi.\delta(E_i - H). \tag{20.189}$$

The contributions from the principal parts vanish in the difference, but the imaginary parts add.

Thus we have

$$T_{fi} - T_{if}^* = -2i\pi \left\{\langle f \left| V\delta(E_i - H)V \right| i\rangle\right\}$$
$$= -2i\pi \left\{ \sum_n \langle f \left| V \right| \psi_n^{(+)}\rangle\langle\psi_n^{(+)} \left| V \right| i\rangle\delta(E_i - E_n) \right\} \tag{20.190}$$

where we have made note of the fact that

$$H|\psi_n^{(+)}\rangle = E_n|\psi_n^{(+)}\rangle. \tag{20.191}$$

The right-hand side of (20.190) is then equal to

$$-2i\pi \sum_n \delta(E_i - E_n) T_{fn} T_{in}^*. \tag{20.192}$$

However, this term exactly cancels the term in the square bracket in the expression (20.184) for $\langle f \left| SS^\dagger \right| i\rangle$. From this we obtain

$$\langle f \left| SS^\dagger \right| i\rangle = \delta_{fi} \tag{20.193}$$

and hence the S-matrix satisfies the unitarity condition

$$SS^\dagger = \mathbf{I}, \tag{20.194}$$

while the T-matrix satisfies

$$T_{if} - T_{if}^* = -2i\pi \sum_n \delta(E_i - E_n) T_{fn} T_{in}^*. \tag{20.195}$$

20.12 Properties of the *T*-matrix and the optical theorem (again)

In the relation (20.195) for the T-matrix we note that $E_f = E_i$. For the case of forward scattering, $f = i$, the relation (20.195) gives

$$\text{Im}\, T_{ii} = -\pi \sum_n \delta(E_i - E_n) \left| T_{in} \right|^2 . \tag{20.196}$$

From the results given by (20.118) we also know that

$$T_{fi} = -\frac{\hbar^2}{4\pi^2 m} f(\theta) \tag{20.197}$$

where θ is the angle between $\mathbf{k}_i$ and $\mathbf{k}_f$. Thus,

$$T_{ii} = -\frac{\hbar^2}{4\pi^2 m} f(0) \tag{20.198}$$

where $f(0)$ corresponds to the forward scattering amplitude. The above relation gives

$$\begin{aligned} -\frac{\hbar^2}{4\pi^2 m}\, \text{Im} f(0) &= -\pi \left(\frac{\hbar^2}{4\pi^2 m} \right)^2 \int d^3k_n \, \delta\left(\frac{\hbar^2 k^2}{2m} - \frac{\hbar^2 k_n^2}{2m} \right) \left| f(\theta') \right|^2 \\ &= -\left(\frac{1}{4\pi} \right) \int \frac{k_n}{2} dk_n^2 \, d\Omega' \delta(k^2 - k_n^2) \left(\frac{\hbar^2}{4\pi^2 m} \right) \left| f(\theta') \right|^2 \end{aligned} \tag{20.199}$$

which gives

$$\text{Im} f(0) = \left(\frac{1}{4\pi} \right) \int k_n \, dk_n^2 \, d\Omega' \delta(k^2 - k_n^2) \left| f(\theta') \right|^2 \tag{20.200}$$

where θ' = polar angle of k_n and $d^3k_n = k_n^2 \, dk_n \, d\Omega'$. From this relation we can once again derive the famous optical theorem

$$\text{Im} f(0) = \frac{k}{4\pi} \sigma \tag{20.201}$$

where σ is the total cross-section given by

$$\sigma = \int d\Omega' \left| f(\theta') \right|^2 . \tag{20.202}$$

20.13 Appendix to Chapter 20

20.13.1 Integrals involved in Green's function calculations

Consider the following integral, where α is a real quantity,

$$I = \int_{-\infty}^{\infty} dx\, f(x) e^{i\alpha x}. \tag{20.203}$$

For $\alpha > 0$, one can express this integral in the complex z-plane as

$$I = \int_{C_1} dz\, f(z) e^{i\alpha z} \tag{20.204}$$

where C_1 is a contour taken in a counter-clockwise direction consisting of the real axis $(-\infty, \infty)$ and the upper-half infinite semi-circle.

For $\alpha < 0$, we will have

$$I = \int_{C_2} dz\, f(z) e^{i\alpha z} \tag{20.205}$$

where C_2 is a contour consisting of the real axis $(-\infty, \infty)$ and the lower-half infinite semi-circle, in a clockwise direction.

The proof is actually quite straightforward. First let us consider, $\alpha > 0$. One can then write for the right-hand side of (20.204),

$$I = \int_{C_1} dz f(z) e^{i\alpha z} = \int_{-\infty}^{\infty} dx f(x) e^{i\alpha x} + \int_{\Gamma} dz f(z) e^{i\alpha z} \tag{20.206}$$

where Γ is an infinite semi-circle in the upper half-plane and where one can write

$$z = \rho e^{i\theta} = \rho \left(\cos\theta + i \sin\theta\right). \tag{20.207}$$

We have

$$\int_{\Gamma} dz f(z) e^{i\alpha z} = \lim_{\rho\to\infty} i\rho \int_{0}^{\pi} d\theta\, e^{i\theta} f(\rho e^{i\theta}) e^{i\alpha\rho\cos\theta} e^{-\alpha\rho\sin\theta}. \tag{20.208}$$

For $\alpha > 0$, the second exponential factor in the integral above vanishes in the limit $\rho \to \infty$ as long as $f(\rho e^{i\theta})$ is a well-behaved function. Thus the integral over Γ vanishes and

$$\int_{-\infty}^{\infty} dx f(x) e^{i\alpha x} = \int_{C_1} dz f(z) e^{i\alpha z}. \tag{20.209}$$

We can prove (20.205) in a similar manner, for $\alpha < 0$, by taking the lower half-plane. We note that we can not take the contour C_1 because then $\exp(i\alpha z)$ will blow up at infinity along the upper-half contour Γ. Thus, once we know the sign of α, we have only one choice.

Consider now the following integral that occurs typically in Green's function calculations,

$$\lim_{\epsilon\to 0}\int_{-\infty}^{\infty} dx\frac{e^{i\alpha x}}{x-(x_0+i\epsilon)} \tag{20.210}$$

where x_0 is a real quantity and ϵ is an infinitesimally small, positive, quantity. To evaluate the integral we use Cauchy's residue theorem

$$\int_C dz\frac{\phi(z)}{z-z_0} = 2\pi i\phi(z_0) \tag{20.211}$$

where z_0 lies inside the contour C.

If $\alpha > 0$ then from (20.204) we can write, with the same definition for C_1,

$$\lim_{\epsilon\to 0}\int_{-\infty}^{\infty} dx\frac{e^{i\alpha x}}{x-(x_0+i\epsilon)} = \lim_{\epsilon\to 0}\int_{C_1} dz\frac{e^{i\alpha z}}{z-(x_0+i\epsilon)} = 2\pi i e^{i\alpha x_0} \tag{20.212}$$

where we have used the Cauchy theorem (20.211) with $z_0 = x_0 + i\epsilon$.

For $\alpha < 0$, we must use the contour C_2 (and not C_1). We then find

$$\lim_{\epsilon\to 0}\int_{-\infty}^{\infty} dx\frac{e^{i\alpha x}}{x-(x_0+i\epsilon)} = \lim_{\epsilon\to 0}\int_{C_2} dz\frac{e^{i\alpha z}}{z-(x_0+i\epsilon)} = 0. \tag{20.213}$$

The integral now vanishes because the pole, $z_0 = x_0 + i\epsilon$, lies outside the contour C_2.

The conclusions will be exactly the opposite, apart from overall sign, for $z_0 = x_0 - i\epsilon$.

20.14 Problems

1. Obtain the low-energy cross-section for the potential given by

$$\begin{aligned} V(r) &= -V_0, \quad \text{for } r < a \\ &= 0, \quad \text{for } r > a \end{aligned}$$

2. Obtain the scattering amplitude and cross-section in the Born approximation for the potential given by $V(\mathbf{r}) = g\delta^{(3)}(\mathbf{r})$.
3. Answer the same question as above for $V(\mathbf{r}) = g\delta(r-a)$.

4. Obtain the scattering amplitude and cross-section in the Born approximation for the potential

$$V(r) = ge^{-r^2/a^2}.$$

5. Consider a scattering of two identical particles of spin ½ through the Yukawa potential given by

$$V(r) = g\frac{e^{-\mu r}}{r}.$$

Obtain the cross-sections in the Born approximation when the particles are in spin-symmetric and in spin-antisymmetric states, respectively.

6. A spin ½ particle scatters off a heavy spin ½ target particle through the potential

$$V(r) = g\boldsymbol{\sigma}_1 \cdot \boldsymbol{\sigma}_2 \frac{e^{-\mu r}}{r}$$

where $\boldsymbol{\sigma}_1$ and $\boldsymbol{\sigma}_2$ are the Pauli matrices representing the spins of the particles. Determine the scattering amplitude for the case when the two particles have total spin $S = 0$ and $S = 1$, respectively. Designate the corresponding scattering amplitudes by $f_0(\theta)$ and $f_1(\theta)$, respectively, and obtain the differential cross-sections in terms of these amplitudes for different spin orientation of individual particles. Also obtain the differential cross-section after summing over final and averaging over the initial spin states.

7. Consider the scattering of a particle by a bound system described by the wavefunction $\phi(\mathbf{r}) = a\exp(-r^2/\beta^2)$. If the interaction potential between the two is given by

$$V(\mathbf{r}) = \lambda\delta^{(3)}(\mathbf{r}),$$

obtain the differential scattering cross-section in the Born approximation.

8. For the potential

$$\begin{aligned} V &= -V_0, \quad r < a \\ &= 0, \qquad r > a \end{aligned}$$

obtain the scattering amplitude, $f_B(\theta)$, for $\theta = 0$ and $\theta = \pi$.

21 Partial wave amplitudes and phase shifts

This chapter is devoted to the consequences of scattering due to the presence of a potential that is spherically symmetric. This leads to solving radial equations in terms of angular momentum and the partial waves. We find that the scattering problem can be described one partial wave at a time in terms of the newly defined quantity, the phase shift.

21.1 Scattering amplitude in terms of phase shifts

We consider the Schrödinger equation in three dimensions in the presence of a spherically symmetric potential, $V(r)$:

$$-\frac{\hbar^2}{2m}\nabla^2 u(\mathbf{r}) + V(r)\, u(\mathbf{r}) = Eu(\mathbf{r}). \tag{21.1}$$

We found in the previous chapter that for $V(r)$ going to zero faster than $1/r^2$, one can write the asymptotic behavior of $u(\mathbf{r})$ in terms of the scattering amplitude $f(\theta)$ as follows:

$$u(\mathbf{r}) \to \frac{1}{(\sqrt{2\pi})^3}\left[e^{ikz} + f(\theta)\frac{e^{ikr}}{r}\right] \tag{21.2}$$

where the incoming particle is traveling along the z-direction. This is described pictorially in Fig. 21.1.

In the following we will consider the same problem in spherical coordinates. First we express $\exp(ikz)$ as

$$e^{ikz} = \sum_{l=0}^{\infty}(2l+1)i^l j_l(kr)P_l(\cos\theta) \tag{21.3}$$

and, similarly,

$$u(\mathbf{r}) = \frac{1}{(\sqrt{2\pi})^3}\sum_{l=0}^{\infty}(2l+1)i^l R_l(r)\, P_l(\cos\theta) \tag{21.4}$$

where R_l is the radial wavefunction.

We will consider the asymptotic behavior of $R_l(r)$ and define the corresponding partial wave scattering amplitude, which in practical situations is often more useful than the amplitude $f(\theta)$.

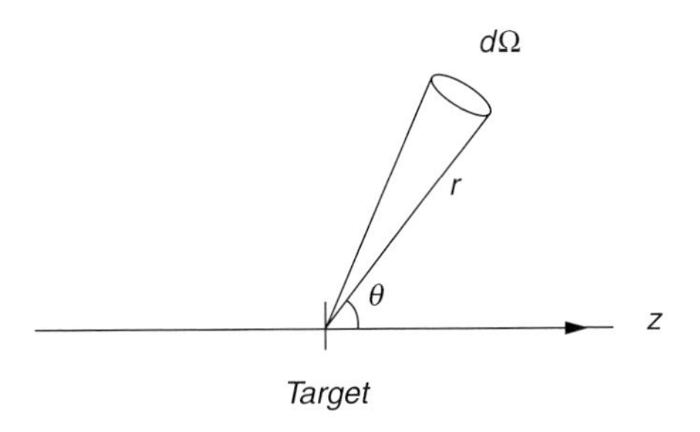

Fig. 21.1

We start with the radial Schrödinger equation,

$$-\frac{\hbar^2}{2m}\frac{1}{r}\frac{d^2(rR_l)}{dr^2}+\left[V(r)+\frac{\hbar^2 l(l+1)}{2mr^2}\right]R_l=ER_l. \tag{21.5}$$

Let us first consider the case where $V(r)=0$ everywhere in $0\leq r\leq\infty$. The equation is then given by the free particle equation

$$-\frac{\hbar^2}{2m}\frac{1}{r}\frac{d^2(rR_l)}{dr^2}+\frac{\hbar^2 l(l+1)}{2mr^2}R_l=ER_l. \tag{21.6}$$

As discussed in Chapter 4 on free particles, there are two possible solutions to this equation: the spherical Bessel function, $j_l(kr)$, and spherical Neumann function, $n_l(kr)$. However, only $j_l(kr)$ is allowed among the two since it is the only function that is finite at $r=0$. Hence we write

$$R_l(r)=j_l(kr). \tag{21.7}$$

From the asymptotic behavior of $j_l(r)$ discussed in Chapter 4 we find

$$R_l(r)\to\frac{1}{kr}\sin\left[kr-\frac{l\pi}{2}\right]\quad \text{as } r\to\infty,\quad \text{for } V(r)=0. \tag{21.8}$$

For the case where $V(r)\neq 0$ but still $V(r)\to 0$ faster than $1/r^2$ as $r\to\infty$, one can neglect the potential in comparison with the angular momentum term. The equation satisfied by R_l in that limit is

$$-\frac{\hbar^2}{2m}\frac{1}{r}\frac{d^2(rR_l)}{dr^2}+\frac{\hbar^2 l(l+1)}{2mr^2}R_l=ER_l\ \ \text{for } r\to\infty, \tag{21.9}$$

which is, once again, an equation for a free particle. Since we are considering this equation for large values of r, the finiteness constraints at $r=0$ no longer apply and, therefore, we are not restricted to having just j_l but must include n_l also. Hence, for large values of r we have

$$R_l(r) = A_l j_l(kr) + B_l n_l(kr). \tag{21.10}$$

From the asymptotic behavior of j_l and n_l discussed in Chapter 4 we obtain, for $r \to \infty$,

$$R_l(r) \to A_l \frac{1}{kr} \sin\left[kr - \frac{l\pi}{2}\right] - B_l \frac{1}{kr} \cos\left[kr - \frac{l\pi}{2}\right]. \tag{21.11}$$

If we write

$$A_l = D_l \cos\delta_l \quad \text{and} \quad B_l = -D_l \sin\delta_l \tag{21.12}$$

then

$$R_l(r) \to D_l \left[j_l(kr)\cos\delta_l - n_l(kr)\sin\delta_l\right]. \tag{21.13}$$

Therefore

$$R_l(r) \to D_l \frac{1}{kr} \sin\left[kr - \frac{l\pi}{2} + \delta_l\right] \quad \text{as } r \to \infty. \tag{21.14}$$

Note, once again, that the asymptotic behaviors of the two solutions (21.8) and (21.14) are different because the solution (21.8) corresponds to $V(r) = 0$ everywhere in $0 \leq r \leq \infty$, while for (21.14), $V(r) \neq 0$, except that $V(r) \to 0$ as $r \to \infty$, faster than $1/r^2$. Comparing the two solutions we conclude that the presence of a potential creates a "phase-shift" of δ_l in the asymptotic forms of the two radial wavefunctions.

From the asymptotic expressions above for $u(\mathbf{r})$ and $\exp(ikz)$ we will now relate the scattering amplitude, $f(\theta)$, to the phase-shift δ_l, assuming that the particle is traveling along the z-direction (i.e., with azimuthal angle, $\phi = 0$). We express $u(\mathbf{r})$ as

$$u(\mathbf{r}) = \frac{1}{(\sqrt{2\pi})^3} \sum_{l=0}^{\infty} C_l R_l(r) P_l(\cos\theta). \tag{21.15}$$

This is the partial wave expansion of the total wavefunction, where C_l is a constant,

$$C_l = (2l+1)i^l. \tag{21.16}$$

In the limit $r \to \infty$ we have, using (21.14) for $R_l(r)$,

$$u(\mathbf{r}) \to \frac{1}{(\sqrt{2\pi})^3} \sum_{l=0}^{\infty} C_l' \frac{1}{kr} \sin\left[kr - \frac{l\pi}{2} + \delta_l\right] P_l(\cos\theta) \tag{21.17}$$

where C_l' is a constant.

$$C_l' = C_l D_l. \tag{21.18}$$

Let us now look at the right-hand side of (21.2). We first write $\exp(ikz)$ in spherical coordinates as given in Chapter 4,

$$e^{ikz} = \sum_{l=0}^{\infty}(2l+1)i^l j_l(kr)P_l(\cos\theta). \tag{21.19}$$

From the knowledge of the asymptotic behavior of $j_l(kr)$, we find that

$$e^{ikz} \to \sum_{l=0}^{\infty}(2l+1)i^l \frac{1}{kr}\sin\left[kr - \frac{l\pi}{2}\right]P_l(\cos\theta). \tag{21.20}$$

We substitute the asymptotic form (21.17) on the left-hand side of (21.2) while in the right-hand side we now include (21.20). Thus we obtain

$$\begin{aligned}&\frac{1}{(\sqrt{2\pi})^3}\sum C_l' \frac{1}{kr}\sin\left[kr - \frac{l\pi}{2} + \delta_l\right]P_l(\cos\theta)\\ &= \frac{1}{(\sqrt{2\pi})^3}\sum(2l+1)i^l\frac{1}{kr}\sin\left[kr - \frac{l\pi}{2}\right]P_l(\cos\theta) + \frac{1}{(\sqrt{2\pi})^3}f(\theta)\frac{e^{ikr}}{r}.\end{aligned} \tag{21.21}$$

Expressing the sine functions in the exponential form we obtain, after canceling the factor $(1/(\sqrt{2\pi})^3)$,

$$\begin{aligned}&\sum C_l' \frac{\exp\left[i\left(kr - \frac{l\pi}{2} + \delta_l\right)\right] - \exp\left[-i\left(kr - \frac{l\pi}{2} + \delta_l\right)\right]}{2ikr}P_l(\cos\theta)\\ &= \sum(2l+1)i^l\frac{\exp\left[i\left(kr - \frac{l\pi}{2}\right)\right] - \exp\left[-i\left(kr - \frac{l\pi}{2}\right)\right]}{2ikr}P_l(\cos\theta) + f(\theta)\frac{e^{ikr}}{r}.\end{aligned} \tag{21.22}$$

First we note that

$$\exp\left(\pm i\frac{l\pi}{2}\right) = i^{\pm l}. \tag{21.23}$$

Then, comparing the coefficients of $\exp(-ikr)$ on both sides of the above equation, we find

$$\sum C_l'\left(\frac{-i^l e^{-i\delta_l}}{2ikr}\right)P_l(\cos\theta) = \sum(2l+1)i^l\left(\frac{-i^l}{2ikr}\right)P_l(\cos\theta). \tag{21.24}$$

Therefore,

$$C_l' = (2l+1)i^l e^{i\delta_l}. \tag{21.25}$$

From the expression for C_l given in (21.16), we obtain

$$D_l = e^{i\delta_l}. \tag{21.26}$$

Hence the asymptotic form for R_l is

$$R_l(r) \to \frac{e^{i\delta_l}}{kr} \sin\left[kr - \frac{l\pi}{2} + \delta_l\right] \quad \text{as } r \to \infty. \tag{21.27}$$

Comparing now the coefficients of $\exp(ikr)$ we obtain

$$\sum C_l' \left(\frac{i^{-l}e^{i\delta_l}}{2ikr}\right) P_l(\cos\theta)$$
$$= \sum (2l+1) i^l \left(\frac{i^{-l}}{2ikr}\right) P_l(\cos\theta) + f(\theta)\frac{1}{r}. \tag{21.28}$$

Substituting the value of C_l' already derived in (21.25), we obtain the scattering amplitude

$$f(\theta) = \sum_{l=0}^{\infty} (2l+1)\left(\frac{e^{2i\delta_l} - 1}{2ik}\right) P_l(\cos\theta). \tag{21.29}$$

We define the "partial wave scattering amplitude," $f_l(k)$ by

$$f_l(k) = \frac{e^{2i\delta_l} - 1}{2ik} = \frac{e^{i\delta_l}\sin\delta_l}{k}. \tag{21.30}$$

The scattering amplitude in terms of $f_l(k)$ can be written as

$$f(\theta) = \sum_{l=0}^{\infty} (2l+1) f_l(k) P_l(\cos\theta). \tag{21.31}$$

This is then the partial wave expansion of $f(\theta)$.

The differential cross-section is given by

$$\frac{d\sigma}{d\Omega} = |f(\theta)|^2 = \left[\sum_{l=0}^{\infty} (2l+1) f_l^*(k) P_l(\cos\theta)\right]\left[\sum_{l'=0}^{\infty} (2l'+1) f_{l'}(k) P_{l'}(\cos\theta)\right]. \tag{21.32}$$

We note that there are interference terms between different partial waves. The cross-section is then

$$\sigma = \int d\Omega\, |f(\theta)|^2 = 2\pi \int_{-1}^{1} d\cos\theta \sum_{l,l'=0}^{\infty} (2l+1)(2l'+1) f_l^*(k) f_{l'}(k)$$
$$\times P_l(\cos\theta) P_{l'}(\cos\theta). \tag{21.33}$$

Since the Legendre polynomials satisfy the orthogonality property

$$\int_{-1}^{+1} d\cos\theta\, P_l(\cos\theta) P_{l'}(\cos\theta) = \frac{2}{2l+1}\delta_{ll'}, \tag{21.34}$$

the interference terms vanish and we obtain

$$\sigma = 4\pi \sum (2l+1) |f_l(k)|^2 = 4\pi \sum (2l+1) \left(\frac{\sin^2 \delta_l}{k^2} \right). \tag{21.35}$$

We define a partial wave cross-section, σ_l as

$$\sigma_l = (2l+1) \frac{4\pi \sin^2 \delta_l}{k^2}. \tag{21.36}$$

Then the total cross-section is expressed in terms of it as

$$\sigma = \sum \sigma_l. \tag{21.37}$$

21.1.1 Comparing R_l for $V = 0$ and $V = V(r) \neq 0$

We note that for $V = 0$, R_l is simply the spherical Bessel function $j_l(kr)$ whose asymptotic behavior is given by

$$R_l(r) \to \frac{1}{kr} \sin \left[kr - \frac{l\pi}{2} \right], \tag{21.38}$$

which we can write as

$$R_l(r) \to \frac{e^{i(kr - l\pi/2)} - e^{-i(kr - l\pi/2)}}{2ikr}. \tag{21.39}$$

This is a combination of an incoming spherical wave given by the $\exp(-ikr)$ term and the outgoing spherical wave given by the $\exp(ikr)$ term. The absolute value of the coefficient of each is the same, which simply reflects the fact that the incoming flux is the same as the outgoing flux, which in turn is a consequence of probability conservation.

In the presence of the potential we found from (21.27) that

$$R_l(r) \to \frac{e^{i\delta_l}}{kr} \sin \left[kr - \frac{l\pi}{2} + \delta_l \right], \tag{21.40}$$

which we can write as

$$R_l(r) \to \frac{e^{2i\delta_l} e^{i(kr - l\pi/2)} - e^{-i(kr - l\pi/2)}}{2ikr}. \tag{21.41}$$

We find that while we have the same incoming wave, the outgoing part acquires a phase given by $\exp(2i\delta_l)$ due to the presence of the scatterer. The absolute values of the two coefficients are still the same, however, since the probabilities are still conserved. This phase term is just the (partial wave) S-matrix, which we denote S_l,

$$S_l = e^{2i\delta_l}. \tag{21.42}$$

Probability conservation tells us that

$$|S_l| = 1, \tag{21.43}$$

which implies that the S-matrix is unitary.

21.2 χ_l, K_l, and T_l

The radial wave equation is given by

$$-\frac{\hbar^2}{2m}\frac{1}{r}\frac{d^2(rR_l)}{dr^2} + \left[V(r) + \frac{\hbar^2 l(l+1)}{2mr^2}\right]R_l = ER_l. \tag{21.44}$$

As we discussed in Chapter 4, it is often more convenient to work with the wavefunction χ_l defined as

$$(kr)R_l = \chi_l. \tag{21.45}$$

If we let

$$k^2 = \frac{2mE}{\hbar^2}, \quad U = \frac{2mV}{\hbar^2} \tag{21.46}$$

and multiply (21.44) by $\left(-2m/\hbar^2\right)$, we obtain the following equation for χ_l:

$$\left[\frac{d^2}{dr^2} - \frac{l(l+1)}{r^2} - U(r) + k^2\right]\chi_l = 0. \tag{21.47}$$

We divide the above equation by k^2 and define

$$\rho = kr. \tag{21.48}$$

A much simpler equation is then found:

$$\left[\frac{d^2}{d\rho^2} - \frac{l(l+1)}{\rho^2} - \frac{U(r)}{k^2} + 1\right]\chi_l = 0. \tag{21.49}$$

The threshold behavior of $\chi_l(\rho)$ is given by

$$\chi_l(\rho) \sim \rho^{l+1} \quad \text{as } \rho \to 0. \tag{21.50}$$

Therefore,

$$\chi_l(0) = 0. \tag{21.51}$$

At this stage it is convenient to introduce the functions s_l and c_l given by

$$s_l(\rho) = \rho j_l(\rho) \tag{21.52}$$

and

$$c_l(\rho) = -\rho n_l(\rho). \tag{21.53}$$

These functions are close to sine and cosine functions, respectively.

From (21.13), (21.26) and (21.45) the asymptotic form of $\chi_l(\rho)$ is found to be

$$\chi_l(\rho) \to e^{i\delta_l}\left[s_l \cos\delta_l + c_l \sin\delta_l\right] \quad \text{as } \rho \to \infty. \tag{21.54}$$

By rearranging the terms we can express the asymptotic expression (20.54) in different forms:

$$\chi_l(\rho) \to A_l'\left[s_l + K_l c_l\right] \tag{21.55}$$

where

$$K_l = \tan\delta_l, \tag{21.56}$$

or as

$$\chi_l(\rho) = \chi_l^{(+)}(\rho) \to B_l'\left[s_l + T_l e_l^{(+)}\right] \tag{21.57}$$

where

$$e_l^{(+)} = c_l + is_l$$

and

$$T_l = e^{i\delta_l}\sin\delta_l, \tag{21.58}$$

where A_l' and B_l' are constants. The K_l and T_l functions introduced above play a very important role in the theory of partial wave scattering, some examples of which will be discussed below.

21.3 Integral relations for χ_l, K_l, and T_l

Below we present a series of integral relations.

Following the same procedure as the one we followed for the three-dimensional wave functions, we can express $\chi_l(r)$ in the Green's function formalism,

$$\chi_l(r) = s_l(kr) + \int_0^\infty dr'\, g_l(r,r')U(r')\chi_l(r'), \tag{21.59}$$

where $g_l(r,r')$ is the radial Green's function which satisfies the equation

$$\left(\frac{d^2}{dr^2} - \frac{l(l+1)}{r^2} + k^2\right) g_l(r,r') = \delta(r - r'). \tag{21.60}$$

This equation can be solved by considering the regions $r > r'$ and $r < r'$ separately. There are several different forms in which the solutions can be written depending on the asymptotic form one chooses for χ_l.

(i) One of the forms is

$$g_l(r,r') = -\frac{1}{k} s_l(kr_<) c_l(kr_>) \tag{21.61}$$

where $r_<$ means the smaller of the two variables r and r', while $r_>$ implies the larger of the two. If we put this in the relation (21.59) and let $r \to \infty$, then we find from the asymptotic form for $c_l(kr_>)$,

$$\chi_l(r) \to s_l(kr) - \frac{1}{k}\left[\int_0^\infty dr'\, s_l(kr')U(r')\chi_l(r')\right] c_l(kr) \tag{21.62}$$

where since $r \to \infty$ we have taken $r_> = r$ and $r_< = r'$.

From the asymptotic behavior given by

$$\chi_l(\rho) \to s_l + K_l c_l \tag{21.63}$$

we obtain

$$K_l = \tan \delta_l = -\frac{1}{k}\int_0^\infty dr'\, s_l(kr')U(r')\chi_l(r'). \tag{21.64}$$

This is an important relation that relates the phase shifts to the potential and to the wave-functions.

(ii) Another form for the Green's function is

$$g_l^{(+)}(r,r') = -\frac{1}{k} s_l(kr_<)\, e_l^{(+)}(kr_>). \tag{21.65}$$

As $r \to \infty$ one now finds

$$\chi_l^{(+)}(r) \to s_l(kr) - \frac{1}{k}\left[\int_0^\infty dr'\, s_l(kr')U(r')\chi_l^{(+)}(r')\right] e_l^{(+)}(kr). \tag{21.66}$$

We use the asymptotic form

$$\chi_l^{(+)}(\rho) \to s_l + T_l e_l^{(+)}, \tag{21.67}$$

which allows one to obtain the integral relation for the partial wave T-matrix

$$T_l = -\frac{1}{k}\int_0^\infty dr'\, s_l(kr')U(r')\chi_l^{(+)}(r'). \tag{21.68}$$

21.4 Wronskian

Let us now construct what is called a Wronskian between two wavefunctions, which plays an important role in solving differential equations.

21.4.1 Same potentials

We write the differential equations for two possible wavefunctions $\chi_l^{(1)}(r)$ and $\chi_l^{(2)}(r)$ with the same potentials as

$$\left(\frac{d^2}{d\rho^2} - \frac{l(l+1)}{\rho^2} - \frac{U(r)}{k^2} + 1\right)\chi_l^{(1)}(r) = 0, \tag{21.69}$$

$$\left(\frac{d^2}{d\rho^2} - \frac{l(l+1)}{\rho^2} - \frac{U(r)}{k^2} + 1\right)\chi_l^{(2)}(r) = 0. \tag{21.70}$$

Multiplying (21.69) by $\chi_l^{(2)}(r)$ and (21.70) by $\chi_l^{(1)}(r)$ and subtracting the first term from the second, we find

$$\chi_l^{(1)}\frac{d^2\chi_l^{(2)}}{d\rho^2} - \chi_l^{(2)}\frac{d^2\chi_l^{(1)}}{d\rho^2} = 0. \tag{21.71}$$

This expression is a perfect differential of the form

$$\frac{d}{d\rho}\left[\chi_l^{(1)}\frac{d\chi_l^{(2)}}{d\rho} - \chi_l^{(2)}\frac{d\chi_l^{(1)}}{d\rho}\right] = 0. \tag{21.72}$$

Let us now define a Wronskian, $W(\rho)$, between $\chi_l^{(1)}$ and $\chi_l^{(2)}$ as

$$W(\rho) = \left[\chi_l^{(1)}\frac{d\chi_l^{(2)}}{d\rho} - \chi_l^{(2)}\frac{d\chi_l^{(1)}}{d\rho}\right]. \tag{21.73}$$

From (21.72) we find that

$$W(\rho) = \text{constant}. \tag{21.74}$$

This result is useful for many derivations. For example, one can rederive the result (21.64),

$$\tan\delta_l = -\frac{1}{k}\int_0^\infty dr\, s_l U(r)\chi_l. \tag{21.75}$$

21.4.2 Different potentials

Let us consider the case where the potentials are different corresponding to two different wavefunctions $\chi_l^{(1)}$ and $\chi_l^{(2)}$,

$$\left(\frac{d^2}{d\rho^2} - \frac{l(l+1)}{\rho^2} - \frac{U_1}{k^2} + 1\right)\chi_l^{(1)}(\rho) = 0, \tag{21.76}$$

$$\left(\frac{d^2}{d\rho^2} - \frac{l(l+1)}{\rho^2} - \frac{U_2}{k^2} + 1\right)\chi_l^{(2)}(\rho) = 0. \tag{21.77}$$

We now find that

$$\chi_l^{(1)}\frac{d^2\chi_l^{(2)}}{d\rho^2} - \chi_l^{(2)}\frac{d^2\chi_l^{(1)}}{d\rho^2} = [U_2(\rho) - U_2(\rho)]\chi_l^{(1)}\chi_l^{(2)}. \tag{21.78}$$

Integrating both sides in the interval $(0, \rho)$, we obtain the following relation with $W(0) = 0$:

$$W(\rho) = \frac{1}{k^2}\int_0^{\rho} d\rho'\,[U_2(\rho') - U_1(\rho')]\chi_l^{(1)}\chi_l^{(2)}. \tag{21.79}$$

By taking $\rho \to \infty$, we obtain

$$\tan\delta_l^{(2)} - \tan\delta_l^{(1)} = \frac{1}{k}\int_0^{\infty} dr\,[U_1(r) - U_2(r)]\chi_l^{(1)}\chi_l^{(2)}. \tag{21.80}$$

21.4.3 Properties of δ_l

We summarize below some of the important properties of δ_l.

Sign of δ_l

From (21.40) and (21.45), the asymptotic behavior of χ_l in the presence of a potential is given by

$$\chi_l \to e^{i\delta_l}\sin\left[kr - \frac{l\pi}{2} + \delta_l\right]. \tag{21.81}$$

In the absence of the potential the behavior is

$$\chi_l^{(0)} \to \sin\left[kr - \frac{l\pi}{2}\right]. \tag{21.82}$$

Comparing the two wavefunctions, for the case of a finite range potential, it is clear that the wavefunction χ_l outside the range is "pulled in" if $\delta_l > 0$. A wavefunction that has this property would normally indicate a potential that is attractive (see Fig. 21.2). Hence

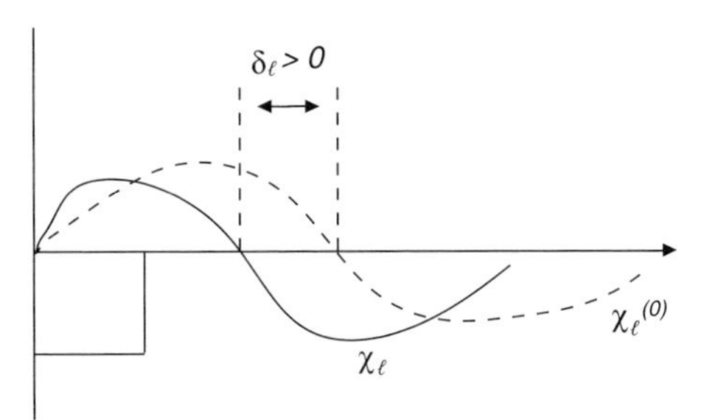

Fig. 21.2

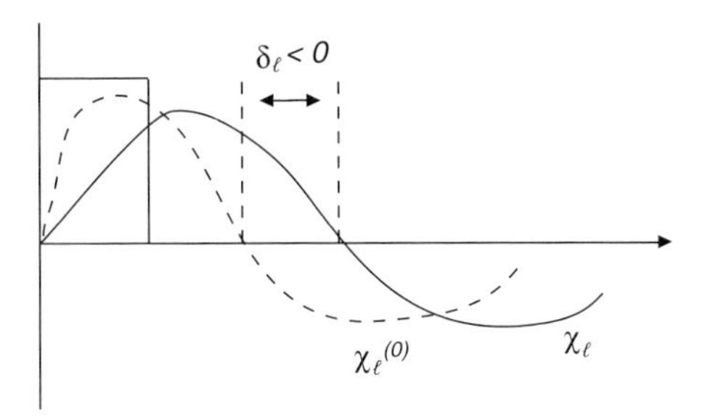

Fig. 21.3

an attractive potential will correspond to a positive phase shift. Similarly $\delta_l < 0$ will correspond to the outside wavefunction being "pushed out" and, therefore, will reflect the presence of a repulsive potential (see Fig. 21.3).

These results can be understood from the expression (21.75), which shows that for an attractive potential, $U(r) < 0$, we have $\delta_l > 0$, with the opposite result for a repulsive potential. In fact, when looking at the difference $(\tan\delta_l^{(1)} - \tan\delta_l^{(2)})$ involving two different potentials, U_1 and U_2, if we take $U_1 \approx U_2$ so that $\chi_l^{(1)} \approx \chi_l^{(2)}$ but with an overall contribution of the difference $(U_1 - U_2) < 0$ then we obtain $\tan\delta_l^{(1)} > \tan\delta_l^{(2)}$, implying $\delta_l^{(1)} > \delta_l^{(2)}$, while for $(U_1 - U_2) > 0$, we have $\delta_l^{(1)} < \delta_l^{(2)}$.

Ramsauer–Townsend effect

As we have found, the wavefunction is pulled in when the potential is attractive corresponding to a positive phase shift. As the strength of the potential increases and becomes large enough, one encounters a situation in which the sinusoidal wavefunction goes through a complete half-cycle. As a result, the outside wavefunction is 180° out of phase with the unperturbed wavefunction. This corresponds to $\delta_0 = \pi$ and $\sin^2\delta_0 = 0$. Thus, the cross-section vanishes and there is perfect transmission (if one ignores the small effects due to higher partial waves). This is called the Ramsauer–Townsend effect and has been observed experimentally.

Dependence of δ_l on l

As $\rho \to 0$, where $\rho = kr$, the functions s_l and χ_l have the behavior

$$s_l \to \frac{\rho^l}{(2l+1)!!} \quad \text{and} \quad \chi_l \to A_l \rho^l \tag{21.83}$$

where A_l depends on the details of the potential. As l increases, this behavior will diminish the contribution of $V(r)$ to the integral (21.75) for small values of r. Since $V(r)$ is expected to be of finite range, this integral and therefore δ_l, will decrease as we increase l. Physically one can interpret this as due to the repulsive centrifugal barrier

$$\frac{\hbar^2 l(l+1)}{2mr^2}, \tag{21.84}$$

which increasingly dilutes the effect of the potential as l increases.

The largest maximum of $s_l(\rho)$ occurs for $\rho \sim l$, that is, at

$$r \sim \frac{l}{k}. \tag{21.85}$$

If R is the range of the potential then only partial waves with $l \lesssim kR$ will give significant values of δ_l. Thus only a finite number of partial waves need be considered in the partial wave expansion, from $l = 0$ to kR. The partial wave analysis and the phase shift treatment will, therefore, be most useful at low energies, since then only a small number of partial waves will come into play. The maximum value of the total cross-section obtained by taking $\sin^2 \delta_l = 1$ is

$$\sigma = \frac{4\pi}{k^2} \sum_0^{kR} (2l+1) \simeq 4\pi R^2. \tag{21.86}$$

Thus, the cross-section is given by the range of interaction, R, with a maximum value that is four times the classical value of πR^2, corresponding to a disk of radius R representing the effective scattering area.

Low-energy behavior of δ_l

The behavior as $k \to 0$ can be easily obtained from (21.75). Since $\tan \delta_l \sim \delta_l$ for small values of δ_l, we obtain

$$\delta_l \to k^{2l+1}. \tag{21.87}$$

Since $k = 0$ is the threshold for the scattering to take place, this relation is called the threshold behavior for the phase shift. Again, one interprets this as due to the dominant centrifugal effect.

High-energy behavior of δ_l

The right-hand side of the integral relation (21.75) for $\tan \delta_l$ vanishes as $k \to \infty$ due to the term $(1/k)$ in front of the integral. In this limit $\chi_l \sim s_l$, therefore, the factor multiplying $V(r)$ in the integrand will be $\sim s_l^2$ which is $\lesssim 1$. Hence,

$$\delta_l \to 0 \quad \text{as} \quad k \to \infty.$$

This is not surprising because in the high-energy limit the kinetic energy term in the Hamiltonian is much larger than the potential. Therefore, in this limit the particle acts like a free particle, which by definition has $\delta_l = 0$. The precise functional form for δ_l for $l = 0$ in this limit will be determined in Chapter 22.

Born approximation

The Born approximation corresponds to keeping only the first-order contribution to the wavefunction. Therefore, we take

$$\chi_l^B(r) = s_l(kr). \tag{21.88}$$

It is a valid approximation if the potentials are weak, in which case we expect the integral (21.75) to be small. For small values of $\tan \delta_l^B$ one can use the approximation

$$\tan \delta_l^B \approx \delta_l^B. \tag{21.89}$$

Therefore,

$$\delta_l^B = -\frac{1}{k}\int_0^\infty dr\, U(r) s_l^2(kr). \tag{21.90}$$

For $k \to 0$, we can obtain δ_l^B explicitly from the behavior of $s_l(\rho)$ if the potential is of finite range,

$$\delta_l^B \simeq -\frac{k^{2l+1}}{[(2l+1)!!]^2}\int_0^R dr\, r^{2l+1} U(r) \tag{21.91}$$

where R is the range of the potential. For infinite-range potentials like the Yukawa potential, $ge^{-\mu r}/r$, one can take $R = \infty$ for the upper limit since the integral will still be convergent. The k-dependence is consistent with what we obtained in (21.87). At high energies $\delta_l^B \to 0$.

Finally, we note that the Born approximation is not valid if the potential is strong enough to form bound states. In the case of bound states one can not use perturbation theory. The Born approximation is quite meaningful if the strength of the potential is weak or if the energies are high.

21.5 Calculation of phase shifts: some examples

21.5.1 Square well

Below we determine the phase shift δ_l for $l = 0$ for the case of a square well potential

$$V = -V_0, \quad r \leq a \tag{21.92}$$

$$= 0, \quad r > a. \tag{21.93}$$

Earlier we considered bound states for which the energy $E < 0$. We are interested in the scattering states corresponding to positive energies, $E > 0$. The radial wavefunction $\chi_0(r)$ for this problem can be calculated in the same manner as for the bound states,

$$\chi_0(r) = A \sin \alpha r, \quad r < a \tag{21.94}$$

where A is an arbitrary constant and

$$\alpha = \sqrt{\frac{2m}{\hbar^2}(E + V_0)}. \tag{21.95}$$

Since for $r > a$, $V(r) = 0$, the wavefunction in this region discussed earlier can be written down in terms of the phase shift, δ_0, as

$$\chi_0(r) = B \sin(kr + \delta_0), \quad r > a \tag{21.96}$$

where B is an arbitrary constant and $k = \sqrt{2mE/\hbar^2}$.

The boundary conditions imply that the two wavefunctions and their derivatives must be equal at $r = a$. Therefore,

$$A \sin \alpha a = B \sin(ka + \delta_0) \tag{21.97}$$

and

$$A\alpha \cos \alpha a = Bk \cos(ka + \delta_0). \tag{21.98}$$

By dividing (21.98) by (21.97) we eliminate A and B and obtain

$$\alpha \cot \alpha a = k \cot(ka + \delta_0). \tag{21.99}$$

One can then easily show that

$$\tan \delta_0 = \frac{k - \alpha \cot(\alpha a) \tan(ka)}{\alpha \cot(\alpha a) + k \tan(ka)}. \tag{21.100}$$

Approximate properties of δ_0

(i) *Behavior as* $k \to 0$.

We find from (21.99) leading order in k,

$$\tan \delta_0 \to \frac{ka\,(1 - \bar{\alpha}a \cot \bar{\alpha}a)}{\bar{\alpha}a \cot \bar{\alpha}a} = ka\gamma \tag{21.101}$$

where

$$\gamma = \frac{1 - \bar{\alpha}a \cot(\bar{\alpha}a)}{\bar{\alpha}a \cot \bar{\alpha}a} = \frac{\tan \bar{\alpha}a}{\bar{\alpha}a} - 1 \tag{21.102}$$

and

$$\bar{\alpha} = \sqrt{\frac{2m}{\hbar^2} V_0}. \tag{21.103}$$

Since $\tan \delta_0$ is found to be proportional to k and, therefore, small, it can be approximated by δ_0. Thus

$$\delta_0 \to ka\gamma \quad \text{as} \quad k \to 0. \tag{21.104}$$

The partial wave cross-section is then

$$\sigma_0 = \frac{4\pi \sin^2 \delta_0}{k^2} \to 4\pi a^2 \left[\frac{\tan \bar{\alpha}a}{\bar{\alpha}a} - 1\right]^2. \tag{21.105}$$

(ii) *Dependence on the strength of the potential.*

We will continue to assume that k is small. If the potential is weak (but attractive), then $\bar{\alpha}a < 1$ and therefore we can expand $\tan \bar{\alpha}a$ in this limit,

$$\tan \bar{\alpha}a = \bar{\alpha}a + \frac{1}{3}(\bar{\alpha}a)^3 + \cdots. \tag{21.106}$$

Hence,

$$\gamma \approx \frac{1}{3}(\bar{\alpha}a)^2 + \cdots. \tag{21.107}$$

Therefore, γ and the phase shift δ_0 are positive.

If the potential is very strong (attractive) then $\tan \bar{\alpha}a > \bar{\alpha}a$, and the phase shift δ_0 is large and positive, until it reaches the value $\pi/2$. If

$$\bar{\alpha}a > \frac{\pi}{2} \tag{21.108}$$

then $\tan \bar{\alpha}a < 0$ and γ is negative. But $\bar{\alpha}a > \pi/2$ corresponds to

$$V_0 > \frac{\pi^2\hbar^2}{8ma^2}. \tag{21.109}$$

Our results in Chapter 8 indicate that this is just the condition that there be a bound state. More appropriately, the condition

$$\frac{\pi^2\hbar^2}{8ma^2} < V_0 < \frac{\pi^2\hbar^2}{2ma^2} \tag{21.110}$$

corresponds to one bound state. Since $\gamma < 0$, for the above potential

$$\tan\delta_0 \to -|\gamma|\,ka \quad \text{as } k \to 0. \tag{21.111}$$

Therefore,

$$at\ \ k = 0, \quad \delta_0 = \pi.$$

This is as predicted by Levinson's theorem, which we will discuss in the next chapter, which states that $\delta_0 \to n\pi$, as $k \to 0$ if there are n bound states. Thus, for $n = 1$ we have $\delta_0 = \pi$.

21.5.2 Rigid sphere

A rigid sphere potential is given by

$$V(r) = 0, \quad r \le a \tag{21.112}$$

$$V(r) = \infty, \quad r > a. \tag{21.113}$$

Once again we consider the S waves. We note that because it is an infinitely repulsive potential the particle will be entirely outside the sphere with the wavefunction given by

$$\chi(r) = A\sin(kr + \delta_0), \quad r > a \tag{21.114}$$

where δ_0 is the S-wave phase shift. The boundary condition implies that the wavefunction must vanish at $r = a$. Hence, $\chi(a) = 0$ and therefore

$$ka + \delta_0 = 0, \quad \text{i.e.,} \quad \delta_0 = -ka. \tag{21.115}$$

Thus the phase shift is negative and stays negative throughout the scattering region, with $\delta_0 \to -\infty$ as $k \to \infty$. In Fig. 21.4 we have plotted a typical behavior of δ_0 as a function of ka and, for comparison, we also have δ_0 for the purely attractive case.

The S-wave cross-section is then given by

$$\sigma_0 = \frac{4\pi\sin^2(ka)}{k^2}, \tag{21.116}$$

which in the limit $k \to 0$, gives

$$\sigma_0 = 4\pi a^2. \tag{21.117}$$

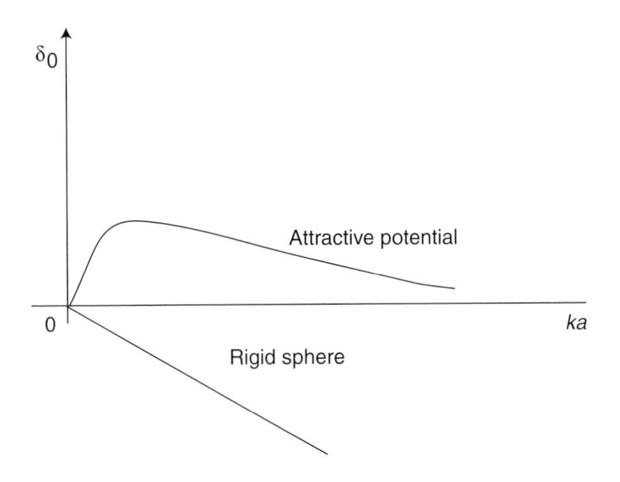

Fig. 21.4

This value for the cross-section is of the right order of magnitude since it corresponds to an area presented by a disk of radius a. The factor of 4 is a consequence of low-energy behavior.

For $l \neq 0$, the phase shift can be determined from the relation

$$\chi_l(r) = D_l\left[s_l \cos\delta_l + c_l \sin\delta_l\right]. \tag{21.118}$$

The condition $\chi_l(a) = 0$ implies that

$$\tan\delta_l = -\frac{s_l(ka)}{c_l(ka)}. \tag{21.119}$$

From the properties of s_l and c_l we find the following behavior near $k = 0$ and at $k = \infty$:

$$\delta_l \to -k^{2l+1} \quad \text{as } k \to 0 \tag{21.120}$$

and

$$\delta_l \to -ka + \frac{l\pi}{a} \quad \text{as } k \to \infty. \tag{21.121}$$

Thus, once again, as with the S waves, the phase shift stays negative and $\delta_l \to -\infty$ as $k \to \infty$.

For finite repulsive potentials one can show that the phase shift is negative near $k = 0$. But as $k \to \infty$, $\delta_0 \to 0$.

21.5.3 Absorption and inelastic amplitudes

We have found that the partial wave scattering amplitude is given by

$$f_l = \frac{e^{2i\delta_\ell} - 1}{2ik}. \tag{21.122}$$

As we saw in our earlier discussions, absorptive or inelastic processes can be incorporated by introducing a complex potential, $V = V_R - iV_I$ with $V_I > 0$. If we include an imaginary part, V_I, in the relation between the phase shift and the potential derived in this chapter, one can easily show that the phase shift itself can be expressed as a sum of real and imaginary parts,

$$\delta_l = \delta_{l(R)} + i\delta_{l(I)} \tag{21.123}$$

where $V_I < 0$ will correspond to $\delta_{l(I)} > 0$.

If we set

$$e^{-2\delta_I} = \eta_l, \tag{21.124}$$

then the partial wave amplitude will be modified to

$$f_l = \frac{\eta_l e^{2i\delta_l} - 1}{2ik} \tag{21.125}$$

and the amplitude $f(\theta)$ to

$$f(\theta) = \sum_{l=0}^{\infty}(2l+1)\left(\frac{\eta_l e^{2i\delta_l} - 1}{2ik}\right) P_l(\cos\theta). \tag{21.126}$$

The elastic scattering cross-section is then given by

$$\sigma_{el} = \int d\Omega\, |f(\theta)|^2 = 4\pi\sum_{l=0}^{\infty}(2l+1)\left|\frac{\eta_l e^{2i\delta_l} - 1}{2ik}\right|^2, \tag{21.127}$$

which gives

$$\sigma_{el} = \frac{\pi}{k^2}\sum_{l=0}^{\infty}(\eta_l^2 + 1 - 2\eta_l \cos 2\delta_l). \tag{21.128}$$

The total cross-section is obtained from the generalized optical theorem,

$$\sigma_T = \frac{4\pi}{k}\,\mathrm{Im} f(0) = \frac{2\pi}{k^2}\sum_{l=0}^{\infty}(-\eta_l \cos\delta_l + 1). \tag{21.129}$$

Therefore, the inelastic cross-section is

$$\sigma_{inel} = \sigma_T - \sigma_{el} = \frac{\pi}{k^2}\sum_{l=0}^{\infty}(1 - \eta_l^2). \tag{21.130}$$

21.6 Problems

1. Determine the $l = 0$ and $l = 1$ partial wave phase shifts in the low-energy limit for the potential

$$\begin{aligned} V(r) &= V_0, \quad \text{for } r \leq R \\ &= 0, \quad \text{for } r > 0. \end{aligned}$$

In terms of these two phase shifts, determine the scattering amplitude, $f(\theta)$, and the total cross-section, σ.

2. For a potential represented by a spherical shell

$$V(r) = \lambda\delta(r - R),$$

determine the phase shift δ_l for $l = 0$. Also obtain the cross-section in the limit $k \to 0$. Consider both signs for λ.

3. In the probability current conservation relation for the time-independent case, express the current in terms of the radial wavefunctions. Show that the outgoing flux must be the same as the incoming flux. From this result show that the partial wave S-matrix, S_l, is unitary.
4. Consider a square-well potential

$$\begin{aligned} V(r) &= -V_0, \quad r \leq a \\ &= 0 \quad\quad\; r > a. \end{aligned}$$

Show that S-wave partial wave amplitude can be expressed as

$$f_0 = \frac{e^{-2ika}}{\alpha \cot \alpha a - ik} - \frac{e^{-ika} \sin ka}{k}$$

where

$$\alpha = \sqrt{\frac{2m(E + V_0)}{\hbar^2}}$$

and the second term in the expression for f_0 is characteristic of the scattering amplitude for a hard sphere.

5. Obtain the upper and lower limits of the ratio of the partial wave cross-sections $(\sigma_{inel}/\sigma_{tot})$ in terms of η_l.
6. Consider the following potential:

$$\begin{aligned} V &= \infty, \quad r < a \\ &= -V_0, \quad a < r < b \\ &= 0, \quad r > b. \end{aligned}$$

Obtain the S-wave phase shift δ_0.

7. For a potential given by

$$\begin{aligned} V(r) &= +V_0, \quad 0 < r < a \\ &= 0, \quad r > a \end{aligned}$$

obtain the phase-shift δ_l if the energy of the particle satisfies $0 < E < V_0$.

22 Analytic structure of the S-matrix

We discuss the properties of the partial wave S-matrix in the complex momentum and complex energy plane. We establish that bound states appear as poles in these planes. The relation between resonances and bound states are also discussed.

22.1 S-matrix poles

22.1.1 Bound states

In this section we discuss how the bound state energies in a given angular momentum state are related to the poles in the complex energy plane of the scattering amplitude in that state. As an illustration we consider a square well potential given by

$$V = -V_0, \quad r \leq a \tag{22.1}$$

$$= 0, \quad r > a \tag{22.2}$$

and consider specifically an S-wave ($l = 0$) bound state. Here we have $E < 0$, so we define

$$E = -E_B \quad \text{with } E_B > 0 \tag{22.3}$$

as the binding energy. For the S-states it is much simpler to consider the function $\chi_0(r)$. It is given by

$$\chi_0(r) = A \sin \alpha r, \quad r \leq a \tag{22.4}$$

$$= Be^{-\beta r}, \quad r > a \tag{22.5}$$

where

$$\alpha = \sqrt{\frac{2m(V_0 - E_B)}{\hbar^2}} \quad \text{and} \quad \beta = \sqrt{\frac{2mE_B}{\hbar^2}}. \tag{22.6}$$

The boundary conditions at $r = a$ give

$$A \sin \alpha a = Be^{-\beta a}, \tag{22.7}$$

$$A\alpha \cos \alpha a = -B\beta e^{-\beta a}. \tag{22.8}$$

Dividing (22.8) by (22.7) we obtain the relation

$$\beta = -\alpha \cot \alpha a. \tag{22.9}$$

Since we will be examining the relation between the bound state and the scattering amplitude in the low energy limit, we take E_B to be small $E_B \ll V_0$ and, in that limit, write the relation (22.9) as

$$\beta = -\alpha_0 \cot \alpha_0 a \tag{22.10}$$

where

$$\alpha_0 = \sqrt{\frac{2mV_0}{\hbar^2}}. \tag{22.11}$$

We note that since $\beta > 0$, a solution of (22.10) will exist if $\cot \alpha_0 a$ is negative, which occurs when $\pi/2 < \alpha_0 a < \pi$, i.e., when

$$\frac{\pi^2\hbar^2}{8ma^2} < V_0 < \frac{\pi^2\hbar^2}{2ma^2}. \tag{22.12}$$

This is, of course, confirmed by the results obtained in Chapter 8.

Now let us consider the scattering state. Here $E > 0$, and the wavefunctions are given by

$$\chi_0(r) = C \sin \kappa r, \qquad r \leq a \tag{22.13}$$

$$= D \sin (kr + \delta_0), \quad r > a \tag{22.14}$$

where δ_0 is the S-wave phase shift with

$$\kappa = \sqrt{\frac{2m(V_0 + E)}{\hbar^2}} \quad \text{and} \quad k = \sqrt{\frac{2mE}{\hbar^2}}. \tag{22.15}$$

From the boundary conditions at $r = a$, we have

$$C \sin \kappa a = D \sin (ka + \delta_0), \tag{22.16}$$

$$C\kappa \cos \kappa a = Dk \cos (ka + \delta_0), \tag{22.17}$$

which leads to the following relation, by eliminating C and D,

$$\kappa \cot \kappa a = k \cot (ka + \delta_0). \tag{22.18}$$

In the low-energy limit, i.e., $k \to 0$ and $E \ll V_0$, relation (22.18) becomes

$$\kappa_0 \cot \kappa_0 a = k \cot \delta_0 \tag{22.19}$$

where

$$\kappa_0 = \sqrt{\frac{2mV_0}{\hbar^2}}. \tag{22.20}$$

The S-wave scattering amplitude is given by

$$f_0 = \frac{e^{i\delta_0} \sin \delta_0}{k} = \frac{1}{k \cot \delta_0 - ik}. \tag{22.21}$$

A pole occurs in the complex k-plane when the denominator of (22.21) vanishes, i.e., when

$$k \cot \delta_0 = ik. \tag{22.22}$$

It is clear that the position of the pole in the k-plane does not correspond to physical values of k, i.e., to real and positive values of k. If it did, then the scattering amplitude would become infinite, which is not allowed.

Replacing $k \cot \delta_0$ on the right-hand side of (22.19) by ik we find

$$k = -i\kappa_0 \cot \kappa_0 a. \tag{22.23}$$

We also note from relations (22.11) and (22.20) that

$$\kappa_0 = \alpha_0. \tag{22.24}$$

Therefore, the pole in the scattering amplitude is located at

$$k = -i\alpha_0 \cot \alpha_0 a, \tag{22.25}$$

which is along the imaginary k-axis. An imaginary k from (22.15) corresponds to $E < 0$. If we take $E = -E_B$ in (22.15) then we find that

$$k = i\beta \tag{22.26}$$

where β is given by (22.6). Hence relation (22.25) gives

$$\beta = -\alpha_0 \cot \alpha_0 a, \tag{22.27}$$

which is precisely the relation (22.10) for the bound-state energy. Therefore, the pole in the scattering amplitude occurs at the position of the bound state. Since β is positive, from (22.26) we find that the pole occurs along the positive imaginary axis of the k-plane.

We conclude, based on this example, that a bound state corresponds to a pole in the scattering amplitude at negative values of E when the energy values of the position of the pole are continued to $E = -E_B$. We have shown this only under special conditions, e.g., at low energies and for S-waves, but it can be shown to be generally true. We will return to this subject when we discuss, albeit very briefly, the so-called Jost functions.

22.1.2 Resonances

In an interaction through a potential $V(r)$, a resonance in a partial wave l is said to occur when the scattering phase shift, δ_l, passes though $\pi/2$ or any odd multiple of it. At that point $\sin\delta_l = 1$ and the partial wave cross-section is given by

$$\sigma_l = \frac{4\pi}{k^2}, \tag{22.28}$$

which is the maximum possible value that σ_l can reach for a given value of k. This is also called the unitarity limit.

We will show, particularly for $l \neq 0$, that if the potential is attractive and sufficiently strong to create a bound state at low energies (i.e., E small and negative), then the scattering amplitude will exhibit a resonance behavior at low energies (i.e., E small and positive) for the same partial wave. We will show this, once again, under the simple situation of a square-well potential.

Bound states for $l \neq 0$

We will consider the same square-well potential that we considered in the previous section,

$$V = -V_0, \quad r \le a \tag{22.29}$$

$$= 0, \quad r > a, \tag{22.30}$$

with the same parameters,

$$E = -E_B \quad \text{with} \quad \alpha = \sqrt{\frac{2m(V_0 - E_B)}{\hbar^2}} \quad \text{and} \quad \beta = \sqrt{\frac{2mE_B}{\hbar^2}}, \tag{22.31}$$

but we revert back to considering the radial wavefunction R_l instead of χ_l. The wavefunctions in the two regions are

$$R_l(r) = A_l j_l(\alpha r), \qquad r \le a \tag{22.32}$$

$$= B_l h_l^{(1)}(i\beta r), \quad r > a. \tag{22.33}$$

The boundary conditions at $r = a$ give

$$\frac{\alpha j_l'(\alpha a)}{j_l(\alpha a)} = \frac{i\beta h_l'^{(1)}(i\beta a)}{h_l^{(1)}(i\beta a)}. \tag{22.34}$$

We will examine the above relation for the case of low energies and strong potentials. These conditions can be incorporated by taking (with α_0 defined in (22.11))

$$\beta a \ll l \quad \text{and} \quad \alpha_0 a \gg l \tag{22.35}$$

which facilitates using the well-known behaviors of $j_l(\alpha_0 a)$ and $h_l^{(1)}(i\beta a)$ under those limits. We have

$$j_l(\rho) \to \frac{\sin(\rho - l\pi/2)}{\rho} \quad \text{for } \rho \gg l, \tag{22.36}$$

while we have

$$h_l^{(1)}(\rho) \to -i(2l-1)!!\rho^{-l-1} \quad \text{for } |\rho| \ll l \tag{22.37}$$

where $(2l-1)!! = 1.3\cdots(2l-1)$. Inserting (22.36) and (22.37) in (22.34) with the conditions given by (22.35), we obtain the following simple relation that must be satisfied for a bound state to occur:

$$\alpha_0 a \cot\left(\alpha_0 a - \frac{l\pi}{2}\right) = -l. \tag{22.38}$$

Dividing both sides of the above equation by $\alpha_0 a$, and remembering that $\alpha_0 a$ is very large, we find $\cot(\alpha a - l\pi/2) \approx 0$, which implies that

$$\alpha_0 a - \frac{l\pi}{2} = \left(n + \frac{1}{2}\right)\pi \tag{22.39}$$

where n is an integer.

Let us consider the scattering state. We once again use the symbols

$$\kappa = \sqrt{\frac{2m(V_0+E)}{\hbar^2}} \quad \text{and} \quad k = \sqrt{\frac{2mE}{\hbar^2}} \tag{22.40}$$

with $E > 0$, and use the general expression for the phase shift, δ_l, derived earlier for square well potentials,

$$\cot\delta_l = \frac{kn_l'(ka)j_l(\kappa_0 a) - \kappa_0 n_l(ka)j_l'(\kappa_0 a)}{kj_l'(ka)j_l(\kappa_0 a) - \kappa_0 j_l(ka)j_l'(\kappa_0 a)} \tag{22.41}$$

where κ_0 has already been defined in (22.20). At a resonance we have $\delta_l = \pi/2$ and, therefore, $\cot\delta_l = 0$. The above relation then implies that

$$kn_l'(ka)j_l(\kappa_0 a) - \kappa_0 n_l(ka)j_l'(\kappa_0 a) = 0. \tag{22.42}$$

In the above relation we use, once again,

$$j_l(\rho) \to \frac{\sin(\rho - l\pi/2)}{\rho} \quad \text{for } \rho \gg l \tag{22.43}$$

where $\rho = \kappa_0 a$ and also

$$n_l(\rho') \to -(2l-1)!!\rho'^{-l-1} \quad \text{for } \rho' \ll l \tag{22.44}$$

where $\rho' = ka$. Note that we have taken the behavior of n_l to be the same as that of $h_l^{(1)}$ given in Chapter 4 since in this region the contribution of j_l to $h_l^{(1)}$ is negligible. The relation (22.42) under these limits reads

$$-(l+2)\sin\left[\kappa_0 a - \frac{l\pi}{2}\right] + \kappa_0 a\cos\left[\kappa_0 a - \frac{l\pi}{2}\right] = 0. \tag{22.45}$$

Dividing both sides of the above equation by $\kappa_0 a$ we obtain

$$\cos\left[\kappa_0 a - \frac{l\pi}{2}\right] = -\left(\frac{l+2}{\kappa_0 a}\right)\sin\left[\kappa_0 a - \frac{l\pi}{2}\right]. \tag{22.46}$$

In the limit when $\kappa_0 a$ is very large, the right-hand side vanishes and, therefore, $\cos\left[\kappa_0 a - l\pi/2\right] = 0$, which implies

$$\kappa_0 a - \frac{l\pi}{2} = \left(n + \frac{1}{2}\right)\pi. \tag{22.47}$$

This is the same result as (22.39), since $\kappa_0 = \alpha_0$.

We thus conclude that the resonance condition is equivalent to the bound-state condition. If the potential is attractive enough to form a bound state, then there is a strong possibility that there will also be resonances. One can, in fact, have a bound state for a given partial wave and find that there is a resonance in the next higher partial wave; it depends on how strong the potential is.

22.1.3 Resonance as complex poles of the partial wave S-matrix

For the partial wave amplitude

$$f_l = \frac{1}{k\cot\delta_l - ik} \tag{22.48}$$

we consider the neighborhood of the point where $\delta_l = \pi/2$. As previously discussed, at $k = 0$ the phase shift δ_l vanishes. As k goes toward the resonance position, δ_l will increase for attractive potentials, through positive values, toward $\pi/2$. From our earlier discussions we found that δ_l is an odd function of k; therefore, $k\cot\delta_l$ is an even function. If a resonance occurs at $k = k_R$, i.e., $\cot\delta_l = 0$ at $k = k_R$, then we can expand $k\cot\delta_l$ near that point and write

$$k\cot\delta_l = \frac{k_R^2 - k^2}{\gamma}. \tag{22.49}$$

The partial wave amplitude can now be expressed as

$$f_l = \frac{\gamma}{k_R^2 - k^2 - i\gamma k}. \tag{22.50}$$

The poles of f_l are complex, given by the solution of

$$k^2 + i\gamma k - k_R^2 = 0. \tag{22.51}$$

The roots are given by $(-i\gamma \pm \sqrt{-\gamma^2 + 4k_R^2})/2$. In the limit $\gamma \ll k_R$, the locations of the poles are

$$k_R - i\frac{\gamma}{2}, \quad -k_R - i\frac{\gamma}{2}. \tag{22.52}$$

Thus, a resonance corresponds to complex poles in the lower half of the complex k-plane, which are symmetric with respect to the imaginary axis.

The term

$$\left|\frac{1}{k_R^2 - k^2 - i\gamma k}\right|^2 = \frac{1}{\left(k_R^2 - k^2\right)^2 + \gamma^2 k^2} \tag{22.53}$$

corresponding to the modulus squared of the scattering amplitude will have a peak at the position of the resonance, $k^2 = k_R^2$. How sharp the peak is depends on the value of γ, which is related to the width of the resonance. A smaller γ will produce a sharper peak.

The term "resonance" makes sense only if the peak is relatively sharp. A broad peak provides no special distinction to the scattering amplitude since it will blend in with the background. Thus, how close the complex poles are to the real axis is very important. The poles cannot lie on the real axis because that would make the scattering amplitude infinite. But the closer they are to the real axis the more they act as the real-axis poles, which instead of giving an infinite contribution give a very sharp peak consistent with the unitarity limit

$$|f_l|^2 = \left|\frac{e^{i\delta_l}\sin\delta_l}{k}\right|^2 \leq \frac{1}{k^2}. \tag{22.54}$$

Let us consider the analytic structure in the complex E-plane, where

$$E = \frac{\hbar^2 k^2}{2m}. \tag{22.55}$$

We note that the upper half of the k-plane will project onto the top sheet (first sheet) of the complex E-plane, which we call the "physical sheet." In terms of k, $\operatorname{Im} k > 0$, i.e., the upper half of the complex k-plane corresponds to the physical sheet. The lower half of the k-plane will project on to the second Riemann sheet of the complex E-plane, which we call the "unphysical sheet." In the complex E-plane the bound state poles appear along the negative real axis as we already found in the previous section. The pair of resonance poles, on the other hand, appear on the second Riemann sheet (the unphysical sheet), underneath the first sheet, one above the real axis and the other just below the real axis. Sitting there, these two poles control the shape of the peak.

22.2 Jost function formalism

We will confine ourselves to the S-waves. The wavefunction $\chi_0(r)$, which we discussed previously, is a well-behaved function as $r \to 0$. As $r \to \infty$ it is described by a linear combination of e^{ikr} and e^{-ikr} for the scattering states, and a decaying behavior, $e^{-\beta r}$, for the bound states. For the purposes of constructing the Jost function, which we designate as $F(k)$, we define a "regular wavefunction" $\phi_0(k, r)$ that is proportional to $\chi_0(r)$ but has the following behavior as $r \to 0$,

$$\phi_0(k,0) = 0, \tag{22.56}$$

$$\phi_0'(k,0) = 1. \tag{22.57}$$

Since the boundary conditions are independent of k, on the basis of the analysis by Poincaré it can be shown that $\phi_0(k,r)$ is an even function of k. As a function of complex k, it can also be shown that $\phi_0(k,r)$ is an entire function of k, i.e., it has no singularities in the complex k-plane.

We now define the so-called "irregular wavefunction," $f_0(k,r)$, which satisfies the boundary condition

$$f_0(k,r) \to e^{-ikr} \quad \text{as } r \to \infty. \tag{22.58}$$

Based on this boundary condition, once again, one can conclude that $f_0(k,r)$ is regular in the lower half of the complex k-plane (with Im $k < 0$) where it is well-defined, as the substitution, $k = \text{Re } k + i \text{ Im } k$, in the above boundary condition shows. Moreover, based on the same boundary condition, we conclude that

$$f_0^*(k,r) - f_0(-k^*,r) \tag{22.59}$$

for complex values of k. We note that $f_0(-k,r)$ which goes like e^{ikr} as $r \to \infty$, is also a solution of the Schrödinger equation. Moreover, because $f_0(k,r)$ is regular in the upper half of the complex k-plane plane, $f_0(-k,r)$ will be regular in the lower half.

Both ϕ_0 and f_0 satisfy the radial Schrödinger equation, with $l = 0$,

$$-\frac{\hbar^2}{2m}\frac{d^2\psi}{dr^2} + V(r)\psi = E\psi, \tag{22.60}$$

which we rewrite as

$$\frac{d^2\psi}{dr^2} - U(r)\psi + k^2\psi = 0 \tag{22.61}$$

where

$$U = \frac{2mV}{\hbar^2} \quad \text{and} \quad k^2 = \frac{2mE}{\hbar^2}. \tag{22.62}$$

If ψ_1 and ψ_2 are two solutions of (22.61) then following the discussion in the previous chapter we define the Wronskian $W[\psi_1, \psi_2]$ between them as follows:

$$W[\psi_1, \psi_2] = \psi_1 \frac{d\psi_2}{dr} - \psi_2 \frac{d\psi_1}{dr}. \tag{22.63}$$

As we have shown in the previous chapter,

$$W[\psi_1, \psi_2] = \text{constant independent of } r. \tag{22.64}$$

Since the functions $\phi_0(k,r)$ and $f_0(\pm k,r)$ are three independent solutions of a second-order differential equation, any one of them can be expressed in terms of the other two. We will write $\phi_0(k,r)$ in terms of $f_0(\pm k,r)$. Taking account of the fact that $\phi_0(k,r)$ is an even function of k, we write

$$\phi_0(k,r) = A(k)f_0(k,r) + A(-k)f_0(-k,r). \tag{22.65}$$

We now obtain the Wronskian $W\left[f_0(-k,r), \phi_0(k,r)\right]$, which from (22.65) gives

$$W\left[f_0(-k,r), \phi_0(k,r)\right] = A(k)W\left[f_0(-k,r), f_0(k,r)\right]. \tag{22.66}$$

Since the Wronskian is independent of r, we evaluate the left-hand side at $r = 0$, which gives

$$W\left[f_0(-k,r), \phi_0(k,r)\right]_{r=0} = f_0(-k,0) \tag{22.67}$$

where we have utilized the boundary conditions (22.57). Using the boundary condition (22.58) we now evaluate the right-hand side of (22.65) at $r = \infty$, which gives

$$W\left[f_0(-k,r), f_0(k,r)\right]_{r=\infty} = 2ik. \tag{22.68}$$

From (22.66), (22.67), and (22.68) we obtain

$$A(k) = \frac{f_0(-k,0)}{2ik}. \tag{22.69}$$

We now define the Jost function $F_0(k)$ as

$$F_0(k) = f_0(k,0). \tag{22.70}$$

Hence from (22.69), $\phi_0(k,r)$ given by (22.65) can be written as

$$\phi_0(k,r) = \frac{1}{2ik}\left[F_0(-k)f_0(k,r) - F_0(k)f_0(-k,r)\right]. \tag{22.71}$$

In particular, as $r \to \infty$,

$$\phi_0(k,r) \to \frac{1}{2ik}\left[F_0(-k)e^{-ikr} - F_0(k)e^{ikr}\right]. \tag{22.72}$$

Since this corresponds to a scattered state we compare it with $\chi_0(r)$, which has the behavior

$$\chi_0(r) \to \frac{S_0(k)e^{ikr} - e^{-ikr}}{2i} \tag{22.73}$$

where $S_0(k)$ is the S-matrix for $l = 0$. Thus we obtain the following two equalities:

$$\chi_0(r) = -\frac{k}{F_0(-k)}\phi_0(k,r) \tag{22.74}$$

and

$$S_0(k) = \frac{F_0(k)}{F_0(-k)}. \tag{22.75}$$

22.2.1 Zeros of $F(k)$

Consider now the situation when $\phi_0(k,r)$ represents a bound state. The wavefunction $\chi_0(r)$ has the asymptotic behavior

$$\chi_0(r) \to Be^{-\beta r} \tag{22.76}$$

where B is the normalization constant,

$$\beta = \sqrt{\frac{2mE_B}{\hbar^2}} \tag{22.77}$$

and E_B is the binding energy, $E_B = -|E|$.

Comparing the behavior given by (22.76) with that of $\phi_0(k,r)$, given by (22.72), we conclude that

$$F_0(-k) = 0 \quad \text{at } k = i\beta \tag{22.78}$$

and from (22.75) we find that

$$S_0(k) \text{ has a pole at } k = i\beta. \tag{22.79}$$

Thus the bound states correspond to poles of the S-matrix, $S_0(k)$, on the positive imaginary axis, the location being determined by the binding energy.

One can come to the same conclusion more directly by observing that the Jost function, $F_0(k)$, is itself a Wronskian of $\phi_0(k,r)$ and $f_0(k,r)$

$$F_0(k) = W\left[f_0(k,r), \phi_0(k,r)\right]. \tag{22.80}$$

If $F_0(k) = 0$ at $k = k_0$ then the Wronskian will vanish at that value and hence $\phi_0(k_0,r)$ will be a multiple of $f_0(k_0,r)$. Thus at

$$F_0(k_0) = 0 \tag{22.81}$$

we have

$$\phi_0(k_0,r) = Cf_0(k_0,r). \tag{22.82}$$

We know that ϕ_0 is well behaved at $r = 0$. If we take

$$k_0 = -i\beta \tag{22.83}$$

with $\beta > 0$, then from the asymptotic behavior of $f_0(k,r)$ we find that ϕ_0 will vanish exponentially at infinity. Thus $\phi_0(k_0,r)$ is regular both at $r = 0$ and at $r = \infty$. This implies that ϕ_0 is square integrable, which is a classic definition of a bound-state wavefunction. Thus, a bound state corresponds to a zero at $k_0 = -i\beta$ in $F_0(k)$ or, equivalently, to a zero at $k_0 = i\beta$ in $F_0(-k)$, given by

$$F_0(-i\beta) = 0. \tag{22.84}$$

22.2.2 Representation of $F_0(k)$ and $S_0(k)$ in terms of zeros and poles

The functions $F_0(k)$ and $F_0(-k)$ inherit the properties of $f_0(k,r)$ and $f_0(-k,r)$ in the complex k-plane. Thus, for example, from (22.59) we conclude that

$$F_0^*(k) = F_0(-k^*). \tag{22.85}$$

Similarly, like $f_0(k,r)$, $F_0(k)$ will be analytic in the lower half-plane while $F_0(-k)$ will be analytic in the upper half.

In order to express the S-matrix in terms of its poles it is important to keep in mind that $S_0(k)$ given by

$$S_0(k) = \frac{F_0(k)}{F_0(-k)} \tag{22.86}$$

is unitary along the positive real axis and in the upper half-plane (physical sheet) with $\operatorname{Im} k > 0$. From the relation (22.85) one can extend the unitarity relation to complex k''s by writing

$$S_0^*(k)S_0(k^*) = 1. \tag{22.87}$$

We will now concentrate only on the bound states and resonances.

(i) Bound states

The energy E is related to k by

$$E = \frac{\hbar^2 k^2}{2m}. \tag{22.88}$$

As we discussed previously, the bound state poles in the complex E-plane will appear along the negative real axis. We have found that the bound state poles of the S-matrix are located at

$$k = i\beta \quad (\beta > 0) \tag{22.89}$$

where β is a real number. Thus we can write, consistently with (22.85), for the case of a bound state

$$F_0(k) = \frac{k + i\beta}{k - i\alpha} \tag{22.90}$$

where α is a real number with $\alpha > 0$.

Since we are only interested in the bound state singularity we will write

$$S_0(k) = \frac{k + i\beta}{k - i\beta}, \tag{22.91}$$

which satisfies the unitarity property.

(ii) Resonances

For the complex poles corresponding to the resonances we will write

$$S_0(k) = \frac{F_0(k)}{F_0(-k)} = \frac{(k-\alpha-i\beta)\,(k+\alpha-i\beta)}{(k+\alpha+i\beta)\,(k-\alpha+i\beta)}. \tag{22.92}$$

This matrix is unitary on the physical sheet, and also satisfies the extended unitarity condition. It has poles on the unphysical sheet with $\operatorname{Im} k < 0$ (note $\beta > 0$).

To compare with section (22.1.3) we make the substitution

$$\alpha = k_R \tag{22.93}$$

with γ and k_0 positive, and obtain

$$S_0(k) = \frac{(k-k_R-i\beta)\,(k+k_R-i\beta)}{(k+k_R+i\beta)\,(k-k_R+i\beta)}. \tag{22.94}$$

The T-matrix is obtained from this through the relation

$$T_0(k) = \frac{S_0(k)-1}{2ik}. \tag{22.95}$$

Substituting (22.94) in (22.95), one finds

$$T_0(k) = \frac{-2\beta}{(k+k_R+i\beta)\,(k-k_R+i\gamma)} = \frac{2\beta}{k_R^2-k^2-2ik\beta} \tag{22.96}$$

where we have ignored the γ^2 term, assumed small, in the denominator of the middle term. This is a classic form for a resonance as we have already discussed. Thus, we reconfirm that a resonance corresponds to complex poles symmetric with respect to the imaginary axis and in the second sheet of the complex E-plane. Figs. 22.1 and 22.2 describe the structure of the S-matrix poles in the complex k-plane and complex E-plane, respectively.

22.2.3 Residue of the pole

The determination of the residue of the bound-state S-matrix pole is rather involved, though the final result is quite simple. We will derive the result by making the simplifying approximation of assuming that the bound-state wavefunction $\chi_0(r)$ is described entirely by the wavefunction for $r > a$ (the "outside" wavefunction), which is of the form $\exp(-\beta r)$. We then write for all values of r,

$$\chi_0(r) = Be^{-\beta r}. \tag{22.97}$$

The normalization constant B is obtained from the relation

$$\int_0^\infty dr\, \chi_0^2(r) = 1. \tag{22.98}$$

Substituting (22.97) in the above integral we obtain

$$B^2 = 2\beta. \tag{22.99}$$

The residue of the S-matrix pole can be determined from the simple formula in the previous section given by

$$S_0(k) = \frac{k + i\beta}{k - i\beta}. \tag{22.100}$$

The residue of the pole at $k = i\beta$ is found to be

$$\text{Res}\, S_0(k = i\beta) = 2i\beta. \tag{22.101}$$

Hence, comparing with (22.99), we conclude that

$$\text{Res}\, S_0(k = i\beta) = iB^2. \tag{22.102}$$

Therefore, the residue of the bound-state pole is given by the normalization of the wavefunction for the bound-state pole. This result turns out to be valid quite generally.

22.2.4 Phase shift $\delta_0(k)$

The Jost function formalism allows us to determine some important properties of the phase shift, $\delta_0(k)$.

We note that, for real values of k,

$$F_0^*(k) = F_0(-k). \tag{22.103}$$

The S-matrix, which is written as

$$S_0(k) = \frac{F_0(k)}{F_0(-k)}, \tag{22.104}$$

is, therefore, unitary for real values of k:

$$|S_0(k)| = 1. \tag{22.105}$$

In view of this unitary nature of the S-matrix, one writes

$$S_0(k) = e^{2i\delta_0(k)} \tag{22.106}$$

where $\delta_0(k)$ is the phase shift. From (22.103), (22.104), and (22.106) we have the following two results. First, from (22.104) we find

$$S_0(k) = \frac{1}{S_0(-k)}. \tag{22.107}$$

Therefore, from (22.106) and (22.107),

$$\delta_0(-k) = -\delta_0(k). \tag{22.108}$$

That is, the phase shift $\delta_0(k)$ is an odd function of k. Secondly, one can write the Jost function in terms of phase shifts as

$$F_0(k) = |F_0(k)|\, e^{i\delta_0(k)}. \tag{22.109}$$

22.3 Levinson's theorem

This is a very interesting theorem that relates the number of bound states in a system to the phase shift $\delta_0(k)$ at $k = 0$. It is assumed that $\delta_0(k) \to 0$ as $k \to \infty$, and that $F_0(0) \neq 0$.

Consider the contour integral

$$\frac{1}{2\pi i}\oint_C dk \frac{F_0'(-k)}{F_0(-k)} \tag{22.110}$$

where C is a closed contour that consists of the entire real axis and a semi-circle at infinity in the upper half of the k-plane, as indicated in Fig. 22.3.

The integrand in (22.110) has poles at the zeros of $F_0(-k)$ which, as we know, correspond to bound states. Since one can write $F_0(-k) = (k - k_0)F_0'(-k)$ near a zero at $k = k_0$, the residue of the pole is unity. Using Cauchy's residue theorem we conclude that the above integral is, therefore, simply equal to the number of bound states, n_B. Thus,

$$n_B = \frac{1}{2\pi i}\oint_C dk\, \frac{F_0'(-k)}{F_0(-k)}. \tag{22.111}$$

The integrand in (22.110) can also be written as

$$-\frac{1}{2\pi i}\oint_C dk\, \frac{d}{dk}\,|\ln F_0(-k)|\,. \tag{22.112}$$

Hence it is a perfect differential. Since

$$F_0(-k) = |F_0(-k)|\, e^{i\delta_0(-k)} = |F_0(-k)|\, e^{-i\delta_0(k)}, \tag{22.113}$$

we have

$$\ln F_0(-k) = \ln|F_0(-k)| - i\delta_0(k). \tag{22.114}$$

The integral, therefore, gives

$$\lim_{\substack{\epsilon\to 0\\ R\to\infty}} \frac{1}{2\pi}\left[\delta_0(\epsilon) - \delta_0(R) + \delta_0(-R) - \delta_0(-\epsilon)\right] = \frac{1}{\pi}\left[\delta_0(\epsilon) - \delta_0(R)\right] \tag{22.115}$$

where ϵ is an infinitesimally small quantity and where we have used the fact that $\delta_0(-k) = -\delta_0(k)$. Since we have assumed that δ_0 goes to zero at infinity, the above integral is

$$\frac{1}{\pi}\delta_0(0). \tag{22.116}$$

Thus,

$$\delta_0(0) = n_B\pi. \tag{22.117}$$

22.4 Explicit calculation of the Jost function for $l = 0$

Once again we consider the square-well potential

$$U(r) = -U_0, \quad r < a \tag{22.118}$$

$$U(r) = 0, \quad r > a \tag{22.119}$$

where, as defined earlier, $U(r) = 2mV(r)/\hbar^2$. We will work entirely with the irregular wavefunction $f_0(k,r)$ since the Jost function is given in terms of it by the relation

$$F_0(k) = f_0(k,0).$$

Since the potential vanishes for $r > a$, $f_0(k,r)$ will be the same as its asymptotic form,

$$f_0(k,r) = e^{-ikr}, \quad r > a. \tag{22.120}$$

For $r < a$, it can be written as a linear combination of the two solutions, $\exp(\pm i\kappa r)$,

$$f_0(k,r) = c_1(k)\, e^{i\kappa r} + c_2(k)\, e^{-i\kappa r}, \quad r < a \tag{22.121}$$

where

$$\kappa = \sqrt{U_0 + k^2}. \tag{22.122}$$

The boundary conditions at $r = a$ give

$$c_1 e^{i\kappa a} + c_2 e^{-i\kappa a} = e^{-ika}, \tag{22.123}$$

$$i\kappa c_1 e^{i\kappa a} - i\kappa c_2\, e^{-i\kappa a} = -ik\, e^{-ika}. \tag{22.124}$$

Therefore, we obtain

$$c_1 = \frac{\kappa - k}{2\kappa} e^{-i(\kappa+k)}, \quad c_2 = \frac{\kappa + k}{2\kappa} e^{-i(\kappa-k)} \tag{22.125}$$

and

$$F_0(k) = f_0(k, 0) = c_1 + c_2. \tag{22.126}$$

Hence, substituting the values of c_1 and c_2 we find the expression for the Jost function to be

$$F_0(k) = e^{-ika}\left[\cos\kappa a + i\frac{k}{\kappa}\sin\kappa a\right]. \tag{22.127}$$

Let us consider the zeros of $F_0(-k)$, which correspond to the bound states. The relation $F_0(-k) = 0$ corresponds to

$$\kappa\cot\kappa a - ik = 0. \tag{22.128}$$

If we consider the low-energy limit $k^2 \ll U_0$, the above relation becomes

$$ka = -i\kappa_0 a\cot\kappa_0 a \tag{22.129}$$

where $\kappa_0 a = \sqrt{U_0 a^2}$. This is the same relation as (22.23).

For

$$0 < \kappa_0 a < \frac{\pi}{2} \tag{22.130}$$

the zero lies along the negative imaginary axis of the complex k-plane. That is, the pole of the S-matrix lies along the negative imaginary axis, and will not correspond to a bound state. This zero moves up toward the origin as $\kappa_0 a$ increases toward $\pi/2$ (see Fig. 22.4). If we write

$$\kappa_0 a = \frac{\pi}{2} + \epsilon \tag{22.131}$$

for infinitesimally small ϵ, then the solution of (22.129) is given by

$$ka = i\frac{\pi}{2}\epsilon. \tag{22.132}$$

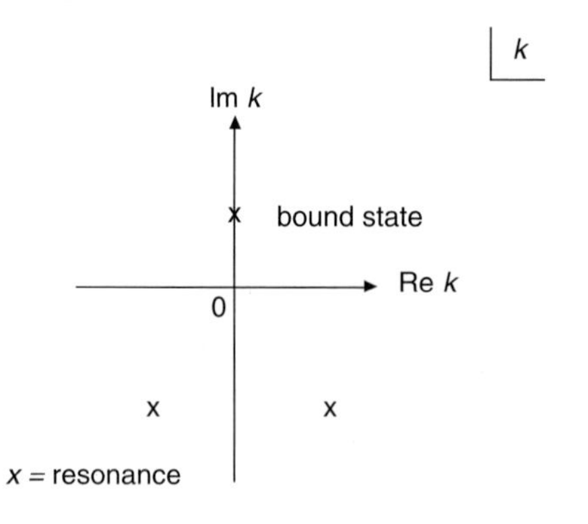

Fig. 22.1

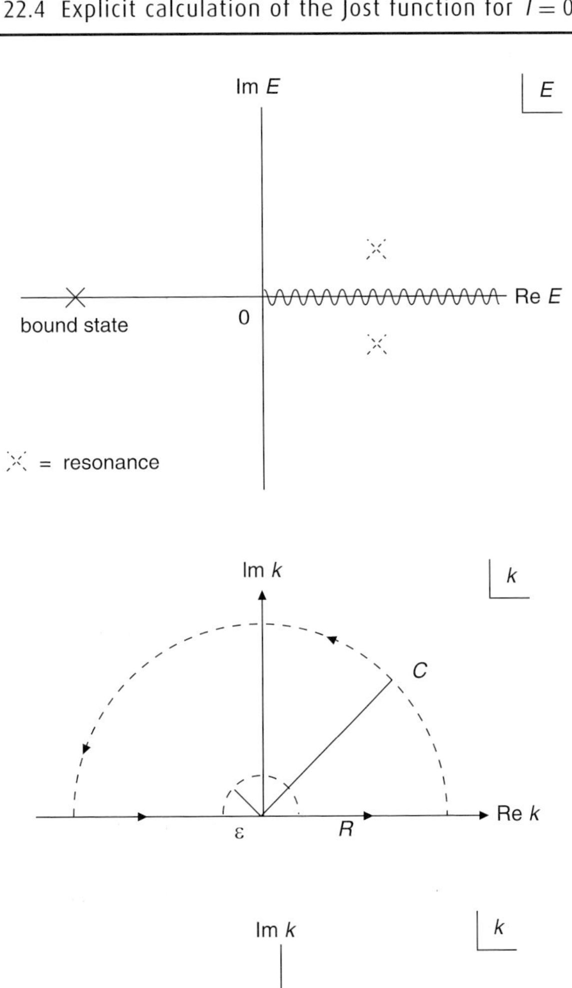

Fig. 22.2

Fig. 22.3

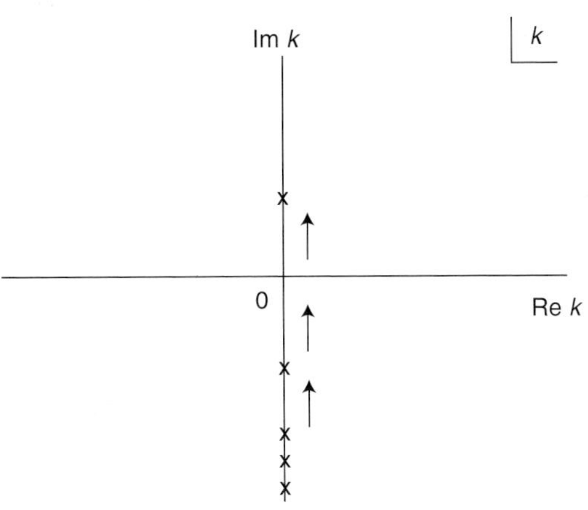

Fig. 22.4

Thus for $\epsilon > 0$ we have a zero in the upper half of the imaginary axis, which will correspond to a bound state, confirming our earlier results. For $\epsilon < 0$ we do not have any bound states since the zero is along the negative imaginary axis, though if ϵ is very small that zero corresponds to a "virtual state" since its contribution to the S-matrix is very large.

As the potential is increased, more zeros of $F_0(-k)$ move up along the imaginary axis and cross the origin to become bound-state zeros (poles of the S-matrix). According to the theory of analytic functions, the zeros cannot be created or destroyed, they pre-exist. For certain types of potentials they are all along the negative imaginary axis with an accumulation at $\operatorname{Im} k = -\infty$. As the strength of the potential increases, these zeros move up along the imaginary axis, one by one, and become bound-state zeros. One may also have pairs of zeros, $\alpha - i\beta, -\alpha - i\beta$. As the strength of the potential increases, both zeros move toward the negative imaginary axis and meet. One of them will continue to move up along the imaginary axis, with the other following it, and so on. Thus, as the potential gets stronger one will have an increasing number of bound states.

22.5 Integral representation of $F_0(k)$

Let us return to the irregular function $f_0(k,r)$ and consider the function

$$g(k,r) = e^{ikr} f_0(k,r), \tag{22.133}$$

which will satisfy the equation

$$\left[\frac{d^2}{dr^2} - 2ik\frac{d}{dr}\right] g(k,r) = U(r)g(k,r). \tag{22.134}$$

We can solve this equation using the Green's function technique. From the asymptotic property of $f_0(k,r)$ discussed earlier we find

$$\lim_{r \longrightarrow \infty} g(k,r) = 1. \tag{22.135}$$

One can write

$$g(k,r) = 1 + \int_0^\infty dr'\, D(k, r' - r) U(r') g(k,r') \tag{22.136}$$

where $D(k, r' - r)$ is the Green's function. One finds that

$$D(k,r') = \frac{1 - e^{-2ikr'}}{2ik} \qquad \text{for } r' \geq 0 \tag{22.137}$$

$$= 0 \qquad \text{for } r' < 0. \tag{22.138}$$

Hence,

$$g(k,r) = 1 + \int_r^\infty dr'\, D(k, r' - r) U(r') g(k,r'). \tag{22.139}$$

From the relation

$$g(k,0) = f_0(k,0) = F_0(k) \tag{22.140}$$

we can, therefore, write down an integral relation for $F_0(k)$:

$$F_0(k) = 1 + \int_0^\infty dr' \, D(k,r')U(r')g(k,r'). \tag{22.141}$$

From this relation one can derive several important properties of the Jost function, including its analytic structure and asymptotic behavior. We will derive an interesting result about the asymptotic behavior of the phase shift in the Born approximation.

The first Born approximation corresponds to replacing $g(k,r')$ in integral (22.141) by its zeroth value, which is 1. Substituting the expression (22.137) and (22.138) for $D(k,r')$ in (22.141), we obtain

$$F_0^B(k) = 1 + \frac{2m}{\hbar^2}\int_0^\infty dr' \left[\frac{1-e^{-2ikr'}}{2ik}\right] V(r'), \tag{22.142}$$

where we have substituted the value of $U(r')$ in terms of $V(r')$. In the limit $k \to \infty$ the exponential in (22.142), because of its rapidly oscillating behavior, will not contribute to the integral. Hence

$$F_0^B(k) \to 1 - i\frac{m}{k\hbar^2}\int_0^\infty dr' \, V(r'). \tag{22.143}$$

Since the second term above is small, one can write the expression for $F_0(k)$ in the approximate form

$$F_0(k) = e^{i\delta_0(k)} \approx 1 + i\delta_0(k). \tag{22.144}$$

Thus,

$$\delta_0^B(k) \to -\frac{m}{k\hbar^2}\int_0^\infty dr' \, V(r') \quad \text{as } k \to \infty, \tag{22.145}$$

which implies that $\delta_0(k) \to 0$ as $k \to \infty$. This is expected since in the high-energy limit the contribution from kinetic energy to the Hamiltonian is much larger than the potential energy and the centrifugal term becomes less important. Therefore, the particle will act like a free particle, which implies $\delta_l \to 0$ as $k \to \infty$.

22.6 Problems

1. Consider the bound-state problem where the wavefunction satisfies the equation

$$\frac{d^2u}{dx^2} - \beta^2 u(x) = \frac{2m}{\hbar^2} V(x) u(x).$$

Express the bound-state wavefunction in terms of the Green's function formalism. Obtain the corresponding Green's function.

2. For the above problem consider an attractive delta function potential

$$\frac{2m}{\hbar^2} V(x) = -\lambda \delta(x).$$

From the consistency of the wavefunction at $x = 0$ show that one must have $\beta = \lambda/2$. Show that this is precisely the point where the scattering solution has a pole in the complex k-plane. Compare your results with those obtained in Chapter 8.

3. Show that the radial wavefunction $R_l(k,r)$ can be expressed, in the Green's function formalism, as

$$R_l(k,r) = j_l(kr) + \int_0^\infty dr' r'^2 G_{0l}(k;r,r') U(r') R_l(k,r')$$

where $U(r) = \frac{2m}{\hbar^2} V(r)$ and

$$G_{0l}(k;r,r') = -ikj_l(kr_<) h_l^{(1)}(kr_>).$$

If there is a bound state at $k = i\alpha$ then express the wavefunction $R_l^B(k, i\alpha)$ in terms of an appropriate integral representation. Show that

$$\lim_{k \to i\alpha} (k - i\alpha) R_l(k,r) = R_l^B(k, i\alpha).$$

Hence, show that $R_l(k,r)$ has a pole at $k = i\alpha$.

4. Consider an attractive potential

$$V(r) = -\lambda \delta(r - a).$$

Show that the radial wavefunction and the partial wave scattering amplitudes have a pole in the complex k-plane. Determine the location of the pole and compare your results with the bound-state solutions in Chapter 8.

23 Poles of the Green's function and composite systems

Here we dig deeper into the properties of the Green's function and find an incredible amount of rich information. We also discover the elegant R-matrix formalism which, at least formally, allows one to describe the Green's function to all orders in perturbation theory. We describe through the R-matrix the resonance scattering involving composite systems.

23.1 Relation between the time-evolution operator and the Green's function

As we discussed in Chapter 2, the time evolution operator $U(t,t')$ is given by

$$U(t,t') = e^{-iH(t-t')/\hbar} \tag{23.1}$$

where H is the Hamiltonian. We define an operator

$$U(t) = U(t,0) \tag{23.2}$$

and take $\hbar = c = 1$. Thus

$$U(t) = e^{-iHt}. \tag{23.3}$$

The time evolution of a state vector $|\psi(t)\rangle$ is given by

$$|\psi(t)\rangle = U(t)|\psi(0)\rangle, \quad t > 0. \tag{23.4}$$

Hence we have, in terms of H,

$$|\psi(t)\rangle = e^{-iHt}|\psi(0)\rangle, \quad t > 0. \tag{23.5}$$

From the properties of the δ-function, we can write

$$e^{-iHt} = \int_{-\infty}^{\infty} dx\ e^{-ixt}\delta(x - H). \tag{23.6}$$

From the relations discussed in Chapter 2 we have

$$\frac{1}{x-H-i\epsilon} = P\left(\frac{1}{x-H}\right) + i\pi\delta(x-H), \tag{23.7}$$

$$\frac{1}{x-H+i\epsilon} = P\left(\frac{1}{x-H}\right) - i\pi\delta(x-H), \tag{23.8}$$

where P stands for the principal part. By subtracting (23.8) from (23.7), we have

$$\delta(x-H) = \frac{1}{2\pi i}\left[\frac{1}{x-H+i\epsilon} - \frac{1}{x-H-i\epsilon}\right]. \tag{23.9}$$

Therefore, (23.6) can be written as

$$e^{-iHt} = \frac{1}{2\pi i}\int_{-\infty}^{\infty} dx\ e^{-ixt}\left[\frac{1}{x-H-i\epsilon} - \frac{1}{x-H+i\epsilon}\right]. \tag{23.10}$$

The integrand has complex poles. Since H is Hermitian, its eigenvalues are real. Hence the real parts of the two poles are given by the eigenvalues of H. The imaginary parts are $\pm i\epsilon$. Thus the poles are off the real axis, while the integration in (23.10) is along the real axis. To evaluate the integral, therefore, one can, through complex integration techniques, utilize Cauchy's theorem. One then needs to add a semi-circle at infinity, either along the upper half of the complex z-plane ($z = x + iy$) or the lower half-plane to construct a closed contour. Since we are interested only in $t > 0$, the presence of the factor e^{-izt} in the complex plane demands that we choose the lower half-plane. Hence only the second term in (23.10) will be relevant since it has poles below the real axis. Discarding the first term in (23.10), which does not contribute to the integral, we write, for $t > 0$

$$e^{-iHt} = -\frac{1}{2\pi i}\int_{-\infty}^{\infty} dx\ \frac{e^{-ixt}}{x-H+i\epsilon}. \tag{23.11}$$

We can rewrite the integral in (23.11) by changing x to E and shifting the line of integration to the line above the real axis. Thus we write

$$e^{-iHt} = \frac{1}{2\pi i}\int_C \frac{dE\ e^{-iEt}}{E-H}, \quad t > 0 \tag{23.12}$$

where the contour C is a line that runs from $\infty + iy_0$ to $-\infty + iy_0$ (thereby removing the negative sign in front of the integral), where y_0 is a positive constant.

However, the total Green's function, $G(E)$, is given by

$$G(E) = \frac{1}{E-H}. \tag{23.13}$$

Therefore, (23.12) can be re-written as

$$U(t) = e^{-iHt} = \frac{1}{2\pi i}\int_C dE\ e^{-iEt}G(E), \quad t > 0 \tag{23.14}$$

and

$$|\psi(t)\rangle = \left[\frac{1}{2\pi i}\int_C dE\ e^{-iEt}G(E)\right]|\psi(0)\rangle,\ t > 0. \tag{23.15}$$

Hence the time evolution operator $U(t)$ is intimately connected to the Green's function $G(E)$. The poles and other singularities of $G(E)$ in the complex E-plane then largely determine $U(t)$, and thereby the state vector $|\psi(t)\rangle$.

23.2 Stable and unstable states

We discuss two important examples of $G(E)$. First we consider

$$G(E) = \frac{1}{E - E_0} \tag{23.16}$$

where E_0 is real. From (23.15) we obtain after contour integration

$$|\psi(t)\rangle = e^{-iE_0t}|\psi(0)\rangle, \quad t > 0, \tag{23.17}$$

which corresponds to an energy eigenstate with eigenvalue E_0. In other words, it will represent a stable particle, e.g., a free particle or a bound state. This result is not surprising since expression (23.16) from the definition (23.10) implies that E_0 is the eigenvalue of H. As a consequence, we find that the state vector (23.17) represents a stable state, e.g., a free particle or a bound state of energy E_0.

As a second example we consider

$$G(E) = \frac{1}{E - E_0 + i\frac{\Gamma}{2}} \tag{23.18}$$

where E_0 is real and Γ is real and positive. This expression appears to imply that the eigenvalue of H is complex and not real as it should be since H is Hermitian. We will return to this apparent contradiction in Section 23.4.

Substituting (23.18) in (23.15) we find

$$|\psi(t)\rangle = e^{-iE_0t}e^{-\frac{\Gamma}{2}t}|\psi(0)\rangle, \quad t > 0, \tag{23.19}$$

which corresponds to a state that "decays" and eventually disappears. If Γ is large then the state $|\psi(t)\rangle$ decays quickly; if it is small then it "lives" for a long time. One calls $1/\Gamma$ the "lifetime" of the state. Thus $|\psi(t)\rangle$ given by (23.19) describes an unstable or a metastable state.

23.3 Scattering amplitude and resonance

The Green's function also appears in scattering problems through the T-matrix as we discussed earlier. The relation between T and G is given by

$$T = V + VGV \tag{23.20}$$

where V is the interaction potential. For illustration purposes let us consider the scattering of two particles A and B,

$$A + B \to A + B. \tag{23.21}$$

For simplicity we assume B to be much heavier than A. If k represents the momentum of A in the center-of-mass system then the kinetic energy of A is given by

$$E = \frac{k^2}{2m}. \tag{23.22}$$

The reaction (23.21) occurs for $k \geq 0$ (i.e., $E \geq 0$). Hence the point $k = E = 0$ is called the "threshold" for the reaction $A + B \to A + B$.

Below we discuss the properties of the T-matrix for the two examples of the Green's function, (23.16) and (23.18).

Consider first the Green's function given by (23.16). Taking the matrix element of T between the initial and final states, $|i\rangle$ and $|f\rangle$ respectively, and inserting (23.16) for G, we obtain

$$\langle f|\, T\, |i\rangle = \langle f|\, V\, |i\rangle + \langle f|\, V \left[\frac{1}{E - E_0}\right] V\, |i\rangle\ . \tag{23.23}$$

This matrix then describes the scattering $A + B \to A + B$ above the threshold $E = 0$. If $E_0 > 0$ then the T-matrix becomes infinite at $E = E_0$ and hence so does the S-matrix. This is untenable since the S-matrix is supposed to be unitary for $E > 0$. Thus E_0 must be negative and outside the scattering region. This result is not surprising because the region $E < 0$ where the infinity in the Green's function occurs is precisely the region where a bound state between the particles A and B will occur.

Consider now the Green's function given by (23.18) and insert it in expression (23.10) for the T-matrix. We obtain

$$\langle f|\, T\, |i\rangle = \langle f|\, V\, |i\rangle + \langle f|\, V \left[\frac{1}{E - E_0 + i\frac{\Gamma}{2}}\right] V\, |i\rangle\, . \tag{23.24}$$

If $E_0 > 0$ (i.e., above the threshold) and Γ is small, then the T-matrix near $E = E_0$ can be approximated as

$$\langle f|\, T\, |i\rangle \simeq \langle f|\, V \left[\frac{1}{E - E_0 + i\frac{\Gamma}{2}}\right] V\, |i\rangle\, . \tag{23.25}$$

The differential cross-section is then proportional to

$$\left| \langle f | V \left[\frac{1}{E - E_0 + i\frac{\Gamma}{2}} \right] V | i \rangle \right|^2 . \tag{23.26}$$

The contribution of the middle square bracket is given by

$$\frac{1}{(E - E_0)^2 + \frac{\Gamma^2}{4}} . \tag{23.27}$$

The scattering cross-section will, therefore, have a peak at $E = E_0$, which will become very sharp as Γ becomes small. This phenomenon is called "resonance" and E_0 and Γ are called the position and the width of the resonance, respectively. The existence of such a peak indicates that the system has almost a bound state, a metastable state, at the position of the peak.

23.4 Complex poles

We note that E_0 is a real quantity both for bound states, where it is negative, and for resonance, where it is positive. Earlier we stated that the poles of the Green's function must be real since the eigenvalues of the Hamiltonian, H, which is a Hermitian operator, must be real. The result for the bound states where G has a pole at $E = E_0$ is consistent with this observation. However, for the resonance the appearance of a complex pole at $E = E_0 - i\Gamma/2$ appears to contradict this assertion. The answer to this apparent contradiction lies not in the complex E-plane, which we have been considering thus far, but in the complex k-plane as we have already discussed in Chapter 22.

The point is that the poles that give rise to the resonances are actually in the second sheet, the so-called unphysical sheet, with $\operatorname{Im} k > 0$, in the complex k-plane and not on the physical sheet. Their presence gives rise to peaks in the S-matrix and similarly in the Green's function, not infinities, and the expression (23.18) then remains compatible with the unitarity of the S-matrix.

23.5 Two types of resonances

Typically, there are two types of resonances. One of them was discussed in Chapter 22 that had as its origin a combination of centrifugal barrier and a strong attractive potential that created a "temporary" (metastable) state. One can write this process as

$$A + B \rightarrow B^* \rightarrow A + B \tag{23.28}$$

where B^* is called a metastable state.

One can view this phenomenon by considering the combination of the potential and the centrifugal:

$$V(r) + \frac{\hbar^2 l(l+1)}{2mr^2}. \tag{23.29}$$

This is called the effective potential, V_{eff}. Suppose $V(r)$ is of the form

$$V(r) = -g\frac{e^{-\mu r}}{r} \tag{23.30}$$

so that it is attractive and of finite range. If it is sufficiently attractive, that is, if g is sufficiently large, then we have a situation which for $l \neq 0$ looks very much like a potential well that would trap a particle giving rise to bound states. However, here $E > 0$, with a barrier height that is finite, which will allow the particle to tunnel through. Therefore, we will "almost" have a bound state, which is sometimes called a "metastable" state.

There is also another possibility where B is a composite state, e.g., an atom in its ground state, where A provides enough energy to B to raise it to an excited state B^* that eventually decays back to the ground state consisting of A and B.

In our discussion on time-dependent perturbation theory in Chapter 18 we considered the case of the interaction between a photon and an atom in its ground state, where the atom "absorbed" the photon, resulting in the atom being kicked to an excited state. This process will describe the reaction $A + B \to B^*$, in other words, "half" the process described by (23.28), where A is the incident photon, B the atom in its ground state, and B^* the atom in its excited state. We obtained the transition probability per unit time for this process, which will clearly be related to the lifetime of B^*. If we designate this probability as w, then we find

$$\Gamma \sim w \tag{23.31}$$

where Γ^{-1}, defined in (23.19), is the lifetime of B^*.

Our calculations in the earlier chapters on bound states, transition probabilities and resonances were confined to the lowest order in perturbation. In the next section we describe a formalism involving composite systems that provides an exact expression for the diagonal element for the Green's function, $G(E)$, and from it the T-matrix under quite general conditions. In the end, in all cases, we express G in a form similar to expression (23.18), in terms of E_0 and Γ.

23.6 The reaction matrix

Even though the formalism of the reaction matrix (R), which we will describe below, has been widely used in nuclear physics, it also has great usefulness in discussing bound states, resonances, and unstable states. We define R as follows:

$$R = V + VP\left(\frac{1}{E - H_0}\right)R \tag{23.32}$$

where H_0 is the unperturbed Hamiltonian, and the symbol P stands for principal part that appears in the relation

$$\frac{1}{E - H_0 + i\epsilon} = P\left(\frac{1}{E - H_0}\right) - i\pi\,\delta\left(E - H_0\right). \tag{23.33}$$

The term corresponding to the principal part is real, while the term with the δ-function is imaginary.

Moving the second term on the right-hand side of (23.32) that involves R and combining it with R on the left and dividing both sides by the factor multiplying R we obtain

$$R = \frac{1}{1 - VP\left(\dfrac{1}{E - H_0}\right)} V. \tag{23.34}$$

Expanding the denominator we then have

$$R = V + VP\left(\frac{1}{E - H_0}\right) V + VP\left(\frac{1}{E - H_0}\right) VP\left(\frac{1}{E - H_0}\right) V + \cdots. \tag{23.35}$$

Returning to the definition given in (23.32) we note that the principal parts normally enter in integrals of the type

$$\int_{-\infty}^{\infty} dx\, \frac{f(x)}{x - x_0 + i\epsilon} \tag{23.36}$$

where one writes

$$\frac{1}{x - x_0 + i\epsilon} = \frac{x - x_0}{(x - x_0)^2 + \epsilon^2} - i\frac{\epsilon}{(x - x_0)^2 + \epsilon^2}. \tag{23.37}$$

As we have discussed earlier, in the limit $\epsilon \to 0$ one has

$$\lim_{\epsilon \to 0} \frac{1}{x - x_0 + i\epsilon} = P\left(\frac{1}{x - x_0}\right) - i\pi\,\delta(x - x_0) \tag{23.38}$$

where

$$P\left(\frac{1}{x - x_0}\right) = \lim_{\epsilon \to 0} \frac{x - x_0}{(x - x_0)^2 + \epsilon^2}. \tag{23.39}$$

This expression vanishes at $x = x_0$ for a fixed ϵ and hence one writes the principal part in (23.36) as

$$P\int_{-\infty}^{\infty} dx\, \frac{f(x)}{x - x_0} = \int_{-\infty}^{x_0 - \delta} dx\, \frac{f(x)}{x - x_0} + \int_{x_0 + \delta}^{\infty} dx\, \frac{f(x)}{x - x_0} \tag{23.40}$$

where δ is a positive, infinitesimal quantity. The principal part integral, therefore, excludes the point $x = x_0$.

The above definitions involve continuous variables but we can extend them to the discrete case. Consider, specifically, the following diagonal matrix element with respect to the eigenstate $|s\rangle$ of H_0:

$$\langle s|VP(\frac{1}{E-H_0})V|s\rangle, \tag{23.41}$$

which we encountered in the expansion for R in (23.35). By inserting a complete set of orthonormal eigenstates $|s'\rangle$ and $|s''\rangle$ we obtain

$$\left\langle s \left| VP\left(\frac{1}{E-H_0}\right)V \right| s\right\rangle = \sum_{s'}\sum_{s''}\langle s|V|s'\rangle\left\langle s' \left| P\left(\frac{1}{E-H_0}\right)\right| s''\right\rangle\langle s''|V|s\rangle. \tag{23.42}$$

First of all we note that E in the denominator in (23.42) multiplies a unit operator, while H_0 is diagonal: $H_0|s\rangle = E_s|s\rangle$, where E_s is the unperturbed energy. Thus $1/(E-H_0)$, and therefore, $P(1/(E-H_0))$ are diagonal. Furthermore, since the principal part implies that the state $|s\rangle$ should be excluded in the sum, we obtain

$$\left\langle s \left| VP\left(\frac{1}{E-H_0}\right)V \right| s\right\rangle = \sum_{s'\neq s}\langle s|V|s'\rangle\frac{1}{E-E_{s'}}\langle s'|V|s\rangle. \tag{23.43}$$

We discuss below an alternative method to accomplish the above result more simply by defining a projection operator, Λ_s, which projects out the state $|s\rangle$,

$$\Lambda_s|s'\rangle = |s\rangle\delta_{ss'}. \tag{23.44}$$

One can also write $\Lambda_s = |s\rangle\langle s|$. It follows that $(1-\Lambda_s)$ projects out all the states except the state $|s\rangle$, i.e.,

$$(1-\Lambda_s)|s\rangle = 0. \tag{23.45}$$

Hence we will write

$$P\left(\frac{1}{E-H_0}\right) = \frac{1-\Lambda_s}{E-H_0} \tag{23.46}$$

whenever the principal part term occurs in a diagonal matrix with respect to the state $|s\rangle$. Inserting (23.46) in the left-hand side of (23.42) we recover (23.43).

Thus the diagonal matrix element of R defined as a series expansion in (23.35) can be written as

$$\langle s\,|R|\,s\rangle = \left\langle s \left| V + V\frac{1-\Lambda_s}{E-H_0}V + V\frac{1-\Lambda_s}{E-H_0}V\frac{1-\Lambda_s}{E-H_0}V + \cdots \right| s\right\rangle. \tag{23.47}$$

Therefore,

$$\langle s|R|s\rangle = \langle s|V|s\rangle + \sum_{s'\neq s} \frac{\langle s|V|s'\rangle\langle s'|V|s\rangle}{E - E_{s^e}} \tag{23.48}$$

$$+\sum_{s'\neq s}\sum_{s''\neq s} \frac{\langle s|V|s'\rangle\langle s'|V|s''\rangle\langle s''|V|s\rangle}{(E - E_{s'})(E - E_{s''})} + \cdots . \tag{23.49}$$

Using (23.46), expression (23.34) can be written as

$$\langle s|R|s\rangle = \left\langle s \left| \frac{1}{1 - V\dfrac{1-\Lambda_s}{E - H_0}} V \right| s \right\rangle. \tag{23.50}$$

By expanding the denominator it is straightforward to show that one can also write

$$\langle s|R|s\rangle = \left\langle s \left| V\frac{1}{1 - \dfrac{1-\Lambda_s}{E - H_0} V} \right| s \right\rangle. \tag{23.51}$$

23.6.1 Relation between *R*- and *T*-matrices

We now turn our attention to obtaining the T-matrix in terms of R by writing it in the form we considered earlier,

$$T = V + V\frac{1}{E - H_0 + i\epsilon}T. \tag{23.52}$$

From (23.33) applied to the middle factor in the second term on the right-hand side, we obtain

$$T = V + VP\left(\frac{1}{E - H_0}\right)T - i\pi V\delta(E - H_0)T. \tag{23.53}$$

In order to relate T and R we now write

$$T = R(1 + \Delta) \tag{23.54}$$

and determine Δ in a self-consistent manner by inserting (23.54) in the second term of the right-hand side of expression (23.53):

$$T = V + VP\left(\frac{1}{E - H_0}\right)R + VP\left(\frac{1}{E - H_0}\right)R\Delta - i\pi V\delta(E - H_0)T. \tag{23.55}$$

Using the definition of R in (23.32) we replace the first two terms in (23.55) by R and rewrite the third as $(R - V)\Delta$ using (23.54). Thus we have

$$T = R + (R - V)\Delta - i\pi V\delta(E - H_0)T, \tag{23.56}$$

$$= R + R\Delta - V\Delta - i\pi V\delta(E - H_0)T. \tag{23.57}$$

Let us take

$$\Delta = -i\pi\delta(E - H_0)T. \tag{23.58}$$

Then from (23.57) we obtain

$$T = R + R\Delta. \tag{23.59}$$

Hence, we recover (23.54) with the value of Δ now determined through (23.58). Inserting Δ in (23.59) we obtain the relation between T and R:

$$T = R - i\pi R\delta(E - H_0)T. \tag{23.60}$$

23.6.2 Relation between R and G

The Green's function, G, is given by

$$G(E) = \frac{1}{E - H_0 - V} \tag{23.61}$$

where the total Hamiltonian is $H = H_0 + V$. The denominator is kept real by removing the $i\epsilon$ term which can easily be restored by considering the function $G(E + i\epsilon)$. From (23.61) we obtain

$$(E - H_0 - V)G = 1. \tag{23.62}$$

Taking the diagonal elements, we have

$$\langle s|(E - H_0 - V)G|s\rangle = 1. \tag{23.63}$$

Since E and H_0 are diagonal operators and $H_0|s\rangle = E_s|s\rangle$, we obtain

$$(E - E_s)\langle s|G|s\rangle - \langle s|VG|s\rangle = 1. \tag{23.64}$$

We determine $\langle s|VG|s\rangle$ by first writing

$$G = \frac{1}{E - H_0 - \Lambda_s V - (1 - \Lambda_s)V} \tag{23.65}$$

where we have added and subtracted the $\Lambda_s V$ term, where Λ_s is the projection operator $(= |s\rangle\langle s|)$ defined earlier for the state $|s\rangle$. Consider now the product

$$\left[1 - \frac{(1-\Lambda_s)V}{E-H_0}\right][E - H_0 - \Lambda_s V] \tag{23.66}$$

$$= E - H_0 - \Lambda_s V - (1-\Lambda_s)V + \frac{(1-\Lambda_s)\Lambda_s V^2}{(E-H_0)} \tag{23.67}$$

$$= E - H_0 - \Lambda_s V - (1-\Lambda_s)V, \tag{23.68}$$

where we have used the relation

$$\Lambda_s(1-\Lambda_s) = 0 \tag{23.69}$$

as it corresponds to the product of a term that allows only the state $|s\rangle$ and, simultaneously, a term that excludes state $|s\rangle$.

The right-hand side of (23.68) is the denominator of G in (23.65). Thus we obtain from (23.65) and (23.68) the following result for G:

$$G = \frac{1}{\left[1 - \frac{(1-\Lambda_s)V}{E-H_0}\right][E - H_0 - \Lambda_s V]}. \tag{23.70}$$

Inserting this expression for G in the matrix element $\langle s\,|VG|\,s\rangle$ in (23.64), we obtain

$$\langle s\,|VG|\,s\rangle = \left\langle s \left| V \frac{1}{\left[1 - \frac{(1-\Lambda_s)V}{E-H_0}\right][E - H_0 - \Lambda_s V]} \right| s \right\rangle \tag{23.71}$$

$$= \sum_{s'} \left\langle s \left| V \frac{1}{\left[1 - \frac{(1-\Lambda_s)V}{E-H_0}\right]} \right| s' \right\rangle \left\langle s' \left| \frac{1}{[E - H_0 - \Lambda_s V]} \right| s \right\rangle \tag{23.72}$$

where we have introduced a complete set of states $|s'\rangle$.

We note that

$$\frac{1}{E - H_0 - \Lambda_s V} \tag{23.73}$$

is a diagonal operator because of the presence of Λ_s and the fact that E and H_0 are diagonal. Therefore, only $s' = s$ contribute in (23.72). The diagonal matrix element of (23.73) is identical to the diagonal matrix element of G, since the diagonal element of $(1-\Lambda_s)$ vanishes in (23.65). Thus,

$$\left\langle s \left| \frac{1}{E - H_0 - \Lambda_s V} \right| s \right\rangle = \langle s\,|G|\,s\rangle\,. \tag{23.74}$$

Hence (23.72) can be written as

$$\langle s\,|VG|\,s\rangle = \left\langle s \left| V \frac{1}{\left[1 - \frac{(1-\Lambda_s)V}{E-H_0}\right]} \right| s \right\rangle \langle s\,|G|\,s\rangle\,. \tag{23.75}$$

The first factor on the right-hand side is identical to $\langle s|R|s\rangle$ given in (23.51). If we write $\langle s\,|G|\,s\rangle = G_{ss}$ and $\langle s|R|s\rangle = R_{ss}$ then the relation (23.75) can be written as

$$\langle s\,|VG|\,s\rangle = R_{ss}G_{ss}. \tag{23.76}$$

Inserting (23.76) in (23.64) we obtain a remarkably simple result for the diagonal matrix element of the Green's function, G_{ss}, as

$$G_{ss}\,(E) = \frac{1}{E - E_s - R_{ss}\,(E)} \tag{23.77}$$

where we have made the dependence on E explicit.

From (23.49) we have

$$R_{ss}\,(E) = \langle s|V|s\rangle + \sum_{s'\neq s} \frac{\langle s|V|s'\rangle\langle s'|V|s\rangle}{E - E_{s'}} + \sum_{s'\neq s}\sum_{s''\neq s} \frac{\langle s|V|s'\rangle\langle s'|V|s''\rangle\langle s''|V|s\rangle}{(E - E_{s'})(E - E_{s''})} + \cdots. \tag{23.78}$$

Inserting this in (23.77) we find that we have reproduced the results of time-independent perturbation theory from a different perspective. We elaborate on this comment: if we designate the energy of a bound state as E_{bound}, which is a pole of $G_{ss}\,(E)$, and note that $E_s, E_{s'}, \ldots$ are the energies of the unperturbed states, which can be rewritten as $E_s^0, E_{s'}^0, \ldots$ to conform to the notations used in the perturbation treatment of Chapter 16 then

$$E_{bound} = E_s^0 + R_{ss}\left(E_s^0\right), \tag{23.79}$$

which is precisely what we derived in Chapter 16. If one isolates the pole then

$$G_{ss}\,(E) \sim \frac{1}{E - E_{bound}}.$$

We have already noted that the poles corresponding to the bound states occurring below the threshold will be real. Therefore, R_{ss} will be real for those cases. For the poles occurring above the threshold, R_{ss} will be complex, which is one of the properties we will discuss below.

23.6.3 Properties of $R_{ss}(E)$

Let us consider $R(E)$ for all values of E including complex values. To simplify things we write

$$P\left(\frac{1}{E-H_0}\right) = P(E, H_0), \tag{23.80}$$

where the principal part, P, can be written as

$$P(E, H_0) = \frac{1}{2}\left[\frac{1}{E-H_0+i\epsilon} + \frac{1}{E-H_0-i\epsilon}\right](1-\Lambda_s). \tag{23.81}$$

The factor $(1-\Lambda_s)$ ensures that the diagonal elements of P with respect to $|s\rangle$ will vanish, a property we have already discussed.

From (23.34) we have

$$R(E+i\sigma) = \frac{1}{1-VP(E+i\sigma, H_0)}V \tag{23.82}$$

where σ is an infinitesimal quantity different from ϵ in (23.81). This result leads to the following:

$$R(E+i\sigma) - R(E-i\sigma) = \left[\frac{1}{1-VP(E+i\sigma, H_0)} - \frac{1}{1-VP(E-i\sigma, H_0)}\right]V \tag{23.83}$$

$$= -V\frac{[P(E-i\sigma, H_0) - P(E+i\sigma, H_0)]}{[1-VP(E+i\sigma, H_0)][1-VP(E-i\sigma, H_0)]}V. \tag{23.84}$$

To obtain $P(E+i\sigma) - P(E-i\sigma)$ let us consider the following difference:

$$\frac{1}{E+i\sigma-H_0\pm i\epsilon} - \frac{1}{E-i\sigma-H_0\pm i\epsilon} = \frac{-2i\sigma}{(E-H_0\pm i\epsilon)^2+\sigma^2}. \tag{23.85}$$

We now take $\epsilon \to 0$ first, keeping σ fixed, and obtain from (23.81)

$$P(E-i\sigma, H_0) - P(E+i\sigma, H_0) = \left[\frac{-2i\sigma}{(E-H_0)^2+\sigma^2}\right](1-\Lambda_s). \tag{23.86}$$

If we now take $\sigma \to 0$, then the square bracket on the right-hand side is $-2\pi i\delta(E-H_0)(1-\Lambda_s)$.

Expression (23.84) in the limit $\sigma \to 0$ will then be of the form

$$R(E+i\sigma) - R(E-i\sigma) = 2\pi i V\frac{\delta(E-H_0)(1-\Lambda_s)V}{[1-VP(E+i\sigma, H_0)][1-VP(E-i\sigma, H_0)]} \tag{23.87}$$

$$= 2\pi i R(E+i\sigma)\delta(E-H_0)(1-\Lambda_s)R(E-i\sigma). \tag{23.88}$$

We use the property of the functions of complex variables that states that if a function $f(z)$ of $z(=x+iy)$ is analytic in a region of the complex z-plane and is real along the real axis, then by analytic continuation the value of the function at z^* is given by $f(z^*) = f^*(z)$. We assume that $R(E)$ satisfies this analyticity property and, since it is real for negative values of E, which is the region of the bound states, one can write

$$R(E - i\sigma) = R^\dagger(E + i\sigma). \tag{23.89}$$

Hence from (23.88) we have in the limit $\sigma \to 0$,

$$R(E + i\sigma) - R^\dagger(E + i\sigma) = 2\pi i R(E + i\sigma)\delta(E - H_0)(1 - \Lambda_s)R^\dagger(E + i\sigma). \tag{23.90}$$

We now consider the region above the threshold where R can be complex. The left-hand side above is $2i \operatorname{Im} R(E + i\sigma)$. Thus, by taking the diagonal component of (23.90) and inserting a complete set of states on the right-hand side we obtain the following:

$$\lim_{\sigma \to 0} \langle s| \operatorname{Im} R(E + i\sigma)|s\rangle = \pi \sum_{s' \neq s} \langle s|R(E_{s'} + i\sigma)|s'\rangle \langle s'|R^\dagger(E_{s'} + i\sigma)|s\rangle \delta(E_s - E_{s'}). \tag{23.91}$$

Therefore,

$$\operatorname{Im}\langle s|R(E + i\sigma)|s\rangle = \pi \sum_{s' \neq s} \left|\langle s'|R(E_{s'} + i\sigma)|s\rangle\right|^2 \delta(E_s - E_{s'}). \tag{23.92}$$

We simplify some of the notations and define

$$R(E + i\sigma) = R^{(+)}(E), \tag{23.93}$$

$$\langle s|R(E + i\sigma)|s\rangle = R_{ss}^{(+)}(E), \tag{23.94}$$

$$\langle s|R(E + i\sigma)|s'\rangle = R_{ss'}^{(+)}(E), \tag{23.95}$$

and

$$\langle s|V|s'\rangle = V_{ss'}. \tag{23.96}$$

For the superscript $(-)$ on the right one must change the sign of σ on the left.

Let us write

$$R_{ss}^{(+)}(E) = D_s(E) - iI_s(E) \tag{23.97}$$

where the real part of $R_s(E)$ will be identical to $R_s(E)$ determined earlier in (23.78) when $\sigma = 0$. Thus,

$$D_s(E) = V_{ss} + \sum_{s' \neq s} \frac{V_{ss'}V_{s's}}{E - E_{s'}} + \cdots . \tag{23.98}$$

The imaginary part is given in (23.91),

$$I_s(E) = \pi \sum_{s' \neq s} \left| R_{ss'}^{(+)}(E_s) \right|^2 \delta(E_s - E_{s'}) \tag{23.99}$$

$$= \frac{1}{2} w_s \tag{23.100}$$

where w_s is the transition probability for transition from a state $|s\rangle$ to a group of final states $|s'\rangle$, with $E_s = E_{s'}$. The first-order perturbation version of w_s has already been derived in the chapter on time-dependent perturbation theory, where

$$w_s = \frac{2\pi}{\hbar} \sum_{s' \neq s} |V_{ss'}(E_s)|^2 \, \delta(E_s - E_{s'}) \tag{23.101}$$

except that we have taken $\hbar = c = 1$ in our discussions here and have written $V_{ss'}$ instead of $H'_{ss'}$. We emphasize that the quantity w_s expressed in (23.100) is given to all orders in perturbation, while (23.101) was only to first order. The diagonal Green's function is then given by

$$G_{ss}(E) = \frac{1}{E - E_s - D_s(E) + \frac{i}{2} w_s}, \tag{23.102}$$

which is of the resonance form with the decay rate given by

$$\Gamma_s = w_s. \tag{23.103}$$

The position of the resonance is given by

$$E_{resonance} = E_s^0 + D_s(E_s^0) \tag{23.104}$$

where as before we have replaced E_s by E_s^0, as both correspond to unperturbed values of the energy eigenstate. It can be shown that $R_s^{(+)}(E)$ can be written in the form of dispersion relations (see Appendix),

$$R_{ss}^{(+)}(E) = V_{ss} - \frac{1}{\pi} \int_{E_0}^{\infty} dE' \frac{I_s(E')}{E' - (E + i\sigma)} \tag{23.105}$$

where I_s is given in (23.100). Hence,

$$G_{ss} = \frac{1}{E - E_s - V_{ss} + \frac{1}{\pi} \int_{\epsilon_0}^{\infty} dE' \dfrac{I_s(E')}{E' - (E + i\sigma)}}. \tag{23.106}$$

Since G is related to the T-matrix and, therefore, to the scattering amplitude, the reaction matrix formalism outlined above is very powerful in establishing a relation that connects the scattering amplitude for a given process to the bound states and resonances in that process to all orders in perturbation.

23.7 Composite systems

Below we consider two important scattering processes in which a particle, A, of sufficiently high energy, such as a photon or an electron, scatters off a composite object, B, such as an atom in its ground state, producing an excited state B^*. That is,

$$A + B \rightarrow B^*. \tag{23.107}$$

The state B^* then goes back to $A + B$ or, for the case of the electron, to $A + B + photon$. These two problems provide an interesting application of the R-matrix approach.

We begin in each case by writing

$$T = V + VGV. \tag{23.108}$$

To apply the R-matrix formalism we need to keep in mind that we are dealing here with a complex system of particles and interactions. However, one finds that for the above processes, under reasonable approximations, one can express T in terms of the diagonal matrix elements of G, which can then be written as

$$G_{ss}(E) = \frac{1}{E - E_s - R_{ss}(E)}. \tag{23.109}$$

We write R_{ss} in the form

$$R_{ss} = D_s(E) - \frac{i}{2}w_s. \tag{23.110}$$

As we saw earlier for a weak potential, $D_s(E) \sim V_{ss}$ which to a good approximation will allow the T-matrix to be written in the form typical of a resonant state:

$$T_{fi} \simeq \frac{1}{E - E_s + \frac{i}{2}w_s}. \tag{23.111}$$

23.7.1 Resonant photon–atom scattering

Let us consider the case in which a photon scatters off an atom (as depicted in Fig. 23.1). The photon here represents an electromagnetic wave of frequency ω described in field-theoretic

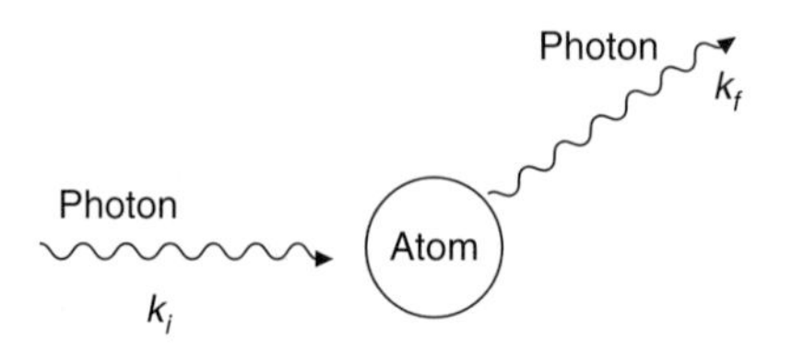

Fig. 23.1

language as a particle with definite energy $\hbar\omega$. The Hamiltonian for this process is given by

$$H = \frac{p^2}{2m} + \hbar\omega + V_0(\mathbf{r}) + V_r \tag{23.112}$$

where V_r is the interaction of radiation (photon) with an electron in the atom. As we discussed earlier, V_r is given by

$$V_r = -\frac{e}{mc}\mathbf{A}\cdot\mathbf{p} \tag{23.113}$$

where $\mathbf{A}$ is the vector potential describing the photon and $\mathbf{p}$ is the momentum of the electron in the atom. The potential $V_0(\mathbf{r})$, which is Coulomb-like, is responsible for the formation of the atomic bound state; $p^2/2m$ is the kinetic energy of the atom, which is assumed to be heavy; and $\hbar\omega$ represents the energy of the photon. We combine the first three terms in (23.112) and designate the sum as H_0. Thus,

$$H = H_0 + V_r. \tag{23.114}$$

The potential V_r is responsible for transitions between the atomic levels. The T-matrix for the scattering of the photon is then given by

$$T = V_r + V_r \frac{1}{E - H + i\epsilon} V_r. \tag{23.115}$$

Let us calculate the matrix element of T,

$$\langle f|T|i\rangle = \langle f|V_r|i\rangle + \left\langle f \left| V_r \frac{1}{E - H + i\epsilon} V_r \right| i \right\rangle \tag{23.116}$$

where $|i\rangle$ and $|f\rangle$ represent the initial and final states, respectively,

$$|i\rangle = |\mathbf{k}_i, E_i\rangle \tag{23.117}$$

where $\mathbf{k}_i$ is the momentum vector of the photon with energy $\hbar\omega_i$ ($c|\mathbf{k}_i| = ck_i = \omega_i$), and E_i is the energy of the atom. We are ignoring the polarization components of the photons and the spin of the electrons. The final state is designated by

$$|f\rangle = |\mathbf{k}_f, E_f\rangle \tag{23.118}$$

with the respective notations.

In Chapter 18 in our calculations involving photon–atom interactions in second-order time-dependent perturbation theory, it was pointed out that

$$\langle f|V_r|i\rangle = 0. \tag{23.119}$$

This result is derived from quantum electrodynamics (QED). It signifies the fact that two photons do not interact via electromagnetic potentials or the potential V_r. Recapitulating our discussion in that chapter, we note that if we write the states in terms of photons, that is, for example, designate $|0\rangle, |1\rangle, |2\rangle$ to correspond to states with 0, 1, 2 photons respectively, then QED tells us that

$$\langle 0|\mathbf{A}|0\rangle = 0 = \langle 1|\mathbf{A}|1\rangle \tag{23.120}$$

while

$$\langle 1|\mathbf{A}|0\rangle \neq 0 \neq \langle 1|\mathbf{A}|2\rangle. \tag{23.121}$$

The point is that in QED A is expressed as a sum of two operators, one of which when operating on a state defined in a multiparticle system destroys (removes) a photon from this state and the other of which creates a new photon in the state. Since $|0\rangle$, $|1\rangle$, and $|2\rangle$ are orthonormal, this explains results (23.120) and (23.121). It also explains that $\langle f|V_r|i\rangle$ will vanish since it is related to $\langle 1|\mathbf{A}|1\rangle$.

The matrix element of T is then given by the second term in (23.116). Thus,

$$\langle f|T|i\rangle = \sum_{mn} \langle f|V_r|n\rangle \left\langle n \left| \frac{1}{E - H + i\epsilon} \right| m \right\rangle \langle m|V_r|i\rangle \tag{23.122}$$

where we have inserted a complete sets of states $|n\rangle$ and $|m\rangle$ that designate states with different energy levels of the atom. These states are assumed not to have any photons.

If we expand the denominator in (23.122) in powers of V_r, we can easily show, based on QED, that since neither $|n\rangle$ nor $|m\rangle$ have photons in them, only even powers of V_r will contribute, in which case only the diagonal elements will be nonzero. Thus the middle matrix in (23.122) will be diagonal, given simply by the diagonal matrix of the Green's function, G_{nn}. One can therefore write

$$\langle f|T|i\rangle = \sum_{n} \langle f|V_r|n\rangle G_{nn} \langle n|V_r|i\rangle. \tag{23.123}$$

We note the since $|i\rangle$ and $|f\rangle$ each contain a single photon, and $|n\rangle$ does not contain any photons, the above matrix elements of V_r will be nonzero.

We can now use the R-matrix formalism and write

$$G_{nn}(E) = \frac{1}{E - E_n - R_{nn}(E + i\epsilon)}. \tag{23.124}$$

Since

$$R_{nn} = D_n(E) - \frac{i}{2} w_n \approx -\frac{i}{2} w_n \tag{23.125}$$

where $D_n(E)$ and w_n have already been defined, we obtain

$$\langle f|T|i\rangle \simeq \sum_{n} \langle f|V_r|n\rangle \langle n|V_r|i\rangle \frac{1}{E - E_n + \frac{i}{2} w_n}. \tag{23.126}$$

This is then in the resonance form.

23.7.2 Resonant electron–atom scattering

Consider an electron scattering off a very heavy atom of energy E_i in its ground state (see Fig. 23.2). If the electron can impart sufficient energy to the atom to raise it to an excited state with energy $E_n > E_i$ then the scattering amplitude can exhibit a resonance. Let us consider such a case. We also assume that the atom after reaching the excited state drops down to a lower level by the emission of a photon. We will derive a formal expression for the T-matrix for this process.

The Hamiltonian for this system can be written as

$$H = H_0 + V. \tag{23.127}$$

The Hamiltonian H_0 includes the energy of the atom, the kinetic energy of the electron, and other interactions that do not come into play in this problem. The interaction between the electron and the atom is represented by V. We divide V into two parts, V_r and V_c:

$$V = V_r + V_c. \tag{23.128}$$

Here V_r is the interaction between the radiation field, represented by the vector potential $\mathbf{A}$, and an electron, of momentum $\mathbf{p}$, in the atom, given by

$$V_r = -\frac{e}{2mc}\mathbf{A}\cdot\mathbf{p}. \tag{23.129}$$

This interaction is responsible for transitions from one atomic level to another as well as for possible emission of a photon, while V_c is the typical Coulomb interaction between the electron and the atom which, by itself, would give rise to Rutherford-type scattering. The T-matrix expressed as $T = V + VGV$ will be of the form

$$T = V + V\frac{1}{E - H + i\epsilon}V. \tag{23.130}$$

Since we are only interested in the possible presence of a resonance, we ignore the first term in (23.130). We then write

$$\frac{1}{E - H + i\epsilon} = \frac{1}{(E - H_0 - V_r + i\epsilon)}\left[\frac{E - H + V_c + i\epsilon}{E - H + i\epsilon}\right] \tag{23.131}$$

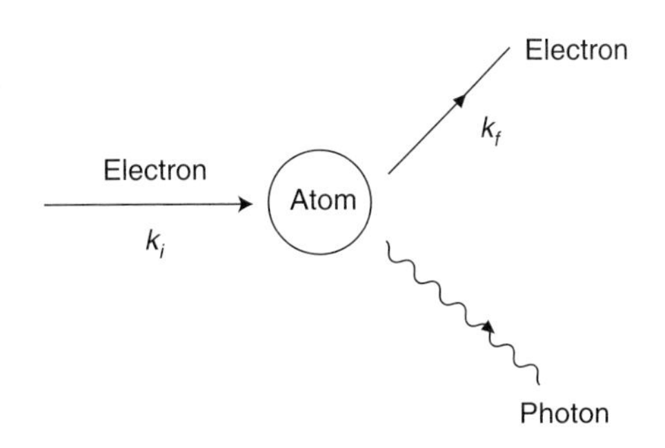

Fig. 23.2

where the first denominator on the right exactly cancels the numerator. After rearranging the terms in the square bracket above we obtain

$$\frac{1}{E - H + i\epsilon} = \frac{1}{(E - H_0 - V_r + i\epsilon)} \left[1 + V_c \frac{1}{E - H + i\epsilon} \right]. \tag{23.132}$$

Hence,

$$T = V \frac{1}{E - H_0 - V_r + i\epsilon} \left[V + V_c \frac{1}{E - H + i\epsilon} V \right]. \tag{23.133}$$

For the terms inside the square bracket, if we ignore the weak V_r term in relation to V_c and replace V by V_c, then the expression will be simply the T-matrix for the electron–atom scattering in the absence of a radiation field. We write this term as T_e, given by

$$T_e = \left[V_c + V_c \frac{1}{E - H_0 - V_c + i\epsilon} V_c \right]. \tag{23.134}$$

The complete T-matrix is then

$$T = V \frac{1}{E - H_0 - V_r + i\epsilon} T_e. \tag{23.135}$$

If we designate $|i\rangle$ and $|f\rangle$ as initial and final states, respectively, and insert complete sets of intermediate states $|m\rangle$ and $|n\rangle$, then the matrix element of T is given by

$$\langle f|T|i\rangle = \sum_{m,n} \langle f\,|V|\,n\rangle \left\langle n \left| \frac{1}{E - H_0 - V_r + i\epsilon} \right| m \right\rangle \langle m\,|T_e|\,i\rangle. \tag{23.136}$$

We assume that the states $|m\rangle$ and $|n\rangle$ correspond to different atomic states, ignoring the presence of any photons, whose contribution is expected to be small. Following the arguments in the previous section, $\langle n\,|V_r|\,m\rangle$ will vanish and the middle matrix in (23.136) will be diagonal. It will be given by the diagonal element, G_{nn}, of the total Green's function. Hence,

$$\langle f|T|i\rangle = \sum_{n} \langle f\,|V|\,n\rangle G_{nn} \langle n\,|T_e|\,i\rangle. \tag{23.137}$$

One can then invoke the R-matrix formalism and write

$$G_{nn}(E) = \frac{1}{E - E_n - R_{nn}(E + i\epsilon)} \simeq \frac{1}{E - E_n + \frac{i}{2} w_n}. \tag{23.138}$$

Since we are considering the emission of a photon in the final state and since the state $|n\rangle$ does not contain a photon,

$$\langle f\,|V|\,n\rangle = \langle f\,|V_r|\,n\rangle. \tag{23.139}$$

However, the external electron will not participate in this process, rather it will contribute to $\langle n | T_e | i \rangle$. If we designate E_i, E_n, and E_f as the initial, intermediate, and final energies of the atom, $\mathbf{e}_i$ and $\mathbf{e}_f$ as the initial and final momenta of the external electron, and $\mathbf{k}_f$ as the momentum of the emitted photon in the final state, then the T-matrix will be given by

$$\langle f | T | i \rangle = \sum_n \frac{\langle \mathbf{k}_f | V_r | E_n \rangle \langle E_n \mathbf{e}_f | T_e | E_i \mathbf{e}_i \rangle}{E - E_n + \frac{i}{2} w_n}. \tag{23.140}$$

This is in the resonance form as expected.

In our derivation of $\langle f | T | i \rangle$ for the photon–atom and electron–atom interactions the matrix elements found in the numerators are by no means trivial to calculate. Our purpose here was not to do detailed calculations for these processes but rather to express the T-matrix in the R-matrix formalisms and show that this formalism is very useful if there are resonances in the interacting system (see also Goldberger and Watson (1964) for a detailed discussion on this subject).

23.8 Appendix to Chapter 23

23.8.1 Dispersion relations

Consider a function $f(z)$ in the complex z-plane that is analytic in the upper half-plane. We assume further that $f(z)$ has a branch point at $z = 0$ (see Fig. 23.3). If $f(z)$ is real along the negative real axis, then along this axis it satisfies the relation

$$f\left(z^*\right) = f^*(z). \tag{23.141}$$

However, z^* corresponds to a point in the lower half-plane if z lies in the upper half-plane where $f(z)$ is defined. Therefore, according to the theorem on analytic functions, $f^*(z)$ is an analytic continuation of $f(z)$ in the lower half-plane.

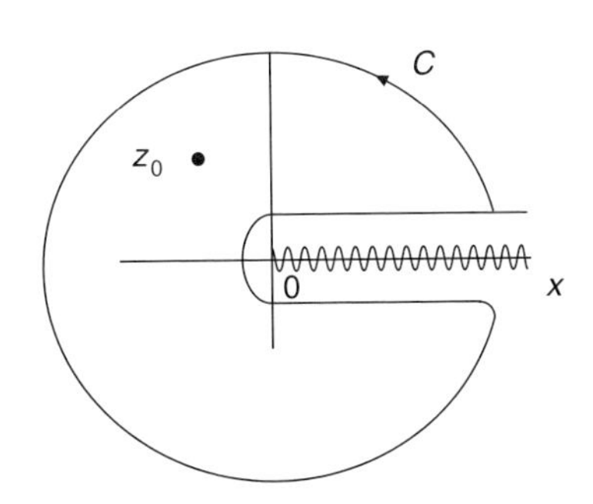

Fig. 23.3

With $z = 0$ being a branch point, it follows from the above-mentioned properties of $f(z)$ that we place the branch cut along the positive real axis. The discontinuity across this branch cut is then defined as

$$\operatorname{disc} f(z) = \frac{f(z) - f(z^*)}{zi}. \tag{23.142}$$

We now use the relation (23.141) for $f(z^*)$. Since the discontinuity across the branch cut along the real axis will correspond to the difference in $f(z)$ just above and below the real axis, we write $z = x + i\epsilon$, where ϵ is an infinitesimal quantity. Thus from (23.142) and (23.141) we write

$$\operatorname{disc} f(x) = \lim_{\epsilon\to 0} \frac{f(x+i\epsilon) - f(x-i\epsilon)}{2i} \tag{23.143}$$

$$= \lim_{\epsilon\to 0} \frac{f(x+i\epsilon) - f^*(x+i\epsilon)}{2i} \tag{23.144}$$

$$= \operatorname{Im} f(x). \tag{23.145}$$

For a point z in the upper half-plane where $f(z)$ is analytic, we utilize Cauchy's theorem to write

$$f(z) = \frac{1}{2\pi i} \oint_C \frac{dz' f(z')}{z' - z} \tag{23.146}$$

where C is a contour inside which $f(z)$ is analytic and z lies inside C. In the above relation one can always include any contributions from isolated poles of $f(z)$.

For our problem we will take C to cover the entire z-plane inside which $f(z)$ is analytic which would, therefore, exclude the positive real axis along which there is a branch cut. The contour is described in Fig. 23.3. We also assume that $f(z)$ vanishes at infinity to ensure that the integral converges. We will now take

$$z = x + i\epsilon. \tag{23.147}$$

Since $f(z)$ vanishes at infinity, the contribution to the integral will come only from $f(z')$ above and below the real axis. The contribution will be in the form of the discontinuity function, which we found to be

$$\operatorname{disc} f(x) = \operatorname{Im} f(x). \tag{23.148}$$

Hence,

$$f(x+i\epsilon) = \frac{1}{\pi} \int_0^\infty \frac{dx' \operatorname{Im} f(x')}{x' - (x+i\epsilon)}. \tag{23.149}$$

The above relation is assumed to be in the limit $\epsilon \to 0$. This is the basic dispersion relation. One can write it slightly differently by using the well-known results

$$\frac{1}{x' - x - i\epsilon} = P\left(\frac{1}{x' - x}\right) + i\pi\,\delta(x' - x) \tag{23.150}$$

and

$$f(x + i\epsilon) = \text{Re} f(x) + i\,\text{Im} f(x), \tag{23.151}$$

again in the limit $\epsilon \to 0$, where the two terms correspond to real and imaginary parts.

Hence we write (23.149) as

$$\text{Re} f(x) = \frac{P}{\pi}\int_0^\infty \frac{dx'\,\text{Im} f(x')}{x' - x}. \tag{23.152}$$

The usefulness of this relation lies in the fact that often one has some information about the imaginary part of $f(x)$ from which one determines the real part. For example, if $f(x)$ represents the scattering amplitude in the forward direction, with x as the energy, then from the optical theorem one relates the imaginary part of $f(x)$ to the total cross-section, which in turn may be obtained from experiments. The crucial remaining question then is to determine whether the integral in (23.152) converges, which is not always simple.

This method has been applied with considerable success in low-energy nuclear physics and in optics, where the term "dispersion relation" arises. In optics it is called the Kramers–Kronig relation.

24 Approximation methods for bound states and scattering

Having looked at solutions that are obtained exactly or through perturbation theory, we now consider cases where we have to resort to some unique approximation schemes, both for scattering and for bound states. In relation to the bound states we introduce the WKB approximation, and the variational method. For scattering at high energies we introduce the eikonal approximation, which is closely allied to the WKB method.

24.1 WKB approximation

24.1.1 Introduction

We begin with the Schrödinger equation in one dimension,

$$-\frac{\hbar^2}{2m}\frac{d^2u}{dx^2} + V(x)u = Eu, \tag{24.1}$$

which we write as

$$\frac{d^2u}{dx^2} + \frac{2m}{\hbar^2}(E - V(x)u = 0. \tag{24.2}$$

If $V(x)$ is a constant, equal to V_0, then in that region one can write the solutions as

$$u = e^{\pm ikx} \quad \text{for } E > V_0 \quad \text{with } k^2 = \frac{2m}{\hbar^2}(E - V_0), \tag{24.3}$$

$$u = e^{\pm \kappa x} \quad \text{for } E < V_0 \quad \text{with } \kappa^2 = \frac{2m}{\hbar^2}(V_0 - E). \tag{24.4}$$

Suppose now that V is not exactly a constant but varies very slowly as a function of x. To be specific, let us first consider $E > V(x)$ and write

$$k^2(x) = \frac{2m}{\hbar^2}(E - V(x)), \quad \text{with } E > V(x). \tag{24.5}$$

The Schrödinger equation is then given by

$$\frac{d^2u}{dx^2} + k^2(x)u = 0. \tag{24.6}$$

We define $\lambda(x) = 2\pi/k(x)$ to be the wavelength at a point x. We will obtain approximate solutions for those cases where

$$\left|\frac{d\lambda}{dx}\right| \ll 1, \quad \text{i.e.,} \quad \left|\frac{k'(x)}{k^2(x)}\right| \ll 1. \tag{24.7}$$

This is basis of the WKB approximation, named after Wentzel, Kramers, and Brillouin who formulated it. There are two situations in which this approximation applies: (i) small E with $V(x)$ varying very slowly, (ii) very large E.

Let us write the solution of (24.6) as

$$u(x) = Ae^{i\phi(x)}. \tag{24.8}$$

Substituting this in equation (24.6), we obtain

$$-\phi'^2 + i\phi'' + k^2(x) = 0, \tag{24.9}$$

which can be written as

$$\phi'^2 = k^2(x) + i\phi''. \tag{24.10}$$

This equation can be solved iteratively by assuming $\phi'' \ll 1$, which corresponds to the approximation (24.7). In the first iteration, obtained by taking $\phi'' = 0$, we find

$$\phi' \approx \pm k(x). \tag{24.11}$$

Thus,

$$\phi'' \approx \pm k'(x). \tag{24.12}$$

Substituting (24.11) into (24.9) gives

$$\phi' \approx \pm\sqrt{k^2(x) \pm ik'(x)} = \pm k(x)\left[1 \pm \frac{ik'(x)}{2k^2(x)}\right], \tag{24.13}$$

using the approximation (24.7). Therefore,

$$\phi' = \pm k(x) + \frac{ik'(x)}{2k(x)} \tag{24.14}$$

and

$$\phi(x) = \pm\int dx\, k(x) + \frac{i}{2}\int dx \frac{k'(x)}{k(x)} + \text{const.} \tag{24.15}$$

$$= \pm\int dx\, k(x) + \frac{i}{2}\log(k(x)). \tag{24.16}$$

Therefore, the solution becomes

$$u(x) = e^{i\left[\pm\int dx\, k(x) + \frac{i}{2}\log(k(x))\right]} \tag{24.17}$$

or, now including also the case $E < V(x)$,

$$u(x) = \left\{ \begin{array}{lll} \frac{1}{\sqrt{k(x)}} e^{\pm i \int dx\, k(x)}, & k(x) = \sqrt{\frac{2m}{\hbar^2}(E - V(x))}, & \text{for } E > V(x) \\ \frac{1}{\sqrt{k(x)}} e^{\pm \int dx \kappa(x)}, & \kappa(x) = \sqrt{\frac{2m}{\hbar^2}(V(x) - E)}, & \text{for } E < V(x) \end{array} \right\}. \tag{24.18}$$

In the region where $E \approx V(x)$ we find that $k(x) \approx 0$ and the condition (24.7) for the applicability of the approximation is violated. The point where

$$E = V(x) \tag{24.19}$$

is called the classical turning point.

Consider the configuration of the potential as given by Fig. 24.1, with the dotted line corresponding to the energy of the particle. We note that to the left of the classical turning point is a region where $E < V(x)$, which is classically forbidden. We call this region 2. A particle in region 1 going to the left will be turned back in classical mechanics, before it enters region 2.

One therefore needs to treat the region near the turning point separately. One can actually obtain the exact solution if the behavior of $k(x)$ is already known or if one can make a reasonably good approximation to it near the turning point, $x = 0$. If we assume

$$k^2(x) = cx^n, \tag{24.20}$$

then one finds the solution in region 1 to be

$$u_1(x) = A\, \xi_1^{\frac{1}{2}} k^{\frac{1}{2}} J_{\pm\, m}(\xi_1), \quad m = \frac{1}{n+2} \tag{24.21}$$

where J is the cylindrical Bessel function, and

$$\xi_1 = \int_0^x dx\, k(x). \tag{24.22}$$

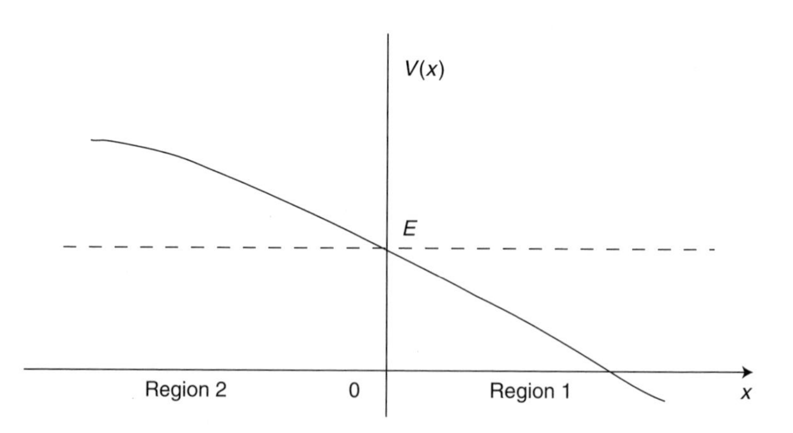

Fig. 24.1

As a simple example, let us take $n = 1$ so that

$$k^2(x) = cx \tag{24.23}$$

near $x = 0$. Then the solutions in the two regions are given by

$$u_1^{\pm}(x) = A_{\pm}\, \xi_1^{\frac{1}{2}} k^{-\frac{1}{2}} J_{\pm\frac{1}{3}}(\xi_1), \tag{24.24}$$

$$u_2^{\pm}(x) = B_{\pm}\, \xi_2^{\frac{1}{2}} \kappa^{-\frac{1}{2}} I_{\pm\frac{1}{3}}(\xi_2), \tag{24.25}$$

where I is a Bessel function with imaginary argument, and

$$\xi_2 = \int_x^0 dx\, \kappa(x). \tag{24.26}$$

One can determine the coefficients $A_{\pm}$ and $B_{\pm}$ by substituting (24.23) into the solutions (24.24) and (24.25) and equating the two functions at the boundary.

The behavior near the boundary $x = 0$ of the Bessel functions is

$$J_{\pm\frac{1}{3}}(\xi_1) \underset{x\to 0}{\longrightarrow} \frac{\left(\frac{1}{2}\xi_1\right)^{\pm\frac{1}{2}}}{\Gamma\left(1 \pm \frac{1}{3}\right)}, \tag{24.27}$$

$$I_{\pm\frac{1}{2}}(\xi_2) \underset{x\to 0}{\longrightarrow} \frac{\left(\frac{1}{2}\xi_2\right)^{\pm\frac{1}{2}}}{\Gamma\left(1 \pm \frac{1}{3}\right)}. \tag{24.28}$$

Inserting these expressions and equating the wavefunctions at the boundary, one finds

$$u_1^+ = u_2^+ \;\Rightarrow\; A_+ = -B_+, \tag{24.29}$$

$$u_1^- = u_2^- \;\Rightarrow\; A_- = B_-. \tag{24.30}$$

We will construct proper linear combinations of the wavefunctions in each region so that they are consistent with the boundary conditions and at the same time have the correct asymptotic behavior. We need to ensure that the wavefunction vanishes in the limit $x \to -\infty$ in region 2. The asymptotic behavior of the Bessel functions is given by

$$J_{\pm\frac{1}{3}}(\xi_1) \underset{x\to\infty}{\longrightarrow} \left(\frac{\pi}{2}\xi_1\right)^{-\frac{1}{2}} \cos\left(\xi_1 \mp \frac{\pi}{6} - \frac{\pi}{4}\right), \tag{24.31}$$

$$I_{\pm\frac{1}{3}}(\xi_2) \underset{x\to -\infty}{\longrightarrow} (2\pi\, \xi_2)^{-\frac{1}{2}} \left[e^{\xi_2} + e^{-\xi_2} e^{-i\left(\frac{1}{2}\pm\frac{1}{3}\right)\pi}\right]. \tag{24.32}$$

For the wavefunctions $u_2^{\pm}(x)$ in region 2 this implies

$$u_2^+(x) \underset{x\to -\infty}{\longrightarrow} -(2\pi\kappa)^{-\frac{1}{2}} \left[e^{\xi_2} + e^{-\xi_2} e^{-\left(\frac{5i\pi}{6}\right)}\right], \tag{24.33}$$

$$u_2^-(x) \underset{x\to -\infty}{\longrightarrow} (2\pi\kappa)^{-\frac{1}{2}} \left[e^{\xi_2} + e^{-\xi_2} e^{-\left(\frac{i\pi}{6}\right)}\right], \tag{24.34}$$

where we have ignored $A_\pm$ after incorporating the relative magnitudes and signs between the coefficients $A_\pm$ and $B_\pm$. We find that the sum of the two wavefunctions $u_2^+(x)$ and $u_2^-(x)$ will vanish as $x \to -\infty$. One must then also take a sum $u_1^+(x) + u_1^-(x)$ to describe the region to the right. We then have the following "connection formula" connecting the asymptotic forms of the left and the right hand sides:

$$\frac{1}{2}\kappa^{-\frac{1}{2}}e^{-\xi_2} \Leftrightarrow k^{-\frac{1}{2}}\cos\left(\xi_1 - \frac{\pi}{4}\right). \tag{24.35}$$

One needs to keep in mind that the functional form on the left has been determined first so that it vanishes strictly as $\exp(-\xi_2)$, since otherwise any fluctuations in the function on the right can introduce the unwanted term $\exp(\xi_2)$ on the left.

24.1.2 Energy levels of a particle trapped inside a potential

Consider a particle trapped inside a one-dimensional potential of the type given by Fig. 24.2. The precise functional dependence of $V(x)$ is left unknown and could be complicated. We show below how WKB approximation allows one to determine the energy eigenvalues, E, of the particle.

From Fig. 24.2 we note that we have two separate regions outside where $E < V(x)$. These are of the type of region 2 discussed in Section 24.1.1 earlier. If x_1 and x_2 are the two turning points then we can determine the wavefunctions in the inside region (region 1) through the connection formula. Thus we find

$$u(x) = \cos\left(\int_{x_1}^{x} dx\, k(x) - \frac{\pi}{4}\right) \qquad x > x_1, \tag{24.36}$$

$$u(x) = \cos\left(\int_{x}^{x_2} dx\, k(x) - \frac{\pi}{4}\right) \qquad x < x_2, \tag{24.37}$$

where x is any point in the inside region. The multiplicative factors will be the same for the two functions but one can adjust the relative signs. Since the two cosines must be equal at

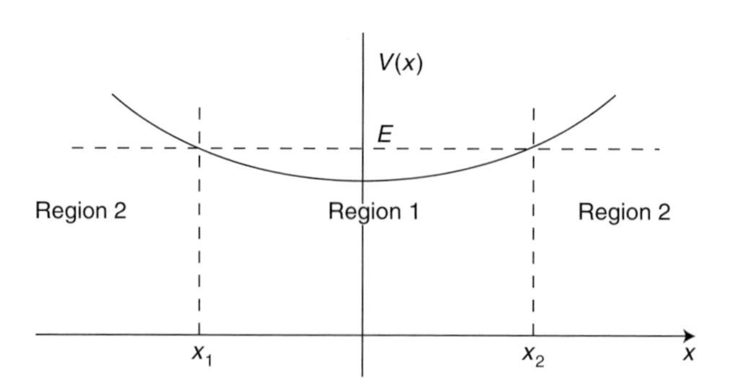

Fig. 24.2

the common point x, arguments of the cosines must be the same apart from the sign and up to a multiple of π. Thus we have

$$\int_{x_1}^{x} k(x)\, dx - \frac{\pi}{4} = \pm \left[\int_{x}^{x_2} dx\, k(x) - \frac{\pi}{4} \right] + n\pi. \tag{24.38}$$

To be valid for all values of x in the inside region, the sign in front of the square bracket must be negative. Therefore, moving the integral term on the right to the left-hand side and collecting all the factors of π on the right we obtain

$$\int_{x_1}^{x_2} dx\, k(x) = \frac{\pi}{2} + n\pi = \left(n + \frac{1}{2}\right)\pi. \tag{24.39}$$

Hence,

$$\int_{x_1}^{x_2} dx \left[\frac{2m}{\hbar^2}(E - V(x))\right]^{\frac{1}{2}} = \left(n + \frac{1}{2}\right)\pi. \tag{24.40}$$

This condition, remarkably, is the same as the old Bohr–Sommerfeld quantization condition. Once $V(x)$ is known, the above formula will yield the eigenvalues, E.

24.1.3 Tunneling through a barrier

One encounters this type of phenomenon in nuclear physics, for example, in α-particle decay. Consider a heavy nucleus of charge Z that contains a bound α-particle of charge $Z_\alpha = 2$. The α-decay corresponds to this particle fragmenting off the heavy nucleus. The particle inside the nucleus is subjected to a very strong and attractive interaction due to the nuclear potential, which is typically given by

$$-\frac{ge^{-r/R}}{r}. \tag{24.41}$$

These are short-range forces of radius R. Outside $r = R$, there are two forces that are very effective since they are of longer range. One is the repulsive Coulomb potential

$$\frac{Z_\alpha Z'}{r} \tag{24.42}$$

where Z' is the residual nuclear charge, and the other is the centrifugal (repulsive) barrier

$$\frac{\hbar^2 l(l+1)}{2mr^2} \tag{24.43}$$

where l is the angular momentum of the particle. These two together constitute a barrier preventing the escape of the α-particle. The configuration of the potentials is described

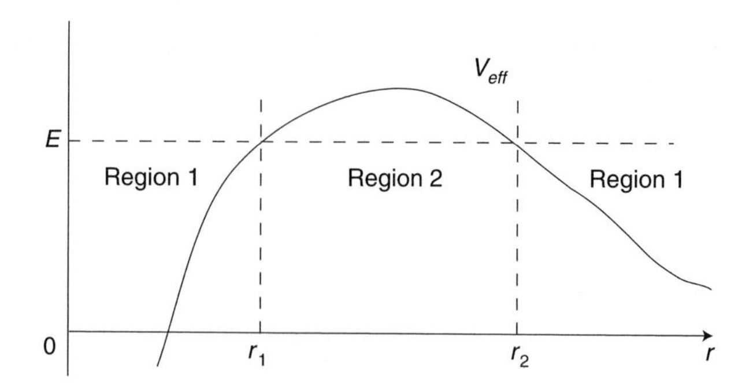

Fig. 24.3

pictorially in Fig. 24.3. The occurrence of α-decay implies that the α-particle has a finite probability of penetrating through this barrier.

The WKB approximation has an interesting application to this problem in determining the barrier penetration factor and, therefore, the decay rate. Our earlier formulation for one dimension can easily be extended to the case of a three-dimensional potential that is a function of a single variable, r. This is in a sense a one-dimensional problem in which one includes the centrifugal barrier by defining an effective potential, V_{eff}, given by

$$V_{eff} = V(r) + \frac{\hbar^2 l(l+1)}{2mr^2}. \tag{24.44}$$

From Fig. 24.3 we note that the middle region will correspond to our region 2 discussed in Section 24.1.2 where we write

$$\kappa(r) = \left[\frac{2m}{\hbar^2}\left(V_{eff} - E\right)\right]^{\frac{1}{2}}. \tag{24.45}$$

If a and b are the turning points and $u(r)$ is the radial wavefunction, then the probability of tunneling through the barrier is the same as the transmission coefficient given by

$$T = \left|\frac{u(b)}{u(a)}\right|^2. \tag{24.46}$$

From our WKB analysis of Section 24.1.2 this is simply proportional to

$$e^{-2\int_a^b dr\kappa(r)} = e^{-2\int_a^b dr\sqrt{\frac{2m}{\hbar^2}(V_{eff}-E)}}. \tag{24.47}$$

Considering the case $l = 0$, the effective potential is

$$V_{eff}(r) = \frac{Z_\alpha Z' e^2}{r} \tag{24.48}$$

where Z' is the number of left-over protons in the nucleus. Since the nuclear potential is very strong and of short range, we can take the end point to be simply given by $a = R$, the range of nuclear interaction. The other turning point, b, is given by the value of r when

$$E = \frac{Z_\alpha Z' e^2}{r}. \tag{24.49}$$

Thus,

$$\int dr\, k(r) = \sqrt{\frac{2m}{\hbar^2}} \int_R^{\frac{Z Z' e^2}{E}} dr \sqrt{\frac{Z Z' e^2}{r} - E} \tag{24.50}$$

$$= \sqrt{\frac{2mE}{\hbar^2}} \int_R^{\frac{Z Z' e^2}{E}} dr \sqrt{\frac{Z Z' e^2}{E} \frac{1}{r} - 1}. \tag{24.51}$$

Let

$$\frac{Z Z' e^2}{E} \frac{1}{r} = \frac{1}{\rho} \quad \text{and} \quad \frac{Z Z' e^2}{E} \frac{1}{R} = \frac{1}{\rho_0}. \tag{24.52}$$

Hence,

$$k(r) = \sqrt{\frac{2mE}{\hbar^2}} \frac{Z Z' e^2}{E} \int_{\rho_0}^{1} d\rho \sqrt{\frac{1}{\rho} - 1}. \tag{24.53}$$

If the energies are low, $E \sim 0$, then $\rho_0 \ll 1$. We can then replace the lower limit of the integral above by zero and write

$$k(r) = \frac{2Z Z' e^2}{\hbar v} \int_0^1 d\rho \sqrt{\frac{1}{\rho} - 1} \tag{24.54}$$

where we have taken the nonrelativistic approximation $E = \frac{1}{2} m v^2$ with v as the velocity.

To evaluate the integral we set

$$\rho = \cos^2 \theta. \tag{24.55}$$

Then

$$d\rho = -2 \cos\theta \sin\theta \, d\theta. \tag{24.56}$$

and

$$\int_0^1 d\rho \sqrt{\frac{1}{\rho} - 1} = \int_0^{\pi} 2 \sin^2 \theta \, d\theta = \pi. \tag{24.57}$$

Hence,

$$\int dr\, k(r) = \frac{2\pi Z\, Z' e^2}{\hbar v}. \tag{24.58}$$

The transmission coefficient for nuclear α decay is then equal to

$$e^{-2\int_a^b drk(r)} = e^{-\frac{4\pi Z\, Z' e^2}{\hbar v}}. \tag{24.59}$$

This is known as the Gamow factor for nuclear alpha decay. The results we derived here are consistent with experiments.

From our derivations above we find that the connection formulas play a secondary role in the WKB approximation, as long as we have a good knowledge of what the wavefunctions should be in the internal regions.

24.2 Variational method

The variational method is most useful in estimating the ground-state energy of a bound system for which the eigenfunctions are not known.

Let $u_n(\mathbf{r})$ be the eigenfunctions of energy with eigenvalues E_n that satisfy the equation

$$Hu_n = E_n u_n \tag{24.60}$$

where H is the Hamiltonian for the system. Consider now an arbitrary function $\phi(\mathbf{r})$. This function can be expanded in terms of the complete set of eigenstates u_n,

$$\phi = \sum_n a_n u_n. \tag{24.61}$$

If u_n's, and, therefore, ϕ, are normalized then

$$\int d^3r\, \phi^* \phi = \sum_n |a_n|^2 = 1. \tag{24.62}$$

Now let us take the expectation value of H with respect to ϕ:

$$\langle H \rangle = \int d^3r\, \phi^* H \phi = \sum_n E_n |a_n|^2 . \tag{24.63}$$

If E_0 is the ground-state energy, then

$$\langle H \rangle \geq \sum_n E_0 |a_n|^2 = E_0 \sum_n |a_n|^2 . \tag{24.64}$$

This implies that

$$E_0 \leq \langle H \rangle = \int d^3r \, \phi^* H \phi. \tag{24.65}$$

If u_n's are not normalized then one writes

$$E_0 \leq \frac{\int d^3r \, \phi^* H \phi}{\int d^3r \, |\phi|^2}. \tag{24.66}$$

The variational method consists in choosing a trial function ϕ with respect to which one obtains $\langle H \rangle$. This trial function will depend on one or more parameters with respect to which $\langle H \rangle$ is minimized to obtain the best upper bound on E_0. Let us now turn to an interesting application of this method.

24.2.1 Helium atom

Helium consists of two electrons in a nucleus of positive charge with $Z = 2$. The total Hamiltonian is given by

$$H = -\frac{\hbar^2}{2m}\left(\nabla_1^2 + \nabla_2^2\right) - 2e^2\left(\frac{1}{r_1} + \frac{1}{r_2}\right) + \frac{e^2}{|\mathbf{r}_1 - \mathbf{r}_2|}. \tag{24.67}$$

The nucleus is assumed to be infinitely heavy. The two electrons are designated by the subscripts 1 and 2. The first term above corresponds to their kinetic energy, the second one to the Coulomb attraction between the nucleus and the individual electrons and the last term to the mutual Coulomb repulsion between the electrons. Were it not for this repulsion the wavefunction for the electrons would be simply a product of their individual hydrogen-like wavefunctions. This product, however, will be an ideal candidate for us to use as a trial function. It is given by

$$\phi_{trial} = \phi\left(\mathbf{r}_1, \mathbf{r}_2\right) = \frac{Z^3}{\pi a_0^3} e^{-(Z/a_0)(r_1 + r_2)} \tag{24.68}$$

where Z is now taken as a parameter with respect to which the Hamiltonian will be minimized.

The expectation values of the first two terms with respect to the trial wavefunction can easily be calculated and are given by

$$\left\langle -\frac{\hbar^2}{2m}\left(\nabla_1^2 + \nabla_2^2\right) \right\rangle = \frac{e^2 Z^2}{a_0}, \tag{24.69}$$

$$\left\langle 2e^2\left(\frac{1}{r_1} + \frac{1}{r_2}\right) \right\rangle = \frac{4e^2 Z}{a_0}. \tag{24.70}$$

The expectation value of the third term is

$$\left\langle \frac{e^2}{|\mathbf{r}_1 - \mathbf{r}_2|} \right\rangle = \int\int d^3r_1\, d^3r_2\, \phi^*(\mathbf{r}_1, \mathbf{r}_2) \frac{e^2}{|\mathbf{r}_1 - \mathbf{r}_2|} \psi(\mathbf{r}_1, \mathbf{r}_2) \tag{24.71}$$

$$= \left(\frac{Z^3}{\pi a_0^3} \right)^2 e^2 \int\int d^3r_1\, d^3r_2 \frac{1}{|\mathbf{r}_1 - \mathbf{r}_2|} e^{-(2Z/a_0)(r_1 + r_2)}. \tag{24.72}$$

To determine this integral we use the result

$$\frac{1}{r} = \frac{1}{2\pi^2} \int \frac{d^3k e^{i\mathbf{k}\cdot\mathbf{r}}}{k^2}. \tag{24.73}$$

This integral is actually a familiar one. It is related to the scattering amplitude in the Born approximation for the Coulomb interaction, which is effectively a Fourier transform of $(1/r)$. The integral above is just an inverse transform of it. We replace r on the left by $|\mathbf{r}_1 - \mathbf{r}_2|$ and $\mathbf{r}$ in the exponent on the right by $(\mathbf{r}_1 - \mathbf{r})$. The integral then becomes a product of two identical integrals. This result when used in (24.72) gives

$$\left\langle \frac{e^2}{|\mathbf{r}_1 - \mathbf{r}_2|} \right\rangle = \left(\frac{Z^3}{\pi a_0^3} \right)^2 e^2 \left(\frac{1}{2\pi^2} \right) \int \frac{d^3k}{k^2} \left[\int d^3r\, e^{-(2Z/a_0)r} e^{i\mathbf{k}\cdot\mathbf{r}} \right]^2. \tag{24.74}$$

We use the integrals

$$\int d^3r\, e^{-(2Z/a_0)r} e^{i\mathbf{k}\cdot\mathbf{r}} = \frac{16\pi Z a_0^3}{\left(k^2 a_0^2 + 4Z^2\right)^2}, \tag{24.75}$$

$$\int_0^\infty \frac{d\kappa}{(\kappa + 1)^4} = \frac{5\pi}{32}, \tag{24.76}$$

to obtain

$$\left\langle \frac{e^2}{|\mathbf{r}_1 - \mathbf{r}_2|} \right\rangle = \frac{5e^2 Z}{8a_0}. \tag{24.77}$$

Thus combining (24.69), (24.70), and (24.77) we find

$$\langle H \rangle = \frac{e^2 Z^2}{a_0} - \frac{4e^2 Z}{a_0} + \frac{5e^2 Z}{8a_0} = \frac{e^2}{a_0} \left(Z^2 - \frac{27}{8} Z \right). \tag{24.78}$$

We minimize $\langle H \rangle$ with respect to the parameter Z,

$$\frac{\partial \langle H \rangle}{\partial Z} = 0. \tag{24.79}$$

The solution is found to be

$$Z = \frac{27}{16}. \tag{24.80}$$

Hence the ground-state energy of the helium atom is estimated to be

$$E_0 = -\left(\frac{27}{16}\right)^2 \frac{e^2}{a_0} = -2.85\frac{e^2}{a_0}. \tag{24.81}$$

The experimental value is found to be close to $-2.90(e^2/a_0)$. Hence the variational method provides an excellent approximation to the actual value.

The value (24.80) obtained for Z can be called the effective charge, which can be understood physically if we write

$$Z = \frac{27}{16} = 2 - \frac{5}{16}. \tag{24.82}$$

The reduction in the original value $Z = 2$ for the nucleus can be interpreted by saying that each electron screens the nucleus and hence reduces its charge by $5/16$, giving us a smaller effective charge of $27/16$.

24.3 Eikonal approximation

The WKB method has been useful for low-energy problems. Let us now turn to the scattering problems in which the particle energies are very high. We will discuss here a closely related approach to WKB called the eikonal approximation.

Consider once again a one-dimensional Schrödinger equation for a scattering problem, where $E > V(x)$,

$$-\frac{\hbar^2}{2m}\frac{d^2u}{dx^2} + V(x)u = Eu. \tag{24.83}$$

We can write it as

$$\frac{d^2u}{dx^2} + k^2(x)u = 0 \tag{24.84}$$

where

$$k^2(x) = \frac{2m}{\hbar^2}(E - V(x)). \tag{24.85}$$

The solution of this equation, as we discussed for the WKB problems, can be written as

$$u(x) = Ae^{i\int^x dx'\, k(x')}. \tag{24.86}$$

Let us consider the case where

$$E \gg V. \tag{24.87}$$

This is the high-energy approximation. One can then make the expansion

$$k(x) = \sqrt{\frac{2mE}{\hbar^2}} \left(1 - \frac{1}{2}\frac{V(x)}{E}\right), \tag{24.88}$$

which can be written as

$$k(x) = k - \frac{1}{\hbar v} V(x) \tag{24.89}$$

where

$$k = \sqrt{\frac{2mE}{\hbar^2}}, \tag{24.90}$$

which is a constant and

$$v = \text{velocity of the particle}, \quad E = \frac{1}{2}mv^2 \quad \text{and} \quad \hbar k = mv. \tag{24.91}$$

Therefore,

$$u(x) = e^{ikx} \, e^{\frac{-i}{\hbar v}\int^x dx' \, V(x')}. \tag{24.92}$$

We extend this to three dimensions by writing the Schrödinger equation in Cartesian coordinates as

$$\frac{d^2u}{dx^2} + \frac{d^2u}{dy^2} + \frac{d^2u}{dz^2} + \frac{2m}{\hbar^2}(E - V(x, y, z))u = 0. \tag{24.93}$$

We assume that the particle is traveling in the z-direction at sufficiently high energies that the effect of the interaction will be to deflect the particle by only small angles. This process can be described more simply by using cylindrical coordinates with the axis of a cylinder being the z-axis. Let us denote the radial vector in the $x-y$ plane by $\mathbf{b}$, which is also called the impact parameter, and the angle by ϕ. The relation between the coordinate systems is given by

$$x = b\cos\phi, \quad y = b\sin\phi, \quad z = z. \tag{24.94}$$

Thus,

$$d^3r = b\, db\, d\phi\, dz. \tag{24.95}$$

The potential can be written as

$$V(x, y, z) = V(b, z) \tag{24.96}$$

assuming symmetry about the z-axis. This process is described in Fig. 24.4.

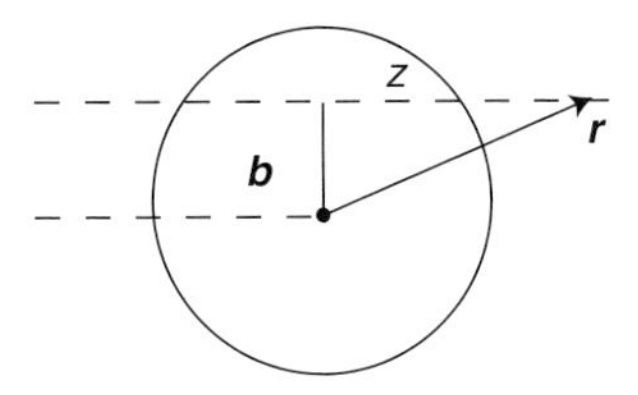

Fig. 24.4

The wavefunction in three dimensions is then given by the extension of the one-dimensional result,

$$u_i(\mathbf{r}) = e^{i\mathbf{k}_i\cdot\mathbf{r}} e^{\frac{-i}{\hbar v}\int_{-\infty}^{z} dz'\, V(b,z')} \tag{24.97}$$

where $\mathbf{k}_i = k\mathbf{e}_3$, corresponding to the motion of the particle in the z-direction. Since this represents the initial or incoming wave, we have introduced the subscript i.

From our results in scattering theory we have found that the scattering amplitude, $f(\theta)$, is given by

$$f(\theta) = -\frac{m}{2\pi\hbar^2}\int d^3r\, e^{-ik_f\cdot r} V(r) u_i(\mathbf{r}) \tag{24.98}$$

where $\mathbf{k}_f$ is the momentum of the particle after scattering, the final momentum. Substituting (24.97) we obtain

$$f(\theta) = -\frac{m}{2\pi\hbar^2}\int d^3r\, V(b,z) e^{-i\mathbf{q}\cdot\mathbf{r}} e^{\frac{-i}{\hbar v}\int_{-\infty}^{z} dz'\, V(b,z')} \tag{24.99}$$

where $\mathbf{q}$ is the momentum transfer

$$\mathbf{q} = \mathbf{k}_f - \mathbf{k}_i. \tag{24.100}$$

For small deflections from the direction of motion, the z-axis, we have $\mathbf{q}_z \approx 0$, and, therefore, $\mathbf{q}$ will be entirely in the x–y plane. Thus,

$$\mathbf{q}\cdot\mathbf{r} = \mathbf{q}\cdot\mathbf{b} = qb\cos\phi \tag{24.101}$$

and

$$f(\theta) = -\frac{m}{2\pi\hbar^2}\int_0^{\infty} b\, db \int_0^{2\pi} d\phi \int_{-\infty}^{\infty} dz\, V(b,z) e^{iqb\cos\phi}\, e^{\frac{-i}{\hbar v}\int_{-\infty}^{z} dz'\, V(b,z')}. \tag{24.102}$$

To evaluate the integral over z, we note that since $V(b,z)$ also appears in the exponent, the integrand, as a function of z, corresponds to a perfect differential. We can then make use of the relation

$$\int dz\, \frac{df(z)}{dz} e^{f(z)} = e^{f(z)}. \tag{24.103}$$

Thus,

$$\int_{-\infty}^{\infty} dz\, V(b,z) e^{\frac{-i}{\hbar v}\int_{-\infty}^{z} dz'\, V(b,z')} = \left[i\hbar v e^{\frac{-i}{\hbar v}\int_{-\infty}^{z} dz'\, V(b,z')} \right]_{z=-\infty}^{z=\infty} \tag{24.104}$$

$$= i\hbar v \left[e^{\frac{-i}{\hbar v}\int_{-\infty}^{z} dz'\, V(b,z')} - 1 \right]. \tag{24.105}$$

The scattering amplitude is then found to be

$$f(\theta) = \frac{ik}{2\pi}\int_0^{\infty} db\, b \int_0^{2\pi} d\phi\, e^{iqb\cos\phi} \left[1 - e^{\frac{-i}{\hbar v}\int_{-\infty}^{\infty} dz'\, V(b,z')} \right]. \tag{24.106}$$

The integral over ϕ is a well-known identity involving the Bessel function $J_0(qb)$,

$$\int_0^{2\pi} d\phi\, e^{iqb\cos\phi} = 2\pi\, J_0(qb). \tag{24.107}$$

Thus the scattering amplitude is found to be

$$f(\theta) = ik\int_0^{\infty} db\, b\, J_0(qb) \left[1 - e^{\frac{-i}{\hbar v}\int_{-\infty}^{\infty} dz'\, V(b,z')} \right]. \tag{24.108}$$

We note that the angular momentum l is related to the impact parameter by

$$l = kb. \tag{24.109}$$

Therefore, the above relation for $f(\theta)$ is related to the partial wave expansion we considered in Chapter 20. Here, since $b = l/k$, the sum over l at high energies is replaced by an integral over the impact parameter b.

24.3.1 Absorption

As we have already discussed in earlier chapters, to account for absorption, albeit phenomenologically, we can replace V by a complex V given by $(V_R - iV_I)$, with positive V_I, and substitute it in (24.108). We obtain

$$f(\theta) = ik\int_0^{\infty} db\, b\, J_0(qb) \left[1 - e^{\frac{-i}{\hbar v}\int_{-\infty}^{\infty} dz'\, V_R(b,z')}\, e^{-\frac{1}{\hbar v}\int_{-\infty}^{\infty} dz'\, V_I(b,z')} \right]. \tag{24.110}$$

We will now call this $f(\theta)$ the elastic scattering amplitude, which will describe elastic process of the type $A + B \to A + B$. The elastic cross-section is then

$$\sigma_{el} = \int d\Omega\, |f(\theta)|^2 . \tag{24.111}$$

The total cross-section is obtained from the optical theorem:

$$\sigma_{total} = \frac{k}{4\pi}\,\mathrm{Im} f(0). \tag{24.112}$$

The absorption cross-section is then the difference

$$\sigma_{abs} = \sigma_{total} - \sigma_{el}. \tag{24.113}$$

24.3.2 Perfect absorption

Of great interest in high-energy scattering processes is the case when there is perfect absorption. The scattering process for this case is also called pure diffraction. It will correspond to having V_I infinite in the region of interaction. Therefore, if the range of interaction is R, perfect absorption will correspond to

$$V = 0, \quad b > R, \tag{24.114}$$

$$V_I = \infty, \quad b < R. \tag{24.115}$$

This is referred to as representing a "black" disk. The scattering amplitude is then given by

$$f(\theta) = ik \int_0^R db\, bJ_0(kb\theta). \tag{24.116}$$

Using a well-known relation involving the Bessel functions, this integral can be calculated exactly and one finds

$$f(\theta) = i\frac{RJ_1(kR\theta)}{\theta} \tag{24.117}$$

where $J_1(kR\theta)$ is the cylindrical Bessel function of the first order. It has the property

$$J_1(x) \to \frac{x}{2} \quad \text{as } x \to 0. \tag{24.118}$$

The quantity $|f(\theta)|^2$ has a sharp peak at $\theta = 0$ with height $(1/4)\,k^2R^4$. The differential cross-section is then given by

$$\frac{d\sigma}{d\Omega} = R^2\left[\frac{J_1(kR\theta)}{\theta}\right]^2 . \tag{24.119}$$

This result is well-known in optics and corresponds to Fraunhofer diffraction by a black sphere.

The elastic cross-section can be calculated. One can write it approximately as

$$\sigma_{el} = \int |f(\theta)|^2 \, d\Omega \approx \pi R^2. \tag{24.120}$$

From the generalized optical theorem the total cross-section is found to be

$$\sigma_T = \frac{4\pi}{k} \operatorname{Im} f(0) = \left(\frac{4\pi}{k}\right)\left(\frac{kR^2}{2}\right) = 2\pi R^2. \tag{24.121}$$

The inelastic cross-section due to a perfect absorber is then

$$\sigma_{inel} = \pi R^2. \tag{24.122}$$

This is once again a consequence of the "shadow" effect, where the black disk takes away from the incident wave an amount equal to the area of the disk.

24.4 Problems

1. Use the WKB approximation to obtain the energy levels of a harmonic oscillator.
2. Consider the following potential,

$$V(x) = \lambda |x| \quad \text{for} \quad -\infty < x < +\infty$$

   with $\lambda > 0$. Use the WKB approximation to obtain the energy eigenvalues.
3. For a one-dimensional harmonic oscillator use the variational method with the trial wavefunction $\exp(-\beta |x|)$ to estimate the ground-state energy.
4. Consider the infinite barrier problem where the potential is given by

$$\begin{aligned} V(x) &= 0, \quad &|x| < a \\ &= \infty, \quad &|x| > a. \end{aligned}$$

   Use the variational method with the following trial wavefunction:

$$\begin{aligned} u(x) &= A(a^2 - x^2), \quad &|x| < a \\ &= 0, \quad &|x| > a \end{aligned}$$

   where A is the normalization constant, to estimate the ground-state energy. Compare your result with the exact result.
5. Consider a three-dimensional isotropic harmonic oscillator for which the Hamiltonian is given by

$$H = \frac{\mathbf{p}^2}{2m} + \frac{1}{2} m\omega^2 \mathbf{r}^2.$$

Use the variational method with the trial function

$$u(\mathbf{r}) = \frac{1}{\left(\pi a^2\right)^{3/4}} \exp(-r^2/2a^2)$$

and obtain E. Minimizing E with respect to a^2, show that the upper bound for the ground-state energy reproduces the exact result for the energy given by

$$a = \sqrt{\frac{\hbar}{m\omega}} \quad \text{and} \quad E_a = \frac{3}{2}\hbar\omega.$$

Substitute the above value of a in the trial function and show that it also reproduces the exact ground-state wavefunction.

6. Consider the Hamiltonian for the hydrogen atom:

$$H = \frac{\mathbf{p}^2}{2m} - \frac{e^2}{r}.$$

Assume $Ne^{-\alpha r}$ as a trial wavefunction where N is the normalization constant (to be determined in terms of α). Use α as the variational parameter to estimate the energy for the ground state, which is an S-state.

7. Use

$$R_l = e^{-\beta r}$$

as the trial wavefunction to obtain the $l = 0$ ground-state energy for the potentials

$$\text{(i)} \;\; V(r) = -\lambda e^{-\mu r},$$

$$\text{(ii)} \;\; V(r) = -\lambda r^n, \quad \text{with } n \geq -1.$$

Compare the results of (ii) with those for Coulomb and for harmonic oscillator potentials.

8. In an equivalent description of the eikonal approximation, express the wavefunction for a particle moving in the z-direction as

$$u(\mathbf{r}) = \frac{1}{\left(\sqrt{2\pi}\right)^3} e^{ikz} w(\mathbf{r}).$$

Show that in the approximation in which $\nabla^2 w(\mathbf{r})$ can be neglected, the Schrödinger equation gives

$$\left[\frac{d}{dz} + \frac{im}{k} V(r)\right] w(\mathbf{r}) = 0$$

with the solution

$$w(\mathbf{b}, z) = \exp\left(-\frac{i}{v} \int_{-\infty}^{z} dz' \, V(\mathbf{b}, z')\right)$$

where $\mathbf{b} = \sqrt{x^2 + y^2}$ is the impact parameter, $v = k/m$ is the velocity.

9. In the previous problem, assuming the scattering to be essentially in the forward direction, use the Born approximation to show that the scattering amplitude is given by

$$f(\theta) = \frac{k}{2\pi i} \int d\mathbf{b} \left[e^{2i\Delta(\mathbf{b})} - 1 \right]$$

where

$$\Delta(\mathbf{b}) = -\frac{1}{2v} \int_{-\infty}^{\infty} dz' V(\mathbf{b}, z').$$

Assume $\theta \approx 0$, and take the magnitude of the momentum transfer $|\mathbf{q}| \approx k\theta$.

10. Determine the ground-state energy of the helium atom using perturbation theory with the Coulomb potential between the two electrons as the perturbing potential. Assume the unperturbed wavefunction to be given by the product of two hydrogen atom wave functions. Compare your result with the variational result.
11. Obtain the ground-state energy using the variational principle with the trial function $\exp(-\alpha r)$ for (i) $V(r) = \lambda/r$ and (ii) $V(r) = \lambda r$.

25 Lagrangian method and Feynman path integrals

The Lagrangian formulation, in essence, describes the dynamics of a particle starting with the action principle. Through the so-called Euler–Lagrange equations it allows one to determine the particle's equations of motion. One of the properties of a Lagrangian is that the conservation laws can be incorporated through it quite simply. It is important to note that the action principle also plays a role in quantum mechanics through Feynman path integral techniques.

In the path integral technique in quantum mechanics, the concept of a unique path for a classical particle is replaced, starting from the superposition principle, by a sum of infinitely many possible paths that a particle can take between any two points. In this formulation the probability amplitude for a given process is calculated and found to be proportional to $\exp[iS/\hbar]$, where S is the action. It is then summed over all paths. The interference between the terms in the summation then determines the motion of the particles. We will discuss all of this in the following.

25.1 Euler–Lagrange equations

In classical mechanics the motion of a point particle is described by a Lagrangian, $L(q_i, \dot{q}_i)$, where $q_i(t)$ $(i = 1, 2, 3)$ are the three space coordinates and $\dot{q}_i = dq_i/dt$. The Lagrangian is defined as

$$L = T - V \tag{25.1}$$

where T is the kinetic energy and V the potential energy given, respectively, by

$$T = \frac{1}{2}m\dot{q}_i^2, \quad V = V(q) \tag{25.2}$$

where $V(q)$ depends only on the magnitude of q. We are using the summation convention whenever the index i is repeated.

The momentum p_i is defined as

$$p_i = \frac{\partial L}{\partial \dot{q}_i} \tag{25.3}$$

and the Hamiltonian is then

$$H = p_i\dot{q}_i - L. \tag{25.4}$$

Thus from (25.1) and (25.2), we find

$$p_i = m\dot{q}_i \tag{25.5}$$

and

$$H = \frac{1}{2}m\dot{q}_i^2 + V(q) = T + V \tag{25.6}$$

as expected.

The basic hypothesis of the Lagrangian formulation is that the motion of a particle is obtained by minimizing the action, S, which is defined as

$$S = \int_{t_1}^{t_2} dt\, L(q_i, \dot{q}_i)\,. \tag{25.7}$$

The minimization then corresponds to

$$\delta S = 0 \tag{25.8}$$

where δS is the variation of S with respect to an arbitrary path with the end points at t_1 and t_2 fixed. The relation (25.8) implies that the path taken by a classical particle will correspond to a minimum of S.

We can calculate δS as follows:

$$\delta S = \int_{t_1}^{t_2} dt \left[\frac{\partial L}{\partial q_i}\delta q_i + \frac{\partial L}{\partial \dot{q}_i}\delta \dot{q}_i\right] \tag{25.9}$$

where a summation of q_i and $\dot{q}_i$ for $i = 1, 2, 3$ is implied. Integrating the second term by parts we get

$$\int_{t_1}^{t_2} dt\, \frac{\partial L}{\partial \dot{q}_i}\delta \dot{q}_i = -\int_{t_1}^{t_2} dt\, \delta q_i \frac{d}{dt}\left(\frac{\partial L}{\partial \dot{q}_i}\right). \tag{25.10}$$

Since the end points are fixed we have taken $\delta q_i(t_1) = 0 = \delta q_i(t_2)$ in doing the partial integration. The relation (25.9) is now given by

$$\delta S = \int_{t_1}^{t_2} dt\, \delta q_i \left[\frac{\partial L}{\partial q_i} - \frac{d}{dt}\left(\frac{\partial L}{\partial \dot{q}_i}\right)\right]. \tag{25.11}$$

Since this relation holds for any arbitrary path between t_1 and t_2, the minimization condition (25.8) implies that the integrand in (25.11) must vanish. Hence,

$$\frac{\partial L}{\partial q_i} - \frac{d}{dt}\left(\frac{\partial L}{\partial \dot{q}_i}\right) = 0. \tag{25.12}$$

These are called Euler–Lagrange equations, which constitute the basic equations of motion in classical mechanics.

If we insert the relations (25.2) in (25.1) and substitute L in (25.12), we reproduce the well-known classical equations of motion,

$$m\ddot{q}_i = -\frac{\partial V}{\partial q_i}. \tag{25.13}$$

A one-dimensional harmonic oscillator provides a simple example for an application of the Lagrangian formulation. Replacing q_i by x and $\dot{q}_i$ by $\dot{x}$, the Lagrangian for this problem is given by

$$L = \frac{1}{2}m\dot{x}^2 - \frac{1}{2}Kx^2 \tag{25.14}$$

where K is the spring constant. From (25.13) we obtain

$$m\ddot{x} = -Kx, \tag{25.15}$$

which is the classical equation of motion for a one-dimensional harmonic oscillator.

25.2 *N* oscillators and the continuum limit

Consider N particles connected along a single dimension by identical springs with the same spring constant, K (see Fig. 25.1). If y_i is the displacement of the ith particle from its equilibrium position, then the Lagrangian will be given by

$$L = \sum_{i=1}^{N} \frac{1}{2}\left[m\dot{y}_i^2 - K\left(y_{i+1} - y_i\right)^2\right]. \tag{25.16}$$

We assume that the equilibrium positions of two neighboring particles are separated by a distance, a, and define the Lagrangian density per unit length as $\mathcal{L}$ so that

$$L = \sum_{i=1}^{N} a\mathcal{L}_i \tag{25.17}$$

where

$$\mathcal{L} = \frac{1}{2}\left[\left(\frac{m}{a}\right)\dot{y}^2 - Ka\left(\frac{y_{i+1} - y_i}{a}\right)^2\right]. \tag{25.18}$$

Fig. 25.1

Let us now go to the continuum limit with infinite degrees of freedom, $N \to \infty$, by taking

$$\frac{m}{a} = \mu = \text{ mass per unit length,} \tag{25.19}$$

$$Y = Ka = \text{Young's modulus,} \tag{25.20}$$

and change

$$\sum_{i=1}^{N} a\mathcal{L} \to \int dx\, \mathcal{L} \tag{25.21}$$

where we have replaced a by dx while the summation over i is replaced by an integral. The Lagrangian density, $\mathcal{L}$, is now given by

$$\mathcal{L} = \frac{1}{2}\left[\mu\dot{\phi}^2 - Y\left(\frac{\partial\phi}{\partial x}\right)^2\right]. \tag{25.22}$$

We have also made the following replacements of discrete quantities by their continuous counterparts in writing the above result:

$$y_i \to \phi(x,t), \tag{25.23}$$

$$\frac{y_{i+1} - y_i}{a} \to \frac{\partial\phi}{\partial x}. \tag{25.24}$$

The Lagrangian density, $\mathcal{L}$, is now a function of ϕ, $\partial\phi/\partial x, \dot{\phi}$.

The Lagrangian, L, is written as

$$L = \int dx\, \mathcal{L}\left(\phi, \dot{\phi}, \frac{\partial\phi}{\partial x}\right) \tag{25.25}$$

and the action S as

$$S = \int dt \int dx\, \mathcal{L}\left(\phi, \dot{\phi}, \frac{\partial\phi}{\partial x}\right). \tag{25.26}$$

The minimization $\delta S = 0$ gives the Euler–Lagrange equations

$$\frac{\partial\mathcal{L}}{\partial\phi} - \frac{\partial}{\partial x}\frac{\partial\mathcal{L}}{\partial\left(\frac{\partial\phi}{\partial x}\right)} - \frac{\partial}{\partial t}\frac{\partial\mathcal{L}}{\partial\left(\dot{\phi}\right)} = 0 \tag{25.27}$$

where as before we have assumed $\delta\phi$ to vanish at the end points of x and t. From $\mathcal{L}$ given by (25.22) we obtain

$$\mu\frac{\partial^2\phi}{\partial t^2} - Y\frac{\partial^2\phi}{\partial x^2} = 0, \tag{25.28}$$

which corresponds to the known wave equation in one dimension for a wave amplitude ϕ traveling with velocity $\sqrt{Y/\mu}$.

In summary, in the discrete case we discussed earlier, we found that once the Lagrangian was known, as it was for the case of the harmonic oscillator given by (25.14), the Lagrangian formulation enabled us to obtain the equation of motion for the displacement variable. In the continuum limit it was shown through the example of a set of harmonic oscillators that the equation of motion for the displacement variable is replaced by a wave equation for ϕ representing the displacement in the continuum limit.

25.3 Feynman path integrals

25.3.1 Time evolution operator and Green's function

Consider a matrix element of the time-evolution operator

$$U(t_2, t_1) = \exp\left[-i\frac{H(t_2 - t_1)}{\hbar}\right]. \tag{25.29}$$

Confining to one space dimension, we define the matrix element

$$K(x_2, t_2; x_1, t_1) = \langle x_2 | \exp\left[-i\frac{H(t_2 - t_1)}{\hbar}\right] | x_1 \rangle \tag{25.30}$$

where H is the Hamiltonian

$$H = \frac{p^2}{2m} + V(x). \tag{25.31}$$

We can write K as

$$K(x_2, t_2; x_1, t_1) = \langle x_2(t_2) | x_1(t_1) \rangle. \tag{25.32}$$

In a sense K propagates the information between the two points, (x_2, t_2) and (x_1, t_1); hence it is called a "propagator" or simply a Green's function. One also refers to it as a "transition amplitude" to go from the state $|x_2(t_2)\rangle$ to the state $|x_1(t_1)\rangle$. We have come across many types of propagators before that generally depend on momentum variables. This depends on space-time. We will write it as

$$K(x_2, t_2; x_1, t_1) = \langle x_2, t_2 | x_1, t_1 \rangle. \tag{25.33}$$

It satisfies the properties

$$K(x_2, t_2; x_1, t_1) = 0, \qquad \text{for } t_2 < t_1, \tag{25.34}$$

$$\lim_{t_2 \to t_1} K(x_2, t_2; x_1, t_1) = \delta(x_2 - x_1). \tag{25.35}$$

We note that for fixed values of x_1 and t_1, K satisfies the time-dependent Schrödinger equation, as is evident from the presence of the time-evolution operator $U(t_2, t_1)$.

Let us assume the two space-time points (x_2, t_2) and (x_1, t_1) to be only infinitesimally apart,

$$x_2 = x_1 + \Delta x, \quad t_2 = t_1 + \epsilon \tag{25.36}$$

where Δx and ϵ are infinitesimal quantities. We then make a linear expansion of the exponential in (25.30) in terms of ϵ, and obtain,

$$K(x_2, t_2; x_1, t_1) = \langle x_2 | \left[1 - \frac{i\epsilon}{\hbar} \left(\frac{p^2}{2m} + V(x) \right) \right] | x_1 \rangle . \tag{25.37}$$

At this stage p and x are operators. However, one can make the following simplification,

$$\langle x_2 | \left[1 - \frac{i\epsilon}{\hbar} V(x) \right] | x_1 \rangle = \left[1 - \frac{i\epsilon}{\hbar} V(x_1) \right] \delta(x_2 - x_1) \tag{25.38}$$

where $V(x_1)$ is now a function that corresponds to the potential at the point x_1, and no longer an operator. Let us now express $\delta(x_2 - x_1)$ in the following familiar form:

$$\delta(x_2 - x_1) = \frac{1}{2\pi} \int dp_1 \, e^{ip_1(x_2 - x_1)}. \tag{25.39}$$

Expression (25.38) is then given by

$$\frac{1}{2\pi} \int dp_1 \left[1 - \frac{i\epsilon}{\hbar} V(x_1) \right] e^{ip_1(x_2 - x_1)}. \tag{25.40}$$

To obtain the matrix element of p^2 in (25.37), we insert a complete set of states $|p_1\rangle$ and $|p_2\rangle$; therefore,

$$\langle x_2 | p^2 | x_1 \rangle = \int dp_1 \int dp_2 \, \langle x_2 | p_2 \rangle \langle p_2 | p^2 | p_1 \rangle \langle p_1 | x_1 \rangle. \tag{25.41}$$

Since p^2 is a diagonal operator, for (25.41) we obtain

$$\frac{1}{2\pi} \int dp_1 \, p_1^2 \, e^{ip_1(x_2 - x_1)} \tag{25.42}$$

where we have written the momentum eigenfunction $\langle x | p \rangle$ as

$$\langle x | p \rangle = \frac{1}{\sqrt{2\pi}} e^{ipx}. \tag{25.43}$$

After combining (25.40) and (25.42), the expression (25.37) can be written as

$$\frac{1}{2\pi} \int dp_1 \left[1 - \frac{i\epsilon}{\hbar} \left[\frac{p_1^2}{2m} + V(x_1) \right] \right] e^{ip_1(x_2 - x_1)}. \tag{25.44}$$

The Hamiltonian is now to be evaluated at specific values x_1 and p_1. To order ϵ, we can revert to writing the term in the square bracket as an exponential and write (25.44) as

$$\frac{1}{2\pi}\int dp_1 e^{ip_1(x_2-x_1)} e^{-\frac{-i\epsilon}{\hbar}\left[\frac{p_1^2}{2m}+V(x_1)\right]}. \tag{25.45}$$

Let us write

$$(x_2 - x_1) = \Delta x = \dot{x}_1 \epsilon. \tag{25.46}$$

Therefore, (25.45) becomes

$$\frac{1}{2\pi}\int dp_1 e^{i\left[p_1\dot{x}_1 - \frac{p_1^2}{2m} - V(x_1)\right]\frac{-\epsilon}{\hbar}}. \tag{25.47}$$

Hence

$$\langle x_2, t_2 | x_1, t_1\rangle = \frac{1}{2\pi}\int dp_1 \exp\left[\frac{-i\epsilon}{\hbar}\left[p_1\dot{x}_1 - H(x_1, p_1)\right]\right]. \tag{25.48}$$

We note that the above simplification was achieved only because we made a linear approximation for the exponential on the grounds that the space-time points for the matrix element are separated by an infinitesimal amount. This allowed us to avoid going to higher powers in $(t_2 - t_1)$, which would have involved the products of x and p, and, therefore, would have brought in the commutator $[x, p]$ $(= i\hbar)$.

The integral (25.47) can be simplified further by writing

$$p_1\dot{x}_1 - \frac{p_1^2}{2m} = -\frac{1}{2m}(p_1 - m\dot{x}_1)^2 + \frac{m\dot{x}_1^2}{2}. \tag{25.49}$$

The integral over the p_1-term in (25.47) is

$$\int dp_1 \exp\frac{-i\epsilon}{2m\hbar}\left[(p_1 - m\dot{x}_1)^2\right] = \sqrt{\frac{2m\pi\hbar}{i\epsilon}}. \tag{25.50}$$

Thus,

$$\langle x_2, t_2 | x_1, t_1\rangle = \sqrt{\frac{m\hbar}{2\pi i\epsilon}} \exp\epsilon\left[\frac{m\dot{x}_1^2}{2} - V(x_1)\right]. \tag{25.51}$$

The right-hand side can be expressed as

$$\frac{1}{w(\epsilon)}\exp\left[-i\frac{L}{\hbar}\Delta t_{21}\right] \tag{25.52}$$

where

$$w(\epsilon) = \sqrt{\frac{2\pi i\epsilon}{m\hbar}} \quad \text{and} \quad \Delta t_{21} = (t_2 - t_1) \tag{25.53}$$

and L is the Lagrangian

$$L = L(x_1, \dot{x}_1) = \frac{m\dot{x}_1^2}{2} - V(x_1). \tag{25.54}$$

Thus,

$$\langle x_2, t_2 | x_1, t_1 \rangle = \frac{1}{w(\epsilon)} \exp\left(-i\left[\frac{L}{\hbar}\Delta t_{21}\right]\right). \tag{25.55}$$

25.3.2 *N*-intervals

In the following we will continue to make the assumption that the separation between all the adjacent space-time points is infinitesimal. In going from a point (x_1, t_1) a finite distance to (x_N, t_N) we will divide the interval into $(N - 1)$ equal parts and take

$$\Delta t_{i+1,i} = \epsilon = \lim_{N\to\infty} \frac{t_{i+1} - t_i}{N}. \tag{25.56}$$

Introducing a complete set of states, the matrix element $\langle x_N, t_N | x_1, t_1 \rangle$ can be written as

$$\langle x_N, t_N | x_1, t_1 \rangle$$
$$= \int dx_{N-1} \int dx_{N-2} \cdots \int dx_2 \, \langle x_N, t_N | x_{N-1}, t_{N-1} \rangle \langle x_{N-1}, t_{N-1} | x_{N-2}, t_{N-2} \rangle \cdots \langle x_2, t_2 | x_1, t_1 \rangle. \tag{25.57}$$

Pictorially this is expressed in the space-time plot in Fig. 25.2. We note that the end points (x_1, t_1), and (x_N, t_N) are fixed, while the transition amplitudes are integrated out over all values of the intermediate space points $x_2 \cdots x_{N-1}$ at each value of time. This implies that we must sum over all possible paths in the space-time plane with fixed end points. This is the essence of the Feynman path integration technique for transition amplitudes. The formalism shows that if we are given the transition amplitude over an infinitesimal interval then through this technique, one can generate, the amplitude for finite distances.

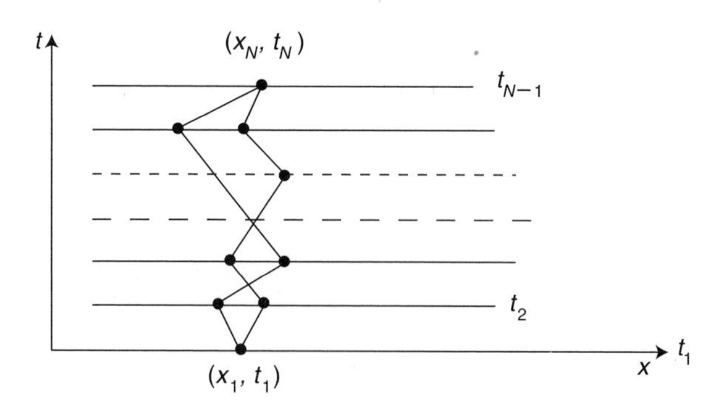

Fig. 25.2

Using (25.48) the above relation can be expressed as

$$\langle x_N, t_N | x_1, t_1 \rangle = \lim_{N\to\infty} \int \prod_{i=2}^{N-1} dx_i \prod_{i=1}^{N-1} \frac{dp_i}{2\pi} \exp\left[\frac{i\epsilon}{\hbar}\left[p_i \dot{x}_1 - H(x_i, p_i)\right]\right]. \tag{25.58}$$

We can also write it, using (25.52), as

$$\langle x_N, t_N | x_1, t_1 \rangle = \lim_{N\to\infty} \frac{1}{w(\epsilon)^N} \int dx_2 \int dx_3 \cdots \int dx_{N-1} \exp\left[-\frac{i}{\hbar}\left[\int_{t_1}^{t_N} L\, dt\right]\right] \tag{25.59}$$

where we have expressed

$$L_1 \Delta t_{21} + L_2 \Delta t_{32} + \cdots + L_N \Delta t_{NN-1} = \int_{t_1}^{t_N} L\, dt \tag{25.60}$$

where L_i is the Lagrangian

$$L_i = \frac{m\dot{x}_i^2}{2} - V(x_i). \tag{25.61}$$

The integral over the Lagrangian is just the action S defined earlier. We write it as

$$S(n, n-1) = \int_{t_{n-1}}^{t_n} L\, dt. \tag{25.62}$$

Hence, one can write

$$\langle x_N, t_N | x_1, t_1 \rangle = \frac{1}{w(\epsilon)^N} \int D(x) \exp\left[-\frac{S(N,1)}{\hbar}\right] \tag{25.63}$$

where $D(x)$ symbolically represents the product of the integrals

$$\int D(x) = \int dx_2 \int dx_3 \cdots \int dx_{N-1}. \tag{25.64}$$

This expression is known as the Feynman path integral (see Fig. 25.3), and it represents the sum over all possible paths, $\sum_{paths}$.

We find from the above formalism that in quantum mechanics all possible paths must play a role. In the classical limit, $\hbar \to 0$, we note from (25.63) that the phases will fluctuate wildly as one goes from one path to another, leading to cancelations between the terms. The exception will occur when the variation in the numerator of the exponent is minimal, leading to the condition

$$\delta \int_{t_{n-1}}^{t_n} L\, dt = 0 \tag{25.65}$$

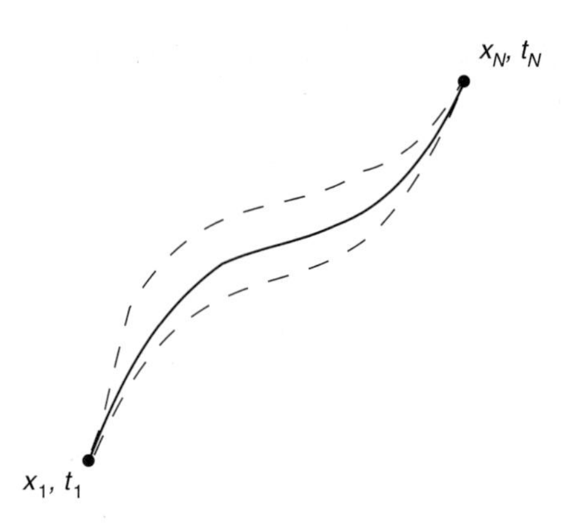

Fig. 25.3

for fixed end points. This will then correspond to the classical path. The above condition is, indeed, the basis for the Lagrangian formulation of classical mechanics.

It is often found that one has to go through very difficult calculations to solve a relatively simple problem when one utilizes the Feynman path techniques. The advantage of this technique is that only functions of variables are involved and not the operators and hence the problems of commutation relations are not encountered. Also, symmetries are easier to incorporate into the theory. One finds that Feynman techniques provide a powerful tool in treating problems in quantum field theory.

25.4 Problems

1. Show that one can write

$$\langle x_N, t_N \,| x_1, t_1 \rangle = \int dx_{N-1}\, F\left(x_N, x_{N-1}; t_N, t_{N-1}\right) \exp\left[-i\frac{V\Delta t}{\hbar}\right] \langle x_{N-1}, t_{N-1} \,| x_1, t_1 \rangle$$

   where $F = K$ for $V = 0$. Take $x_N = x$, $x_{N-1} = x - \Delta x$, $t_N = t + \Delta t$, and $t_{N-1} = t$ and expand the matrix elements on both sides of the equation, as well as the exponential on the right-hand side, in terms of Δx and Δt, keeping only the lowest-order nonvanishing terms. Compare the coefficients of Δt on both sides and show that $\langle x, t \,| x_1, t_1 \rangle$ satisfies the time-dependent Schrödinger equation.
2. A particle is in a ground state at $t = 0$ bound through a harmonic oscillator potential to a force center. For $t > 0$ the force center is suddenly removed. Determine the wavefunction for $t > 0$.
3. A particle trapped within two infinite walls is described at $t = 0$ by

$$\psi(x, 0) = a\delta(x).$$

   The walls are removed for $t > 0$. Determine $\psi(x, t)$.

26 Rotations and angular momentum

We will, temporarily, abandon calculations of dynamical quantities and concentrate on the symmetry properties of physical systems beginning with rotations. We demonstrate the direct connection between rotation and angular momentum. This time we also include spin. Based on rotation of the coordinate systems, we establish the rotation properties of state vectors and operators.

26.1 Rotation of coordinate axes

Consider a point P with coordinates (x, y, z) in a three-dimensional space. If O is the origin then we will call OP the coordinate vector. Let the z axis be the axis of rotation, ρ the magnitude of the projection of OP on the x–y plane, and α the angle made by this projection with the x-axis; then

$$x = \rho \cos \alpha, \tag{26.1}$$

$$y = \rho \sin \alpha. \tag{26.2}$$

We note that a counter-clockwise rotation of the axes keeping the coordinate vector fixed is equivalent to a clockwise rotation of the coordinate vector keeping the axes fixed. Specifically, for a counter-clockwise rotation of the axes by an angle ϕ about the z-axis, which we will consider below, the new coordinates (x', y', z') and the old coordinates (x, y, z) are related as follows (see Fig. 26.1):

$$x' = \rho \cos (\phi + \alpha), \tag{26.3}$$

$$y' = \rho \sin (\phi + \alpha), \tag{26.4}$$

$$z' = z. \tag{26.5}$$

Expanding the trigonometric functions in (26.3) and (26.4) and substituting the relations involving x and y in (26.1) and (26.2), we obtain

$$x' = x \cos \phi - y \sin \phi, \tag{26.6}$$

$$y' = x \sin \phi + y \cos \phi, \tag{26.7}$$

$$z' = z. \tag{26.8}$$

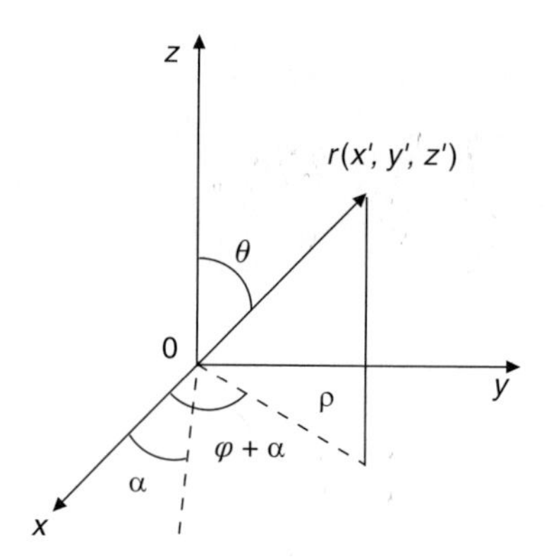

Fig. 26.1

Writing this relation in a matrix form we have

$$\begin{pmatrix} x' \\ y' \\ z' \end{pmatrix} = \begin{pmatrix} \cos\phi & -\sin\phi & 0 \\ \sin\phi & \cos\phi & 0 \\ 0 & 0 & 1 \end{pmatrix} \begin{pmatrix} x \\ y \\ z \end{pmatrix}. \tag{26.9}$$

The matrix

$$R_z(\phi) = \begin{pmatrix} \cos\phi & -\sin\phi & 0 \\ \sin\phi & \cos\phi & 0 \\ 0 & 0 & 1 \end{pmatrix} \tag{26.10}$$

is called the rotation matrix or rotation operator for rotations around the z-axis. The above expression represents the relation between the original coordinate vector $\mathbf{r} = (x, y, z)$ and the rotated vector $\mathbf{r}' = (x', y', z')$.

We can write the relation (26.9) as

$$\begin{pmatrix} x' \\ y' \\ z' \end{pmatrix} = R_z(\phi) \begin{pmatrix} x \\ y \\ z \end{pmatrix}. \tag{26.11}$$

Since the distance OP remains unchanged whether we use the original coordinate system or the rotated one, the following relation is satisfied:

$$x'^2 + y'^2 + z'^2 = x^2 + y^2 + z^2. \tag{26.12}$$

In terms of (26.11) the left-hand side of (26.12) gives

$$\begin{pmatrix} x' & y' & z' \end{pmatrix} \begin{pmatrix} x' \\ y' \\ z' \end{pmatrix} = \widetilde{R}_z(\phi) R_z(\phi) \begin{pmatrix} x & y & z \end{pmatrix} \begin{pmatrix} x \\ y \\ z \end{pmatrix}. \tag{26.13}$$

The relation (26.13) implies that

$$\widetilde{R}_z(\phi)R_z(\phi) = \mathbf{I} \tag{26.14}$$

where $\mathbf{I}$ is a unit matrix. This relation is, of course, satisfied by the explicit expression for $R_z(\phi)$ we obtained in (26.10). Furthermore, we also find from (26.10) that

$$\det R_z(\phi) = 1. \tag{26.15}$$

The significance of the relations (26.14) and (26.15) will be elaborated upon in some of the later sections.

The transformation of a general vector $\mathbf{V} = \left(V_x, V_y, V_z\right)$ can be written in a similar manner as (26.9). We obtain

$$\begin{pmatrix} V_{x'} \\ V_{y'} \\ V_{z'} \end{pmatrix} = \begin{pmatrix} \cos\phi & -\sin\phi & 0 \\ \sin\phi & \cos\phi & 0 \\ 0 & 0 & 1 \end{pmatrix} \begin{pmatrix} V_x \\ V_y \\ V_z \end{pmatrix}. \tag{26.16}$$

26.1.1 Infinitesimal transformations

Under the transformations with $\phi = \epsilon$, where ϵ is an infinitesimal quantity, we have $\cos\phi \simeq 1$ and $\sin\phi \simeq \epsilon$. The rotation matrix is then

$$R_z(\epsilon) = \begin{pmatrix} 1 & -\epsilon & 0 \\ \epsilon & 1 & 0 \\ 0 & 0 & 1 \end{pmatrix}. \tag{26.17}$$

The relation (26.16) reads

$$\begin{pmatrix} V'_x \\ V'_y \\ V'_z \end{pmatrix} = \begin{pmatrix} 1 & -\epsilon & 0 \\ \epsilon & 1 & 0 \\ 0 & 0 & 1 \end{pmatrix} \begin{pmatrix} V_x \\ V_y \\ V_z \end{pmatrix} = \left[\begin{pmatrix} 1 & 0 & 0 \\ 0 & 1 & 0 \\ 0 & 0 & 1 \end{pmatrix} + \epsilon \begin{pmatrix} 0 & -1 & 0 \\ 1 & 0 & 0 \\ 0 & 0 & 0 \end{pmatrix}\right] \begin{pmatrix} V_x \\ V_y \\ V_z \end{pmatrix}. \tag{26.18}$$

In order to apply this result to quantum-mechanical problems, we divide and multiply the second term by $\hbar$, and write this relation as

$$\begin{pmatrix} V'_x \\ V'_y \\ V'_z \end{pmatrix} = \left(\mathbf{1} - i\frac{\epsilon}{\hbar}S_z\right) \begin{pmatrix} V_x \\ V_y \\ V_z \end{pmatrix} \tag{26.19}$$

where S_z is given by

$$S_z = i\hbar \begin{pmatrix} 0 & -1 & 0 \\ 1 & 0 & 0 \\ 0 & 0 & 0 \end{pmatrix}. \tag{26.20}$$

The operator S_z is called the generator of $R_z(\epsilon)$, i.e., the generator of infinitesimal rotations about the z-axis.

Rotation by a finite angle ϕ about the z-axis is accomplished by substituting $\epsilon = \phi/N$ in the factor $(\mathbf{1} - i\epsilon S_z/\hbar)$ in (26.19), and taking the product of $(\mathbf{1} - i\phi S_z/N\hbar)$ repeatedly N times with $N \to \infty$. We then use the well-known identity

$$\lim_{N\to\infty}\left(1+\frac{\alpha}{N}\right)^N = e^{\alpha} \tag{26.21}$$

with $\alpha = -i\phi S_z/\hbar$ and obtain the following result for finite transformations:

$$\begin{pmatrix} V'_x \\ V'_y \\ V'_z \end{pmatrix} = e^{-i\phi S_z/\hbar}\begin{pmatrix} V_x \\ V_y \\ V_z \end{pmatrix}. \tag{26.22}$$

We have thus written the rotation operator in terms of its generator.

Following a similar procedure, generators for counter-clockwise rotations along the x- and y-axes can be obtained and are given by

$$S_x = i\hbar\begin{pmatrix} 0 & 0 & 0 \\ 0 & 0 & -1 \\ 0 & 1 & 0 \end{pmatrix}, \quad S_y = i\hbar\begin{pmatrix} 0 & 0 & 1 \\ 0 & 0 & 0 \\ -1 & 0 & 0 \end{pmatrix}. \tag{26.23}$$

It is easy to verify that the matrices S_x, S_y, S_z, satisfy the relation

$$(S_i)_{jk} = -i\hbar\varepsilon_{ijk}, \tag{26.24}$$

where i,j,k can be taken interchangeably as $1,2,3$ or x,y,z. They also satisfy the commutation relations

$$\left[S_i, S_j\right] = i\hbar\sum_k \varepsilon_{ijk}S_k = (i\hbar)\,\varepsilon_{ijk}S_k \tag{26.25}$$

where ε_{ijk} is the totally antisymmetric tensor. On the right-hand side above we have removed the summation sign, $\sum$, in order to follow the summation convention in which any repeated index is automatically summed. We will follow this convention throughout the rest of this book, except where it is essential to use the explicit form.

To generalize the above results to rotations about an arbitrary axis we note that, as we stated earlier, rotating coordinate axes counter-clockwise while keeping the vector $\mathbf{V}$ fixed is equivalent to rotating the vector $\mathbf{V}$ clockwise, to $\mathbf{V}'$ while keeping the coordinate axes fixed. For infinitesimal clockwise rotations, the relation between $\mathbf{V}$ and $\mathbf{V}'$ is particularly simple. It is given by

$$\mathbf{V}' = \mathbf{V} + \epsilon\mathbf{n}\times\mathbf{V} \tag{26.26}$$

where $\mathbf{n}$ is a unit vector in the direction of the rotation axis. The ith component of $\mathbf{V}'$ is then given by

$$V_i' = V_i + \epsilon\,(\mathbf{n}\times\mathbf{V})_i = V_i + \epsilon\left[\varepsilon_{ijk}n_jV_k\right] = V_i - \frac{i\epsilon}{\hbar}\left[n_j\left(S_j\right)_{ik}V_k\right]$$
$$= V_i - \frac{i\epsilon}{\hbar}\left[(\mathbf{n}\cdot\mathbf{S})_{ik}V_k\right] = \left[\delta_{ik} - \frac{i\epsilon}{\hbar}(\mathbf{n}\cdot\mathbf{S})_{ik}\right]V_k. \tag{26.27}$$

Following the procedure outlined above, the rotation of a vector $\mathbf{V}$ by a finite angle χ along a direction $\mathbf{n}$ can be obtained. We find

$$\begin{pmatrix} V_x' \\ V_y' \\ V_z' \end{pmatrix} = e^{-i\chi\mathbf{n}\cdot\mathbf{S}/\hbar}\begin{pmatrix} V_x \\ V_y \\ V_z \end{pmatrix}. \tag{26.28}$$

We will be using the following convention for the angle of rotation: whenever we have a rotation about the z-axis we will denote the angle of rotation by ϕ, which is an obvious choice since we generally denote an azimuthal angle by ϕ. On the other hand, if the rotation is about an arbitrary direction $\mathbf{n}$ then the angle of rotation about $\mathbf{n}$ will be denoted by χ, as we have done above.

26.2 Scalar functions and orbital angular momentum

Let us consider the transformation properties of a function $g(x,y,z)$. Under infinitesimal counter-clockwise rotations of the coordinates about the z-axis we have from (26.6), (26.7), and (26.8),

$$x' = x - \epsilon y, \tag{26.29}$$

$$y' = \epsilon x + y, \tag{26.30}$$

$$z' = z. \tag{26.31}$$

The function $g(x,y,z)$ transforms as

$$g(x,y,z) \to g(x',y',z') = g\left(x-\epsilon y, y+\epsilon x, z\right). \tag{26.32}$$

We expand the right-hand side in a Taylor expansion to obtain

$$g\left(x-\epsilon y, y+\epsilon x, z\right) = g\left(x,y,z\right) - \epsilon y\frac{\partial g\left(x,y,z\right)}{\partial x} + \epsilon x\frac{\partial g\left(x,y,z\right)}{\partial y} + \cdots \tag{26.33}$$

$$= \left[1 - \epsilon\left(y\frac{\partial}{\partial x} - x\frac{\partial}{\partial y}\right) + \cdots\right]g(x,y,z). \tag{26.34}$$

We recall the relation $\mathbf{p} = -i\hbar\nabla$ between the momentum and the space coordinates, which is given by

$$p_x = -i\hbar\frac{\partial}{\partial x}, \quad p_y = -i\hbar\frac{\partial}{\partial y}, \quad p_z = -i\hbar\frac{\partial}{\partial z}. \tag{26.35}$$

Thus,

$$y\frac{\partial}{\partial x} - x\frac{\partial}{\partial y} = \frac{i}{\hbar}\left(yp_x - xp_y\right) = -\frac{iL_z}{\hbar} \tag{26.36}$$

where on the right-hand side we have used the relation for the angular momentum $\mathbf{L} = \mathbf{r}\times\mathbf{p}$, so that

$$L_z = \left(xp_y - yp_x\right) = -i\hbar\left(x\frac{\partial}{\partial y} - y\frac{\partial}{\partial x}\right). \tag{26.37}$$

Hence, under infinitesimal rotations, we have to order ϵ,

$$g\left(x - \epsilon y, y + \epsilon x, z\right) = \left(1 + \frac{i\epsilon L_z}{\hbar}\right) g(x,y,z). \tag{26.38}$$

The rotation by a finite angle ϕ about the z-axis can be obtained, once again, through (26.21),

$$g(x',y',z') = e^{i\phi L_z/\hbar} g(x,y,z). \tag{26.39}$$

One can, in a similar manner, express rotations about the x-axis and y-axis through the angular momentum operators L_x and L_y given by

$$L_x = -i\hbar\left(y\frac{\partial}{\partial z} - z\frac{\partial}{\partial y}\right), \quad L_y = -i\hbar\left(z\frac{\partial}{\partial x} - x\frac{\partial}{\partial z}\right). \tag{26.40}$$

It is easy to check that the operators L_x, L_y, and L_z satisfy the following commutators:

$$\left[L_x, L_y\right] = i\hbar L_z, \quad \left[L_y, L_z\right] = i\hbar L_x, \quad \left[L_z, L_x\right] = i\hbar L_y. \tag{26.41}$$

We can combine these three relations into a single relation and write

$$\left[L_i, L_j\right] = i\hbar\varepsilon_{ijk}L_k, \quad i,j,k = 1,2,3 \tag{26.42}$$

where we have followed the summation convention.

We can generalize to rotations about arbitrary axes by noting, once again, from (26.26),

$$\mathbf{r}' = \mathbf{r} + \epsilon\mathbf{n}\times\mathbf{r} \tag{26.43}$$

where $\mathbf{r}$ $(=(x,y,z))$ and $\mathbf{r}'(=(x',y',z'))$ are the coordinate vectors. Hence, through Taylor expansion, we obtain to order ϵ,

$$g(\mathbf{r}') = g(\mathbf{r} + \epsilon\mathbf{n} \times \mathbf{r}) = g(\mathbf{r}) + \epsilon\,(\mathbf{n} \times \mathbf{r}) \cdot \nabla g(\mathbf{r}). \tag{26.44}$$

Using the vector identity $(\mathbf{n} \times \mathbf{r}) \cdot \nabla = \mathbf{n} \cdot (\mathbf{r} \times \nabla)$, we can write, taking $\mathbf{p} = -i\hbar\nabla$ and $\mathbf{L} = \mathbf{r} \times \mathbf{p}$, the relation

$$g(\mathbf{r}') = \left(1 + i\epsilon\frac{\mathbf{n} \cdot \mathbf{L}}{\hbar}\right) g(\mathbf{r}). \tag{26.45}$$

For finite transformations we obtain, as before,

$$g(\mathbf{r}') = e^{i\chi\mathbf{n}\cdot\mathbf{L}/\hbar} g(\mathbf{r}) \tag{26.46}$$

for rotations by an angle χ about the direction $\mathbf{n}$.

In the above discussions we have written the angular momentum operators in Cartesian coordinates. However, one can also express them in spherical coordinates by writing (x,y,z) in terms of (r,θ,ϕ) as follows:

$$x = r\sin\theta\cos\phi, \tag{26.47}$$

$$y = r\sin\theta\sin\phi, \tag{26.48}$$

$$z = r\cos\theta, \tag{26.49}$$

to give

$$L_x = i\hbar\left(\sin\phi\frac{\partial}{\partial\theta} + \cot\theta\cos\phi\frac{\partial}{\partial\phi}\right), \tag{26.50}$$

$$L_y = i\hbar\left(-\cos\phi\frac{\partial}{\partial\theta} + \cot\theta\sin\phi\frac{\partial}{\partial\phi}\right), \tag{26.51}$$

$$L_z = -i\hbar\frac{\partial}{\partial\phi}, \tag{26.52}$$

$$\mathbf{L}^2 = L_x^2 + L_y^2 + L_z^2 = -\hbar^2\left[\frac{1}{\sin\theta}\frac{\partial}{\partial\theta}\left(\sin\theta\frac{\partial}{\partial\theta}\right) + \frac{1}{\sin^2\theta}\frac{\partial^2}{\partial\phi^2}\right]. \tag{26.53}$$

26.3 State vectors

We now consider the rotation of the state vectors $|x,y,z\rangle$. Under rotation by infinitesimal angle $\phi = \epsilon$ about the z-axis, the state $|x,y,z\rangle$ behaves as

$$|x,y,z\rangle \to |x',y',z'\rangle = |x - \epsilon y, y + \epsilon x, z\rangle \tag{26.54}$$

where we have used (26.29), (26.30), and (26.31). One can, once again, express the right-hand side in a Taylor expansion and write, to order ϵ,

$$|x-\epsilon y, y+\epsilon x, z\rangle = \left[|x,y,z\rangle - \epsilon\left(y\frac{\partial}{\partial x} - x\frac{\partial}{\partial y}\right)\right]|x,y,z\rangle = \left(1+\frac{i\epsilon L_z}{\hbar}\right)|x,y,z\rangle . \tag{26.55}$$

For finite rotations we obtain as before

$$\left|x',y',z'\right\rangle = e^{i\phi L_z/\hbar}\,|x,y,z\rangle . \tag{26.56}$$

If $|\psi\rangle$ describes the state of a particle then its wavefunctions in the (x',y',z') system and in the (x,y,z) system can easily be related:

$$\psi'(x,y,z) = \left\langle x',y',z'|\psi\right\rangle = e^{-i\phi L_z/\hbar}\,\langle x,y,z|\psi\rangle = e^{-i\phi L_z/\hbar}\psi(x,y,z) . \tag{26.57}$$

Writing in terms of vectors $\mathbf{r}$ and $\mathbf{r}'$, the above relation becomes

$$\psi'(\mathbf{r}) = \left\langle \mathbf{r}'|\psi\right\rangle = e^{-i\phi L_z/\hbar}\psi(\mathbf{r}) . \tag{26.58}$$

In terms of the ket vectors we write this in a short-hand form:

$$|\psi\rangle' = \left|\psi'\right\rangle = e^{-i\phi L_z/\hbar}\,|\psi\rangle . \tag{26.59}$$

For rotations about an axis in the direction $\mathbf{n}$, we obtain the following transformation of the ket vector $\left|\mathbf{r}'\right\rangle$ to order ϵ,

$$\left|\mathbf{r}'\right\rangle = |\mathbf{r}+\epsilon\mathbf{n}\times\mathbf{r}\rangle \tag{26.60}$$

$$= |\mathbf{r}\rangle + \epsilon\mathbf{n}\times\mathbf{r}\cdot\nabla\,|\mathbf{r}\rangle$$

$$= \left(1+i\epsilon\frac{\mathbf{n}\cdot\mathbf{L}}{\hbar}\right)|\mathbf{r}\rangle . \tag{26.61}$$

A finite transformation by an angle χ then gives

$$\left|\mathbf{r}'\right\rangle = e^{i\chi\mathbf{n}\cdot\mathbf{L}/\hbar}\,|\mathbf{r}\rangle \tag{26.62}$$

and hence

$$\left\langle\mathbf{r}'|\psi\right\rangle = e^{-i\chi\mathbf{n}\cdot\mathbf{L}/\hbar}\,\langle\mathbf{r}|\psi\rangle \tag{26.63}$$

where the components of $\mathbf{L} = (L_x, L_y, L_z)$ are already given above. In terms of ket vectors we write this as

$$|\psi\rangle' = \left|\psi'\right\rangle = e^{-i\chi\mathbf{n}\cdot\mathbf{L}/\hbar}\,|\psi\rangle . \tag{26.64}$$

Finally, to obtain the transformation of a vector wavefunction $\psi_i(\mathbf{r})$ with $i = 1,2,3$, we follow a two-step process: transform the column matrix formed by $\psi_i(\mathbf{r})$, through the rotation matrix **S**, in the same manner as we transformed the column matrix V_i in Section 26.1.1; then make a transformation involving **L** of the coordinate vectors **r** in $\psi_i(\mathbf{r})$ as we did in this section. We obtain the following result for a rotation by an angle χ along the direction **n**:

$$\begin{pmatrix} \psi_1'(\mathbf{r}) \\ \psi_2'(\mathbf{r}) \\ \psi_3'(\mathbf{r}) \end{pmatrix} = e^{-i\chi \frac{\mathbf{n}\cdot\mathbf{L}}{\hbar}} e^{-i\chi \frac{\mathbf{n}\cdot\mathbf{S}}{\hbar}} \begin{pmatrix} \psi_1(\mathbf{r}) \\ \psi_2(\mathbf{r}) \\ \psi_3(\mathbf{r}) \end{pmatrix} = e^{-i\chi \frac{\mathbf{n}\cdot\mathbf{J}}{\hbar}} \begin{pmatrix} \psi_1(\mathbf{r}) \\ \psi_2(\mathbf{r}) \\ \psi_3(\mathbf{r}) \end{pmatrix} \tag{26.65}$$

where $\mathbf{J} = \mathbf{L} + \mathbf{S}$. The last step follows from the fact that **L** and **S** commute as they are in different spaces, **L** is a differential operator that operates on the coordinates, and **S** is a matrix operator.

While **S** is referred to as "spin angular momentum" or simply "spin" and **L** as "orbital angular momentum," the operator **J** is then the so-called "total angular momentum" operator.

26.4 Transformation of matrix elements and representations of the rotation operator

As we saw previously in Section 26.1.1 for a rotation operator R characterized by a 3×3 matrix, one can associate an operator $D(R)$ in the ket space by writing

$$|\psi\rangle' = |\psi\rangle_R = D(R)\,|\psi\rangle\,. \tag{26.66}$$

For example, the operator for rotations about the z-axis is given by

$$R_z(\phi) = \begin{pmatrix} \cos\phi & -\sin\phi & 0 \\ \sin\phi & \cos\phi & 0 \\ 0 & 0 & 0 \end{pmatrix}, \tag{26.67}$$

for which we have

$$D(R) = D(\phi) = e^{-i\phi L_z/\hbar} \tag{26.68}$$

as the corresponding operator on the ket vectors.

Let us consider now the commutators of the operators x, y, z with the operators L_x, L_y, L_z. In particular, we consider $[x, L_z]$:

$$\begin{aligned} [x, L_z]\,|\psi\rangle &= -\left\{\left[x\,(i\hbar)\left(x\frac{\partial}{\partial y} - y\frac{\partial}{\partial x}\right)\right] - \left[(i\hbar)\left(x\frac{\partial}{\partial y} - y\frac{\partial}{\partial x}\right)x\right]\right\}|\psi\rangle \\ &= -i\hbar y\,|\psi\rangle \end{aligned} \tag{26.69}$$

where we note that

$$\frac{\partial}{\partial x_i}x_j\,|\psi\rangle = \left(\delta_{ij} + x_j\frac{\partial}{\partial x_i}\right)|\psi\rangle\,. \tag{26.70}$$

The above commutation relation can be generalized to

$$\left[x_i, L_j\right] = i\hbar\varepsilon_{ijk}x_k \tag{26.71}$$

where we have used the summation convention. Thus x_i satisfies the same commutation relations as we obtained for L_i given by

$$\left[L_i, L_j\right] = i\hbar\varepsilon_{ijk}L_k. \tag{26.72}$$

Consider now the following matrix element under rotations about the z-axis:

$$\langle\psi\,|x|\,\psi\rangle' = \left\langle\psi'\,|x|\,\psi'\right\rangle = \left\langle\psi\,\left|e^{i\phi L_z/\hbar}xe^{-i\phi L_z/\hbar}\right|\,\psi\right\rangle. \tag{26.73}$$

For infinitesimal transformations we take $\phi = \varepsilon$,

$$\langle\psi\,|x|\,\psi\rangle' = \left\langle\psi\,\left|x + \frac{i\varepsilon}{\hbar}\left[L_z, x\right]\right|\,\psi\right\rangle \tag{26.74}$$

$$= \langle\psi\,|x - \varepsilon y|\,\psi\rangle \tag{26.75}$$

$$= \langle\psi\,|x|\,\psi\rangle - \varepsilon\,\langle\psi\,|y|\,\psi\rangle \tag{26.76}$$

where we have made use of the relation (26.69). Thus we obtain precisely the same relations as we did for the coordinates in (26.29). If we write a general rotation

$$x_i' = R_{ij}x_j \tag{26.77}$$

for the coordinate vectors then the corresponding matrix elements will have the same behavior:

$$\langle\psi\,|x_i|\,\psi\rangle' = R_{ij}\left\langle\psi\,\left|x_j\right|\,\psi\right\rangle. \tag{26.78}$$

From the relation (26.64) we can write this equation in the operator form:

$$e^{i\chi\mathbf{n}\cdot\mathbf{L}/\hbar}x_ie^{-i\chi\mathbf{n}\cdot\mathbf{L}/\hbar} = R_{ij}x_j. \tag{26.79}$$

This relation is then of the form

$$D^\dagger(R)x_iD(R) = R_{ij}x_j \tag{26.80}$$

where

$$D(R) = D(\chi) = e^{-i\chi\mathbf{n}\cdot\mathbf{L}/\hbar}. \tag{26.81}$$

For a vector V_i we can similarly write

$$D^\dagger(R)V_iD(R) = R_{ij}V_j \tag{26.82}$$

where we replace $\mathbf{L}$ by $\mathbf{S}$ in $D(R)$ since the transformation here involves only the components of the vectors $\mathbf{V}$ not the coordinate vectors.

26.5 Generators of infinitesimal rotations: their eigenstates and eigenvalues

If we designate R to represent any one of the rotation matrices we came across earlier, e.g., $\exp(-i\chi\mathbf{n}\cdot\mathbf{S}/\hbar), \exp(-i\chi\mathbf{n}\cdot\mathbf{L}/\hbar)$, or $\exp(-i\chi\mathbf{n}\cdot\mathbf{J}/\hbar)$, we notice that they all satisfy the relation

$$R^\dagger R = \mathbf{1}, \tag{26.83}$$

which implies that they are all unitary operators. If the matrix elements of R are real then we replace the Hermitian conjugate $R^\dagger$ by the transpose $\widetilde{R}$. For rotations about the imaginary axis one can write

$$R = e^{-i\theta_i J_i/\hbar} \tag{26.84}$$

where J_i is a generic notation for S_i or L_i or their sum, J_i. We are not using the summation convention here since we are considering a specific direction. Below we discuss some of the important properties of the generator J_i and obtain its eigenvectors and eigenvalues.

For infinitesimal angles ($\theta_i = \epsilon$), we have, upon expanding the exponential in (26.84),

$$R = \mathbf{1} - i\epsilon\frac{J_i}{\hbar}, \quad R^\dagger = \mathbf{1} + i\epsilon\frac{J_i^\dagger}{\hbar}. \tag{26.85}$$

Hence

$$R^\dagger R = \mathbf{1} \Rightarrow \left(\mathbf{1} + i\epsilon\frac{J_i^\dagger}{\hbar}\right)\left(\mathbf{1} - i\epsilon\frac{J_i}{\hbar}\right) = \mathbf{1}. \tag{26.86}$$

Keeping just the leading terms in ϵ on the left hand side, we obtain

$$\mathbf{1} + i\frac{\epsilon}{\hbar}\left(J_i^\dagger - J_i\right) = \mathbf{1}. \tag{26.87}$$

Therefore,

$$J_i^\dagger = J_i. \tag{26.88}$$

Hence the generators are Hermitian. We also note from (26.15) that

$$\det(R) = 1. \tag{26.89}$$

Going to infinitesimal angles we obtain

$$\det\left(\mathbf{1} - i\epsilon\frac{J_i}{\hbar}\right) = 1. \tag{26.90}$$

Keeping only the leading term in ϵ in the calculation for the determinant we obtain

$$1 - i\frac{\epsilon}{\hbar}\text{Tr}\,(J_i) = 1. \tag{26.91}$$

where "Tr" means trace of the matrix J_i. Therefore,

$$\text{Tr}\,(J_i) = 0. \tag{26.92}$$

Hence the matrix representation of the generator must be traceless.

From the properties of S_i and L_i that we previously deduced, one can write

$$\left[J_i, J_j\right] = (i\hbar)\varepsilon_{ijk}\, J_k \tag{26.93}$$

where the summation convention is used. Writing explicitly in terms of the generators J_x, J_y, J_z, we obtain

$$\left[J_x, J_y\right] = i\hbar J_z, \quad \left[J_y, J_z\right] = i\hbar J_x, \quad \left[J_z, J_x\right] = i\hbar J_y. \tag{26.94}$$

Let us define the total angular momentum operator $\mathbf{J}$ as

$$\mathbf{J}^2 = J_x^2 + J_y^2 + J_z^2 \tag{26.95}$$

and determine $\left[\mathbf{J}^2, J_x\right]$. Since $\left[J_x^2, J_x\right] = 0$, we can write

$$\left[\mathbf{J}^2, J_x\right] = \left[J_x^2 + J_y^2 + J_z^2, J_x\right] = \left[J_y^2, J_x\right] + \left[J_z^2, J_x\right]. \tag{26.96}$$

Utilizing the commutator relations (26.94) we find the terms on the right-hand side above to be

$$\left[J_y^2, J_x\right] = J_y\left[J_y, J_x\right] + \left[J_y, J_x\right]J_y = -i\hbar\left(J_yJ_z + J_zJ_y\right), \tag{26.97}$$

$$\left[J_z^2, J_x\right] = J_z\left[J_z, J_x\right] + \left[J_z, J_x\right]J_z = i\hbar\left(J_zJ_y + J_yJ_z\right). \tag{26.98}$$

Hence $\left[\mathbf{J}^2, J_x\right] = 0$. A similar relation will hold for J_y and J_z. Therefore, we obtain the following general result:

$$\left[\mathbf{J}^2, J_i\right] = 0. \tag{26.99}$$

Thus, an eigenstate of J_i will also be an eigenstate of $\mathbf{J}^2$.

Since J_i's do not commute among themselves they cannot have a common eigenstate. We specifically pick J_z among them even though the other two are equally eligible; then J_z and $\mathbf{J}^2$ will have a common eigenstate which we designate as $|j, m\rangle$ with the properties

$$J_z\,|j, m\rangle = m\hbar\,|j, m\rangle \tag{26.100}$$

and

$$\mathbf{J}^2 |j,m\rangle = n_j \hbar^2 |j,m\rangle \tag{26.101}$$

where the states $|j,m\rangle$ satisfy the orthogonality property

$$\langle j',m' |j,m\rangle = \delta_{jj'}\delta_{mm'}. \tag{26.102}$$

The quantum numbers m and n_j are still to be determined. For that purpose we consider the expectation value

$$\langle j,m \left| \left(\mathbf{J}^2 - J_z^2 \right) \right| j,m\rangle = \left(n_j - m^2 \right) \hbar^2. \tag{26.103}$$

One can write

$$\mathbf{J}^2 - J_z^2 = J_x^2 + J_y^2 = \frac{1}{2} \left[\left(J_x + iJ_y \right) \left(J_x - iJ_y \right) + \left(J_x - iJ_y \right) \left(J_x + iJ_y \right) \right]. \tag{26.104}$$

We note that the two products on the right-hand side above are each of the form $AA^\dagger$ and hence each is positive definite. Therefore, from (26.100) and (26.101), we can write

$$\left(n_j - m^2 \right) \hbar^2 \geq 0. \tag{26.105}$$

Hence,

$$-\sqrt{n_j} \leq m \leq \sqrt{n_j}. \tag{26.106}$$

Therefore, the eigenvalue m is bounded from above and below.

Let us define new operators

$$J_+ = J_x + iJ_y, \quad J_- = J_x - iJ_y \tag{26.107}$$

and consider $J_+ |j,m\rangle$. Since from (26.99)

$$\left[\mathbf{J}^2, J_+ \right] = 0, \tag{26.108}$$

we find that $J_+ |j,m\rangle$ is also an eigenstate of $\mathbf{J}^2$, i.e.,

$$\mathbf{J}^2 \left(J_+ |j,m\rangle \right) = J_+ \left(\mathbf{J}^2 |j,m\rangle \right) = n_j \hbar^2 \left(J_+ |j,m\rangle \right). \tag{26.109}$$

Furthermore,

$$J_z J_+ = J_z \left(J_x + iJ_y \right) = \left[\left(J_x J_z + i\hbar J_y \right) + i \left(J_y J_z - i\hbar J_x \right) \right] \tag{26.110}$$

where we have used the commutation relations involving J_z with J_x and J_y. Combining the appropriate terms we obtain

$$J_z J_+ = \left(J_x + iJ_y \right) \left(J_z + \hbar \right) = J_+ \left(J_z + \hbar \right). \tag{26.111}$$

Hence,

$$J_z \left(J_+ \left|j,m\right\rangle\right) = J_+ \left(J_z + \hbar\right) \left|j,m\right\rangle = (m+1)\,\hbar \left(J_+ \left|j,m\right\rangle\right). \tag{26.112}$$

Thus $J_+ \left|j,m\right\rangle$ corresponds to a state with eigenvalue $(m+1)\,\hbar$ for the operator J_z and eigenvalue $n_j\hbar^2$ for the operator $\mathbf{J}^2$. Hence it is proportional to $\left|j,m+1\right\rangle$, i.e.,

$$J_+ \left|j,m\right\rangle = a_j\hbar \left|j,m+1\right\rangle \tag{26.113}$$

where a_j is a constant yet to be determined. The above results show that J_+ acts as a "raising" operator. It increases the eigenvalue of J_z by one unit. In other words, the values of m change by one unit as one goes from one state to the next higher state. The m-values are, therefore, separated by integers, i.e., the difference Δm between the m-quantum numbers satisfies the relation

$$\Delta m = \text{integer}\,. \tag{26.114}$$

Multiplying (26.113) by its Hermitian conjugate, we get

$$\left(\left\langle j,m\right| J_+^\dagger\right) \left(J_+ \left|j,m\right\rangle\right) = \left|a_j\right|^2 \hbar^2 \left\langle j,m+1 \left|j,m+1\right\rangle\right. = \left|a_j\right|^2 \hbar^2. \tag{26.115}$$

Therefore,

$$\left\langle j,m\right| J_+^\dagger J_+ \left|j,m\right\rangle = \left|a_j\right|^2 \hbar^2. \tag{26.116}$$

Since $J_+^\dagger = J_- = \left(J_x - iJ_y\right)$, we can write the operator term on the left-hand side as

$$J_+^\dagger J_+ = \left(J_x - iJ_y\right)\left(J_x + iJ_y\right) = J_x^2 + J_y^2 + i\left[J_x, J_y\right]. \tag{26.117}$$

Using the fact that $J_x^2 + J_y^2 = \mathbf{J}^2 - J_z^2$, and $\left[J_x, J_y\right] = i\hbar J_z$, we obtain

$$J_+^\dagger J_+ = \mathbf{J}^2 - J_z^2 - \hbar J_z. \tag{26.118}$$

Thus relation (26.116) can be written as

$$\left\langle j,m\right| \left(J^2 - J_z^2 - \hbar J_z\right) \left|j,m\right\rangle = \left|a_j\right|^2 \hbar^2. \tag{26.119}$$

Therefore,

$$\left|a_j\right|^2 = \left(n_j - m^2 - m\right). \tag{26.120}$$

Since m is bounded from above, we take m_2 to be the highest value of m, then

$$J_+ \left|j,m_2\right\rangle = 0. \tag{26.121}$$

Hence $a_j = 0$ for this state and, as a result, from (26.120) we obtain

$$n_j = m_2(m_2+1). \tag{26.122}$$

Similarly, for the state $J_- |j, m\rangle$ we can write

$$J_- |j, m\rangle = b_j \hbar |j, m-1\rangle \tag{26.123}$$

and obtain

$$|b_j|^2 = n_j - m(m-1). \tag{26.124}$$

If m_1 is the lowest value of m then we have

$$J_- |j, m_1\rangle = 0, \tag{26.125}$$

and hence $b_j = 0$ for this state, which leads to

$$n_j = m_1(m_1 - 1). \tag{26.126}$$

Therefore, since both (26.122) and (26.126) equal the same quantity, n_j, we have

$$m_2(m_2 + 1) = m_1(m_1 - 1). \tag{26.127}$$

This equation has two solutions:

$$m_1 = -m_2 \quad \text{and} \quad m_1 = m_2 + 1. \tag{26.128}$$

The second solution is not allowed since, by definition, $m_1 < m_2$. Hence,

$$m_1 = -m_2. \tag{26.129}$$

Since the values of m are separated by integers, as stated in (26.114) we also have the relation

$$m_2 - m_1 = \text{ integer.} \tag{26.130}$$

From (26.129) we then have

$$2m_2 = \text{integer.} \tag{26.131}$$

If we write

$$m_2 = j \tag{26.132}$$

then we have

$$2j = \text{integer.} \tag{26.133}$$

Also, since $m_1 < m < m_2$, we have

$$-j \leq m \leq j. \tag{26.134}$$

The possible values of j are therefore

$$j = \text{integer or half-integer}. \tag{26.135}$$

Furthermore, from (26.122) and (26.132) we obtain

$$n_j = j(j+1). \tag{26.136}$$

Thus,

$$|a_j|^2 = j(j+1) - m(m+1) = (j-m)(j+m+1), \tag{26.137}$$

$$|b_j|^2 = j(j+1) - m(m-1) = (j+m)(j-m+1). \tag{26.138}$$

Summarizing the above results, we have

$$J_z |j, m\rangle = m\hbar |j, m\rangle, \qquad \text{with } -j \le m \le j, \tag{26.139}$$

$$\mathbf{J}^2 |j, m\rangle = j(j+1)\hbar^2 |j, m\rangle \quad \text{with } j = \tfrac{1}{2}, 1, \tfrac{3}{2}, \ldots, \tag{26.140}$$

$$J_+ |j, m\rangle = \sqrt{(j-m)(j+m+1)}\hbar\ |j, m+1\rangle, \tag{26.141}$$

$$J_- |j, m\rangle = \sqrt{(j+m)(j-m+1)}\hbar\, |j, m-1\rangle. \tag{26.142}$$

The operators J_+ and J_- are often called "ladder" operators.

26.6 Representations of J^2 and J_i for $j = \frac{1}{2}$ and $j = 1$

We start with $j = \frac{1}{2}$, for which m takes the values $\frac{1}{2}$ and $-\frac{1}{2}$. We write the eigenstates $|j, m\rangle$ as column matrices,

$$\left\{\left|\tfrac{1}{2}, m\right\rangle\right\} = \begin{pmatrix} \left|\frac{1}{2}, \frac{1}{2}\right\rangle \\ \left|\frac{1}{2}, -\frac{1}{2}\right\rangle \end{pmatrix}. \tag{26.143}$$

The operators J_i will then be 2×2 matrices. From the relations given in (26.139) through (26. 142) we obtain

$$J_z = \begin{pmatrix} \frac{1}{2} & 0 \\ 0 & -\frac{1}{2} \end{pmatrix}\hbar, \quad J_+ = \begin{pmatrix} 0 & 1 \\ 0 & 0 \end{pmatrix}\hbar, \quad J_- = \begin{pmatrix} 0 & 0 \\ 1 & 0 \end{pmatrix}\hbar, \tag{26.144}$$

from which we find

$$J_x = \frac{1}{2}\begin{pmatrix} 0 & 1 \\ 1 & 0 \end{pmatrix}\hbar, \quad J_y = \frac{1}{2}\begin{pmatrix} 0 & -i \\ i & 0 \end{pmatrix}\hbar. \tag{26.145}$$

If we define

$$\mathbf{J} = \frac{\hbar}{2}\boldsymbol{\sigma}, \tag{26.146}$$

then we have

$$\sigma_x = \begin{pmatrix} 0 & 1 \\ 1 & 0 \end{pmatrix}, \quad \sigma_y = \begin{pmatrix} 0 & -i \\ i & 0 \end{pmatrix}, \quad \sigma_z = \begin{pmatrix} 1 & 0 \\ 0 & -1 \end{pmatrix}. \tag{26.147}$$

These are just the Pauli matrices we discussed in Chapter 2.

Next we consider $j = 1$, for which $m = 1, 0, -1$. The eigenstates $|j, m\rangle$ are then given by

$$\{|1, m\rangle\} = \begin{pmatrix} |1, 1\rangle \\ |1, 0\rangle \\ |1, -1\rangle \end{pmatrix}. \tag{26.148}$$

We then obtain

$$J_z = \begin{pmatrix} 1 & 0 & 0 \\ 0 & 0 & 0 \\ 0 & 0 & -1 \end{pmatrix}, \quad J_+ = \begin{pmatrix} 0 & \sqrt{2} & 0 \\ 0 & 0 & \sqrt{2} \\ 0 & 0 & 0 \end{pmatrix}, \quad J_- = \begin{pmatrix} 0 & 0 & 0 \\ \sqrt{2} & 0 & 0 \\ 0 & \sqrt{2} & 0 \end{pmatrix}, \tag{26.149}$$

and hence

$$J_x = \frac{1}{\sqrt{2}} \begin{pmatrix} 0 & 1 & 0 \\ 1 & 0 & 1 \\ 0 & 1 & 0 \end{pmatrix} \quad \text{and} \quad J_y = -\frac{i}{\sqrt{2}} \begin{pmatrix} 0 & -1 & 0 \\ 1 & 0 & -1 \\ 0 & 1 & 0 \end{pmatrix}. \tag{26.150}$$

Clearly, one can go on to higher representations by following the same procedure we have outlined above.

26.7 Spherical harmonics

26.7.1 Ladder operators and eigenfunctions

Let us consider the eigenstates $|jm\rangle$ discussed in Section 26.5 that are relevant to orbital angular momentum. We designate these states as $|lm\rangle$ where l and m correspond to the orbital angular momentum quantum numbers. Their representatives in the coordinate space with state vectors $|\theta\phi\rangle$, where θ and ϕ are the angular coordinates of a point (r, θ, ϕ), are written as

$$\langle\theta\phi|lm\rangle = Y_{lm}(\theta, \phi) \tag{26.151}$$

where $Y_{lm}(\theta, \phi)$ are called spherical harmonics. The normalization condition

$$\langle l'm'|lm\rangle = \delta_{ll'}\delta_{mm'} \tag{26.152}$$

gives, after putting a complete set of intermediate states $|\theta\phi\rangle$,

$$\int_0^{2\pi}\int_0^{\pi} d\Omega \,\langle l'm'|\theta\phi\rangle\langle\theta\phi|lm\rangle = \delta_{ll'}\delta_{mm'}. \tag{26.153}$$

Since $d\Omega = \sin\theta \, d\theta \, d\phi$, we obtain the following normalization condition for the spherical harmonics:

$$\int_0^{2\pi}\int_0^{\pi} Y^*_{l'm'}Y_{l'm}(\theta,\phi)Y_{lm}(\theta,\phi) \, \sin\theta \, d\theta \, d\phi = \delta_{ll'}\delta_{mm'}. \tag{26.154}$$

The state $|lm\rangle$ will satisfy the same properties as did $|jm\rangle$, where we replace J_i by the orbital angular momentum operators L_i and in place of the equations (26.139)–(26.142), write the following:

$$\mathbf{L}^2|lm\rangle = l(l+1)\hbar^2|lm\rangle,$$

$$L_z|lm\rangle = m\hbar|lm\rangle, \tag{26.155}$$

$$L_+Y_{lm}(\theta,\phi) = \sqrt{(l-m)(l+m+1)}\hbar Y_{lm+1}(\theta,\phi),$$

$$L_-Y_{lm}(\theta,\phi) = \sqrt{(l+m)(l-m+1)}\hbar Y_{lm-1}(\theta,\phi). \tag{26.156}$$

Multiplying equation (26.155) by $\langle\theta\phi|$ we obtain

$$\mathbf{L}^2 Y_{lm}(\theta,\phi) = l(l+1)\hbar^2 Y_{lm}(\theta,\phi),$$

$$L_z Y_{lm}(\theta,\phi) = m\hbar Y_{lm}(\theta,\phi) \text{ with } -l < m < l. \tag{26.157}$$

The operators L_x, L_y, L_z, and $\mathbf{L}^2$ were defined previously and are given by

$$L_x = i\hbar\left(\sin\phi\frac{\partial}{\partial\theta} + \cot\theta\cos\phi\frac{\partial}{\partial\phi}\right), \tag{26.158}$$

$$L_y = i\hbar\left(-\cos\phi\frac{\partial}{\partial\theta} + \cot\theta\sin\phi\frac{\partial}{\partial\phi}\right), \tag{26.159}$$

$$L_z = -i\hbar\frac{\partial}{\partial\phi}, \tag{26.160}$$

$$\mathbf{L}^2 = -\hbar^2\left[\frac{1}{\sin\theta}\frac{\partial}{\partial\theta}\left(\sin\theta\frac{\partial}{\partial\theta}\right) + \frac{1}{\sin^2\theta}\frac{\partial^2}{\partial\phi^2}\right]. \tag{26.161}$$

From these we can determine the ladder operators

$$L_+ = L_x + iL_y = \hbar e^{i\phi}\left[\frac{\partial}{\partial\theta} + i\cot\theta\frac{\partial}{\partial\phi}\right], \tag{26.162}$$

$$L_- = L_x - iL_y = -\hbar e^{-i\phi}\left[\frac{\partial}{\partial\theta} - i\cot\theta\frac{\partial}{\partial\phi}\right]. \tag{26.163}$$

26.7.2 Explicit expression for $Y_{lm}(\theta, \phi)$

We now wish to obtain an explicit expression for $Y_{lm}(\theta, \phi)$. To that end, let us consider the equation for the operator L_z.

$$-i\hbar \frac{\partial}{\partial \phi} Y_{lm}(\theta, \phi) = m\hbar Y_{lm}(\theta, \phi). \tag{26.164}$$

Based on the separation of variables technique we can write

$$Y_{lm}(\theta, \phi) = e^{im\phi} T_{lm}(\theta) \tag{26.165}$$

where $T_{lm}(\theta)$ will depend on θ alone. Let us operate L_+ on Y_{lm} for $m = l$. Since $m = l$ corresponds to the highest state, we have

$$L_+ Y_{ll}(\theta, \phi) = L_+(e^{il\phi} T_{ll}(\theta)) = 0. \tag{26.166}$$

One can write $L_+(e^{il\phi} T_{ll}(\theta))$ as

$$L_+(e^{il\phi} T_{ll}(\theta)) = (L_+ e^{il\phi}) T_{ll}(\theta) + e^{il\phi} L_+ T_{ll}(\theta). \tag{26.167}$$

To evaluate (26.167) we use relation (26.162)

$$L_+ e^{il\phi} = \left(-l\hbar \cot\theta e^{i\phi}\right) e^{il\phi} \tag{26.168}$$

and

$$L_+ T_{ll}(\theta) = \hbar e^{i\phi} \left(\frac{\partial}{\partial \theta} T_{ll}(\theta)\right). \tag{26.169}$$

Taking account of the relations (26.168) and (26.169) we find that (26.166) gives

$$-l\hbar \cot\theta e^{i\phi} e^{il\phi} T_{ll} + e^{il\phi} \hbar e^{i\phi} \frac{\partial}{\partial \theta} T_{ll} = 0. \tag{26.170}$$

Thus,

$$\frac{dT_{ll}}{d\theta} = l \cot\theta T_{ll} \tag{26.171}$$

where we have replaced the partial derivative by the total derivative in θ because T_{ll} depends on θ alone. This equation can be written as

$$\frac{dT_{ll}}{T_{ll}} = l \frac{d \sin\theta}{\sin\theta}. \tag{26.172}$$

Hence,

$$T_{ll} = c_l (\sin\theta)^l \tag{26.173}$$

where c_l is a constant. Substituting this result in (26.165) we obtain

$$Y_{ll}(\theta, \phi) = c_l e^{il\phi} (\sin\theta)^l. \tag{26.174}$$

Let us now operate L_- on $Y_{lm}(\theta, \phi)$. From (26.163) we obtain

$$\begin{aligned} L_- \left[e^{im\phi} T_{lm}(\theta) \right] &= -\hbar e^{i\phi} \left[\frac{\partial}{\partial\theta} - i \cot\theta \frac{\partial}{\partial\phi} \right] e^{im\phi} T_{lm}(\theta) \\ &= -\hbar e^{-i\phi} e^{im\phi} \left[\frac{d}{d\theta} + m \cot\theta \right] T_{lm}(\theta). \end{aligned} \tag{26.175}$$

In order to calculate the right-hand side above we note that for a function $f(\theta)$,

$$\frac{d}{d\cos\theta} \left[(\sin\theta)^m f(\theta) \right] = -(\sin\theta)^{m-1} \left[m \cot\theta f + \frac{df}{d\theta} \right] \tag{26.176}$$

where we have used the relation $d\cos\theta = -\sin\theta \, d\theta$. Hence,

$$\left[\frac{d}{d\theta} + m \cot\theta \right] f = -(\sin\theta)^{1-m} \frac{d}{d\cos\theta} \left[(\sin\theta)^m f(\theta) \right]. \tag{26.177}$$

Substituting the above relation in (26.175) we obtain

$$\begin{aligned} L_- \left[e^{im\phi} T_{lm}(\theta) \right] &= \hbar e^{i(m-1)\phi} (\sin\theta)^{1-m} \frac{d}{d\cos\theta} \left[(\sin\theta)^m T_{lm}(\theta) \right] \\ &= e^{i(m-1)\phi} T'_{lm}(\theta) \end{aligned} \tag{26.178}$$

where

$$T'_{lm}(\theta) = (\sin\theta)^{1-m} \frac{d}{d\cos\theta} \left[(\sin\theta)^m T_{lm}(\theta) \right]. \tag{26.179}$$

Through repeated application of L_- we find

$$(L_-)^N \left[e^{im\phi} T_{lm}(\theta) \right] = e^{i(m-N)\phi} (\sin\theta)^{N-m} \left(\frac{d}{d\cos\theta} \right)^N [(\sin\theta)^m T_{lm}(\theta)]. \tag{26.180}$$

Taking $m = l$ we obtain

$$(L_-)^N \left[e^{il\phi} T_{ll}(\theta) \right] = e^{i(l-N)\phi} (\sin\theta)^{N-l} \left(\frac{d}{d\cos\theta} \right)^N [(\sin\theta)^l T_{ll}(\theta)]. \tag{26.181}$$

From (26.165) and (26.173) we find

$$(L_-)^N \left[Y_{ll}(\theta, \phi) \right] = c_l e^{i(l-N)\phi} (\sin\theta)^{N-l} \left(\frac{d}{d\cos\theta} \right)^N [(\sin\theta)^{2l}]. \tag{26.182}$$

However, since L_- is a lowering operator, lowering $Y_{ll}(\theta,\phi)$ N times corresponds to writing the above relation as

$$Y_{l,l-N}(\theta,\phi) = c_l'' e^{i(l-N)\phi} (\sin\theta)^{N-l} \left(\frac{d}{d\cos\theta}\right)^N \left[(\sin\theta)^{2l}\right] \tag{26.183}$$

where c_l'' is a constant.

To obtain $Y_{lm}(\theta,\phi)$ we now simply take $l - N = m$. Hence,

$$Y_{lm}(\theta,\phi) = c_l'' e^{im\phi} (\sin\theta)^{-m} \left(\frac{d}{d\cos\theta}\right)^{l-m} \left[(\sin\theta)^{2l}\right]. \tag{26.184}$$

From the normalization condition (26.154) we can obtain c_l''. One finds

$$Y_{lm}(\theta,\phi) = \frac{(-1)^l}{2^l l!} \sqrt{\frac{(2l+1)}{4\pi}\frac{(l+m)!}{(l-m)!}} e^{im\phi} (\sin\theta)^{-m} \left(\frac{d}{d\cos\theta}\right)^{l-m} \left[(\sin\theta)^{2l}\right]. \tag{26.185}$$

26.7.3 Properties of $Y_{lm}(\theta,\phi)$

Having determined the functional form of $Y_{lm}(\theta,\phi)$, let us examine its properties under rotations. Continuing with the discussions in Section 26.4 we note that under rotations

$$e^{i\chi\frac{\mathbf{n}\cdot\mathbf{j}}{\hbar}} x_i e^{-i\chi\frac{\mathbf{n}\cdot\mathbf{j}}{\hbar}} = \sum_j R_{ij} x_i \tag{26.186}$$

where $i, j = 1, 2, 3$. We can construct linear combinations of x_i, e.g., $(x + iy)$, $(x - iy)$, and z and find that under rotations about the z-axis

$$e^{i\frac{\phi}{\hbar}J_z}(x+iy)e^{-i\frac{\phi}{\hbar}J_z} = e^{i\phi}(x+iy), \tag{26.187}$$

$$e^{i\frac{\phi}{\hbar}J_z}(x-iy)e^{-i\frac{\phi}{\hbar}J_z} = e^{-i\phi}(x+iy), \tag{26.188}$$

$$e^{i\frac{\phi}{\hbar}J_z} z e^{-i\frac{\phi}{\hbar}J_z} = z. \tag{26.189}$$

From the functional form of $Y_{1m}(\theta,\phi)$ given in the previous section for $l = 1$ and $m = \pm 1, 0$, we find that the above three relations imply

$$e^{i\frac{\phi}{\hbar}J_z} Y_{1m}(\theta,\phi) e^{-i\frac{\phi}{\hbar}J_z} = e^{im\phi} Y_{1m}(\theta,\phi).$$

This implies that for a general rotation

$$e^{i\chi\frac{\mathbf{n}\cdot\mathbf{j}}{\hbar}} Y_{1m}(\theta,\phi) e^{-i\chi\frac{\mathbf{n}\cdot\mathbf{j}}{\hbar}} = \sum_{m'} D^1_{mm'}(\chi) Y_{1m'}(\theta,\phi). \tag{26.190}$$

We note that for rotations about the z-axis by angle ϕ,

$$D^1_{mm'}(\phi) = e^{im\phi}\delta_{mm'}. \tag{26.191}$$

Thus, under rotations Y_{1m} transform among themselves with $l = 1$ fixed and $m = 0, \pm 1$. This property is preserved for any l. In general one can write the following relation for a general spherical harmonic $Y_{lm}(\theta, \phi)$:

$$e^{i\chi\frac{\mathbf{n}\cdot\mathbf{j}}{\hbar}} Y_{lm}(\theta,\phi) e^{-i\chi\frac{\mathbf{n}\cdot\mathbf{j}}{\hbar}} = \sum_{m'} D^l_{mm'}(\chi) Y_{lm'}(\theta,\phi), \tag{26.192}$$

i.e.,

$$D^{\dagger}(\chi) Y_{lm}(\theta,\phi) D(\chi) = \sum_{m'} D^l_{mm'}(\chi) Y_{lm'}(\theta,\phi). \tag{26.193}$$

We can also discuss the properties of $Y_{lm}(\theta,\phi)$ from a different perspective. Based on our earlier results we write

$$|jm\rangle' = e^{-i\chi\frac{\mathbf{n}\cdot\mathbf{J}}{\hbar}} |jm\rangle = \sum_{m'} c_{jm'} |jm'\rangle. \tag{26.194}$$

This relation follows from the fact that when one expands $\exp(-i\chi\mathbf{n}\cdot\mathbf{J}/\hbar) = 1 - i\chi\mathbf{n}\cdot\mathbf{J}/\hbar + \cdots$ the operators J_i contained in each term do not change the eigenvalue j when operating on $|jm\rangle$ but may change the eigenvalues of m. Therefore, the right-hand side above will involve a summation only over m'. We can evaluate $c_{jm'}$ by using orthogonality of the eigenstates $|jm'\rangle$ to obtain

$$c_{jm'} = \langle jm'| e^{-i\chi\frac{\mathbf{n}\cdot\mathbf{J}}{\hbar}} |jm\rangle = D^j_{m'm}(\chi). \tag{26.195}$$

Hence,

$$|jm\rangle' = \sum_{m'} D^j_{m'm}(\chi) |jm'\rangle. \tag{26.196}$$

For $j = l$ we consider its representative in the (θ, ϕ) space and obtain

$$\langle\theta\phi|lm\rangle' = \sum_{m'} D^l_{m'm}(\chi)\langle\theta\phi|lm'\rangle. \tag{26.197}$$

Hence,

$$Y_{lm}\left(\theta',\phi'\right) = \sum_{m'} D^l_{m'm}(\chi)\, Y_{lm'}(\theta,\phi) \tag{26.198}$$

where θ' and $\phi'$are the angles in the rotated system. We note that l in (26.197) and (26.198) remains the same on both sides of the equation.

We can carry through with the same arguments as above by replacing x_i, which are the components of a vector $\mathbf{r}$, with the components V_i of a vector $\mathbf{V}$. That is,

$$Y_{lm}(\theta, \phi) \rightarrow Y_{lm}(\mathbf{V}). \tag{26.199}$$

In that case one would be dealing with the so called spherical tensors $T(k, q)$, with k replacing l and q replacing m.

We will follow up on this when we consider irreducible tensors.

26.8 Problems

1. Consider a state that has angular momentum quantum numbers $l = 1$ and $m = 0$. If this state is rotated by an angle θ, determine the probability that the new state has $m = 1$. Answer the same question if the state had $l = 2$ and $m = 0$.
2. If a system is in an eigenstate of J_z, show that $\langle J_x \rangle = 0 = \langle J_y \rangle$.
3. For an eigenstate of J_z given by $|jm\rangle$ show that for another operator $J_{z'}$ where z' is pointing in the direction making an angle θ with respect to z, the expectation value is $\langle jm | J_{z'} | jm \rangle = m \cos\theta$.
4. Obtain the unitary matrix that relates the generators S_x, S_y, S_z of rotations of Cartesian coordinates, the generators of the corresponding spherical harmonics with $l = 1$.
5. A normalized wavefunction is given by $u(\mathbf{r}) = A(x+y+z)f(r)$, where $f(r)$ is a function only of the radial coordinate r. (i) Determine the following for this case: $\langle L_x \rangle, \langle L_y \rangle, \langle L_z \rangle$, and $\langle \mathbf{L}^2 \rangle$. (ii) What is the probability that $u(\mathbf{r})$ corresponds to $L_x = +\hbar$?

27 Symmetry in quantum mechanics and symmetry groups

From rotations we now go to the general symmetry properties of physical systems, including parity and time reversal invariance. In some of the previous chapters we have considered the consequences of symmetries without saying so explicitly. In Chapter 8 we discussed one-dimensional potentials which were symmetric about the origin and concluded that there are two types of solutions depending on their properties under reflection. That was an example of the consequences of symmetry under parity transformations. In the previous chapter as well as in Chapter 4 on angular momentum we described eigenfunctions in terms of the quantum numbers l and m. That was an example of invariance under rotations. We discuss these questions within the framework of general group theory. We also consider the properties of linear algebra and the groups $O(2)$, $O(3)$, and $SU(2)$.

We need now to introduce the Hamiltonian and discuss its properties under various transformations. Let us start with rotations.

27.1 Rotational symmetry

The rotation operator, as we discussed in the previous chapter, is unitary. A Hamiltonian in three dimensions that includes a spherically symmetric potential is given by

$$H = \frac{\mathbf{p}^2}{2m} + V(r). \tag{27.1}$$

Since $\mathbf{p}^2 = p_x^2 + p_y^2 + p_z^2$ and $r = \sqrt{x^2 + y^2 + z^2}$, which are both invariant under rotations, H will also be invariant. Thus, if R is the rotation operator then

$$R^\dagger H R = H \tag{27.2}$$

and hence

$$HR = RH. \tag{27.3}$$

27.1.1 Consequences of rotational symmetry

1. If $|\alpha_n\rangle$ is an eigenstate of H with eigenvalue α_n then we have

$$H\,|\alpha_n\rangle = \alpha_n\,|\alpha_n\rangle\,. \tag{27.4}$$

We also write

$$R|\alpha_n\rangle = \left|\alpha_n^R\right\rangle. \tag{27.5}$$

Multiplying each side of the equality (27.3) on the right by $|\alpha_n\rangle$ we have

$$H(R|\alpha_n\rangle) = R(H|\alpha_n\rangle). \tag{27.6}$$

Therefore, from (27.4) and (27.5) we obtain the relation

$$H\left|\alpha_n^R\right\rangle = \alpha_n\left|\alpha_n^R\right\rangle. \tag{27.7}$$

Hence, $\left|\alpha_n^R\right\rangle$ is also an eigenstate of H with the same eigenvalue α_n. States obtained through rotation are degenerate if the Hamiltonian is invariant under the transformation. Degeneracy of energy levels is an essential outcome of symmetry principles.

2. We found in the previous chapter that we can write the expression for the rotation operator around a unit vector $\mathbf{n}$ as

$$R(\theta,\phi) = e^{-i\mathbf{n}\cdot\mathbf{L}\chi/\hbar} \tag{27.8}$$

where $\mathbf{L}$ is called the generator, which is Hermitian and, as we found, is identical to the angular momentum operator. For infinitesimal transformations one writes

$$R = \mathbf{1} - i\frac{\epsilon}{\hbar}\mathbf{n}\cdot\mathbf{L}. \tag{27.9}$$

From the unitary property of R it follows that

$$[\mathbf{L}, H] = 0. \tag{27.10}$$

From the Heisenberg relation the above result implies that

$$\frac{d\mathbf{L}}{dt} = 0. \tag{27.11}$$

Thus $\mathbf{L}$ is a constant of the motion. One can also show that $\mathbf{L}^2$ is also a constant of the motion.

3. Furthermore, we found in the last chapter that

$$\left[\mathbf{L}^2, L_i\right] = 0 \tag{27.12}$$

but

$$\left[L_i, L_j\right] \neq 0. \tag{27.13}$$

Taking $\mathbf{L}^2$ and L_z as the commuting operators,

$$\left[\mathbf{L}^2, L_z\right] = 0, \tag{27.14}$$

we can designate the eigenstates as $|l,m\rangle$ such that

$$\mathbf{L}^2 |l,m\rangle = l(l+1)\hbar^2 |l,m\rangle \tag{27.15}$$

and

$$L_z |l,m\rangle = m\hbar |l,m\rangle . \tag{27.16}$$

Let us assume that the above relation is valid at $t=0$ and substitute $|l,m(0)\rangle$ for $|l,m\rangle$. The time evolution of $|l,m\rangle$ is then given by

$$|l,m(t)\rangle = e^{-iHt/\hbar} |l,m(0)\rangle . \tag{27.17}$$

Since L_z commutes with H,

$$L_z |l,m(t)\rangle = e^{-iHt/\hbar} L_z |l,m(0)\rangle = m\hbar |l,m(t)\rangle . \tag{27.18}$$

Thus the eigenvalues remain unchanged as a function of time. A similar result will hold for the quantum number l of the operator $\mathbf{L}^2$.

Let us now designate the eigenstates of H as $|n;l,m\rangle$ satisfying the relation

$$H |n;l.m\rangle = E_n |n;l,m\rangle . \tag{27.19}$$

The following relation will hold for the matrix elements of H between degenerate states, assuming $E_n \neq 0$,

$$\langle n;l.m| [L_z,H] |n;l.m'\rangle = (m-m') \langle n;l.m| n;l.m'\rangle E_n = 0. \tag{27.20}$$

Therefore,

$$\text{either} \quad m = m' \tag{27.21}$$

or

$$\text{if } m \neq m' \text{ then } \langle n;l,m| n;l,m'\rangle = 0. \tag{27.22}$$

Similar results will be obtained for the states with the same E_n and different values of l. Thus, for example, the degeneracy in the hydrogen atom levels with the same principal quantum number, n, and different values of m and l is a simple consequence of the invariance of the Hamiltonian under rotations.

Finally, since

$$\langle \theta,\phi |l,m\rangle = Y_{lm}(\theta,\phi). \tag{27.23}$$

The orthogonality of the $Y_{lm}(\theta,\phi)$ is a natural outcome of rotational symmetry.

27.2 Parity transformation

Parity transformation, P, is the same as reflection defined by

$$P\left|\mathbf{r}'\right\rangle = \left|-\mathbf{r}'\right\rangle. \tag{27.24}$$

Multiplying by P one more time we find

$$P^2 = \mathbf{1}. \tag{27.25}$$

Thus, P is unitary with eigenvalues ± 1. It is a discrete transformation for which only finite values are possible, unlike rotations where the angle of rotation can take continuous values.

Let us define

$$P\left|\alpha\right\rangle = \left|\alpha^P\right\rangle. \tag{27.26}$$

If a state has a definite parity then we will write

$$P\left|\alpha\right\rangle = \left|\alpha^P\right\rangle = \pi_\alpha \left|\alpha\right\rangle \tag{27.27}$$

where π_α is $+1$ or -1.

27.2.1 Consequences of parity transformation

1. The transformation of the operator $\mathbf{r}$ will be defined as

$$\left\langle\alpha^P\right|\mathbf{r}\left|\alpha^P\right\rangle = -\left\langle\alpha\right|\mathbf{r}\left|\alpha\right\rangle, \tag{27.28}$$

which implies

$$P\mathbf{r}P^{-1} = -\mathbf{r}. \tag{27.29}$$

If an operator A has a definite parity then its matrix element is given by

$$\left\langle\alpha^P\right|A\left|\alpha^P\right\rangle = \pi_A \left\langle\alpha\right|A\left|\alpha\right\rangle. \tag{27.30}$$

That is,

$$PAP^{-1} = \pi_A A. \tag{27.31}$$

The matrix element between states $|\alpha\rangle$ and $|\beta\rangle$ with parities π_α and π_β respectively will be given by

$$\left\langle\alpha^P\right|A\left|\beta^P\right\rangle = \pi_\alpha \pi_\beta \left\langle\alpha\right|A\left|\beta\right\rangle. \tag{27.32}$$

However, we also have the following:

$$\left\langle \alpha^P \middle| A \middle| \beta^P \right\rangle = \pi_A \langle \alpha | A | \beta \rangle . \tag{27.33}$$

Hence $\pi_\alpha \pi_\beta = \pi_A$. Or since $\pi_A^2 = 1$, we have

$$\pi_A \pi_\alpha \pi_\beta = 1. \tag{27.34}$$

If A is invariant under parity transformation then

$$\pi_\alpha \pi_\beta = \pi_A = 1. \tag{27.35}$$

The conditions outlined above correspond to "selection rules" that allow only certain types of processes to occur.

2. Let us consider the eigenstates of H,

$$H |n\rangle = E_n |n\rangle . \tag{27.36}$$

If H is invariant under parity then $[P, H] = 0$. Taking the matrix element of this between two different eigenstates $|n; \alpha\rangle$ and $|n; \beta\rangle$ we obtain

$$\langle n; \beta | [P, H] | n; \alpha \rangle = 0, \tag{27.37}$$

which gives

$$E_n \left(\pi_\alpha - \pi_\beta \right) = 0. \tag{27.38}$$

Discarding $E_n = 0$ as the possibility we conclude that the two states must have the same parity.

3. We can construct the following parity eigenstates of H:

$$|n_+\rangle = \frac{1}{2}(1 + \pi) |n\rangle , \tag{27.39}$$

$$|n_-\rangle = \frac{1}{2}(1 - \pi) |n\rangle . \tag{27.40}$$

These are positive and negative parity eigenstates respectively since

$$P |n_+\rangle = + |n_+\rangle , \tag{27.41}$$

$$P |n_-\rangle = - |n_-\rangle . \tag{27.42}$$

From these two equations we get

$$\langle n_- | P^2 | n_+ \rangle = - \langle n_- | n_+ \rangle . \tag{27.43}$$

However, since $P^2 = 1$ the above relation leads to

$$\langle n_- | n_+ \rangle = - \langle n_- | n_+ \rangle , \tag{27.44}$$

which implies $\langle n_- | n_+ \rangle = 0$. Hence $|n_+\rangle$ and $|n_-\rangle$ are orthogonal states.

27.3 Time reversal

Invariance under time reversal, more correctly referred to as reversal of the direction of motion, is not as straightforward as the previous two invariances we have considered. Let us start with the time-dependent Schrödinger equation with real, and time-independent, potential,

$$i\hbar\frac{\partial\psi(\mathbf{r},t)}{\partial t} = \left(-\frac{\hbar^2}{2m}\nabla^2 + V\right)\psi(\mathbf{r},t). \tag{27.45}$$

Under time reversal given by

$$t \to -t \tag{27.46}$$

the equation becomes

$$-i\hbar\frac{\partial\psi(\mathbf{r},-t)}{\partial t} = \left(-\frac{\hbar^2}{2m}\nabla^2 + V\right)\psi(\mathbf{r},-t). \tag{27.47}$$

The new wavefunction does not satisfy the original equation (27.45). Let us, however, take the complex conjugate of (27.45),

$$-i\hbar\frac{\partial\psi^*(\mathbf{r},t)}{\partial t} = \left(-\frac{\hbar^2}{2m}\nabla^2 + V\right)\psi^*(\mathbf{r},t). \tag{27.48}$$

This equation looks similar to (27.47) and so one is inclined to propose that the wavefunction must transform as

$$\psi(\mathbf{r},t) \to \psi^*(\mathbf{r},t). \tag{27.49}$$

Thus the time reversal operator cannot be a linear operator. However, this is not the complete story, as we demonstrate below.

Let T designate the time reversal operator. Since $\mathbf{r}$ does not change sign under T,

$$T\left[\mathbf{r}\,|\alpha\rangle\right] = \mathbf{r}\left[T\,|\alpha\rangle\right]. \tag{27.50}$$

Therefore,

$$T\mathbf{r}T^{-1}\left[T\,|\alpha\rangle\right] = \mathbf{r}\left[T\,|\alpha\rangle\right]. \tag{27.51}$$

Since $|\alpha\rangle$ is any arbitrary state, we have

$$T\mathbf{r}T^{-1} = \mathbf{r}. \tag{27.52}$$

The velocity $v(= d\mathbf{r}/dt)$ reverses sign under time reversal; therefore, the momentum $(\mathbf{p} = m\mathbf{v})$ satisfies the relation

$$T\mathbf{p}T^{-1} = -\mathbf{p}. \tag{27.53}$$

Consider now the fundamental commutator,

$$\left[x_i, p_j\right] |\alpha\rangle = i\hbar\delta_{ij} |\alpha\rangle . \tag{27.54}$$

Applying the time reversal operator we have

$$\left[x_i, -p_j\right] T |\alpha\rangle = i\hbar\delta_{ij} T |\alpha\rangle . \tag{27.55}$$

Therefore,

$$\left[x_i, p_j\right] T |\alpha\rangle = \left(-i\hbar\delta_{ij}\right) T |\alpha\rangle . \tag{27.56}$$

The time-reversed states do not satisfy the correct commutation relation. It is clear that T must include an operator that changes $i \to -i$.

27.3.1 Correct form of T

Let us write

$$T|\alpha\rangle = |\alpha^T\rangle \tag{27.57}$$

and require that

$$T(a|\alpha\rangle + b|\beta\rangle) = a^*|\alpha^T\rangle + b^*|\alpha^T\rangle. \tag{27.58}$$

This establishes the fact that T is antilinear. Let us then write

$$T = UK \tag{27.59}$$

where U is a unitary operator of the type we have used earlier, and K has the following property:

$$Ka|\alpha\rangle = a^*K|\alpha\rangle. \tag{27.60}$$

We emphasize that K changes a complex coefficient to its complex conjugate. If $|\alpha\rangle$ is a base ket then K does not affect it. We elaborate on this by expressing $|\alpha\rangle$ in terms of a complete set of states

$$|\alpha\rangle = \sum_n |a_n\rangle\langle a_n|\alpha\rangle. \tag{27.61}$$

We apply the time-reversal operator:

$$|\alpha^T\rangle = UK|\alpha\rangle = \sum_n \langle a_n|\alpha\rangle^* UK|a_n\rangle = \sum_n \langle a_n|\alpha\rangle^* U|a_n\rangle. \tag{27.62}$$

We note that when K operates on $\langle a_n|\alpha\rangle$ it changes it to its complex conjugate but does not affect the basis kets $|a_n\rangle$. We simplify the above and write

$$|\alpha^T\rangle = \sum_n \langle\alpha|a_n\rangle U|a_n\rangle. \tag{27.63}$$

Similarly, if we write

$$|\beta\rangle = \sum_n |a_n\rangle\langle a_n|\beta\rangle, \tag{27.64}$$

then

$$|\beta^T\rangle = \sum_m \langle a_m|\beta\rangle^* U|a_m\rangle. \tag{27.65}$$

Taking the complex conjugate,

$$\langle\beta^T| = \sum_m \langle a_m|\beta\rangle\langle a_m|U^\dagger. \tag{27.66}$$

Taking the scalar product we find

$$\langle\beta^T|\alpha^T\rangle = \sum_m\sum_n \langle\alpha|a_n\rangle\langle a_m|\beta\rangle\langle a_m|U^\dagger U|a_n\rangle. \tag{27.67}$$

Since

$$U^\dagger U = 1, \qquad \langle a_m|a_n\rangle = \delta_{mn} \tag{27.68}$$

we have

$$\langle\beta^T|\alpha^T\rangle = \sum_n \langle\alpha|a_n\rangle\langle a_n|\beta\rangle = \langle\alpha|\beta\rangle = \langle\beta|\alpha\rangle^*. \tag{27.69}$$

This result implies that instead of imposing the equality of the scalar products we require that their absolute values be the same. This requirement still preserves the invariance condition of the probability:

$$\left|\langle\beta^T|\alpha^T\rangle\right|^2 = |\langle\beta|\alpha\rangle|^2. \tag{27.70}$$

Let us return to our earlier results and confirm that the difficulties we found in defining time reversal have been resolved. First, as for the properties of the wavefunction as a function of **r**, we find

$$|\alpha^T\rangle = T|\alpha\rangle = \int d^3r'\, U|\mathbf{r}'\rangle K\langle\mathbf{r}'|\alpha\rangle = \int d^3r'\, |\mathbf{r}'\rangle\langle\mathbf{r}'|\alpha\rangle^*. \tag{27.71}$$

Its representative in the **r**-space is given by

$$\langle \mathbf{r}|\alpha^T\rangle = \int d^3r' \, \langle \mathbf{r}|\mathbf{r}'\rangle\langle \mathbf{r}'|\alpha\rangle^* = \int d^3r' \, \delta^{(3)}\left(\mathbf{r}-\mathbf{r}'\right)\langle \mathbf{r}'|\alpha\rangle^* = \langle \mathbf{r}|\alpha\rangle^*. \tag{27.72}$$

Thus we reproduce the required wavefunction given by (27.49).

Next we consider the fundamental commutator

$$\left[x_i, p_j\right]|\alpha\rangle = i\hbar\delta_{ij}|\alpha\rangle. \tag{27.73}$$

In terms of the time-reversed state we can write this as

$$T\left[x_i, p_j\right]T^{-1}|\alpha^T\rangle = T\left(i\hbar\delta_{ij}\right)T^{-1}|\alpha^T\rangle, \tag{27.74}$$

$$\left[x_i, Up_jU^{-1}\right]|\alpha^T\rangle = K\left(i\hbar\delta_{ij}\right)K^{-1}|\alpha^T\rangle. \tag{27.75}$$

K will now change i to $-i$ and U will change p_j to $-p_j$, and we recover the same commutator condition

$$\left[x_i, p_j\right] = i\hbar\delta_{ij}. \tag{27.76}$$

27.3.2 Consequences of time reversal

Orbital angular momentum

The orbital angular momentum is defined as $\mathbf{L} = \mathbf{r} \times \mathbf{p}$. Thus we require that

$$T^{-1}\mathbf{L}T = -\mathbf{L}. \tag{27.77}$$

Accordingly, the direction of **L** must change sign. The angular momentum state $|l.m\rangle$ must then change to $|l. - m\rangle$. Thus, the spherical harmonics defined by

$$Y_{lm}(\theta, \phi) = \langle \theta, \phi\, | l, m\rangle \tag{27.78}$$

must transform as

$$Y_{lm}(\theta, \phi) \rightarrow Y_{l-m}(\theta, \phi). \tag{27.79}$$

Spin $\frac{1}{2}$

A state with spin directed along a unit vector **n** with polar angle α and azimuthal angle β is given in terms of the spin-up state $| + z\rangle$ by

$$|\hat{\mathbf{n}}; +\rangle = e^{-i\sigma_z\alpha/2\hbar}e^{-i\sigma_y\beta/2\hbar}| + z\rangle, \tag{27.80}$$

while

$$|\hat{\mathbf{n}};-\rangle = e^{-i\sigma_z\alpha/2\hbar}e^{-i\sigma_y(\beta+\pi)/2\hbar}|+z\rangle. \tag{27.81}$$

Under time reversal we expect

$$T|\hat{\mathbf{n}};+\rangle = \left|\hat{\mathbf{n}};-\right\rangle. \tag{27.82}$$

The relation (27.59) is then given by

$$T = e^{-i\pi\sigma_y/2\hbar}K. \tag{27.83}$$

From this result we find that the T-operator when acting on the basis states $|+\rangle$ and $|-\rangle$ gives

$$T|+\rangle = +|-\rangle, \qquad T|-\rangle = -|+\rangle. \tag{27.84}$$

Thus, when acting on a linear combination we have

$$T(a|+\rangle + b|-\rangle) = a^*|-\rangle - b^*|+\rangle, \tag{27.85}$$

which implies that

$$T^2(a|+\rangle + b|-\rangle) = -(a|+\rangle + b|-\rangle). \tag{27.86}$$

Hence for spin $\frac{1}{2}$ states we have the most usual property

$$T^2 = -1. \tag{27.87}$$

It can be shown generally that for total angular momentum states $\mathbf{J} = \mathbf{L} + \mathbf{S}$,

$$T^2 = -1 \quad \text{for } j = 1/2, 3/2, \ldots, \tag{27.88}$$

$$T^2 = +1 \quad \text{for } j = 1, 2, \ldots. \tag{27.89}$$

27.4 Symmetry groups

27.4.1 Definition of a group

A group is a collection of elements such that the product of any two of its elements is also a member of the group. Here "product" is an abstract concept. It is defined in the context of the type of group one is considering, which does not necessarily imply ordinary multiplication. For example, all integers, including zero, form a group "under addition" since a sum of any two integers is also an integer. Here the abstract word "product" actually means ordinary sum.

The basic definition of a group as follows:

(i) If a and b are any two elements of a group and

$$ab = c \tag{27.90}$$

then c is also a member of the group.

(ii) There is an identity element, e, such that for any element a,

$$ae = a. \tag{27.91}$$

(iii) For every element a there is an inverse, a^{-1}, such that

$$aa^{-1} = a^{-1}a = e. \tag{27.92}$$

(iv) The elements of the group obey the associative law such that for any three elements a, b, c,

$$a\,(bc) = (ab)\,c. \tag{27.93}$$

A group for which the elements satisfy the relation $ab = ba$ is called an Abelian group.

Thus, in our example of the integers, which form a group under addition, the identity element is 0, and the inverse of an integer is the negative of that integer. It is also an Abelian group.

If there are only a finite number of elements in the group then it is called a finite or a discrete group. Parity transformation and time reversal are examples of finite, Abelian groups.

A group can also be formed if an element is a function of a continuous variable. For example, $g(\alpha)$ as a function of a continuous variable α can form a group where

$$g(\alpha_1)g\,(\alpha_2) = g(\alpha_1 + \alpha_2), \tag{27.94}$$

$$g(0) = \text{identity}, \tag{27.95}$$

$$g(-\alpha) = \text{inverse of } g(\alpha). \tag{27.96}$$

These are called Lie groups. The above is an example of a function of a single variable, α. It is called a one-dimensional group. But g can also be a function of more than one variable, $g\,(\alpha, \beta, \ldots)$. If there are n such variables then it is called a Lie group of n dimensions.

Some important examples

1. Rotation in three dimensions is a classic example of a three-dimensional Lie group.
2. A set of $n \times n$ matrices, U, that are unitary,

$$U^{\dagger}U = \mathbf{1}, \tag{27.97}$$

form a unitary group called $U(n)$. A group formed by phase transformations $\exp(i\theta)$ where θ is a continuous variable form the group $U(1)$.

3. If the $n \times n$ unitary matrices also have a unit determinant then it is called the $SU(n)$ group where S stands for "special."
4. If a set of $n \times n$ matrices, A, are orthogonal,

$$\tilde{A}A = \mathbf{1}, \tag{27.98}$$

then they form the group $SO(n)$ where SO stands for "special orthogonal."

5. A mapping of the elements of an abstract group to a set of matrices gives rise to their matrix representation. Thus,

$$a \to D(a) \tag{27.99}$$

with

$$ab = c \to D(a)D(b) = D(c) \tag{27.100}$$

where D's are the matrix representations.

Lie algebra

Transformations in quantum mechanics corresponding to continuous unitary operators form Lie groups. For infinitesimal transformations they can be written in the form

$$a(\boldsymbol{\alpha}) = \mathbf{1} - i\frac{\boldsymbol{\epsilon} \cdot \mathbf{X}}{\hbar} \tag{27.101}$$

where $\boldsymbol{\alpha} = (\alpha, \beta, \ldots)$ depending on the number of dimensions and similarly $\boldsymbol{\epsilon} = \left(\epsilon_\alpha, \epsilon_\beta, \ldots\right)$. The operators $\mathbf{X} = \left(X_\alpha, X_\beta, \ldots\right)$ are called the generators of the group. Since $a(\boldsymbol{\alpha})$'s are unitary, they satisfy the relation

$$\left[X_\alpha, X_\beta\right] = i\hbar C_{\alpha\beta\gamma} X_\gamma \tag{27.102}$$

where we have used the summation convention. The quantities $C_{\alpha\beta\gamma}$ are called the structure constants. For rotations, we have found that they are simply the totally antisymmetric tensor ϵ_{ijk}. The generators are then said to satisfy the Lie algebra.

For finite values of $\boldsymbol{\alpha}$ one then writes

$$a(\boldsymbol{\alpha}) = \exp\left[-i\frac{\boldsymbol{\alpha} \cdot \mathbf{X}}{\hbar}\right]. \tag{27.103}$$

The matrix representation of the group can be written as

$$D\left[a(\boldsymbol{\alpha})\right] = \exp\left[-i\frac{\boldsymbol{\alpha} \cdot \mathbf{T}}{\hbar}\right] \tag{27.104}$$

where T's will satisfy the same relations as the X's,

$$\left[T_\alpha, T_\beta\right] = i\hbar C_{\alpha\beta\gamma} T_\gamma. \tag{27.105}$$

Below we consider two well-known examples of groups in quantum mechanics.

27.5 $D^j(R)$ for $j = \frac{1}{2}$ and $j = 1$: examples of $SO(3)$ and $SU(2)$ groups

Let us consider the representation of the rotation operator $D(R)$ discussed in Section 26.4,

$$D^j(R) = e^{-i\chi \frac{\mathbf{n}.\mathbf{J}}{\hbar}} \tag{27.106}$$

for specific values of j where, as discussed earlier, $\mathbf{n}$ is a unit operator in the direction of the axis of rotation, χ is the rotation angle around the axis $\mathbf{n}$, and J_i's are the generators already defined.

27.5.1 Spin $j = \frac{1}{2}$

Consider now the case of spin $\frac{1}{2}$ rotations where

$$\mathbf{J} = \frac{\hbar}{2}\boldsymbol{\sigma} \tag{27.107}$$

where σ_i's are the Pauli matrices:

$$\sigma_x = \begin{pmatrix} 0 & 1 \\ 1 & 0 \end{pmatrix}, \quad \sigma_y = \begin{pmatrix} 0 & -i \\ i & 0 \end{pmatrix}, \quad \sigma_z = \begin{pmatrix} 1 & 0 \\ 0 & -1 \end{pmatrix}. \tag{27.108}$$

The representation of the rotation matrix can be written as

$$D^{\frac{1}{2}}(R) = e^{-i\chi \frac{\mathbf{n}\cdot\boldsymbol{\sigma}}{2}}. \tag{27.109}$$

$D^{\frac{1}{2}}(R)$ depends on three parameters: the angle χ and two out of the three components of the vector $\mathbf{n}$, the third component being determined by the fact that $\mathbf{n}$ is a unit vector, and, therefore, satisfies,

$$n_x^2 + n_y^2 + n_z^2 = 1. \tag{27.110}$$

For an infinitesimal χ $(= \epsilon)$ one obtains

$$D^{\frac{1}{2}}(R) = \mathbf{1} - i\frac{\mathbf{n}\cdot\boldsymbol{\sigma}}{2}\epsilon. \tag{27.111}$$

Inserting the Pauli matrices and expressing the unit vector $\mathbf{n}$ in terms of its components (n_x, n_y, n_z), we find

$$D^{\frac{1}{2}}(R) = \begin{bmatrix} 1 - i\dfrac{n_z\epsilon}{2} & -i\dfrac{(n_x - in_y)\,\epsilon}{2} \\ -i\dfrac{(n_x + in_y)\,\epsilon}{2} & 1 + i\dfrac{n_z\epsilon}{2} \end{bmatrix}. \tag{27.112}$$

$D^{\frac{1}{2}}(R)$ is, therefore, a 2×2 matrix with matrix elements that are complex. Furthermore, $\det D^{\frac{1}{2}}(R)$ and the magnitude of $\left[D^{\frac{1}{2}}(R)\right]^{\dagger} D^{\frac{1}{2}}(R)$ have the form $1+o(\epsilon^2)$. If higher powers are kept, then all the terms involving ϵ will cancel, giving us

$$\left[D^{\frac{1}{2}}(R)\right]^{\dagger} D^{\frac{1}{2}}(R) = \mathbf{1}, \tag{27.113}$$

$$\det D^{\frac{1}{2}}(R) = 1. \tag{27.114}$$

Matrix $D^{\frac{1}{2}}(R)$ belongs to an $SU(2)$ group, where the lowest dimension of the matrix is 2 and where, as we stated earlier, S stands for "special" group, signifying that it has a unit determinant as written in (27.114). The letter U implies that the matrix is unitary as indicated by (27.113).

Actually one can write $D^{\frac{1}{2}}(R)$ in a closed form,

$$D^{\frac{1}{2}}(R) = \mathbf{1}\cos\left(\frac{\chi}{2}\right) - i\frac{\mathbf{n}\cdot\boldsymbol{\sigma}}{2}\sin\left(\frac{\chi}{2}\right) \tag{27.115}$$

which gives

$$D^{\frac{1}{2}}(R) = \begin{bmatrix} \cos\left(\frac{\chi}{2}\right) - in_z\sin\left(\frac{\chi}{2}\right) & -i\left(n_x - in_y\right)\sin\left(\frac{\chi}{2}\right) \\ -i\left(n_x + in_y\right)\sin\left(\frac{\chi}{2}\right) & \cos\left(\frac{\chi}{2}\right) + in_z\sin\left(\frac{\chi}{2}\right) \end{bmatrix}. \tag{27.116}$$

27.5.2 Spin $j = 1$

Let us consider the case with $j = 1$, where the J_i's, obtained in Chapter 26 are given as follows:

$$J_x = \frac{1}{\sqrt{2}}\begin{pmatrix} 0 & 1 & 0 \\ 1 & 0 & 1 \\ 0 & 1 & 0 \end{pmatrix}, \quad J_y = -\frac{i}{\sqrt{2}}\begin{pmatrix} 0 & -1 & 0 \\ 1 & 0 & -1 \\ 0 & 1 & 0 \end{pmatrix}, \quad J_z = \begin{pmatrix} 1 & 0 & 0 \\ 0 & 0 & 0 \\ 0 & 0 & -1 \end{pmatrix}. \tag{27.117}$$

We note that the J_i's defined above and the S_i's obtained in Section 26.1.1 correspond to the same three-dimensional rotation, except that the matrices above were obtained for the case when J_z is diagonal. The representation matrix $D^j(R)$ for $j = 1$ is given by

$$D^1(R) = e^{-i\chi\frac{\mathbf{n}\cdot\mathbf{J}}{\hbar}}. \tag{27.118}$$

For small angles, $\chi = \epsilon$, the matrix $D^1(R)$ can be written as

$$D^1(R) = \mathbf{1} - i\frac{\mathbf{n}\cdot\mathbf{J}}{\hbar}\epsilon. \tag{27.119}$$

Substituting (27.117), we obtain for infinitesimal ϵ,

$$D^1(R) = \begin{bmatrix} 1 & -n_z\epsilon & n_y\epsilon \\ n_z\epsilon & 1 & -n_x\epsilon \\ -n_y\epsilon & n_x\epsilon & 1 \end{bmatrix}. \tag{27.120}$$

The matrix elements here are real. If $\widetilde{D}^1(R)$ denotes the transpose of $D^1(R)$ then it is easy to verify that $\det D^1(R)$ and the magnitude of $\widetilde{D}^1(R)D^1(R)$ are of the form $1 + o(\epsilon^2)$. If one keeps all the higher powers of ϵ then one finds that

$$\widetilde{D}^1(R)D^1(R) = \mathbf{1}, \tag{27.121}$$

$$\det D^1(R) = 1. \tag{27.122}$$

We note that $D^1(R)$, and therefore R, which are the same for $j = 1$, are 3×3 matrices that depend on three parameters: the angle χ and two out of the three components of the vector $\mathbf{n}$, the third component being determined by the fact that $\mathbf{n}$ is a unit vector. They represent the so-called $SO(3)$ group, where the O in $SO(3)$, as indicated earlier, stands for the fact that $D^1(R)$ and R are orthogonal satisfying the property (27.121) and S stands for "special" transformation, corresponding to $\det R = 1$.

A major difference between $SO(3)$ and $SU(2)$ is that $D^1(R) \to D^1(R)$ as $\chi \to \chi + 2\pi$, while for $D^{\frac{1}{2}}(R)$ one needs to go through twice the amount to return to the same form, i.e., $D^{\frac{1}{2}}(R) \to D^{\frac{1}{2}}(R)$ as $\chi \to \chi + 4\pi$. Keeping this difference in mind, we say that $SO(3)$ and $SU(2)$ are "isomorphic" to each other.

Finally, the rotations that we have considered are called "proper" rotations, which implies that the transformations are achieved by continuous rotations. If we had included reflection, for example, that is, if the transformation for the three-dimensional case were of the product form

$$\begin{bmatrix} -1 & 0 & 0 \\ 0 & -1 & 0 \\ 0 & 0 & -1 \end{bmatrix} D^1(R), \tag{27.123}$$

for which the determinant is -1, then it would not be a proper transformation. It then carries the designation $O(3)$.

27.6 Problems

1. Consider the Hamiltonian for the hydrogen atom,

$$H = \frac{\mathbf{p}^2}{2m} - \frac{Ze^2}{r}.$$

Show that the operator

$$A = \frac{\hbar}{2}[\mathbf{L} \times \mathbf{p} - \mathbf{p} \times \mathbf{L}] + Ze^2 m \frac{\mathbf{r}}{r}$$

commutes with H, where A is called the Runge–Lenz vector. Furthermore, show that

$$\left[L_i, L_j\right] = i\epsilon_{ijk}L_k, \quad \left[L_i, M_j\right] = i\epsilon_{ijk}M_k \quad \text{and} \quad \left[M_i, M_j\right] = i\epsilon_{ijk}L_k$$

where

$$\mathbf{M} = \frac{1}{\sqrt{-2mE/\hbar^2}}\mathbf{A}$$

where E is the energy eigenvalue. Construct the following linear combinations:

$$\mathbf{J} = \frac{\mathbf{L} + \mathbf{M}}{2}, \quad \mathbf{K} = \frac{\mathbf{L} - \mathbf{M}}{2}$$

and show that

$$\left[J_i, J_j\right] = i\epsilon_{ijk}J_k, \quad \left[K_i, K_j\right] = i\epsilon_{ijk}K_k, \quad \text{and} \quad \left[J_i, K_j\right] = 0.$$

Thus we have two mutually commuting operators both of which have the same properties as the angular momentum and both of which also commute with the Hamiltonian. Thus one can designate the common eigenstate as $|j, m_j; k, m_k; E\rangle$.
One can now show that

$$\mathbf{J}^2 = \mathbf{K}^2,$$

whose common value can be written as $j(j+1)$. Finally, returning to $\mathbf{L}$ and $\mathbf{M}$, show that

$$\mathbf{L}^2 + \mathbf{M}^2 + 1 = -\frac{mZ^2e^2}{2E\hbar^2}.$$

From this and the above relations involving the operators $\mathbf{J}$ and $\mathbf{K}$, show that the energy eigenvalues are given by

$$E = -\frac{Z^2\alpha^2 mc^2}{2n^2}$$

where

$$n = j + \frac{1}{2}.$$

28 Addition of angular momenta

In the chapter on rotation we discussed methods for obtaining the eigenvalues and eigenstates of the operator related to angular momentum. These obviously refer to single-particle states. However, often one has a more complex system containing two or more particles, as happens for two-electron systems, the electron–proton state of the atom, or many-nucleon systems in nuclei. In this chapter we tackle some of these problems involving addition of angular momenta.

28.1 Combining eigenstates: simple examples

Let us consider the problem of expressing the product of two states $|j_1, m_1\rangle$, $|j_2, m_2\rangle$ in terms of a state that is an eigenstate of $\mathbf{J}^2$ and J_z, where

$$\mathbf{J} = \mathbf{J}_1 + \mathbf{J}_2 \quad \text{and} \quad J_z = J_{1z} + J_{2z}. \tag{28.1}$$

The operators $\mathbf{J}_1$ and $\mathbf{J}_2$ commute since they operate on different sets of eigenstates. If we write this product state simply as $|j_1, m_1\rangle\, |j_2, m_2\rangle$, and keep in mind that $\mathbf{J}_i^2$ and J_{iz} operate only on states $|j_i, m_i\rangle$ for $i = 1, 2$, then we obtain the following

$$J_z\, |j_1, m_1\rangle\, |j_2, m_2\rangle = (J_{1z} + J_{2z})\, |j_1, m_1\rangle\, |j_2, m_2\rangle \tag{28.2}$$

$$= (J_{1z}\, |j_1, m_1\rangle)\, |j_2, m_2\rangle + |j_1, m_1\rangle\, (J_{2z}\, |j_2, m_2\rangle) \tag{28.3}$$

$$= (m_1 + m_2)\, |j_1, m_1\rangle\, |j_2, m_2\rangle\, . \tag{28.4}$$

Therefore $|j_1, m_1\rangle\, |j_2, m_2\rangle$ is an eigenstate of J_z with eigenvalues $m = m_1 + m_2$.

However, such a product state need not be an eigenstate of $\mathbf{J}^2$ because, from (28.1),

$$\mathbf{J}^2 = (\mathbf{J}_1 + \mathbf{J}_2)^2 = \mathbf{J}_1^2 + \mathbf{J}_2^2 + 2\mathbf{J}_1 \cdot \mathbf{J}_2 \tag{28.5}$$

$$= \mathbf{J}_1^2 + \mathbf{J}_2^2 + J_{1+}\, J + J_{1-}\, J_{2+} + 2J_{1z}J_{2z} \tag{28.6}$$

where the operators $J_{i\pm}$ for $i = 1, 2$ are the ladder operators defined before as

$$J_{i\pm} = J_{ix} \pm iJ_{iy}. \tag{28.7}$$

The presence of these operators will not allow the product state $|j_1, m_1\rangle\, |j_2, m_2\rangle$ to be an eigenstate of $\mathbf{J}^2$ as these operators change the m_i values by ± 1 (for $i = 1, 2$). The only exceptions will be when the ladder operators give vanishing contributions.

A simple procedure to construct the eigenstates of $\mathbf{J}^2$ and J_z will be illustrated below with a specific example where $j_1 = \frac{1}{2}$ and $j_2 = \frac{1}{2}$.

28.1.1 $j_1 = \frac{1}{2}$, $j_2 = \frac{1}{2}$

We start with the product of the two "highest" states, with $m_1 = \frac{1}{2} = m_2$,

$$\left|\tfrac{1}{2},\tfrac{1}{2}\right\rangle\left|\tfrac{1}{2},\tfrac{1}{2}\right\rangle. \tag{28.8}$$

From (28.4) we deduce that

$$J_z\left|\tfrac{1}{2},\tfrac{1}{2}\right\rangle\left|\tfrac{1}{2},\tfrac{1}{2}\right\rangle = 1\left|\tfrac{1}{2},\tfrac{1}{2}\right\rangle\left|\tfrac{1}{2},\tfrac{1}{2}\right\rangle. \tag{28.9}$$

From (28.6) we find

$$\mathbf{J}^2\left|\tfrac{1}{2},\tfrac{1}{2}\right\rangle\left|\tfrac{1}{2},\tfrac{1}{2}\right\rangle = \left(\mathbf{J}_1^2 + \mathbf{J}_2^2 + J_{1+}J_{2-} + J_{1-}J_{2+} + 2J_{1z}J_{2z}\right)\left|\tfrac{1}{2},\tfrac{1}{2}\right\rangle\left|\tfrac{1}{2},\tfrac{1}{2}\right\rangle. \tag{28.10}$$

Taking note of the relations involving raising and lowering operators we obtain

$$\mathbf{J}^2\left|\tfrac{1}{2},\tfrac{1}{2}\right\rangle\left|\tfrac{1}{2},\tfrac{1}{2}\right\rangle = \left[\frac{3}{4} + \frac{3}{4} + (0)(1) + (1)(0) + 2\left(\frac{1}{2}\right)\left(\frac{1}{2}\right)\right]\left|\tfrac{1}{2},\tfrac{1}{2}\right\rangle\left|\tfrac{1}{2},\tfrac{1}{2}\right\rangle = 2\left|\tfrac{1}{2},\tfrac{1}{2}\right\rangle\left|\tfrac{1}{2},\tfrac{1}{2}\right\rangle. \tag{28.11}$$

Thus $\left|\frac{1}{2},\frac{1}{2}\right\rangle\left|\frac{1}{2},\frac{1}{2}\right\rangle$ is, indeed, an eigenstate of J_z and $\mathbf{J}^2$. Since the eigenvalues of these operators for a state $|j,m\rangle$ are m and $j(j+1)$ respectively, we conclude that this product state has $j = 1$ and $m = 1$. Therefore, we can write

$$\left|\tfrac{1}{2},\tfrac{1}{2}\right\rangle\left|\tfrac{1}{2},\tfrac{1}{2}\right\rangle = a\,|1,1\rangle. \tag{28.12}$$

The right-hand side corresponds to the eigenstate $|j,m\rangle$ for $j = 1$ and $m = 1$. Since both sides are normalized, we have $a = 1$ and, therefore,

$$|1,1\rangle = \left|\tfrac{1}{2},\tfrac{1}{2}\right\rangle\left|\tfrac{1}{2},\tfrac{1}{2}\right\rangle. \tag{28.13}$$

From the above, we can construct other states through the operation of the lowering operator J_- $(= J_{1-} + J_{2-})$ on both sides of the equation (28.13):

$$J_-\left|\tfrac{1}{2},\tfrac{1}{2}\right\rangle\left|\tfrac{1}{2},\tfrac{1}{2}\right\rangle = J_-\,|1,1\rangle. \tag{28.14}$$

Using the relation $J_-\,|j,m\rangle = \sqrt{(j+m)(j-m+1)}\,|j,m-1\rangle$ in (28.14) we have for the left-hand side of (28.14),

$$J_-\left|\tfrac{1}{2},\tfrac{1}{2}\right\rangle\left|\tfrac{1}{2},\tfrac{1}{2}\right\rangle = (J_{1-} + J_{2-})\left|\tfrac{1}{2},\tfrac{1}{2}\right\rangle\left|\tfrac{1}{2},\tfrac{1}{2}\right\rangle = \left|\tfrac{1}{2},-\tfrac{1}{2}\right\rangle\left|\tfrac{1}{2},\tfrac{1}{2}\right\rangle + \left|\tfrac{1}{2},\tfrac{1}{2}\right\rangle\left|\tfrac{1}{2},-\tfrac{1}{2}\right\rangle. \tag{28.15}$$

For the right-hand side of (28.14) we obtain

$$J_- |1,1\rangle = \sqrt{2}\,|1,0\rangle\,. \tag{28.16}$$

Equating (28.15) and (28.16) we get

$$|1,0\rangle = \frac{1}{\sqrt{2}}\left(\left|\tfrac{1}{2},-\tfrac{1}{2}\right\rangle\left|\tfrac{1}{2},\tfrac{1}{2}\right\rangle + \left|\tfrac{1}{2},\tfrac{1}{2}\right\rangle\left|\tfrac{1}{2},-\tfrac{1}{2}\right\rangle\right). \tag{28.17}$$

By applying J_- again, we get

$$|1,-1\rangle = \left|\tfrac{1}{2},-\tfrac{1}{2}\right\rangle\left|\tfrac{1}{2},-\tfrac{1}{2}\right\rangle. \tag{28.18}$$

This is clearly the lowest state and there will be no more states possible.

Thus (28.13), (28.17), and (28.18) constitute a triplet of states for the total angular momentum $j = 1$ and $m = 1, 0, -1$ made up from the product of states with $j_1 = \frac{1}{2}$, $m_1 = \pm\,\frac{1}{2}$ and $j_2 = \frac{1}{2}$, $m_2 = \pm\,\frac{1}{2}$.

There is one more state that is orthogonal to the state $|1,0\rangle$ given in (28.17), and that is

$$|0,0\rangle = \frac{1}{\sqrt{2}}\left(\left|\tfrac{1}{2},-\tfrac{1}{2}\right\rangle\left|\tfrac{1}{2},\tfrac{1}{2}\right\rangle - \left|\tfrac{1}{2},\tfrac{1}{2}\right\rangle\left|\tfrac{1}{2},-\tfrac{1}{2}\right\rangle\right). \tag{28.19}$$

By operating with $\mathbf{J}^2$ and J_z it is easy to check that the state given above has $j = 0$ and $m = 0$.

In summary, if we symbolically designate

$$\left|\tfrac{1}{2},\tfrac{1}{2}\right\rangle = \uparrow,\quad \left|\tfrac{1}{2},-\tfrac{1}{2}\right\rangle = \downarrow \tag{28.20}$$

then we have the following categories of product states:

Triplet states

$$|1,1\rangle = \uparrow\uparrow,\quad |1,0\rangle = \frac{1}{\sqrt{2}}\left(\uparrow\downarrow + \downarrow\uparrow\right),\quad |1,-1\rangle = \downarrow\downarrow\,. \tag{28.21}$$

Singlet state

$$|0,0\rangle = \frac{1}{\sqrt{2}}\left(\uparrow\downarrow - \downarrow\uparrow\right). \tag{28.22}$$

These constitute four mutually orthogonal states.

One can symbolically also write the four states contained in (28.21) and (28.22) as

$$\left(j_1 = \tfrac{1}{2}\right) \times \left(j_2 = \tfrac{1}{2}\right) = (j = 1) + (j = 0) \tag{28.23}$$

where the left-hand side corresponds to the product states $|j_1, m_1\rangle\,|j_2, m_2\rangle$ with $j_1 = \frac{1}{2}$ and $j_2 = \frac{1}{2}$, while the right-hand side corresponds to three types of states in (28.21) with $j = 1$ and one state in (28.22) with $j = 0$.

In terms of the number of orthogonal states we can write, symbolically once again,

$$\mathbf{2} \times \mathbf{2} = \mathbf{3} + \mathbf{1} \tag{28.24}$$

where the left-hand side refers to the product states $|1/2, m_1\rangle\,|1/2, m_2\rangle$ where $\mathbf{2}$ corresponds to the doublet, $|1/2, 1/2\rangle$ and $|1/2, -1/2\rangle$, formed by each individual state, $|1/2, m_i\rangle$. On the right-hand side, $\mathbf{3}$ and $\mathbf{1}$ correspond respectively to the triplet states with $j = 1$ and the singlet state with $j = 0$ as we have already discussed in (28.21) and (28.22). One can also express (28.23) as

$$(\text{doublet}) \times (\text{doublet}) = (\text{triplet}) + (\text{singlet})\,. \tag{28.25}$$

We make the following remarks regarding these results.

(1) The Pauli principle, which we discussed in Chapter 2, says that the electrons, or any fermions with spin $\frac{1}{2}$, must be in a state that is antisymmetric. It refers to the combined wavefunction that includes the spatial dependence and spin dependence. We will assume the spatial part to be symmetric and thus concentrate entirely on the spin part.

Thus two electrons designated by subscripts "1" and "2" and described by the states $\left|\frac{1}{2}, m_1\right\rangle_1$ and $\left|\frac{1}{2}, m_2\right\rangle_2$ respectively cannot be in any of the product states $|1, 1\rangle$, $|1, 0\rangle$, or $|1, -1\rangle$ since these are symmetric under particle exchange ($1 \leftrightarrows 2$). Hence the two electrons can only be in the state

$$|0, 0\rangle = \frac{1}{\sqrt{2}}\left(\left|\tfrac{1}{2}, \tfrac{1}{2}\right\rangle_1 \left|\tfrac{1}{2}, -\tfrac{1}{2}\right\rangle_2 - \left|\tfrac{1}{2}, -\tfrac{1}{2}\right\rangle_1 \left|\tfrac{1}{2}, \tfrac{1}{2}\right\rangle_2\right) \tag{28.26}$$

which is antisymmetric since under the particle interchange $1 \to 2$ one finds $|0, 0\rangle \to -\,|0, 0\rangle$.

(2) A complete wavefunction for a particle of "spin j_1" and projection m_1 is

$$|j_1, m_1\rangle\, \phi\left(\mathbf{r}\right) \tag{28.27}$$

where $|j_1, m_1\rangle$ is the spin-wave part (= column matrix), and $\phi\left(\mathbf{r}\right)$ is the space-dependent part. For example, as we discussed in Chapter 2, a free-electron wavefunction with spin up is

$$\text{Spin-up wavefunction} = \begin{pmatrix} 1 \\ 0 \end{pmatrix} e^{i\mathbf{k}\cdot\mathbf{r}} \tag{28.28}$$

and

$$\text{Spin-down wavefunction} = \begin{pmatrix} 0 \\ 1 \end{pmatrix} e^{i\mathbf{k}\cdot\mathbf{r}}. \tag{28.29}$$

One can similarly obtain products of different spin combinations. A very helpful tool to accomplish all of that, however, is through the so called Clebsch–Gordan coefficients, which we will discuss below.

28.2 Clebsch–Gordan coefficients and their recursion relations

In the previous section, in discussing the product states, we considered a simple case with $j_1 = \frac{1}{2}$ and $j_2 = \frac{1}{2}$. Let us now tackle the problem involving arbitrary j_1 and j_2. We start with the product states defined as follows:

$$|j_1 m_1\rangle |j_2 m_2\rangle = |j_1 j_2 m_1 m_2\rangle. \tag{28.30}$$

In terms of these states we now form a state that we denote as $|j_1 j_2 jm\rangle$, with a definite total angular momentum, j, and the z-component m by using completeness:

$$|j_1 j_2 jm\rangle = \sum_{m_1' m_2'} \langle j_1 j_2 m_1' m_2' |j_1 j_2, jm\rangle | |j_1 j_2 m_1' m_2'\rangle \tag{28.31}$$

where m_1' goes from $-j_1$ to j_1 and m_2' varies from $-j_2$ to j_2. Since j_1 and j_2 are fixed quantities throughout our calculations, we can simplify things by removing them from the notations. Thus, we write

$$|j_1 j_2 jm\rangle = |jm\rangle, \quad |j_1 j_2 m_1 m_2\rangle = |m_1 m_2\rangle \quad \text{and} \quad \langle j_1 j_2 m_1' m_2' |j_1 j_2, jm\rangle | = \langle m_1' m_2' |jm\rangle. \tag{28.32}$$

Equation (28.31) then reads

$$|jm\rangle = \sum_{m_1' m_2'} \langle m_1' m_2' |jm\rangle |m_1' m_2'\rangle. \tag{28.33}$$

The coefficients $\langle m_1' m_2' |jm\rangle$ are called the Clebsch–Gordan coefficients.

First of all, since $J_z = J_{1z} + J_{2z}$, by operating J_z on the left-hand side of (28.33) and $(J_{1z} + J_{2z})$ on the right-hand side we find that the following relation must be satisfied:

$$m = m_1' + m_2'. \tag{28.34}$$

Let us now operate J_+ on equation (28.33). Since $J_+ = J_{1+} + J_{2+}$ we obtain

$$\sqrt{(j-m)(j+m+1)}|j, m+1\rangle$$
$$= \sum_{m_1' m_2'} \langle m_1' m_2' |jm\rangle \left[\sqrt{(j_1 - m_1')(j_1 + m_1' + 1)}|m_1' + 1, m_2'\rangle \right.$$
$$\left. + \sqrt{(j_2 - m_2')(j_2 + m_2' + 1)}|m_1', m_2' + 1\rangle \right] \tag{28.35}$$

where J_+ is applied to the left-hand side of (28.33) and J_{1+} and J_{2+} are applied to the appropriate terms on the right. Let us now multiply the above equation by $\langle m_1 m_2|$ to obtain

$$\sqrt{(j-m)(j+m+1)}\langle m_1 m_2|j, m+1\rangle$$
$$= \sqrt{(j_1-m_1+1)(j_1+m_1)}\langle m_1-1, m_2|jm\rangle$$
$$+ \sqrt{(j_2-m_2+1)(j_2+m_2)}\langle m_1, m_2-1|jm\rangle \tag{28.36}$$

where we have considered the fact that because of orthonormality of the ket vectors, we have $m_1 = m_1' + 1$ (i.e., $m_1' = m_1 - 1$), and $m_2 = m_2'$ in the first term on the right-hand side of (28.35), while in the second term we have $m_1 = m_1'$ and $m_2 = m_2' + 1$ (i.e., $m_2' = m_2 - 1$). The above matrix elements are nonzero only when $m_1 + m_2 = m + 1$.

Similarly, if we operate the equation (28.33) with J_- and then multiply the resulting equation by $\langle m_1 m_2|$ we obtain

$$\sqrt{(j+m)(j-m+1)}\langle m_1, m_2|jm-1\rangle$$
$$= \sqrt{(j_1-m_1)(j_1+m_1+1)}\langle m_1+1, m_2|jm\rangle$$
$$+ \sqrt{(j_2-m_2)(j_2+m_2+1)}\langle m_1, m_2+1|jm\rangle. \tag{28.37}$$

The above matrix elements are nonzero only when $m_1 + m_2 = m - 1$.

Let us take $m = j - 1$ and $m_1 = j_1$ in (28.36), then $m_2 = j - j_1$. We obtain

$$\sqrt{2j}\langle j_1, j-j_1|jj\rangle = \sqrt{2j_1}\langle j_1-1, j-j_1|j, j-1\rangle$$
$$+ \sqrt{(j_2+j-j_1)(j_2-j+j_1+1)}\langle j_1, j-j_1-1|j, j-1\rangle \tag{28.38}$$

and take $m = j$ and $m_1 = j_1$ in (28.37), then $m_2 = j - j_1 - 1$. We obtain

$$\sqrt{2j}\langle j_1, j-j_1-1|j, j-1\rangle = \sqrt{(j_2-j+j_1+1)(j_2+j-j_1)}\langle j_1, j-j_1|j, j\rangle. \tag{28.39}$$

We now see the beginnings of a recursion relation: if $\langle j_1, j-j_1|jj\rangle$ is known then $\langle j_1, j-j_1-1|j, j-1\rangle$ can be determined from (28.39). This in turn allows us to obtain $\langle j_1-1, j-j_1|j, j-1\rangle$ from the previous equation (28.38). Thus all the Clebsch–Gordan coefficients can be determined in terms of a single quantity

$$\langle j_1, j-j_1|jj\rangle. \tag{28.40}$$

The above matrix element refers to $\langle m_1 m_2|jj\rangle$ and, therefore, corresponds to $m_2 = j - j_1$. However, since m_2 lies between $-j_2$ and j_2, we have

$$-j_2 \le j - j_1 \le j_2. \tag{28.41}$$

Therefore,

$$j_1 - j_2 \le j \le j_1 + j_2. \tag{28.42}$$

If we had started the process with $m_2 = j_2$, instead of $m_1 = j_1$, then through the recursion relations we would have determined all the Clebsch–Gordan coefficients in terms of

$$\langle j - j_2, j_2 | jj \rangle. \tag{28.43}$$

This would imply that

$$-j_1 \leq j - j_2 \leq j_1. \tag{28.44}$$

Hence

$$j_2 - j_1 \leq j \leq j_1 + j_2. \tag{28.45}$$

This result combined with (28.45) leads to the triangle condition

$$|j_1 - j_2| \leq j \leq j_1 + j_2. \tag{28.46}$$

28.3 Combining spin ½ and orbital angular momentum l

Generally, obtaining Clebsch–Gordan coefficients can be quite cumbersome, and there may be no alternative but to refer to the tables that are provided in various specialized books. However, in certain specific cases one can resort to simpler methods. One such case involves combining spin and orbital angular momentum when the spin is $\frac{1}{2}$. Here

$$j_1 = l, \quad \text{and} \quad j_2 = \frac{1}{2}. \tag{28.47}$$

Because we have $j_2 = \frac{1}{2}$ there are only two possible values for m_2, $+\frac{1}{2}$ or $-\frac{1}{2}$ and hence there can be only two product states $|m_1, m_2\rangle$, where m_1 is the z-component of the orbital angular momentum. Thus if the total angular momentum is j and its z-component is m, one can write

$$|jm\rangle = a_1 \left|m_1, \tfrac{1}{2}\right\rangle + a_2 \left|m_1 + 1, -\tfrac{1}{2}\right\rangle. \tag{28.48}$$

From the first term on the right-hand side of the above equation we have $m = m_1 + \frac{1}{2}$ with $m_2 = \frac{1}{2}$. In the second term, since $m_2 = -\frac{1}{2}$, the z-component of the orbital angular momentum must be $m_1 + 1$ in order to satisfy the relation $m = m_1 + \frac{1}{2}$. From the previous section, we note that there are two possibilities for j: $j = l + \frac{1}{2}$ and $j = l - \frac{1}{2}$.

Let us consider the case where $j = l + \frac{1}{2}$. The constants a_1 and a_2 correspond to the Clebsch–Gordan coefficients and, since $|jm\rangle$ is normalized, we have

$$|a_1|^2 + |a_2|^2 = 1. \tag{28.49}$$

All we now need is another relation between a_1 and a_2 so that we can determine their magnitudes (though not necessarily the phase). We note that the state (28.48) as written is automatically an eigenstate of

$$J_z = L_z + S_z. \tag{28.50}$$

However, it is not automatically an eigenstate of $\mathbf{J}^2$, a condition that needs to be imposed, providing us with the second relation involving a_1 and a_2. Since

$$\mathbf{J} = \mathbf{L} + \mathbf{S}, \tag{28.51}$$

we have

$$\mathbf{J}^2 = \mathbf{L}^2 + \mathbf{S}^2 + 2\mathbf{L}\cdot\mathbf{S}. \tag{28.52}$$

The states on the right-hand side of (28.48) are eigenstates of $\mathbf{L}^2$ and $\mathbf{S}^2$ with eigenvalues $l(l+1)$ and $s(s+1)(=\frac{3}{4})$ with $s=\frac{1}{2}$, respectively, while the state on the left is, of course, an eigenstate of $\mathbf{J}^2$ with eigenvalue $j(j+1)$. The operators in (28.52) upon operating on $|jm\rangle$ and regrouping the terms give

$$\left[j(j+1) - l(l+1) - \frac{3}{4}\right]|jm\rangle = 2\mathbf{L}\cdot\mathbf{S}|jm\rangle. \tag{28.53}$$

For $j = l + \frac{1}{2}$, the factor on the left-hand side of the above equation is simply l. Hence we have

$$l|jm\rangle = 2\mathbf{L}\cdot\mathbf{S}|jm\rangle. \tag{28.54}$$

Using the ladder operators $L_\pm$ and $S_\pm$ one can show that

$$2\mathbf{L}\cdot\mathbf{S} = (L_+S_- + L_-S_+ + 2L_zS_z)\,. \tag{28.55}$$

Hence, from (28.55) we have

$$2\mathbf{L}\cdot\mathbf{S}|jm\rangle = (L_+S_- + L_-S_+ + 2L_zS_z)\left[a_1\left|m_1, \tfrac{1}{2}\right\rangle + a_2\left|m_1+1, -\tfrac{1}{2}\right\rangle\right]. \tag{28.56}$$

Using the familiar relations

$$L_\pm|m_1, m_2\rangle = \sqrt{(l\mp m_1)(l\pm m_1+1)}|m_1\pm 1, m_2\rangle, \tag{28.57}$$

$$S_\pm|m_1, m_2\rangle = \sqrt{\left(\frac{1}{2}\mp m_2\right)\left(\frac{1}{2}\pm m_2+1\right)}\,\left|m_1, m_2\pm\tfrac{1}{2}\right\rangle, \tag{28.58}$$

and

$$L_zS_z|m_1, m_2\rangle = m_1m_2|m_1m_2\rangle, \tag{28.59}$$

the right-hand side of (28.56) is found to be

$$\left[a_1 m_1 + a_2\sqrt{(l+m_1+1)(l-m_1)}\right]\left|m_1, \tfrac{1}{2}\right\rangle$$
$$+\left[-(m_1+1)\,a_2 + a_1\sqrt{(l-m_1)(l+m_1+1)}\right]\left|m_1+1, -\tfrac{1}{2}\right\rangle. \tag{28.60}$$

Hence, combining all of the above relations, we obtain

$$\left[(l-m_1)a_1 - \sqrt{(l+m_1+1)}a_2\right]\left|m_1, \tfrac{1}{2}\right\rangle$$
$$+\left[(l+m_1+1)\,a_2 - \sqrt{(l-m_1)(l+m_1+1)}a_1\right]\left|m_1+1, -\tfrac{1}{2}\right\rangle = 0. \tag{28.61}$$

Since $\left|m_1, \frac{1}{2}\right\rangle$ and $\left|m_1+1, -\frac{1}{2}\right\rangle$ are orthonormal, the terms inside each square bracket vanish giving, in each case, the result

$$\frac{a_1}{a_2} = \sqrt{\frac{l+m_1+1}{l-m_1}}. \tag{28.62}$$

Since $m_1 = m - \frac{1}{2}$, we obtain the following using the normalized result:

$$a_1 = \sqrt{\frac{l+m+\frac{1}{2}}{2l+1}}, \quad a_2 = \sqrt{\frac{l-m+\frac{1}{2}}{2l+1}}. \tag{28.63}$$

Hence

$$|jm\rangle = \sqrt{\frac{l+m+\frac{1}{2}}{2l+1}}\left|m-\tfrac{1}{2}, \tfrac{1}{2}\right\rangle + \sqrt{\frac{l-m+\frac{1}{2}}{2l+1}}\left|m+\tfrac{1}{2}, -\tfrac{1}{2}\right\rangle \tag{28.64}$$

where $j = l + \frac{1}{2}$. Taking the representatives of these states in the (θ, ϕ) system and taking into account the fact that the state $|m_1 m_2\rangle$ is a product state $|lm_1\rangle|\frac{1}{2}m_2\rangle$, we have

$$\langle\theta, \phi|m_1 m_2\rangle = \langle\theta, \phi|l, m_1\rangle\left|\tfrac{1}{2}, m_2\right\rangle = Y_{lm_1}(\theta, \phi)\left|\tfrac{1}{2}, m_2\right\rangle \tag{28.65}$$

where $Y_{lm_1}(\theta, \phi)$ is the spherical harmonic function. Hence (28.64) gives

$$\mathcal{Y}_{l+\frac{1}{2},m}(\theta, \phi) = \sqrt{\frac{l+m+\frac{1}{2}}{2l+1}}Y_{l,m-\frac{1}{2}}(\theta, \phi)\chi_+ + \sqrt{\frac{l-m+\frac{1}{2}}{2l+1}}Y_{l,m+\frac{1}{2}}(\theta, \phi)\chi_- \tag{28.66}$$

where we denote

$$\langle\theta\phi|jm\rangle = \left\langle\theta\phi|l+\tfrac{1}{2}, m\right\rangle = \mathcal{Y}_{l+\frac{1}{2},m}(\theta, \phi), \tag{28.67}$$

$$\left|\tfrac{1}{2}, \tfrac{1}{2}\right\rangle = \chi_+, \tag{28.68}$$

$$\left|\tfrac{1}{2}, -\tfrac{1}{2}\right\rangle = \chi_-. \tag{28.69}$$

Following the same procedure we obtain, for $j = l - \frac{1}{2}$,

$$\mathcal{Y}_{l-\frac{1}{2},m}(\theta,\phi) = -\sqrt{\frac{l-m+\frac{1}{2}}{2l+1}}\,Y_{l,m-\frac{1}{2}}(\theta,\phi)\chi_+ + \sqrt{\frac{l+m+\frac{1}{2}}{2l+1}}\,Y_{l,m+\frac{1}{2}}(\theta,\phi)\chi_-. \quad (28.70)$$

This example illustrates that one need not always go through the recursion relations to obtain the Clebsch–Gordan coefficients. If the circumstances are right, as they were in this example, where there were only two possible spin states, one can follow the type of procedure just outlined.

28.4 Appendix to Chapter 28

28.4.1 Table of Clebsch–Gordan coefficients

$J_1 = \frac{1}{2},\ J_2 = \frac{1}{2}$

		1	0	0	1
		1	0	0	-1
$\frac{1}{2}$	$\frac{1}{2}$	1			
$\frac{1}{2}$	$-\frac{1}{2}$		$\sqrt{\frac{1}{2}}$	$\sqrt{\frac{1}{2}}$	
$-\frac{1}{2}$	$\frac{1}{2}$		$\sqrt{\frac{1}{2}}$	$-\sqrt{\frac{1}{2}}$	
$-\frac{1}{2}$	$-\frac{1}{2}$				1

$J_1 = 1,\ J_2 = 1$

		$\frac{3}{2}$	$\frac{3}{2}$	$\frac{1}{2}$	$\frac{3}{2}$	$\frac{1}{2}$	$\frac{3}{2}$
		$\frac{3}{2}$	$\frac{1}{2}$	$\frac{1}{2}$	$-\frac{1}{2}$	$-\frac{1}{2}$	$-\frac{1}{2}$
1	$\frac{1}{2}$	1					
1	$-\frac{1}{2}$		$\sqrt{\frac{1}{3}}$	$\sqrt{\frac{2}{3}}$			
0	$\frac{1}{2}$		$\sqrt{\frac{2}{3}}$	$-\sqrt{\frac{1}{3}}$			
0	$-\frac{1}{2}$				$\sqrt{\frac{2}{3}}$	$\sqrt{\frac{1}{3}}$	
-1	$\frac{1}{2}$				$\sqrt{\frac{1}{3}}$	$-\sqrt{\frac{2}{3}}$	
1	$-\frac{1}{2}$						1

$J_1 = 1,\ J_2 = 1$

$$
\begin{array}{cc|ccccccccc}
 & & 2 & 2 & 1 & 2 & 1 & 0 & 2 & 1 & 2 \\
 & & 2 & 1 & 1 & 0 & 0 & 0 & -1 & -1 & -2 \\
\hline
1 & 1 & 1 & & & & & & & & \\
1 & 0 & & \sqrt{\frac{1}{2}} & \sqrt{\frac{1}{2}} & & & & & & \\
0 & 1 & & \sqrt{\frac{1}{2}} & -\sqrt{\frac{1}{2}} & & & & & & \\
1 & -1 & & & & \sqrt{\frac{1}{6}} & \sqrt{\frac{1}{2}} & \sqrt{\frac{1}{3}} & & & \\
0 & 0 & & & & \sqrt{\frac{2}{3}} & 0 & -\sqrt{\frac{1}{3}} & & & \\
-1 & 1 & & & & \sqrt{\frac{1}{6}} & -\sqrt{\frac{1}{2}} & \sqrt{\frac{1}{3}} & & & \\
0 & -1 & & & & & & & \sqrt{\frac{1}{2}} & \sqrt{\frac{1}{2}} & \\
-1 & 0 & & & & & & & \sqrt{\frac{1}{2}} & -\sqrt{\frac{1}{2}} & \\
-1 & -1 & & & & & & & & & 1
\end{array}
$$

28.5 Problems

1. Consider the spin triplet and singlet states χ_1 and χ_0, respectively, which are constructed out of two spin $\frac{1}{2}$ particles. If $\boldsymbol{\sigma}_1$ and $\boldsymbol{\sigma}_2$ correspond to Pauli operators for the particles, then show that

$$(\boldsymbol{\sigma}_1 \cdot \boldsymbol{\sigma}_2)\, \chi_1 = +\chi_1$$

and

$$(\boldsymbol{\sigma}_1 \cdot \boldsymbol{\sigma}_2)\, \chi_0 = -3\chi_0.$$

Obtain the eigenvalues of $(\boldsymbol{\sigma}_1 \cdot \boldsymbol{\sigma}_2)^n$.

2. If $\boldsymbol{\sigma} = \boldsymbol{\sigma}_1 + \boldsymbol{\sigma}_2$, then show that

$$(\boldsymbol{\sigma} \cdot \mathbf{r})^2 = 2[(\sigma_1 \cdot \mathbf{r})\, (\sigma_2 \cdot \mathbf{r}) + 1].$$

3. A system consisting of two spin $\frac{1}{2}$ particles is subjected to a time-dependent perturbation

$$
\begin{aligned}
H' &= \lambda \boldsymbol{\sigma}_1 \cdot \boldsymbol{\sigma}_2, \quad && t \geq 0 \\
&= 0, && t < 0.
\end{aligned}
$$

The initial state $|\psi(t=0)\rangle$ of the system is given by $|-+\rangle$. Express H' in terms of the total spin in triplet and singlet states through the Pauli operator with the total spin $\boldsymbol{\sigma}$ $(= \boldsymbol{\sigma}_1 + \boldsymbol{\sigma}_2)$. Obtain $|\psi(t)\rangle$ for $t > 0$ and determine the probability that it is in $|++\rangle$, $|--\rangle$, and $|++\rangle$ states.

29 Irreducible tensors and Wigner–Eckart theorem

Irreducible spherical tensors are like spherical harmonics that under rotations transform into one another. A typical irreducible spherical tensor depends on two quantities, usually designated as k and q, which play a role parallel to j and m of the angular momentum. The matrix elements of these tensors between different angular momentum states j', m' and j, m are often quite complicated but they can be simplified considerably due to a theorem by Wigner and Eckart. This allows one to separate the matrix element into two factors: one which is just the Clebsch–Gordan coefficient involved in combining k with j to give j'. This is called the geometrical factor and the other is a single term that is characteristic of the spherical tensor and is independent of m, m', and q. All this will be explored in this chapter.

29.1 Irreducible spherical tensors and their properties

Consider the following property of the spherical harmonics $Y_{lm}(\theta, \phi)$:

$$\begin{aligned} J_z \left[Y_{lm}(\theta,\phi) f(\theta,\phi)\right] &= f(\theta,\phi) J_z Y_{lm} + Y_{lm} J_z f(\theta,\phi) \\ &= m\hbar f(\theta,\phi) Y_{lm} + Y_{lm} J_z f(\theta,\phi) \end{aligned} \tag{29.1}$$

where $f(\theta, \phi)$ is an arbitrary function of θ and ϕ. We have used J_z in a generic sense, since in this case $J_z = L_z = -i\hbar\partial/\partial\phi$. Since Y_{lm} is an eigenstate of J_z we have used the property

$$J_z Y_{lm} = m\hbar Y_{lm} \tag{29.2}$$

in obtaining the relation (29.1). Thus (29.1) gives

$$[J_z, Y_{lm}] f = m\hbar f \, Y_{lm}. \tag{29.3}$$

Therefore, since f is arbitrary, we have

$$[J_z, Y_{lm}] = m\hbar \, Y_{lm}. \tag{29.4}$$

By following similar steps we can also obtain the following commutator relations for the ladder operators:

$$[J_+, Y_{lm}] = \sqrt{(l-m)(l+m+1)}\hbar Y_{l,m+1}, \tag{29.5}$$

$$[J_-, Y_{lm}] = \sqrt{(l+m)(l-m+1)}\hbar Y_{l,m-1}. \tag{29.6}$$

We can generalize the above example of the spherical harmonics to define spherical tensors, $T(k,q)$, where

$$k = 0, \frac{1}{2}, 1, \frac{3}{2}, \ldots \quad \text{and} \quad q = -k, -k+1, \ldots, 0, \ldots, k. \tag{29.7}$$

Thus k replaces l in Y_{lm} and q replaces m and has $2k + 1$ values. We note that k and, therefore, q can take integral as well as half-integral values.

The spherical tensors are simple generalizations of spherical harmonics in that, while the angles θ and ϕ in $Y_{lm}(\theta, \phi)$ refer to the coordinate vector $\mathbf{r}$, in spherical tensors they refer to a general vector $\mathbf{V}$ of which $\mathbf{r}$ is a special case. Furthermore, no reference is made to angles in writing down $T(k,q)$. For example, corresponding to

$$Y_{11}(\theta, \phi) = -\sqrt{\frac{3}{8\pi}} \sin\theta e^{i\phi} = -\sqrt{\frac{3}{8\pi}} \frac{(x+iy)}{r}, \tag{29.8}$$

$$Y_{10}(\theta, \phi) = \sqrt{\frac{3}{4\pi}} \cos\theta = \sqrt{\frac{3}{4\pi}} \frac{z}{r}, \tag{29.9}$$

$$Y_{1-1}(\theta, \phi) = \sqrt{\frac{3}{8\pi}} \sin\theta e^{-i\phi} = \sqrt{\frac{3}{8\pi}} \frac{(x-iy)}{r}, \tag{29.10}$$

we define the spherical tensors for an arbitrary vector, $\mathbf{V}$, as

$$T(1,1) = -\frac{\left(V_x + iV_y\right)}{\sqrt{2}} = V_{+1}, \tag{29.11}$$

$$T(1,0) = V_z = V_0, \tag{29.12}$$

$$T(1,-1) = \frac{\left(V_x - iV_y\right)}{\sqrt{2}} = V_{-1}, \tag{29.13}$$

where, the terms $V_{\pm1}$, V_0 are alternative ways of describing the tensors.

We define a set of irreducible tensors $T(k,q)$, which include spherical harmonics Y_{lm} with the range of values of k and q defined above, as tensors that satisfy the same type of relations as satisfied by Y_{lm}:

$$[J_z, T(k,q)] = q\hbar\, T(k,q), \tag{29.14}$$

$$[J_+, T(k,q)] = \sqrt{(k-q)(k+q+1)}\hbar T(k, q+1), \tag{29.15}$$

$$[J_-, T(k,q)] = \sqrt{(k+q)(k-q+1)}\hbar T(k, q-1). \tag{29.16}$$

The fact that these relations are similar to the relations satisfied by the eigenstates $|j,m\rangle$ when operated by J_z and $J_\pm$ leads one to ask whether the T's also follow the Clebsch–Gordan combination law we discussed in Chapter 28, where we wrote down the combination $|j_1m_1\rangle|j_2m_2\rangle$ in terms of $|jm\rangle$ as

$$|jm\rangle = \sum_{m_1'=-j_1}^{j_1}\sum_{m_2'=-j_2}^{j_2} \langle m_1'm_2'|jm\rangle\, |j_1m_1'\rangle|j_2m_2'\rangle, \tag{29.17}$$

$$|j_1m_1\rangle|j_2m_2\rangle = \sum_{j=|j_1-j_2|}^{j_1+j_2}\sum_{m'=-j}^{j} \langle jm'|m_1m_2\rangle|jm'\rangle, \tag{29.18}$$

where in (29.17) $m_1'+m_2'=m$ while in (29.18) $m_1+m_2=m'$. In other words, we need to investigate whether we can make the following replacements in (29.17):

$$|jm\rangle \to T(k,q), \tag{29.19}$$

$$|j_1m_1\rangle \to T_1(k_1,q_1), \tag{29.20}$$

$$|j_2m_2\rangle \to T_2(k_2,q_2), \tag{29.21}$$

and obtain

$$T(k,q) = \sum_{q_1q_2}\langle q_1q_2|kq\rangle T_1(k_1,q_1)T_2(k_2,q_2) \tag{29.22}$$

where $\langle q_1q_2|kq\rangle$ is the appropriate Clebsch–Gordan coefficient, and q_1 and q_2 sum from $-k_1$ to k_1, and $-k_2$ to k_2, respectively.

Using (29.22) we will show that if T_1 and T_2 are irreducible spherical tensors then so is T. In other words, if $T_1(k_1,q_1)$ and $T_2(k_2,q_2)$ satisfy the relations (29.14), (29.15), and (29.16) with the operators $\mathbf{J}_1$ and $\mathbf{J}_2$, then $T(k,q)$ will have the same properties with $\mathbf{J}=\mathbf{J}_1+\mathbf{J}_2$.

We demonstrate this first for the $J_z(=J_{1z}+J_{2z})$ operator.

$$[J_z,T(k,q)] = \sum_{q_1q_2}\langle q_1q_2|kq\rangle\{[J_{1z},T_1(k_1,q_1)]T_2(k_2,q_2)+T_1(k_1,q_1)[J_{2z},T_2(k_2,q_2)]\}$$

$$= \sum_{q_1q_2}\langle q_1q_2|kq\rangle\,(q_1+q_2)\,T_1(k_1,q_1)T_2(k_2,q_2)$$

$$= qT(k,q) \tag{29.23}$$

where in (29.23) we have used the relation (29.14) for T_1 and T_2, and we have used the fact that $\langle q_1q_2|kq\rangle = 0$ unless $q_1+q_2=q$. Thus, $T(k,q)$ satisfies (29.14).

Let us now consider the raising operator by taking the commutator of (29.22) with $\mathbf{J}_+ (= \mathbf{J}_{1+}+\mathbf{J}_{2+})$,

$$[J_+,T(k,q)] = \sum_{q_1q_2}\langle q_1q_2|kq\rangle\,\{[J_{1+},T_1(k_1,q_1)]T_2(k_2,q_2)+T_1(k_1,q_1)[J_{2+},T_2(k_2,q_2)]\}. \tag{29.24}$$

Using (29.15), the right-hand side in (29.24) is given by

$$\sum_{q_1 q_2} \langle q_1 q_2 | kq \rangle \left\{ \sqrt{(k_1 - q_1)(k_1 + q_1 + 1)} T_1(k_1, q_1 + 1) T_2(k_2, q_2) \right.$$
$$\left. + \sqrt{(k_2 - q_2)(k_2 + q_2 + 1)} T_1(k_1, q_1) T_2(k_2, q_2 + 1) \right\}. \tag{29.25}$$

In the above expression, taking $q_1 = q_1' - 1$ in the first term and $q_2 = q_2' - 1$ in the second term, we obtain from (29.24)

$$[J_+, T(k,q)] = \sum_{q_1 q_2} \langle q_1' - 1, q_2 | kq \rangle \left\{ \sqrt{(k_1 - q_1' + 1)(k_1 + q_1')} T_1(k_1, q_1') T_2(k_2, q_2) \right.$$
$$\left. + \langle q_1, q_2' - 1 | kq \rangle \sqrt{(k_2 - q_2' + 1)(k_2 + q_2')} T_1(k_1, q_1) T_2(k_2, q_2') \right\}. \tag{29.26}$$

Relabeling q_1' and q_2' in the two terms of (29.25) as $q_1$and q_2, respectively, we obtain

$$[J_+, T(k,q)] = \sum_{q_1 q_2} \sqrt{(k_1 - q_1 + 1)(k_1 + q_1)} \langle q_1 - 1, q_2 | kq \rangle$$
$$+ \sqrt{(k_2 - q_2 + 1)(k_2 + q_2)} \} \langle q_1, q_2 - 1 | kq \rangle T_1(k_1, q_1) T_2(k_2, q_2). \tag{29.27}$$

However, from the previous chapter we found that

$$\sqrt{(k_1 - q_1 + 1)(k_1 + q_1)} \langle q_1 - 1, q_2 | kq \rangle + \sqrt{(k_2 - q_2 + 1)(k_2 + q_2)} \langle q_1, q_2 - 1 | kq \rangle$$
$$= \sqrt{(k - q)(k + q + 1)} \langle q_1, q_2 | k, q + 1 \rangle. \tag{29.28}$$

Hence,

$$[J_+, T(k,q)] = \sqrt{(k - q)(k + q + 1)} \sum_{q_1 q_2} \langle q_1, q_2 | kq + 1 \rangle T_1(k_1, q_1) T_2(k_2, q_2), \tag{29.29}$$

which from (29.22) can be written as

$$[J_+, T(k,q)] = \sqrt{(k - q)(k + q + 1)} T(k, q + 1). \tag{29.30}$$

Similarly, one can show that

$$[J_-, T(k,q)] = \sqrt{(k + q)(k - q + 1)} T(k, q - 1). \tag{29.31}$$

Thus from (29.14), (29.15), and (29.16) we conclude that $T(k,q)$ satisfies the properties of an irreducible tensor if $T_1(k_1, q_1)$ and $T_2(k_2, q_2)$ do.

Just as (29.18) is obtained by inverting (29.17), we can write the following by inverting (29.22),

$$T_1(k_1,q_1)T_2(k_2,q_2) = \sum_k \sum_{q'} \langle kq'|q_1q_2\rangle T(k,q') \tag{29.32}$$

where k runs from $|k_1 - k_2|$ to $(k_1 + k_2)$ and q' runs from $-k$ to $+k$.

This shows that the product of two irreducible spherical tensors is a linear combination of irreducible tensors.

29.2 The irreducible tensors: $Y_{lm}(\theta,\phi)$ and $D^j(\chi)$

In the context of irreducible tensors let us discuss certain important properties of the spherical harmonics $Y_{lm}(\theta,\phi)$ and the matrix element of the rotation operator, $D^j(\chi)$. We note from our previous calculations that

$$|jm\rangle' = e^{-i\chi\mathbf{n}\cdot\mathbf{J}/\hbar}|jm\rangle = \sum_{m'} c^j_{m'm}|jm'\rangle. \tag{29.33}$$

By taking an infinitesimal value $\chi = \epsilon$ and expanding $e^{-i\chi\mathbf{n}\cdot\mathbf{J}/\hbar}$ we note that we will have terms of the type $J_i|jm\rangle$ in the above equation, which will result in eigenstates with different m-values but the value of j will not change. Hence the right-hand side of (29.33) will contain only a sum over m'. If we define

$$D^j_{m'm}(\chi) = \langle jm'|e^{-i\chi\mathbf{n}\cdot\mathbf{J}/\hbar}|jm\rangle, \tag{29.34}$$

then using the orthogonality of $|jm\rangle$ we find

$$c^j_{m'm} = \langle jm'|e^{-i\chi\mathbf{n}\cdot\mathbf{J}/\hbar}|jm\rangle = D^j_{m'm}(\chi) \tag{29.35}$$

and write (29.33) as

$$|jm\rangle' = \sum_{m'} D^j_{m'm}(\chi)|jm'\rangle. \tag{29.36}$$

We take $j = l$, the orbital angular momentum, in (29.36) and in the θ,ϕ space write

$$\langle\theta\phi|lm\rangle' = \sum_{m'} D^l_{m'm}(\chi)\langle\theta\phi|lm'\rangle. \tag{29.37}$$

Hence,

$$Y_{lm}\left(\theta',\phi'\right) = \sum_{m'} D^l_{m'm}(\chi)\, Y_{lm'}(\theta,\phi) \tag{29.38}$$

where θ',ϕ' are the rotated angles.

Let us now demonstrate the orthogonality property of $D^j_{m'm}(\chi)$ by starting with the identity

$$e^{i\chi\mathbf{n}\cdot\mathbf{J}/\hbar}e^{-i\chi\mathbf{n}\cdot\mathbf{J}/\hbar} = 1. \tag{29.39}$$

Sandwiching the above operator between states $|jm\rangle$ and $|jm''\rangle$ and inserting a complete set of states we find

$$\sum_{m'} \langle jm''|e^{i\chi\mathbf{n}\cdot\mathbf{J}/\hbar}|jm'\rangle\langle jm'|e^{-i\chi\mathbf{n}\cdot\mathbf{J}/\hbar}|jm\rangle = \delta_{m''m}. \tag{29.40}$$

Thus, from (29.34) we have

$$\sum_{m'} D^{*j}_{m''m}(\chi)D^j_{m'm}(\chi) = \delta_{m''m}, \tag{29.41}$$

which establishes the orthogonality. Multiplying both sides of (29.38) by $D^{*l}_{mm'}$ and using this orthogonality property of D^j we can invert relation (29.38) to obtain

$$Y_{lm}(\theta,\phi) = \sum_m D^{*l}_{mm'}(\chi)\, Y_{lm'}(\theta',\phi'). \tag{29.42}$$

If we let $\theta' = 0$ and note that

$$Y_{lm'}(0,\phi') = \sqrt{\frac{2l+1}{4\pi}}\delta_{m'0} \tag{29.43}$$

then relation (29.40) yields

$$Y_{lm}(\theta,\phi) = D^{*l}_{m0}(\chi)\sqrt{\frac{2l+1}{4\pi}}. \tag{29.44}$$

This is an important relation connecting $Y_{lm}(\theta,\phi)$ and $D^{*l}_{m0}(\chi)$.

Consider now a combination of momentum operators $\mathbf{J}_1$ and $\mathbf{J}_2$,

$$\mathbf{J} = \mathbf{J}_1 + \mathbf{J}_2, \tag{29.45}$$

and write

$$e^{-i\chi\mathbf{n}\cdot\mathbf{J}/\hbar} = e^{-i\chi\mathbf{n}\cdot\mathbf{J}_1/\hbar}e^{-i\chi\mathbf{n}\cdot\mathbf{J}_2/\hbar}. \tag{29.46}$$

We take the matrix elements of both sides of the above with respect to

$$|j_1j_2jm\rangle = |jm\rangle. \tag{29.47}$$

The left-hand side of (29.46) will then give

$$D^j_{m'm}(\chi) = \langle jm'|e^{-i\chi\mathbf{n}\cdot\mathbf{J}/\hbar}|jm\rangle. \tag{29.48}$$

On the right-hand side we use the result

$$|jm\rangle = \sum_{m_1 m_2} \langle m_1 m_2 | jm\rangle |m_1 m_2\rangle, \tag{29.49}$$

$$|jm'\rangle = \sum_{m_1' m_2'} \langle m_1' m_2' | jm'\rangle |m_1' m_2'\rangle, \tag{29.50}$$

to obtain

$$\begin{aligned}\sum_{m_1' m_2'} \sum_{m_1 m_2} &\langle jm' | m_1' m_2'\rangle \langle m_1' m_2' | e^{-i\chi \mathbf{n}\cdot\mathbf{J}_1/\hbar} e^{-i\chi \mathbf{n}\cdot\mathbf{J}_2/\hbar} |m_1 m_2\rangle \langle m_1 m_2 | jm\rangle \\ &= \sum_{m_1' m_2'} \sum_{m_1 m_2} \langle jm' | m_1' m_2'\rangle \langle j_1 m_1' | e^{-i\chi \mathbf{n}\cdot\mathbf{J}_1/\hbar} | j_1 m_1\rangle \langle j_2 m_2' | e^{-i\chi \mathbf{n}\cdot\mathbf{J}_2/\hbar} | j_2 m_2\rangle \langle m_1 m_2 | jm\rangle\end{aligned} \tag{29.51}$$

where in the last step we have used the relations

$$|m_1 m_2\rangle = |j_1 m_1\rangle |j_2 m_2\rangle, \tag{29.52}$$

$$|m_1' m_2'\rangle = |j_1 m_1'\rangle |j_2 m_2'\rangle. \tag{29.53}$$

We thus obtain from (29.48)

$$D^j_{m'm}(\chi) = \sum_{m_1' m_2'} \sum_{m_1 m_2} \langle jm' | m_1' m_2'\rangle D^{j_1}_{m_1' m_1}(\chi)\, D^{j_2}_{m_2' m_2}(\chi)\, \langle m_1 m_2 | jm\rangle, \tag{29.54}$$

which is the counterpart of a similar relation for the spherical tensors $T(k,q)$ discussed in the previous section.

Similarly, one can write

$$\begin{aligned}\langle m_1' m_2' | &e^{-i\chi \mathbf{n}\cdot\mathbf{J}_1/\hbar} e^{-i\chi \mathbf{n}\cdot\mathbf{J}_2/\hbar} |m_1 m_2\rangle \\ &= \langle m_1' m_2' | e^{-i\chi \mathbf{n}\cdot\mathbf{J}/\hbar} |m_1 m_2\rangle \\ &= \sum_{jmm'} \langle m_1' m_2' | jm'\rangle \langle jm' | e^{-i\chi \mathbf{n}\cdot\mathbf{J}/\hbar} | jm\rangle \langle jm | m_1 m_2\rangle.\end{aligned} \tag{29.55}$$

Hence,

$$D^{j_1}_{m_1' m_1}(\chi)\, D^{j_2}_{m_2' m_2}(\chi) = \sum_{jmm'} \langle m_1' m_2' | jm'\rangle D^j_{m'm}(\chi)\, \langle jm | m_1 m_2\rangle \tag{29.56}$$

which is a counterpart of the relation for the spherical tensors $T(k,q)$.

To obtain a relation between the spherical harmonics that are also irreducible tensors, we take account of the relation (29.44) and put $m_1 = 0$ and $m_2 = 0$ in (29.56) and find, after replacing j_1, j_2 and j by $l_1 l_2$ and l,

$$D^{l_1}_{m'_1 0}(\chi)\, D^{l_2}_{m'_2 0}(\chi) = \sum_{lmm'} \langle m'_1 m'_2 | lm' \rangle D^{l}_{m'}(\chi)\, \langle lm|00\rangle. \tag{29.57}$$

Changing $m'_1, m'_2 \to m_1, m_2$ and $m' \to m = m_1 + m_2$, we obtain from (29.57)

$$Y_{l_1 m_1}(\theta,\phi)\, Y_{l_2 m_2}(\theta,\phi)$$

$$= \sum_l \sqrt{\frac{(2l_1+1)(2l_2+1)}{4\pi(2l+1)}} \langle m_1 m_2 | l, m_1+m_2\rangle Y_{l,m_1+m_2}(\theta,\phi)\, \langle lm|00\rangle. \tag{29.58}$$

29.3 Wigner–Eckart theorem

Let us consider the matrix element of expression (29.30),

$$\langle j'm' |\, [J_+, T(k,q)]\, | jm \rangle = \sqrt{(k-q)(k+q+1)} \langle j'm' |\, T(k,q+1)\, | jm \rangle. \tag{29.59}$$

To determine the left-hand side we note that

$$J_+ \, |jm\rangle = \sqrt{(j-m)(j+m+1)}\, |jm+1\rangle \tag{29.60}$$

and

$$\left[\langle j'm' |\, J_+\right] = \left[J_- \, |j'm'\rangle\right]^\dagger = \sqrt{(j'+m')(j'-m'+1)} \langle j'm'-1 |. \tag{29.61}$$

Thus, after bringing all the terms to the left, (29.59) can be written as

$$\sqrt{(j'+m')(j'-m'+1)} \langle j', m'-1 |\, T(k,q)\, | jm\rangle$$

$$- \sqrt{(j-m)(j+m+1)} \langle j'm' |\, T(k,q)\, | j, m+1\rangle$$

$$- \sqrt{(k-q)(k+q+1)} \langle j'm' |\, T(k,q+1)\, | jm\rangle = 0. \tag{29.62}$$

Consider now the recursion relation for the Clebsch–Gordan coefficients given by (28.37), which we write as follows, after taking the complex conjugate,

$$\sqrt{(j+m)(j-m+1)} \langle jm-1 | m_1, m_2\rangle - \sqrt{(j_1-m_1)(j_1+m_1+1)} \langle jm | m_1+1, m_2\rangle$$

$$- \sqrt{(j_2-m_2)(j_2+m_2+1)} \langle jm | m_1, m_2+1\rangle = 0. \tag{29.63}$$

We make the following replacements in (29.63) in order to compare with (29.62).

$$j \to j', \quad m \to m', \quad m_1 \to m, \quad j_1 \to j, \quad m_2 \to q, \quad j_2 \to k. \tag{29.64}$$

The relation (29.63) then becomes

$$\sqrt{(j'+m')(j'-m'+1)}\langle j'm'-1|m,q\rangle - \sqrt{(j-m)(j+m+1)}\langle j'm'|m+1,q\rangle$$

$$-\sqrt{(k-q)(k+q+1)}\langle j'm'|m,q+1\rangle = 0. \quad (29.65)$$

If we now compare (29.62) and (29.65), we note that they are both homogeneous equations of the form

$$A_1x + A_2y + A_3z = 0, \quad (29.66)$$

$$B_1x + B_2y + B_3z = 0. \quad (29.67)$$

We take

$$B_i = \lambda A_i \quad \text{for} \quad i = 1, 2, 3 \quad (29.68)$$

where λ is a constant independent of i. Substituting B_i in (29.67) we obtain the equation (29.66). Hence (29.68) provides a possible solution to both equations.

Applying this result to (29.62) and (29.65), where the A_i and B_i terms differ in their dependence on m', q, and m, we obtain

$$\langle j'm'-1|\, T(k,q)\, |jm\rangle = \lambda\langle j'm'-1|m,q\rangle, \quad (29.69)$$

$$\langle j'm'|\, T(k,q)\, |jm+1\rangle = \lambda\langle j'm'|m+1,q\rangle, \quad (29.70)$$

$$\langle j'm'|\, T(k,q+1)\, |jm\rangle = \lambda\langle j'm'|m,q+1\rangle, \quad (29.71)$$

where λ does not depend on m', q, or m.

We can summarize the above results by writing

$$\langle j'm'|\, T(k,q)\, |jm\rangle = \lambda\langle j'm'|mq\rangle \quad (29.72)$$

where m', q, or m take different values as indicated by equations (29.69), (29.70), and (29.71). One generally writes

$$\lambda = \langle j'\|T(k)\|j\rangle \quad (29.73)$$

where $\langle j'\|T(k))\|j\rangle$ is called the reduced matrix element of the spherical tensor $T(k,q)$, which is independent of m', q, or m. Hence,

$$\langle j'm'|\, T(k,q)\, |jm\rangle = \langle j'\|T(k)\|j\rangle\, \langle j'm'|mq\rangle. \quad (29.74)$$

This is the statement of the Wigner–Eckart theorem, according to which the matrix element $\langle j'm'|\, T(k,q)\, |jm\rangle$ is expressed as a product of two terms, the reduced matrix element that depends only on j' and j and therefore describes the overall physical properties of the spherical tensor, and (the already known) Clebsch–Gordan coefficient $\langle j'm'|mq\rangle$ that describes the geometrical content of the matrix element since it depends on the quantum numbers m', q, and m that determine the orientation of the operators.

Equation (29.74) tells us, therefore, that, for a given j', j, and k, if we know the matrix element $\langle j'm' | T(k,q) | jm \rangle$ for one specific combination of m', m, and q then the rest of the matrix elements are determined by the Clebsch–Gordan coefficients $\langle j'm' | mq \rangle$. It allows considerable simplifications in calculating the matrix elements and leads to many interesting selection rules.

29.4 Applications of the Wigner–Eckart theorem

Below we consider some simple applications of the Wigner–Eckart theorem.

(i) Because the Clebsch–Gordan coefficients $\langle j'm' | mq \rangle$ vanish unless $\mathbf{J}' = \mathbf{J} + \mathbf{k}$ and $m' = m + q$, we have

$$\langle j'm' | T(k,q) | jm \rangle = 0 \tag{29.75}$$

unless

$$|j' - j| \leq k \leq j' + j \quad \text{and} \quad q = m' - m. \tag{29.76}$$

(ii) Specifically, for a scalar operator $T(0,0) = S$ with $k = 0$ and $q = 0$,

$$\langle j'm' | S | jm \rangle = 0 \tag{29.77}$$

unless

$$\nabla j = j' - j = 0 \quad \text{and} \quad \nabla m = m' - m = 0. \tag{29.78}$$

(iii) For a vector operator $T(1,q) = V_q$ with $k = 1$ and $q = 0, \pm 1$,

$$\langle j'm' | V_q | jm \rangle = 0 \tag{29.79}$$

unless

$$\nabla j = j' - j = 0, \pm 1 \quad \text{and} \quad \nabla m = m' - m = 0, \pm 1. \tag{29.80}$$

For the case $j' = j = 0$, however, $\langle j'm' | V_q | jm \rangle$ vanishes identically because the relation $\mathbf{J}' = \mathbf{J} + \mathbf{k}$ is not satisfied as $|\mathbf{k}| = 1$.

(iv) If $Y_{kq}(\theta, \phi)$ is a spherical harmonic then

$$\langle j'm' | Y_{kq}(\theta, \phi) | jm \rangle = 0 \tag{29.81}$$

unless

$$|j' - j| \leq k \leq j' + j \quad \text{and} \quad |m' - m| \leq q \leq m' + m. \tag{29.82}$$

A state with total angular moment j that has a 2^k-electric or magnetic moment gives rise to the matrix element

$$\langle jm|\, Y_{kq}(\theta,\phi)\, |jm\rangle\,. \tag{29.83}$$

The above result then shows that one must have $k \leq 2j$. Thus, a scalar particle, with $j=0$, and therefore $k=0$, cannot have a magnetic dipole moment.

We consider some examples that involve the matrix elements of $\mathbf{J}$.

(i) In particular we evaluate $\langle j\|\mathbf{J}\|j\rangle$. According to the Wigner–Eckart theorem,

$$\langle jm'|\, J_0\, |jm\rangle = \langle j\|\mathbf{J}\|j\rangle\, \langle jm'|m0\rangle \tag{29.84}$$

where J_0 is the spherical component that equals J_z. Let us take $m'=m=j$, then

$$\langle jj|\, J_0\, |jj\rangle = \langle j\|\mathbf{J}\|j\rangle\, \langle jj|j0\rangle. \tag{29.85}$$

The Clebsch–Gordan term is given by

$$\langle jj|j0\rangle = \sqrt{\frac{j}{j+1}}. \tag{29.86}$$

Also,

$$J_0\, |jj\rangle = J_z\, |jj\rangle = j\, |jj\rangle\,. \tag{29.87}$$

Hence,

$$\langle j\|\mathbf{J}\|j\rangle = \sqrt{j\,(j+1)}. \tag{29.88}$$

The reduced matrix element of the angular momentum operator is thus determined.

(ii) Let us now consider a general vector operator $\mathbf{V}$. Since both $\mathbf{V}$ and $\mathbf{J}$ are vectors, they will have the same factor for the Clebsch–Gordan coefficient when their matrix elements are expressed in the form dictated by the Wigner–Eckart theorem. Hence

$$\frac{\langle jm'|\, V_q\, |jm\rangle}{\langle jm'|\, J_q\, |jm\rangle} = \frac{\langle j\|\mathbf{V}\|j\rangle}{\langle j\|\mathbf{J}\|j\rangle} = \frac{\langle j\|\mathbf{V}\|j\rangle}{\sqrt{j\,(j+1)}}. \tag{29.89}$$

In the last step we have utilized the result in (29.88).

Next let us consider the matrix element $\langle jm'|\, \mathbf{J}.\mathbf{V}\, |jm\rangle$ written in terms of spherical components,

$$\langle jm'|\, \mathbf{J}.\mathbf{V}\, |jm\rangle = \langle jm'|\, (J_0V_0 - J_{+1}V_{-1} - J_{-1}V_{+1})\, |jm\rangle\,. \tag{29.90}$$

Since we already know how to evaluate $\langle jm'|\, J_q$, the above matrix element will be a linear combination of $\langle jm'|\, V_q\, |jm\rangle$ with $q = 0, \pm 1$, and where, according to the Wigner–Eckart theorem,

$$\langle jm'|\, V_q\, |jm\rangle = \langle j\|\mathbf{V}\|j\rangle\, \langle jm'|mq\rangle. \tag{29.91}$$

Hence in (29.90), $\langle j\|\mathbf{V}\|j\rangle$ will be a common term and can be factored out. We note, however, that $\mathbf{J.V}$ is a scalar operator and as such its matrix element will not depend on the nature of $\mathbf{V}$. Thus, the term that will multiply $\langle j\|\mathbf{V}\|j\rangle$ in (29.90) must be independent of m', m, and q, leading to the relation

$$\langle jm'|\,\mathbf{J.V}\,|jm\rangle = c_j\,\langle j\|\mathbf{V}\|j\rangle\,. \tag{29.92}$$

To evaluate c_j we take $\mathbf{V} = \mathbf{J}$ in (29.92) and obtain

$$\langle jm'|\,\mathbf{J}^2\,|jm\rangle = c_j\,\langle j\|\mathbf{J}\|j\rangle\,. \tag{29.93}$$

Since $\mathbf{J}^2\,|jm\rangle = j\,(j+1)\,|jm\rangle$, and since $\langle j\|\mathbf{J}\|j\rangle$ has already been found in (29.88) to be $\sqrt{j\,(j+1)}$, we obtain

$$c_j = \sqrt{j\,(j+1)}. \tag{29.94}$$

The relation (29.92) can then be written as

$$\langle jm'|\,\mathbf{J.V}\,|jm\rangle = \sqrt{j\,(j+1)}\,\langle j\|\mathbf{V}\|j\rangle\,. \tag{29.95}$$

Expression (29.89), after substituting for $\langle j\|\mathbf{V}\|j\rangle$ from (29.95), can be written as

$$\langle jm'|\,V_q\,|jm\rangle = \frac{\langle jm'|\,\mathbf{J.V}\,|jm\rangle}{j(j+1)}\,\langle jm'|\,J_q\,|jm\rangle\,. \tag{29.96}$$

This is called the projection theorem, which has applications in spectroscopy.

(ii) We consider, as an example, a magnetic moment operator

$$\boldsymbol{\mu} = g_1\mathbf{J}_1 + g_2\mathbf{J}_2 \quad \text{where } \mathbf{J} = \mathbf{J}_1 + \mathbf{J}_2. \tag{29.97}$$

Let us calculate the diagonal matrix element $\langle jj|\,\mu_q\,|jj\rangle$. Specifically, we take $q = 0$. We then have, from the projection theorem,

$$\langle jj|\,\mu_0\,|jj\rangle = \frac{\langle jj|\,\mathbf{J}.\boldsymbol{\mu}\,|jj\rangle}{j(j+1)}\,\langle jj|\,J_0\,|jj\rangle = \frac{\langle jj|\,g_1\mathbf{J.J}_1 + g_2\mathbf{J.J}_2\,|jj\rangle}{j(j+1)}\,\langle jj|\,J_0\,|jj\rangle\,. \tag{29.98}$$

From the relation $\mathbf{J} = \mathbf{J}_1 + \mathbf{J}_2$, we obtain

$$\mathbf{J.J}_1 = \frac{1}{2}\left(\mathbf{J}^2 + \mathbf{J}_1^2 - \mathbf{J}_2^2\right), \quad \mathbf{J.J}_2 = \frac{1}{2}\left(\mathbf{J}^2 - \mathbf{J}_1^2 + \mathbf{J}_2^2\right). \tag{29.99}$$

If $|jj\rangle$ are eigenstates of $\mathbf{J}_1^2$ and $\mathbf{J}_2^2$ in addition to $\mathbf{J}^2$, then substituting (29.95) in (29.98) and using the fact that $\langle jj|\,J_0\,|jj\rangle = j$, we obtain

$$\langle jj|\,\mu_0\,|jj\rangle = \frac{j(j+1)\,(g_1+g_2) + j_1(j_1+1)\,(g_1-g_2) + j_2(j_2+1)\,(g_2-g_1)}{2(j+1)}. \tag{29.100}$$

29.5 Appendix to Chapter 29: $SO(3)$, $SU(2)$ groups and Young's tableau

29.5.1 Constructing irreducible tensors

In Chapter 27 we discussed the groups $SO(3)$ and $SU(2)$ in connection with the representation of the rotation operators. We return to these two groups and their relation to the irreducible tensors, which we will define below.

Tensors belong to a set of elements that transform in a specific way. The vectors V_i $(i = 1, 2, 3)$ are a special case of tensors. They are often referred to as first-rank tensors. Higher-rank tensors are written as T_{ij}, T_{ijk} . . ., with $i, j, k = 1, 2, 3$.

$SO(3)$

The rotation matrix $R_z(\phi)$ that we described in Section 5.1 belongs to a class of transformations that have certain unique properties. We discuss these below in the context of the abstract group theory.

Let us consider a 3×3 transformation matrix, denoted by $\{\alpha\}$, that has real matrix elements and satisfies the properties

$$\alpha\tilde{\alpha} = \mathbf{1}, \tag{29.101}$$

$$\det \alpha = 1. \tag{29.102}$$

We note that the rotation matrices described in Chapter 26 exhibit the same properties. The transformation of a vector with components A_i $(i = 1, 2, 3)$ will then be written as

$$A'_i = \alpha_{ij}A_j, \qquad i, j = 1, 2, 3 \tag{29.103}$$

where A_i's are assumed to be real numbers and where we use the convention that repeated indices imply summation. Thus, a summation over j is implied in (29.103). A product of two vectors A_iB_j transforms as

$$A'_iB'_j = \alpha_{ik}A_k\alpha_{jl}B_l = \alpha_{ik}\alpha_{jl}A_kB_l. \tag{29.104}$$

Defining a second-rank tensor $T_{ij} = A_iB_j$ as a product of two vectors, the above relation becomes

$$T'_{ij} = \alpha_{ik}\alpha_{jl}T_{kl}. \tag{29.105}$$

One can form higher-dimensional tensors by constructing products of vectors, e.g., $T_{ijk\ldots} = A_iB_jC_k \ldots$ and their transformation is given by

$$T'_{ijk\ldots} = \alpha_{ia}\alpha_{jb}\alpha_{kc} \ldots T_{abc\ldots}. \tag{29.106}$$

In a second-rank tensor, T_{ij}, a total of nine elements are involved since i and j take on values $1, 2, 3$. However, we will show that certain linear combinations of these elements transform among themselves. These combinations are the irreducible tensors.

To obtain these linear combinations we first consider some special tensors that we have defined in the previous chapters: δ_{ij}, the Kronecker delta, and ϵ_{ijk}, the totally antisymmetric tensor. Their transformations are given from (29.103) as follows:

$$\delta'_{ij} = \alpha_{ik}\alpha_{jl}\delta_{kl} = \alpha_{ik}\alpha_{jk} = (\alpha\tilde{\alpha})_{ij}\,. \tag{29.107}$$

However, since α_{ij} satisfies orthonormality, $(\alpha\tilde{\alpha})_{ij} = \delta_{ij}$, we have

$$\delta'_{ij} = \delta_{ij}. \tag{29.108}$$

Thus δ_{ij} is "form invariant," i.e., its matrix elements remain unchanged after the transformation.

As for ϵ_{ijk}, the transformation is given by

$$\epsilon'_{ijk} = \alpha_{il}\alpha_{jm}\alpha_{kn}\epsilon_{lmn}. \tag{29.109}$$

From a rather simple but slightly laborious calculation, it is possible to show that the right-hand side in (29.109) is $(\det\alpha)\,\epsilon_{ijk}$. From (29.102), since $\det\alpha = 1$, we find that

$$\epsilon'_{ijk} = \epsilon_{ijk}. \tag{29.110}$$

Thus ϵ_{ijk} is also "form invariant."

We can now use δ_{ij} and ϵ_{ijk} to form special linear combinations that have unique transformation properties.

If we now consider the tensor A_iB_j and multiply the product by δ_{ij} and sum over the indices i and j, we obtain

$$S = \delta_{ij}A_iB_j \tag{29.111}$$

where the summation convention is used so that both i and j are summed over. This sum, denoted by S, transforms back to itself as we show below,

$$S' = \delta'_{ij}A'_iB'_j = \delta_{ij}A'_iB'_j = A'_iB'_i = (\tilde{\alpha}\alpha)_{kj}\,A_kB_j \tag{29.112}$$

where we have used the transformation property (29.103). Since $(\tilde{\alpha}\alpha)_{kj} = \delta_{kj}$, we obtain

$$S' = A'_iB'_i = A_jB_j = S. \tag{29.113}$$

Hence the linear combination $\delta_{ij}A_iB_j$ transforms back to itself, i.e.,

$$A'_1B'_1 + A'_2B'_2 + A'_3B'_3 = A_1B_1 + A_2B_2 + A_3B_3. \tag{29.114}$$

The linear combination, S, is thus said to form a scalar under the group transformation (29.112). Similarly, one can show that $V_i = \epsilon_{ijk}A_jB_k$ transforms as a vector with the transformation property given by (29.103). In the context of ordinary three-dimensional vectors, S corresponds to the dot product $\mathbf{A} \cdot \mathbf{B}$ and V_i corresponds to the ith component of the cross-product $\mathbf{A} \times \mathbf{B}$.

Thus δ_{ij} and ϵ_{ijk} act as "reducing agents" for this group by reducing the original dimensionality to a lower one, so that out of the nine elements of the tensor T_{ij} one linear combination transforms as a scalar and three linear combinations transform as the three components of a vector. The remaining terms of the tensor A_iB_j must be such that they cannot be reduced further by δ_{ij} or ϵ_{ijk}. That is, for these terms $\delta_{ij}A_iB_j$ and $\epsilon_{ijk}A_jB_k$ must vanish:

$$\delta_{ij}A_iB_j = 0 \quad \text{and} \quad \epsilon_{ijk}A_jB_k = 0. \tag{29.115}$$

Hence the matrix formed by the remaining terms of T_{ij} must be traceless and symmetric, which accounts for the five terms.

Thus the nine elements of A_iB_j in $SO(3)$ can be expressed in terms of tensors that cannot be reduced further, called "irreducible tensors," made up of a scalar, a vector, and a symmetric traceless matrix with the number of components (i.e., the number of linear combinations) 1, 3, and 5 respectively. This result is written symbolically as

$$\mathbf{3} \otimes \mathbf{3} = \mathbf{1} \oplus \mathbf{3} \oplus \mathbf{5}. \tag{29.116}$$

The irreducible tensors we have determined are called "spherical tensors" since they correspond to spherical harmonics with $l = 0, 1, 2$, where for each l there are $(2l+1)$ independent terms.

If we had started with $SO(2)$ with a simpler 2×2 transformation matrix with vectors A_i $(i = 1, 2)$, the only reducing agent available would be δ_{ij} since ϵ_{ijk} would have required a minimum of three dimensions. In this case it is easy to check that

$$\mathbf{2} \otimes \mathbf{2} = \mathbf{1} \oplus \mathbf{3}. \tag{29.117}$$

These would correspond to $l = 0, 1$ of the spherical harmonics.

$SU(2)$

Here our basic transformation matrix $\{\alpha\}$ is a unitary 2×2 matrix with complex matrix elements,

$$\alpha\alpha^\dagger = \mathbf{1}. \tag{29.118}$$

The transformation of a vector is given by

$$A_i' = \alpha_{ij}A_j, \qquad i,j = 1, 2. \tag{29.119}$$

We take the complex conjugates of both sides and write

$$A_i^{*\prime} = \alpha_{ij}^* A_j^*. \tag{29.120}$$

However, because of their unitary nature, the α's satisfy

$$\alpha_{ij}^* = \left(\alpha^{-1}\right)_{ji}. \tag{29.121}$$

To simplify notations, let us write complex conjugates of the vectors as

$$A_i^* = A^i. \tag{29.122}$$

Thus the transformation (29.120) can be written as

$$A'^i = \left(\alpha^{-1}\right)_{ji} A^j = A^j \left(\alpha^{-1}\right)_{ji}. \tag{29.123}$$

The last step above facilitates writing the product as a matrix multiplication.

Let us now consider the tensor

$$T_j^i = A^i B_j. \tag{29.124}$$

We will call it a mixed tensor because it has one superscript and one subscript. It will transform as

$$T_j'^i = \left(\alpha^{-1}\right)_{ki} \alpha_{jl} T_l^k. \tag{29.125}$$

To obtain irreducible combinations let us consider, once again, the Kronecker delta, δ_j^i, which is constructed as a mixed tensor and transforms as

$$\delta_j'^i = \left(\alpha^{-1}\right)_{ki} \alpha_{jl} \delta_l^k. \tag{29.126}$$

The right-hand side above simply becomes $\left(\alpha^{-1}\right)_{ki} \alpha_{jk} = \alpha_{jk} \left(\alpha^{-1}\right)_{ki} = \left(\alpha\alpha^{-1}\right)_{ji}$ and, since $\alpha\alpha^{-1} = \mathbf{1}$, it behaves like a Kronecker delta. Hence,

$$\delta_j'^i = \delta_j^i. \tag{29.127}$$

Thus, δ_j^i is form invariant. As with $SO(3)$, one can then show that $\delta_i^j A^i B_j$ transforms as a scalar. There is no counterpart of ϵ_{ijk} in the 2×2 system because i, j, and k can only take on the values 1 or 2 so at least two of the three indices of ϵ_{ijk} will be the same, in which case ϵ_{ijk} vanishes identically.

Thus the irreducible elements of $A^i B_j$ are obtained by taking a linear combination that is a scalar with the remaining elements that satisfy $\delta_i^j A^i B_j = 0$ and which are, therefore, matrix elements of a traceless matrix. Here for $SU(2)$ with complex elements one writes symbolically

$$\mathbf{2} \times \mathbf{2}^* = \mathbf{1} + \mathbf{3}. \tag{29.128}$$

For $SU(3)$ with complex elements $A^i B_j$ $(i,j = 1,2,3)$ the only reducing agent available is δ_i^j. One cannot construct a tensor of the type ϵ_{ijk} that could act as a reducing agent. Hence one finds

$$\mathbf{3} \times \mathbf{3}^* = \mathbf{1} + \mathbf{8}. \tag{29.129}$$

29.5.2 Young's tableau

This method provides a simple pictorial way to construct irreducible combinations. We apply it to $SU(2)$ systems. Since $SO(3)$, the group that concerns us, is isomorphic to $SU(2)$, we can use the results we outline below (without proof) for $SO(3)$ as well.

The lowest-dimensional matrices, the "building blocks" of $SU(2)$, are 2×2 matrices, one of the representations of which is $D^{\frac{1}{2}}(R)$, which we have already discussed for spin $\frac{1}{2}$ particles. Let us symbolize a spin $\frac{1}{2}$ system as simple squares:

$$\square\,.$$

This is then a doublet. It consist of spin-up and spin-down states. Below we describe the process by which higher symmetries are built.

29.5.3 Two spin ½ particles

We start by making the designations

$$\boxed{1} = \text{spin-up} \qquad \boxed{2} = \text{spin-down}$$

and consider a system with two boxes. The combinations $\square\ \square$ are made up of spin-up and spin-down states. Specifically, they are made up of the following combinations:

$$\boxed{1}\ \boxed{1}$$
$$\boxed{1}\ \boxed{2}$$
$$\boxed{2}\ \boxed{2}\,.$$

The combination $\boxed{1}\ \boxed{2}$ is assumed to be symmetric. Thus $\boxed{2}\ \boxed{1}$ is considered redundant. This leads to one of the rules of the Young's tableau.

Rule 1: The numbers from left to right cannot decrease.

We therefore have three possibilities as indicated above. All three possibilities are indicated by a single pair of boxes

$$\square\ \square$$

without any numbers inserted. This is then a triplet with $j = 1$. Another combination of boxes gives, according to the rules,

$$\begin{array}{c}\boxed{1}\\ \boxed{2}\end{array}\,.$$

This is an antisymmetric state. The rule here is that

Rule 2: Numbers must increase in group from the top to the bottom. Since there is only one possibility, this is a singlet state with $j = 0$.

To obtain the combination of two boxes, which we write as $\square \otimes \square$, one can move boxes around and make newer systems as long as they are consistent with the rules indicated above. We then find

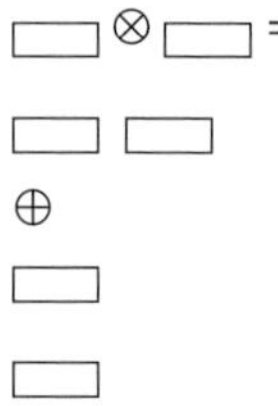

$$\square \otimes \square =$$

$$\square\,\square \tag{29.130}$$

$$\oplus$$

$$\begin{array}{c}\square\\ \square\end{array} \tag{29.131}$$

which is of the form

$$\left(j = \frac{1}{2}, \text{doublet}\right) \otimes \left(j = \frac{1}{2}, \text{doublet}\right) = (j = 1, \text{triplet}) \oplus (j = 0, \text{singlet}),$$

or

$$2 \otimes 2 = 3 \oplus 1.$$

We note that for the addition of two doublets in $SU(2)$, which we are considering here, there are only two rows possible, since

$$\begin{array}{c}\square\\ \square\\ \square\end{array}$$

will imply antisymmetrization between three spin $\frac{1}{2}$ states, which is impossible. For example, in terms of $\boxed{1}$ and $\boxed{2}$ we note that the only way to have the blocks in a vertical array is to have terms like

$$\begin{array}{c}\boxed{1}\\ \boxed{2}\\ \boxed{2}\end{array}$$

which is zero since antisymmetrization between $\boxed{2}$ and $\boxed{2}$ is impossible.

Thus there are the following additional rules.

Rule 3: The number of blocks as one goes down the rows cannot be larger than the number of blocks in the row immediately above.

Rule 4: The dimensionality of any diagram is

$$D(\lambda) = 1 + \lambda \tag{29.132}$$

where λ is the number of blocks in the top row minus the number of blocks in the bottom row. Since the dimensionality of the system is also $2j+1$, we have

$$D(\lambda) = 2j + 1. \tag{29.133}$$

The rule about the dimensionality, once again, is applicable only to $SU(2)$. For higher systems such as $SU(3), \ldots, SU(n)$ this particular rule becomes more involved. Thus,

$$\square\ \square \qquad D(\lambda) = 3,\ j = 1$$

$$\begin{matrix}\square \\ \square .\end{matrix} \qquad D(\lambda) = 1,\ j = 0$$

29.5.4 Higher systems

Let us consider the addition of $j_1 = 1$ and $j_2 = 1$, each being denoted by $\square\ \square$. In forming the product states, we note that the individual blocks, $\square$, in the combination can be moved around, consistent with the above rules. That is, even though we have

$$\square\ \square \otimes \square\ \square$$

the individual single blocks can be moved around. We then get the following:

(i) All four blocks in the same row;

$$\square\ \square\ \square\ \square .$$

This is a completely symmetric state, it has $\lambda = 4$ and hence $D(\lambda) = 5$. Therefore, $2j+1 = 5$ and $j = 2$.

(ii) Take one of the blocks and move it down just below the first block:

$$\begin{matrix}\square & \square & \square \\ \square . & & \end{matrix}$$

For this system, $\lambda = 3 - 1 = 2$ and hence $D(\lambda) = 3$. That is, $2j + 1 = 3$ and $j = 1$.

(iii) Take the last block in the top row of (ii) and move it down just below the second block:

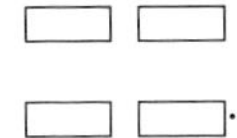

Here $\lambda = 2 - 2 = 0$ and hence $D(\lambda) = 1$. This corresponds to $j = 0$, a singlet. Thus we have

$$\square\,\square \otimes \square\,\square =$$

$$\square\,\square\,\square\,\square$$

$$\oplus$$

$$\begin{array}{l}\square\,\square\,\square\\ \square\end{array}$$

$$\oplus$$

$$\begin{array}{l}\square\,\square\\ \square\,\square\end{array}$$

which expresses the decomposition

$$(j = 1, \text{triplet}) \otimes (j = 1, \text{triplet}) = (j = 2, \text{quintuplet}) \oplus (j = 1, \text{triplet}) \oplus (j = 0, \text{singlet}),$$

or

$$3 \otimes 3 = 5 \oplus 3 \oplus 1.$$

What we have described above are independent irreducible tensors. We find that the number of irreducible tensors is equal to the number of independent tableaus. One can now proceed with other spin combinations by following the above rules.

29.6 Problems

1. Express xy, yz, and xz as components of an irreducible tensor of rank 2.
2. Consider three vector operators $\mathbf{A}$, $\mathbf{r}$, and $\mathbf{p}$, where $\mathbf{A}$ satisfies the commutation relations

$$\left[A_i, A_j\right] = i\epsilon_{ijk}A_k, \qquad \left[A_i, r_j\right] = i\epsilon_{ijk}r_k, \qquad \text{and} \qquad \left[A_i, p_j\right] = i\epsilon_{ijk}p_k.$$

   Express the three operators as spherical tensors.
3. Consider two operators, $\mathbf{A}$ and $\mathbf{L}$ (the angular momentum operator), which satisfy the commutation relation

$$\left[L_i, A_j\right] = i\epsilon_{ijk}A_k.$$

   Determine the matrix elements of A_i in the representation in which $\mathbf{L}^2$ and L_z are diagonal. Use the Wigner–Eckart theorem.

30 Entangled states

Entanglement is a remarkable quantum phenomenon in which a system of two or more particles are linked closely together so that the description of one of the objects determines the state of the others. As we know, for a spin $\frac{1}{2}$ particle the spin will be found to be either up or down. If a sufficient number of spin measurements are made, one will find 50% of the time the spin is up and 50% of the time the spin is down. If we have two spin $\frac{1}{2}$ particles, one would generally find that each particle independently of the other will have the same probability profile as the single particle – except when these two particles are entangled. In that case, if one of the particles is found to have the spin in a certain direction then the spin orientation of the other is predicted even when the physical separation between the two particles is very large. We discuss all this below.

30.1 Definition of an entangled state

Consider the following two spin $\frac{1}{2}$ states $|\chi_1\rangle$ and $|\chi_2\rangle$ representing two different particles:

$$|\chi_1\rangle = a_1 |+\rangle + b_1 |-\rangle, \tag{30.1}$$

$$|\chi_2\rangle = a_2 |+\rangle + b_2 |-\rangle, \tag{30.2}$$

where $|a_i|^2 + |b_i|^2 = 1$ for $i = 1, 2$, and where $|\pm\rangle$ designate states with the quantization axes $\pm z$. A state describing both particles combined will be given by

$$|\chi_1\rangle \cdot |\chi_2\rangle = |\chi_1 \otimes \chi_2\rangle = a_1 a_2 |+ \otimes +\rangle + a_1 b_2 |+ \otimes -\rangle + b_1 a_2 |- \otimes +\rangle + b_1 b_2 |- \otimes -\rangle \tag{30.3}$$

where we have used the symbol $\otimes$ for a product state. If we write the above product state as

$$|\psi\rangle = \alpha |+ \otimes +\rangle + \beta |+ \otimes -\rangle + \gamma |- \otimes +\rangle + \delta |- \otimes -\rangle \tag{30.4}$$

then, comparing the coefficients of (30.3) and (30.4), we find that they satisfy

$$\alpha\delta = \beta\gamma. \tag{30.5}$$

We define an entangled state describing two particles as a state which, when written as (30.4), satisfies the property

$$\alpha\delta \neq \beta\gamma. \tag{30.6}$$

A state with this property is thus not a simple product of two separate states, but is "entangled." In contrast, the states that satisfy (30.5) are separable.

The state

$$|\phi\rangle = \frac{|+\otimes-\rangle - |-\otimes+\rangle}{\sqrt{2}} \tag{30.7}$$

is a classic example of an entangled state. We show below that $|\phi\rangle$ is invariant under rotations.

Let $|\pm n\rangle$ correspond to the rotated axes in the direction $\pm\mathbf{n}$ with polar angle θ (and azimuthal angle $\phi = 0$), then as we showed in Chapter 5 we obtain

$$|+n\rangle = \cos\left(\frac{\theta}{2}\right)|+z\rangle + \sin\left(\frac{\theta}{2}\right)|-z\rangle, \tag{30.8}$$

$$|-n\rangle = -\sin\left(\frac{\theta}{2}\right)|+z\rangle + \cos\left(\frac{\theta}{2}\right)|-z\rangle. \tag{30.9}$$

Therefore,

$$\begin{aligned}|+n\otimes-n\rangle = &-\cos\left(\frac{\theta}{2}\right)\sin\left(\frac{\theta}{2}\right)|+\otimes+\rangle + \cos^2\left(\frac{\theta}{2}\right)|+\otimes-\rangle\\ &-\sin^2\left(\frac{\theta}{2}\right)|-\otimes+\rangle + \sin\left(\frac{\theta}{2}\right)\cos\left(\frac{\theta}{2}\right)|-\otimes-\rangle.\end{aligned} \tag{30.10}$$

Similarly,

$$\begin{aligned}|-n\otimes+n\rangle = &-\sin\left(\frac{\theta}{2}\right)\cos\left(\frac{\theta}{2}\right)|+\otimes+\rangle - \sin^2\left(\frac{\theta}{2}\right)|+\otimes-\rangle\\ &+\cos^2\left(\frac{\theta}{2}\right)|-\otimes+\rangle + \cos\left(\frac{\theta}{2}\right)\sin\left(\frac{\theta}{2}\right)|-\otimes-\rangle.\end{aligned} \tag{30.11}$$

Taking the difference, we find

$$\frac{|+n\otimes-n\rangle - |-n\otimes+n\rangle}{\sqrt{2}} = \frac{|+\otimes-\rangle - |-\otimes+\rangle}{\sqrt{2}}. \tag{30.12}$$

Therefore, $|\phi\rangle$ is invariant under rotations.

30.2 The singlet state

Let us consider a state consisting of two identical particles of spin $\frac{1}{2}$ that are traveling in opposite directions with equal momenta. An example of such a situation is the decay of a neutral meson into an electron and a positron, i.e., the rare process

$$\pi^0 \to e^+ + e^-. \tag{30.13}$$

Suppose that there are two observers A and B positioned to measure the spin components of each particle when the particles are far apart and unable to interact with each other, as described in Fig. 30.1. Let us call the particle moving to the left particle 1 and the particle moving to the right particle 2. The observer A using a Stern–Gerlach device measures the component along an arbitrary direction, we call it the **a** direction, of the spin of particle 1, and B similarly measures the component of the spin in another direction, the **b** direction of particle 2.

First let us assume that the two axes **a** and **b** are the same, both coinciding with the z-axis. Consider the two-electron system to be in a spin singlet state and hence with a total spin of zero. The state of the system is described by the state vector (30.7),

$$|\phi\rangle = \frac{|+\otimes -\rangle - |-\otimes +\rangle}{\sqrt{2}}. \tag{30.14}$$

When A measures the spin component of particle 1 and finds it to be $+\hbar/2$, B will find the spin component of particle 2 to be $-\hbar/2$, and vice versa. The point to note is the following: once A makes the measurement, B will automatically find that the spin component of particle 2 is exactly opposite to that measurement. The results for each pair of measurements are perfectly anticorrelated.

What is remarkable is that they need not have both chosen the z-axis; the same result would occur if they had chosen the x-axis since the singlet state is invariant under rotation as expression (30.12) makes clear. Going a step further, we note that if A measures the spin component along the z-axis and B along the x-axis, then if A finds the spin component to be $+\hbar/2$ along the z-axis, B – who in the previous experiment found the spin to be in the $-z$ direction – will now find the spin component along the x-axis to be $+\hbar/2$ or $-\hbar/2$ with equal probability since $|z-\rangle = \left(1/\sqrt{2}\right)[|x+\rangle + |x-\rangle]$.

The fact is that this kind of anticorrelation persists even when the particles are far apart and no longer interacting and with the observers A and B who are long way apart with no

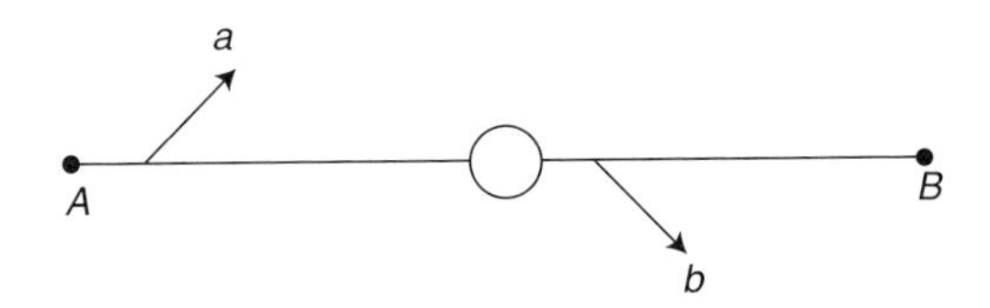

Fig. 30.1

possibility of communication. The observer B may decide how to orient the Stern–Gerlach device long after the two particles have separated, but the outcome is already known. There is, therefore, a certain degree of certainty what B will measure without disturbing particle 2.

Einstein, Podolsky, and Rosen stated with reference to these types of measurements that if the value of a physical property can be predicted with certainty without disturbing the system in any way then there is an element of reality associated with this property.

A question then arises whether there is a "super theory" beyond quantum mechanics which is deterministic and in which there are "hidden variables" describing the system of π^0, e^+, e^- in the reaction (30.13). Thus, e^+ and e^- inside π^0 may already have spin correlations that persist as they leave and fly apart. What one observes then is just a statistical outcome. The hidden variables here are the initial spin, velocity, etc. of e^+ and e^-.

This situation would be akin to the case in which one tosses coins into the air a large number of times and, as the coins fall on the ground, discover that they fall 50% of the time heads and 50% tails. This is a statistical outcome, but the system is actually deterministic. If, for example, one knew the specific details about the process, e.g., the exact velocity of the coin upon release, the wind velocity, etc. one could predict how the coin would fall. However, the process is normally sufficiently complex that one is satisfied with the statistical outcome. The variables in this illustration are, of course, classical and known and not hidden.

30.3 Differentiating the two approaches

Consider two measurement axes **a** and **b** to lie in the x–z plane perpendicular to the direction of propagation, which we take to be the y-axis. We shall use $A(\mathbf{a})$ and $B(\mathbf{b})$ as the results of measurements of the projections $\boldsymbol{\sigma} \cdot \mathbf{a}$ and $\boldsymbol{\sigma} \cdot \mathbf{b}$ of particles 1 and 2 with possible values

$$A(\mathbf{a}) = \epsilon_a = \pm 1, \qquad B(\mathbf{b}) = \epsilon_b = \pm 1. \tag{30.15}$$

Let $E(\mathbf{a}, \mathbf{b})$ be the expectation value that determines the correlation between the spin measurements of the observers A and B, who use axes **a** and **b**, respectively. We can write it as

$$E(\mathbf{a}, \mathbf{b}) = \lim_{N \to \infty} \frac{1}{N} \sum A_n(\mathbf{a}) B_n(\mathbf{b}), \tag{30.16}$$

which corresponds to N different measurements of the pairs $A_n(\mathbf{a})$, $B_n(\mathbf{b})$. A typical term in the product is

$$\epsilon_a \epsilon_b = \pm 1. \tag{30.17}$$

Hence,

$$|E(\mathbf{a}, \mathbf{b})| \leq 1. \tag{30.18}$$

Quantum-mechanically, the expectation value will be

$$E(\mathbf{a},\mathbf{b}) = \left\langle \phi \left| \boldsymbol{\sigma}^{(1)} \cdot \mathbf{a} \boldsymbol{\sigma}^{(2)} \cdot \mathbf{b} \right| \phi \right\rangle, \tag{30.19}$$

where $|\phi\rangle$ is the entangled state representing the singlet spin state, and $\sigma^{(1)}$ and $\sigma^{(2)}$ are the Pauli spin operators corresponding to the particles 1 and 2. To evaluate the matrix element we first consider

$$\left\langle \phi \left| \boldsymbol{\sigma}^{(1)} + \boldsymbol{\sigma}^{(2)} \right| \phi \right\rangle. \tag{30.20}$$

Since $|\phi\rangle$ is invariant under rotations and $\left(\boldsymbol{\sigma}^{(1)} + \boldsymbol{\sigma}^{(2)}\right)$ is a vector, this matrix element vanishes. Hence,

$$\boldsymbol{\sigma}^{(1)} |\phi\rangle = -\boldsymbol{\sigma}^{(2)} |\phi\rangle. \tag{30.21}$$

We first multiply both the terms by $\mathbf{b}$ and then by $\boldsymbol{\sigma}^{(1)} \cdot \mathbf{a}$. We obtain

$$\boldsymbol{\sigma}^{(1)} \cdot \mathbf{a} \boldsymbol{\sigma}^{(1)} \cdot \mathbf{b} |\phi\rangle = -\boldsymbol{\sigma}^{(1)} \cdot \mathbf{a} \boldsymbol{\sigma}^{(2)} \cdot \mathbf{b} |\phi\rangle. \tag{30.22}$$

The two $\boldsymbol{\sigma}^{(1)}$'s on the left-hand side are in the same space, so that we can utilize the relation

$$\boldsymbol{\sigma}^{(1)} \cdot \mathbf{a} \boldsymbol{\sigma}^{(1)} \cdot \mathbf{b} = \mathbf{a} \cdot \mathbf{b} + \mathbf{i} \boldsymbol{\sigma}^{(1)} \cdot (\mathbf{a} \times \mathbf{b}). \tag{30.23}$$

The expectation value with respect to $|\phi\rangle$ of the second term on the right-hand side above can be shown to vanish. From (30.22) and (30.23) we therefore obtain

$$\left\langle \phi \left| \boldsymbol{\sigma}^{(1)} \cdot \mathbf{a} \boldsymbol{\sigma}^{(2)} \cdot \mathbf{b} \right| \phi \right\rangle = -\mathbf{a} \cdot \mathbf{b} = -\cos\theta \tag{30.24}$$

where θ is the angle between the axes $\mathbf{a}$ and $\mathbf{b}$. Thus, quantum-mechanically we find

$$|E(\mathbf{a},\mathbf{b})| \leq 1, \tag{30.25}$$

which gives the same conclusion as the classical or the hidden theory.

In the absence of any quantitative resolution of the differences, the debates between the proponents of quantum theory and hidden-variables theory remained essentially philosophical in nature. Things changed in 1964, however, when J. S. Bell proposed a quantitative way in which one could determine the difference.

30.4 Bell's inequality

Consider the following situation: observers A and B use a pair of axes $\mathbf{a}$ and $\mathbf{a}'$ and a pair of axes $\mathbf{b}$ and $\mathbf{b}'$, respectively for their spin measurements. We construct the following combination using the notations we have already defined in the previous section:

$$X_n = A_n(\mathbf{a})B_n(\mathbf{b}) + A_n(\mathbf{a})B_n'(\mathbf{b}') + A_n'(\mathbf{a}')B_n'(\mathbf{b}') - A_n'(\mathbf{a}')B_n(\mathbf{b}). \tag{30.26}$$

A typical result of the measurement for X_n will be

$$\epsilon_a\epsilon_b + \epsilon_a\epsilon'_b + \epsilon'_a\epsilon'_b - \epsilon'_a\epsilon_b. \tag{30.27}$$

We can write this as

$$\epsilon_a(\epsilon_b + \epsilon'_b) + \epsilon'_a(\epsilon'_b - \epsilon_b). \tag{30.28}$$

Since the values of the ϵ's are ± 1, we have then two possibilities.

(i) $\epsilon_b = \epsilon'_b$. In this case we have

$$X_n = 2\epsilon_a\epsilon_b = \pm 2. \tag{30.29}$$

(ii) $\epsilon_b = -\epsilon'_b$. In this case we have

$$X_n = 2\epsilon'_a\epsilon'_b = \pm 2. \tag{30.30}$$

Hence we have

$$|E(\mathbf{a},\mathbf{b})| = \left| \lim_{N\to\infty} \frac{1}{N} \sum X_n \right| \leq 2. \tag{30.31}$$

Quantum-mechanically, we have

$$E(\mathbf{a},\mathbf{b}) = -\,\mathbf{a}\cdot\mathbf{b} - \mathbf{a}\cdot\mathbf{b}' - \mathbf{a}'\cdot\mathbf{b}' + \mathbf{a}'\cdot\mathbf{b}. \tag{30.32}$$

Let us now consider the specific case as given by the Fig. 30.2 with the angles between $\mathbf{a}, \mathbf{b}, \mathbf{a}'$, and $\mathbf{b}'$ as prescribed. We find

$$E(\mathbf{a},\mathbf{b}) = -\,\cos\left(\frac{\pi}{4}\right) - \cos\left(\frac{\pi}{4}\right) - \cos\left(\frac{\pi}{4}\right) + \cos\left(\frac{3\pi}{4}\right). \tag{30.33}$$

Since $\cos(\pi/4) = 1/\sqrt{2}$, and $\cos(3\pi/4) = -1/\sqrt{2}$, we obtain

$$E(\mathbf{a},\mathbf{b}) = -\,2\sqrt{2}. \tag{30.34}$$

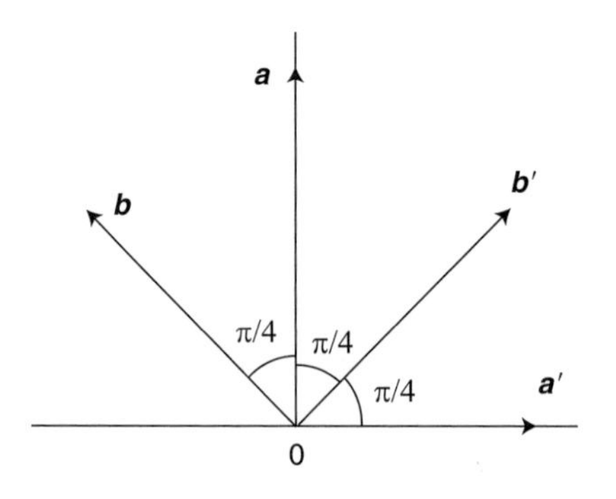

Fig. 30.2

We thus see that there is a clear difference between the classical and quantum-mechanical results for the correlation.

Experimental evidence points squarely in the direction of quantum mechanics.

Entanglement has a number of applications, e.g., in quantum computing, quantum communications, quantum state teleportation, and many aspects of information theory. Evidently, this is an exciting, modern subject. What we have done in this chapter is just to scratch the surface.

30.5 Problems

1. Consider a spin singlet state made up of two spin $\frac{1}{2}$ states. The two particles – we call them particle 1 and particle 2 – are traveling in opposite directions. An observer, A, measures the spin of particle 1 and another observer, B, measures the spin of particle 2. What is the probability that A finds the spin of particle 1 to be pointing in the positive z-direction when **B** makes no measurement? What if B made a measurement and found the spin of particle 2 to be in the positive x-direction?

31 Special theory of relativity: Klein–Gordon and Maxwell's equations

We now enter the domain of relativity. First we modify and extend the nonrelativistic theories to be compatible with the postulates of the special theory of relativity. We express the previously defined physical quantities in a four-dimensional language and introduce the concept of covariance so that the equations look the same in any inertial frame. Then we introduce the Klein–Gordon equation, which is a relativistic extension of the Schrödinger equation. We discover that the solutions of this equation can have negative energies and the probability density is not necessarily positive definite. We also consider Maxwell's equation and discuss the consequences of gauge invariance.

31.1 Lorentz transformation

The postulates of the special theory of relativity state in the simplest terms that:

(i) If a system of coordinates exists so that, in relation to it, physical laws hold in their simplest form, the same laws hold good in relation to any other system of coordinates moving uniformly with respect to it. These are the inertial frames.
(ii) The velocity of light is the same in all inertial frames.

Consider two inertial frames with origins at O and O', respectively, moving with respect to each other with a uniform velocity. We designate the frames by their origins as the O-frame and the O'-frame. If a point in the O-frame is described by four coordinates (x, y, z, ct), where c is the velocity of light, then the coordinates of the corresponding point in the O'-frame will be denoted by (x', y', z', ct').

We assume that the two frames coincide at time $t = 0$, in which case we must have $x' = x$, $y' = y$, $z' = z$, at $t' = t = 0$. According to the postulates of the special theory of relativity, the speed of light remains the same when measured in any two inertial frames. This implies that the trajectory of light emitted at $t' = t = 0$ from the common origin of the two frames must satisfy the following relation

$$c^2t'^2 - x'^2 - y'^2 - z'^2 = c^2t^2 - x^2 - y^2 - z^2. \tag{31.1}$$

Consider now the motion of a particle along the x-direction. One can write the following relations between the two coordinate-systems:

$$x' = a_1x + a_2t, \tag{31.2}$$

$$t' = b_1x + b_2t, \tag{31.3}$$

$$y' = y, \tag{31.4}$$

$$z' = z. \tag{31.5}$$

We assume that the O'-frame is moving with uniform velocity v in the x-direction with respect to the O-frame. Thus the point O' will travel, in time t, a distance vt along the x-direction with respect to O. Hence $x' = 0$ will correspond to $x = vt$ since they represent the coordinates of the same point, O', in the two frames. With the initial conditions as outlined, we now substitute (31.2)–(31.5) in (31.1). The following result is obtained:

$$x' = \gamma\,(x - vt)\,, \tag{31.6}$$

$$t' = \gamma\left(t - \frac{vx}{c^2}\right), \tag{31.7}$$

$$y' = y, \tag{31.8}$$

$$z' = z, \tag{31.9}$$

where

$$\gamma = \frac{1}{\sqrt{1 - (v/c)^2}}. \tag{31.9a}$$

The above relations correspond to Lorentz transformation along the x-axis. In the discussions to follow we will confine ourselves entirely to the transformation along the x-axis.

31.2 Contravariant and covariant vectors

We now define the following four-component vector, called a contravariant vector:

$$x^{\mu} = (x, y, z, ct)\,. \tag{31.10}$$

The Lorentz transformation relations (31.6)–(31.9) can be written in terms of these four-vectors as

$$x'^1 = \gamma\left(x^1 - \beta x^4\right), \tag{31.11}$$

$$x'^4 = \gamma\left(x^4 - \beta x^1\right), \tag{31.12}$$

$$x'^2 = x^2, \tag{31.13}$$

$$x'^3 = x^3, \tag{31.14}$$

where γ is defined in (31.9a) and

$$\beta = \frac{v}{c}. \tag{31.15}$$

A covariant vector, x_μ, is defined as

$$x_\mu = (-x, -y, -z, ct) . \tag{31.16}$$

The invariance condition (31.1) can be written as

$$x'_\mu x'^\mu = x_\rho x^\rho \tag{31.17}$$

where we adopt the convention that whenever the same index appears in opposite locations then that index is to be summed over, from 1 to 4. This will remain our summation convention. The important thing to note is that the indices must be in opposite locations for the summation convention to apply.

We note further that in (31.17) there is no free index left after the summation, signifying that it is an invariant quantity that will remain unchanged under the Lorentz transformation. Thus, terms with no free indices will be the invariants, also called scalars, while those with one index free will be four-vectors, either contravariant or covariant. The objects with two or more indices will be called tensors, following the convention in three-dimensional vector algebra.

One can relate x^μ and x_μ through the tensor relations

$$x_\mu = g_{\mu\nu} x^\nu \tag{31.18}$$

as well as by

$$x^\mu = g^{\mu\nu} x_\nu \tag{31.19}$$

where $g_{\mu\nu}$ and $g^{\mu\nu}$ are each called the metric tensor. Their matrix representations are given by

$$\{g^{\mu\nu}\} = \{g_{\mu\nu}\} = \begin{bmatrix} -1 & 0 & 0 & 0 \\ 0 & -1 & 0 & 0 \\ 0 & 0 & -1 & 0 \\ 0 & 0 & 0 & 1 \end{bmatrix} . \tag{31.20}$$

We note, that after the summation, the right-hand sides of (31.18) and (31.19) have one free index remaining, which is at the same location as the one on the left-hand side. The metric tensor thus acts like a raising or a lowering operator in changing a contravariant vector to a covariant vector or vice versa. It is also called a "reducing agent."

The invariance relation (31.1) can be expressed as

$$g_{\mu\nu} x'^\mu x'^\nu = g_{\rho\sigma} x^\rho x^\sigma = \text{invariant}. \tag{31.21}$$

We now write the Lorentz transformation in a matrix form as

$$x'^\mu = L^\mu_{.\nu} x^\nu \tag{31.22}$$

where

$$\{L^{\mu}_{.\nu}\} = \begin{bmatrix} \gamma & 0 & 0 & -\gamma\beta \\ 0 & 1 & 0 & 0 \\ 0 & 0 & 1 & 0 \\ -\gamma\beta & 0 & 0 & \gamma \end{bmatrix}. \tag{31.23}$$

By writing the matrix as $L^{\mu}_{.\nu}$ we follow the rules of matrix multiplication in that the first index, μ, is the row index and the second index, ν, is the column index. The dot separates the two indices.

To obtain the Lorentz transformation for a covariant vector, x'_{ρ}, we multiply (31.22) on both sides on the left by $g_{\rho\mu}$ and then sum over μ,

$$x'_{\rho} = g_{\rho\mu}x'^{\mu} = g_{\rho\mu}L^{\mu}_{.\nu}x^{\nu} = g_{\rho\mu}L^{\mu}_{.\nu}g^{\nu\sigma}x_{\sigma} = M_{\rho}^{.\sigma}x_{\sigma} \tag{31.24}$$

where $M_{\rho}^{.\sigma}$ is a matrix related to $L^{\mu}_{.\nu}$:

$$M_{\rho}^{.\sigma} = g_{\rho\mu}L^{\mu}_{.\nu}g^{\nu\sigma}. \tag{31.25}$$

For the specific case of Lorentz transformations along the x-axis for which $L^{\mu}_{.\nu}$ is given by (31.23), $M_{\rho}^{.\sigma}$ is given by

$$\{M_{\rho}^{.\sigma}\} = \begin{bmatrix} \gamma & 0 & 0 & \gamma\beta \\ 0 & 1 & 0 & 0 \\ 0 & 0 & 1 & 0 \\ \gamma\beta & 0 & 0 & \gamma \end{bmatrix}. \tag{31.26}$$

Here, again, the first index in $M_{\rho}^{.\sigma}$ is ρ, which is the row index, and the second index, σ, is the column index. Thus we can write the relation (31.25) in a compact matrix form as follows:

$$M = gLg. \tag{31.27}$$

Below we will obtain a simpler relation between M and L.

The relation (31.21) can be written as

$$g_{\mu\nu}x'^{\mu}x'^{\nu} = g_{\mu\nu}\left(L^{\mu}_{.\rho}x^{\rho}\right)\left(L^{\nu}_{.\sigma}x^{\sigma}\right) = g_{\mu\nu}L^{\mu}_{.\rho}L^{\nu}_{.\sigma}x^{\rho}x^{\sigma}. \tag{31.28}$$

Equating this to the equation on the right-hand side of (31.21) we obtain

$$g_{\mu\nu}L^{\mu}_{.\rho}L^{\nu}_{.\sigma} = g_{\rho\sigma}. \tag{31.29}$$

The above relation can also be written as

$$\left(\widetilde{L}\right)_{\rho}^{\cdot\mu} g_{\mu\nu} L^{\nu}_{.\sigma} = g_{\rho\sigma}, \tag{31.30}$$

i.e.,

$$\widetilde{L} g L = g \tag{31.31}$$

where $\widetilde{L}$ corresponds to the transpose of the matrix L. From (31.29) and (31.21) we find

$$\widetilde{L} M = \widetilde{L} g L g = g^2 = \mathbf{1} \tag{31.32}$$

where $\mathbf{1}$ is a unit matrix. Hence from (31.27), (31.31), and (31.32) we find

$$M = \left(\widetilde{L}\right)^{-1}. \tag{31.33}$$

Thus, M is not an independent matrix but is related to L.

By starting with the product of covariant vectors, $g^{\mu\nu} x'_{\mu} x'_{\nu}$, we can follow the same steps that led to (31.31), and obtain

$$\widetilde{M} g M = g. \tag{31.34}$$

From (31.33), one can write the above relation as

$$L^{-1} g \left(\widetilde{L}\right)^{-1} = g. \tag{31.35}$$

Multiplying both sides of (31.35) on the left by L and on the right by $\widetilde{L}$, we obtain

$$L g \widetilde{L} = g. \tag{31.36}$$

31.3 An example of a covariant vector

A familiar example of a covariant vector is the derivative operator $\partial/\partial x^{\mu}$. Consider a function $f(x, y, z, ct)$ that is invariant under Lorentz transformation (e.g., expression (31.1) is such a function). As we stated earlier, this type of function is called a scalar function. Let us expand it in a Taylor series around the origin.

$$f(x, y, z, ct) = f(0,0,0,0) + \left(x\frac{\partial}{\partial x} + y\frac{\partial}{\partial y} + z\frac{\partial}{\partial z} + (ct)\frac{\partial}{\partial (ct)}\right) f(0,0,0,0) + \cdots. \tag{31.37}$$

Since the left-hand side is a scalar function, each term on the right side must also be a scalar (i.e., invariant). Therefore, since $f(0,0,0,0)$ is a scalar we conclude

$$\left(x\frac{\partial}{\partial x} + y\frac{\partial}{\partial y} + z\frac{\partial}{\partial z} + (ct)\frac{\partial}{\partial (ct)}\right) = \text{invariant}. \tag{31.38}$$

Since (x, y, z, ct) are components of a contravariant vector x^μ, the derivatives must be part of a covariant vector so that the product is of the form $A_\rho B^\rho$ (e.g., expression (31.17)). To make this explicit we define

$$\partial_\mu = \frac{\partial}{\partial x^\mu} = \left(\frac{\partial}{\partial x^1}, \frac{\partial}{\partial x^2}, \frac{\partial}{\partial x^3}, \frac{\partial}{\partial x^4}\right) = \left(\frac{\partial}{\partial x}, \frac{\partial}{\partial y}, \frac{\partial}{\partial z}, \frac{\partial}{c\partial t}\right), \tag{31.39}$$

$$\partial^\mu = \frac{\partial}{\partial x_\mu} = \left(\frac{\partial}{\partial x_1}, \frac{\partial}{\partial x_2}, \frac{\partial}{\partial x_3}, \frac{\partial}{\partial x_4}\right) = \left(-\frac{\partial}{\partial x}, -\frac{\partial}{\partial y}, -\frac{\partial}{\partial z}, \frac{\partial}{c\partial t}\right). \tag{31.40}$$

We can express (31.38) as

$$x^\mu \partial_\mu = x_\mu \partial^\mu = \text{invariant}. \tag{31.41}$$

31.4 Generalization to arbitrary tensors

We can now generalize the results obtained above from the coordinate vectors to arbitrary vectors and tensors. A contravariant vector A^μ and a covariant vector A_μ are defined as

$$A^\mu = \left(A_x, A_y, A_z, A^4\right), \tag{31.42}$$

$$A_\mu = \left(-A_x, -A_y, -A_z, A_4\right), \tag{31.43}$$

where A_x, A_y, A_z are the ordinary space components and A^4 is the fourth (time) component with $A_4 = A^4$. The corresponding Lorentz transformation properties are

$$A'^\mu = L^\mu_{.\nu} A^\nu, \tag{31.44}$$

$$A'_\mu = M_\mu^{.\nu} A_\nu. \tag{31.45}$$

One can show that

$$A^\mu B_\mu \tag{31.46}$$

is an invariant through the following steps.

$$A'^\mu B'_\mu = \left(L^\mu_{.\nu} A^\nu\right)\left(M_\mu^{.\rho} B_\rho\right) = A^\nu \left(\widetilde{L}M\right)_\nu^{.\rho} B_\rho = A^\nu (\mathbf{1})_\nu^{.\rho} B_\rho = A^\nu B_\nu. \tag{31.47}$$

One often writes the above product as a dot-product

$$A^\mu B_\mu = A \cdot B. \tag{31.48}$$

A (totally) contravariant tensor, $T^{\mu\nu\cdots}$, can be thought of as having the same transformation property as the product $A^\mu B^\nu \ldots$. Thus,

$$T'^{\mu\nu\cdots} = L^\mu_{.\rho} L^\nu_{.\sigma} \cdots T^{\rho\sigma\cdots}. \tag{31.49}$$

In a similar fashion the mixed tensor $T^{\mu\nu\ldots}_{\ldots\ldots\ldots\rho\sigma}$, and totally covariant tensor $T_{\mu\nu\ldots.}$ will transform as

$$T'^{\mu\nu\ldots}_{\ldots\ldots\ldots\rho\sigma} = L^{\mu}_{.a}L^{\nu}_{.b}M^{.c}_{\rho}M^{.d}_{\sigma}\cdots T^{ab\ldots}_{\ldots\ldots\ldots cd} \tag{31.50}$$

and

$$T'_{\mu\nu\ldots..} = M^{.\rho}_{\mu}M^{.\sigma}_{\nu}\ldots T_{\rho\sigma}. \tag{31.51}$$

We note that, according to our summation convention, a product

$$T^{\mu\nu}_{..\rho\sigma}B^{\sigma}C_{\nu} \tag{31.52}$$

transforms as a mixed tensor $S^{\mu}_{.\rho}$. This follows from the fact that the repeated indices ν and σ are situated in opposite locations and, are, therefore, summed over. However, in products like

$$A^{\mu}B^{\mu}, \tag{31.53}$$

since the index μ is not situated in opposite locations in the two terms, there is no summation involved. The above term designates the $\mu\mu$ diagonal component of the tensor $T^{\mu\nu} = A^{\mu}B^{\nu}$.

For second-rank tensors the transformations can be written quite simply in terms of 4×4 matrices. Using the usual interpretation of the row index and column index in matrices one follows the matrix multiplication rules. So, for example, one can write

$$T'^{\mu\nu} = L^{\mu}_{.\rho}L^{\nu}_{.\sigma}T^{\rho\sigma} = L^{\mu}_{.\rho}T^{\rho\sigma}\widetilde{L}^{.\nu}_{\sigma} \tag{31.54}$$

as

$$T' = LT\widetilde{L} \quad \text{(contravariant)}. \tag{31.55}$$

Similarly,

$$T' = LT\widetilde{M} \quad \text{(mixed)}, \tag{31.56}$$

$$T' = MT\widetilde{M} \quad \text{(covariant)}. \tag{31.57}$$

One can easily verify that the metric tensor is "form-invariant," i.e.,

$$g'^{\mu\nu} = g^{\mu\nu}. \tag{31.58}$$

This follows after using the transformation (31.36). We can also write

$$g' = Lg\widetilde{L} = g. \tag{31.59}$$

Furthermore, if one assumes the identity matrix as a mixed tensor then it is also a form-invariant matrix. This can be shown if we write

$$\{\delta^{\mu}_{.\nu}\} = \begin{bmatrix} 1 & 0 & 0 & 0 \\ 0 & 1 & 0 & 0 \\ 0 & 0 & 1 & 0 \\ 0 & 0 & 0 & 1 \end{bmatrix}. \tag{31.60}$$

The transformed matrix, $\delta'^{\mu}_{.\nu}$, because of its mixed character, can be written as

$$\delta'^{\mu}_{.\nu} = \left(L\mathbf{1}\widetilde{M}\right)^{\mu}_{.\nu} = \delta^{\mu}_{.\nu}. \tag{31.61}$$

The last equality follows from (31.32).

Finally, we emphasize that writing a matrix without specifying its transformation properties is meaningless because the same matrix will transform differently depending on the type of tensor it represents. For example if we took the matrix (31.60) as a contravariant tensor,

$$\{T^{\mu\nu}\} = \begin{bmatrix} 1 & 0 & 0 & 0 \\ 0 & 1 & 0 & 0 \\ 0 & 0 & 1 & 0 \\ 0 & 0 & 0 & 1 \end{bmatrix} \tag{31.62}$$

then from (31.54)

$$T'^{\mu\nu} = \begin{bmatrix} \gamma^2(1+\beta^2) & 0 & 0 & -2\beta\gamma^2 \\ 0 & 1 & 0 & 0 \\ 0 & 0 & 1 & 0 \\ -2\beta\gamma^2 & 0 & 0 & \gamma^2(1+\beta^2) \end{bmatrix}. \tag{31.63}$$

This is a totally different result compared to (31.62).

Finally, just as we did for rotations, it is important to discuss infinitesimal Lorentz transformations. This we have done in the Appendix at the end of this chapter.

31.5 Relativistically invariant equations

According to the special theory of relativity, the laws of physics must remain the same in any two inertial frames. This statement implies that both sides of an equation describing a physical phenomenon must transform the same way under Lorentz transformations. For example, an equation of the type

$$A^{\mu} = B^{\mu} \tag{31.64}$$

where both sides transform as contravariant vectors will lead to an equation that looks exactly the same in another inertial frame. This can be shown by taking the Lorentz transform of the left-hand side of (31.64),

$$A'^{\mu} = L^{\mu}_{.\nu} A^{\nu} = L^{\mu}_{.\nu} B^{\nu} = B'^{\mu}. \tag{31.65}$$

Thus the same equation holds in the primed-frame. These types of equations are called **covariant equations**.

Our task in the topics to follow is to cast different physical equations into a covariant form.

31.5.1 Electromagnetism

First let us consider charge and current densities in classical electromagnetism, ρ and $\mathbf{j}$. They satisfy the charge conservation relation

$$\frac{\partial \rho}{\partial t} + \nabla \cdot \mathbf{j} = 0. \tag{31.66}$$

We can cast this equation into a covariant form by defining a "four-vector current"

$$j^{\mu} = (\mathbf{j}, c\rho) . \tag{31.67}$$

The charge conservation relation can now be expressed as

$$\partial_{\mu} j^{\mu} = 0. \tag{31.68}$$

The left-hand side of this equation is invariant under Lorentz transformations, as is the right-hand side. This is then a covariant equation that will be valid in any inertial frame.

We have thus established $\mathbf{j}$ and ρ, which were originally thought of as separate entities, as being parts of the same four-current j^{μ} which, therefore, transform into each other under Lorentz transformations. Below we will find similar results for other physical quantities.

Let us consider the scalar and vector potentials ϕ and $\mathbf{A}$, which satisfy equations

$$\nabla^2 \phi - \frac{\partial^2}{c^2 \partial t^2} \phi = -4\pi \rho, \tag{31.69}$$

$$\nabla^2 \mathbf{A} - \frac{\partial^2}{c^2 \partial t^2} \mathbf{A} = -\frac{4\pi}{c} \mathbf{j}. \tag{31.70}$$

A covariant equation can be achieved by taking

$$A^{\mu} = (\mathbf{A}, \phi) . \tag{31.71}$$

Using the fact that

$$\partial_{\mu} \partial^{\mu} = -\nabla^2 + \frac{\partial^2}{c^2 \partial t^2}, \tag{31.72}$$

the two equations (31.69) and (31.70) can now be written in a covariant form as

$$\left(\partial_\mu \partial^\mu\right) A^\nu = \frac{4\pi}{c} j^\nu. \tag{31.73}$$

In electromagnetic theory the electric and magnetic fields are expressed in terms of the vector and scalar potentials, $\mathbf{A}$ and ϕ, respectively, as follows:

$$\mathbf{E} = -\nabla\phi - \frac{\partial \mathbf{A}}{c\partial t} \quad \text{and} \quad \mathbf{B} = \nabla \times \mathbf{A}. \tag{31.74}$$

Taking the x-components of these fields we note that

$$E_x = -\frac{\partial \phi}{\partial x} - \frac{\partial A_x}{c\partial t} = \partial^1 A^4 - \partial^4 A^1. \tag{31.75}$$

Similarly,

$$B_x = \frac{\partial A_z}{\partial y} - \frac{\partial A_y}{\partial z} = -\partial^2 A^3 + \partial^3 A^2. \tag{31.76}$$

One can easily generalize these results by writing a second rank tensor

$$F^{\mu\nu} = \partial^\mu A^\nu - \partial^\nu A^\mu. \tag{31.77}$$

This is an antisymmetric tensor, called the Maxwell tensor, whose matrix elements are related to the components of the $\mathbf{E}$ and $\mathbf{B}$ fields. For example, from the relations (31.75) and (31.76) we find

$$F^{14} = E_x, \qquad F^{23} = -B_x. \tag{31.78}$$

The Maxwell tensor in a matrix form is then

$$\{F^{\mu\nu}\} = \begin{bmatrix} 0 & -B_z & B_y & E_x \\ B_z & 0 & -B_x & E_y \\ -B_y & B_x & 0 & E_z \\ -E_x & -E_y & -E_z & 0 \end{bmatrix}. \tag{31.79}$$

Finally, if we consider

$$\partial_\mu F^{\mu\nu} = \partial_\mu \left(\partial^\mu A^\nu - \partial^\nu A^\mu\right) = \partial_\mu \partial^\mu A^\nu - \partial_\mu \partial^\nu A^\mu, \tag{31.80}$$

then using (31.73) we obtain

$$\partial_\mu F^{\mu\nu} = \frac{4\pi}{c} j^\nu \tag{31.81}$$

provided the following condition holds:

$$\partial_\mu A^\mu = 0. \tag{31.82}$$

This is, of course, the famous Lorentz condition

$$\nabla \cdot \mathbf{A} + \frac{\partial \phi}{c\partial t} = 0. \tag{31.83}$$

Taking $\nu = 1, 2, 3$ in (31.81) we obtain

$$\nabla \times \mathbf{B} = \frac{1}{c}\frac{\partial \mathbf{E}}{\partial t} + \frac{4\pi}{c}\mathbf{j}, \quad \nabla \cdot \mathbf{E} = 4\pi\rho. \tag{31.84}$$

These are two of the four Maxwell's equations that involve the source terms.

The remaining two of Maxwell's equations are homogeneous. These are obtained by noting that because the Maxwell tensor (31.77) is antisymmetric it satisfies

$$\partial^\mu F^{\rho\sigma} + \partial^\rho F^{\sigma\mu} + \partial^\sigma F^{\mu\rho} = 0 \tag{31.85}$$

where μ, ρ, and σ appear in cyclic order. Relation (31.85) can be proved very simply by substituting the relation (31.77) for the Maxwell tensor for the three terms above. Taking $\mu = 1$, $\rho = 2$, and $\sigma = 3$ in (31.85) we obtain

$$\nabla \cdot \mathbf{B} = 0. \tag{31.86}$$

Similarly, if we take $\mu = 1$, $\rho = 2$, and $\sigma = 4$ we obtain

$$\frac{\partial E_y}{\partial x} - \frac{\partial E_x}{\partial y} = -\frac{1}{c}\frac{\partial B_z}{\partial t}. \tag{31.87}$$

Similar equations result by taking different values of μ and ρ with $\sigma = 4$. These equations are found to be different components of the last of the four Maxwell's equations:

$$\nabla \times \mathbf{E} = -\frac{1}{c}\frac{\partial \mathbf{B}}{\partial t}. \tag{31.88}$$

Thus, we have successfully converted Maxwell's equations to a covariant form through the equations (31.81) and (31.85). We conclude that **E** and **B** are not two separate entities but are part of the same tensor, the Maxwell tensor $F^{\mu\nu}$.

31.5.2 Classical mechanics

In classical mechanics, momentum and kinetic energy are given by

$$\mathbf{p} = m\mathbf{v} \tag{31.89}$$

and

$$E = \frac{1}{2}mv^2, \tag{31.90}$$

respectively, where the velocity vector is

$$\mathbf{v} = \frac{d\mathbf{r}}{dt}. \tag{31.91}$$

We wish to generalize these definitions to the relativistic case by expressing them in a four-vector language. The relation

$$v^{\mu} = \frac{dx^{\mu}}{dt} \tag{31.92}$$

looks like an obvious candidate for the velocity four-vector. However, it will not work since, in the relativistic context, t is a fourth component which makes the term on the right-hand side behave as a tensor. We need to investigate the possibility of replacing dt with an equivalent quantity which transforms as a scalar. This is, indeed, possible from the invariant condition

$$c^2\, dt'^2 - dx'^2 - dy'^2 - dz'^2 = c^2 dt^2 - dx^2 - dy^2 - dz^2 = \text{invariant}. \tag{31.93}$$

Let the O'-frame be a frame in which the particle is at rest. This type of frame is called the "rest frame" or "proper frame" for the particle, in which

$$dx' = dy' = dz' = 0. \tag{31.94}$$

One often designates dt' in the proper frame as $d\tau$. Thus from (31.93) and (31.94) we obtain

$$c^2\, d\tau^2 = c^2 dt^2 - dx^2 - dy^2 - dz^2 = \text{invariant}. \tag{31.95}$$

By replacing dt in (31.92) by $d\tau$ we write.

$$v^{\mu} = \frac{dx^{\mu}}{d\tau}. \tag{31.96}$$

Since $d\tau$ is a scalar, this is a covariant relation in which both sides transform as a four-vector. We can write this relation in terms of the time, t, as measured in the O-frame in which all the quantities like v^{μ} are measured. We note that since the O'-frame is the proper frame with respect to which the particle is at rest, the particle velocity in the O-frame is the same as the velocity, v, of the O'-frame. Therefore,

$$\sqrt{dx^2 + dy^2 + dz^2} = v\, dt. \tag{31.97}$$

From relation (31.95) we obtain the following

$$d\tau = \sqrt{1 - \beta^2}\, dt \tag{31.98}$$

where, as defined earlier, $\beta = v/c$.

We can then write (31.96) in the form

$$v^{\mu} = \frac{dx^{\mu}}{\sqrt{1-\beta^2}dt} = \gamma \frac{dx^{\mu}}{dt} \tag{31.99}$$

where, as defined earlier, $\gamma = 1/\sqrt{1-\beta^2}$. In the nonrelativistic limit, $\beta \ll 1$, we recover (31.91) as we should. The momentum vector is

$$p^{\mu} = m_0 v^{\mu} = \gamma m_0 \frac{dx^{\mu}}{dt} \tag{31.100}$$

where we take the mass term m_0 to be a scalar to maintain the four-vector character of p^{μ}. The space components of p^{μ} are simply related to the classical momentum vector $m\mathbf{v}$ by

$$\mathbf{p} = \gamma m_0 \mathbf{v}. \tag{31.101}$$

In the limit $\beta \ll 1$ we recover the nonrelativistic result

$$\mathbf{p} \approx m_0 \mathbf{v}. \tag{31.102}$$

Let us consider the fourth component of p^{μ}, in particular cp^4,

$$cp^4 = \gamma m_0 c \frac{dx^4}{dt} = \gamma m_0 c^2 = \frac{m_0 c^2}{\sqrt{1-\beta^2}}. \tag{31.103}$$

In the nonrelativistic limit,

$$\frac{m_0 c^2}{\sqrt{1-\beta^2}} = m_0 c^2 (1-\beta^2)^{-\frac{1}{2}} \approx m_0 c^2 + \frac{1}{2} m v^2. \tag{31.104}$$

Because of the presence of $\frac{1}{2}mv^2$, which is the kinetic energy in classical mechanics, one should identify cp^4 with energy, E, in relativistic mechanics. We then note that we have an extra term, m_0c^2, whose presence is entirely a consequence of relativity. This quantity is called "rest energy" or "rest mass" since it exists even when the particle is at rest, that is when $\mathbf{v} = 0$. The concept of rest mass is a fundamental consequence of the special theory of relativity whose manifestation we are already familiar with in radioactivity and nuclear fission and, of course, in particle physics. Thus we write

$$E = cp^4. \tag{31.105}$$

If we define

$$m = \gamma m_0 = \frac{m_0}{\sqrt{1-\beta^2}} \tag{31.106}$$

we obtain the well-known mass–energy equivalence relation

$$E = mc^2. \tag{31.107}$$

In summary, we can write

$$p^{\mu} = \left(\mathbf{p}, \frac{E}{c}\right), \tag{31.108}$$

with

$$\mathbf{p} = \gamma m_0 \mathbf{v}, \tag{31.109}$$

$$E = \gamma m_0 c^2. \tag{31.110}$$

Furthermore, one can construct the following invariant quantity after substituting the value of γ:

$$p_{\mu} p^{\mu} = -\mathbf{p}^2 + \frac{E^2}{c^2} = -\gamma^2 m_0^2 v^2 + \gamma^2 m_0^2 c^2 = m_0^2 c^2. \tag{31.111}$$

This gives us the result

$$E^2 = c^2 \mathbf{p}^2 + m_0^2 c^4. \tag{31.112}$$

There are two solutions to the above equation for E, $+\sqrt{c^2\mathbf{p}^2 + m_0^2 c^4}$, and $-\sqrt{c^2\mathbf{p}^2 + m_0^2 c^4}$. The positive solution corresponds to the physical case in (relativistic) mechanics. The question of negative energies will not be resolved until we get to quantum field theory in Chapter 37, where these particles will be identified as antiparticles with positive energy.

31.6 Appendix to Chapter 31

31.6.1 Infinitesimal Lorentz transformations: rotations and "pure" Lorentz transformations

We write the Lorentz transformation operator $L^{\mu}_{.\nu}$ under infinitesimal transformations as

$$L^{\mu}_{.\nu} = \delta^{\mu}_{.\nu} + \epsilon e^{\mu}_{.\nu} \tag{31.113}$$

where ϵ is an infinitesimal quantity. We will determine $e^{\mu}_{.\nu}$ from the relation

$$\tilde{L} g L = g, \tag{31.114}$$

which can be written as

$$\tilde{L}^{\mu}_{.\nu} g^{\nu\alpha} L^{.\beta}_{\alpha} = g^{\mu\beta}. \tag{31.115}$$

Substituting (31.113) in the relation above we obtain

$$e^{\beta\mu} + e^{\mu\beta} = 0 \tag{31.116}$$

where

$$e^{\mu\beta} = e^{\mu}_{.\upsilon} g^{\upsilon\beta}. \tag{31.117}$$

Let us now look at the Lorentz transformation

$$x'^{\mu} = L^{\mu}_{.\upsilon} x^{\upsilon}. \tag{31.118}$$

The transformation along the x-axis, which we will be considering below, affects only the x and t coordinates, therefore, instead of writing a 4×4 matrix transformation matrix we will just use the relevant 2×2 matrix:

$$\begin{bmatrix} x'^{1} \\ x'^{4} \end{bmatrix} = \begin{bmatrix} \gamma & -\gamma\beta \\ -\gamma\beta & \gamma \end{bmatrix} \begin{bmatrix} x^{1} \\ x^{4} \end{bmatrix}. \tag{31.119}$$

An infinitesimal transformation will correspond to β being small therefore, $\gamma \approx 1$. Let us take

$$\beta = \epsilon. \tag{31.120}$$

The above transformation will correspond to

$$\begin{bmatrix} x'^{1} \\ x'^{4} \end{bmatrix} = \begin{bmatrix} 1 & -\epsilon \\ -\epsilon & 1 \end{bmatrix} \begin{bmatrix} x^{1} \\ x^{4} \end{bmatrix}. \tag{31.121}$$

We write it in the form

$$x'^{\mu} = L^{\mu}_{.\upsilon} x^{\upsilon} \tag{31.122}$$

for a submatrix formed with $\mu = 4$ and $\mu = 1$,

$$\begin{bmatrix} x'^{1} \\ x'^{4} \end{bmatrix} = \begin{bmatrix} L^{1}_{.1} & L^{1}_{.4} \\ L^{4}_{.1} & L^{4}_{.4} \end{bmatrix} \begin{bmatrix} x^{1} \\ x^{4} \end{bmatrix}, \tag{31.123}$$

which under infinitesimal transformation will correspond to

$$\begin{bmatrix} x'^{1} \\ x'^{4} \end{bmatrix} = \begin{bmatrix} 1 & \epsilon e^{1}_{.4} \\ \epsilon e^{4}_{.1} & 1 \end{bmatrix} \begin{bmatrix} x^{1} \\ x^{4} \end{bmatrix}. \tag{31.124}$$

For the off-diagonal terms, keeping only the nonzero terms we have

$$e^{1}_{.4} = e^{1\alpha} g_{\alpha 4} = e^{14} g_{44} = e^{14} \tag{31.125}$$

and

$$e^{4}_{.1} = e^{4\beta} g_{\beta 1} = e^{41} g_{11} = -e^{41}. \tag{31.126}$$

If we take $e^{41} = 1$, then from the antisymmetry property we have $e^{14} = -e^{41}$. Therefore, the transformation relation (31.124) will be of the form

$$\begin{bmatrix} x'^1 \\ x'^4 \end{bmatrix} = \begin{bmatrix} 1 & -\epsilon \\ -\epsilon & 1 \end{bmatrix} \begin{bmatrix} x^1 \\ x^4 \end{bmatrix}, \tag{31.127}$$

which reproduces (31.121).

We refer to the Lorentz transformation matrix $L^i_{.\upsilon}$ with $i = 1, 2, 3$ and $\nu = 4$ as a "pure" Lorentz transformation in which only the space coordinates get a "boost." The results we have derived above for infinitesimal transformations can be shown to be valid for pure Lorentz transformations:

$$\begin{bmatrix} x'^i \\ x'^4 \end{bmatrix} = \begin{bmatrix} L^i_{.i} & L^i_{.4} \\ L^4_{.i} & L^4_{.4} \end{bmatrix} \begin{bmatrix} x^i \\ x^4 \end{bmatrix} \tag{31.128}$$

with $i = 1, 2, 3$.

We should note that Lorentz transformations contain rotations if $L^\mu_{.\upsilon}$ corresponds to $\mu, \nu = 1, 2, 3$. For rotations we already know, for example, that a rotation about the z-axis by an angle θ corresponds to

$$x' = x\cos\theta + y\sin\theta, \tag{31.129}$$

$$y' = -x\sin\theta + y\cos\theta, \tag{31.130}$$

with $z' = z$. An infinitesimal rotation will correspond to

$$\theta = \epsilon \tag{31.131}$$

and, therefore,

$$x' = x + \epsilon y, \tag{31.132}$$

$$y' = -\epsilon x + y, \tag{31.133}$$

which we can write in terms of the relevant 2×2 matrix as

$$\begin{bmatrix} x' \\ y' \end{bmatrix} = \begin{bmatrix} 1 & \epsilon \\ -\epsilon & 1 \end{bmatrix} \begin{bmatrix} x \\ y \end{bmatrix}. \tag{31.134}$$

In terms of $L^\mu_{.\upsilon}$ the rotation is given by

$$\begin{bmatrix} x'^1 \\ x'^2 \end{bmatrix} = \begin{bmatrix} L^1_{.1} & L^1_{.2} \\ L^2_{.1} & L^2_{.2} \end{bmatrix} \begin{bmatrix} x^1 \\ x^2 \end{bmatrix}, \tag{31.135}$$

which under infinitesimal transformation (31.113) is

$$\begin{bmatrix} x'^1 \\ x'^2 \end{bmatrix} = \begin{bmatrix} 1 & \epsilon e^1_{.2} \\ \epsilon e^2_{.1} & 1 \end{bmatrix} \begin{bmatrix} x^1 \\ x^2 \end{bmatrix}. \tag{31.136}$$

Now,

$$e^{1}_{.2} = e^{1\alpha} g_{\alpha 2} = e^{12} g_{22} = -e^{12} \tag{31.137}$$

and

$$e^{2}_{.1} = e^{2\alpha} g_{\alpha 1} = e^{21} g_{11} = -e^{21}. \tag{31.138}$$

If we take $e^{12} = -1$, then from the antisymmetric property we have $e^{21} = 1$. Therefore,

$$\begin{bmatrix} x'^1 \\ x'^2 \end{bmatrix} = \begin{bmatrix} 1 & \epsilon \\ -\epsilon & 1 \end{bmatrix} \begin{bmatrix} x^1 \\ x^2 \end{bmatrix}, \tag{31.139}$$

which confirms the relation (31.134). Hence the relation derived checks out both types of transformations.

The infinitesimal transformation as given by (31.113), is, therefore, a correct expression for Lorentz transformation including rotations.

Finite, pure, Lorentz transformation can easily be derived through a series expansion, and instead of the sine and cosine functions we have hyperbolic sines and cosines. Thus one finds

$$\begin{bmatrix} x'^1 \\ x'^4 \end{bmatrix} = \begin{bmatrix} \cosh \chi & -\sinh \chi \\ -\sinh \chi & \cosh \chi \end{bmatrix} \begin{bmatrix} x^1 \\ x^4 \end{bmatrix} \tag{31.140}$$

where

$$\tanh \chi = \beta. \tag{31.141}$$

Writing it out explicitly in terms of β, we obtain

$$\begin{bmatrix} x'^1 \\ x'^4 \end{bmatrix} = \begin{bmatrix} \gamma & -\gamma\beta \\ -\gamma\beta & \gamma \end{bmatrix} \begin{bmatrix} x^1 \\ x^4 \end{bmatrix} \tag{31.142}$$

where $\gamma = 1/\sqrt{1-\beta^2}$. This is, of course, the matrix (31.119) we started with.

31.7 Problems

1. For the following 4 × 4 matrices, condensed into a 2 × 2 form, consider a Lorentz transformation along the x-axis.
   (a) If

$$T^{\mu}_{.\nu} = \begin{bmatrix} a & 0 \\ 0 & b \end{bmatrix},$$

   obtain a and b so that $T'^{\mu}_{.\nu} = T^{\mu}_{.\nu}$.

(b) If

$$T^{\mu}_{.\nu} = \begin{bmatrix} -1 & 0 \\ 0 & 1 \end{bmatrix},$$

obtain $T'^{\mu}_{.\nu}$.

(c) Answer the same questions as (a) and (b) above if in each of the relations the tensor on the left-hand side, $T^{\mu}_{.\nu}$, is replaced by $T^{\mu\nu}$.

2. Which of the following are covariant equations (give brief explanations)?
   (a) $\dfrac{\partial \phi}{\partial x^{\mu}} = A^{\mu}$
   (b) $\dfrac{\partial \phi}{\partial x_{\mu}} = a\left(c^2 t^2 - r^2\right)$
   (c) $T^{\mu\nu} A^{\nu} = B^{\mu}$
   (d) $A^{\mu} = B^{\mu\nu} C_{\mu}$
   (e) $\partial_{\mu} A^{\mu} = C_{\mu}$
   (f) $\partial_{\mu} T^{\rho\sigma\nu} = 1$.
3. Indicate whether the following quantities transform as scalar, vector or a specific type of a tensor under Lorentz transformations.
   (a) $A^{\mu} B_{\upsilon}$
   (b) $A^{\mu} B^{\mu\nu}$
   (c) $\partial A_{\nu} / \partial x_{\mu}$
   (d) $T^{\mu\upsilon} T_{\nu\mu\rho}$
   (e) $A^{\mu} \partial_{\mu} B_{\nu}$
   (f) $A^{\mu} T^{\rho\sigma\nu} C_{\sigma}$.
4. Show that d^4x is invariant under Lorentz transformations.
5. Consider a stick at rest in the O'-system (proper frame) placed along the x'-axis. The two end points are x_1' and x_2'. The same stick is measured in another inertial system, the O-system, by recording x_1 and x_2 of the two end points at the same time $t_1 = t_2$. If L and L' are the lengths in the two systems, obtain the relation between them. Which is longer?
6. Consider a clock at rest in the O'-system placed on the x'-axis. The time recorded by this clock is also measured in another inertial system, the O-system. If Δt and $\Delta t'$ are the times recorded in each system, obtain the relation between then. Which clock is slower?
7. Determine how the velocities transform under Lorentz transformation. That is, if (u_x, u_y, u_z) corresponds to the velocity of a particle in the O-frame, then in terms of it obtain the velocity, (u_x', u_y', u_z'), in the O'-frame.
8. From Problem 7 show that the velocity of light remains unchanged under Lorentz transformation.
9. If $\mathbf{E} = (E_x, E_y, 0)$ and $\mathbf{B} = (0, B_y, B_z)$, obtain $\mathbf{E}'$ and $\mathbf{B}'$.
10. Light coming vertically down from a distant star is registered on earth. Assume the star to be stationary and representing the O-frame. Assume the earth to be moving in the x-direction and representing the O'-frame. Taking the velocity of light to have only the

y-component in the O-frame, determine the components of the velocity and the total velocity of light in the O'-frame.

11. A particle is traveling with a velocity u in the y-direction in a given frame (O-frame). What is its velocity (i.e., u'_x, u'_y, u'_z) in a frame (O'-frame) that is moving with respect to the O-frame with a uniform velocity, v, in the x-direction? If $u = c$ (the velocity of light), determine u'.

32 Klein–Gordon and Maxwell's equations

32.1 Covariant equations in quantum mechanics

Let us now continue our task of obtaining relativistically covariant equations, this time concentrating on quantum mechanics.

The quantum-mechanical operators for energy and momentum are given by

$$E \to i\hbar\frac{\partial}{\partial t}, \tag{32.1}$$

$$\mathbf{p} \to -i\hbar\boldsymbol{\nabla}. \tag{32.2}$$

These are reproduced by the covariant relation

$$p^{\mu} \to i\hbar\partial^{\mu}. \tag{32.3}$$

The wavefunction, $\phi(x^{\mu})$, now is a function of four variables, $x^{\mu} = (x, y, z, ct)$. We will write it simply as $\phi(x)$. The equation satisfied by $\phi(x)$ describing a free particle with definite energy, E and momentum $\mathbf{p}$ is

$$i\hbar\partial^{\mu}\phi = p^{\mu}\phi, \tag{32.4}$$

i.e.,

$$i\hbar\frac{\partial\phi}{\partial x_{\mu}} = p^{\mu}\phi, \tag{32.5}$$

whose solution is

$$\phi\,(x) = C\exp\left(-\frac{i}{\hbar}p\cdot x\right) = Ce^{i\frac{\mathbf{p}\cdot\mathbf{r}}{\hbar}}e^{-i\frac{Et}{\hbar}} \tag{32.6}$$

where we have written the four-vector product $p \cdot x$ in the component form $(Et - \mathbf{p}.r)$. Thus, the expression for the free particle wavefunction is unchanged from the nonrelativistic case. What is different now is the relation between the energy and momentum.

We will return in Section 32.4 to the question of the normalization constant C, since we are now in the relativistic regime.

32.2 Klein–Gordon equations: free particles

The relativistic energy–momentum relation is given by

$$p_\mu p^\mu = m_0^2 c^2, \tag{32.7}$$

which can be converted into a quantum-mechanical equation through (32.3). We find,

$$-\hbar^2 \partial_\mu \partial^\mu \phi = m_0^2 c^2 \phi. \tag{32.8}$$

In terms of space and time variables it can be written as

$$\nabla^2 \phi - \frac{1}{c^2}\frac{\partial^2 \phi}{\partial t^2} = \frac{m_0^2 c^2}{\hbar^2}\phi. \tag{32.9}$$

This is the so-called Klein–Gordon equation, also called the relativistic Schrödinger equation. The solution $\phi(x)$ has already been obtained. This solution will correspond to a free particle with a definite energy E and momentum $\mathbf{p}$.

32.2.1 Current conservation and negative energies

To construct the probability density and probability current density for the Klein–Gordon case, let us first recall the results in Chapter 1 where we considered the nonrelativistic Schrödinger equation,

$$i\hbar\frac{\partial \psi}{\partial t} + \frac{\hbar^2}{2m}\nabla^2 \psi = 0. \tag{32.10}$$

Taking the complex conjugate we find

$$-i\hbar\frac{\partial \psi^*}{\partial t} + \frac{\hbar^2}{2m}\nabla^2 \psi^* = 0. \tag{32.11}$$

We multiply (32.10) on the left by $\psi^*$and (32.11) on the right by ψ and make the following subtraction,

$$i\hbar\left(\psi^*\frac{\partial \psi}{\partial t} + \frac{\partial \psi^*}{\partial t}\psi\right) + \frac{\hbar^2}{2m}\left(\psi^*\left(\nabla^2\psi\right) - \left(\nabla^2\psi^*\right)\psi\right) = 0 \tag{32.12}$$

which is expressed as

$$i\hbar\frac{\partial}{\partial t}\left(\psi^*\psi\right) + \frac{\hbar^2}{2m}\nabla\cdot\left[\psi^*\left(\nabla\psi\right) - \left(\nabla\psi^*\right)\psi\right] = 0 \tag{32.13}$$

and written in the form

$$\frac{\partial \rho}{\partial t} + \nabla.\mathbf{j} = 0. \tag{32.14}$$

This is the nonrelativistic version of the charge–current conservation relation where ρ, the probability density, and $\mathbf{j}$, the probability current density are given by

$$\rho = \psi^* \psi, \tag{32.15}$$

$$\mathbf{j} = \frac{\hbar}{2im} \left[\psi^* (\nabla \psi) - (\nabla \psi^*) \psi \right]. \tag{32.16}$$

If we substitute the free particle solution (32.6), which is also a solution of the Schrödinger equation, we get

$$\rho = |C|^2 \quad \text{and} \quad \mathbf{j} = \frac{\mathbf{p}}{m} |C|^2 . \tag{32.17}$$

Let us now return to the Klein–Gordon equation and write it in the following form:

$$\frac{1}{c^2} \frac{\partial^2 \phi}{\partial t^2} - \nabla^2 \phi = -\frac{m_0^2 c^2}{\hbar^2} \phi. \tag{32.18}$$

We now determine the corresponding charge and current densities. Repeating the same process of subtraction as in the Schrödinger case we obtain

$$\frac{1}{c^2} \frac{\partial}{\partial t} \left[\phi^* \frac{\partial \phi}{\partial t} - \frac{\partial \phi^*}{\partial t} \phi \right] - \nabla \cdot \left[\phi^* \ (\nabla \phi) - (\nabla \phi^*) \phi \right] = 0. \tag{32.19}$$

The second term on the left-hand side is the same type of term as the second term of (32.13) for the Schrödinger equation if we multiply the equation by $(-\hbar/2im)$,

$$-\frac{\hbar}{2im} \frac{\partial}{c^2 \partial t} \left[\phi^* \frac{\partial \phi}{\partial t} - \frac{\partial \phi^*}{\partial t} \phi \right] + \frac{\hbar}{2im} \nabla \cdot \left[\phi^* \ (\nabla \phi) - (\nabla \phi^*) \phi \right] = 0. \tag{32.20}$$

We write this equation as

$$\frac{\partial}{\partial t} \rho + \nabla \cdot \mathbf{j} = 0. \tag{32.21}$$

Hence for the Klein–Gordon equation we have

$$\rho = -\frac{\hbar}{2imc^2} \left[\phi^* \frac{\partial \phi}{\partial t} - \frac{\partial \phi^*}{\partial t} \phi \right], \tag{32.22}$$

$$\mathbf{j} = \frac{\hbar}{2im} \nabla \cdot \left[\phi^* \ (\nabla \phi) - (\nabla \phi^*) \phi \right]. \tag{32.23}$$

The current density $\mathbf{j}$ is described by the same expression as in the Schrödinger case, but ρ is quite different. Indeed, if we substitute the free particle wavefunction we obtain

$$\rho = \frac{E}{mc^2} |C|^2 , \tag{32.24}$$

which is a different expression from (32.17), while $\mathbf{j}$ remains the same,

$$\mathbf{j} = \frac{\mathbf{p}}{m} |C|^2 . \tag{32.25}$$

The presence of E in ρ is not surprising since ρ is the fourth component of the current four-vector, $j^{\mu} = (\mathbf{j}, c\rho)$, just as E is the fourth component of the momentum four-vector, $p^{\mu} = (\mathbf{p}, E/\mathbf{c})$. Indeed, one can write

$$j^{\mu} = \frac{p^{\mu}}{m} |C|^2 . \tag{32.26}$$

However, because in the relativistic case we have both positive- and negative-energy solutions,

$$E = \pm\sqrt{c^2\mathbf{p}^2 + m_0^2 c^4}, \tag{32.27}$$

the probability density ρ is no longer positive definite. The problem of negative energies persists.

32.3 Normalization of matrix elements

In the nonrelativistic case the scalar product $\langle\phi_1|\phi_2\rangle$ of two state vectors is given by

$$\langle\phi_1|\phi_2\rangle = \int d^3r\, \phi_1^*(x)\, \phi_2(x) \tag{32.28}$$

where ϕ's are normalized wavefunctions, and the probability is written as

$$P = \int d^3r\, \rho(x) \tag{32.29}$$

where ρ is the probability density

$$\rho = \phi^*(x)\phi(x) . \tag{32.30}$$

In the relativistic case the situation is different, as we learned from the previous section. From our discussion on current conservation for the Klein–Gordon equation we found that the probability density is not given by (32.30) but is instead given by

$$\phi^* \left(\frac{\partial \phi}{\partial t}\right) - \left(\frac{\partial \phi^*}{\partial t}\right)\phi. \tag{32.31}$$

In fact, unlike ρ in the nonrelativistic case, which is a scalar, in this case it was found that it is proportional to energy and hence proportional to the fourth component of a four-vector.

In order to write (32.31) more compactly we introduce the symbol $\partial_{(t)}$ defined by

$$\phi_1^*\partial_{(t)}\phi_2 = \phi_1^* \left(\frac{\partial \phi_2}{\partial t}\right) - \left(\frac{\partial \phi_1^*}{\partial t}\right)\phi_2. \tag{32.32}$$

To maintain the fact that the matrix elements follow the same transformation properties as the probability density defined above, the product $\langle\phi_1|\phi_2\rangle$ for the Klein–Gordon wavefunctions will not be defined by (32.28) but by

$$\langle\phi_1|\phi_2\rangle = i\int d^3r\,\phi_1^*(x)\,\overleftrightarrow{\partial}_{(t)}\phi_2(x)\,. \tag{32.33}$$

This definition can be carried over to matrix elements of operators. For example, if A is an operator then

$$\langle\phi_1|A|\phi_2\rangle = i\int d^3r\,\phi_1^*\overleftrightarrow{\partial}_{(t)}(A\phi_2)\,. \tag{32.34}$$

If we write the wavefunction $\phi(x)$ in the same manner as we did for the nonrelativistic case, then the expression for $\phi(x)$ will be

$$\phi(x) = \frac{1}{\left(\sqrt{2\pi}\right)^3}e^{ik\cdot x} \tag{32.35}$$

where $k^\mu = (\omega_k/c, \mathbf{k})$, with ω_k as the energy of the particle $(= \sqrt{c^2\mathbf{k}^2 + m_0^2c^4})$. The product $\langle\phi_1|\phi_2\rangle$ from (32.33) is then given by

$$\langle\phi_1|\phi_2\rangle = 2\omega_k\delta^{(3)}(\mathbf{k}_1 - \mathbf{k}_2)\,. \tag{32.36}$$

On the other hand, a wavefunction normalized to a δ-function will have the form

$$\phi(x) = \frac{1}{\sqrt{2\omega_k}}\frac{1}{\left(\sqrt{2\pi}\right)^3}e^{-ik\cdot x} \tag{32.37}$$

which gives

$$\langle\phi_1|\phi_2\rangle = \delta^{(3)}(\mathbf{k}_1 - \mathbf{k}_2)\,. \tag{32.38}$$

The standard convention for normalization is (32.37), containing the factor $\left(1/\sqrt{2\omega_k}\right)$, which we will use in what follows.

In the Appendix at the end of this chapter we consider the problem of energy levels for the hydrogen atom for a relativistic electron, which now follows the Klein–Gordon equation. The electron will not be a spin ½ particle in this calculation but effectively a spin 0 particle.

32.4 Maxwell's equations

32.4.1 Gauge invariance and zero mass

In view of our discussion on the Klein–Gordon equation it is interesting to return to Maxwell's equation. We start with the definition of the Maxwell tensor, $F_{\mu\nu}$, in terms of the potential, A_μ,

$$F_{\mu\nu} = \partial_\mu A_\nu - \partial_\nu A_\mu. \tag{32.39}$$

As we have pointed out in the earlier chapters, there is a certain arbitrariness in the definition involving A_μ because, for example, one could change

$$A_\mu \to A'_\mu = A_\mu - g\partial_\mu \chi(x) \tag{32.40}$$

while $F_{\mu\nu}$ remains the same. The transformation (32.40) is called a gauge transformation under which $F_{\mu\nu}$ is invariant. Since χ depends on the space-time point x, this is called a "local" gauge transformation.

Maxwell's equation in the absence of sources is given by

$$\partial_\mu F^{\mu\nu} = 0. \tag{32.41}$$

This equation is also gauge invariant since $F^{\mu\nu}$ is invariant.

The equation for A^μ obtained from (32.41) through the definition (31.77) is given by

$$\Box A^\mu + \partial^\mu(\partial_\nu A^\nu) = 0 \tag{32.42}$$

where, as before,

$$\Box = \partial_\nu \partial^\nu. \tag{32.43}$$

We can take advantage of the arbitrariness given by (32.40) and eliminate the second term in (32.42) by imposing the condition

$$\partial_\nu A^\nu = 0. \tag{32.44}$$

Under gauge transformation (32.40) we find

$$\partial_\nu A^\nu \to \partial_\nu A'^\nu = \partial_\nu A^\nu - \partial_\nu \partial^\nu \chi. \tag{32.45}$$

To keep the condition (32.44) invariant under this transformation we choose χ such that

$$\Box \chi = 0. \tag{32.46}$$

The particular choice (32.44) and (32.46) corresponds to what is called the Lorentz gauge. Thus A_μ satisfies the equation

$$\Box A^\mu = 0 \tag{32.47}$$

provided the relations (32.44) and (32.46) are imposed.

Equation (32.47) is now the wave equation for the electromagnetic field. There are actually four equations since A^μ is a four-vector. The solution of (32.47), following the result for the Klein–Gordon equation, is simply

$$A^\mu(x) = \frac{a_0^\mu}{\sqrt{2\omega_k}} \frac{1}{\left(\sqrt{2\pi}\right)^3} e^{-ik\cdot x} \tag{32.48}$$

where a_0^μ is a constant four-vector, and k^μ $(= (\mathbf{k}, \omega_k/c))$ satisfies the equation

$$k^\mu k_\mu = 0 = \mathbf{k}^2 - \frac{\omega_k^2}{c^2}.$$

Therefore, $\omega_k = c\,|\mathbf{k}|$.

Let us compare the above equation for A^μ with the Klein–Gordon equation,

$$\Box\phi = -\frac{m_0^2 c^2}{\hbar^2}\phi. \tag{32.49}$$

We note that (32.47) corresponds to the Klein–Gordon equation for each of the four components of A^μ but with zero mass. As we will observe in our discussions on second quantization, Maxwell's equation for the electromagnetic field can be interpreted as an equation for a particle of zero mass, the photon.

If a mass term is added to (32.47) in the same manner as it appears in the Klein–Gordon equation, then the equation, for A^μ becomes

$$\Box A^\mu + m^2 A^\mu = 0. \tag{32.50}$$

We note that (32.50) is no longer invariant under gauge transformations. The presence of a mass term destroys the gauge invariance. We can restate this situation slightly differently and say that because of gauge invariance the photon is massless, as given by (32.47). We will return to this subject in some of the later chapters.

32.5 Propagators

32.5.1 Maxwell's equation

Consider Maxwell's equation in the presence of a source term given by an external current $J^\mu(x)$,

$$\Box A^\mu(x) = J^\mu(x) \tag{32.51}$$

with

$$\Box = \frac{1}{c^2}\frac{\partial^2}{\partial t^2} - \nabla^2. \tag{32.52}$$

We can write the solution of this equation using the Green's function technique,

$$A^\mu(x) = A_0^\mu(x) + \int d^4x' D_F(x - x')J^\mu(x') \tag{32.53}$$

where $A_0^\mu(x)$ is a solution of the homogeneous equation

$$\Box A_0^\mu(x) = 0 \tag{32.54}$$

and $D_F(x-x')$ is the Green's function, which from (32.51), (32.53), and (32.54) satisfies the equation

$$\Box D_F\left(x-x'\right)=\delta^{(4)}\left(x-x'\right) \tag{32.55}$$

where $\delta^{(4)}\left(x-x'\right)$ is a four-dimensional δ-function. The subscript "F" in D_F refers to a specific type of Green's function which we will discuss below. Since the right-hand side of (32.55) is a function of $\left(x-x'\right)$, D_F will also be a function of $\left(x-x'\right)$. In order to simplify our task, therefore, we will first take $x'=0$. After completing the calculation we will replace x by $\left(x-x'\right)$ to obtain the final solution. Thus we consider the equation

$$\Box D_F\left(x\right)=\delta^{(4)}\left(x\right). \tag{32.56}$$

We will solve equation (32.56) by writing $D_F\left(x\right)$ as a Fourier integral,

$$D_F\left(x\right)=-\int d^4p\, D_F\left(p\right)e^{-ip\cdot x} \tag{32.57}$$

where the differential element is

$$d^4p=dp^1dp^2dp^3dp^4=\frac{1}{c}d^3p\,dE. \tag{32.58}$$

The Dirac δ-function in four-dimensions is also written as a Fourier integral

$$\delta^{(4)}\left(x\right)=\frac{1}{(2\pi)^4}\int d^4p\, e^{-ip\cdot x}. \tag{32.59}$$

The left-hand side of equation (32.56) in terms of the Fourier integral is given by

$$\Box D_F\left(x\right)=\int d^4p\,p^2D_F\left(p\right)e^{-ip\cdot x} \tag{32.60}$$

where $p^2=\left(p^4\right)^2-\mathbf{p}^2$. The right-hand side of (32.56) is given by (32.59). Equating the two, we find that the functional form of $D_F\left(p\right)$ will be

$$\frac{1}{p^2}. \tag{32.61}$$

If we substitute this in (32.57), we notice that the integral will diverge because (32.61) has a pole at $p^2=0$. To obtain a finite result, however, we shift the pole off the real axis and define

$$D_F\left(p\right)=\lim_{\epsilon\to 0}\frac{1}{p^2+i\epsilon}. \tag{32.62}$$

The subscript "F" refers to this specific prescription for the pole. We then obtain

$$D_F\left(x\right)=-\frac{1}{(2\pi)^4}\lim_{\epsilon\to 0}\int d^4p\,\frac{e^{-ip\cdot x}}{p^2+i\epsilon}. \tag{32.63}$$

Replacing x by $(x - x')$, we have

$$D_F\left(x - x'\right) = -\frac{1}{(2\pi)^4} \lim_{\epsilon \to 0} \int d^4p \frac{e^{-ip\cdot(x-x')}}{p^2 + i\epsilon}. \tag{32.64}$$

Before we calculate this function, let us go to the next section on the Klein–Gordon equation and determine the propagator there as well.

32.5.2 Klein–Gordon equation

Let us consider the Klein–Gordon equation in the presence of a source term, $f(x)$,

$$\left(\Box + m^2\right) \phi(x) = f(x). \tag{32.65}$$

The solution for $\phi(x)$ in terms of the Green's function, $\Delta_F\left(x - x'\right)$, is given by

$$\phi(x) = \phi_0(x) + \int d^4x' \, \Delta_F\left(x - x'\right) f\left(x'\right) \phi\left(x'\right) \tag{32.66}$$

where $\phi_0(x)$ and $\Delta_F\left(x - x'\right)$ satisfy the equations

$$\left(\Box + m^2\right) \phi_0(x) = 0 \tag{32.67}$$

and

$$\left(\Box + m^2\right) \Delta_F\left(x - x'\right) = \delta^{(4)}\left(x - x'\right). \tag{32.68}$$

Taking $x' = 0$ again, as we did for the Maxwell case, the above equation reads

$$\left(\Box + m^2\right) \Delta_F(x) = \delta^{(4)}(x). \tag{32.69}$$

Writing $\Delta_F(x)$ in terms of a Fourier integral

$$\Delta_F(x) = \int d^4p \, \Delta_F(p) \, e^{-ip\cdot x} \tag{32.70}$$

and following the same steps as in the Maxwell case, we obtain

$$\Delta_F(p) = -\lim_{\epsilon \to 0} \frac{1}{(2\pi)^4} \frac{1}{p^2 - m^2 + i\epsilon} \tag{32.71}$$

where we have used the same "$+i\epsilon$" as we did for $D_F(x)$. Hence,

$$\Delta_F(x) = -\frac{1}{(2\pi)^4} \lim_{\epsilon \to 0} \int d^4p \, \frac{e^{-ip\cdot x}}{p^2 - m^2 + i\epsilon} \tag{32.72}$$

and

$$\Delta_F\left(x-x'\right) = -\frac{1}{(2\pi)^4}\lim_{\epsilon\to 0}\int d^4p\,\frac{e^{-ip\cdot(x-x')}}{p^2-m^2+i\epsilon}. \tag{32.73}$$

Let us consider the integral in (32.72) in some detail. We first write $d^4p = d^3p\,dp_4$ and consider the poles in the complex p_4-plane. They are given by

$$p_4 = \pm\sqrt{\mathbf{p}^2+m^2+i\epsilon} = \pm\left(\omega_p + i\epsilon'\right) \tag{32.74}$$

where $\omega_p = \sqrt{\mathbf{p}^2+m^2}$ with $\epsilon' = \epsilon/2\omega_p$, which is an infinitesimal quantity. We then write

$$\frac{1}{p^2-m^2+i\epsilon} = \frac{1}{(p_4-\omega_p-i\epsilon')(p_4+\omega_p+i\epsilon')} \tag{32.75}$$

$$= \frac{1}{2\omega_p}\left[\frac{1}{(p_4-\omega_p-i\epsilon')} - \frac{1}{(p_4+\omega_p+i\epsilon')}\right]. \tag{32.76}$$

Thus,

$$\int d^4p\,\frac{e^{-ip\cdot x}}{p^2-m^2+i\epsilon} = \int d^3p\,\frac{e^{i\mathbf{p}\cdot\mathbf{r}}}{2\omega_p}\left[\int_{-\infty}^{\infty} dp_4\,\frac{e^{-ip_4t}}{p_4-\omega_p+i\epsilon'} - \int_{-\infty}^{\infty} dp_4\,\frac{e^{-ip_4t}}{p_4+\omega_p-i\epsilon'}\right]. \tag{32.77}$$

The second term in (32.77) corresponds to a negative-energy pole: $p_4 = -\omega_p$. The space-time dependence of this term is obtained upon integration over p_4 for $t > 0$ and is found to be $e^{i\mathbf{p}\cdot\mathbf{r}}e^{i\omega_p t}$. It represents a plane wave that travels backward in time; that is, as we change $t \to t + \Delta t$ one must have $\mathbf{r} \to \mathbf{r}-\Delta\mathbf{r}$ for the motion of the wavefront. To integrate the second term and keep the motion in the forward direction, we make the change in variables $\mathbf{p} \to -\mathbf{p}$ and $p_4 \to -p_4$, and obtain, after changing the limits,

$$\int d^4p\,\frac{e^{-ip\cdot x}}{p^2-m^2+i\epsilon} = \int d^3p\,\frac{e^{i\mathbf{p}\cdot\mathbf{r}}}{2\omega_p}\int_{-\infty}^{\infty} dp_4\,\frac{e^{-ip_4t}}{p_4-\omega_p+i\epsilon'}$$

$$+\int d^3p\,\frac{e^{-i\mathbf{p}\cdot\mathbf{r}}}{2\omega_p}\int_{-\infty}^{\infty} dp_4\,\frac{e^{ip_4t}}{p_4-\omega_p+i\epsilon'}$$

$$= (-2\pi i)\left[\theta(t)\int d^3p\,\frac{e^{-ip\cdot x}}{2\omega_p} + \theta(-t)\int d^3p\,\frac{e^{ip\cdot x}}{2\omega_p}\right]$$

with

$$p\cdot x = \omega_p t - \mathbf{p}\cdot\mathbf{r} \tag{32.78}$$

where we have made note of the fact that, because the poles corresponding to $p_4 = \omega_p - i\epsilon'$ are in the lower half-plane, we have

$$\int_{-\infty}^{\infty} dp_4 \frac{e^{-ip_4 t}}{p_4 - \omega_p + i\epsilon'} = (-2\pi i)\, e^{-i\omega_p t}\theta(t), \tag{32.79}$$

$$\int_{-\infty}^{\infty} dp_4 \frac{e^{ip_4 t}}{p_4 - \omega_p + i\epsilon'} = (-2\pi i)\, e^{i\omega_p t}\theta(-t). \tag{32.80}$$

Hence $\Delta_F\left(x - x'\right)$ reads

$$\Delta_F\left(x - x'\right) = \frac{i}{(2\pi)^3}\left[\theta\left(t - t'\right)\int d^3p \frac{e^{-ip\cdot(x-x')}}{2\omega_p} + \theta\left(-t + t'\right)\int d^3p \frac{e^{ip\cdot(x-x')}}{2\omega_p}\right]. \tag{32.81}$$

Let us now define the wavefunction of a free particle state with momentum $\mathbf{p}_n$ and energy $\omega_{p_n} = \sqrt{\mathbf{p}_n^2 + m^2}$ as

$$\phi_{n0}^{(-)}(x) = \frac{1}{\left(\sqrt{2\pi}\right)^3}\frac{1}{2\omega_{p_n}} e^{-ip_n.x}. \tag{32.82}$$

The negative sign for the superscript is conventional and indicates a negative sign of the exponent. Note that this state actually corresponds to a positive-energy state, the first term in (32.76) and (32.77). We also define for the positive sign of the exponent

$$\phi_{n0}^{(+)}(x) = \frac{1}{\left(\sqrt{2\pi}\right)^3}\frac{1}{2\omega_{p_n}} e^{ip_n.x}. \tag{32.83}$$

We note that this wavefunction corresponds to a negative-energy state, the second term in (32.76) and (32.77).

Let us rewrite (32.81) as a summation rather than an integral, so that the expression looks like a completeness sum:

$$\Delta_F\left(x - x'\right) = i\left[\theta\left(t - t'\right)\sum_n \psi_{0n}^{(-)}(x)\,\psi_{0n}^{*(-)}\left(x'\right) + \theta\left(t' - t\right)\sum_n \psi_{0n}^{(+)}(x)\,\psi_{0n}^{*(+)}\left(x'\right)\right]. \tag{32.84}$$

This is a time-ordered product, which indicates that the positive-energy terms contribute when $t > t'$, while the negative-energy terms contribute when $t < t'$.

32.6 Virtual particles

In the integral equation (32.66) for $\phi(x)$ in terms of the source, $f(x')$ at x', we note that the link between the two points is provided by $\Delta_F(x-x')$. In a sense, this function carries the "cause" originating at the source of interaction at x' to the point x where the "effect" of the interaction is felt. Hence the name "propagator" is attached to $\Delta_F(x-x')$.

To give a particle interpretation to this process, we note that its Fourier transform, $\Delta_F(p)$, is proportional to $1/(p^2-m^2+i\epsilon)$, where m is the mass of the Klein–Gordon particle. One could say, therefore, from this expression that a "virtual" particle of mass m propagates to provide the link between the two points. It cannot be a real particle for which $p^2=m^2$, though at this point $\Delta_F(p)$ is very large, but not infinite due to the presence of $i\epsilon$. Similarly, one could interpret $D_F(p)$ as representing the propagation of a zero-mass "virtual" particle, a photon.

32.7 Static approximation

We have so far presented our results in a fully four-dimensional space, but often situations arise (e.g., in nonrelativistic calculations), where the time dependence can be ignored. This is what is called the "static approximation." We examine the two equations we have studied, the Klein–Gordon and Maxwell's equations, in the static approximation.

32.7.1 Klein-Gordon equation

Let $\phi(\mathbf{r})$ represent the scalar wavefunction, which now depends only on $\mathbf{r}$. If $f(\mathbf{r})$ is the source then the equation reads (with m replaced by μ)

$$\left(\nabla^2-\mu^2\right)\phi(\mathbf{r})=f(\mathbf{r}). \tag{32.85}$$

The solution is then given by

$$\phi(\mathbf{r})=\phi_0(\mathbf{r})+\int d^3r'\,G(\mathbf{r}-\mathbf{r}')f(\mathbf{r}') \tag{32.86}$$

where $\phi_0(\mathbf{r})$ is the homogeneous solution, and G is the Green's function that satisfies the equation

$$\left(\nabla^2-\mu^2\right)G(\mathbf{r}-\mathbf{r}')=\delta^{(3)}(\mathbf{r}-\mathbf{r}'). \tag{32.87}$$

One can write G in the form of a Fourier transform,

$$G\left(\mathbf{r}-\mathbf{r}'\right)=\frac{1}{(\sqrt{2\pi})^3}\int d^3k\,G(\mathbf{k})e^{i\mathbf{k}.(\mathbf{r}-\mathbf{r}')}. \tag{32.88}$$

From (32.88) $G(\mathbf{k})$ is found to be

$$G(\mathbf{k}) = \frac{1}{\mathbf{k}^2 + \mu^2}. \tag{32.89}$$

Reiterating the arguments made earlier, one could consider $G(\mathbf{k})$ to represent the propagation of a particle of virtual mass μ.

32.7.2 Maxwell's equation

Here by static approximation we also mean that we will ignore the vector field, $\mathbf{A}$. What is left then is just the electrostatic field, which we designate as $\phi_e(\mathbf{r})$. It satisfies the electrostatic equation

$$\nabla^2 \phi_e(\mathbf{r}) = 4\pi e \rho_{el}(\mathbf{r}), \tag{32.90}$$

whose solution is

$$\phi_e(\mathbf{r}) = \phi_0(\mathbf{r}) + 4\pi e \int d^3r' \, D\left(\mathbf{r} - \mathbf{r}'\right) \rho_{el}(\mathbf{r}') \tag{32.91}$$

where $\phi_0(\mathbf{r})$ is the homogeneous solution and D is the Green's function that satisfies the equation

$$\nabla^2 D\left(\mathbf{r} - \mathbf{r}'\right) = \delta^{(3)}\left(\mathbf{r} - \mathbf{r}'\right). \tag{32.92}$$

We write $D\left(\mathbf{r} - \mathbf{r}'\right)$ in the form of a Fourier transform,

$$D\left(\mathbf{r} - \mathbf{r}'\right) = \frac{1}{(\sqrt{2\pi})^3} \int d^3k \, D(\mathbf{k}) e^{i\mathbf{k}.(\mathbf{r}-\mathbf{r}')} \tag{32.93}$$

where

$$D(\mathbf{k}) = \frac{1}{\mathbf{k}^2}. \tag{32.94}$$

$D(\mathbf{k})$ then corresponds to the propagation of a zero-mass particle, the photon.

32.8 Interaction potential in nonrelativistic processes

32.8.1 Coulomb potential

In a typical interaction in electrostatics of an electron of charge $-e$ and a nucleus of charge Ze, the Coulomb potential, $V(r)$, felt by the electron is defined as

$$V(r) = -e\phi_e(\mathbf{r}) \tag{32.95}$$

where $\phi_e(\mathbf{r})$ is the electrostatic field whose source is Ze. It satisfies the equation

$$\nabla^2\phi_e(\mathbf{r}) = 4\pi Ze\delta^{(3)}(\mathbf{r}). \tag{32.96}$$

The solution is obtained by first integrating both sides over a spherical volume of radius r. We find

$$\int d^3r\, \nabla^2\phi_e(\mathbf{r}) = 4\pi Ze. \tag{32.97}$$

We write

$$\nabla^2\phi_e(\mathbf{r}) = (\nabla \cdot \nabla)\, \phi_e(\mathbf{r}). \tag{32.98}$$

Using the divergence theorem to convert the volume integral to a surface integral, we obtain

$$4\pi r^2 \frac{d\phi_e(\mathbf{r})}{dr} = 4\pi Ze. \tag{32.99}$$

Hence, with boundary condition $\phi_e(\infty) = 0$, we find

$$\phi_e(\mathbf{r}) = \frac{Ze}{r}, \tag{32.100}$$

which gives

$$V(r) = -\frac{Ze^2}{r}. \tag{32.101}$$

This is indeed the well-known Coulomb potential between two point charges.

Thus the interaction potential and the electrostatic field of the source are intimately connected.

32.8.2 Yukawa potential

Following essentially the same steps as in the Coulomb case we consider a (static) nuclear field $\phi_Y(\mathbf{r})$ and define the interaction potential between two nuclear charges $-g$ and g as

$$V(r) = -g\phi_Y(\mathbf{r}) \tag{32.102}$$

where $\phi_Y(\mathbf{r})$ is the nuclear field, whose source is the nuclear charge g, which has the same properties as the (scalar) Klein–Gordon wavefunction corresponding to a mass μ. For a point source with charge density $g\delta^{(3)}(\mathbf{r})$, $\phi_Y(\mathbf{r})$ then satisfies the equation

$$\left(\nabla^2 - \mu^2\right)\phi_Y(\mathbf{r}) = 4\pi g\delta^{(3)}(\mathbf{r}). \tag{32.103}$$

The solution is obtained by first making the transformation

$$\phi_Y = \phi_Y' e^{-\mu r} \tag{32.104}$$

in order that the equation (32.103) is exactly like in the Coulomb case (note: $\exp(\mu r)\delta^{(3)}(\mathbf{r}) = \delta^{(3)}(\mathbf{r})$),

$$\nabla^2 \phi'_Y = 4\pi g \delta^{(3)}(\mathbf{r}) \tag{32.105}$$

with the solution

$$\phi_Y(\mathbf{r}) = g\frac{e^{-\mu r}}{r}. \tag{32.106}$$

The potential $V(r)$ is then

$$V(r) = -g^2 \frac{e^{-\mu r}}{r},$$

which is the famous (attractive) Yukawa potential. Once again we confirm an intimate connection between the potential and the field.

32.9 Scattering interpreted as an exchange of virtual particles

We found in the nonrelativistic processes that for a particle scattering off a heavy force center located at the origin, the scattering amplitude is given by

$$f(\theta) = -\left(\frac{m\sqrt{2\pi}}{\hbar^2}\right)\int d^3r'\, e^{-i\mathbf{k}_f \cdot \mathbf{r}'} V(r') u(\mathbf{r}'), \tag{32.107}$$

which in the Born approximation, i.e., in lowest-order perturbation theory, is given by

$$f_B(\theta) = -\frac{m}{2\pi\hbar^2} \int d^3r'\, e^{-i\mathbf{q}\cdot\mathbf{r}'} V(r') \tag{32.108}$$

where

$$\mathbf{q} = \mathbf{k}_f - \mathbf{k}_i \tag{32.109}$$

is the momentum transfer with $\mathbf{k}_f$ and $\mathbf{k}_i$ as the final and initial momenta of the particle.

In the discussion below we will concentrate only on the Born approximation for two typical potentials, the Yukawa and Coulomb potentials, where

$$q^2 = |\mathbf{q}|^2 = \left|\mathbf{k}_f\right|^2 + |\mathbf{k}_i|^2 - 2\mathbf{k}_f \cdot \mathbf{k}_i \cos\theta \tag{32.110}$$

with θ the scattering angle. For elastic scattering, which we are considering,

$$\left|\mathbf{k}_f\right| = |\mathbf{k}_i| = k. \tag{32.111}$$

Therefore,

$$q^2 = 2k^2 (1 - \cos\theta). \tag{32.112}$$

We note that q^2 is positive definite.

32.9.1 Rutherford scattering

Rutherford scattering due to the Coulomb potential corresponds to the scattering of an electron off a heavy charge. If this charge, say Ze, is located at the origin, then $V(r)$ is the familiar Coulomb potential

$$V(r) = -\frac{Ze^2}{r}. \tag{32.113}$$

The scattering amplitude in the Born approximation is then found to be

$$f_B(\theta) = \left(\frac{2mZe^2}{\hbar^2}\right)\frac{1}{\mathbf{q}^2} \tag{32.114}$$

where, repeating the results from the previous section, once again, $\mathbf{q}$ is the momentum transfer and

$$|\mathbf{q}|^2 = \left|\mathbf{k}_f\right|^2 + |\mathbf{k}_i|^2 - 2\mathbf{k}_f \cdot \mathbf{k}_i \cos\theta \tag{32.115}$$

with θ the scattering angle. For elastic scattering, which we are considering,

$$\left|\mathbf{k}_f\right| = |\mathbf{k}_i| = k \tag{32.116}$$

therefore,

$$\mathbf{q}^2 = 2k^2 (1 - \cos\theta). \tag{32.117}$$

We note that the scattering amplitude has the term $(1/\mathbf{q}^2)$, which is reminiscent of the Green's function, in the momentum representation, in Maxwell's equation. Below we will make this apparent connection more concrete.

32.9.2 Rutherford scattering as due to the exchange of a zero-mass particle (photon)

Let us go back to the relation between $V(r)$ and the electrostatic field ϕ_e,

$$V(r) = -e\phi_e(\mathbf{r}) \tag{32.118}$$

and hence

$$f_B(\theta) = \frac{em}{2\pi\hbar^2}\int d^3r'\, e^{-i\mathbf{q}.\mathbf{r}'}\phi_e(\mathbf{r}').$$

The field $\phi_e(\mathbf{r})$ satisfies the equation

$$\nabla^2\phi_e(\mathbf{r}) = 4\pi\,\rho_{el}(\mathbf{r}) \tag{32.119}$$

with $\rho_{el}(\mathbf{r})$ is the charge density of the source. The solution for $\phi_e(\mathbf{r})$ can be written in the Green's function formalism as

$$\phi_e(\mathbf{r}) = \phi_0(\mathbf{r}) + 4\pi \int d^3r'\, D\left(\mathbf{r} - \mathbf{r}'\right) \rho_{el}(\mathbf{r}') \tag{32.120}$$

where $\phi_0(\mathbf{r})$ is the homogeneous solution and D is the Green's function.

Since ϕ_e involved in our calculation for the scattering process is due to the source ρ_{el} and is not a free field, we will be interested only in the second term of (32.120). Therefore, we have

$$\phi_e(\mathbf{r}) = 4\pi \int d^3r'\, D\left(\mathbf{r} - \mathbf{r}'\right) \rho_{el}(\mathbf{r}'). \tag{32.121}$$

For a point charge Ze as the source,

$$\rho_{el}(\mathbf{r}') = Ze\delta^{(3)}(\mathbf{r}'), \tag{32.122}$$

which gives

$$\phi_e(\mathbf{r}) = ZeD\,(\mathbf{r})\,. \tag{32.123}$$

Substituting this in the expression for $f_B\,(\theta)$ we obtain

$$f_B\,(\theta) = \frac{Ze^2m}{2\pi\hbar^2} \int d^3r'\, e^{-i\mathbf{q}.\mathbf{r}'} D\,(\mathbf{r})\,. \tag{32.124}$$

The integral above is simply the Fourier transform $D(\mathbf{q})$, of $D(\mathbf{r})$. Therefore,

$$f_B\,(\theta) = \frac{Ze^2m}{2\pi\hbar^2} D\,(\mathbf{q})\,. \tag{32.125}$$

Substituting the $D(\mathbf{q})$ we have already obtained in (32.94),

$$D\,(\mathbf{q}) = \frac{1}{\mathbf{q}^2}, \tag{32.126}$$

we recover the result (32.114) for $f_B\,(\theta)$. Hence we find that the denominator term $(1/\mathbf{q}^2)$ that appears in $f_B\,(\theta)$ in (32.114) is actually the Green's function involved in the solution to the electrostatic problem.

We recall from our discussions in the previous chapter that G corresponds to the propagation of a "virtual" photon. Thus in the Coulomb amplitude what appeared to be simply an outcome of a mathematical integration has a deeper physical meaning. In fact, it implies that Rutherford scattering can be interpreted as the process in which an electron scatters off a heavy charge through the exchange of a virtual photon.

32.9.3 Yukawa scattering as due to the exchange of a massive particle

For a Yukawa potential given by

$$V(r) = -g\frac{e^{-\mu r}}{r} \tag{32.127}$$

the scattering amplitude is found to be

$$f_B(\theta) = \frac{2mg}{\hbar^2}\left(\frac{1}{\mu^2 + \mathbf{q}^2}\right). \tag{32.128}$$

One can follow the same steps as for the Coulomb case, but instead of the electric charge we have a nuclear charge g. We then obtain

$$f_B(\theta) = \frac{2mg^2}{\hbar^2}G(\mathbf{p}) \tag{32.129}$$

where $G(\mathbf{p})$ is the appropriate Green's function given by

$$G(\mathbf{p}) = \frac{1}{\mathbf{p}^2 + \mu^2}. \tag{32.130}$$

Substituting this in (32.129) we find that the $f_B(\theta)$ we obtained above is the same as we obtained directly from the Yukawa potential.

We conclude then that the Yukawa scattering can be viewed as the process in which a particle with nuclear charge scatters off another nuclear particle by the exchange of a virtual (Yukawa) particle of mass μ.

Rutherford and Yukawa scattering are interpreted pictorially as particle exchange processes in Fig. 32.1, where the wavy line corresponds to the virtual particle that is exchanged. The solid lines correspond to the particles that undergo scattering. We might add that higher-order perturbation terms correspond to multiple exchanges of virtual particles. The particle interpretation and the general connection between particles and fields will be firmed up when we go to the quantum field theory.

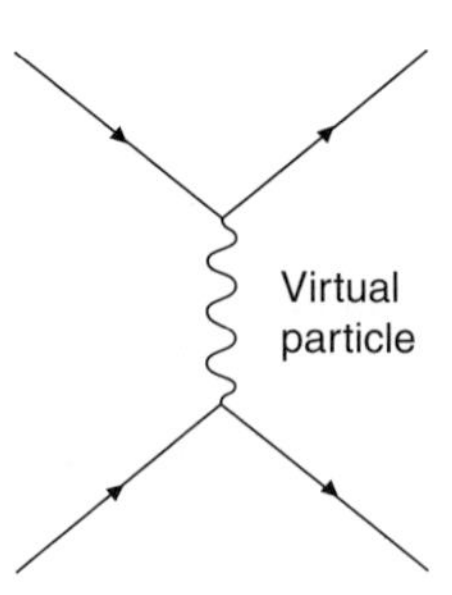

Fig. 32.1

32.10 Appendix to Chapter 32

32.10.1 Hydrogen atom

As we discussed previously, one can easily take into account the presence of electromagnetic interactions by the following transformations of the energy E and momentum $\mathbf{p}$:

$$E \to E - e\phi \quad \text{and} \quad \mathbf{p} \to \mathbf{p} - \frac{e}{c}\mathbf{A} \tag{32.131}$$

where $\mathbf{A}$ is the vector potential and ϕ is the scalar potential. Therefore, the energy–momentum relation for a free particle given by

$$E^2 = c^2\mathbf{p}^2 + m^2c^4 \tag{32.131a}$$

now becomes

$$(E - e\phi)^2 = (c\mathbf{p} - e\mathbf{A})^2 + m^2c^4 \tag{32.131b}$$

We keep in mind the operator relations

$$E \to i\hbar\frac{\partial}{\partial t} \quad \text{and} \quad \mathbf{p} \to -i\hbar\boldsymbol{\nabla}, \tag{32.132}$$

so that (32.131b) implies an operator relation on the wavefunction $\psi(\mathbf{r}, t)$.

Since we are considering the hydrogen atom problem, we need only be concerned with the electrostatic Coulomb interaction. Therefore,

$$\mathbf{A} = \mathbf{0}, \quad \phi = -\frac{Ze}{r}. \tag{32.133}$$

Using the separation of variables technique we write

$$\psi(\mathbf{r}, t) = u(\mathbf{r})\, e^{-iEt/\hbar}. \tag{32.133a}$$

The Klein–Gordon equation corresponding to (32.131b) can, therefore, be written as

$$[E - e\phi(r)]^2\, u(\mathbf{r}) = \left(-\hbar^2c^2\nabla^2 + m^2c^4\right) u(\mathbf{r}) \tag{32.133b}$$

where E is now the energy eigenvalue. We rewrite (32.133b) as

$$\nabla^2 u = \frac{[E - e\phi(r)]^2 - m^2c^4}{\hbar^2c^2} u. \tag{32.134}$$

Following the same procedure that was used for the nonrelativistic hydrogen atom, we do a further separation of variables into radial coordinates and the spherical harmonics expressed in terms of the angular momentum quantum numbers, l and m,

$$u(r, \theta, \phi) = R_l(r)Y_{lm}(\theta, \phi). \tag{32.135}$$

The ∇^2 operator can, once again, be expressed in terms of angular momentum so, in a manner very similar to the previous hydrogen atom problem, we will obtain the following radial equation:

$$\left[-\frac{1}{r^2}\frac{d}{dr}\left(r^2\frac{d}{dr}\right)+\frac{l(l+1)}{r^2}\right]R_l=\left[\frac{(E-e\phi)^2-m^2c^4}{\hbar^2c^2}\right]R_l. \tag{32.136}$$

Using (32.133), the right-hand side of the above equation can be written as

$$RHS=\frac{1}{\hbar^2c^2}\left[E^2-m^2c^4+\frac{2e^2ZE}{r}+\frac{Z^2e^4}{r^2}\right]. \tag{32.137}$$

To simplify the equations let us introduce the following notations:

$$\gamma=\frac{Ze^2}{\hbar c},\quad \alpha^2=\frac{4\left(m^2c^4-E^2\right)}{\hbar^2c^2},\quad \lambda=\frac{2E\gamma}{\hbar c\alpha},\quad \rho=\alpha r. \tag{32.138}$$

From these relations we can easily derive the following expression for E,

$$E=mc^2\left(1+\frac{\gamma^2}{\lambda^2}\right)^{-\frac{1}{2}}. \tag{32.139}$$

The radial equation now reads

$$\frac{1}{\rho^2}\frac{d}{d\rho}\left(\rho^2\frac{dR}{d\rho}\right)+\left[\frac{\lambda}{\rho}-\frac{1}{4}-\frac{l(l+1)-\gamma^2}{\rho^2}\right]R=0. \tag{32.140}$$

The eigenvalue equation for the nonrelativistic hydrogen atom that we discussed in Chapter 8 is given by

$$\frac{d^2R_l}{d\rho^2}+\frac{2}{\rho}\frac{dR_l}{d\rho}+\left[\frac{\lambda}{\rho}-\frac{1}{4}-\frac{l(l+1)}{\rho^2}\right]R_l=0 \tag{32.141}$$

where

$$\lambda=\frac{2mZe^2}{\hbar^2\alpha}\quad\text{and}\quad\frac{2mE_B}{\hbar^2\alpha^2}=\frac{1}{4}. \tag{32.142}$$

We found the bound-state energy eigenvalues to be

$$E_B=\frac{mZ^2e^4}{2\hbar^2\lambda^2}. \tag{32.143}$$

Apart from the presence of γ^2 in the numerator of the third term in the square bracket in (32.140), the two equations are the same. One can write

$$R_l=\rho^sL(\rho)\,e^{-\frac{1}{2}\rho} \tag{32.144}$$

where L is expressed as a power series,

$$L = a_0 + a_1\rho + \cdots + a_{n'}\rho^{n'}. \tag{32.145}$$

Substituting (32.145) and (32.144) in (32.110) it is found that

$$s(s+1) = l(l+1) - \gamma^2. \tag{32.146}$$

Proceeding as we did in the nonrelativistic case, one finds

$$\lambda = n' + s + 1 = n \tag{32.147}$$

where n is the principal quantum number, and n' is the radial quantum number as in the nonrelativistic case. The solution of (32.146) is given by

$$s = -\frac{1}{2} \pm \frac{1}{2}\left[(2l+1)^2 - 4\gamma^2\right]^{\frac{1}{2}}. \tag{32.148}$$

Substituting (32.148) in (32.147) we obtain

$$\lambda = n' + \frac{1}{2} + \left[\left(l + \frac{1}{2}\right)^2 - \gamma^2\right]^{\frac{1}{2}}. \tag{32.149}$$

Substituting λ given above in the expression for E in (32.139) we obtain

$$E = \frac{mc^2}{\left[1 + \dfrac{\gamma^2}{n' + \frac{1}{2} + \sqrt{\left[\left(l+\frac{1}{2}\right)^2 - \gamma^2\right]}}\right]^{\frac{1}{2}}}. \tag{32.150}$$

One can express the above result as a power series in γ which is proportional to e^2. We find, up to power γ^4,

$$E = mc^2\left[1 - \frac{\gamma^2}{2n^2} - \frac{\gamma^4}{2n^4}\left(\frac{n}{l + \frac{1}{2}} - \frac{3}{4}\right)\right]. \tag{32.151}$$

The first term corresponds to the rest energy, mc^2, of the electron. The second term is given by

$$-\frac{mc^2\gamma^2}{2n^2} = -\frac{mZ^2e^4}{2\hbar^2n^2} \tag{32.152}$$

which is the same as in the nonrelativistic hydrogen atom.

The third term, proportional to γ^4, is the new term which is a consequence of relativity. It removes the degeneracy between different values of l. This is the "fine" structure that separates the nonrelativistic from the relativistic hydrogen atom. However, this contribution is found to be inconsistent with the experimental results. As we will see when we come to the Dirac equation, an electron considered as a Dirac particle with spin ½ gives correct results.

33 The Dirac equation

Below we outline the formulation of the Dirac equation incorporating momentum and energy as linear terms in the Hamiltonian. We find that it leads to the prediction of particles with spin ½.

33.1 Basic formalism

We found in our discussion on the subject of relativistic quantum mechanics in Chapter 31 that, as a consequence of the quadratic nature of the energy–momentum relation, negative energy solutions must exist and that the probability density is not positive definite. To solve these problems, Dirac set out to write a quantum-mechanical equation that was linear in energy and momentum based on the following Hamiltonian for a free particle,

$$H = c\boldsymbol{\alpha} \cdot \mathbf{p} + \beta m_0 c^2, \tag{33.1}$$

with m_0 the rest mass, $\mathbf{p}$ the momentum of the particle and the unknown $\boldsymbol{\alpha}$ and β to be determined from relativistic considerations. We have temporarily reinstated $\hbar$ and c. Below we discuss the consequences of Dirac's postulate.

In the four-dimensional x-space the free-particle Hamiltonian has the form

$$H = -i\hbar c\boldsymbol{\alpha} \cdot \boldsymbol{\nabla} + \beta m_0 c^2 \tag{33.2}$$

where we have replaced $\mathbf{p}$ by the operator $-i\hbar\boldsymbol{\nabla}$. The eigenvalue equations for momentum and energy are given by the usual relations

$$-i\hbar\boldsymbol{\nabla}\psi(x) = \mathbf{p}\psi(x) \tag{33.3}$$

and

$$i\hbar\frac{\partial\psi(x)}{\partial t} = H\psi(x) = E\psi(x) \tag{33.4}$$

where $\psi(x)$ is the eigenfunction. The solution for a free particle with energy, E, and momentum, $\mathbf{p}$, is obtained in the same manner as in Chapter 12. It is of the form

$$\psi(x) = u(p)\, e^{-\frac{ip\cdot x}{\hbar}} \tag{33.5}$$

where we have replaced the normalization constant, C, by $u(p)$, which is independent of x but depends on p. We will discuss the functional form and the normalization of $u(p)$ in Chapter 35.

In the following we will take $\hbar = c = 1$, and $m_0 = m$ in order to simplify the calculations. The Dirac equation is then given by

$$(E - \boldsymbol{\alpha} \cdot \mathbf{p} - \beta m)\, u(p) = 0. \tag{33.6}$$

We note that E in the above equation multiplies a unit operator. The energy–momentum relation for a particle is given by the quadratic formula $E^2 - \mathbf{p}^2 - m^2 = 0$. Therefore, besides (33.6), $u(p)$ must also satisfy the equation

$$\left(E^2 - \mathbf{p}^2 - m^2\right) u = 0. \tag{33.7}$$

In order to relate (33.6) and (33.7) we multiply (33.6) on the left by $(E + \boldsymbol{\alpha} \cdot \mathbf{p} + \beta m)$ and obtain a quadratic relation between E and $\mathbf{p}$, which we compare with (33.7). We need, therefore, to satisfy the relation

$$(E + \boldsymbol{\alpha} \cdot \mathbf{p} + \beta m)\,(E - \boldsymbol{\alpha} \cdot \mathbf{p} - \beta m) = \left(E^2 - \mathbf{p}^2 - m^2\right), \tag{33.8}$$

which implies that

$$E^2 - \sum_i (\alpha_i p_i)(\sum_j \alpha_j p_j) - \sum_i (\alpha_i \beta + \beta \alpha_i) p_i - \beta^2 m^2 = E^2 - \mathbf{p}^2 - m^2 \tag{33.9}$$

where we have kept the order of multiplication intact and where we write the dot product $\boldsymbol{\alpha} \cdot \mathbf{p}$ in the expanded form $\sum_i \alpha_i p_i$, with i taking on the values $1, 2, 3$. We take $\mathbf{p} = (p_1, p_2, p_3)$ with $\mathbf{p}^2 = p_1^2 + p_2^2 + p_3^2$.

Comparing both sides in (33.9), we have the following relations for α_i and β with $i,j = 1, 2, 3$:

$$\alpha_i^2 = \mathbf{1} = \beta^2, \tag{33.10}$$

$$\alpha_i \beta + \beta \alpha_i = 0, \tag{33.11}$$

$$\alpha_i \alpha_j + \alpha_j \alpha_i = 0 \quad (i \neq j). \tag{33.12}$$

First of all, it is clear that the above relations are incompatible if β and the α_i's are numbers, otherwise we would have to conclude from (33.10) that $\alpha_i = \pm 1$ and $\beta = \pm 1$, a result that is inconsistent with (33.11) and (33.12). Thus they must be matrices and as such they will not necessarily commute. The following properties then hold for α_i and β.

(A) *Square matrices*

Since the Hamiltonian must be Hermitian,

$$H^\dagger = H. \tag{33.13}$$

Therefore, since **p** is Hermitian, we must have

$$\alpha_i^\dagger = \alpha_i \tag{33.14}$$

and

$$\beta^\dagger = \beta. \tag{33.15}$$

Considering β for example, if β_{kl} is a (kl) matrix element, then (33.15) implies that there must be an (lk) element, which then satisfies the relation $\beta_{lk} = \beta_{kl}^*$. Hence β must be a square matrix. Similar arguments apply to the α_i's. Therefore, α_i and β must be square matrices. They will be matrices of dimensions $2 \times 2, 3 \times 3, 4 \times 4, \ldots$.

(B) *Traceless*
From (33.11) we obtain

$$\alpha_i = -\beta\alpha_i\beta^{-1}. \tag{33.16}$$

Taking the trace of both sides, and using the cyclic property of the trace of a product of matrices we get

$$\mathrm{Tr}(\alpha_i) = -\mathrm{Tr}\left(\beta\alpha_i\beta^{-1}\right) = -\mathrm{Tr}\left(\beta^{-1}\beta\alpha_i\right) = -\mathrm{Tr}\left(\alpha_i\right). \tag{33.17}$$

Hence $\mathrm{Tr}(\alpha_i)$ must vanish. A similar result will hold for $\mathrm{Tr}(\beta)$. Thus,

$$\mathrm{Tr}(\alpha_i) = 0 = \mathrm{Tr}(\beta). \tag{33.18}$$

(C) *Even rank*
We have the freedom to choose one of the four matrices, α_i and β, to be diagonal. Let us take β to be diagonal. From (33.10) we note that the eigenvalues (i.e., the diagonal values) of β must be ± 1. We can arrange them in a block form such that all the $+1$'s are separated from the -1's. However, because of the tracelessness condition (33.18), one must have the same number of $+1$'s as -1's. Hence β must be even-dimensional. The same argument can be applied to the α_i's. Thus α_i and β must be $2 \times 2, 4 \times 4, \ldots$.

(D) *At least* 4×4
The two-dimensional matrices that satisfy anticommutation relations of the type (33.11) are the Pauli matrices σ_i $(i = 1, 2, 3)$, which we have discussed in Chapter 2 and have the property

$$\sigma_i\sigma_j + \sigma_j\sigma_i = 0 \quad (i \neq j), \tag{33.19}$$

$$\sigma_i^\dagger = \sigma_i, \tag{33.20}$$

$$\sigma_i^2 = 1. \tag{33.21}$$

They have the representation

$$\sigma_1 = \begin{pmatrix} 0 & 1 \\ 1 & 0 \end{pmatrix}, \quad \sigma_2 = \begin{pmatrix} 0 & -i \\ i & 0 \end{pmatrix}, \quad \sigma_3 = \begin{pmatrix} 1 & 0 \\ 0 & -1 \end{pmatrix}. \tag{33.22}$$

As we discussed in Chapter 2, there are only four independent 2×2 matrices. These can be taken as the three σ_i's and a unit matrix. Among them, only three, the σ_i's, anticommute. In the case of the Dirac equation, however, we have four matrices all of which anticommute. Hence these matrices cannot be 2×2. A simple way to confirm this is to take $\alpha_i = \sigma_i$ and

$$\beta = \begin{pmatrix} a & b \\ c & d \end{pmatrix}. \tag{33.23}$$

Imposing condition (33.11) leads to $a = 0 = b = c = d$, i.e, $\beta = 0$, which is untenable.

Thus the minimal dimensionality of the matrices α_i and β must be 4×4.

33.2 Standard representation and spinor solutions

From Section 33.1 we found that the matrices α_i and β have the properties that they are: (i) Hermitian, (ii) traceless, and (iii) even-dimensional with a minimum dimensionality of 4×4.

Let us consider a 4×4 representation. Following the above discussions we will take β as a diagonal matrix and write it as

$$\beta = \begin{bmatrix} 1 & 0 & 0 & 0 \\ 0 & 1 & 0 & 0 \\ 0 & 0 & -1 & 0 \\ 0 & 0 & 0 & -1 \end{bmatrix}. \tag{33.24}$$

We will write this matrix in a condensed 2×2 block form where the individual entries are themselves 2×2 matrices,

$$\beta = \begin{bmatrix} 1 & 0 \\ 0 & -1 \end{bmatrix} \tag{33.25}$$

where by 1 we mean a two-dimensional unit matrix. To determine the matrix representation of α_i we write it in the form

$$\alpha_i = \begin{bmatrix} \mathbf{a}_i & \mathbf{b}_i \\ \mathbf{c}_i & \mathbf{d}_i \end{bmatrix} \tag{33.26}$$

where the individual matrix elements are 2×2 matrices. Substituting this into (33.11) with the matrix representation β given in (33.25) we find

$$\begin{bmatrix} 2\mathbf{a}_i & 0 \\ 0 & -2\mathbf{d}_i \end{bmatrix} = 0. \tag{33.27}$$

Thus,

$$\mathbf{a}_i = 0 = \mathbf{d}_i. \tag{33.28}$$

From (33.12) we obtain

$$\mathbf{b}_i\mathbf{c}_j + \mathbf{b}_j\mathbf{c}_i = 0. \tag{33.29}$$

Moreover, since $\alpha_i^\dagger = \alpha_i$, and $\alpha_i^2 = \mathbf{1}$ we have $\mathbf{b}_i^\dagger = \mathbf{c}_i$ and $\mathbf{b}_i^2 = \mathbf{1} = \mathbf{c}_i^2$. We note, from the relations (33.10), (33.11), and (33.12) that the matrices satisfying the above requirements are the Pauli matrices, σ_i, themselves. Hence, we take

$$\mathbf{b}_i = \sigma_i = \mathbf{c}_i. \tag{33.30}$$

In summary, then, we have the following representation:

$$\beta = \begin{bmatrix} 1 & 0 \\ 0 & -1 \end{bmatrix}, \tag{33.31}$$

$$\alpha_i = \begin{bmatrix} 0 & \sigma_i \\ \sigma_i & 0 \end{bmatrix}. \tag{33.32}$$

This is called the "standard representation" for the matrices α_i and β.

Since α_i and β are 4×4 matrices, the function u will be a column matrix,

$$u(p) = \begin{bmatrix} \checkmark \\ \checkmark \\ \checkmark \\ \checkmark \end{bmatrix}. \tag{33.33}$$

Following the prescriptions for α_i and β, we also write $u(p)$ with two entries,

$$u(p) = \begin{bmatrix} u_1 \\ u_2 \end{bmatrix} \tag{33.34}$$

where u_1 and u_2 are two-dimensional column matrices. In the next section we obtain some of the important properties of $u(p)$.

33.3 Large and small components of $u(p)$

We determine $u(p)$ by first writing the Dirac equation in (33.6) in a 2×2 form,

$$\left(E\begin{bmatrix} 1 & 0 \\ 0 & 1 \end{bmatrix} - \begin{bmatrix} 0 & \boldsymbol{\sigma}.\mathbf{p} \\ \boldsymbol{\sigma}.\mathbf{p} & 0 \end{bmatrix} - \begin{bmatrix} 1 & 0 \\ 0 & -1 \end{bmatrix} m\right)\begin{bmatrix} u_1 \\ u_2 \end{bmatrix} = 0, \tag{33.35}$$

which simplifies to

$$\begin{bmatrix} E-m & -\boldsymbol{\sigma}\cdot\mathbf{p} \\ -\boldsymbol{\sigma}\cdot\mathbf{p} & E+m \end{bmatrix}\begin{bmatrix} u_1 \\ u_2 \end{bmatrix} = 0. \tag{33.36}$$

We then have two coupled equations:

$$(E-m)\,u_1 - \boldsymbol{\sigma}\cdot\mathbf{p}u_2 = 0, \tag{33.37}$$

$$-\boldsymbol{\sigma}\cdot\mathbf{p}u_1 + (E+m)\,u_2 = 0. \tag{33.38}$$

A solution to the coupled equations exist provided the determinant of the 2×2 matrix formed by the coefficients of the matrix in (33.36) vanish, i.e., if

$$E^2 - \mathbf{p}^2 - m^2 = 0 \tag{33.39}$$

where we have used the relation $\boldsymbol{\sigma}\cdot\mathbf{p}\boldsymbol{\sigma}\cdot\mathbf{p} = \mathbf{p}^2$. Equation (33.39) is the same as the relativistic energy–momentum relation with positive- and negative-energy solutions $E = \pm\sqrt{\mathbf{p}^2 + m^2}$.

33.3.1 Positive-energy solution

First let us consider the positive-energy solution

$$E = +\sqrt{\mathbf{p}^2 + m^2} = |E|\,. \tag{33.40}$$

We note that in the nonrelativistic limit, $|\mathbf{p}| \ll m$, we have $|E| \approx m$. Therefore, in equation (33.38), in the nonrelativistic limit, the coefficient of u_2 is much larger than the coefficient of u_1. This equation can be satisfied only if

$$u_2 \ll u_1. \tag{33.41}$$

We thus designate u_2 as the "small component," and u_1 as the "large component," and write them as

$$u_1 = u_L, \tag{33.42}$$

$$u_2 = u_S. \tag{33.43}$$

Hence,

$$u = \begin{bmatrix} u_L \\ u_S \end{bmatrix}. \tag{33.44}$$

Equations (33.37) and (33.38) then correspond to

$$(|E| - m)\,u_L - \boldsymbol{\sigma}\cdot\mathbf{p}u_S = 0, \tag{33.45}$$

$$-\boldsymbol{\sigma}\cdot\mathbf{p}u_L + (|E| + m)\,u_S = 0, \tag{33.46}$$

where we have written $E = |E| = \sqrt{\mathbf{p}^2 + m^2}$ for reasons that will be clear later when we consider negative-energy solutions.

From (33.46) we obtain

$$u_S = \frac{\boldsymbol{\sigma} \cdot \mathbf{p}}{(|E| + m)} u_L. \tag{33.47}$$

Substituting this result in (33.45) gives

$$\left[(|E| - m) - \boldsymbol{\sigma} \cdot \mathbf{p} \frac{\boldsymbol{\sigma} \cdot \mathbf{p}}{(|E| + m)} \right] u_L = 0. \tag{33.48}$$

Since

$$(\boldsymbol{\sigma} \cdot \mathbf{p})\, (\boldsymbol{\sigma} \cdot \mathbf{p}) = \mathbf{p}^2 \tag{33.49}$$

and since in the nonrelativistic limit, $|E| \approx m$, we can replace the denominator in the second term on the left by $2m$, we can write (33.48) as

$$\left(\frac{\mathbf{p}^2}{2m} + m \right) u_L = E u_L. \tag{33.50}$$

We can write this equation as $Hu_L = Eu_L$, where H is a 2×2 matrix and u_L is a column matrix with two entries. This is just the restatement of the fact that the total energy in the nonrelativistic limit is the sum of the kinetic energy and the rest mass.

We note that $\sigma_3 (= \sigma_z)$ commutes with this H corresponding to (33.50). Hence we take u_L to be also an eigenstate of σ_3. We can then express u_L's as "spin-up" and "spin-down" matrices given by

$$u_L^+ = \begin{bmatrix} 1 \\ 0 \end{bmatrix} \quad \text{and} \quad u_L^- = \begin{bmatrix} 0 \\ 1 \end{bmatrix}. \tag{33.51}$$

Correspondingly, expressing the Pauli matrices σ_i's in the 2×2 form, we have from (33.47)

$$u_S^+ = \frac{\boldsymbol{\sigma} \cdot \mathbf{p}}{(|E| + m)} u_L^+ = \frac{1}{(|E| + m)} \begin{bmatrix} p_z & p_x - ip_y \\ p_x + ip_y & -p_z \end{bmatrix} \begin{bmatrix} 1 \\ 0 \end{bmatrix} \tag{33.52}$$

$$= \frac{1}{(|E| + m)} \begin{bmatrix} p_z \\ p_x + ip_y \end{bmatrix}. \tag{33.53}$$

Similarly, we have

$$u_S^- = \frac{\boldsymbol{\sigma} \cdot \mathbf{p}}{(|E| + m)} u_L^- = \frac{1}{(|E| + m)} \begin{bmatrix} p_z & p_x - ip_y \\ p_x + ip_y & -p_z \end{bmatrix} \begin{bmatrix} 0 \\ 1 \end{bmatrix} \tag{33.54}$$

$$= \frac{1}{(|E| + m)} \begin{bmatrix} p_x - ip_y \\ -p_z \end{bmatrix}. \tag{33.55}$$

In summary, the positive-energy solutions are

$$u^{+}(p) = C\begin{bmatrix} u_L^{+} \\ \dfrac{\boldsymbol{\sigma}\cdot\mathbf{p}}{(|E|+m)}u_L^{+} \end{bmatrix} = C\begin{bmatrix} 1 \\ 0 \\ \dfrac{p_z}{(|E|+m)} \\ \dfrac{(p_x+ip_y)}{(|E|+m)} \end{bmatrix}, \tag{33.56}$$

$$u^{-}(p) = C\begin{bmatrix} u_L^{-} \\ \dfrac{\boldsymbol{\sigma}\cdot\mathbf{p}}{(|E|+m)}u_L^{-} \end{bmatrix} = C\begin{bmatrix} 0 \\ 1 \\ \dfrac{(p_x-ip_y)}{(|E|+m)} \\ -\dfrac{p_z}{(|E|+m)} \end{bmatrix}, \tag{33.57}$$

where C is a constant to be determined from the normalization condition of the u's.

33.3.2 Negative-energy solutions

For negative energies, $E = -|E|$, examining (33.37) and (33.38) it is clear that the roles of the small and large components are reversed. Hence we have

$$u_1 = u_S, \tag{33.58}$$

$$u_2 = u_L, \tag{33.59}$$

and

$$u = \begin{bmatrix} u_S \\ u_L \end{bmatrix}, \tag{33.60}$$

with equations

$$(-|E| - m)\,u_S - \boldsymbol{\sigma}\cdot\mathbf{p}u_L = 0, \tag{33.61}$$

$$-\boldsymbol{\sigma}\cdot\mathbf{p}u_S + (-|E| + m)\,u_L = 0. \tag{33.62}$$

From (33.61) we obtain

$$u_S = -\frac{\boldsymbol{\sigma}\cdot\mathbf{p}}{(|E|+m)}u_L. \tag{33.63}$$

Following arguments similar to the positive-energy case, we obtain

$$u^{+}(p) = C\begin{bmatrix} -\dfrac{\boldsymbol{\sigma}\cdot\mathbf{p}}{(|E|+m)}u_L^{+} \\ u_L^{+} \end{bmatrix} = C\begin{bmatrix} -\dfrac{p_z}{(|E|+m)} \\ -\dfrac{(p_x+ip_y)}{(|E|+m)} \\ 1 \\ 0 \end{bmatrix}, \tag{33.64}$$

$$u^{-}(p) = C\begin{bmatrix} -\dfrac{\boldsymbol{\sigma}\cdot\mathbf{p}}{(|E|+m)}u_L^{-} \\ u_L^{-} \end{bmatrix} = C\begin{bmatrix} -\dfrac{(p_x-ip_y)}{(|E|+m)} \\ \dfrac{p_z}{(|E|+m)} \\ 0 \\ 1 \end{bmatrix}. \tag{33.65}$$

We will defer the discussion on the normalization constant to Chapter 35.

33.4 Probability conservation

Let us consider the time-dependent equation

$$i\frac{\partial\psi}{\partial t} = H\psi = -i\boldsymbol{\alpha}\cdot\nabla\psi + \beta m\psi \tag{33.66}$$

where ψ is a column matrix. The Hermitian conjugate of this equation is given by

$$-i\frac{\partial\psi^{\dagger}}{\partial t} = i\left(\nabla\psi^{\dagger}\right)\cdot\boldsymbol{\alpha} + \beta m\psi^{\dagger}. \tag{33.67}$$

where we have made use of the fact that $\boldsymbol{\alpha}^{\dagger} = \boldsymbol{\alpha}$, $\beta^{\dagger} = \beta$. We multiply (33.66) on the left by $\psi^{\dagger}$ and subtract from the product the product obtained by multiplying expression (33.67) on the right by ψ. We obtain the following:

$$\frac{\partial(\psi^{\dagger}\psi)}{\partial t} = -\nabla\cdot\left(\psi^{\dagger}\boldsymbol{\alpha}\psi\right). \tag{33.68}$$

This relation can be expressed as

$$\frac{\partial\rho}{\partial t} + \nabla\cdot\mathbf{j} = 0 \tag{33.69}$$

where we define

$$\rho = \psi^{\dagger}\psi \quad \text{and} \quad \mathbf{j} = \psi^{\dagger}\boldsymbol{\alpha}\psi. \tag{33.70}$$

This is then the probability conservation relation, where ρ is the probability density and, as we will confirm below by going to the nonrelativistic limit, $\mathbf{j}$ has the properties of a current density.

We arrive at a very important result that in contrast to the Klein–Gordon case, ρ for the Dirac equation is positive definite. The vector $\mathbf{j}$ will be identified as the current density by first writing it in terms of large and small components for positive-energy solutions. From (33.70) and from the relations for u given in (33.60) along with the definition of $\boldsymbol{\alpha}$ given in (33.32), we obtain

$$\mathbf{j} = \left[u_L^\dagger \boldsymbol{\sigma} u_S + u_S^\dagger \boldsymbol{\sigma} u_L\right]. \tag{33.71}$$

To determine (33.71) let us consider a positive-energy particle with spin up traveling in the z-direction, which in the nonrelativistic approximation, with $|E| \simeq m$, gives

$$u_L = C\begin{pmatrix} 1 \\ 0 \end{pmatrix}, \quad u_S = \frac{Cp_z}{2m}\begin{pmatrix} 1 \\ 0 \end{pmatrix}. \tag{33.72}$$

Substituting these in the expression for $\mathbf{j}$ in (33.71), we find that the only nonzero component of $\mathbf{j}$ is the z-component, given by

$$j_z = \frac{p_z}{m}|C|^2. \tag{33.73}$$

Since p_z/m is the velocity, the current is proportional to the velocity as expected.

We will come to the same conclusion, without going into the nonrelativistic approximation, if we consider the Heisenberg equation

$$i\hbar\frac{dz}{dt} = [z, H]. \tag{33.74}$$

Only the term $\boldsymbol{\alpha} \cdot \mathbf{p}$ in H will contribute to this commutator, so that

$$[z, H] = [z, \boldsymbol{\alpha} \cdot \mathbf{p}] = [z, p_z]\alpha_z = i\hbar\alpha_z \tag{33.75}$$

where we have taken into account the fact that z commutes with p_x and p_y, and have taken $\alpha_3 = \alpha_z$. Equation (33.74), therefore, implies

$$\frac{dz}{dt} = \alpha_z. \tag{33.76}$$

Hence, $j_z = u^\dagger\alpha_z u$ is proportional to the velocity $v_z = dz/dt$ and therefore $\mathbf{j}$, defined by (33.70), has the properties of a current. Thus, equation (33.69) correctly signifies the probability conservation relation.

The result that the probability density, ρ, is positive definite is a fundamental breakthrough achieved by Dirac.

We now go to the next section in which we discover another extraordinary property of this equation – that the Dirac particle has an intrinsic angular momentum, the "spin", of ½.

33.5 Spin ½ for the Dirac particle

The time derivative of the angular momentum operator, **L**, in the Heisenberg representation is given by

$$i\hbar\frac{dL_i}{dt} = [L_i, H], \quad i = 1, 2, 3. \tag{33.77}$$

Writing the relation $\mathbf{L} = \mathbf{r} \times \mathbf{p}$ in the component form we obtain

$$L_i = \epsilon_{ijk} r_j p_k \tag{33.78}$$

where ϵ_{ijk} is the totally antisymmetric tensor and where we adopt the convention that a repeated index implies summation over that index. The fundamental commutator relation between **r** and **p** can be written as

$$\left[r_j, p_k\right] = i\hbar\delta_{jk}. \tag{33.79}$$

First let us consider a nonrelativistic Hamiltonian with a spherically symmetric potential, $V(r)$.

$$H = \frac{\mathbf{p}^2}{2m} + V(r). \tag{33.80}$$

In terms of components p_l one can write

$$\mathbf{p}^2 = p_l p_l \tag{33.81}$$

where, again, a summation over l is implied. To simplify writing, however, we write p_l^2 in place of $p_l p_l$. Thus the commutator in (33.77) is given by

$$[L_i, H] = \left[L_i, \frac{p_l^2}{2m}\right] + [L_i, V(r)]. \tag{33.82}$$

We have shown previously that this commutator vanishes, but we will rederive it, nevertheless, since it is relevant to the steps that follow pertaining to the Dirac equation. Let us consider each term separately. Using (33.78) for L_i we have

$$\left[L_i, \frac{p_l^2}{2m}\right] = \epsilon_{ijk}\frac{1}{2m}[(r_j p_k)\, p_l^2 - p_l^2\,(r_j p_k)] = \epsilon_{ijk}\frac{1}{2m}\left[r_j, p_l^2\right] p_k \tag{33.83}$$

where we have taken out p_k as a common factor on the right since it commutes with p_l. The commutator on the right can be simplified by making use of (33.79). We obtain

$$\left[r_j, p_l^2\right] = 2i\hbar\delta_{jl} p_l. \tag{33.84}$$

Inserting this on the right-hand side of (33.83) we have

$$\left[L_i, \frac{p_l^2}{2m}\right] = i\epsilon_{ijk}\frac{\hbar}{m}p_j p_k. \tag{33.85}$$

Since ϵ_{ijk} is an antisymmetric term in j and k while $p_j p_k$ is symmetric, we conclude that

$$\left[L_i, \frac{p_l^2}{2m}\right] = 0. \tag{33.86}$$

Hence, the angular momentum operator commutes with the kinetic energy term, a result we are familiar with.

Let us now consider the second term in (33.82),

$$[L_i, V(r)] = \epsilon_{ijk}\left(r_j p_k V(r) - V(r) r_j p_k\right). \tag{33.87}$$

We employ the definition of p_k as a derivative operator,

$$p_k = -i\hbar\partial_k, \tag{33.88}$$

and note that when operating on a function like $V(r)$, which depends on $r = |\mathbf{r}|$, we have

$$\partial_k V(r) = \frac{r_k}{r}\left(\frac{\partial V(r)}{\partial r}\right). \tag{33.89}$$

Keeping in mind that (33.87) involves operators that operate on a wavefunction $\psi(\mathbf{r})$, we find the following for the right-hand side of (33.87)

$$\epsilon_{ijk}\left(r_j p_k V(r) - V(r) r_j p_k\right)\psi = -i\hbar\epsilon_{ijk}(r_j\partial_k V(r)\psi - V(r) r_j \partial_k\psi) \tag{33.90}$$

where we have used (33.88). Since

$$\partial_k\left(V(r)\psi\right) = \left(\partial_k V(r)\right)\psi + V(r)\left(\partial_k\psi\right) = \frac{r_k}{r}\left(\frac{\partial V(r)}{\partial r}\right)\psi(\mathbf{r}) + V(r)\left(\partial_k\psi\right), \tag{33.91}$$

we find the right-hand side of (33.87) to be simply

$$-i\hbar\epsilon_{ijk} r_j \frac{r_k}{r}\left(\frac{\partial V(r)}{\partial r}\right) \tag{33.92}$$

after removing ψ. Once again we have a product of an antisymmetric term ϵ_{ijk} in j and k, and a symmetric term $r_j r_k$. Thus,

$$[L_i, V(r)] = 0. \tag{33.93}$$

Hence,

$$[L_i, H] = 0. \tag{33.94}$$

That is, angular momentum is conserved for the nonrelativistic Hamiltonian. As stated before, this is a well-known result.

We turn now to the Dirac equation, with the Hamiltonian that includes the potential $V(r)$,

$$H = \boldsymbol{\alpha} \cdot \mathbf{p} + \beta m + V(r). \tag{33.95}$$

Let us consider the commutator, $[L_i, H]$, which is given by

$$[L_i, H] = [L_i, \boldsymbol{\alpha} \cdot \mathbf{p}] + [L_i, \beta m] + [L_i, V(r)]. \tag{33.96}$$

Since β is a matrix that does not involve the coordinate variables, the second term above vanishes and, from (33.93), the third term above also vanishes, leaving just the first term for us to consider. Using the summation convention, we write

$$[L_i, \boldsymbol{\alpha} \cdot \mathbf{p}] = \alpha_l [L_i, p_l]. \tag{33.97}$$

Using the relation $L_i = \epsilon_{ijk} r_j p_k$, we find

$$[L_i, p_l] = \epsilon_{ijk} \left(r_j p_k p_l - p_l r_j p_k \right). \tag{33.98}$$

Since p_k and p_l commute we can take p_k out as a common factor and write

$$[L_i, p_l] = \epsilon_{ijk} \left(r_j p_l - p_l r_j \right) p_k = \epsilon_{ijk} \left(i\hbar \delta_{jl} \right) p_k \tag{33.99}$$

where in the last equality we have used the commutator relation (33.79). Therefore, for (33.96) we have

$$[L_i, H] = \epsilon_{ijk} \alpha_l \left(i\hbar \delta_{jl} \right) p_k = i\hbar \epsilon_{ijk} \alpha_j p_k. \tag{33.100}$$

In the vector notation one can write

$$[\mathbf{L}, H] = i\hbar \left(\boldsymbol{\alpha} \times \mathbf{p} \right). \tag{33.101}$$

Therefore, $\mathbf{L}$ does not commute with H for the Dirac particle.

To proceed further let us now define a new 4×4 matrix written in the 2×2 form as

$$\boldsymbol{\sigma}' = \begin{bmatrix} \boldsymbol{\sigma} & 0 \\ 0 & \boldsymbol{\sigma} \end{bmatrix}. \tag{33.102}$$

We evaluate its commutator with H:

$$\left[\sigma_i', H\right] = \left[\sigma_i', \boldsymbol{\alpha} \cdot \mathbf{p}\right] + \left[\sigma_i', \beta m\right] + \left[\sigma_i', V(r)\right]. \tag{33.103}$$

The second term above vanishes since β is diagonal in our representation. The third term also vanishes since the quantities involved in the commutator are in different spaces. Thus,

$$\left[\sigma_i', H\right] = \left[\sigma_i', \boldsymbol{\alpha} \cdot \mathbf{p}\right] = \left[\sigma_i', \alpha_l p_l\right] = (\sigma_i' \alpha_l - \alpha_l \sigma_i') p_l. \tag{33.104}$$

Writing this result in an explicit matrix form we obtain

$$\left[\sigma_i', H\right] = \begin{bmatrix} 0 & \sigma_i\sigma_l - \sigma_l\sigma_i \\ \sigma_i\sigma_l - \sigma_l\sigma_i & 0 \end{bmatrix} p_l \tag{33.105}$$

where we have used the standard representation for α_l. We now use the identity mentioned in Chapter 5,

$$\sigma_i\sigma_l = \delta_{il} + i\epsilon_{ilk}\sigma_k, \tag{33.106}$$

and obtain

$$\left[\sigma_i', H\right] = 2i\epsilon_{ilk} \begin{bmatrix} 0 & \sigma_k \\ \sigma_k & 0 \end{bmatrix} p_l = 2i\epsilon_{ilk}\alpha_k p_l \tag{33.107}$$

using the standard representation for α_k. The above relation can be written in the vector form as

$$\left[\boldsymbol{\sigma}', H\right] = 2i\left(\mathbf{p} \times \boldsymbol{\alpha}\right). \tag{33.108}$$

Combining (33.101) and (33.108), we derive the following result:

$$\left[\left(\mathbf{L} + \frac{1}{2}\hbar\boldsymbol{\sigma}'\right), H\right] = 0. \tag{33.109}$$

Thus, it is not $\mathbf{L}$ but the combination $\left(\mathbf{L} + \frac{1}{2}\hbar\boldsymbol{\sigma}'\right)$ that commutes with H. This implies that the Dirac equation has an additional degree of freedom in the category of angular momentum, which one calls the "spin angular momentum," $\mathbf{S}$, defined as

$$\mathbf{S} = \frac{1}{2}\hbar\boldsymbol{\sigma}'. \tag{33.110}$$

This result tells us that a Dirac particle has spin ½. It is then the "total angular momentum" $\mathbf{J}$ given by

$$\mathbf{J} = \mathbf{L} + \mathbf{S} \tag{33.111}$$

that commutes with H:

$$[\mathbf{J}, H] = 0. \tag{33.112}$$

Thus $\mathbf{J}$ is a constant of the motion in Dirac theory.

34 Dirac equation in the presence of spherically symmetric potentials

We introduce interaction into the Dirac framework by considering a spherically symmetric potential. Consequences that result in obtaining the spin–orbit contribution and energy levels of the hydrogen atom are then discussed and compared with their nonrelativistic counterparts. It is found that the predictions of the Dirac equation are much closer to the experimental results.

34.1 Spin-orbit coupling

The Dirac equation for a particle with energy eigenvalue E in the presence of a radial potential $V(r)$ is given by

$$[E - \boldsymbol{\alpha} \cdot \mathbf{p} - \beta m - V(r)]\, \phi(\mathbf{r}) = 0 \tag{34.1}$$

where we have taken $\psi(x) = \phi(\mathbf{r}) \exp(-iEt)$. It is understood that $\mathbf{p}$ is represented by the operator $-i\nabla$. We express (34.1) in a 2×2 compact matrix form where E and V multiply unit matrices, while $\boldsymbol{\alpha}$ and β are given by the standard representation, and $\phi(\mathbf{r})$ is a column matrix represented by ϕ_L, the large component, and ϕ_S, the small component. Equation (34.1) reads

$$\begin{pmatrix} E - m - V & -\boldsymbol{\sigma} \cdot \mathbf{p} \\ -\boldsymbol{\sigma} \cdot \mathbf{p} & E + m - V \end{pmatrix} \begin{pmatrix} \phi_L \\ \phi_S \end{pmatrix} = 0. \tag{34.2}$$

Therefore,

$$(E - m - V)\, \phi_L - \boldsymbol{\sigma} \cdot \mathbf{p} \phi_S = 0, \tag{34.3}$$

$$-\boldsymbol{\sigma} \cdot \mathbf{p} \phi_L + (E + m - V)\, \phi_S = 0. \tag{34.4}$$

We assume $|V| \ll m$ and consider the nonrelativistic approximation, $E \approx m$, with

$$E - m = E_T \tag{34.5}$$

where E_T is the kinetic energy. From (34.3) we obtain

$$(E_T - V)\phi_L = \boldsymbol{\sigma} \cdot \mathbf{p} \phi_S. \tag{34.6}$$

Since both E_T and $V \ll m$, equation (34.4) gives

$$(2m + E_T - V)\phi_S = \boldsymbol{\sigma} \cdot \mathbf{p}\phi_L. \tag{34.7}$$

We can write this relation as

$$\phi_S = \frac{1}{(2m + E_T - V)}\boldsymbol{\sigma} \cdot \mathbf{p}\phi_L \tag{34.8}$$

$$\simeq \frac{1}{2m}\left[1 - \left(\frac{E_T - V}{2m}\right)\right]\sigma \cdot \mathbf{p}\phi_L \tag{34.9}$$

$$\simeq \left[\frac{\boldsymbol{\sigma} \cdot \mathbf{p}}{2m} - \left(\frac{E_T - V}{4m^2}\right)\boldsymbol{\sigma} \cdot \mathbf{p}\right]\phi_L. \tag{34.10}$$

We need to point out that, ϕ_S is $\sim (v/c)\phi_L$, given by the first term on the right in (34.10); the next order term is $\sim (v/c)^2\phi_L$ is given by the second term. It is, therefore, essential that all the calculations are carried out to this order. This has been done in the Appendix.

It is shown in the Appendix that if we define a new wavefunction ψ such that

$$\phi_L = \left(1 - \frac{\mathbf{p}^2}{8m^2}\right)\psi, \tag{34.11}$$

then ψ satisfies the eigenvalue equation

$$H\psi = E_T\psi \tag{34.12}$$

where the Hamiltonian H is given by

$$H = \frac{\mathbf{p}^2}{2m} - \frac{\mathbf{p}^4}{8m^3} + V - \frac{1}{8m^2}\left(V\mathbf{p}^2 + \mathbf{p}^2V\right) + \frac{\boldsymbol{\sigma} \cdot \mathbf{p}\ V\ \boldsymbol{\sigma} \cdot \mathbf{p}}{4m^2}. \tag{34.13}$$

We can simplify this expression by noting that

$$\boldsymbol{\sigma} \cdot \mathbf{p}\ V\ \boldsymbol{\sigma} \cdot \mathbf{p} = -i\boldsymbol{\sigma} \cdot (\nabla V)\boldsymbol{\sigma} \cdot \mathbf{p} + V\mathbf{p}^2. \tag{34.14}$$

Hence,

$$H = \frac{\mathbf{p}^2}{2m} - \frac{\mathbf{p}^4}{8m^3} + V - \frac{1}{8m^2}\left(\mathbf{p}^2V - V\mathbf{p}^2\right) - \frac{i\boldsymbol{\sigma} \cdot (\nabla V)\boldsymbol{\sigma} \cdot \mathbf{p}}{4m^2}. \tag{34.15}$$

Since H operates on the wavefunction ψ, we have, for example,

$$\mathbf{p}^2V\psi = \nabla^2(V\psi) = -\left[(\nabla^2V)\psi + 2\nabla V \cdot (\nabla\psi) + V(\nabla^2\psi)\right]. \tag{34.16}$$

Thus,

$$\mathbf{p}^2V - V\mathbf{p}^2 = -(\nabla^2V)\psi - 2\nabla V \cdot (\nabla\psi) = -(\nabla^2V)\psi + i2\nabla V \cdot (\mathbf{p}\psi). \tag{34.17}$$

Also,

$$\sigma \cdot (\nabla V)\sigma \cdot \mathbf{p} = \nabla V \cdot \mathbf{p} + i\sigma \cdot (\nabla V \times \mathbf{p}). \tag{34.18}$$

Putting together the relations (34.13)–(34.18) we find

$$H = \frac{\mathbf{p}^2}{2m} - \frac{\mathbf{p}^4}{8m^3} + V + \frac{1}{8m^2}\nabla^2 V + \frac{\sigma \cdot (\nabla V \times \mathbf{p})}{4m^2}. \tag{34.19}$$

The fourth term on the right-hand side of (34.19) is designated as

$$\frac{1}{8m^2}\nabla^2 V = \text{ Darwin term.} \tag{34.20}$$

If V corresponds to the Coulomb potential due to a unit point charge then

$$\nabla^2 V = 4\pi\delta^3(\mathbf{r}). \tag{34.21}$$

With regard to the fifth term on the right, since $V = V(r)$ is spherically symmetric, we can write

$$\nabla V = \mathbf{r}\frac{1}{r}\frac{dV}{dr}. \tag{34.22}$$

Therefore,

$$\sigma \cdot (\nabla V \times \mathbf{p}) = \frac{1}{r}\frac{dV}{dr}\sigma \cdot (\mathbf{r} \times \mathbf{p}) = \frac{1}{r}\frac{dV}{dr}\sigma \cdot \mathbf{L} \tag{34.23}$$

where $\mathbf{L}$ is the angular momentum operator.

In summary, the Hamiltonian is expressed as

$$H' = \frac{\mathbf{p}^2}{2m} - \frac{\mathbf{p}^4}{8m^3} + V + \frac{1}{8m^2}\nabla^2 V + \frac{1}{2m^2 r}\frac{dV}{dr}\mathbf{L} \cdot \mathbf{S} \tag{34.24}$$

where $\mathbf{S}$ is the spin-operator $(= (1/2)\sigma)$. The last term corresponds to "spin–orbit coupling" which is a well-known term observed for spherically symmetric potentials. See also our discussion in Chapter 6.

34.2 *K*-operator for the spherically symmetric potentials

We consider the Hamiltonian with a spherically symmetric potential, $V(r)$,

$$H = \boldsymbol{\alpha} \cdot \mathbf{p} + \beta m + V(r). \tag{34.25}$$

We have found previously that for spherically symmetric potentials the total angular momentum $\mathbf{J}$ commutes with the Hamiltonian. There is another operator that also commutes with **H** and that allows one to distinguish the direction of the 4×4 spin operator $\boldsymbol{\sigma}'$ with respect to **L**. It is defined as

$$K = \beta(1 + \boldsymbol{\sigma}' \cdot \mathbf{L}) = \beta\left(\boldsymbol{\sigma}' \cdot \mathbf{J} - \frac{1}{2}\right) \tag{34.26}$$

where $\mathbf{J} = \mathbf{L} + \frac{1}{2}\boldsymbol{\sigma}'$, and $\left(\boldsymbol{\sigma}'\right)^2 = 3$.

In the Appendix we show that K commutes with H, that is,

$$[K, H] = 0. \tag{34.27}$$

We note that $\mathbf{J}^2$ and J_z besides K commute with H, but $\mathbf{L}^2$, L_z, and S_z do **not**. If $|\phi\rangle$ is an eigenstate of H then it will also be an eigenstate of these three operators. In particular, for the operator K we write

$$K\,|\phi\rangle = -\kappa\,|\phi\rangle. \tag{34.28}$$

Let us now determine the eigenvalue κ. From the definition (34.26) of K, we find

$$K^2 = \beta\left(1 + \boldsymbol{\sigma}' \cdot \mathbf{L}\right)\beta\left(1 + \boldsymbol{\sigma}' \cdot \mathbf{L}\right) = 1 + 2\boldsymbol{\sigma}' \cdot \mathbf{L} + \left(\boldsymbol{\sigma}' \cdot \mathbf{L}\right)^2. \tag{34.29}$$

The last term above can be simplified through the following steps:

$$\boldsymbol{\sigma}' \cdot \mathbf{L}\boldsymbol{\sigma}' \cdot \mathbf{L} = \sigma_i' \sigma_j' L_i L_j = \left[\delta_{ij} + i\epsilon_{ijk}\sigma_k'\right]\left[L_i L_j\right] = \mathbf{L}^2 + i\boldsymbol{\sigma}' \cdot (\mathbf{L} \times \mathbf{L}) = \mathbf{L}^2 - \boldsymbol{\sigma}' \cdot \mathbf{L} \tag{34.30}$$

where in the last step we have used the relation

$$\mathbf{L} \times \mathbf{L} = i\mathbf{L}. \tag{34.31}$$

Therefore,

$$K^2 = 1 + \boldsymbol{\sigma}' \cdot \mathbf{L} + \mathbf{L}^2 = \left(\mathbf{L} + \frac{1}{2}\boldsymbol{\sigma}'\right)^2 + \frac{1}{4} \tag{34.32}$$

since $\boldsymbol{\sigma}'^2 = 3$. And so

$$K^2 = \mathbf{J}^2 + \frac{1}{4}, \tag{34.33}$$

or, in terms of eigenvalues,

$$\kappa^2 = j\,(j + 1) + \frac{1}{4} = \left(j + \frac{1}{2}\right)^2. \tag{34.34}$$

Hence,

$$j + \frac{1}{2} = |\kappa|. \tag{34.35}$$

Since on the right-hand side of (34.35) we have a positive definite quantity and since $j \geq 1/2$, κ cannot be zero, and its minimum value will be $|\kappa| = 1$.

34.2.1 Nonrelativistic limit

Let us determine the properties of K in the nonrelativistic limit. Since in this limit $\beta \to 1$, we have

$$K = \beta \left(1 + \boldsymbol{\sigma}' \cdot \mathbf{L}\right) \to \left(1 + \boldsymbol{\sigma} \cdot \mathbf{L}\right). \tag{34.36}$$

Hence,

$$K = \left(1 + \boldsymbol{\sigma} \cdot \mathbf{L}\right) = \left(1 + 2\mathbf{L} \cdot \mathbf{S}\right) = \left(1 + \mathbf{J}^2 - \mathbf{L}^2 - \mathbf{S}^2\right) \tag{34.37}$$

where we have used the relation

$$\mathbf{J}^2 = (\mathbf{L} + \mathbf{S})^2 = \mathbf{L}^2 + \mathbf{S}^2 + 2\mathbf{L} \cdot \mathbf{S}. \tag{34.38}$$

Therefore, in terms of eigenvalues, the relation (34.37) gives

$$-\kappa = \left[1 + j\left(j+1\right) - l\left(l+1\right) - \frac{3}{4}\right]. \tag{34.39}$$

In the nonrelativistic approximation, $\mathbf{J}^2$, $\mathbf{L}^2$, $\mathbf{S}^2$, J_z as well as K commute with each other and with the Hamiltonian containing the spin–orbit coupling,

$$H = V + \frac{1}{2m^2}\frac{1}{r}\frac{dV}{dr}\mathbf{L} \cdot \mathbf{S}. \tag{34.40}$$

However, L_z and S_z do not commute with H. Now from (34.35) we have

$$j + \frac{1}{2} = |\kappa| \tag{34.41}$$

and from (34.39)

$$-\kappa = \left[\frac{1}{4} + j\left(j+1\right) - l\left(l+1\right)\right]. \tag{34.42}$$

Let us consider both positive and negative κ. If $\kappa < 0$ or $\kappa = -|\kappa|$, then from (34.42),

$$|\kappa| = \frac{1}{4} + j\left(j+1\right) - l\left(l+1\right). \tag{34.43}$$

Hence,

$$j + \frac{1}{2} = \frac{1}{4} + j^2 + j - l\left(l+1\right), \tag{34.44}$$

which gives the result

$$l = j - \frac{1}{2}. \tag{34.45}$$

If $\kappa > 0$ or $\kappa = |\kappa|$, then

$$-\left(j + \frac{1}{2}\right) = \frac{1}{4} + j^2 + j - l\,(l+1) \tag{34.46}$$

and

$$l = j + \frac{1}{2}. \tag{34.47}$$

The tabulation (34.48) below shows the relationship between various eigenvalues in the nonrelativistic approximation along with the spectral designation L_j for a state with the orbital angular momentum L and total angular momentum j:

k	j	l	State
-1	$\frac{1}{2}$	0	$S_{\frac{1}{2}}$
$+1$	$\frac{1}{2}$	1	$P_{\frac{1}{2}}$
-2	$\frac{3}{2}$	1	$P_{\frac{3}{2}}$
$+2$	$\frac{3}{2}$	2	$D_{\frac{3}{2}}$

(34.48)

One can obtain the wavefunction using the separation of variables technique as follows:

$$\psi(\mathbf{r}) = \psi(r,k)\,\chi(k,M,\theta,\phi) \tag{34.49}$$

where

$$\chi = \sum_m C\left(l, \frac{1}{2}, j; M, M-m\right) Y_{lm}(\theta,\phi)\chi_{\frac{1}{2}M-m} \tag{34.50}$$

and where Y_{lm} is a spherical harmonic function, χ is the spin wavefunction, and C is the Clebsch–Gordan coefficient.

34.3 Hydrogen atom

We have already studied the hydrogen atom nonrelativistically and have obtained energy levels and bound-state wavefunctions for the electron. We found that the binding energies closely reproduced the experimental values. Assuming the electron to be a spin 0 particle, we also considered the relativistic Klein–Gordon equation but found that the fine structure due to relativistic corrections was inconsistent with experiments. We now know that the electron is actually a spin ½ particle and, therefore, we expect that the Dirac equation will correctly describe the energy spectrum. This is, indeed, what we will find below.

Since the Coulomb potential given by

$$V(r) = -\frac{Ze^2}{r} \tag{34.51}$$

is spherically symmetric, then, as we discussed earlier, k and M are good quantum numbers. The Dirac equation (with units $\hbar = c = 1$) is given by

$$[E - \boldsymbol{\alpha} \cdot \mathbf{p} - \beta m - V]\,\psi = 0. \tag{34.52}$$

Replacing $\mathbf{p}$ by the operator $-i\boldsymbol{\nabla}$, we write

$$[E + i\boldsymbol{\alpha} \cdot \boldsymbol{\nabla} - \beta m - V]\,\psi = 0. \tag{34.53}$$

In order to make the above equation more transparent we consider the following trick in which we use a well-known identity for a triple cross-product involving an operator $\mathbf{A}$:

$$\mathbf{r} \times (\mathbf{r} \times \mathbf{A}) = \mathbf{r}\,(\mathbf{r} \cdot \mathbf{A}) - r^2\mathbf{A}. \tag{34.54}$$

Therefore,

$$\mathbf{A} = \frac{1}{r^2}\,[\mathbf{r}\,(\mathbf{r} \cdot \mathbf{A}) - \mathbf{r} \times (\mathbf{r} \times \mathbf{A})]. \tag{34.55}$$

Replacing $\mathbf{A}$ by $\boldsymbol{\nabla}$ we obtain

$$\boldsymbol{\nabla} = \frac{1}{r^2}\,[\mathbf{r}\,(\mathbf{r} \cdot \boldsymbol{\nabla}) - \mathbf{r} \times (\mathbf{r} \times \boldsymbol{\nabla})] \tag{34.56}$$

$$= \frac{1}{r^2}\left[r\mathbf{r}\frac{\partial}{\partial r} - i\mathbf{r} \times \mathbf{L}\right] \tag{34.57}$$

$$= \frac{1}{r}\left[\mathbf{r}\frac{\partial}{\partial r} - i\frac{\mathbf{r} \times \mathbf{L}}{r}\right] \tag{34.58}$$

where we have used the relations

$$\mathbf{r} \cdot \boldsymbol{\nabla} = r\frac{\partial}{\partial r} \quad \text{and} \quad \mathbf{L} = \mathbf{r} \times \mathbf{p} = -i\mathbf{r} \times \boldsymbol{\nabla}. \tag{34.59}$$

Let us now consider the term $\boldsymbol{\alpha} \cdot \boldsymbol{\nabla}$ in the Dirac equation (34.53) and carry out the following simplifications:

$$\boldsymbol{\alpha} \cdot \boldsymbol{\nabla} = \frac{1}{r}\left[\boldsymbol{\alpha} \cdot \mathbf{r}\frac{\partial}{\partial r} - \frac{i}{r}\boldsymbol{\alpha} \cdot (\mathbf{r} \times \mathbf{L})\right]. \tag{34.60}$$

Furthermore, from our earlier calculations we have the relation

$$\boldsymbol{\alpha} \cdot \mathbf{A}\boldsymbol{\alpha} \cdot \mathbf{B} = \mathbf{A} \cdot \mathbf{B} + i\sigma' \cdot (\mathbf{A} \times \mathbf{B}). \tag{34.61}$$

Previously, we introduced the off-diagonal matrix γ_5 given by

$$\gamma_5 = \begin{bmatrix} 0 & 1 \\ 1 & 0 \end{bmatrix}. \tag{34.62}$$

Multiplying (34.61) by γ_5 on the right and noting that $\boldsymbol{\alpha}\gamma_5 = \boldsymbol{\sigma}'$ and $\boldsymbol{\sigma}'\gamma_5 = \boldsymbol{\alpha}$, we obtain

$$\boldsymbol{\alpha} \cdot \mathbf{A}\boldsymbol{\sigma}' \cdot \mathbf{B} = \mathbf{A} \cdot \mathbf{B}\gamma_5 + i\boldsymbol{\alpha}\left(\mathbf{A} \times \mathbf{B}\right). \tag{34.63}$$

For $\mathbf{A} = \mathbf{r}$, and $\mathbf{B} = \mathbf{L}$, the above relation gives

$$\boldsymbol{\alpha} \cdot \mathbf{r}\boldsymbol{\sigma}' \cdot \mathbf{L} = i\boldsymbol{\alpha} \cdot (\mathbf{r} \times \mathbf{L}) \tag{34.64}$$

where we have used the result $\mathbf{r}\cdot\mathbf{L} = \mathbf{0}$.

On the basis of the results obtained above, we carry out the following steps:

$$\boldsymbol{\alpha} \cdot \nabla = \frac{1}{r}\left[\boldsymbol{\alpha} \cdot \mathbf{r}\frac{\partial}{\partial r} - \frac{\boldsymbol{\alpha} \cdot \mathbf{r}}{r}\boldsymbol{\sigma}' \cdot \mathbf{L}\right] \tag{34.65}$$

$$= \frac{\boldsymbol{\alpha} \cdot \mathbf{r}}{r}\left[\frac{\partial}{\partial r} - \frac{1}{r}\boldsymbol{\sigma}' \cdot \mathbf{L}\right] = \alpha_r\left[\frac{\partial}{\partial r} - \frac{1}{r}\boldsymbol{\sigma}' \cdot \mathbf{L}\right] \tag{34.66}$$

$$= \alpha_r\left[\frac{\partial}{\partial r} + \frac{1}{r} - \frac{1}{r}\left(1 + \boldsymbol{\sigma}' \cdot \mathbf{L}\right)\right] = \alpha_r\left[\left(\frac{\partial}{\partial r} + \frac{1}{r}\right) - \frac{\beta K}{r}\right] \tag{34.67}$$

where the operator K has already been defined in (34.26) and

$$\alpha_r = \boldsymbol{\alpha} \cdot \mathbf{r}. \tag{34.68}$$

We point out that by expressing the results directly in terms of the angular momentum operator, **L**, we have basically sidestepped the question of that part of the wavefunction that depends on the angles, as the differential equation for these functions are related to **L**. We will discuss these functions which involve the spherical harmonics later in this section but, for the moment, we discuss only the radial part of the equation.

34.4 Radial Dirac equation

The equation in radial coordinates is now

$$H\phi(\mathbf{r}) = E\phi(\mathbf{r}) \tag{34.69}$$

with H given by

$$H = \alpha_r p_r + i\frac{\alpha_r \beta}{r}K + \beta m + V \tag{34.70}$$

where the operator K has already been defined and

$$p_r = -i\left(\frac{\partial}{\partial r} + \frac{1}{r}\right). \tag{34.71}$$

One can easily show that p_r is a Hermitian operator. From the properties of $\boldsymbol{\alpha}$ and β one can also show that

$$\alpha_r^2 = \beta^2 = 1 \quad \text{and} \quad \alpha_r\beta + \beta\alpha_r = 0. \tag{34.72}$$

Since we now only have two operators, α_r and β, and not four, we can start afresh and take the following 2×2 representation for β and α_r:

$$\beta = \begin{pmatrix} 1 & 0 \\ 0 & -1 \end{pmatrix}, \quad \alpha_r = \begin{pmatrix} 0 & -i \\ i & 0 \end{pmatrix}. \tag{34.73}$$

This representation satisfies (34.72). We can now write (34.70) as

$$\left[E - \alpha_r p_r + i\frac{\alpha_r\beta}{r}\kappa - \beta m - V\right]\psi = 0 \tag{34.74}$$

where $-\kappa$ is the eigenvalue of K.

Let us express ψ as a column matrix given by

$$\psi = \begin{pmatrix} \frac{1}{r}F \\ \frac{1}{r}G \end{pmatrix}. \tag{34.75}$$

Furthermore, the following results will be useful:

$$p_r\left(\frac{1}{r}f\right) = -i\frac{1}{r}\frac{df}{dr}, \tag{34.76}$$

$$\alpha_r p_r\left(\frac{1}{r}f\right) = \begin{pmatrix} 0 & -i \\ i & 0 \end{pmatrix}\left(-i\frac{1}{r}\frac{df}{dr}\right) = \begin{pmatrix} 0 & -1 \\ 1 & 0 \end{pmatrix}\frac{1}{r}\frac{df}{dr}, \tag{34.77}$$

$$i\alpha_r\beta\kappa\left(\frac{1}{r}f\right) = \begin{pmatrix} 0 & -1 \\ -1 & 0 \end{pmatrix}\left(\frac{\kappa}{r}f\right). \tag{34.78}$$

Equation (37.74), with the help of (34.73) and (34.75), is found to be

$$\left.\begin{aligned} (E - m - V)F + \frac{dG}{dr} - \frac{\kappa}{r}G = 0, \\ (E + m - V)G - \frac{dF}{dr} - \frac{\kappa}{r}F = 0. \end{aligned}\right\} \tag{34.79}$$

We define the following constants:

$$\begin{aligned}\alpha_1 &= m+E, \quad \alpha_2 = m-E \\ \alpha &= \sqrt{\alpha_1\alpha_2} = \sqrt{m^2-E^2}, \quad \rho = \alpha r\end{aligned} \tag{34.80}$$

and write

$$\gamma = Ze^2, \quad V = -\frac{\alpha\gamma}{\rho}. \tag{34.81}$$

Then the equations (34.79) are of the form

$$\left(\frac{d}{d\rho} - \frac{\kappa}{\rho}\right)G - \left(\frac{\alpha_2}{\alpha} - \frac{\gamma}{\rho}\right)F = 0, \tag{34.82}$$

$$\left(\frac{d}{d\rho} + \frac{\kappa}{\rho}\right)F - \left(\frac{\alpha_1}{\alpha} + \frac{\gamma}{\rho}\right)G = 0. \tag{34.83}$$

First we determine the asymptotic behaviors of F and G. Equations (34.82) and (34.83) in the limit $\rho \to \infty$ are

$$\frac{dG}{d\rho} - \frac{\alpha_2}{\alpha}F = 0, \tag{34.84}$$

$$\frac{dF}{d\rho} - \frac{\alpha_1}{\alpha}G = 0. \tag{34.85}$$

Taking the derivative with respect to ρ of equation (34.84) and substituting in it the expression for $dF/d\rho$ in (34.85), we obtain

$$\frac{d^2G}{d\rho^2} - G = 0. \tag{34.86}$$

A solution for G that is convergent at infinity is found to be $e^{-\rho}$, the same result holds for F.

Let us write

$$F = f(\rho)\,e^{-\rho} \quad \text{and} \quad G = g(\rho)\,e^{-\rho}. \tag{34.87}$$

Then,

$$\frac{d}{d\rho}\begin{pmatrix} G \\ F \end{pmatrix} = \begin{pmatrix} \dfrac{dg}{d\rho} - g \\ \dfrac{df}{d\rho} - f \end{pmatrix} e^{-\rho}. \tag{34.88}$$

Thus, equations (34.84) and (34.85) are of the form

$$g' - g - \frac{\kappa}{\rho}g - \left(\frac{\alpha_2}{\alpha} - \frac{\gamma}{\rho}\right)f = 0, \tag{34.89}$$

$$f' - f + \frac{\kappa}{\rho}f - \left(\frac{\alpha_1}{\alpha} + \frac{\gamma}{\rho}\right)g = 0. \tag{34.90}$$

Let us now write a power series expansion for g and f as follows:

$$g = \rho^s \left(a_0 + a_1\rho + \cdots + a_\nu \rho^\nu\right), \tag{34.91}$$

$$f = \rho^s \left(b_0 + b_1\rho + \cdots + b_\nu \rho^\nu\right). \tag{34.92}$$

Substituting these expansions in (34.89) and (34.90) and comparing the coefficients of $\rho^{s+\nu-1}$, we obtain

$$(s+\nu)\, a_\nu - a_{\nu-1} - \kappa a_\nu - \left(\frac{\alpha_2}{\alpha} b_{\nu-1} - \gamma b_\nu\right) = 0, \tag{34.93}$$

$$(s+\nu)\, b_\nu - b_{\nu-1} + \kappa b_\nu - \left(\frac{\alpha_1}{\alpha} a_{\nu-1} + \gamma a_\nu\right) = 0. \tag{34.94}$$

Consider first $\nu = 0$ in the above two equations; then since $a_{\nu-1}$ and $b_{\nu-1}$ do not exist for this value of ν, we obtain

$$(s-\kappa)\, a_0 + \gamma b_0 = 0, \tag{34.95}$$

$$(s+\kappa)\, b_0 - \gamma a_0 = 0. \tag{34.96}$$

Combining the two we find

$$s^2 - \kappa^2 = -\gamma^2. \tag{34.97}$$

Therefore,

$$s = \pm\sqrt{\kappa^2 - \gamma^2}. \tag{34.98}$$

We take the + sign, $s = \sqrt{\kappa^2 - \gamma^2}$, in order to keep the wavefunction finite at $\rho = 0$.

Following the same arguments as for the nonrelativistic case, we conclude that the series for f and g must terminate in order to keep the wavefunction finite at infinity. We assume that the series terminates at $\nu = n'$, i.e.,

$$a_{n'+1} = 0 = b_{n'+1}. \tag{34.99}$$

To determine the consequences of this, we write down the following relation obtained by multiplying (34.94) by α and subtracting from it (34.93) multiplied by α_1:

$$b_\nu \left[\alpha\,(s+\nu+\kappa) - \alpha_1\gamma\right] = a_\nu \left[\alpha_1\,(s+\nu-\kappa) + \gamma\alpha\right]. \tag{34.100}$$

Let us take $\nu = n' + 1$ in relation (34.94) then we find

$$b_{n'} = -\frac{\alpha_1}{\alpha} a_{n'}, \quad n' = 0, 1, 2, \ldots. \tag{34.101}$$

Also, by taking $\nu = n'$ in (34.100) we obtain

$$b_{n'} \left[\alpha\left(s + n' + \kappa\right) - \alpha_1\gamma\right] = a_{n'} \left[\alpha_1\left(s + n' - \kappa\right) + \gamma\alpha\right]. \tag{34.102}$$

Substituting $b_{n'}$ from the relation in the above equation we find

$$-\left[\alpha\left(s+n'+\kappa\right)-\alpha_1\gamma\right]=\left[\alpha\left(s+n'-\kappa\right)+\alpha_2\gamma\right], \tag{34.103}$$

which gives

$$2\alpha\left(s+n'\right)=\gamma\left(\alpha_1-\alpha_2\right)=2E\gamma. \tag{34.104}$$

We therefore obtain the following result for E:

$$E^2=m^2\left[1+\frac{\gamma^2}{(s+n')^2}\right]^{-1}. \tag{34.105}$$

Thus, the energy levels are given by

$$E=m\left[1+\frac{\gamma^2}{(s+n')^2}\right]^{-\frac{1}{2}}. \tag{34.106}$$

In terms of κ it has the form

$$E=m\left[1+\frac{\gamma^2}{\left(\sqrt{\kappa^2-\gamma^2}+n'\right)^2}\right]^{-\frac{1}{2}}, \qquad \begin{array}{l} n'=0,1,\ldots, \\ k=\mp 1,\mp 2,\ldots, \end{array} \tag{34.107}$$

while in terms of j it can be written as

$$E=m\left[1+\frac{\gamma^2}{\left(\sqrt{\left(j+\frac{1}{2}\right)^2-\gamma^2}+n'\right)^2}\right]^{-\frac{1}{2}}, \qquad \begin{array}{l} n'=0,1,\ldots, \\ j=\frac{1}{2},\frac{3}{2},\ldots. \end{array} \tag{34.108}$$

Thus E depends on n' and j. If $n'=0$, then $k\neq 1$. Let us now expand the above result in powers of γ, which is proportional to the Coulomb interaction strength Ze^2. The parameter s can be written as

$$s=|\kappa|\left(1-\frac{\gamma^2}{\kappa^2}\right)^{\frac{1}{2}}\simeq|\kappa|\left(1-\frac{1}{2}\frac{\gamma^2}{\kappa^2}\right). \tag{34.109}$$

To write the expansion for E we note that

$$s+n'\simeq n'+|\kappa|-\frac{1}{2}\frac{\gamma^2}{|\kappa|}=n-\frac{1}{2}\frac{\gamma^2}{|\kappa|} \tag{34.110}$$

where we have defined

$$n'+|\kappa|=n. \tag{34.111}$$

As we will find below, n is the same as the principal quantum number we defined for the hydrogen atom. Thus,

$$E \simeq m\left[1+\frac{\gamma^2}{\left(n-\frac{1}{2}\frac{\gamma^2}{|\kappa|}\right)^2}\right]^{-\frac{1}{2}}. \quad (34.112)$$

Expanding this expression up to the power γ^4, we obtain

$$E \simeq m\left[1-\frac{\gamma^2}{2n^2}-\frac{\gamma^4}{2n^4}\left(\frac{n}{|\kappa|}-\frac{3}{4}\right)\right]. \quad (34.113)$$

In terms of j, this result can be written as

$$E \simeq m\left[1-\frac{\gamma^2}{2n^2}-\frac{\gamma^4}{2n^4}\left(\frac{n}{\left(j+\frac{1}{2}\right)}-\frac{3}{4}\right)\right]. \quad (34.114)$$

We notice that here, once again, as in the Klein–Gordon case, we find a fine structure in the energy levels given by the third term in (34.114). The denominator of this term is $(j + 1/2)$ in contrast to the K-G equation where it was $(l + 1)$. The results found for the Dirac case are consistent with experiments. We note that the states with the same n and j are still degenerate. This will be removed by the so-called Lamb shift, which we will discuss in Chapter 45.

34.5 Hydrogen atom states

Below we tabulate the states denoted in spectroscopic notation. In doing so we note that, in the nonrelativistic limit, the large components (i.e., the upper components) of the wavefunction $\phi(\mathbf{r})$ are dominant and satisfy the relations

$$l = \begin{cases} = -\kappa - 1 = j - \dfrac{1}{2} & \text{for } \kappa < 0, \\ = \kappa \qquad\quad\; = j + \dfrac{1}{2} & \text{for } \kappa > 0. \end{cases} \quad (34.115)$$

It is in terms of these angular momenta that the states are designated in spectroscopic notations. Also, we have already shown that one cannot have $\kappa = 0$. As for the states with $n' = 0$, we find that

$$(s+\kappa) = \gamma\frac{a_0}{b_0} = -\gamma\frac{\alpha}{\alpha_1} \quad (34.116)$$

where we have used the relation (34.102). The right-hand side above is negative, which implies that κ cannot be a positive integer otherwise the wavefunction would be infinite at $\rho = 0$. With these observations we present the following tabulation (34.117).

(nL_j) State	n'	κ	$n = n' + \|\kappa\|$	j	l
$1S_{\frac{1}{2}}$	0	-1	1	$\frac{1}{2}$	0
$2S_{\frac{1}{2}}$	1	-1	2	$\frac{1}{2}$	0
$2P_{\frac{1}{2}}$	1	1	2	$\frac{1}{2}$	1
$2P_{\frac{3}{2}}$	0	-2	2	$\frac{3}{2}$	1

(34.117)

We note that the $2S_{\frac{1}{2}}$ and $2P_{\frac{1}{2}}$ are degenerate but not $2S_{\frac{1}{2}}$ and $2P_{\frac{3}{2}}$. Thus, the degeneracy between states with the same values of n and l is partially removed.

34.6 Hydrogen atom wavefunction

The angular dependence of a wavefunction is given by the spherical harmonics, $Y_{lm}(\theta, \phi)$, which we designate as Y_l^m to simplify writing. We already know the orbital angular momentum, l, of the upper component in terms of the total angular momentum, j. To get the relation for the lower two components, we note that the operator $K = \beta(1 + \boldsymbol{\sigma}' \cdot \mathbf{L})$ will have a contribution from the lower two diagonal components of β. Therefore,

$$K \to -(1 + \boldsymbol{\sigma}' \cdot \mathbf{L}). \tag{34.118}$$

We have already shown that in the nonrelativistic limit we have the relation

$$l = j + \frac{1}{2}. \tag{34.119}$$

Let us write the wavefunction in the full four-component form as

$$\psi = \begin{pmatrix} \psi_1 \\ \psi_2 \\ \psi_3 \\ \psi_4 \end{pmatrix} \tag{34.120}$$

where the following relations are satisfied:

$$J_z \psi = M\psi, \tag{34.121}$$

$$\mathbf{J} = \mathbf{L} + \frac{1}{2}\boldsymbol{\sigma}', \tag{34.122}$$

$$M = m + \frac{1}{2}\sigma_z', \tag{34.123}$$

$$\sigma_z' \begin{pmatrix} \psi_1 \\ \psi_2 \\ \psi_3 \\ \psi_4 \end{pmatrix} = \begin{pmatrix} \psi_1 \\ -\psi_2 \\ \psi_3 \\ -\psi_4 \end{pmatrix}, \tag{34.124}$$

and, therefore,

$$\psi_1 = \frac{1}{r} F(r)\, \lambda_1 Y_{j-\frac{1}{2}}^{M-\frac{1}{2}}, \tag{34.125}$$

$$\psi_2 = \frac{1}{r} F(r)\, \lambda_2 Y_{j-\frac{1}{2}}^{M+\frac{1}{2}}, \tag{34.126}$$

$$\psi_3 = \frac{1}{r} G(r)\, \lambda_3 Y_{j+\frac{1}{2}}^{M-\frac{1}{2}}, \tag{34.127}$$

$$\psi_4 = \frac{1}{r} G(r)\, \lambda_4 Y_{j+\frac{1}{2}}^{M+\frac{1}{2}}. \tag{34.128}$$

For ease of writing, as we mentioned earlier we have changed the notation for the spherical harmonics from Y_{lm} to Y_l^m.

The upper two components of (34.120) can be written as

$$\begin{pmatrix} \psi_1 \\ \psi_2 \end{pmatrix} = \psi_1 \begin{pmatrix} 1 \\ 0 \end{pmatrix} + \psi_2 \begin{pmatrix} 0 \\ 1 \end{pmatrix} \tag{34.129}$$

$$= \frac{1}{r} F(r) \left[\lambda_1 Y_{j-\frac{1}{2}}^{M-\frac{1}{2}} \begin{pmatrix} 1 \\ 0 \end{pmatrix} + \lambda_2 Y_{j-\frac{1}{2}}^{M+\frac{1}{2}} \begin{pmatrix} 0 \\ 1 \end{pmatrix} \right] \tag{34.130}$$

where the λ_i's can be written as the Clebsch–Gordan coefficients

$$\lambda_1 = C\left(j, l, \frac{1}{2}; M - \frac{1}{2} m \frac{1}{2}\right) = \sqrt{\frac{j+M}{2j}}, \tag{34.131}$$

$$\lambda_2 = C\left(j, l, \frac{1}{2}; M + \frac{1}{2} m \frac{1}{2}\right) = \sqrt{\frac{j-M}{2j}}. \tag{34.132}$$

Similarly,

$$\begin{pmatrix} \psi_3 \\ \psi_4 \end{pmatrix} = \psi_3 \begin{pmatrix} 1 \\ 0 \end{pmatrix} + \psi_4 \begin{pmatrix} 0 \\ 1 \end{pmatrix} \tag{34.133}$$

$$= \frac{1}{r} g(r) \left[\lambda_3 Y_{j+\frac{1}{2}}^{M-\frac{1}{2}} \begin{pmatrix} 1 \\ 0 \end{pmatrix} + \lambda_4 Y_{j+\frac{1}{2}}^{M+\frac{1}{2}} \begin{pmatrix} 0 \\ 1 \end{pmatrix} \right] \tag{34.134}$$

with

$$\lambda_3 = C\left(j, l, \frac{1}{2}; M - \frac{1}{2}m, \frac{1}{2}\right) = \sqrt{\frac{j - M + 1}{2j + 2}}, \tag{34.135}$$

$$\lambda_4 = C\left(j, l, \frac{1}{2}; M + \frac{1}{2}m, \frac{1}{2}\right) = -\sqrt{\frac{j + M + 1}{2j + 2}}. \tag{34.136}$$

One can then write the compact relation

$$\psi(\mathbf{r}) = \begin{bmatrix} \frac{1}{r} F(r)\, \mathcal{Y}^M_{j,j-\frac{1}{2}} \\ \frac{1}{r} G(r)\, \mathcal{Y}^M_{j,j+\frac{1}{2}} \end{bmatrix} \tag{34.137}$$

where

$$\mathcal{Y}^M_{j,j-\frac{1}{2}} = \left[\sqrt{\frac{j+M}{2j}}\, Y^{M-\frac{1}{2}}_{j-\frac{1}{2}} \begin{pmatrix} 1 \\ 0 \end{pmatrix} + \sqrt{\frac{j-M}{2j}}\, Y^{M+\frac{1}{2}}_{j-\frac{1}{2}} \begin{pmatrix} 0 \\ 1 \end{pmatrix}\right], \tag{34.138}$$

$$\mathcal{Y}^M_{j,j+\frac{1}{2}} = \left[\sqrt{\frac{j-M+1}{2j+2}}\, Y^{M-\frac{1}{2}}_{j+\frac{1}{2}} \begin{pmatrix} 1 \\ 0 \end{pmatrix} - \sqrt{\frac{j+M+1}{2j+2}}\, Y^{M+\frac{1}{2}}_{j+\frac{1}{2}} \begin{pmatrix} 0 \\ 1 \end{pmatrix}\right]. \tag{34.139}$$

The radial wavefunction can be obtained as in the Schrödinger case, keeping in mind that for the $r = 0$ behavior, given by r^s, the exponent s will no longer be an integer. We will not pursue this calculation further.

34.7 Appendix to Chapter 34

34.7.1 The commutator $[K,H]$

From the definition

$$K = \beta(1 + \boldsymbol{\sigma}' \cdot \mathbf{L}) = \beta\left(\boldsymbol{\sigma}' \cdot \mathbf{J} - \frac{1}{2}\right) \tag{34.140}$$

and

$$H = \boldsymbol{\alpha} \cdot \mathbf{p} + \beta m + V(r), \tag{34.141}$$

we note that

$$[K,H] = \left[\boldsymbol{\beta}\boldsymbol{\sigma}' \cdot \mathbf{J}, H\right] - \frac{1}{2}\left[\boldsymbol{\beta}, H\right]. \tag{34.142}$$

First we write below two of the results we have already derived:

$$\left[\boldsymbol{\sigma}',H\right]=2i\hbar\mathbf{p}\times\boldsymbol{\alpha},\quad [\mathbf{L},H]=i\hbar\boldsymbol{\alpha}\times\mathbf{p}. \tag{34.143}$$

These relations were derived in Section 33.5 and led to the result $[\mathbf{J},H]=0$. To obtain the commutator $[K,H]$, let us first consider $[\beta,H]$, where H is given by (34.141). Since β commutes with the second and third terms, we need only consider the first term of H,

$$[\beta,H]=[\beta,\boldsymbol{\alpha}\cdot\mathbf{p}]=[\beta,\alpha_i]p_i=-2\alpha_i\beta p_i=-2\boldsymbol{\alpha}\cdot\mathbf{p}\beta \tag{34.144}$$

where we have used the anticommutation relation $\{\beta,\alpha_i\}=0$.

In order to evaluate the commutator involving $\boldsymbol{\beta\sigma}'\cdot\mathbf{J}$, we note that the commutator $[A_1A_2,B]$, where A_1, A_2, and B are operators, can be written as

$$[A_1A_2,B]=A_1\,[A_2,B]+[A_1,B]\,A_2. \tag{34.145}$$

Hence,

$$\left[\beta\boldsymbol{\sigma}'\cdot\mathbf{J},H\right]=\beta\left[\boldsymbol{\sigma}'\cdot\mathbf{J},H\right]+[\beta,H]\,\boldsymbol{\sigma}'\cdot\mathbf{J} \tag{34.146}$$

$$=\beta\left[\boldsymbol{\sigma}'\cdot\mathbf{J},H\right]-2\boldsymbol{\alpha}\cdot\mathbf{p}\beta\boldsymbol{\sigma}'\cdot\mathbf{J} \tag{34.147}$$

where we have used the result (34.144) for the second term above. We can simplify the first term in (34.147) further by using (34.145) once again:

$$\beta\left[\boldsymbol{\sigma}'\cdot\mathbf{J},H\right]=\beta\left\{\boldsymbol{\sigma}'\cdot[\mathbf{J},H]+\left[\boldsymbol{\sigma}',H\right]\cdot\mathbf{J}\right\}. \tag{34.148}$$

The first term vanishes because $\mathbf{J}$ commutes with H, and for the second term we use (34.143). Thus,

$$\beta\left[\boldsymbol{\sigma}'\cdot\mathbf{J},H\right]=2i\beta\,(\mathbf{p}\times\boldsymbol{\alpha})\cdot\mathbf{J}. \tag{34.149}$$

From this result we obtain

$$\left[\beta\boldsymbol{\sigma}'\cdot\mathbf{J},H\right]=2\beta\left\{i\,(\mathbf{p}\times\boldsymbol{\alpha})\cdot\mathbf{J}+\boldsymbol{\alpha}\cdot\mathbf{p}\boldsymbol{\sigma}'\cdot\mathbf{J}\right\}. \tag{34.150}$$

In order to obtain the second term in (34.150) we multiply the well-known relation

$$\boldsymbol{\sigma}'\cdot\mathbf{A}\boldsymbol{\sigma}'\cdot\mathbf{B}=\mathbf{A}\cdot\mathbf{B}+i\boldsymbol{\sigma}'\cdot(\mathbf{A}\times\mathbf{B}) \tag{34.151}$$

on both sides by the γ_5 matrix, which we have already introduced and which in the standard representation can be written in a 2×2 form as

$$\gamma_5=\begin{pmatrix}0&1\\1&0\end{pmatrix}. \tag{34.152}$$

We obtain

$$\boldsymbol{\alpha}\cdot\mathbf{A}\boldsymbol{\sigma}'\cdot\mathbf{B}=\gamma_5\mathbf{A}\cdot\mathbf{B}+i\boldsymbol{\alpha}\cdot(\mathbf{A}\times\mathbf{B}). \tag{34.153}$$

Using this result we find

$$\boldsymbol{\alpha}\cdot\mathbf{p}\boldsymbol{\sigma}'\cdot\mathbf{J}=\gamma_5\mathbf{p}\cdot\mathbf{J}+i\boldsymbol{\alpha}\cdot(\mathbf{p}\times\mathbf{J}) \tag{34.154}$$

$$=\frac{1}{2}\mathbf{p}\cdot\boldsymbol{\alpha}+i\boldsymbol{\alpha}\cdot(\mathbf{p}\times\mathbf{J}) \tag{34.155}$$

$$=\frac{1}{2}\mathbf{p}\cdot\boldsymbol{\alpha}+i\,(\boldsymbol{\alpha}\times\mathbf{p})\cdot\mathbf{J} \tag{34.156}$$

where we have taken $\mathbf{J}=\mathbf{L}+\frac{1}{2}\boldsymbol{\sigma}'$, with $\mathbf{p}\cdot\mathbf{L}=0$, and the identity $\mathbf{A}\cdot(\mathbf{B}\times\mathbf{C})=(\mathbf{A}\times\mathbf{B})\cdot\mathbf{C}$. Thus (34.150) gives

$$\left[\beta\boldsymbol{\sigma}'\cdot\mathbf{J},H\right]=\beta\boldsymbol{\alpha}\cdot\mathbf{p}. \tag{34.157}$$

Combining (34.142), (34.144), and (34.157) we obtain

$$[K,H]=\beta\boldsymbol{\alpha}\cdot\mathbf{p}+\boldsymbol{\alpha}\cdot\mathbf{p}\beta=0 \tag{34.158}$$

because β and $\boldsymbol{\alpha}$ anticommute. Thus we complete the proof that K commutes with the Hamiltonian.

34.7.2 Derivation of the spin-orbit term

We start by substituting ϕ_S on the right-hand side of (34.6) and obtain

$$(E_T-V)\phi_L=\frac{1}{2m}\boldsymbol{\sigma}\cdot\mathbf{p}\left[1-\left(\frac{E_T-V}{2m}\right)\right]\boldsymbol{\sigma}\cdot\mathbf{p}\phi_L \tag{34.159}$$

$$=\left\{\frac{\boldsymbol{\sigma}\cdot\mathbf{p}\boldsymbol{\sigma}\cdot\mathbf{p}}{2m}-\frac{\boldsymbol{\sigma}\cdot\mathbf{p}\,E_T\,\boldsymbol{\sigma}\cdot\mathbf{p}}{4m^2}+\frac{\boldsymbol{\sigma}\cdot\mathbf{p}\,V\,\boldsymbol{\sigma}\cdot\mathbf{p}}{4m^2}\right\}\phi_L. \tag{34.160}$$

Therefore,

$$(E_T-V)\phi_L=\left\{\frac{\mathbf{p}^2}{2m}-\frac{\mathbf{p}^2}{4m^2}E_T+\frac{\boldsymbol{\sigma}\cdot\mathbf{p}\,V\,\boldsymbol{\sigma}\cdot\mathbf{p}}{4m^2}\right\}\phi_L \tag{34.161}$$

where we have used the result

$$\boldsymbol{\sigma}\cdot\mathbf{p}\ \boldsymbol{\sigma}\cdot\mathbf{p}=\mathbf{p}^2. \tag{34.162}$$

We note that equation (34.161) cannot be cast as an eigenvalue equation of the type $H\phi_L=E_T\phi_L$ because the right-hand side of (34.161) itself contains E_T. Also, since the second term in (34.161) is of the order of $(v/c)^2$ we need to make sure that the approximate wavefunctions we are using are also normalized to that order. The normalization condition, however, reads

$$\int d^3r\,\phi^\dagger(\mathbf{r})\phi(\mathbf{r})=\int d^3r\left[|\phi_L|^2+|\phi_S|^2\right]=1 \tag{34.163}$$

which contains ϕ_S. Keeping the leading term proportional to v/c in the first term in (34.10) relating ϕ_S to ϕ_L we find

$$\int d^3r \left[|\phi_L|^2 + \frac{\mathbf{p}^2}{4m^2} |\phi_L|^2 \right] = 1 \quad \text{to order} \quad \frac{v^2}{c^2}. \tag{34.164}$$

Hence,

$$\int d^3r \left[1 + \frac{\mathbf{p}^2}{4m^2} \right] |\phi_L|^2 = 1. \tag{34.165}$$

Let us define a new wavefunction

$$\psi = \sqrt{1 + \frac{\mathbf{p}^2}{4m^2}} \phi_L, \tag{34.166}$$

which has the normalization

$$\int d^3r \ |\psi|^2 = 1. \tag{34.167}$$

Expanding ψ and keeping the leading term, we write

$$\psi = \left(1 + \frac{\mathbf{p}^2}{8m^2} \right) \phi_L. \tag{34.168}$$

In (34.161), after bringing the E_T term from the right to the left, we rewrite it as

$$E_T \left(1 + \frac{\mathbf{p}^2}{4m^2} \right) \phi_L = \left\{ \frac{\mathbf{p}^2}{2m} + V + \frac{\boldsymbol{\sigma} \cdot \mathbf{p} \, V \, \boldsymbol{\sigma} \cdot \mathbf{p}}{4m^2} \right\} \phi_L. \tag{34.169}$$

Our aim is to write the eigenvalue relation in terms of ψ. From (34.168) we write ϕ_L in terms of ψ to leading order in $(v/c)^2$,

$$\phi_L = \left(1 - \frac{\mathbf{p}^2}{8m^2} \right) \psi. \tag{34.170}$$

Substituting this in (34.169) we obtain, to leading order in $(v/c)^2$,

$$E_T \left(1 + \frac{\mathbf{p}^2}{4m^2} \right) \left(1 - \frac{\mathbf{p}^2}{8m^2} \right) \psi = E_T \left(1 + \frac{\mathbf{p}^2}{8m^2} \right) \psi \tag{34.171}$$

$$= \left\{ \frac{\mathbf{p}^2}{2m} + V + \frac{\boldsymbol{\sigma} \cdot \mathbf{p} \, V \, \boldsymbol{\sigma} \cdot \mathbf{p}}{4m^2} \right\} \left(1 - \frac{\mathbf{p}^2}{8m^2} \right) \psi. \tag{34.172}$$

Multiplying (34.172) on the left by $\left(1 - p^2/8m^2\right)$ and neglecting terms of order $(v/c)^4$, we have

$$E_T\psi = \left(1 - \frac{\mathbf{p}^2}{8m^2}\right)\left\{\frac{\mathbf{p}^2}{2m} + V + \frac{\boldsymbol{\sigma}\cdot\mathbf{p}\,V\,\boldsymbol{\sigma}\cdot\mathbf{p}}{4m^2}\right\}\left(1 - \frac{\mathbf{p}^2}{8m^2}\right)\psi. \tag{34.173}$$

We write the above equation in terms of ψ as follows:

$$E_T\psi = H\psi. \tag{34.174}$$

To order $(v/c)^2$ the Hamiltonian H is found to be

$$H = \frac{\mathbf{p}^2}{2m} - \frac{\mathbf{p}^4}{8m^3} + V - \frac{1}{8m^2}\left(V\mathbf{p}^2 + \mathbf{p}^2V\right) + \frac{\boldsymbol{\sigma}\cdot\mathbf{p}\,V\,\boldsymbol{\sigma}\cdot\mathbf{p}}{4m^2}. \tag{34.175}$$

35 Dirac equation in a relativistically invariant form

The Dirac equation is re-expressed in a covariant form so that the equation is manifestly invariant under Lorentz transformations. The matrices $\boldsymbol{\alpha}$ and β are replaced by the γ^{μ} matrices and the spinor wavefunctions are appropriately defined. The Lorentz transformation of spinors and their bilinear products are obtained. We also derive the Gordon decomposition rule.

35.1 Covariant Dirac equation

Let us express the Dirac equation in a fully relativistic form. We start with the equation in the momentum space,

$$(E\mathbf{1} - \boldsymbol{\alpha} \cdot \mathbf{p} - \beta m)\, u = 0, \tag{35.1}$$

and multiply it on the left by β. Since $\beta^2 = \mathbf{1}$, we have

$$(\beta E - \beta\boldsymbol{\alpha} \cdot \mathbf{p} - m)\, u = 0. \tag{35.2}$$

We introduce the matrices

$$\gamma^{\mu} = \left(\gamma^{i}, \gamma^{4}\right), \quad i = 1, 2, 3, \tag{35.3}$$

which we define in terms β and α_i as follows:

$$\gamma^{i} = \beta\alpha_i, \quad \gamma^4 = \beta. \tag{35.4}$$

The first two terms of (35.2) can be expressed as $\gamma^4 E - \boldsymbol{\gamma} \cdot \mathbf{p}$. Hence the complete Dirac equation in the four-vector form is then

$$\left(\gamma^{\mu} p_{\mu} - m\right) u = 0, \tag{35.5}$$

which we can write as

$$(\gamma \cdot p - m)\, u = 0. \tag{35.6}$$

Replacing $p_{\mu} \to i\partial_{\mu}$, we can express the Dirac equation in the x-space as

$$\left(i\gamma^{\mu}\partial_{\mu} - m\right) u = 0 \tag{35.7}$$

or

$$(i\gamma \cdot \partial - m)\, u = 0. \tag{35.8}$$

Equations (35.5)–(35.8) express the Dirac equation in a covariant form.

35.2 Properties of the γ-matrices

To determine the properties of the γ-matrices we recapitulate below the properties of the $\boldsymbol{\alpha}$ and β matrices (with $i,j = 1, 2, 3$):

$$\boldsymbol{\alpha}^2 = \mathbf{1} = \beta^2, \tag{35.9}$$

$$\alpha_i \beta + \beta \alpha_i = 0, \tag{35.10}$$

$$\alpha_i \alpha_j + \alpha_j \alpha_i = 0, \quad i \neq j. \tag{35.11}$$

Therefore, for the corresponding γ-matrices, we have

$$\left(\gamma^4\right)^2 = (\beta)^2 = 1. \tag{35.12}$$

Similarly,

$$\left(\gamma^i\right)^2 = \beta\alpha_i\beta\alpha_i = \beta\left(-\beta\alpha_i\right)\alpha_i = -\beta^2\alpha_i^2 = -1. \tag{35.13}$$

If we multiply (35.10) on the left by β we obtain

$$(\beta)(\beta\alpha_i) + (\beta)(\alpha_i\beta) = (\beta)(\beta\alpha_i) + (\beta\alpha_i)(\beta) = 0. \tag{35.14}$$

Using the relation (35.4) this equation can be written as

$$\gamma^4\gamma^i + \gamma^i\,\gamma^4 = 0. \tag{35.15}$$

We then consider the relation (35.11) and multiply it on the left and on the right by β to obtain

$$\beta\alpha_i\alpha_j\beta + \beta\alpha_j\alpha_i\beta = -\left(\beta\alpha_i\right)\left(\beta\alpha_j\right) - \left(\beta\alpha_j\right)\left(\beta\alpha_i\right) = 0, \quad \text{for } i \neq j, \tag{35.16}$$

which can be reduced to

$$\gamma^i\gamma^j + \gamma^j\,\gamma^i = 0, \quad \text{for} \quad i \neq j. \tag{35.17}$$

We can write the relations (35.12), (35.13), (35.15), and (35.17) in a compact form as a single relation,

$$\left\{\gamma^\mu, \gamma^\nu\right\} = 2g^{\mu\nu}, \tag{35.18}$$

with $\mu, \nu = 1, 2, 3, 4$, where $g^{\mu\nu}$ is the metric tensor defined earlier, and $\{\gamma^\mu, \gamma^\nu\}$ is the anticommutator $(\gamma^\mu\gamma^\nu + \gamma^\nu\gamma^\mu)$. The anticommutation relation (35.18) is now expressed in a covariant form.

The standard representation we had defined for $\boldsymbol{\alpha}$ and β can now be written for the γ-matrices as follows:

$$\gamma^4 = \begin{bmatrix} 1 & 0 \\ 0 & -1 \end{bmatrix}, \tag{35.19}$$

$$\gamma^i = \begin{bmatrix} 0 & \sigma_i \\ -\sigma_i & 0 \end{bmatrix}. \tag{35.20}$$

35.2.1 Hermitian conjugate of γ^μ

We first note the following relations regarding Hermitian conjugates of the γ-matrices. From the definitions $\gamma^i = \beta\alpha_i$ and $\gamma^4 = \beta$, we obtain, since β is Hermitian,

$$\left(\gamma^4\right)^\dagger = \beta^\dagger = \beta = \gamma^4, \tag{35.21}$$

$$\left(\gamma^i\right)^\dagger = (\beta\alpha_i)^\dagger = \alpha_i\beta = -\beta\alpha_i = -\gamma^i, \tag{35.22}$$

where we have used the anticommutation properties of α_i and β. Furthermore, we note that

$$\gamma^4\left(\gamma^i\right)^\dagger\gamma^4 = -\gamma^4\gamma^i\gamma^4 = \gamma^i\left(\gamma^4\right)^2 = \gamma^i \tag{35.23}$$

where we have used (35.22) and the anticommutation relation between γ^i and γ^4. Also, trivially,

$$\gamma^4\left(\gamma^4\right)^\dagger\gamma^4 = \gamma^4. \tag{35.24}$$

Hence we can write

$$\gamma^4\left(\gamma^\mu\right)^\dagger\gamma^4 = \gamma^\mu \tag{35.25}$$

or

$$\left(\gamma^\mu\right)^\dagger = \gamma^4\gamma^\mu\gamma^4. \tag{35.26}$$

This is an important relation involving the Hermitian conjugate of γ^μ.

35.3 Charge-current conservation in a covariant form

Previously we obtained the probability current conservation relation

$$\frac{\partial\rho}{\partial t} + \nabla\cdot\mathbf{j} = 0 \tag{35.27}$$

where

$$\rho = \psi^\dagger \psi, \quad \mathbf{j} = \psi^\dagger \boldsymbol{\alpha} \psi. \tag{35.28}$$

Using these relations we can express ρ and $\mathbf{j}$ in terms of the γ-matrices as follows:

$$j_i = \psi^\dagger \alpha_i \psi = \psi^\dagger \beta \gamma^i \psi = \psi^\dagger \gamma^4 \gamma^i \psi. \tag{35.29}$$

We define a new quantity $\bar{\psi}$ given by

$$\bar{\psi} = \psi^\dagger \gamma^4 \tag{35.30}$$

and obtain

$$j_i = \bar{\psi} \gamma^i \psi. \tag{35.31}$$

Similarly, since $\left(\gamma^4\right)^2 = 1$,

$$\rho = \psi^\dagger \left(\gamma^4\right)^2 \psi = \left(\psi^\dagger \gamma^4\right) \gamma^4 \psi = \bar{\psi} \gamma^4 \psi. \tag{35.32}$$

The four-vector current j^μ is then

$$j^\mu = \left(j^\alpha, j^4\right) \tag{35.33}$$

where $\alpha = 1, 2, 3$

$$j^\alpha = \bar{\psi} \gamma^\alpha \psi, \quad j^4 = \bar{\psi} \gamma^4 \psi. \tag{35.34}$$

Hence

$$j^\mu = \bar{\psi} \gamma^\mu \psi. \tag{35.35}$$

Condition (35.27) can then be rewritten as

$$\partial_\mu j^\mu = 0. \tag{35.36}$$

The probability current conservation relation is now expressed as a covariant equation.

35.3.1 Derivation directly from the Dirac equation

Let us derive (35.36) directly from the Dirac equation rather than through (35.27). The equation is given by

$$i\gamma^\mu \left(\partial_\mu \psi\right) - m\psi = 0. \tag{35.37}$$

We take the Hermitian conjugate of (35.37),

$$-i\left(\partial_\mu \psi^\dagger\right)(\gamma^\mu)^\dagger - m\psi^\dagger = 0. \tag{35.38}$$

Let us now multiply (35.38) on the right by γ^4 and insert $\left(\gamma^4\right)^2$ in the first term of (35.38):

$$-i\left(\partial_\mu \psi^\dagger\right)\gamma^4\left(\gamma^4(\gamma^\mu)^\dagger\gamma^4\right) - m\psi^\dagger\gamma^4 = 0. \tag{35.39}$$

From the definition (35.30) and the result (35.24) we have

$$-i\left(\partial_\mu \bar{\psi}\right)\gamma^\mu - m\bar{\psi} = 0. \tag{35.40}$$

We now multiply (35.37) on the left by $\bar{\psi}$ and (35.40) on the right by ψ, and make a subtraction as follows:

$$\bar{\psi}\gamma^\mu\left(\partial_\mu\psi\right) + \left(\partial_\mu\bar{\psi}\right)\gamma^\mu\psi = 0. \tag{35.41}$$

This leads to

$$\partial_\mu j^\mu = 0 \tag{35.42}$$

where

$$j^\mu = \bar{\psi}\gamma^\mu\psi. \tag{35.43}$$

The two results (35.42) and (35.43) are the same as (35.36) and (35.35) derived earlier.

35.4 Spinor solutions: $u_r(p)$ and $v_r(p)$

Let $W_r(\mathbf{p}, E)$ be the solution of the Dirac equation

$$(\boldsymbol{\gamma}\cdot p - m)\,W_r(\mathbf{p}, E) = \left(\gamma_4 E - \boldsymbol{\gamma}\cdot\mathbf{p} - m\right)W_r(\mathbf{p}, E) = 0 \tag{35.44}$$

where E can be positive or negative and for each sign of E we have $r = 1, 2$ depending on whether we have a spin-up or spin-down state. The W's are related to the solutions we obtained previously.

For positive energies, $E = E_p = \sqrt{\mathbf{p}^2 + \mathbf{m}^2}$ we write

$$W_r\left(\mathbf{p}, E_p\right) = u_r(\mathbf{p}). \tag{35.45}$$

From (35.44) we then have

$$(\boldsymbol{\gamma}\cdot p - m)\,u_r(\mathbf{p}) = 0; \quad p^\mu = \left(\mathbf{p}, E_p\right). \tag{35.46}$$

For negative energies $E = -E_p = -\sqrt{\mathbf{p}^2 + \mathbf{m}^2}$, the equation satisfied by W_r is

$$\left(-\gamma_4 E_p - \boldsymbol{\gamma} \cdot \mathbf{p} - m\right) W_r\left(\mathbf{p}, -E_p\right) = 0. \tag{35.47}$$

We now reverse the sign of $\mathbf{p}$ in (35.47) to obtain

$$\left(-\gamma_4 E_p + \boldsymbol{\gamma} \cdot \mathbf{p} - m\right) W_r\left(-\mathbf{p}, -E_p\right) = 0 \tag{35.48}$$

and identify

$$W_r\left(-\mathbf{p}, -E_p\right) = v_r(\mathbf{p}), \quad r = 1, 2. \tag{35.49}$$

Factoring out the minus sign in (35.48) we obtain

$$(\boldsymbol{\gamma} \cdot p + m)\, v_r(\mathbf{p}) = 0; \quad p^\mu = \left(\mathbf{p}, E_p\right). \tag{35.50}$$

Thus we have separated the positive and negative solutions into two types of spinors, $u_r(\mathbf{p})$ and $v_r(\mathbf{p})$ each with spin-up ($r = 1$) and spin-down ($r = 2$) states.

35.5 Normalization and completeness condition for $u_r(p)$ and $v_r(p)$

We discuss now the question of normalization of $u_r(\mathbf{p})$ and $v_r(\mathbf{p})$. From our earlier discussion of the spinors we write below the positive-energy solution for spin-up ($r = 1$),

$$r = 1: \qquad W_r(\mathbf{p}, E) = u_r(\mathbf{p}) = C \begin{bmatrix} 1 \\ 0 \\ \dfrac{p_z}{(|E| + m)} \\ \dfrac{\left(p_x + ip_y\right)}{(|E| + m)} \end{bmatrix}. \tag{35.51}$$

The Hermitian conjugate $\bar{u}_r(\mathbf{p})$ is then

$$\bar{u}_r(\mathbf{p}) = u_r^\dagger(\mathbf{p})\gamma^4 = C^* \left(1, 0, \frac{p_z}{(|E| + m)}, \frac{\left(p_x - ip_y\right)}{(|E| + m)}\right) \begin{pmatrix} 1 & 0 & 0 & 0 \\ 0 & 1 & 0 & 0 \\ 0 & 0 & -1 & 0 \\ 0 & 0 & 0 & -1 \end{pmatrix}$$

$$= C^* \left(1, 0, -\frac{p_z}{E_p + m}, -\frac{\left(p_x - ip_y\right)}{\left(E_p + m\right)}\right). \tag{35.52}$$

We will demonstrate in the later sections that $\bar{u}_r(\mathbf{p})u_r(\mathbf{p})$ transforms as a scalar. Hence we use the normalization $\bar{u}_r(\mathbf{p})u_r(\mathbf{p}) = 1$ and find

$$1 = |C|^2 \left(1, 0, -\frac{p_z}{E_p + m}, -\frac{(p_x - ip_y)}{(E_p + m)}\right) \begin{bmatrix} 1 \\ 0 \\ \dfrac{p_z}{(|E| + m)} \\ \dfrac{(p_x + ip_y)}{(|E| + m)} \end{bmatrix}$$

$$= |C|^2 \left| 1 - \frac{p_z^2 + p_x^2 + p_y^2}{(E_p + m)^2} \right| = |C|^2 \left[1 - \frac{\left(E_p^2 - m^2\right)}{(E_p + m)^2} \right]$$

$$= |C|^2 \frac{2m}{E_p + m}. \tag{35.53}$$

The normalization constant is, therefore,

$$C = \sqrt{\frac{E_p + m}{2m}}. \tag{35.54}$$

Thus, we have

$$u_r(\mathbf{p}) = \sqrt{\frac{E_p + m}{2m}} \begin{bmatrix} 1 \\ 0 \\ \dfrac{p_z}{(|E| + m)} \\ \dfrac{(p_x + ip_y)}{(|E| + m)} \end{bmatrix}, \quad \text{for } r = 1. \tag{35.55}$$

We find the value of C for $r = 2$ to be the same as above.

$$u_r(\mathbf{p}) = \sqrt{\frac{E_p + m}{2m}} \begin{bmatrix} 0 \\ 1 \\ \dfrac{(p_x - ip_y)}{(|E| + m)} \\ -\dfrac{p_z}{(|E| + m)} \end{bmatrix}, \quad \text{for } r = 2. \tag{35.56}$$

Keeping the same normalization constant, we write for the negative energies,

$$W_r\left(-\mathbf{p}, -E_p\right) = v_r(\mathbf{p}) = \sqrt{\frac{E_p + m}{2m}} \begin{bmatrix} \dfrac{p_z}{(E_p + m)} \\ \dfrac{(p_x + ip_y)}{(E_p + m)} \\ 1 \\ 0 \end{bmatrix}, \quad \text{for } r = 1. \tag{35.57}$$

Similarly,

$$v_r(\mathbf{p}) = \sqrt{\frac{E_p + m}{2m}} C \begin{bmatrix} \frac{(p_x - ip_y)}{(|E| + m)} \\ -\frac{p_z}{(|E| + m)} \\ 0 \\ 1 \end{bmatrix}, \quad \text{for } r = 2. \tag{35.58}$$

We now obtain $\bar{v}_r(\mathbf{p})$ for $r = 1$, where

$$\bar{v}_r(\mathbf{p}) = v_r^\dagger(\mathbf{p})\gamma^4,$$

$$\bar{v}_r(\mathbf{p}) = \sqrt{\frac{E_p + m}{2m}} \left(\frac{p_z}{E_p + m}, \frac{(p_x - ip_y)}{E_p + m}, 1, 0 \right) \begin{pmatrix} 1 & 0 & 0 & 0 \\ 0 & 1 & 0 & 0 \\ 0 & 0 & -1 & 0 \\ 0 & 0 & 0 & -1 \end{pmatrix}$$

$$= \sqrt{\frac{E_p + m}{2m}} \left(\frac{p_z}{E_p + m}, \frac{(p_x - ip_y)}{E_p + m}, -1, 0 \right), \quad \text{for } r = 1. \tag{35.59}$$

Thus, the normalization of $v_r(\mathbf{p})$ is given by

$$\bar{v}_r(\mathbf{p})v_r(\mathbf{p}) = \frac{E_p + m}{2m} \left(\frac{p_z}{E_p + m}, \frac{(p_x - ip_y)}{E_p + m}, -1, 0 \right) \begin{bmatrix} \frac{p_z}{(E_p + m)} \\ \frac{(p_x + ip_y)}{(E_p + m)} \\ 1 \\ 0 \end{bmatrix}$$

$$= \frac{E_p + m}{2m} \left(\frac{p_z^2 + p_x^2 + p_y^2}{(E_p + m)^2} - 1 \right) = -1, \quad \text{for } r = 1. \tag{35.60}$$

One can similarly calculate the product $\bar{v}_r(\mathbf{p})v_r(\mathbf{p})$ for $r = 2$, as well as other bilinear products.

We summarize all the results compactly as follows:

$$\bar{u}_r(\mathbf{p})u_s(\mathbf{p}) = \delta_{rs}, \quad \bar{v}_r(\mathbf{p})v_s(\mathbf{p}) = -\delta_{rs}, \quad r, s = 1, 2, \tag{35.61}$$

$$\bar{u}_r(\mathbf{p})v_s(\mathbf{p}) = 0 = \bar{v}_r(\mathbf{p})u_s(\mathbf{p}), \quad r, s = 1, 2. \tag{35.62}$$

The completeness relation is found to be of the form

$$\sum_{r=1}^{2} \left[u_r(\mathbf{p}) \bar{u}_r(\mathbf{p}) - v_r(\mathbf{p}) \bar{v}_r(\mathbf{p}) \right] = \mathbf{1} \tag{35.63}$$

where each term on the left-hand side is a matrix while the right-hand side is a unit matrix.

35.5.1 Projection operators

Let us define what are known as projection operators, $\Lambda_\pm(\mathbf{p})$, as follows:

$$\Lambda_+(\mathbf{p}) = \sum_{r=1}^{2} u_r(\mathbf{p})\,\bar{u}_r(\mathbf{p}), \quad \Lambda_-(\mathbf{p}) = -\sum_{r=1}^{2} v_r(\mathbf{p})\,\bar{v}_r(\mathbf{p}). \tag{35.64}$$

The completeness relation (35.63) gives

$$\Lambda_+ + \Lambda_- = \mathbf{1}. \tag{35.65}$$

Furthermore, we find

$$\Lambda_+ u_s = u_s, \quad \Lambda_+ v_s = 0 \tag{35.66}$$

and

$$\Lambda_- v_s = v_s, \quad \Lambda_- u_s = 0. \tag{35.67}$$

Thus Λ_+ projects out positive-energy solutions, u_s, while Λ_- projects out the negative-energy solutions v_s. Moreover, using the orthogonality properties of $u_s(\mathbf{p})$ we find

$$\Lambda_+^2 = \left[\sum_{r=1}^{2} u_r(\mathbf{p})\,\bar{u}_r(\mathbf{p})\right]\left[\sum_{s=1}^{2} u_s(\mathbf{p})\,\bar{u}_s(\mathbf{p})\right] \tag{35.68}$$

$$= \sum_{r=1}^{2}\sum_{s=1}^{2} u_r(\mathbf{p})\,\delta_{rs}\bar{u}_s(\mathbf{p}) \tag{35.69}$$

$$= \sum_{r=1}^{2} u_r(\mathbf{p})\,\bar{u}_r(\mathbf{p}) = \Lambda_+. \tag{35.70}$$

We also obtain the following:

$$\Lambda_-^2 = \Lambda_-, \quad \Lambda_+\Lambda_- = 0. \tag{35.71}$$

Thus we have demonstrated from the above results that the operators $\Lambda_\pm$ act as projection operators.

We now determine the specific forms of the operators $\Lambda_\pm$. Since $\Lambda_+ v_r(\mathbf{p}) = 0$, then from the equation (35.50) satisfied by $v_r(\mathbf{p})$ we can write

$$\Lambda_+ = \lambda\,(m + \gamma \cdot p) \tag{35.72}$$

where λ is a constant. Operating on $u_r(\mathbf{p})$ we find, using (35.46),

$$\Lambda_+ u_r = \lambda 2m u_r. \tag{35.73}$$

Since $\Lambda_+ u_r = u_r$, we obtain

$$\lambda = \frac{1}{2m}. \tag{35.74}$$

Hence,

$$\Lambda_+ = \frac{m + \gamma \cdot p}{2m}. \tag{35.75}$$

From (35.65) we have

$$\Lambda_- = 1 - \Lambda_+ = \frac{m - \gamma \cdot p}{2m}. \tag{35.76}$$

The two relations (35.75) and (35.76) satisfy the conditions (35.66)–(35.71). Thus, in summary,

$$\sum_{r=1}^{2} u_r(\mathbf{p})\, \bar{u}_r(\mathbf{p}) = \Lambda_+(\mathbf{p}) = \frac{m + \gamma \cdot p}{2m} \tag{35.77}$$

and

$$-\sum_{r=1}^{2} v_r(\mathbf{p})\, \bar{v}_r(\mathbf{p}) = \Lambda_-(\mathbf{p}) = \frac{m - \gamma \cdot p}{2m}. \tag{35.78}$$

35.6 Gordon decomposition

We will derive below a very important relation called the Gordon decomposition. We start with the equation

$$(\gamma \cdot p_1 - m)u(p_1) = 0 \tag{35.79}$$

where we have suppressed the index r in u_r. Taking the Hermitian conjugate, we obtain

$$\bar{u}(p_1)\,(\gamma \cdot p_1 - m) = 0. \tag{35.80}$$

We multiply the right-hand side of (35.80) first by $\gamma \cdot A$ followed by $u(p_2)$, where A is an arbitrary four-vector

$$\bar{u}(p_1)(\gamma \cdot p_1 - m)\gamma \cdot A u(p_2) = 0. \tag{35.81}$$

We then consider the same equation as (35.79) but replace p_1 by p_2,

$$(\gamma \cdot p_2 - m)u(p_2) = 0, \tag{35.82}$$

and multiply (35.82) on the left first by $\gamma \cdot A$ followed by $\bar{u}(p_1)$

$$\bar{u}(p_1)\gamma \cdot A(\gamma \cdot p_2 - m)u(p_2) = 0. \tag{35.83}$$

Adding (35.81) and (35.83) we obtain

$$\bar{u}(p_1)\left[\gamma \cdot p_1 \gamma \cdot A + \gamma \cdot A\gamma \cdot p_2 - 2m\gamma \cdot A\right]u(p_2) = 0. \tag{35.84}$$

We can write (35.84) as

$$\bar{u}(p_1)\left[\gamma^\mu \gamma^\nu p_{1\mu} + \gamma^\nu \gamma^\mu p_{2\mu} - 2m\gamma^\nu\right]A_\nu u(p_2) = 0 \tag{35.85}$$

where we have factored out A_ν. Since A_ν is arbitrary, we can remove it to obtain the following relation

$$\bar{u}(p_1)\left[\gamma^\mu \gamma^\nu p_{1\mu} + \gamma^\nu \gamma^\mu p_{2\mu} - 2m\gamma^\nu\right]u(p_2) = 0. \tag{35.86}$$

One can write

$$\gamma^\mu \gamma^\nu = \frac{\gamma^\mu \gamma^\nu + \gamma^\nu \gamma^\mu}{2} + \frac{\gamma^\mu \gamma^\nu - \gamma^\nu \gamma^\mu}{2}. \tag{35.87}$$

The first term above can be simplified because of the relation

$$\gamma^\mu \gamma^\nu + \gamma^\nu \gamma^\mu = 2g^{\mu\nu}. \tag{35.88}$$

We define

$$\sigma^{\mu\nu} = \frac{\gamma^\mu \gamma^\nu - \gamma^\nu \gamma^\mu}{2i}. \tag{35.89}$$

Hence (35.87) can be re-expressed as

$$\gamma^\mu \gamma^\nu = g^{\mu\nu} + i\sigma^{\mu\nu}. \tag{35.90}$$

Concentrating only on the square bracket in (35.86), after inserting (35.90) we obtain

$$g^{\mu\nu}(p_{1\mu} + p_{2\mu}) + i\sigma^{\mu\nu}p_{1\mu} + i\sigma^{\nu\mu}p_{2\mu} - 2m\gamma^\nu = 0. \tag{35.91}$$

Since $\sigma^{\mu\nu} = -\sigma^{\nu\mu}$, we have

$$(p_1^\nu + p_2^\nu) + i\sigma^{\mu\nu}(p_{1\mu} - p_{2\mu}) - 2m\gamma^\nu = 0. \tag{35.92}$$

That is,

$$\gamma^\nu = \frac{1}{2m}(p_1^\nu + p_2^\nu) + i\sigma^{\mu\nu}\frac{1}{2m}(p_{1\mu} - p_{2\mu}). \tag{35.93}$$

This is the Gordon decomposition, where it is understood that this relation is sandwiched between $\bar{u}(p_1)$ and $u(p_2)$.

35.7 Lorentz transformation of the Dirac equation

Let us consider the properties of the Dirac wavefunction under Lorentz transformations. The equation is given by

$$\left(i\gamma_\mu \partial^\mu - m\right)\psi(x) = 0. \tag{35.94}$$

In another Lorentz frame, which we will designate as the primed frame with coordinates x'^μ, the equation will be of the form

$$\left(i\gamma_\mu \partial'^\mu - m\right)\psi'(x') = 0 \tag{35.95}$$

where the Lorentz transformation relates the two coordinate systems by

$$x'^\mu = L^\mu_{.\nu} x^\nu. \tag{35.96}$$

Let

$$\psi'(x') = S\psi(x). \tag{35.97}$$

Then, since $\partial'^\mu = L^\mu_{.\nu}\partial^\nu$, we write equation (35.95) as

$$\left(i\gamma_\mu L^\mu_{.\nu}\partial^\nu - m\right)S\psi(x) = 0. \tag{35.98}$$

Multiplying the above equation on the left by S^{-1} and comparing it with equation (35.94), we find that the following relation must be satisfied:

$$S^{-1}\gamma_\mu L^\mu_{.\nu} S = \gamma_\nu. \tag{35.99}$$

We rewrite this relation as

$$\gamma_\mu L^\mu_{.\nu} = S\gamma_\nu S^{-1}. \tag{35.100}$$

We now consider infinitesimal transformations of both S and $L^\mu_{.\nu}$. First we write

$$S = 1 + \varepsilon\Sigma \tag{35.101}$$

where ε is an infinitesimal quantity. The infinitesimal transformation of $L^\mu_{.\nu}$ has already been considered in Chapter 31, where it was found that one can write

$$L^\mu_{.\nu} = \delta^\mu_{.\nu} + \varepsilon e^\mu_{.\nu} \tag{35.102}$$

where, if we define

$$e^\mu_{.\nu} = e^{\mu\alpha} g_{\alpha\nu}, \tag{35.103}$$

then $e^{\mu\alpha}$ is an antisymmetric tensor:

$$e^{\mu\alpha} = -e^{\alpha\mu}. \tag{35.104}$$

To obtain Σ we substitute (35.102) into (35.100) and obtain

$$\gamma^{\mu}\left(\delta_{\mu}^{\nu} + \varepsilon e_{\mu}^{\nu}\right) = (1 + \varepsilon\Sigma)\,\gamma^{\nu}\,(1 - \varepsilon\Sigma). \tag{35.105}$$

Therefore, to first order in ε,

$$\gamma^{\nu} + \varepsilon e_{\mu}^{\nu}\gamma^{\mu} = \gamma^{\nu} + \varepsilon\left[\Sigma, \gamma^{\nu}\right], \tag{35.106}$$

which leads to the relation

$$\left[\Sigma, \gamma^{\nu}\right] = e_{\mu}^{\nu}\gamma^{\mu}. \tag{35.107}$$

The solution of this is given by

$$\Sigma = \frac{1}{4}\gamma^{\alpha}\gamma^{\beta}e_{\alpha\beta}. \tag{35.108}$$

Once again through the process of infinitesimal transformations one can show that

$$S^{-1} = \gamma_4 S^{\dagger}\gamma_4. \tag{35.109}$$

35.7.1 Bilinear covariant terms

Starting with

$$\psi'(x') = S\psi(x) \tag{35.110}$$

we obtain the following relations for $\bar{\psi}'(x')$:

$$\bar{\psi}'\left(x'\right) = \psi'^{\dagger}\gamma^4 = \psi^{\dagger}S^{\dagger}\gamma^4 = \psi^{\dagger}\gamma^4\left(\gamma^4 S^{\dagger}\gamma^4\right). \tag{35.111}$$

Substituting (35.109) we find

$$\bar{\psi}'(x') = \bar{\psi}(x)S^{-1}. \tag{35.112}$$

Let us consider the following bilinear products:

(i)
$$\bar{\psi}(x)\,\psi(x) \tag{35.113}$$

Under Lorentz transformations

$$\bar{\psi}(x)\,\psi(x) \to \bar{\psi}'\left(x'\right)\psi\left(x'\right) = \bar{\psi}(x)\,S^{-1}S\psi(x) = \bar{\psi}(x)\,\psi(x). \tag{35.114}$$

Hence $\bar{\psi}(x)\psi(x)$ is invariant, i.e., it transforms as a scalar.

(ii)
$$\bar{\psi}(x)\,\gamma^{\mu}\psi(x)\,. \tag{35.115}$$
$$\bar{\psi}'(x')\,\gamma^{\mu}\psi'(x') = \bar{\psi}(x)\,S^{-1}\gamma^{\mu}S\psi(x)\,. \tag{35.116}$$

However,

$$S^{-1}\gamma^{\mu}S = L^{\mu}{}_{\cdot\rho}\,\gamma^{\rho}. \tag{35.117}$$

Thus,

$$\bar{\psi}(x')\,\gamma^{\mu}\psi'(x') = L^{\mu}{}_{\cdot\rho}\,\bar{\psi}(x)\,\gamma^{\rho}\psi(x)\,. \tag{35.118}$$

Hence, $\bar{\psi}(x)\,\gamma^{\mu}\psi(x)$ transforms as a four-vector.

Similarly, one can derive properties of other bilinear products of the form $\bar{\psi}(x)\Gamma\psi(x)$ where Γ is a product of γ-matrices.

35.8 Appendix to Chapter 35

35.8.1 Further properties of γ-matrices

Below we consider some important properties involving the traces of γ-matrices.

Product of even number of γ-matrices

(i) $\text{Tr}\,(\gamma^{\mu}\gamma^{\nu})$

To evaluate this we use the relation

$$\gamma^{\mu}\gamma^{\nu} + \gamma^{\nu}\gamma^{\mu} = 2g^{\mu\nu}\cdot 1. \tag{35.119}$$

Taking the trace of both sides and using the cyclic property of the traces,

$$\text{Tr}\,(abc\cdots z) = \text{Tr}\,(bc\cdots za)\,, \tag{35.120}$$

we find

$$\text{Tr}\left(\gamma^{\mu}\gamma^{\nu}\right) = g^{\mu\nu}\text{Tr}\,(\mathbf{1}) = 4g^{\mu\nu} \tag{35.121}$$

where we have used the fact that $\text{Tr}(\mathbf{1}) = 4$.

(ii) $\text{Tr}\,(\gamma^{\mu}\gamma^{\nu}\gamma^{\rho}\gamma^{\sigma})$

Here we use, as in (i), the relation (35.119) at each stage as we move γ^{μ} from the left all the way to the right, and, once it reaches the extreme right, we use the cyclic property (35.120),

$$\mathrm{Tr}\left(\gamma^{\mu}\gamma^{\nu}\gamma^{\rho}\gamma^{\sigma}\right) = \mathrm{Tr}\left[\left(2g^{\mu\nu} - \gamma^{\nu}\gamma^{\mu}\right)\gamma^{\rho}\gamma^{\sigma}\right] \tag{35.122}$$

$$= 2g^{\mu\nu}\mathrm{Tr}\left(\gamma^{\rho}\gamma^{\sigma}\right) - \mathrm{Tr}\left(\gamma^{\nu}\gamma^{\mu}\gamma^{\rho}\gamma^{\sigma}\right) \tag{35.123}$$

$$= \left(2g^{\mu\nu}\right)\left(g^{\rho\sigma}\right)\cdot 4 - \ \mathrm{Tr}\left(\gamma^{\nu}\gamma^{\mu}\gamma^{\rho}\gamma^{\sigma}\right). \tag{35.124}$$

Now continuing moving γ^{μ} we find for the second term above,

$$\mathrm{Tr}\left(\gamma^{\nu}\gamma^{\mu}\gamma^{\rho}\gamma^{\sigma}\right) = \mathrm{Tr}\left[\gamma^{\nu}\left(2g^{\mu\rho} - \gamma^{\rho}\gamma^{\mu}\right)\gamma^{\sigma}\right] \tag{35.125}$$

$$= 2g^{\mu\rho}\mathrm{Tr}\left(\gamma^{\nu}\gamma^{\sigma}\right) - \mathrm{Tr}\left(\gamma^{\nu}\gamma^{\rho}\gamma^{\mu}\gamma^{\sigma}\right) \tag{35.126}$$

$$= 2g^{\mu\rho}\left(g^{\nu\sigma}\right)\cdot 4 - \mathrm{Tr}\left(\gamma^{\nu}\gamma^{\rho}\gamma^{\mu}\gamma^{\sigma}\right), \tag{35.127}$$

and continuing further for the second term,

$$\mathrm{Tr}\left(\gamma^{\nu}\gamma^{\rho}\gamma^{\mu}\gamma^{\sigma}\right) = \mathrm{Tr}\left[\gamma^{\nu}\gamma^{\rho}\left(2g^{\mu\sigma} - \gamma^{\sigma}\gamma^{\mu}\right)\right] \tag{35.128}$$

$$= 2g^{\mu\sigma}\left(g^{\nu\rho}\right)\cdot 4 - \mathrm{Tr}\left(\gamma^{\nu}\gamma^{\rho}\gamma^{\sigma}\gamma^{\mu}\right) \tag{35.129}$$

$$= 2g^{\mu\sigma}\left(g^{\nu\rho}\right)\cdot 4 - \mathrm{Tr}\left(\gamma^{\mu}\gamma^{\nu}\gamma^{\rho}\gamma^{\sigma}\right). \tag{35.130}$$

In the last step we have used the cyclic property (35.120), so that the second term on the right of (35.130) is the same as the term on the left side of (35.124).

Hence we obtain

$$\mathrm{Tr}\left(\gamma^{\mu}\gamma^{\nu}\gamma^{\rho}\gamma^{\sigma}\right) = 4\left(g^{\mu\nu}g^{\rho\sigma} - g^{\mu\rho}g^{\nu\sigma} + g^{\mu\sigma}g^{\nu\rho}\right). \tag{35.131}$$

One can similarly work out the traces of six or higher numbers of even products.

Product of odd number of γ-matrices

The trace of a product of an odd-number of γ-matrices can be obtained by using the γ_5-trick which we explain below.

(i) $\mathrm{Tr}\left(\gamma^{\mu}\right)$

From the definition of γ^{μ} we know that

$$\mathrm{Tr}\left(\gamma^{\mu}\right) = 0. \tag{35.132}$$

(ii) $\mathrm{Tr}\left(\gamma^{\mu}\gamma^{\nu}\gamma^{\rho}\right)$

Let us use the trick of using the γ_5-matrix which in the standard representation is given by

$$\gamma_5 = \begin{pmatrix} 0 & 1 \\ 1 & 0 \end{pmatrix}; \quad \gamma_5^2 = \begin{pmatrix} 1 & 0 \\ 0 & 1 \end{pmatrix} = 1. \tag{35.133}$$

One can also write

$$\gamma_5 = i\gamma^1\gamma^2\gamma^3\gamma^4. \tag{35.134}$$

It can easily be shown that γ_5 anticommutes with γ^μ,

$$\{\gamma_5, \gamma^\mu\} = 0. \tag{35.135}$$

We will use the properties (35.133) and (35.135) to determine $\mathrm{Tr}(\gamma^\mu\gamma^\nu\gamma^\rho)$, which can be written as

$$\mathrm{Tr}\left(\gamma^\mu\gamma^\nu\gamma^\rho\right) = \mathrm{Tr}\left(\gamma_5\gamma_5\gamma^\mu\gamma^\nu\gamma^\rho\right) \tag{35.136}$$

$$= -\mathrm{Tr}\left(\gamma_5\gamma^\mu\gamma^\nu\gamma^\rho\gamma_5\right) \tag{35.137}$$

where we have moved one of the γ_5's from the left to the right, at each stage using the property (35.135). If we use the cyclic property of the traces then

$$\mathrm{Tr}\left(\gamma_5\gamma^\mu\gamma^\nu\gamma^\rho\gamma_5\right) = \mathrm{Tr}\left(\gamma^\mu\gamma^\nu\gamma^\rho\gamma_5\gamma_5\right) = \mathrm{Tr}\left(\gamma^\mu\gamma^\nu\gamma^\rho\right). \tag{35.138}$$

Hence, from (35.137) and (35.138),

$$\mathrm{Tr}\left(\gamma^\mu\gamma^\nu\gamma^\rho\right) = 0. \tag{35.139}$$

Similarly, we can show that

$$\mathrm{Tr}\left(\gamma^\mu\gamma^\nu \cdots \text{odd number}\right) = 0. \tag{35.140}$$

35.8.2 Trace of products of the form $(\gamma \cdot a_1\gamma \cdot a_2 \cdots)$

We summarize the following properties involving the traces derived from the results of the previous section:

$$\mathrm{Tr}\left(\gamma \cdot a_1\right) = 0, \tag{35.141}$$

$$\mathrm{Tr}\left(\gamma \cdot a_1\gamma \cdot a_2\right) = 4\left(a_1 a_2\right) = 4a_1 \cdot a_2, \tag{35.142}$$

$$\mathrm{Tr}\left(\gamma \cdot a_1\gamma \cdot a_2 \cdots \gamma \cdot a_n\right) = 4[(a_1 \cdot a_2)(a_3 \cdots a_n) - (a_1 \cdot a_3)(a_2 \cdots a_n) + \cdots + (a_1 \cdot a_n)(a_2 \cdots a_{n-1})] \tag{35.143}$$

$$\mathrm{Tr}(\gamma \cdot a_1\gamma \cdot a_2 \cdots \gamma \cdot a_n) = 0, \quad \text{for any odd } n. \tag{35.144}$$

36 Interaction of a Dirac particle with an electromagnetic field

We introduce an electromagnetic field into the Dirac equation in a fully relativistic form and examine the consequences by comparing, for example, the Dirac electromagnetic current with the current in the Klein–Gordon equation. We calculate the propagator for the Dirac particle and obtain the S-matrix involved in electromagnetic scattering. Specifically, we calculate the Rutherford scattering amplitude and compare it with the nonrelativistic result.

36.1 Charged particle Hamiltonian

As we have already discussed, electromagnetic interaction can easily be incorporated into the Hamiltonian by making the substitutions

$$\mathbf{p} \to \mathbf{p} - e\mathbf{A}, \tag{36.1}$$

$$E \to E - e\phi \tag{36.2}$$

where ϕ and $\mathbf{A}$ are scalar and vector potentials.

The nonrelativistic energy momentum relation in the presence of electromagnetic interaction will then be of the form

$$(E - e\phi) - \frac{(\mathbf{p} - e\mathbf{A})^2}{2m} = 0, \tag{36.3}$$

while the relativistic form given by the Klein–Gordon equation will be

$$(E - e\phi)^2 - (\mathbf{p} - e\mathbf{A})^2 - m^2 = 0. \tag{36.4}$$

The Dirac equation in the presence of electromagnetic interactions can, therefore, be written as

$$[E - e\phi - \boldsymbol{\alpha} \cdot (\mathbf{p} - e\mathbf{A}) - \beta m]\, u = 0. \tag{36.5}$$

In order that this equation be consistent with the relativistic relation (36.4) we multiply the above equation by

$$E - e\phi + \boldsymbol{\alpha} \cdot (\mathbf{p} - e\mathbf{A}) + \beta m \tag{36.6}$$

and obtain

$$\begin{aligned} &[(E - e\phi)^2 - \boldsymbol{\alpha} \cdot (\mathbf{p} - e\mathbf{A})\, \boldsymbol{\alpha} \cdot (\mathbf{p} - e\mathbf{A}) - m^2 \\ &\quad + (E - e\phi)\, \boldsymbol{\alpha} \cdot (\mathbf{p} - e\mathbf{A}) - \boldsymbol{\alpha} \cdot (\mathbf{p} - e\mathbf{A})\, (E - e\phi)]u = 0. \end{aligned} \tag{36.7}$$

This equation can be simplified by using the standard representation for $\boldsymbol{\alpha}$. We find

$$\boldsymbol{\alpha}\cdot\mathbf{B}\cdot\boldsymbol{\alpha}\cdot\mathbf{C}=\begin{bmatrix}0 & \boldsymbol{\sigma}\cdot\mathbf{B}\\ \boldsymbol{\sigma}\cdot\mathbf{B} & 0\end{bmatrix}\begin{bmatrix}0 & \boldsymbol{\sigma}\cdot\mathbf{C}\\ \boldsymbol{\sigma}\cdot\mathbf{C} & 0\end{bmatrix}=\begin{bmatrix}\boldsymbol{\sigma}\cdot\mathbf{B}\boldsymbol{\sigma}\cdot\mathbf{C} & 0\\ 0 & \boldsymbol{\sigma}\cdot\mathbf{B}\boldsymbol{\sigma}\cdot\mathbf{C}\end{bmatrix}. \tag{36.8}$$

We have already derived the following property of the 2×2 Pauli matrices,

$$\boldsymbol{\sigma}\cdot\mathbf{B}\boldsymbol{\sigma}\cdot\mathbf{C}=\mathbf{B}\cdot\mathbf{C}+i\boldsymbol{\sigma}\cdot(\mathbf{B}\times\mathbf{C}), \tag{36.9}$$

and from it we have obtained the relation

$$\boldsymbol{\alpha}\cdot\mathbf{B}\boldsymbol{\alpha}\cdot\mathbf{C}=\mathbf{B}\cdot\mathbf{C}+i\boldsymbol{\sigma}'\cdot(\mathbf{B}\times\mathbf{C}) \tag{36.10}$$

where the 4×4 matrix, $\boldsymbol{\sigma}'$, has been defined earlier. Thus,

$$\boldsymbol{\alpha}\cdot(\mathbf{p}-e\mathbf{A})\;\boldsymbol{\alpha}\cdot(\mathbf{p}-e\mathbf{A})=(\mathbf{p}-e\mathbf{A})^2+i\boldsymbol{\sigma}'\cdot(\mathbf{p}-e\mathbf{A})\times(\mathbf{p}-e\mathbf{A}) \tag{36.11}$$

$$=(\mathbf{p}-e\mathbf{A})^2+ie\boldsymbol{\sigma}'\cdot(\mathbf{p}\times\mathbf{A}+\mathbf{A}\times\mathbf{p}) \tag{36.12}$$

where we have taken into account the fact that $\mathbf{p}$ is an operator that operates on the wavefunction $\psi(x)=u(p)\exp(-ip\cdot x)$. Specifically, writing $\mathbf{p}=-\mathbf{i}\nabla$, we have, after including the wavefunction ψ,

$$(\mathbf{p}\times\mathbf{A}+\mathbf{A}\times\mathbf{p})\,\psi=-i\,[\nabla\times(\mathbf{A}\psi)+(\mathbf{A}\times\nabla\psi)] \tag{36.13}$$

$$=-i\,[(\nabla\times\mathbf{A})\,\psi+(\nabla\psi\times\mathbf{A})+(\mathbf{A}\times\nabla\psi)] \tag{36.14}$$

$$=-i\,(\nabla\times\mathbf{A})\,\psi. \tag{36.15}$$

Since $\nabla\times\mathbf{A}=\mathbf{H}$, where $\mathbf{H}$ is the magnetic field, we obtain

$$\boldsymbol{\alpha}\,(\mathbf{p}-e\mathbf{A})\;\boldsymbol{\alpha}\cdot(\mathbf{p}-e\mathbf{A})=(\mathbf{p}-e\mathbf{A})^2-e\hbar\boldsymbol{\sigma}'\cdot\;(\nabla\times\mathbf{A}) \tag{36.16}$$

$$=(\mathbf{p}-e\mathbf{A})^2-e\hbar\boldsymbol{\sigma}'\cdot\mathbf{H}. \tag{36.17}$$

Similarly the following two terms can be simplified by writing E and $\mathbf{p}$ in the operator forms $E=i\partial/\partial t$ and $\mathbf{p}=-i\nabla$, respectively:

$$[(E-e\phi)\,\boldsymbol{\alpha}\cdot(\mathbf{p}-e\mathbf{A})-\boldsymbol{\alpha}\cdot(\mathbf{p}-e\mathbf{A})\,(E-e\phi)]\,\psi \tag{36.18}$$

$$=[-e\boldsymbol{\alpha}\cdot(E\mathbf{A}-\mathbf{A}E)-e\,\boldsymbol{\alpha}\cdot(\phi\mathbf{p}-\mathbf{p}\phi)]\,\psi \tag{36.19}$$

$$=-ie\boldsymbol{\alpha}\cdot\left(\frac{\partial}{\partial t}(\mathbf{A}\psi)-\mathbf{A}\frac{\partial\psi}{\partial t}\right)+ie\boldsymbol{\alpha}\cdot\phi\nabla\psi-\;\nabla(\phi\psi). \tag{36.20}$$

The last term above can be rewritten as

$$ie\boldsymbol{\alpha}\cdot\left(-\frac{\partial\mathbf{A}}{\partial t}-\nabla\phi\right)=ie\boldsymbol{\alpha}\cdot\mathbf{E} \tag{36.21}$$

where we have used the relation $\mathbf{E} = (-\partial \mathbf{A}/\partial \mathbf{t} - \nabla\phi)$ for the electric field $\mathbf{E}$. Hence, relation (36.7) is found to be

$$\left[(E - e\phi)^2 - (\mathbf{p} - e\mathbf{A})^2 - m^2 + e\boldsymbol{\sigma}' \cdot \mathbf{H} + ie\boldsymbol{\alpha} \cdot \mathbf{E}\right] u = 0 \tag{36.22}$$

where we have returned to writing the equation in terms of $u(p)$. This equation is reminiscent of the equation in the nonrelativistic case with a magnetic field, except that we have a 4×4 theory and have the additional $\boldsymbol{\alpha} \cdot \mathbf{E}$ term.

In order to discuss the nonrelativistic limit we write

$$E = E_T + m \tag{36.23}$$

where, in the nonrelativistic limit, E_T can be considered as the kinetic energy with $E_T \ll m$. We also replace u by u_L.

We then express

$$(E - e\phi)^2 - m^2 = (E - e\phi - m)(E - e\phi + m) \approx (E_T - e\phi)\, 2m \tag{36.24}$$

where in the last step we have also assumed $e\phi \ll m$. Thus the nonrelativistic form of the the Dirac equation is given by

$$\left(E_T - e\phi - \frac{1}{2m}(\mathbf{p} - e\mathbf{A})^2 + \frac{e}{2m}\boldsymbol{\sigma}' \cdot \mathbf{H} + \frac{ie}{2m}\boldsymbol{\alpha} \cdot \mathbf{E}\right) u = 0. \tag{36.25}$$

The fourth term in the above equation is of the form $\boldsymbol{\mu} \cdot \mathbf{H}$, which is the interaction energy due to a magnetic dipole moment in an electromagnetic field. The Dirac particle, therefore, acts as if it has a magnetic dipole moment given by

$$\boldsymbol{\mu} = \frac{e\hbar}{2mc}\boldsymbol{\sigma}' \tag{36.26}$$

where we have temporarily re-instated $\hbar$ and c. The quantity $e\hbar/2m$ is called the Bohr magneton. We have already discussed all of this in Chapter 6.

The last term in the equation (36.25) can be estimated by comparing it with $e\phi$,

$$\left|\frac{(e\hbar/2mc)\boldsymbol{\alpha} \cdot \mathbf{E}}{e\phi}\right| \approx \frac{v^2}{c^2} \tag{36.27}$$

where we have made the following assumptions. (i) Since the electric field $\mathbf{E}$ is related to the potential ϕ by $\mathbf{E} = -\nabla\boldsymbol{\phi}$, we have $|\partial\phi/\partial r| = |\mathbf{E}|$, and hence $|\phi| \approx |\mathbf{E}|\, a$, where a is a characteristic length. (ii) The uncertainty relation gives $pa \sim \hbar$, where in the nonrelativistic limit we write the momentum, p, in terms of velocity, v, as $p = mv$. (iii) The relation $\langle\boldsymbol{\alpha}\rangle \sim v/c$. Combining all these factors leads to (36.27). Hence $\boldsymbol{\alpha} \cdot \mathbf{E}$ in (36.25) can be neglected. We then have in place of (36.25) the following

$$\left(E_T - e\phi - \frac{1}{2m}(\mathbf{p} - e\mathbf{A})^2 + \frac{e\hbar}{2m}\boldsymbol{\sigma} \cdot \mathbf{H}\right) u_L = 0. \tag{36.28}$$

We can write the above equation in the x-space as

$$i\hbar\frac{\partial\psi_L}{\partial t} = \left[\frac{1}{2m}(ih\nabla + e\mathbf{A})^2 - \frac{e\hbar}{2m}\boldsymbol{\sigma}\cdot\mathbf{H}+e\phi\right]\psi_L \tag{36.29}$$

where $\psi_L(x) = u_L(p)\exp(-ip\cdot x)$, where in $p\cdot x$ we have taken $p^\mu = (\mathbf{p}, E_T)$ and $u_L(p)$ corresponds to the large component, which is a column matrix with just two entries.

36.2 Deriving the equation another way

We can derive (36.29) by going directly to the large component as follows. We write the Dirac equation (36.5) as

$$\begin{pmatrix} E - e\phi - m & -\sigma\cdot(\mathbf{p} - e\mathbf{A}) \\ -\sigma\cdot(\mathbf{p} - e\mathbf{A}) & E - e\phi + m \end{pmatrix}\begin{pmatrix} u_L \\ u_S \end{pmatrix} = 0, \tag{36.30}$$

which gives

$$(E - e\phi - m)\,u_L - \boldsymbol{\sigma}\cdot(\mathbf{p} - e\mathbf{A})\,u_S = 0, \tag{36.31}$$

$$(E - e\phi + m)\,u_S - \boldsymbol{\sigma}\cdot(\mathbf{p} - e\mathbf{A})\,u_L = 0. \tag{36.32}$$

From (36.32) in the nonrelativistic approximation $e\phi \ll m$ and $E \approx m$, we obtain

$$u_S = \frac{\boldsymbol{\sigma}\cdot(\mathbf{p} - e\mathbf{A})}{E - e\phi + m}u_L \approx \frac{\boldsymbol{\sigma}\cdot(\mathbf{p} - e\mathbf{A})}{2m}u_L. \tag{36.33}$$

Putting this into (36.32) we get

$$(E - e\phi - m)\,u_L - \frac{\boldsymbol{\sigma}\cdot(\mathbf{p} - e\mathbf{A})\,\boldsymbol{\sigma}\cdot(\mathbf{p} - e\mathbf{A})}{2m}u_L = 0. \tag{36.34}$$

As discussed above,

$$\boldsymbol{\sigma}\cdot(\mathbf{p} - e\mathbf{A})\,\boldsymbol{\sigma}\cdot(\mathbf{p} - e\mathbf{A}) = (\mathbf{p} - e\mathbf{A})^2 - e\frac{\boldsymbol{\sigma}\cdot\mathbf{H}}{2m}. \tag{36.35}$$

Once again writing $E = E_T + m$, we get from (36.34)

$$\left(E_T - e\phi - \frac{1}{2m}(\mathbf{p} - e\mathbf{A})^2 + \frac{e\hbar}{2m}\boldsymbol{\sigma}\cdot\mathbf{H}\right)u_L = 0 \tag{36.36}$$

which is the same equation as (36.28).

36.3 Gordon decomposition and electromagnetic current

The interaction of a Dirac particle with an electromagnetic field is given by the Hamiltonian

$$H' = ej^{\mu}A_{\mu} \tag{36.37}$$

where, as we derived earlier, the current, j^{μ} is given by

$$j^{\mu} = \bar{\psi}(x)\gamma^{\mu}\psi(x) \tag{36.38}$$

and A_{μ} is the vector potential. Let us relate this to the expressions we obtained for the current in the Klein–Gordon case and in the Schrödinger case.

We will construct an electromagnetic current $\bar{\psi}\partial^{\mu}\psi$ of the Klein–Gordon type and write it in a form involving two spinors $\psi_1$and ψ_2. We denote it

$$\begin{aligned} j^{\mu(1)} &= i\frac{1}{2m}\left[\bar{\psi}_1(x)\left(\partial^{\mu}\psi_2(x)\right) - \left(\partial^{\mu}\bar{\psi}_1(x)\right)\psi_2(x)\right] \\ &= \frac{1}{2m}\left[\bar{u}(\mathbf{p}_1)\left[p_2^{\mu} + p_1^{\mu}\right]u(\mathbf{p}_2)e^{i(p_1-p_2)\cdot x}\right] \end{aligned} \tag{36.39}$$

where we have taken

$$\psi_i(x) = u(\mathbf{p}_i)e^{-ip_i\cdot x}. \tag{36.40}$$

Let us define another current, $j^{\mu(2)}$, as

$$j^{\mu(2)} = \frac{1}{2m}\frac{\partial}{\partial x^{\nu}}\left(\bar{\psi}_1\sigma^{\mu\nu}\psi_2\right) = \frac{1}{2m}i(p_1 - p_2)_{\nu}\left[\bar{u}(\mathbf{p}_1)\sigma^{\mu\nu}u(\mathbf{p}_2)\right]e^{i(p_1-p_2)\cdot x}. \tag{36.41}$$

Combining (36.39) and (36.41) we obtain

$$j^{\mu(1)} + j^{\mu(2)} = \bar{u}(\mathbf{p}_1)\left[\frac{p_1^{\mu} + p_2^{\mu}}{2m} + i\frac{\sigma^{\mu\nu}}{2m}(p_1 - p_2)_{\nu}\right]u(p_2)e^{i(p_1-p_2)\cdot x}. \tag{36.42}$$

From Gordon decomposition applied to the right-hand side we find

$$j^{\mu(1)} + j^{\mu(2)} = \bar{u}(\mathbf{p}_1)\gamma^{\mu}u(\mathbf{p}_2)e^{i(p_1-p_2)\cdot x}, \tag{36.43}$$

which is the same as the Dirac current j^{μ} defined previously, which we now write as

$$j^{\mu} = j^{\mu(1)} + j^{\mu(2)}. \tag{36.44}$$

Hence,

$$H' = ej^{\mu(1)}A_{\mu} + j^{\mu(2)}A_{\mu}. \tag{36.45}$$

Thus the total Dirac current, j^{μ}, is divided into two parts, one of which, $j^{\mu(1)}$, is the normal current, devoid of any γ-matrices, which one finds in the Klein–Gordon equation or (in the nonrelativistic limit) in the Schrödinger equation. The current $j^{\mu(2)}$, however, which depends on γ-matrices, is related to the spin of the electron as we show below.

We consider the following relation involving the vector potential, A_μ:

$$\frac{\partial}{\partial x^\nu}\left[\bar{\psi}_1\sigma^{\mu\nu}\psi_2 A_\mu\right] = \frac{\partial}{\partial x^\nu}\left[\bar{\psi}_1\sigma^{\mu\nu}\psi_2\right]A_\mu + \bar{\psi}_1\sigma^{\mu\nu}\psi_2\frac{\partial A_\mu}{\partial x^\nu}. \tag{36.46}$$

We note that the first term on the right-hand side is of the form $j^{\mu(2)}A_\mu$. As for the left-hand side, we note that if we integrate the equation over a large volume then, since the left hand is a perfect differential, one can use the divergence theorem to show that it vanishes if the fields themselves go to zero at infinity. Hence we will ignore the left-hand side and write relation (36.46) as

$$j^{\mu(2)}A_\mu = -\bar{\psi}_1\sigma^{\mu\nu}\psi_2\frac{\partial A_\mu}{\partial x^\nu}. \tag{36.47}$$

We consider the following relation, which will simplify the right-hand side above,

$$\sigma^{\mu\nu}\frac{\partial A_\mu}{\partial x^\nu} = \frac{1}{2}\left[\sigma^{\mu\nu}\frac{\partial A_\mu}{\partial x^\nu} + \sigma^{\nu\mu}\frac{\partial A_\nu}{\partial x^\mu}\right]. \tag{36.48}$$

If we use the antisymmetry property, $\sigma^{\nu\mu} = -\sigma^{\mu\nu}$, we find

$$\sigma^{\mu\nu}\frac{\partial A_\mu}{\partial x^\nu} = \frac{1}{2}\sigma^{\mu\nu}\left[\partial_\nu A_\mu - \partial_\mu A_\nu\right] = \frac{1}{2}\sigma^{\mu\nu}F_{\nu\mu} \tag{36.49}$$

where we have used the relation $F_{\nu\mu} = \partial_\nu A_\mu - \partial_\mu A_\nu$, which connects the Maxwell tensor, $F_{\nu\mu}$, to the vector potential, A_μ. Thus,

$$j^{\mu(2)}A_\mu = -\frac{1}{2}\sigma^{\mu\nu}F_{\nu\mu}. \tag{36.50}$$

To simplify the right-hand side of the above equation, we write the following relation in terms of the individual components:

$$\sigma^{\mu\nu}F_{\nu\mu} = \sigma^{i4}F_{i4} + \sigma^{4i}F_{i4} + \sigma^{ji}F_{ij}, \quad i,j = 1,2,3 \tag{36.51}$$

where we have taken into account the fact that, because of their antisymmetry properties, $\sigma^{\alpha\beta} = 0 = F_{\alpha\beta}$ if $\alpha = \beta$. Therefore, from (36.51),

$$\sigma^{\mu\nu}F_{\nu\mu} = 2\sigma^{i4}F_{4i} + \sigma^{ji}F_{ij}. \tag{36.52}$$

From the relations discussed in Chapter 31 connecting the Maxwell field, $F_{\mu\nu}$, to the electric field, **E**, and magnetic field, **B**, we have

$$F_{4i} = -F^{4i} = E_i \tag{36.53}$$

and

$$F_{ji} = -\epsilon_{jik}B_k. \tag{36.54}$$

Thus we can write (36.52) as

$$\sigma^{\mu\nu}F_{\nu\mu} = 2\sigma^{i4}E_i - \epsilon_{jik}\sigma^{ji}B_k. \tag{36.55}$$

From the properties of $\sigma^{\mu\nu}$ we find

$$\sigma^{i4} = \frac{i}{2}\left[\gamma^i\gamma^4 - \gamma^4\gamma^i\right] = i\gamma^i\gamma^4 = i(\beta\alpha_i)(-\beta) = i\alpha_i \tag{36.56}$$

where we have used the relations connecting the $\boldsymbol{\alpha}$ and β matrices to the γ-matrices. Furthermore,

$$\sigma^{ji} = \frac{i}{2}\left[\gamma^j\gamma^i - \gamma^i\gamma^j\right] = -i\begin{pmatrix} \sigma_j\sigma_i & 0 \\ 0 & \sigma_j\sigma_i \end{pmatrix} \tag{36.57}$$

in the standard representation. Since

$$\sigma_j\sigma_i = i\epsilon_{jik}\sigma_k, \tag{36.58}$$

we have

$$\sigma^{ji} = \epsilon_{jik}\sigma_k. \tag{36.59}$$

Hence, we obtain

$$\sigma^{\mu\nu}F_{\nu\mu} = 2\left[i\boldsymbol{\alpha}\cdot\mathbf{E} + \boldsymbol{\sigma}\cdot\mathbf{B}\right]. \tag{36.60}$$

Therefore,

$$j^{\mu(2)}A_\mu = -\left[i\boldsymbol{\alpha}\cdot\mathbf{E} + \boldsymbol{\sigma}\cdot\mathbf{B}\right]. \tag{36.61}$$

The complete Hamiltonian is then

$$H' = j^{\mu(1)}A_\mu + j^{\mu(2)}A_\mu. \tag{36.62}$$

The first term corresponds to the Klein–Gordon type interaction and the second is pure Dirac type.

36.4 Dirac equation with electromagnetic field and comparison with the Klein–Gordon equation

As we have discussed previously, electromagnetic interactions can be included in the Dirac equation quite simply by the substitution

$$p_\mu \to p_\mu - ieA_\mu. \tag{36.63}$$

If we define

$$D_\mu = p_\mu - ieA_\mu, \tag{36.64}$$

then the Dirac equation in the presence of an electromagnetic field reads

$$\left(i\gamma^\mu D_\mu - m\right)\psi = 0. \tag{36.65}$$

To express this equation in a quadratic form as we did earlier for the free Dirac equation, we multiply (36.65) by $(i\gamma^{\nu}D_{\nu}+m)$ on the left,

$$\left(i\gamma^{\nu}D_{\nu}+m\right)\left(i\gamma^{\mu}D_{\mu}-m\right)\psi=0, \tag{36.66}$$

which gives

$$\left(\gamma^{\mu}\gamma^{\nu}D_{\mu}D_{\nu}+m^{2}\right)\psi=0. \tag{36.67}$$

From (35.90) we can write

$$\gamma^{\mu}\gamma^{\nu}=g^{\mu\nu}+i\sigma^{\mu\nu}. \tag{36.68}$$

Substituting this in (36.67) we obtain

$$\left(g^{\mu\nu}D_{\mu}D_{\nu}+i\sigma^{\mu\nu}D_{\mu}D_{\nu}+m^{2}\right)\psi=0. \tag{36.69}$$

We can write

$$\begin{aligned}\sigma^{\mu\nu}D_{\mu}D_{\nu}&=\frac{1}{2}\left[\sigma^{\mu\nu}D_{\mu}D_{\nu}+\sigma^{\nu\mu}D_{\nu}D_{\mu}\right]\\&=\frac{1}{2}\sigma^{\mu\nu}\left[D_{\mu},D_{\nu}\right]\end{aligned} \tag{36.70}$$

where we have used the relation, $\sigma^{\nu\mu}=-\sigma^{\mu\nu}$. Let us now evaluate $\left[D_{\mu},D_{\nu}\right]$. We find

$$\left[D_{\mu},D_{\nu}\right]=\left[\left(\partial_{\mu}+ieA_{\mu}\right),\left(\partial_{\nu}+ieA_{\nu}\right)\right] \tag{36.71}$$

$$=ie\left[\partial_{\mu},A_{\nu}\right]+ie\left[A_{\mu},\partial_{\nu}\right]. \tag{36.72}$$

Since these are operators that operate on a wavefunction, we explicitly include the wavefunction, ψ. We find

$$\left[\partial_{\mu},A_{\nu}\right]\psi=\partial_{\mu}(A_{\nu}\psi)-A_{\nu}\left(\partial_{\mu}\psi\right) \tag{36.73}$$

$$=\partial_{\mu}A_{\nu}\psi. \tag{36.74}$$

Hence,

$$\left[\partial_{\mu},A_{\nu}\right]=\partial_{\mu}A_{\nu}. \tag{36.75}$$

Similarly,

$$\left[A_{\mu},\partial_{\nu}\right]=-\partial_{\nu}A_{\mu}. \tag{36.76}$$

Thus,

$$\left[D_{\mu},D_{\nu}\right]=ie(\partial_{\mu}A_{\nu}-\partial_{\nu}A_{\mu})=ieF_{\mu\nu} \tag{36.77}$$

where $F_{\mu\nu}$ is the Maxwell tensor. Equation (36.69) now reads

$$\left[D^{\mu}D_{\mu}-e\sigma^{\mu\nu}F_{\mu\nu}+m^{2}\right]\psi=0. \tag{36.78}$$

The Klein–Gordon relation in the presence of electromagnetic interactions will be

$$\left[D^{\mu}D_{\mu}+m^{2}\right]\psi=0. \tag{36.79}$$

The difference between the two equations comes about entirely due to the spin of the electron.

36.5 Propagators: the Dirac propagator

We consider the Dirac equation with a specific source term that is due to electromagnetic interactions. The Hamiltonian for this case, once again, can be written simply by changing $p^{\mu}\to p^{\mu}-eA^{\mu}$, that is, changing

$$i\gamma\cdot\partial-m\to i\gamma\cdot(\partial-eA)-m. \tag{36.80}$$

The Dirac equation now reads

$$(i\gamma\cdot\partial-m)\,\psi\,(x)=e\gamma\cdot A\psi(x). \tag{36.81}$$

The solution in terms of the Green's function $S_F(x-x')$ can be written as

$$\psi\,(x)=\psi_0\,(x)+e\int d^4x_0'\,S_F\left(x-x'\right)\gamma\cdot A\left(x'\right)\psi(x') \tag{36.82}$$

where as before $\psi_0(x)$ is the homogeneous solution

$$(i\gamma\cdot\partial-m)\,\psi_0\,(x)=0, \tag{36.83}$$

and taking $x'=0$, $S_F(x)$ satisfies the equation

$$(i\gamma\cdot\partial-m)\,S_F\,(x)=\delta^{(4)}(x). \tag{36.84}$$

We can obtain S_F from Δ_F if we consider the product $(i\gamma\cdot\partial-m)\,(i\gamma\cdot\partial+m)$, which can be expressed as

$$(i\gamma\cdot\partial-m)\,(i\gamma\cdot\partial+m)=-\left(\gamma^{\mu}\partial_{\mu}\right)\left(\gamma^{\nu}\partial_{\nu}\right)-m^{2}. \tag{36.85}$$

Since μ and ν are dummy variables, we can write

$$\left(\gamma^{\mu}\partial_{\mu}\right)\left(\gamma^{\nu}\partial_{\nu}\right)=\left(\gamma^{\nu}\partial_{\nu}\right)\left(\gamma^{\mu}\partial_{\mu}\right)=\frac{1}{2}\left[\gamma^{\mu}\gamma^{\nu}+\gamma^{\nu}\gamma^{\mu}\right]\partial_{\mu}\partial_{\nu}=g^{\mu\nu}\partial_{\mu}\partial_{\nu} \tag{36.86}$$

where in the last term we have used the anticommutation relation for the γ-matrices. We write

$$g^{\mu\nu}\partial_{\mu}\partial_{\nu}=-\Box \tag{36.87}$$

and, therefore,

$$(i\gamma \cdot \partial - m)(i\gamma \cdot \partial + m) = -\left(\Box + m^2\right). \tag{36.88}$$

From (36.88), we find, after comparing (36.84) with the propagator equation involving Δ_F in Chapter 32, that one can write

$$S_F(x) = -[i\gamma \cdot \partial + m]\,\Delta_F(x). \tag{36.89}$$

Therefore, from the properties of Δ_F,

$$S_F(x) = \lim_{\epsilon \to 0} \frac{1}{(2\pi)^4} \int d^4p \frac{\gamma \cdot p + m}{p^2 - m^2 + i\epsilon} e^{-ip\cdot x}. \tag{36.90}$$

To obtain the integral in (36.90) we write

$$I = \int d^4p \frac{\gamma \cdot p + m}{p^2 - m^2 + i\epsilon} e^{-ip\cdot x} \tag{36.91}$$

$$= \int d^3p\, e^{i\mathbf{p}\cdot\mathbf{r}} \int dp_4 \frac{(\gamma_4 p_4 - \boldsymbol{\gamma} \cdot \mathbf{p} + m)\, e^{-ip_4 t}}{\left[p_4 - \sqrt{\mathbf{p}^2 + m^2} + i\epsilon\right]\left[p_4 + \sqrt{\mathbf{p}^2 + m^2} - i\epsilon\right]}. \tag{36.92}$$

If we let $E_p = \sqrt{\mathbf{p}^2 + m^2}$, then I becomes, following the same steps as in the Klein–Gordon case,

$$I = \int d^3p e^{i\mathbf{p}\cdot\mathbf{r}} \left[\int_{-\infty}^{\infty} dp_4 \frac{(\gamma_4 p_4 - \boldsymbol{\gamma} \cdot \mathbf{p} + m)}{p_4 - E_p + i\epsilon} e^{-ip_4 t} - \int_{-\infty}^{\infty} dp_4 \frac{(\gamma_4 p_4 - \boldsymbol{\gamma} \cdot \mathbf{p} + m)}{p_4 + E_p - i\epsilon} e^{-ip_4 t}\right] \tag{36.93}$$

$$= \int d^3p \frac{e^{i\mathbf{p}\cdot\mathbf{r}}}{2E_p} \int_{-\infty}^{\infty} dp_4 \frac{(\gamma_4 p_4 - \boldsymbol{\gamma} \cdot \mathbf{p} + m)}{p_4 - E_p + i\epsilon} e^{-ip_4 t}$$

$$+ \int d^3p \frac{e^{-i\mathbf{p}\cdot\mathbf{r}}}{2E_p} \int_{-\infty}^{\infty} dp_4 \frac{(-\gamma_4 p_4 + \boldsymbol{\gamma} \cdot \mathbf{p} + m)}{p_4 + E_p - i\epsilon} e^{-ip_4 t} \tag{36.94}$$

$$= \int d^3p \frac{e^{i\mathbf{p}\cdot\mathbf{r}}}{2E_p} (\gamma \cdot p + m)\, e^{-ip_4 t} (-2\pi i)\, \theta(t)$$

$$+ \int d^3p \frac{e^{-i\mathbf{p}\cdot\mathbf{r}}}{2E_p} (-\gamma \cdot p + m)\, e^{ip_4 t} (-2\pi i)\, \theta(-t). \tag{36.95}$$

We defined the following operators earlier:

$$\Lambda_+(\mathbf{p}) = \frac{1}{2m}(m + \gamma \cdot p) \tag{36.96}$$

$$\Lambda_-(\mathbf{p}) = \frac{1}{2m}(m - \gamma \cdot p). \tag{36.97}$$

As we showed in Chapter 35, $\Lambda_+(\mathbf{p})$ and $\Lambda_-(\mathbf{p})$ are projection operators for the positive- and negative-energy states, which can be written as

$$\Lambda_+(\mathbf{p}) = \sum_{r=1}^{2} w_r(\mathbf{p})\overline{w}_r(\mathbf{p}) \tag{36.98}$$

$$\Lambda_-(\mathbf{p}) = \sum_{r=3}^{4} w_r(-\mathbf{p})\overline{w}_r(-\mathbf{p}) \tag{36.99}$$

where the w's are the spinors and $r = 1, 2$ correspond to the spin-up and spin-down states for the positive-energy states, while $r = 3, 4$ are the corresponding states for negative energy. One can then write

$$S_F\left(x - x'\right) = -i\sum_n \theta\left(t - t'\right)\left(\sum_{r=1}^{2} \psi_{0n}^{(-)r}(x)\,\bar{\psi}_{0n}^{(-)r}\left(x'\right)\right) + i\sum_n \theta\left(t' - t\right)\left(\sum_{r=3}^{4} \psi_{0n}^{(+)r}(x)\,\bar{\psi}_{0n}^{(+)r}\left(x'\right)\right) \tag{36.100}$$

where

$$\psi_{n0}^{(-)}(x) = \frac{1}{\left(\sqrt{2\pi}\right)^3}\sqrt{\frac{m}{E_p}}\, w_r(\mathbf{p})\; e^{-ip_n.x}, \tag{36.101}$$

$$\psi_{n0}^{(+)}(x) = \frac{1}{\left(\sqrt{2\pi}\right)^3}\sqrt{\frac{m}{E_p}}\, w_r(-\mathbf{p})\; e^{ip_n.x}. \tag{36.102}$$

36.6 Scattering

36.6.1 Rutherford scattering

We return now to Rutherford scattering, which we considered earlier, but this time considering the electron as a Dirac particle. The propagator corresponding to the electron, which is a positive-energy particle traveling forward in time, is

$$S_F\left(x - x', m\right) \to -i\sum_n\sum_r \psi_{0n}^{(-)r}(x)\bar{\psi}_{0n}^{(-)r}(x') = -i\int d^3p \sum_{r=1}^{2} \psi_{0p}^{(-)r}(x)\bar{\psi}_{0p}^{(-)r}(x') \tag{36.103}$$

where we have kept only the terms with $(-)$ in the superscript, corresponding to positive-energy particles.

The wavefunction as $t \to \infty$ is given by

$$\lim_{t\to\infty} \psi_i(x) = \psi_{0i}(x) - ie\int d^4x' \int d^3p \sum_r \psi_{0p}^{(-)r}(x)\bar{\psi}_{0p}^{(-)r}(x')\gamma \cdot A(x')\psi_i(x'). \tag{36.104}$$

Substituting this in the S-matrix gives

$$S_{fi} = \lim_{t\to\infty} \int d^3r \, \overset{*}{\psi}_{0f}(x)\psi_i(x) \tag{36.105}$$

$$= \delta_{fi} - ie \int d^4x \, \bar{\psi}_{0f}(x)\gamma \cdot A(x)\psi_i(x). \tag{36.106}$$

For the scattering process to first order perturbation in e we obtain

$$S_{fi} = -ie \int d^4 x \bar{\psi}_{0f}(x) \, \gamma \cdot A(x) \, \psi_{0i}(x), \quad \text{for} \quad f \neq i. \tag{36.107}$$

Normalizing in a box, we will take the initial and final free particle wavefunctions as

$$\psi_{0i}(x) = \frac{1}{\sqrt{V}}\sqrt{\frac{m}{E_i}} u_{S_i}(\mathbf{p}_i) e^{-ip_i \cdot x} \tag{36.108}$$

and

$$\psi_{0f}(x) = \frac{1}{\sqrt{V}}\sqrt{\frac{m}{E_f}} u_{S_f}(\mathbf{p}_f) e^{-ip_f \cdot x}. \tag{36.109}$$

We will consider a nonrelativistic approximation for the vector potential, so only the Coulomb potential is involved:

$$A_4(x) = -\frac{Ze}{4\pi r}, \quad \mathbf{A}(x) = 0. \tag{36.110}$$

Substituting this in S_{fi} we obtain

$$S_{fi} = \frac{1}{V}\frac{iZe^2}{4\pi}\sqrt{\frac{m^2}{E_f E_i}} \bar{u}(\mathbf{p}_f)\gamma_4 u_{S_i}(\mathbf{p}_i) \int \frac{d^4x}{r} e^{i(p_f - p_i)\cdot x}. \tag{36.111}$$

The above integration can be carried out to give

$$\int \frac{d^4x}{r} e^{i(p_f - p_i)\cdot x} = \int dt e^{i(E_f - E_i)t} \int \frac{d^3r}{r} e^{-i(\mathbf{p}_f - \mathbf{p}_i)\cdot \mathbf{r}} \tag{36.112}$$

$$= 2\pi\delta\left(E_f - E_i\right) \int \frac{d^3r}{r} e^{-i(\mathbf{p}_f - \mathbf{p}_i)\cdot \mathbf{r}} \tag{36.113}$$

$$= 2\pi\delta\left(E_f - E_i\right) \frac{4\pi}{\left|\mathbf{p}_f - \mathbf{p}_i\right|^2}. \tag{36.114}$$

We have already assumed that the scattering center is very heavy (e.g., a heavy nucleus) so that the electron is deflected in its motion without changing its energy (i.e., $E_f = E_i$). We note in passing that the propagator term can also be written as

$$\frac{1}{\left|\mathbf{p}_f - \mathbf{p}_i\right|^2} = -\frac{1}{\left(E_f - E_i\right)^2 - \left(\mathbf{p}_f - \mathbf{p}_i\right)^2}, \quad \text{for } E_f = E_i \tag{36.115}$$

$$= -\frac{1}{\left(p_f - p_i\right)^2}, \tag{36.116}$$

which is just the propagator for the photon of four-momentum $(p_f - p_i)$.

The S-matrix is given by

$$S_{fi} = iZe^2 \frac{1}{V} \sqrt{\frac{m^2}{E_f E_i}} \frac{\bar{u}_{S_f}(\mathbf{p}_f) \gamma_4 u_{S_i}(\mathbf{p}_i)}{|\mathbf{q}|^2} 2\pi \delta(E_f - E_i) \tag{36.117}$$

where $\mathbf{q} = \mathbf{p}_f - \mathbf{p}_i$ is the momentum transfer.

The scattering of an electron off a heavy nucleus is written in terms of the number of final free particle states given by

$$dn = \frac{V\, d^3 p_f}{(2\pi)^3}. \tag{36.118}$$

The transition probability is then

$$|S_{fi}|^2 \frac{V}{(2\pi)^3} d^3 p_f = \frac{1}{V} \left(Ze^2\right)^2 \frac{m^2}{E_f E_i} \frac{\left|\bar{u}_{S_f}(\mathbf{p}_f) \gamma_4 u_{S_i}(\mathbf{p}_i)\right|^2}{|q|^4} \frac{d^3 p_f}{(2\pi)^3} \left[2\pi \delta(E_f - E_i)\right]^2. \tag{36.119}$$

We can express the last factor as follows:

$$\left[2\pi \delta(E_f - E_i)\right]^2 = 2\pi \delta(E_f - E_i) \int_{-T/2}^{T/2} dt\, e^{i(E_f - E_i)} = 2\pi \delta(E_f - E_i)\, T \tag{36.120}$$

where we have replaced the second factor of $2\pi\delta(E_f - E_i)$ by its integral representation with the assumption that T is very large.

The transition probability per unit time, λ_{fi}, is obtained by dividing (36.119) by T. Thus we have

$$\lambda_{fi} = \frac{1}{V} \left(Ze^2\right)^2 \frac{m^2}{E_f E_i} \frac{\left|\bar{u}_{S_f}(\mathbf{p}_f) \gamma_4 u_{S_i}(\mathbf{p}_i)\right|^2}{|q|^4} \frac{d^3 p_f}{(2\pi)^3} 2\pi \delta(E_f - E_i). \tag{36.121}$$

We note that the definition of a cross-section is

$$\text{Cross-section} = \text{transition probability per unit time/flux}. \tag{36.122}$$

The flux for the incident particles has already been calculated and is given by

$$\text{Flux} = \frac{\text{no. of particles}}{\text{per unit area} \times \text{time}} = \frac{v_i}{V}. \tag{36.123}$$

Therefore, the cross-section is found to be

$$d\sigma = \int \frac{1}{v_i} \left(Ze^2\right)^2 \frac{m^2}{E_f E_i} \frac{\left|\bar{u}_{S_f}(\mathbf{p}_f) \gamma_4 u_{S_i}(\mathbf{p}_i)\right|^2}{|q|^4} \frac{d^3 p_f}{(2\pi)^3} 2\pi \delta(E_f - E_i). \tag{36.124}$$

We write

$$d^3p_f = p_f^2\, dp_f\, d\Omega, \tag{36.125}$$

$$p_f dp_f = E_f\, dE_f, \tag{36.126}$$

where we have used the relativistic relation $E^2 - \mathbf{p}^2 = m^2$ for the electron. Since $E_f = E_i$ we have

$$v_i = \frac{|\mathbf{p}_i|}{E_i} = \frac{|\mathbf{p}_f|}{E_f}. \tag{36.127}$$

The differential cross-section is given by

$$\frac{d\sigma}{d\Omega} = 4\left(\frac{Ze^2}{4\pi}\right)^2 m^2 \frac{\left|\bar{u}_{S_f}(\mathbf{p}_f)\,\gamma_4 u_{S_i}(\mathbf{p}_i)\right|^2}{|q|^4}. \tag{36.128}$$

A sum over the final spins and averaging over initial spins gives

$$\frac{4\,(Z\alpha)^2\, m^2}{|q|^2}\frac{1}{2}\sum_{S_f,S_i}\left|\bar{u}_{S_f}(\mathbf{p}_f)\,\gamma_4 u_{S_i}(\mathbf{p}_i)\right|^2. \tag{36.129}$$

We make use of the trace relation derived in Chapters 35 and 36,

$$\sum_s\sum_{s'}\left|\bar{u}_s(\mathbf{p})\,Q u_{s'}(\mathbf{p}')\right|^2 = \mathrm{Tr}\left[\Lambda_+(\mathbf{p})\,Q\Lambda_+(\mathbf{p}')\,\gamma_4 Q^\dagger\gamma_4\right] \tag{36.130}$$

where for our case $Q=\gamma_4$, $\mathbf{p}=\mathbf{p}_f$, $\mathbf{p}'=\mathbf{p}_i$. We then obtain

$$\sum_{S_f,S_i}\left|\bar{u}_{S_f}(\mathbf{p}_f)\,\gamma_4 u_{S_i}(\mathbf{p}_i)\right|^2 = \mathrm{Tr}\left[\Lambda_+(\mathbf{p}_f)\,\gamma_4\Lambda_+(\mathbf{p}_i)\,\gamma_4\gamma_4^\dagger\gamma_4\right] \tag{36.131}$$

$$= \frac{1}{4m^2}\left\{\mathrm{Tr}\left[\gamma\cdot p_f\gamma_4\gamma\cdot p_i\gamma_4\right] + m^2 Tr\left[\gamma_4^2\right]\right\} \tag{36.132}$$

where we have used the relation

$$\Lambda_+(\mathbf{p}) = \frac{\gamma\cdot p + m}{2m}. \tag{36.133}$$

From the trace relations, we have

$$\sum_{S_f,S_i}\left|\bar{u}_{S_f}(\mathbf{p}_f)\,\gamma_4 u_{S_i}(\mathbf{p}_i)\right|^2 = \frac{2\left[2E_iE_f - p_i\cdot p_f + m^2\right]}{m^2}, \tag{36.134}$$

which can be simplified by writing

$$p_i\cdot p_f = E_iE_f - \mathbf{p}_i\cdot\mathbf{p}_f = E^2 - |\mathbf{p}|^2\cos\theta \tag{36.135}$$

where $E_i = E_f = E$, $|\mathbf{p}_i| = |\mathbf{p}_f| = |\mathbf{p}|$, with θ being the angle (the scattering angle) between $\mathbf{p}_i$ and $\mathbf{p}_f$. Putting everything together we find

$$\frac{2E^2\left[2E_iE_f - p_i\cdot p_f + m^2\right]}{m^2} = \frac{2E^2\left[1-\beta^2\sin^2\left(\frac{\theta}{2}\right)\right]}{m^2} \tag{36.136}$$

where we have used the relation $\cos\theta = 1 - 2\sin^2(\theta/2)$. In the nonrelativistic limit we approximate $E \approx m$ and obtain for the differential cross-section

$$\frac{d\sigma}{d\Omega} = \frac{Z^2\alpha^2}{4|\mathbf{p}|^2\sin^4(\theta/2)}\left[1-\beta^2\sin^2\left(\frac{\theta}{2}\right)\right]. \tag{36.137}$$

This is the same expression as the one we obtained in the nonrelativistic calculation except now we have a correction term of order β^2, which is due to the relativistic effects.

36.7 Appendix to Chapter 36

36.7.1 Trace properties of matrix elements and summation over spins

In calculating the cross-section one needs to sum over final spin states and average over initial spin states. As we will see below, this involves calculating the trace of a product of γ-matrices.

The trace of a matrix is given by the sum of the diagonal elements,

$$\mathrm{Tr}(A) = \sum_{i=1}^{4} A_{ii} \tag{36.138}$$

where to avoid confusion in the future discussions we have written the summation index explicitly. The trace of a product is similarly given by

$$\mathrm{Tr}(AB) = \sum_{i=1}^{4}\sum_{k=1}^{4} A_{ik}B_{ki}. \tag{36.139}$$

Let us consider the matrix element, $\bar{u}_s(\mathbf{p}_f)Qu_{s'}(\mathbf{p}_i)$, where Q is a matrix consisting of γ-matrices. In particular we consider the following sum over the spin indices s and s'.

$$\sum_{s=1}^{2}\sum_{s'=1}^{2}\left|\bar{u}_s(\mathbf{p})\,Qu_{s'}(\mathbf{p}')\right|^2$$
$$= \sum_{s}\sum_{s'}\left[\bar{u}_s(\mathbf{p})\,Qu_{s'}(\mathbf{p}')\right]\left[\bar{u}_s(\mathbf{p})\,Qu_{s'}(\mathbf{p}')\right]^{\dagger}. \tag{36.140}$$

We now use the relations $\bar{u} = u_r^{\dagger}\gamma_4$, $\gamma_4^{\dagger} = \gamma_4$, and $\gamma_4^2 = \gamma_4$, to write the above sum

$$\sum_s \sum_{s'} \left[\bar{u}_s(\mathbf{p})\, Q u_{s'}(\mathbf{p}')\right]\left[u_{s'}^{\dagger}(\mathbf{p}')\, Q^{\dagger}\gamma_4 u_s(\mathbf{p})\right]$$

$$= \sum_s \sum_{s'} \left[\bar{u}_s(\mathbf{p})\, Q u_{s'}(\mathbf{p}')\right]\left[u_{s'}^{\dagger}(\mathbf{p}')\, Q^{\dagger}\gamma_4 u_s(\mathbf{p})\right].$$

Writing in terms of the individual matrix elements we obtain

$$= \sum_{\alpha\beta ab}\sum_s\sum_{s'} \left[(\bar{u}_s)_\alpha (Q)_{\alpha\beta} (u_{s'})_\beta\right]\left[(\bar{u}_{s'})_a \left(\gamma_4 Q^{\dagger}\gamma_4\right)_{ab} (u_s)_b\right]$$

$$= \sum_{\alpha\beta ab}\left[\sum_{s'} (u_{s'})_\beta (\bar{u}_{s'})_a\right]\left(\gamma_4 Q^{\dagger}\gamma_4\right)_{ab}\left[\sum_s (u_s)_b (\bar{u}_s)_\alpha\right](Q)_{\alpha\beta}$$

$$= \sum_{\alpha\beta ab} (\Lambda_+)_{\beta a}\left(\gamma_4 Q^{\dagger}\gamma_4\right)_{ab} (\Lambda_+)_{b\alpha} (Q)_{\alpha\beta}$$

$$= \mathrm{Tr}\left[\Lambda_+(\mathbf{p})\gamma_4 Q^{\dagger}\gamma_4\Lambda_+ Q(\mathbf{p}')\right] \tag{36.141}$$

where we have used the definition for the projection operator, $\Lambda_+(\mathbf{p}) = \sum_{r=1}^{2} u_r\bar{u}_r$. Therefore,

$$\sum_{s=1}^{2}\sum_{s'=1}^{2} \left|\bar{u}_s(\mathbf{p})\, Q u_{s'}(\mathbf{p}')\right|^2 = \mathrm{Tr}\left[\Lambda_+(\mathbf{p})\gamma_4 Q^{\dagger}\gamma_4\Lambda_+(\mathbf{p}')Q\right]. \tag{36.142}$$

Similarly, one can show that

$$\sum_s\sum_{s'} \left|\bar{u}_s(\mathbf{p})\, Q u_{s'}(\mathbf{p}')\right|^2 = \mathrm{Tr}\left[\Lambda_+(\mathbf{p})\, Q\Lambda_+(\mathbf{p}')\,\gamma_4 Q^{\dagger}\gamma_4\right], \tag{36.143}$$

$$\sum_s\sum_{s'} \left|\bar{v}_s(\mathbf{p})\, Q v_{s'}(\mathbf{p}')\right|^2 = \mathrm{Tr}\left[\Lambda_-(\mathbf{p})\, Q\Lambda_-(\mathbf{p}')\,\gamma_4 Q^{\dagger}\gamma_4\right], \tag{36.144}$$

$$\sum_s\sum_{s'} \left|\bar{u}_s(\mathbf{p})\, Q v_{s'}(\mathbf{p}')\right|^2 = \mathrm{Tr}\left[\Lambda_+(\mathbf{p})\, Q\Lambda_-(\mathbf{p}')\,\gamma_4 Q^{\dagger}\gamma_4\right]. \tag{36.145}$$

The following relations are useful when they appear in Q:

$$\gamma_4\gamma_\mu^{\dagger}\gamma_4 = \gamma_\mu \quad \text{and} \quad \gamma_4\gamma_5^{\dagger}\gamma_4 = -\gamma_5. \tag{36.146}$$

37 Multiparticle systems and second quantization

The quantum mechanics of single particles is extended to multiparticle systems and single-particle operators are defined in terms of this multiparticle space. Creation and destruction operators are incorporated with commutation relations that correspond to whether the particles are fermions or bosons. These operators are much like the raising and lowering operators for the harmonic oscillator which allow one to add or subtract the number of particles in a multiparticle state. We use this path to introduce second quantization. We then write Klein–Gordon, Dirac, and Maxwell fields in second quantization and define the negative-energy solutions as corresponding to positive-energy antiparticles. The photon as a quantum of electromagnetic radiation naturally emerges from this formalism. The question of vacuum fluctuations, a characteristic aspect of second quantization, is also explored in the context of the Casimir effect.

37.1 Wavefunctions for identical particles

Consider two noninteracting particles with momenta p_α and p_β, respectively, that are described by the wavefunctions $u_\alpha(\mathbf{r})$ and $u_\beta(\mathbf{r})$, respectively. A wavefunction describing both the particles, one at the point $\mathbf{r}_1$ and the other at $\mathbf{r}_2$ is then given by the product

$$\phi(\mathbf{r}_1, \mathbf{r}_2) = u_\alpha(\mathbf{r}_1)u_\beta(\mathbf{r}_2). \tag{37.1}$$

If the particles are identical, however, then the wavefunction

$$\phi(\mathbf{r}_2, \mathbf{r}_1) = u_\alpha(\mathbf{r}_2)u_\beta(\mathbf{r}_1) \tag{37.2}$$

will also describe the same system. From these two wavefunctions we can form the following symmetric and antisymmetric (normalized) combinations respectively, under the interchange $\mathbf{r}_1 \leftrightarrow \mathbf{r}_2$

$$\phi_s(\mathbf{r}_1, \mathbf{r}_2) = \frac{\phi(\mathbf{r}_1, \mathbf{r}_2) + \phi(\mathbf{r}_2, \mathbf{r}_1)}{\sqrt{2}} \tag{37.3}$$

and

$$\phi_a(\mathbf{r}_1, \mathbf{r}_2) = \frac{\phi(\mathbf{r}_1, \mathbf{r}_2) - \phi(\mathbf{r}_2, \mathbf{r}_1)}{\sqrt{2}} \tag{37.4}$$

In the symmetric case it is possible to have more than two particles in the same state, e.g., $\alpha = \beta$. In the antisymmetric case, however, this is impossible since the wavefunction,

$\phi_a(\mathbf{r}_1, \mathbf{r}_2)$, vanishes. The number of particles in an antisymmetric state, therefore, can only be either 1 or 0.

Let us now generalize the above discussion to a multiparticle system. Consider n identical particles out of which n_1 particles are in state λ_1, n_2 particles in state λ_2, and so on. The symmetric wavefunction for this system will be

$$\phi_s(\mathbf{r}_1, \mathbf{r}_2, ...) = \frac{1}{\sqrt{n!}} \frac{1}{\sqrt{n_1!}} \frac{1}{\sqrt{n_2!}} \cdots \sum_P \phi(\mathbf{r}_1, \mathbf{r}_2, ...) \tag{37.5}$$

where $\sum_P$ denotes all possible permutations among the $\mathbf{r}_i$'s, while $\phi(\mathbf{r}_1, \mathbf{r}_2, ...)$ denotes a simple product of one particle wavefunctions of the type considered in (37.1), and n is the total number of particles, $n = n_1 + n_2 + \cdots$.

For the antisymmetric case, as we noted earlier, the only possibilities are $n_i = 0, 1$. One can express the wavefunction for the antisymmetric state as

$$\phi_a(\mathbf{r}_1, \mathbf{r}_2, ...) = \frac{1}{\sqrt{n!}} \sum_P \delta_P \phi(\mathbf{r}_1, \mathbf{r}_2, ...) \tag{37.6}$$

where δ_P is $+$ for even permutations among the $\mathbf{r}_i$'s and $-$ for odd permutations. This wavefunction can also be written in the form of a determinant, called the Slater determinant:

$$\phi_a(\mathbf{r}_1, \mathbf{r}_2, ...) = \frac{1}{\sqrt{n!}} \det \begin{vmatrix} u_{\lambda_1}(\mathbf{r}_1) & u_{\lambda_2}(\mathbf{r}_1) & \cdot & \cdot \\ u_{\lambda_1}(\mathbf{r}_2) & u_{\lambda_2}(\mathbf{r}_2) & \cdot & \cdot \\ \cdot & \cdot & \cdot & \cdot \\ \cdot & \cdot & \cdot & \cdot \end{vmatrix}. \tag{37.7}$$

We note that $\phi_a(\mathbf{r}_1, \mathbf{r}_2, ...) = 0$ if $\lambda_i = \lambda_j$ or if $\mathbf{r}_i = \mathbf{r}_j$.

One can write a symmetric state given by (37.5) in a form similar to (37.7) except in this case the determinant, which we will write as $(\det)_+$, will have each term with positive sign:

$$\phi_s(\mathbf{r}_1, \mathbf{r}_2, ...) = \frac{1}{\sqrt{n!}} \frac{1}{\sqrt{n_1!}} \frac{1}{\sqrt{n_2!}} \cdots (\det)_+ \begin{vmatrix} u_{\lambda_1}(\mathbf{r}_1) & u_{\lambda_2}(\mathbf{r}_1) & \cdot & \cdot \\ u_{\lambda_1}(\mathbf{r}_2) & u_{\lambda_2}(\mathbf{r}_2) & \cdot & \cdot \\ \cdot & \cdot & \cdot & \cdot \\ \cdot & \cdot & \cdot & \cdot \end{vmatrix}. \tag{37.8}$$

Unlike (37.7) we now have the terms $\sqrt{n_i!}$ in the denominator ($i = 1, 2, ...$). We note that for the antisymmetric case $\sqrt{n_i!} = 1$, since $n_i = 0$ or 1.

37.2 Occupation number space and ladder operators

We now define the following ket vector,

$$|n_1, n_2, ... n_k, ...\rangle , \tag{37.9}$$

as representing a multiparticle state with n_1 particles in state λ_1 (e.g., momentum $\mathbf{p}_1$), n_2 particles in state λ_2 (e.g., momentum $\mathbf{p}_2$), and so on. In other words, the n_i's represent the number of particles that "occupy" a state λ_i $(i = 1, 2, \ldots)$. This representation is, therefore, called the "occupation number representation." As a natural generalization of one-particle quantum mechanics these states satisfy orthonormality

$$\langle n_1', n_2', ...n_k', ... |...n_k, ...n_2, n_1\rangle = \delta_{n_1 n_1'} \delta_{n_2 n_2'} \cdots \delta_{n_k n_k'} \cdots \tag{37.10}$$

and completeness

$$\sum_{n_1=0}^{\infty} ... \sum_{n_k=0}^{\infty} ... |n_1, n_2, ...n_k, ...\rangle \langle ...n_k ...n_2, n_1| = 1. \tag{37.11}$$

A multiparticle state which has no particles in it is called a "vacuum" state, designated as $|0\rangle$,

$$|0\rangle = |0, 0, ... 0, ...\rangle . \tag{37.12}$$

We define a number operator N_k for a state k which has the property

$$N_k |n_1, n_2, ... n_k, ...\rangle = n_k |n_1, n_2, ... n_k, ...\rangle . \tag{37.13}$$

Thus N_k's are operators whose eigenvalues are the numbers n_k in states k. One can express them as

$$N_k = n_k |n_1, n_2, ... n_k, ...\rangle \langle ...n_k ...n_2, n_1| . \tag{37.14}$$

A (total) number operator N will be defined as

$$N = \sum_{(n)} N_k = \sum_{(n)} n_k |n_1, n_2, ...n_k, ...\rangle \langle ...n_k ...n_2, n_1| \tag{37.15}$$

where

$$\sum_{(n)} = \sum_{n_1=0}^{\infty} ... \sum_{n_k=0}^{\infty} ... \quad \text{for the symmetric case,} \tag{37.16}$$

$$\sum_{(n)} = \sum_{n_1=0}^{1} ... \sum_{n_k=0}^{1} ... \quad \text{for the antisymmetric case.} \tag{37.17}$$

From now on, we will use the short-hand notation $|...n_k...\rangle$ for a general state vector that has n_i particles in state i with $i = 1, 2, \ldots, k \ldots$, and write all our relations accordingly. For example the number operators will be written as

$$N_k = n_k |...n_k...\rangle \langle ...n_k...| , \tag{37.18}$$

$$N = \sum_{(n)} n_k |...n_k...\rangle \langle ...n_k...| . \tag{37.19}$$

We now consider the symmetric and antisymmetric cases separately.

37.3 Creation and destruction operators

37.3.1 Symmetric case

Let us define the following operator,

$$a_k = c_k(n) \left|...n_k - 1...\right\rangle \left\langle ...n_k...\right|, \tag{37.20}$$

which, when it operates on $|...n_k...\rangle$, gives a state vector in which one particle is removed from the state k. Hence it is called a "destruction operator." The Hermitian conjugate of this operator is given by

$$a_k^\dagger = c_k^*(n) \left|...n_k...\right\rangle \left\langle ...n_k - 1...\right|, \tag{37.21}$$

which adds a particle to the state k. It is called a "creation operator." Taking the product of a_k and $a_k^\dagger$ we obtain

$$a_k^\dagger a_k = |c_k(n)|^2 \left|...n_k...\right\rangle \left\langle ...n_k...\right|. \tag{37.22}$$

This has the same form as the number operator N_k. Equating this expression to (37.18) for N_k we obtain $|c_k(n)|^2 = n_k$ and

$$a_k^\dagger a_k = n_k \left|...n_k...\right\rangle \left\langle ...n_k...\right| = N_k \tag{37.23}$$

where we have assumed $c_k(n)$ to be real. Thus,

$$a_k = \sqrt{n_k} \left|...n_k - 1...\right\rangle \left\langle ...n_k...\right| \tag{37.24}$$

and

$$a_k^\dagger = \sqrt{n_k} \left|...n_k...\right\rangle \left\langle ...n_k - 1...\right|, \tag{37.25}$$

which we can rewrite as

$$a_k^\dagger = \sqrt{n_k + 1} \left|...n_k + 1...\right\rangle \left\langle ...n_k...\right|. \tag{37.26}$$

Thus, from (37.24),

$$a_k \left|...n_k...\right\rangle = \sqrt{n_k} \left|...n_k - 1...\right\rangle. \tag{37.27}$$

and from (37.26)

$$a_k^\dagger \left|...n_k...\right\rangle = \sqrt{n_k + 1} \left|...n_k + 1...\right\rangle. \tag{37.28}$$

The product $a_k a_k^\dagger$ is given by

$$a_k a_k^\dagger = n_k \left|...n_k - 1...\right\rangle \left\langle ...n_k - 1...\right| = (n_k + 1) \left|...n_k...\right\rangle \left\langle ...n_k...\right|. \tag{37.29}$$

Comparing with (37.18) we can express the above relation as

$$a_k a_k^\dagger = N_k + \mathbf{1}. \tag{37.30}$$

Subtracting (37.23) from (37.30), we obtain the following commutation relation between a_k and $a_k^\dagger$:

$$\left[a_k, a_k^\dagger\right] = \mathbf{1}. \tag{37.31}$$

One can easily show, using the definitions (37.27) and (37.28), that

$$\left[a_k, a_l^\dagger\right] = \delta_{kl}, \qquad [a_k, a_l] = 0, \qquad \left[a_k^\dagger, a_l^\dagger\right] = 0. \tag{37.32}$$

One can also show that

$$\left[N_k, a_k^\dagger\right] = a_k^\dagger, \tag{37.33}$$

$$[N_k, a_k] = -a_k. \tag{37.34}$$

We note that a multiparticle state that represents a single-particle state k can be obtained from the vacuum by applying $a_k^\dagger$,

$$a_k^\dagger \left|0\right\rangle = \left|0, 0, ..., (1)_k, 0, ...\right\rangle . \tag{37.35}$$

It can be easily shown that a state $\left|0, 0, ..., n_k, 0, ...\right\rangle$ is generated by repeated applications of $a_k^\dagger$ on the vacuum state. From (37.28) one finds

$$\left|0, 0, ..., n_k, 0, ...\right\rangle = \frac{(a_k^\dagger)^{n_k}}{\sqrt{n_k!}} \left|0\right\rangle . \tag{37.36}$$

Finally, we note from (37.32) that

$$a_k^\dagger a_l^\dagger \left|0\right\rangle = a_l^\dagger a_k^\dagger \left|0\right\rangle . \tag{37.37}$$

Thus, the corresponding wavefunctions are symmetric. Particles with symmetric wavefunctions follow what is known as Bose–Einstein statistics.

37.3.2 Antisymmetric case

Particles with antisymmetric wavefunctions follow what is known as Fermi–Dirac statistics.

Here the eigenvalues of N_k are 0, 1. The previous definitions of N_k, a_k, and $a_k^\dagger$ remain the same, and

$$N_k = a_k^\dagger a_k. \tag{37.38}$$

However, we now impose the condition that a_k and $a_k^\dagger$ satisfy anticommutation relations:

$$\left\{a_k, a_l^\dagger\right\} = \delta_{kl}, \tag{37.39}$$

$$\left\{a_k^\dagger, a_l^\dagger\right\} = 0 = \{a_k, a_l\} , \tag{37.40}$$

where the anticommutator { } is defined as

$$\{A, B\} = AB + BA. \tag{37.41}$$

We note, for example, that if we take $l = k$ then (37.40) implies

$$a_k^\dagger a_k^\dagger = 0, \tag{37.42}$$

which is simply the statement that one can not have two identical particles in the same state.

The following properties of N_k are obtained:

$$N_k^2 = \left(a_k^\dagger a_k\right)\left(a_k^\dagger a_k\right) = a_k^\dagger \left(a_k a_k^\dagger\right) a_k = a_k^\dagger (\mathbf{1} - a_k^\dagger a_k) a_k = a_k^\dagger a_k - a_k^\dagger a_k^\dagger a_k a_k. \tag{37.43}$$

But $a_k^\dagger a_k^\dagger = 0$ as we have just shown. Thus,

$$N_k^2 = N_k, \tag{37.44}$$

i.e.,

$$N_k \left(N_k - 1\right) = 0. \tag{37.45}$$

Hence the eigenvalues of N_k are $n_k = 0, 1$, confirming once again that no more than one particle can occupy a given state.

We define a_k and $a_k^\dagger$ for the antisymmetric case differently from the symmetric case. We write

$$a_k = n_k(-1)^{\Sigma_k} \left|...1 - n_k...\right\rangle \left\langle ...n_k...\right|, \tag{37.46}$$

$$a_k^\dagger = (1 - n_k)(-1)^{\Sigma_k} \left|...1 - n_k...\right\rangle \left\langle ...n_k...\right|, \tag{37.47}$$

where the exponent is given by

$$\Sigma_k = \sum_{i=1}^{k} n_i. \tag{37.48}$$

By taking the Hermitian conjugates of (37.46) and (37.47) we can also write

$$a_k^\dagger = n_k(-1)^{\Sigma_k} \left|...n_k...\right\rangle \left\langle ...1 - n_k...\right|, \tag{37.49}$$

$$a_k = (1 - n_k)(-1)^{\Sigma_k} \left|...n_k...\right\rangle \left\langle ...1 - n_k...\right|. \tag{37.50}$$

We have replaced the number $(n_k - 1)$, which was present in the symmetric case, by $1 - n_k$, which automatically restricts the occupation number for the state k to be either 0 or 1.

Let us elaborate on this choice. From (37.46) we can write

$$a_k \left|...n_k...\right\rangle = n_k(-1)^{\Sigma_k} \left|...1 - n_k...\right\rangle. \tag{37.51}$$

Also, from (37.50),

$$a_k \left|...1 - n_k...\right\rangle = (1 - n_k)(-1)^{\Sigma_k} \left|...n_k...\right\rangle. \tag{37.52}$$

Similarly, from (37.47) we can write

$$a_k^\dagger \left|...n_k...\right\rangle = (1 - n_k)(-1)^{\Sigma_k} \left|...1 - n_k...\right\rangle \tag{37.53}$$

and

$$a_k^\dagger \left|...1 - n_k...\right\rangle = n_k(-1)^{\Sigma_k} \left|...n_k...\right\rangle . \tag{37.54}$$

From (37.51) and (37.54) we obtain

$$a_k^\dagger a_k \left|...n_k...\right\rangle = \left[n_k(-1)^{\Sigma_k}\right] a_k^\dagger \left|...n_k - 1...\right\rangle = \left[n_k(-1)^{\Sigma_k}\right]\left[n_k(-1)^{\Sigma_k}\right] \left|...n_k...\right\rangle \tag{37.55}$$

$$= n_k^2 \left|...n_k...\right\rangle . \tag{37.56}$$

Since $n_k^2 = n_k$, we write

$$a_k^\dagger a_k \left|...n_k...\right\rangle = n_k \left|...n_k...\right\rangle . \tag{37.57}$$

Similarly, we find

$$a_k a_k^\dagger \left|...n_k...\right\rangle = (1 - n_k) \left|...n_k...\right\rangle . \tag{37.58}$$

Thus,

$$a_k^\dagger a_k + a_k a_k^\dagger = 1. \tag{37.59}$$

One also finds

$$a_k a_k \left|...n_k...\right\rangle = n_k(1 - n_k) \left|...n_k...\right\rangle = 0 \tag{37.60}$$

because n_k can only be 0 or 1. In the same manner one can show that

$$a_k^\dagger a_k^\dagger \left|...n_k...\right\rangle = n_k(1 - n_k) \left|...n_k...\right\rangle = 0. \tag{37.61}$$

We have thus proved the relation (37.39) and (37.40) for $k = l$.

To consider the relation for $k \neq l$, we follow the same procedure as outlined above and find for $k > l$

$$a_l a_k \left|...n_k, ..., n_l...\right\rangle = n_k n_l (-1)^{\Sigma_k + \Sigma_l} \left|...1 - n_l, ..., 1 - n_k...\right\rangle , \tag{37.62}$$

while

$$a_k a_l \left|...n_k, ..., n_l...\right\rangle = -n_k n_l (-1)^{\Sigma_k + \Sigma_l} \left|...1 - n_l, ..., 1 - n_k...\right\rangle , \tag{37.63}$$

which leads to

$$(a_l a_k + a_k a_l) = 0. \tag{37.64}$$

The same relation will be obtained if one takes $k < l$. The rest of the relations in (37.39) and (37.40) follow in a similar fashion.

For the case of a two-particle antisymmetric state one can define the Fermi statistics quite simply in terms of the following 2×2 matrices for the operators and the states:

$$a_k = \begin{bmatrix} 0 & 1 \\ 0 & 0 \end{bmatrix}, \quad a_k^\dagger = \begin{bmatrix} 0 & 0 \\ 1 & 0 \end{bmatrix}, \quad N_k = \begin{bmatrix} 0 & 0 \\ 0 & 1 \end{bmatrix} \tag{37.65}$$

with the vacuum and one-particle states described by the column matrices

$$|0\rangle = \begin{bmatrix} 1 \\ 0 \end{bmatrix}, \quad |1\rangle = \begin{bmatrix} 0 \\ 1 \end{bmatrix}. \tag{37.66}$$

The commutation relations we have defined for a_k and $a_k^\dagger$ will lead to what is called second quantization, the first quantization being $[x_i, p_j] = i\hbar\delta_{ij}$. We will discuss this more fully in the following sections.

Finally, following our terminology for the harmonic oscillators, we call $a_k, a_k^\dagger$ the "ladder" operators.

37.4 Writing single-particle relations in multiparticle language: the operators, *N*, *H*, and *P*

The superposition principle implies that a single-particle wavefunction can be expressed in terms of a complete set of (orthonormal) eigenstates $u_k(\mathbf{r})$ as follows:

$$\psi(\mathbf{r}) = \sum_k c_k u_k(\mathbf{r}). \tag{37.67}$$

If the wavefunction is normalized then

$$\int d^3r\psi^*(\mathbf{r})\psi(\mathbf{r}) = \sum_k |c_k|^2 = 1. \tag{37.68}$$

In the multiparticle language, $\psi(\mathbf{r})$ and c_k are replaced by operators $\boldsymbol{\psi}(\mathbf{r})$ and a_k in the multiparticle occupation number space,

$$\psi(\mathbf{r}) \to \boldsymbol{\psi}(\mathbf{r}) \quad \text{and} \quad c_k \to a_k \tag{37.69}$$

where a_k is the destruction operator we have already defined in the previous section. The relation corresponding to (37.67) is then written as

$$\boldsymbol{\psi}(\mathbf{r}) = \sum_k a_k u_k(\mathbf{r}) \tag{37.70}$$

where we use bold letters to signify operators in the multiparticle Hilbert space. The normalization condition now reads

$$\int d^3r\, \boldsymbol{\psi}^\dagger(\mathbf{r})\boldsymbol{\psi}(\mathbf{r}) = \sum_k a_k^\dagger a_k = \sum_k \mathbf{N}_k = \mathbf{N}. \tag{37.71}$$

Thus the operator $\boldsymbol{\psi}(\mathbf{r})$ which we call the field operator or simply the "field" is normalized to the number operator $\mathbf{N}_k$ of particles in state k summed over all k. The sum corresponds to the total number operator, $\mathbf{N}$.

The Hamiltonian operator **H** is defined in terms of the single-particle operator H as

$$\mathbf{H} = \int d^3r \psi^\dagger(\mathbf{r}) H(\mathbf{r}) \psi(\mathbf{r}). \tag{37.72}$$

Substituting (37.70) and noting that a_k's are independent of the coordinate variables, we can write

$$\mathbf{H} = \sum_k \sum_l a_k^\dagger a_l \langle k | H | l \rangle \tag{37.73}$$

where $\langle k | H | l \rangle$ is a matrix element in the single-particle space given by

$$\langle k | H | l \rangle = \int d^3r \, u_k^*(\mathbf{r}) H(\mathbf{r}) u_l(\mathbf{r}) \tag{37.74}$$

where $|k\rangle$, and $|l\rangle$ are single-particle states. If H is diagonal in the single-particle space then

$$\langle k | H | l \rangle = E_l \delta_{kl} \tag{37.75}$$

where E_l is the eigenvalue of H given by $H |l\rangle = E_l |l\rangle$. The Hamiltonian operator in the occupation number space is then given by

$$\mathbf{H} = \sum_k a_k^\dagger a_k E_k = \sum_k \mathbf{N}_k E_k. \tag{37.76}$$

Similar expressions can be written for other operators in terms of the number operator $\mathbf{N}_k$. For example, if **P** is defined as the momentum operator in the occupation space then

$$\mathbf{P} = \sum_k \mathbf{N}_k \mathbf{p}_k \tag{37.77}$$

where $\mathbf{p}_k$ is the momentum of the kth state (not the kth component of the vector $\mathbf{p}$).

The above results hold irrespective of whether the particles satisfy Bose or Fermi statistics. In the latter case, of course, the quantum numbers of $\mathbf{N}_k$ are 0, 1.

37.5 Matrix elements of a potential

If one is considering a bilinear matrix involving two particle states, e.g., the operator corresponding to the potential energy in a two-particle scattering, then following (37.71) and (37.76) the matrix element will involve the product of two number operators, $N_k N_l$. However, in order not to double count, this term must be divided by a factor of 2. Furthermore, one must also include the term in which k and l are the same. This contribution must be of the form $N_k (N_k - 1)$.

Combining the two types of terms we obtain

$$\frac{1}{2} \sum_{k \neq l} N_k N_l + \frac{1}{2} N_k (N_k - 1). \tag{37.78}$$

The above sum can simply be written as

$$\frac{1}{2}\sum_k\sum_l a_k^\dagger a_l^\dagger a_l a_k. \tag{37.79}$$

To prove this, let us consider the term $a_k^\dagger a_l^\dagger a_l a_k$. We note that for $k \neq l$ one can write

$$a_k^\dagger a_l^\dagger a_l a_k = a_k^\dagger a_k a_l^\dagger a_l. \tag{37.80}$$

This is accomplished by making two interchanges on the left-hand side of (37.80) to bring a_k from the fourth place to the second place. Since two interchanges are involved, the result is the same whether one is using Bose or Fermi statistics. Thus we recover the first term in (37.78).

When $k = l$ we need to consider only the second term in (37.78). This term vanishes for the Fermi statistics since, for this case, $N_k^2 = N_k$. For Bose statistics a typical term in (37.79) is of the form $a_k^\dagger a_k^\dagger a_k a_k$ which can be written, using the commutation relations, as

$$a_k^\dagger a_k^\dagger a_k a_k = a_k^\dagger\left[-1 + a_k a_k^\dagger\right] a_k = \mathbf{N}_k\left(\mathbf{N}_k - 1\right), \tag{37.81}$$

recovering the second term in (37.78). Thus (37.78) and (37.79) are equivalent.

We can now express an operator $\mathbf{V}$ for the two-particle potential energy as

$$\mathbf{V} = \frac{1}{2}\int d^3r_1 \int d^3r_2 \psi^\dagger(\mathbf{r}_1)\psi^\dagger(\mathbf{r}_2)V(\mathbf{r}_1,\mathbf{r}_2)\psi(\mathbf{r}_1)\psi(\mathbf{r}_2) \tag{37.82}$$

$$= \frac{1}{2}\sum_k\sum_l a_k^\dagger a_m^\dagger a_l a_n \int d^3r_1 \int d^3r_2 u_k^*(\mathbf{r}_1)u_m^*(\mathbf{r}_2)V(\mathbf{r}_1,\mathbf{r}_2)u_l(\mathbf{r}_1)u_n(\mathbf{r}_2). \tag{37.83}$$

In general, one writes

$$\mathbf{V} = \frac{1}{2}\sum_k\sum_l a_k^\dagger a_m^\dagger a_n a_l \left\langle km\left|V\right|ln\right\rangle \tag{37.84}$$

where

$$\left\langle km\left|V\right|ln\right\rangle = \int d^3r_1 \int d^3r_2\, u_k^*(\mathbf{r}_1)u_m^*(\mathbf{r}_2)V(\mathbf{r}_1,\mathbf{r}_2)u_l(\mathbf{r}_1)u_n(\mathbf{r}_2) \tag{37.85}$$

is the matrix element that involves wavefunctions in the single-particle space.

In the problems we will consider in some of the subsequent chapters, $V(\mathbf{r}_1, \mathbf{r}_2)$ will be assumed to depend only on the magnitude of the distance between the particles, $V(|\mathbf{r}_1 - \mathbf{r}_2|)$.

37.6 Free fields and continuous variables

In this section we consider fields that correspond to free particles. Since we will be concerned with continuous variables we will rewrite the commutation relations as follows.

Bose statistics

$$\left[a(\mathbf{p}),a^{\dagger}(\mathbf{q})\right]=\delta^{(3)}(\mathbf{p}-\mathbf{q});\quad [a(\mathbf{p}),a(\mathbf{q})]=0;\quad \left[a^{\dagger}(\mathbf{p}),a^{\dagger}(\mathbf{q})\right]=0. \tag{37.86}$$

Fermi statistics

$$\{a(\mathbf{p}),a^{\dagger}(\mathbf{q})\}=\delta^{(3)}(\mathbf{p}-\mathbf{q});\quad \{a(\mathbf{p}),a(\mathbf{q})\}=0;\quad \{a^{\dagger}(\mathbf{p}),a^{\dagger}(\mathbf{q})\}=0. \tag{37.87}$$

For both cases,

$$a^{\dagger}(\mathbf{p})a(\mathbf{p})=N(\mathbf{p}). \tag{37.88}$$

Nonrelativistic case

Even though we are considering the nonrelativistic case we will express our results in terms of the four-vector $x^{\mu}=(\mathbf{r},t)\,;p^{\mu}=(\mathbf{p},E)$ with $\hbar=c=1$. A single-particle energy eigenfunction can be expressed as a superposition of free particle wavefunctions written in the form of a Fourier transform,

$$\psi(x)=\frac{1}{\left(\sqrt{2\pi}\right)^{3}}\int d^{4}p\,\delta(E-E_{p})g(p)e^{-ip\cdot x} \tag{37.89}$$

where $g(p)=g(\mathbf{p},E)$, $p\cdot x=Et-\mathbf{p}\cdot\mathbf{r}$ and $d^{4}p=dE\,d^{3}p$. We have inserted the term $\delta(E-E_{p})$, where

$$E_{p}=\frac{\mathbf{p}^{2}}{2m}, \tag{37.90}$$

to ensure that $\psi(x)$ is an energy eigenfunction with eigenvalue E_{p} and satisfies the free particle equation

$$i\frac{\partial\psi}{\partial t}=E_{p}\psi. \tag{37.91}$$

Integrating out dE in the above integral, we obtain

$$\psi(x)=\frac{1}{\left(\sqrt{2\pi}\right)^{3}}\int d^{3}p\,g(p)e^{-ip\cdot x}. \tag{37.92}$$

We note that because of the δ-function E is now replaced by E_{p}; thus,

$$p\cdot x=E_{p}t-\mathbf{p}\cdot\mathbf{r}\quad\text{and}\quad g(p)=g(\mathbf{p},E_{p}). \tag{37.93}$$

Since E_{p} is function of $\mathbf{p}$ and no longer an independent variable, $g(p)$ is a function of $\mathbf{p}$.

To express the above relation in a multiparticle language, we follow the same procedure as in the discrete case and make the substitutions

$$g(\mathbf{p})\rightarrow a(\mathbf{p}) \tag{37.94}$$

and

$$\psi(x)\rightarrow\boldsymbol{\psi}(x) \tag{37.95}$$

where $a(\mathbf{p})$ is the destruction operator defined earlier, satisfying Bose statistics, and $\psi(\mathbf{x})$ is an operator in the multiparticle space,

$$\psi(x) = \frac{1}{\left(\sqrt{2\pi}\right)^3} \int d^3p\, a(\mathbf{p}) e^{-ip\cdot x}. \tag{37.96}$$

With these definitions we obtain

$$\int d^3r \psi^\dagger(x)\psi(x) = \frac{1}{(2\pi)^3} \int d^3r \int d^3p\, a^\dagger(\mathbf{p}) e^{ip\cdot x} \int d^3q\, a(\mathbf{q}) e^{-iq\cdot x}. \tag{37.97}$$

However,

$$\begin{aligned} \frac{1}{(2\pi)^3} \int d^3r\, e^{i(p-q)\cdot x} &= \frac{1}{(2\pi)^3} \int d^3r\, e^{-i(\mathbf{p}-\mathbf{q})\cdot \mathbf{r}} e^{i(E_p - E_q)t} = \delta^{(3)}(\mathbf{p}-\mathbf{q})\, e^{i(E_p-E_q)t} \\ &= \delta^{(3)}(\mathbf{p}-\mathbf{q}). \end{aligned} \tag{37.98}$$

The last step follows since $\mathbf{p} = \mathbf{q}$, as dictated by the δ-function, which implies that $E_p = E_q$. Thus,

$$\int d^3r\, \psi^\dagger(x)\psi(x) = \int d^3p\, a^\dagger(\mathbf{p}) a(\mathbf{p}) = \int d^3p\, N(\mathbf{p}). \tag{37.99}$$

Thus, as expected, the fields are normalized to the number of particles.

Similarly, the Hamiltonian is given by

$$\mathbf{H} = \int d^3r\, \psi^\dagger(x) \left(i\frac{\partial}{\partial t}\right) \psi(x), \tag{37.100}$$

which we write as

$$\mathbf{H} = \frac{1}{(2\pi)^3} \int d^3r \int d^3p\, a^\dagger(\mathbf{p}) e^{ip\cdot x} \left(i\frac{\partial}{\partial t}\right) \int d^3q\, a(\mathbf{q}) e^{-iq\cdot x} = \int d^3p\, E_p a^\dagger(\mathbf{p}) a(\mathbf{p}), \tag{37.101}$$

which is, again, a result that was expected.

37.7 Klein–Gordon/scalar field

We write the single particle Klein–Gordon wavefunction as

$$\phi(x) = \frac{1}{(\sqrt{2\pi})^3} \int d^4p\, \delta\left(p^2 - m^2\right) g(p)\, e^{-ip\cdot x}. \tag{37.102}$$

The δ-function is inserted in order that $\phi(x)$ satisfies the Klein–Gordon equation,

$$\nabla^2\phi - \frac{\partial^2\phi}{\partial t^2} = m^2\phi. \tag{37.103}$$

One can simplify the δ-function term by using the relation

$$\delta(p^2 - m^2) = \frac{\delta\left(p_4 - \sqrt{\mathbf{p}^2 + m^2}\right) + \delta\left(p_4 + \sqrt{\mathbf{p}^2 + m^2}\right)}{2\sqrt{\mathbf{p}^2 + m^2}}. \tag{37.104}$$

Let $\omega_p = \sqrt{\mathbf{p}^2 + m^2}$, which corresponds to the energy of the particle with momentum $\mathbf{p}$. Integrating out dp_4 we find

$$\phi(x) = \frac{1}{(\sqrt{2\pi})^3} \int \frac{d^3p}{2\omega_p} \left[g(\mathbf{p}, \omega_p)e^{-i\omega_p t}e^{i\mathbf{p}\cdot\mathbf{r}} + g(\mathbf{p}, -\omega_p)e^{i\omega_p t}e^{i\mathbf{p}\cdot\mathbf{r}}\right]. \tag{37.105}$$

We note that the second term above comes from the second term in the expression (37.104) for $\delta(p^2 - m^2)$. It corresponds to the negative-energy particle, traveling backward in time, which we came across in our discussion of the Klein–Gordon equation. We will examine below how the negative-energy problem is handled in field theory.

We change $\mathbf{p} \to -\mathbf{p}$ in the second term in (37.105) and obtain

$$\phi(x) = \frac{1}{(\sqrt{2\pi})^3} \int \frac{d^3p}{2\omega_p} [g(\mathbf{p}, \omega_p)e^{-ip\cdot x} + g(-\mathbf{p}, -\omega_p)e^{ip\cdot x}] \tag{37.106}$$

$$= \frac{1}{(\sqrt{2\pi})^3} \int \frac{d^3p}{2\omega_p} [g(p)e^{-ip\cdot x} + g(-p)e^{ip\cdot x}] \tag{37.107}$$

where $p^\mu = (\mathbf{p}, \omega_p)$, and $p \cdot x = \omega_p t - \mathbf{p} \cdot \mathbf{r}$.

37.7.1 Second quantization

In order to convert (37.107) to an expression that is appropriate for a field-theoretic description, let

$$\frac{g(p)}{\sqrt{2\omega_p}} = a(\mathbf{p}), \quad \frac{g(-p)}{\sqrt{2\omega_p}} = a'(\mathbf{p}) \tag{37.108}$$

where $a(\mathbf{p})$ is a destruction operator. We will determine $a'(\mathbf{p})$ based on the properties of the Klein–Gordon field, $\Phi(x)$, which we write as

$$\Phi(x) = \frac{1}{(\sqrt{2\pi})^3} \int \frac{d^3p}{\sqrt{2\omega_p}} [a(\mathbf{p})e^{-ip\cdot x} + a'(\mathbf{p})e^{ip\cdot x}] \tag{37.109}$$

and its Hermitian conjugate as

$$\Phi^\dagger(x) = \frac{1}{(\sqrt{2\pi})^3} \int \frac{d^3p}{\sqrt{2\omega_p}} [a^\dagger(\mathbf{p})e^{ip\cdot x} + a'^\dagger(\mathbf{p})e^{-ip\cdot x}]. \tag{37.110}$$

Thus, we have moved from the classical field description for the Klein–Gordon wavefunction to the quantum field description of the Klein–Gordon field.

If we assume that $\Phi(x)$ is Hermitian, $\Phi(x) = \Phi^\dagger(x)$, then comparing the coefficients of the exponentials, we find

$$a'(\mathbf{p}) = a^\dagger(\mathbf{p}). \tag{37.111}$$

Therefore,

$$\Phi(x) = \frac{1}{(\sqrt{2\pi})^3} \int \frac{d^3p}{\sqrt{2\omega_p}} [a(\mathbf{p})e^{-ip\cdot x} + a^\dagger(\mathbf{p})e^{ip\cdot x}]. \tag{37.112}$$

The second term above, which was originally a negative-energy term traveling backward in time, now involves a creation operator for a positive-energy particle, which we will designate as an antiparticle, which is now moving forward in time but with the frequency appearing with an opposite sign.

We impose the following relations on the ladder operators a and $a^\dagger$:

$$\left[a(\mathbf{p}), a^\dagger(\mathbf{q})\right] = \delta^{(3)}(\mathbf{p} - \mathbf{q}); \quad [a(\mathbf{p}), a(\mathbf{q})] = 0; \quad \left[a^\dagger(\mathbf{p}), a^\dagger(\mathbf{q})\right] = 0, \tag{37.113}$$

reflecting the fact that the Klein–Gordon field satisfies Bose statistics. This condition then implements second quantization.

We now try to determine the operator for charge, which, as we saw in our discussion for the Klein–Gordon wave equation, is obtained through the matrix element representing the scalar product of the Klein–Gordon wavefunctions ϕ_1 and ϕ_2. It is defined as follows:

$$\int d^3\mathbf{r}(\phi_1^* \overleftrightarrow{\partial_t} \phi_2) = i \int d^3\mathbf{r} \left[\phi_1^* \frac{\partial \phi_2}{\partial t} - \frac{\partial \phi_1^*}{\partial t}\phi_2\right]. \tag{37.114}$$

We abbreviate this matrix element as $\langle \phi_1 | \phi_2 \rangle$. Thus,

$$\langle \phi_1 | \phi_2 \rangle = i \int d^3\mathbf{r} \left[\phi_1^* \frac{\partial \phi_2}{\partial t} - \frac{\partial \phi_1^*}{\partial t}\phi_2\right]. \tag{37.115}$$

Similarly, for the matrix element of an operator A we have

$$\langle \phi_1 | A\phi_2 \rangle = i \int d^3\mathbf{r}\, (\phi_1^* \overleftrightarrow{\partial_t} A\phi_2). \tag{37.116}$$

To elaborate further, let us separate the field $\Phi(x)$ into two terms $\Phi^{(+)}(x)$ and $\Phi^{(-)}(x)$ defined as

$$\Phi^{(+)}(x) = \frac{1}{(\sqrt{2\pi})^3} \int \frac{d^3p}{\sqrt{2\omega_p}} a(\mathbf{p})e^{-ip\cdot x} \text{ and } \Phi^{(-)}(x) = \frac{1}{(\sqrt{2\pi})^3} \int \frac{d^3p}{\sqrt{2\omega_p}} a^\dagger(\mathbf{p})e^{ip\cdot x} \tag{37.117}$$

where the positive (negative) sign for the superscript of Φ indicates positive (negative) frequency and involves the destruction (creation) operator $a(\mathbf{p})(a^\dagger(\mathbf{p}))$.

We define the charge operator as

$$\begin{aligned} Q &= \langle \Phi(x) | \Phi(x) \rangle \\ &= \langle \Phi^{(+)}(x) \left| \Phi^{(+)}(x) \right\rangle + \langle \Phi^{(+)}(x) \left| \Phi^{(-)}(x) \right\rangle + \langle \Phi^{(-)}(x) \left| \Phi^{(+)}(x) \right\rangle + \langle \Phi^{(-)}(x) \left| \Phi^{(-)}(x) \right\rangle. \end{aligned} \tag{37.118}$$

We first obtain $\left\langle\Phi^{(+)}(x)|\Phi^{(+)}(x)\right\rangle$,

$$\left\langle\Phi^{(+)}(x)|\Phi^{(+)}(x)\right\rangle = i\int d^3\mathbf{r}\,(\Phi^{(+)\dagger}(x)\overleftrightarrow{\partial_t}\,\Phi^{(+)}(x))$$

$$= \frac{1}{(2\pi)^3}\int \frac{d^3pd^3q}{\sqrt{4\omega_p\omega_q}}a^\dagger(\mathbf{p})a(\mathbf{q})\left[e^{ip\cdot x}\overleftrightarrow{\partial_t}\,e^{-iq\cdot x}\right]. \tag{37.119}$$

It is then easy to show after integration that

$$\left\langle\Phi^{(+)}(x)|\Phi^{(+)}(x)\right\rangle = \int d^3p\,a^\dagger(\mathbf{p})a(\mathbf{p}). \tag{37.120}$$

Also, one finds

$$\left\langle\Phi^{(+)}(x)|\Phi^{(-)}(x)\right\rangle = \left\langle\Phi^{(-)}(x)|\Phi^{(+)}(x)\right\rangle = 0. \tag{37.121}$$

Similarly

$$\left\langle\Phi^{(-)}(x)|\Phi^{(-)}(x)\right\rangle = -\int d^3p\,a(\mathbf{p})\,a^\dagger(\mathbf{p}). \tag{37.122}$$

This term, with the creation operator on the right, when it operates on the vacuum will give an infinite contribution; that is, the charge of the vacuum will be infinite, a result that does not have a physical significance. Since experimental measurements only determine quantities as differences from their vacuum values, we prevent the infinity by defining a so-called "normal product" so that the destruction operator is always to the right of the creation operator. In the process of interchanging we follow Bose statistics for the Klein–Gordon (K-G) field and assume that the sign does not change. A normal product $N[...]$ is, therefore, defined such that

$$N[a(\mathbf{p})a^\dagger(\mathbf{p})] = a^\dagger(\mathbf{p})a(\mathbf{p}). \tag{37.123}$$

Thus,

$$N[\left\langle\Phi^{(-)}(x)|\Phi^{(-)}(x)\right\rangle] = -\int d^3p\,a^\dagger(\mathbf{p})a(\mathbf{p}). \tag{37.124}$$

Therefore, the antiparticles contribute a negative sign to the charge eigenvalue.

We emphasize that this is a crucial result in quantum field theory. The probability density we encountered earlier for the K-G wavefunctions in relativistic quantum mechanics, which had a negative sign corresponding to negative energies, now contributes a negative sign to the charge. The probability density in field theory given by the number remains positive, however.

The plus and minus charges add up to zero. The total charge then vanishes.

$$Q = N[\langle\Phi(x)\,|\Phi(x)\rangle] = 0. \tag{37.125}$$

Hence the Hermitian field $\Phi^\dagger(x) = \Phi(x)$ corresponds to a neutral particle. The field $\Phi(x)$ then represents a neutral scalar particle.

The total energy given by the Hamiltonian operator is written as a normal product,

$$\mathbf{H} = N[\langle\Phi(x)|H|\Phi(x)\rangle] = N[\left\langle\Phi^{(+)}(x)|H|\Phi^{(+)}(x)\right\rangle] + N[\left\langle\Phi^{(-)}(x)|H|\Phi^{(-)}(x)\right\rangle] \tag{37.126}$$

where the cross-terms once again vanish. One obtains

$$N[\left\langle \Phi^{(+)}(x)|H|\Phi^{(+)}(x)\right\rangle] = \left\langle \Phi^{(+)}(x)|H|\Phi^{(+)}(x)\right\rangle$$

$$= \frac{1}{(2\pi)^3}\int \frac{d^3p}{\sqrt{4\omega_p\omega_q}} a^\dagger(\mathbf{p})\, a(\mathbf{q}) \left[e^{ip\cdot x}\overleftrightarrow{\partial_t}\,\omega_q e^{-iq\cdot x}\right]$$

$$= \int d^3p\, \omega_p a^\dagger(\mathbf{p})\, a(\mathbf{p}) \tag{37.127}$$

where we have used (37.116). If we write $a^\dagger(\mathbf{p})\, a(\mathbf{p}) = \mathbf{N}(\mathbf{p})$, the number operator, then

$$N[\left\langle \Phi^{(+)}(x)|H|\Phi^{(+)}(x)\right\rangle] = \int d^3p\, \omega_p \mathbf{N}(\mathbf{p}), \tag{37.128}$$

which corresponds to the energy eigenvalue of each state times the number of particles in that state integrated over the momenta. Similarly,

$$N[\left\langle \Phi^{(-)}(x)|H|\Phi^{(-)}(x)\right\rangle] = \int d^3p\, \omega_p \mathbf{N}(\mathbf{p}). \tag{37.129}$$

Note that the sign is now positive. Thus the antiparticles contribute the same amount to the energy as the particles. Hence, the total energy is twice this value:

$$\mathbf{H} = 2\int d^3p\, \omega_p \mathbf{N}(\mathbf{p}). \tag{37.130}$$

We note that in the absence of any quantum numbers that differentiate the two particles, the antiparticle is the same as the particle.

The difference between particles and antiparticles will become more clear when we go to the complex scalar field, which gives rise to particles and antiparticles having opposite charges.

37.8 Complex scalar field

Let us now consider a complex field $\Phi(x)$ defined as

$$\Phi(x) = \Phi_1 + i\Phi_2, \tag{37.131}$$

which is made up of two Hermitian fields:

$$\Phi_1^\dagger = \Phi_1, \quad \Phi_2^\dagger = \Phi_2. \tag{37.132}$$

In the same manner as for the Hermitian field we discussed previously we write

$$\Phi_1(x) = \frac{1}{(\sqrt{2\pi})^3}\int \frac{d^3p}{\sqrt{2\omega_p}}[a_1(\mathbf{p})e^{-ip\cdot x} + a_1^\dagger(\mathbf{p})e^{ip\cdot x}], \tag{37.133}$$

$$\Phi_2(x) = \frac{1}{(\sqrt{2\pi})^3} \int \frac{d^3p}{\sqrt{2\omega_p}} [a_2(\mathbf{p}) e^{-ip\cdot x} + a_2^\dagger(\mathbf{p}) e^{ip\cdot x}]. \tag{37.134}$$

Thus,

$$\Phi(x) = \frac{1}{(\sqrt{2\pi})^3} \int \frac{d^3p}{\sqrt{2\omega_p}} [(a_1 + ia_2) e^{-ip\cdot x} + (a_1^\dagger + ia_2^\dagger) e^{ip\cdot x}]. \tag{37.135}$$

We define

$$a_1 + ia_2 = a, \quad a_1 - ia_2 = b. \tag{37.136}$$

Hence

$$\Phi(x) = \frac{1}{(\sqrt{2\pi})^3} \int \frac{d^3p}{\sqrt{2\omega_p}} [a(\mathbf{p}) e^{-ip\cdot x} + b^\dagger(\mathbf{p}) e^{ip\cdot x}]. \tag{37.137}$$

The second term now corresponds to an antiparticle traveling forward in time replacing the negative-energy solution.

We impose the following commutation relations on a and b:

$$\left[a(\mathbf{p}), a^\dagger(\mathbf{p})\right] = \delta^{(3)}(\mathbf{p} - \mathbf{q}); \quad \left[b(\mathbf{p}), b^\dagger(\mathbf{p})\right] = \delta^{(3)}(\mathbf{p} - \mathbf{q}). \tag{37.138}$$

All other commutators between a, $a^\dagger$, b, and $b^\dagger$ vanish.

Once again, as for the neutral scalar case, we separate $\Phi(x)$ into two parts:

$$\Phi^{(+)}(x) = \frac{1}{(\sqrt{2\pi})^3} \int \frac{d^3p}{\sqrt{2\omega_p}} a(\mathbf{p}) e^{-ip\cdot x} \quad \text{and} \quad \Phi^{(-)}(x) = \frac{1}{(\sqrt{2\pi})^3} \int \frac{d^3p}{\sqrt{2\omega_p}} b^\dagger(\mathbf{p}) e^{ip\cdot x}. \tag{37.139}$$

The charge operator is defined by

$$\begin{aligned} Q &= \langle \Phi(x) \,|\Phi(x)\rangle \\ &= \langle \Phi^{(+)}(x) \left|\Phi^{(+)}(x)\right\rangle + \langle \Phi^{(+)}(x) \left|\Phi^{(-)}(x)\right\rangle + \langle \Phi^{(-)}(x) \left|\Phi^{(+)}(x)\right\rangle + \langle \Phi^{(-)}(x) \left|\Phi^{(-)}(x)\right\rangle. \end{aligned} \tag{37.140}$$

Carrying out the calculations in the same manner as for the Hermitian case, we obtain

$$\left\langle \Phi^{(+)}(x) | \Phi^{(+)}(x)\right\rangle = \int d^3p \, a^\dagger(\mathbf{p}) a(\mathbf{p}), \tag{37.141}$$

$$\left\langle \Phi^{(+)}(x) | \Phi^{(-)}(x)\right\rangle = \left\langle \Phi^{(-)}(x) | \Phi^{(+)}(x)\right\rangle = 0, \tag{37.142}$$

and

$$\left\langle \Phi^{(-)}(x) | \Phi^{(-)}(x)\right\rangle = -\int d^3p \, b(\mathbf{p}) b^\dagger(\mathbf{p}). \tag{37.143}$$

We define the normal products as

$$N[a(\mathbf{p}) a^\dagger(\mathbf{p})] = a^\dagger(\mathbf{p}) a(\mathbf{p}), \tag{37.144}$$

$$N[b(\mathbf{p}) b^\dagger(\mathbf{p})] = b^\dagger(\mathbf{p}) b(\mathbf{p}). \tag{37.145}$$

Thus,

$$N[\langle \Phi^{(-)}(x)|\Phi^{(-)}(x)\rangle] = -\int d^3p\, b^\dagger(\mathbf{p})b(\mathbf{p}). \tag{37.146}$$

If we define

$$\mathbf{N}_+ = \int d^3p\, \mathbf{N}_+(\mathbf{p}) \quad \text{and} \quad \mathbf{N}_- = \int d^3p\, \mathbf{N}_-(\mathbf{p}), \tag{37.147}$$

where

$$\mathbf{N}_+(\mathbf{p}) = a^\dagger(\mathbf{p})a(\mathbf{p}) \text{ and } \mathbf{N}_-(\mathbf{p}) = b^\dagger(\mathbf{p})b(\mathbf{p}), \tag{37.148}$$

then

$$\mathbf{Q} = N[\langle \Phi(x)\, | \Phi(x)\rangle] = \mathbf{N}_+ - \mathbf{N}_-. \tag{37.149}$$

Thus, the antiparticle states have a negative charge while the particles have a positive charge.
Similarly, the total Hamiltonian is given by

$$\mathbf{H} = \int d^3p \left[\mathbf{N}_+(\mathbf{p})\, \omega_p + \mathbf{N}_-(\mathbf{p})\omega_p\right]. \tag{37.150}$$

In recapitulating the above results we find that second quantization provides a solution to the troublesome negative-energy particles by predicting a new set of particles that are called antiparticles and have quantum numbers, such as charge, that have opposite sign to those of the particles but have positive energy and have the same mass as the particles.

37.9 Dirac field

The Dirac wavefunction can be expressed in a similar fashion to the Klein–Gordon case,

$$\psi(x) = \sum_{s=1}^{2} \frac{1}{(\sqrt{2\pi})^3} \int d^4p\, \delta\left(p^2 - m^2\right) w_s(\mathbf{p}, E) g_s(p) e^{-ip\cdot x}, \tag{37.151}$$

where $\psi(x)$ satisfies the free particle equation $(i\gamma \cdot \partial - m)\, \psi(x) = 0$ and, in the momentum representation, $w_s(\mathbf{p}, E)$ satisfies

$$(\gamma \cdot p - m)\, w_s(\mathbf{p}, E) = 0. \tag{37.152}$$

The index **s** corresponds to the spin of the particle. Simplifying the δ-function we write

$$\psi(x) = \sum_{s=1}^{2} \frac{1}{(\sqrt{2\pi})^3} \int d^4p\, \frac{\left[\delta\left(p_4 - \sqrt{\mathbf{p}^2 + m^2}\right) + \delta\left(p_4 + \sqrt{\mathbf{p}^2 + m^2}\right)\right]}{2\sqrt{\mathbf{p}^2 + m^2}} \times w_s(\mathbf{p}, E) g_s(\mathbf{p}, E) e^{-ip\cdot x}. \tag{37.153}$$

Using the properties of the δ-function we obtain

$$\psi(x) = \sum_{s=1}^{2} \frac{1}{(\sqrt{2\pi})^3} \int \frac{d^3p}{2E_p} \left[w_s(\mathbf{p}, E_p) g_s(\mathbf{p}, E_p) e^{-ip\cdot x} + w_s(\mathbf{p}, -E_p) g_s(\mathbf{p}, -E_p) e^{i\mathbf{p}\cdot\mathbf{r}} e^{iE_p t}\right] \tag{37.154}$$

where $E_p = \sqrt{\mathbf{p}^2 + m^2}$. Changing $\mathbf{p} \to -\mathbf{p}$ in the second term we obtain

$$\psi(x) = \sum_{s=1}^{2} \frac{1}{(\sqrt{2\pi})^3} \int \frac{d^3p}{2E_p} \left[w_s(\mathbf{p}, E_p) g_s(\mathbf{p}, E_p) e^{-ip\cdot x} + w_s(-\mathbf{p}, -E_p) g_s(-\mathbf{p}, -E_p) e^{ip\cdot x} \right]. \tag{37.155}$$

We define

$$w_s(\mathbf{p}, E_p) = \sqrt{\frac{m}{E_p}} u_s(\mathbf{p}) \quad \text{and} \quad w_s(-\mathbf{p}, -E_p) = \sqrt{\frac{m}{E_p}} v_s(\mathbf{p}) \tag{37.156}$$

where $u_s(\mathbf{p})$ and $v_s(\mathbf{p})$ are the same spinors that were introduced in Chapter 35. They satisfy the equations

$$(\gamma \cdot p - m) u_s(\mathbf{p}) = 0 \quad \text{with} \quad p_4 = E_p = +\sqrt{\mathbf{p}^2 + m^2} \tag{37.157}$$

and

$$(\gamma \cdot p + m)\, v_s(\mathbf{p}) = 0 \quad \text{with} \quad p_4 = E_p = +\sqrt{\mathbf{p}^2 + m^2} \tag{37.158}$$

where we use the same normalization as in Chapter 35,

$$u_s^\dagger u_{s'} = \frac{E_p}{m} \delta_{ss'}, \quad v_s^\dagger v_{s'} = \frac{E_p}{m} \delta_{ss'}, \quad u_s^\dagger v_{s'} = v_s^\dagger u_{s'} = 0. \tag{37.159}$$

.

Substituting the above results we obtain

$$\psi(x) = \sum_{s=1}^{2} \frac{1}{(\sqrt{2\pi})^3} \int \frac{d^3p}{2E_p} \left[\sqrt{\frac{m}{E_p}} u_s(\mathbf{p}) g_s(p) e^{-ip\cdot x} + \sqrt{\frac{m}{E_p}} v_s(\mathbf{p}) g_s(-p) e^{ip\cdot x} \right]. \tag{37.160}$$

To convert the above relation to the occupation number space, let us write, in a manner similar to that for the complex Klein–Gordon fields,

$$\frac{g_s(p)}{2E_p} = a_s(\mathbf{p}) \quad \text{and} \quad \frac{g_s(-p)}{2E_p} = b_s^\dagger(\mathbf{p}) \tag{37.161}$$

where $a_s(\mathbf{p})$ now becomes the destruction operator for the particle fields and $b_s^\dagger(\mathbf{p})$ the creation operator for the antiparticles. We also replace $\psi(x)$ by the field operator $\boldsymbol{\psi}(x)$. Thus, we have

$$\boldsymbol{\psi}(x) = \sum_{s=1}^{2} \frac{1}{(\sqrt{2\pi})^3} \int d^3p \sqrt{\frac{m}{E_p}} \left[u_s(\mathbf{p}) a_s(\mathbf{p})\, e^{-ip\cdot x} + v_s(\mathbf{p}) b_s^\dagger(\mathbf{p}) e^{ip\cdot x} \right] \tag{37.162}$$

where, in order that the Dirac fields satisfy Fermi statistics, we assume the following anticommutator relations

$$\left\{ a_s(\mathbf{p}), a_{s'}^{+}(\mathbf{q}) \right\} = \delta_{ss'} \delta^{(3)}(\mathbf{p} - \mathbf{q}), \tag{37.163}$$

$$\left\{ b_s(\mathbf{p}), b_{s'}^{+}(\mathbf{q}) \right\} = \delta_{ss'} \delta^{(3)}(\mathbf{p} - \mathbf{q}), \tag{37.164}$$

with all other anticommutation relations being zero.

The charge operator in Dirac theory is defined as

$$\int d^3r\, \psi^\dagger(x)\psi(x) = \mathbf{Q}. \tag{37.165}$$

As in previous discussions, to obtain $\mathbf{Q}$ let us write

$$\psi(x) = \psi^{(+)}(x) + \psi^{(-)}(x) \tag{37.166}$$

where

$$\psi^{(+)}(x) = \sum_{s=1}^{2} \frac{1}{(\sqrt{2\pi})^3} \int d^3p \sqrt{\frac{m}{E_p}} u_s(\mathbf{p}) a_s(\mathbf{p}) e^{-ip\cdot x} \tag{37.167}$$

and

$$\psi^{(-)}(x) = \sum_{s=1}^{2} \frac{1}{(\sqrt{2\pi})^3} \int d^3p \sqrt{\frac{m}{E_p}} v_s(\mathbf{p}) b^\dagger(\mathbf{p}) e^{ip\cdot x}. \tag{37.168}$$

Since

$$u_s^\dagger u_{s'} = \frac{E_p}{m}\delta_{ss'}, \quad v_s^\dagger v_{s'} = \frac{E_p}{m}\delta_{ss'}, \quad \text{and} \quad u_s^\dagger v_{s'} = v_s^\dagger u_{s'} = 0, \tag{37.169}$$

it is easy to show that

$$\int d^3r\, \psi^{(+)\dagger}(x)\psi^{(-)}(x) = 0 = \int d^3r\, \psi^{(-)\dagger}(x)\psi^{(+)}(x) \tag{37.170}$$

and

$$\int d^3r\, \psi^{(+)\dagger}(x)\psi^{(+)}(x) = \int d^3p\, a^\dagger(\mathbf{p})a(\mathbf{p}) \tag{37.171}$$

and

$$\int d^3r\, \psi^{(-)\dagger}(x)\psi^{(-)}(x) = \int d^3p\, b(\mathbf{p})b^\dagger(\mathbf{p}). \tag{37.172}$$

In the second relation above, because we have a creation operator on the right we again invoke the concept of the normal product but this time, in interchanging the operators to bring the creation operator to the left, we use anticommutation relations since the Dirac field obeys Fermi statistics. Thus,

$$N\left[\int d^3r\, \psi^{(-)\dagger}(x)\psi^{(-)}(x)\right] = -\int d^3p\, b^\dagger(\mathbf{p})b(\mathbf{p}). \tag{37.173}$$

If we define the number operators as

$$\mathbf{N}_+ = \int d^3p \mathbf{N}_+(\mathbf{p}) \quad \text{and} \quad \mathbf{N}_- = \int d^3p \mathbf{N}_-(\mathbf{p}) \tag{37.174}$$

where

$$\mathbf{N}_+(\mathbf{p}) = a^\dagger(\mathbf{p})a(\mathbf{p}) \quad \text{and} \quad \mathbf{N}_-(\mathbf{p}) = b^\dagger(\mathbf{p})b(\mathbf{p}), \tag{37.175}$$

then

$$\mathbf{Q} = \mathbf{N}_+ - \mathbf{N}_-. \tag{37.176}$$

Hence, the antiparticles have opposite charge to the particles. In our case, if we identify the particle as an electron then the antiparticle is called the positron. One of the great successes of quantum field theory is that the positrons have been discovered with the same mass as an electron but with opposite charge.

Based on the single-particle results, the Hamiltonian in the Dirac field is given by

$$\mathbf{H} = \int d^3r\, \boldsymbol{\psi}^\dagger(\mathbf{x})\,(i\boldsymbol{\alpha}\cdot\nabla - \beta m)\,\boldsymbol{\psi}(\mathbf{x}). \tag{37.177}$$

We rewrite this as

$$\mathbf{H} = \int d^3r\, \boldsymbol{\psi}^\dagger(\mathbf{x})\,\gamma_4\,(-i\beta\boldsymbol{\alpha}\cdot\nabla + m)\,\boldsymbol{\psi}(\mathbf{x}) \tag{37.178}$$

where

$$\gamma_4 = -\beta \quad \text{and} \quad \beta^2 = 1. \tag{37.179}$$

Since $\beta\boldsymbol{\alpha} = \boldsymbol{\gamma}$, then, as in the single particle case, defining $\boldsymbol{\psi}^\dagger\gamma_4 = \overline{\boldsymbol{\psi}}(x)$,

$$\mathbf{H} = \int d^3r\, \overline{\boldsymbol{\psi}}(x)\,(-i\boldsymbol{\gamma}\cdot\nabla + m)\,\boldsymbol{\psi}(\mathbf{x}). \tag{37.180}$$

Inserting the free field relation (37.162) into the relation for the Hamiltonian, we find

$$\mathbf{H} = \int d^3p\, E_p \sum_s \left[a_s^\dagger(\mathbf{p})\,a_s(\mathbf{p}) - b_s(\mathbf{p})\,b_s^\dagger(\mathbf{p})\right]. \tag{37.181}$$

Using the normal product for the second term and the anticommutation relation in the interchange of the ladder operators, we obtain

$$\mathbf{H} = \int d^3p\, E_p \sum_s \left[a_s^\dagger(\mathbf{p})\,a_s(\mathbf{p}) + b_s^\dagger(\mathbf{p})\,b_s(\mathbf{p})\right]. \tag{37.182}$$

Hence,

$$\mathbf{H} = \int d^3p\left[\mathbf{N}_+E_p + \mathbf{N}_-E_p\right]. \tag{37.183}$$

Thus the antiparticles contribute with a positive sign to the energy.

37.10 Maxwell field

We will consider the Maxwell field, $A_\mu(x)$, in the same manner as the Klein–Gordon field, $\phi(x)$, except that it is a four-vector. We will first look at the field in the transverse gauge given by

$$A_4(x) = 0, \quad \text{and} \quad \nabla \cdot \mathbf{A}(x) = 0. \tag{37.184}$$

If we take the three-vector, $\mathbf{A}$, to be of the form

$$\boldsymbol{\epsilon} e^{i\mathbf{k}\cdot\mathbf{r}} e^{-i\omega t} \tag{37.185}$$

where $\boldsymbol{\epsilon}$ is the polarization vector, then the transversality condition (37.184) implies

$$\mathbf{k} \cdot \mathbf{A}(x) = 0. \tag{37.186}$$

We will thus consider only the two components of $\mathbf{A}$, and therefore of $\boldsymbol{\epsilon}$, that are perpendicular to $\mathbf{k}$.

We write

$$\mathbf{A}(x) = \frac{1}{\left(\sqrt{2\pi}\right)^3} \sum_{\alpha=1}^{2} \int d^4k\, \delta\left(k^2\right) \boldsymbol{\epsilon}_\alpha g_\alpha(k) e^{-ik\cdot x} \tag{37.187}$$

where $\boldsymbol{\epsilon}_\alpha$ is the polarization vector with $\alpha = 1, 2$. It is a real, unit vector and satisfies the orthogonality condition

$$\boldsymbol{\epsilon}_\alpha \cdot \boldsymbol{\epsilon}_{\alpha'} = \delta_{\alpha\alpha'}. \tag{37.188}$$

The term $\delta\left(k^2\right)$ is inserted so that $\mathbf{A}$ satisfies Maxwell's wave equation:

$$\nabla^2 \mathbf{A} - \frac{1}{c^2} \frac{\partial^2 \mathbf{A}}{\partial t^2} = 0. \tag{37.189}$$

We express $\delta\left(k^2\right)$ as follows:

$$\delta(k^2) = \delta\left(k_4^2 - \mathbf{k}^2\right) = \frac{\delta\left(k_4 - |\mathbf{k}|\right) + \delta\left(k_4 + |\mathbf{k}|\right)}{2\,|\mathbf{k}|}. \tag{37.190}$$

If ω_k is the frequency of the electromagnetic wave, then

$$|\mathbf{k}| = \omega_k. \tag{37.191}$$

In the following we will take $\mathbf{A}(x)$ to be real, so that

$$\mathbf{A}^\dagger(x) = \mathbf{A}(x). \tag{37.192}$$

Following the same steps as for the K-G field, we have

$$\mathbf{A}(x) = \frac{1}{\left(\sqrt{2\pi}\right)^3} \sum_{\alpha=1}^{2} \int \frac{d^3k}{\sqrt{2\omega_k}} \boldsymbol{\epsilon}_\alpha \left[a_\alpha(\mathbf{k}) e^{-ik\cdot x} + a_\alpha^\dagger(\mathbf{k}) e^{ik\cdot x}\right]. \tag{37.193}$$

The operators $a_\alpha(\mathbf{k})$ and $a_\alpha^\dagger(\mathbf{k})$ are the usual destruction and creation operators for $\alpha = 1, 2$ and satisfy the following commutation relations characteristic of the Bose–Einstein statistics that Maxwell's field satisfies:

$$\left[a_\alpha(\mathbf{k}), a_{\alpha'}^\dagger\left(\mathbf{k}'\right)\right] = \delta_{\alpha\alpha'}\delta^{(3)}\left(\mathbf{k} - \mathbf{k}'\right). \tag{37.194}$$

All other commutators vanish. This is then the second quantization condition.

Following the same procedure as for the K-G field, we find that the total Hamiltonian is given by

$$H = \sum_k N_k \hbar\omega_k \tag{37.195}$$

where

$$N_k = \sum_{\alpha=1}^{2} a_\alpha^\dagger(\mathbf{k})\, a_\alpha(\mathbf{k}). \tag{37.196}$$

Even though we have worked with the units $\hbar = c = 1$, we have introduced $\hbar$ explicitly, but only temporarily in (37.195), to signify a very important fact that we now have the electromagnetic field quantized and represented by particles, the photons, each with an energy $\hbar\omega_k$.

Let us confirm that the same expression for energy is achieved for the classical electromagnetic Hamiltonian after the imposition of the second quantization condition. The classical Hamiltonian is given by

$$H = \frac{1}{2}\int d^3r \left(|\mathbf{E}|^2 + |\mathbf{B}|^2\right) \tag{37.197}$$

where **E** and **B** are electric and magnetic fields, respectively, which are given in terms of **A** as

$$\mathbf{E} = -\frac{\partial \mathbf{A}}{\partial t}, \qquad \mathbf{B} = \nabla \times \mathbf{A}. \tag{37.198}$$

To simplify our calculations we first change $\mathbf{k} \to -\mathbf{k}$ in the second term in the expression for $\mathbf{A}(x)$ given by (37.193). We then have, after taking out the common factor $\exp(i\mathbf{k}\cdot\mathbf{r})$,

$$\mathbf{A}(x) = \frac{1}{\left(\sqrt{2\pi}\right)^3}\sum_{\alpha=1}^{2}\int \frac{d^3k}{\sqrt{2\omega_k}}\boldsymbol{\epsilon}_\alpha e^{i\mathbf{k}\cdot\mathbf{r}}\left[a_\alpha(\mathbf{k})\, e^{-i\omega_k t} + a_\alpha^\dagger(-\mathbf{k})\, e^{i\omega_k t}\right]. \tag{37.199}$$

Taking the time-derivative of **A** we obtain

$$-\frac{\partial \mathbf{A}}{\partial t} = \frac{1}{\left(\sqrt{2\pi}\right)^3}\sum_{\alpha=1}^{2}\int \frac{d^3k}{\sqrt{2\omega_k}}\boldsymbol{\epsilon}_\alpha e^{i\mathbf{k}\cdot\mathbf{r}}\,(i\omega_k)\left[a_\alpha(\mathbf{k})\, e^{-i\omega_k t} \quad a_\alpha^\dagger(-\mathbf{k})\, e^{i\omega_k t}\right]. \tag{37.200}$$

We write relations (37.199) and (37.200) as follows:

$$\mathbf{A}(x) = \frac{1}{\left(\sqrt{2\pi}\right)^3}\sum_{\alpha=1}^{2}\int \frac{d^3k}{\sqrt{2\omega_k}}\boldsymbol{\epsilon}_\alpha e^{i\mathbf{k}\cdot\mathbf{r}} G_\alpha^{(+)}(\mathbf{k}) \tag{37.201}$$

and

$$\frac{\partial \mathbf{A}(x)}{\partial t} = -\frac{1}{\left(\sqrt{2\pi}\right)^3}\sum_{\alpha=1}^{2}\int \frac{d^3k}{\sqrt{2\omega_k}}\boldsymbol{\epsilon}_\alpha e^{i\mathbf{k}\cdot\mathbf{r}}\,(i\omega_k)\, G_\alpha^{(-)}(\mathbf{k}) \tag{37.202}$$

where

$$G_{\alpha}^{(+)}(\mathbf{k}) = a_{\alpha}(\mathbf{k})\, e^{-i\omega_k t} + a_{\alpha}^{\dagger}(-\mathbf{k})\, e^{i\omega_k t}, \tag{37.203}$$

$$G_{\alpha}^{(-)}(\mathbf{k}) = a_{\alpha}(\mathbf{k})\, e^{-i\omega_k t} - a_{\alpha}^{\dagger}(-\mathbf{k})\, e^{i\omega_k t}. \tag{37.204}$$

To evaluate

$$\int d^3r\, |\mathbf{E}|^2 = \int d^3r \left(\frac{\partial \mathbf{A}^{\dagger}}{\partial t} \cdot \frac{\partial \mathbf{A}}{\partial t} \right). \tag{37.205}$$

we insert the relation (37.202) and, after several steps of integration, find

$$\int d^3r\, |\mathbf{E}|^2 = \frac{1}{2} \sum_{\alpha=1}^{2} \int d^3k\ \omega_k G_{\alpha}^{(-)\dagger}(\mathbf{k})\, G_{\alpha}^{(-)}(\mathbf{k}). \tag{37.206}$$

Similarly, we write

$$\int d^3r\, |\mathbf{B}|^2 = \int d^3r \left(\nabla \times \mathbf{A}^{\dagger} \right) \cdot (\nabla \times \mathbf{A}). \tag{37.207}$$

We insert the relation (37.201) for $\mathbf{A}$ and note that $\nabla \times \mathbf{A} \sim \mathbf{k} \times \boldsymbol{\epsilon}_{\alpha}$. Using the relation

$$(\mathbf{k} \times \boldsymbol{\epsilon}_{\alpha}) \cdot \left(\mathbf{k}' \times \boldsymbol{\epsilon}_{\alpha'} \right) = \left(\mathbf{k} \cdot \mathbf{k}' \right) (\boldsymbol{\epsilon}_{\alpha} \cdot \boldsymbol{\epsilon}_{\alpha'}) - \left(\mathbf{k}' \cdot \boldsymbol{\epsilon}_{\alpha'} \right) \left(\mathbf{k}' \cdot \boldsymbol{\epsilon}_{\alpha} \right) \tag{37.208}$$

we find that the second term in (37.208) will not contribute in view of the presence of $\delta^{(3)}\left(\mathbf{k} - \mathbf{k}'\right)$ that arises after the integration over d^3r in (37.207), and the transversality condition $\mathbf{k} \cdot \boldsymbol{\epsilon}_{\alpha} = 0$. Hence, since $|\mathbf{k}|^2 = \omega_k^2$, we have

$$\int d^3r\, |\mathbf{B}|^2 = \frac{1}{2} \sum_{\alpha=1}^{2} \int d^3k\, \omega_k G_{\alpha}^{(+)\dagger}(\mathbf{k})\, G_{\alpha}^{(+)}(\mathbf{k}). \tag{37.209}$$

When expressions (37.206) and (37.207) are inserted into the relation (37.197) for the Hamiltonian, we find that it will involve the term

$$G_{\alpha}^{(-)\dagger}(\mathbf{k})\, G_{\alpha}^{(-)}(\mathbf{k}) + G_{\alpha}^{(+)\dagger}(\mathbf{k})\, G_{\alpha}^{(+)}(\mathbf{k}) = 2\left[a_{\alpha}^{\dagger}(\mathbf{k})\, a_{\alpha}(\mathbf{k}) + a_{\alpha}(-\mathbf{k})\, a_{\alpha}^{\dagger}(-\mathbf{k}) \right]. \tag{37.210}$$

Since $\mathbf{k}$ is integrated over all its values, the integral $\int d^3k$ of the second term in (37.210) is the same as what one obtains by replacing $-\mathbf{k}$ by $\mathbf{k}$. Hence,

$$H = \frac{1}{2} \sum_{\alpha=1}^{2} \int d^3k\, \omega_k \left[a_{\alpha}^{\dagger}(\mathbf{k})\, a_{\alpha}(\mathbf{k}) + a_{\alpha}(\mathbf{k})\, a_{\alpha}^{\dagger}(\mathbf{k}) \right]. \tag{37.211}$$

We note that the vacuum energy is given by

$$\langle 0\, |H|\, 0 \rangle = \frac{1}{2} \int d^3k\, \hbar\omega_k. \tag{37.212}$$

One refers to this as the zero-point energy or vacuum "fluctuations" of the electromagnetic field which is, of course, infinite. We will return to this subject when we discuss the Casimir effect at the end of this chapter.

Introducing the normal product as we did for the K-G equation we obtain, once again inserting the factor $\hbar$ explicitly,

$$H = \int d^3k \, N_k \hbar\omega_k \tag{37.213}$$

where

$$N_k = \sum_{\alpha=1}^{2} a_k^\dagger (\mathbf{k})\, a_\alpha (\mathbf{k}) . \tag{37.214}$$

Thus the classical formula for the Hamiltonian will reproduce the same result after second quantization as we obtained previously.

37.11 Lorentz covariance for Maxwell field

By going to the transverse gauge we simplified our calculations considerably for the Maxwell field, but it was done at the expense of Lorentz covariance. For example, even though condition $\partial^\mu A_\mu = 0$ is a covariant relation, the conditions $A_4 = 0$ and $\nabla \cdot \mathbf{A} = 0$ we used will not necessarily hold in other Lorentz frames. In particular, the relation (37.194)

$$\left[a_\alpha (\mathbf{k}) , a_{\alpha'}^\dagger (\mathbf{k}')\right] = \delta_{\alpha\alpha'} \delta^{(3)} \left(\mathbf{k} - \mathbf{k}'\right) , \tag{37.215}$$

is clearly not a covariant relation since it involves only two components $\alpha, \alpha' = 1, 2$ and not all four. One could generalize this relation to

$$\left[a_\mu (\mathbf{k}) , a_\nu^\dagger \left(\mathbf{k}'\right)\right] = -g_{\mu\nu} \delta^{(3)} \left(\mathbf{k} - \mathbf{k}'\right) \tag{37.216}$$

to include all four components. This will reproduce (37.215) for space-like components, but since $g_{44} = 1$ for $\mu = \nu = 4$ we will have a negative sign in (37.216), creating the possibility that the state $a_\mu^\dagger |0\rangle$ can have a negative norm. These problems are solved within the so-called Gupta–Bleuler formalism through appropriate subsidiary conditions on the state vectors, e.g., $k^\mu a_\mu |\psi\rangle = 0$. This keeps the Lorentz covariance of (37.216) intact while avoiding the states with negative norms appearing in the formalism. We will not pursue this subject any further since it is beyond the scope of this book, but we will start using (37.215) in the discussions to follow. Specifically, from now on, we will write

$$A_\mu (x) = \frac{1}{\left(\sqrt{2\pi}\right)^3} \int \frac{d^3k}{\sqrt{2\omega_k}} \left[a_\mu (\mathbf{k})\, e^{-ik\cdot x} + a_\mu^\dagger (\mathbf{k})\, e^{ik\cdot x}\right] \tag{37.217}$$

where it is understood that $a_\mu (\mathbf{k}) = \epsilon_\mu a(\mathbf{k})$.

37.12 Propagators and time-ordered products

37.12.1 Scalar field

Consider a time-ordered product of two scalar fields, $\Phi(x)$ and $\Phi(x')$, defined as

$$T\left(\Phi(x)\,\Phi(x')\right) = \theta\left(t-t'\right)\left(\Phi(x)\,\Phi(x')\right) + \theta\left(t'-t\right)\left(\Phi(x')\,\Phi(x)\right). \tag{37.218}$$

We show below that the vacuum expectation value of this time-ordered product given by

$$\left\langle 0\left|T\left(\Phi(x)\,\Phi(x')\right)\right|0\right\rangle \tag{37.219}$$

is directly related to the propagator function $\Delta_F(x-x')$ that we considered for the Klein–Gordon equation.

The product $\Phi(x)\,\Phi(x')$ can be written as

$$\Phi(x)\,\Phi(x') = \frac{1}{(2\pi)^3}\int \frac{d^3p\,d^3q}{\sqrt{4\omega_p\omega_q}}\left[a(\mathbf{p})\,e^{-ip\cdot x} + a^\dagger(\mathbf{p})\,e^{ip\cdot x}\right]\left[a(\mathbf{q})\,e^{-iq\cdot x'} + a^\dagger(\mathbf{q})\,e^{iq\cdot x}\right]. \tag{37.220}$$

Among the four terms involved in the product above, only one term, the term with the creation operator to the right, will survive when we take the vacuum expectation value

$$\left\langle 0\left|\Phi(x)\,\Phi(x')\right|0\right\rangle = \frac{1}{(2\pi)^3}\int \frac{d^3p\,d^3q}{\sqrt{4\omega_p\omega_q}}\left\langle 0\left|a(\mathbf{p})\,a^\dagger(\mathbf{q})\right|0\right\rangle e^{-(ip-q)\cdot x}. \tag{37.221}$$

We write

$$\left\langle 0\left|a(\mathbf{p})\,a^\dagger(\mathbf{q})\right|0\right\rangle = \left\langle 0\left|\left[a(\mathbf{p}),a^\dagger(\mathbf{q})\right]\right|0\right\rangle + \left\langle 0\left|a^\dagger(\mathbf{q})\,a(\mathbf{p})\right|0\right\rangle. \tag{37.222}$$

The second term on the right-hand side vanishes since $a(\mathbf{p})\,|0\rangle = 0$. Using the commutator relation

$$\left[a(\mathbf{p}),a^\dagger(\mathbf{q})\right] = \delta^{(3)}(\mathbf{p}-\mathbf{q}) \tag{37.223}$$

in (37.221), after integrating d^3q, we obtain

$$\left\langle 0\left|\Phi(x)\,\Phi(x')\right|0\right\rangle = \frac{1}{(2\pi)^3}\int \frac{d^3p}{2\omega_p}e^{-ip\cdot(x-x')}. \tag{37.224}$$

By the interchange $x \leftrightarrow x'$ we obtain $\left\langle 0\left|\Phi(x')\,\Phi(x)\right|0\right\rangle$ from (37.221). Substituting our results in (37.218), we find

$$\begin{aligned}\left\langle 0\left|T\left(\Phi(x)\,\Phi(x')\right)\right|0\right\rangle = {}& \theta\left(t-t'\right)\frac{1}{(2\pi)^3}\int \frac{d^3p}{2\omega_p}e^{-ip\cdot(x-x')}\\ &+\theta\left(t'-t\right)\frac{1}{(2\pi)^3}\int \frac{d^3p}{2\omega_p}e^{ip\cdot(x-x')}.\end{aligned} \tag{37.225}$$

The right-hand side is related to the expression for $\Delta_F(x-x')$ that we derived in Chapter 32. The relation we obtain is

$$\langle 0 \left| T\left(\Phi(x)\, \Phi\left(x'\right)\right) \right| 0 \rangle = -i\Delta_F\left(x - x'\right). \tag{37.226}$$

Thus the time-ordered product of two scalar fields is related to the propagator.

37.12.2 Maxwell field

Here we proceed in the same manner as in the Klein–Gordon case to evaluate

$$\langle 0 \left| T\left(A^{\mu}(x)\, A^{\nu}\left(x'\right)\right) \right| 0 \rangle. \tag{37.227}$$

In place of (37.222) in the previous section we have

$$\left[a_{\mu}(\mathbf{p}), a_{\nu}(\mathbf{q})\right] = -g_{\mu\nu}\delta^{(3)}(\mathbf{p}-\mathbf{q}) \tag{37.228}$$

and obtain

$$\langle 0 \left| T\left(A^{\mu}(x)\, A^{\nu}\left(x'\right)\right) \right| 0 \rangle = -ig_{\mu\nu}D_F\left(x - x'\right) \tag{37.229}$$

where D_F is the photon propagator.

37.12.3 Dirac field

The time-ordered product for fermions is defined as

$$T\left(\psi(x)\,\bar{\psi}(x')\right) = \theta\left(t-t'\right)\psi(x)\,\bar{\psi}\left(x'\right) - \theta\left(t'-t\right)\bar{\psi}\left(x'\right)\psi(x) \tag{37.230}$$

where we note that we have a negative sign between the two terms.

$$\psi(x) = \frac{1}{\left(\sqrt{2\pi}\right)^3}\sum_{s=1}^{2}\int d^3p\sqrt{\frac{m}{E_p}}\left[u_s(\mathbf{p})\, a_s(\mathbf{p})\, e^{-ip\cdot x} + v_s(\mathbf{p})\, b_s^{\dagger}(\mathbf{p})\, e^{ip\cdot x}\right]. \tag{37.231}$$

$$\bar{\psi}(x) = \frac{1}{\left(\sqrt{2\pi}\right)^3}\sum_{s=1}^{2}\int d^3p\sqrt{\frac{m}{E_p}}\left[\bar{u}_s(\mathbf{p})\, a_s^{\dagger}(\mathbf{p})\, e^{ip\cdot x} + \bar{v}_s(\mathbf{p})\, b_s(\mathbf{p})\, e^{-ip\cdot x}\right]. \tag{37.232}$$

In calculating the vacuum expectation value of the product $\psi(x)\,\bar{\psi}\left(x'\right)$, the only term that will survive is given by

$$\langle 0 \left| \psi(x)\,\bar{\psi}\left(x'\right) \right| 0 \rangle$$
$$= \frac{1}{(2\pi)^3}\sum_{s=1}^{2}\sum_{s'=1}^{2}\int d^3p\, d^3q\sqrt{\frac{m^2}{E_pE_q}}u_s(\mathbf{p})\,\bar{u}_s(\mathbf{q})\left\langle 0\left|a_s(\mathbf{p})\, a_{s'}^{\dagger}(\mathbf{q})\right|0\right\rangle e^{-i(p\cdot x - q\cdot x')}. \tag{37.233}$$

Using the anticommutation relations we obtain

$$\left\langle 0 \left| a_s(\mathbf{p})\, a_{s'}^{\dagger}(\mathbf{q}) \right| 0 \right\rangle = \left\langle 0 \left| \left\{ a_s(\mathbf{p})\, a_{s'}^{\dagger}(\mathbf{q}) \right\} \right| 0 \right\rangle - \left\langle 0 \left| a_{s'}^{\dagger}(\mathbf{q})\, a_s(\mathbf{p}) \right| 0 \right\rangle \tag{37.234}$$

where only the first term will contribute, to give

$$\langle 0 \left| \psi(x)\, \bar{\psi}(x') \right| 0 \rangle = \frac{1}{(2\pi)^3} \sum_{s=1}^{2} \int d^3p \left(\frac{m}{E_p} \right) u_s(\mathbf{p})\, \bar{u}_s(\mathbf{p})\, e^{-ip\cdot(x-x')}. \tag{37.235}$$

However,

$$\sum_s u_s(\mathbf{p})\, \bar{u}_s(\mathbf{p}) = \Lambda_+(\mathbf{p}) = \frac{m + \gamma \cdot p}{2m}. \tag{37.236}$$

Hence,

$$\langle 0 \left| \psi(x)\, \bar{\psi}(x') \right| 0 \rangle = \frac{1}{(2\pi)^3} \int d^3p \left(\frac{m}{E_p} \right) \left(\frac{m + \gamma \cdot p}{2m} \right) e^{-ip\cdot(x-x')}. \tag{37.237}$$

Similarly,

$$\langle 0 \left| \bar{\psi}(x')\, \psi(x) \right| 0 \rangle = \frac{1}{(2\pi)^3} \sum_{s=1}^{2} \int d^3p \left(\frac{m}{E_p} \right) v_s(\mathbf{p})\, \bar{v}_s(\mathbf{p})\, e^{ip\cdot(x-x')}. \tag{37.238}$$

However,

$$\sum_s v_s(\mathbf{p})\, \bar{v}_s(\mathbf{p}) = \frac{(m - \gamma \cdot p)}{2m}. \tag{37.239}$$

Hence,

$$\langle 0 \left| \bar{\psi}(x')\, \psi(x) \right| 0 \rangle = \frac{1}{(2\pi)^3} \int d^3p \left(\frac{m}{E_p} \right) \Lambda_-(\mathbf{p})\, e^{ip\cdot(x-x')}. \tag{37.240}$$

Thus, the vacuum expectation of the time-ordered product is related to the fermion propagator S_F. The exact relationship is found to be

$$\langle 0 \left| T\left(\psi(x)\, \bar{\psi}(x')\right) \right| 0 \rangle = -iS_F\left(x - x', m\right). \tag{37.241}$$

37.13 Canonical quantization

The commutation relations between the creation and destruction operators in the occupation number space result in second quantization as we have already discussed. We note that the first quantization was based on the commutation relations postulated between the canonical variables X_i and P_j, e.g.,

$$\left[X_i, P_j\right] = i\hbar\delta_{ij} \tag{37.242}$$

with all other commutators involving X_i and P_j vanishing.

One can also couch the postulate of second quantization in terms of canonical quantities, but this time it involves the commutation relation between the field operator and its canonical conjugate. This is best done in the Lagrangian formulation, where the canonical conjugate to a field Φ is defined by

$$\Pi(\mathbf{r},t) = \frac{\partial \mathcal{L}}{\partial \dot{\mathbf{\Phi}}(r,t)}, \tag{37.243}$$

with $\dot{\mathbf{\Phi}} = d\mathbf{\Phi}/dt$, where $\mathcal{L}$ is the Lagrangian density and where we have written the space and time components separately. It is then postulated that $\mathbf{\Phi}$ and Π satisfy the following equal time commutation relation

$$[\mathbf{\Phi}(r,t), \Pi(\mathbf{r}',t)] = i\delta^{(3)}(\mathbf{r}-\mathbf{r}'). \tag{37.244}$$

We note that in our units $\hbar = c = 1$.

The Lagrangian density for the scalar (Klein–Gordon) particle is given by

$$\mathcal{L} = \frac{1}{2}\partial_\mu \mathbf{\Phi}\partial^\mu \mathbf{\Phi} - \frac{1}{2}m^2\mathbf{\Phi}^2. \tag{37.245}$$

The canonical conjugate field is then found to be

$$\Pi(\mathbf{r},t) = \dot{\mathbf{\Phi}}(r,t). \tag{37.246}$$

Hence the postulate for canonical quantization corresponds to

$$[\mathbf{\Phi}(\mathbf{r},t), \dot{\mathbf{\Phi}}(\mathbf{r}',t)] = i\delta^{(3)}(\mathbf{r}-\mathbf{r}'). \tag{37.247}$$

We show below through a series of steps that this is, indeed, satisfied in our formulation where $\mathbf{\Phi}(\mathbf{r},t)$ is given by (37.112) and the commutation relations for a and $a^\dagger$ are already defined.

$$\begin{aligned}
&[\mathbf{\Phi}(\mathbf{r},t), \dot{\mathbf{\Phi}}(\mathbf{r}',\mathbf{t})] \\
&= \int \frac{d^3p}{\sqrt{(2\pi)^3\,2\omega_p}} \int \frac{i\omega_{p'}d^3p'}{\sqrt{(2\pi)^3\,2\omega_{p'}}} \\
&\quad \left[\left(a(p)\,e^{-ip\cdot x} + a^\dagger(p)\,e^{ip\cdot x}\right), \left(-a(p')\,e^{-ip'\cdot x'} + a^\dagger(p')\,e^{ip'\cdot x'}\right)\right] \\
&= \int \frac{d^3p}{\sqrt{(2\pi)^3\,2\omega_p}} \int \frac{i\omega_{p'}d^3p'}{\sqrt{(2\pi)^3\,2\omega_{p'}}}\delta^{(3)}(\mathbf{p}-\mathbf{p}')\left(e^{-ip\cdot x+ip'\cdot x'} + e^{ip\cdot x-ip'\cdot x'}\right) \\
&= i\int \frac{d^3p}{2\,(2\pi)^3}\left(e^{i\mathbf{p}\cdot(\mathbf{r}-\mathbf{r}')} + e^{-i\mathbf{p}\cdot(\mathbf{r}-\mathbf{r}')}\right).
\end{aligned} \tag{37.248}$$

By changing variables from $\mathbf{p} \to -\mathbf{p}$, we find the right-hand side to be

$$i\int \frac{d^3p}{(2\pi)^3}e^{i\mathbf{p}\cdot(\mathbf{r}-\mathbf{r}')}. \tag{37.249}$$

This is just the integral representation of $i\delta^{(3)}(\mathbf{r}-\mathbf{r}')$. Hence we obtain (37.246). Thus, the commutation relation between the ladder operators leads to the canonical quantization condition.

For the complex scalar field $\mathbf{\Phi}$, the Lagrangian density is given by

$$\mathcal{L} = -\frac{1}{2}\partial^{\mu}\mathbf{\Phi}^{\dagger}\partial_{\mu}\mathbf{\Phi} - \frac{\mu^2}{2}(\mathbf{\Phi}^{\dagger}\mathbf{\Phi}). \tag{37.250}$$

The canonical conjugate field from (37.243) is then found to be

$$\Pi(\mathbf{r},t) = \frac{\partial\mathcal{L}}{\partial\dot{\mathbf{\Phi}}(\mathbf{r},t)} = \dot{\mathbf{\Phi}}^{\dagger}(\mathbf{r,t}), \tag{37.251}$$

leading to the quantization rule

$$[\mathbf{\Phi}(\mathbf{r},t), \dot{\mathbf{\Phi}}^{\dagger}(\mathbf{r}',t)] = i\delta^{(3)}(\mathbf{r}-\mathbf{r}'). \tag{37.252}$$

This can also be proved in a manner similar to that for the real scalar field we considered earlier.

Similarly for the Dirac field, where the Lagrangian density is given by

$$\mathcal{L} = i\overline{\Psi}\boldsymbol{\gamma}^{\mu}\partial_{\mu}\Psi - m\overline{\Psi}\Psi. \tag{37.253}$$

The canonical field operator is

$$\Pi(\mathbf{r},t) = \frac{\partial\mathcal{L}}{\partial\dot{\Psi}(\mathbf{r},t)} = i\overline{\Psi}\boldsymbol{\gamma}^{4} = i\mathbf{\Psi}^{\dagger}, \tag{37.254}$$

giving rise to the quantization rule which now, since we are dealing with the Dirac particle, involves anticommutators,

$$\{\Psi_i(\mathbf{r},t),\Pi_j(\mathbf{r}',t)\} = i\delta_{ij}\delta^{(3)}(\mathbf{r}-\mathbf{r}') \tag{37.255}$$

therefore,

$$\{\Psi_i(\mathbf{r},t), \mathbf{\Psi}_{\mathbf{j}}^{\dagger}(\mathbf{r}',t)\} = \boldsymbol{\delta}_{ij}\delta^{(3)}(\mathbf{r}-\mathbf{r}'). \tag{37.256}$$

Again, this result is reproduced through the anticommutation relations satisfied by the ladder operators.

Finally, as we discussed in Section 37.11, for the Maxwell field, we have the complication of satisfying Lorentz invariance and transversality conditions. As we pointed out these, then follow the formalism of Gupta and Bleuler when writing the canonical quantization rule

$$\left[A_{\mu}(\mathbf{r},t), \Pi_{\nu}(\mathbf{r}',t)\right] = -ig_{\mu\nu}\delta^{(3)}(\mathbf{r}-\mathbf{r}') \tag{37.257}$$

where the Lagrangian density and Π_{ν} are given by

$$\mathcal{L} = -\frac{1}{4}F_{\mu\nu}F^{\mu\nu} \tag{37.258}$$

and

$$\Pi^{\nu} = \frac{\partial \mathcal{L}}{\partial \dot{A}_{\nu}} = \dot{A}^{\nu}. \tag{37.259}$$

One can show from (37.193) that (37.257) is satisfied.

37.14 Casimir effect

As we learned from Section 37.10, the zero-point energy of the electromagnetic radiation is infinite. We asserted, however, that this quantity is unobservable and that what matters experimentally are the differences from this energy. The incredible successes of quantum field theory based on this assumption provide some basis for this assertion. But we can actually go closer to the core of the problem and do a quantitative calculation of such a difference and put the result to experimental tests. This is what the Casimir effect is concerned with. Here one measures the difference between the zero-point energy of the electromagnetic radiation in free space and that between two parallel conductors. This difference can be calculated, and it is found to be finite. It leads to a remarkable prediction, which is experimentally verified, that there is an attractive force between the conductors that is inversely proportional to the fourth power of the distance between them!

Let us start with the energy in free space, E_0, which is simply given by

$$E_0 = \sum_i \frac{1}{2}\hbar\omega_i = c\hbar \int_0^{\infty} dn_x \int_0^{\infty} dn_y \int_0^{\infty} dn_z \sqrt{k_x^2 + k_y^2 + k_z^2} \tag{37.260}$$

where we have taken account of the fact that there are two modes. We will take L as the length of the sides in the x- and y-directions but will select the z-axis as the special direction and take its dimension to be a, with $a \ll L$ (see Fig. 37.1). We will eventually take the limit $L \to \infty$ with a finite. Using boundary conditions typical of electromagnetism we have the relations

$$k_x L = n_x \pi, \quad k_y L = n_y \pi, \quad k_z a = n_z \pi. \tag{37.261}$$

Therefore,

$$E_0 = c\hbar \left(\frac{L}{2\pi}\right)^2 \left(\frac{a}{2\pi}\right) \int_{-\infty}^{\infty} dk_x \int_{-\infty}^{\infty} dk_y \int_{-\infty}^{\infty} dk_z \sqrt{k_x^2 + k_y^2 + k_z^2}. \tag{37.262}$$

We note that the integrals are divergent, which is what we expected.

Introducing spherical coordinates in the k-space with

$$\sqrt{k_x^2 + k_y^2 + k_z^2} = \kappa, \tag{37.263}$$

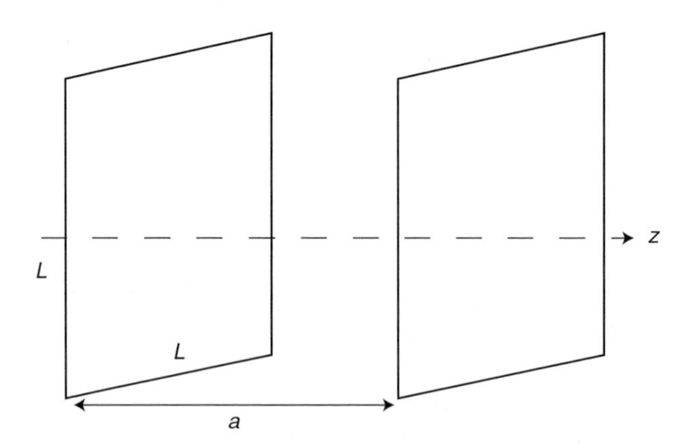

Fig. 37.1

E_0 can be expressed as

$$E_0 = c\hbar\left(\frac{L}{2\pi}\right)^2\left(\frac{a}{2\pi}\right)4\pi\int_0^\infty d\kappa\kappa^3, \tag{37.264}$$

which diverges at infinity as κ^4. To achieve a convergent result we introduce a simple exponential cut-off $\exp(-\lambda\kappa)$ so that

$$\int_0^\infty d\kappa\kappa^3 \to \int_0^\infty d\kappa\kappa^3 e^{-\lambda\kappa} = \frac{1}{\lambda^4}\int_0^\infty dy\, y^3 e^{-y} = \frac{6}{\lambda^4}. \tag{37.265}$$

The vacuum energy is then

$$E_0 = \frac{3c\hbar L^2 a}{\pi^2}\frac{1}{\lambda^4}. \tag{37.266}$$

Thus, the quadratic divergence in κ at infinity is reflected in the quadratic divergence in the limit $\lambda \to 0$. This is the vacuum energy inside the cube of dimensions L, L, and a.

Since we are treating the z-direction separately, we write

$$\frac{a}{2\pi}dk_x\, dk_y\, dk_z = d^2k\, dn_z \tag{37.267}$$

where $k = \sqrt{k_x^2 + k_y^2}$, and E_0 given by (37.262), without the cut-off, is written as

$$E_0 = c\hbar\left(\frac{L}{2\pi}\right)^2\int d^2k\int_0^\infty dn_z\sqrt{k^2 + k_z^2}. \tag{37.268}$$

Let us now consider the situation where two parallel perfect conductors of dimensions $L \times L$ are inserted at a distance a apart in the z-direction, replacing the two walls. Boundary conditions due to Maxwell's equation imply that only discrete numbers of states are allowed. Hence, we make the replacement

$$\int_0^\infty dn_z \rightarrow \sum_{n=0}^\infty. \tag{37.269}$$

It is found that only one mode is allowed for the ground-state energy with $n = 0$, but two for $n \geq 1$. We denote by E the energy of the electromagnetic radiation between the conducting planes and write

$$E = \frac{c\hbar}{2}\left(\frac{L}{2\pi}\right)^2 \int d^2k \left[k + 2\sum_{n=1}^\infty \sqrt{k^2 + \left(\frac{n\pi}{a}\right)^2}\right].$$

Since,

$$d^2k = 2\pi k\, dk, \tag{37.270}$$

we have

$$E = \left(\frac{c\hbar}{2}\right)\left(\frac{L^2}{2\pi}\right)\left[\int_0^\infty dk\, k^2 + 2\sum_{n=1}^\infty \int_0^\infty dk\, k\sqrt{k^2 + \left(\frac{n\pi}{a}\right)^2}\right]. \tag{37.271}$$

We will be interested in obtaining the energy difference $(E - E_0)$ per unit area L^2. First, however, let us convince ourselves that the infinities in the two energies do, indeed, cancel. This is to confirm that E has the same behavior as E_0 as $\lambda \rightarrow 0$. The first integral in (37.121) in terms of the cut-off is found to be

$$\int_0^\infty dk\, k^2 \rightarrow \int_0^\infty dk\, k^2 e^{-\lambda k} = \frac{1}{\lambda^3}\int_0^\infty dy\, y^2 e^{-y} = \frac{2}{\lambda^3}, \tag{37.272}$$

which goes like $1/\lambda^3$ rather than $1/\lambda^4$ of E_0. Let us take a look at the second integral in the presence of a cut-off. It is given by

$$\int_0^\infty dk\, k\sqrt{k^2 + \left(\frac{n\pi}{a}\right)^2} \rightarrow \int_0^\infty dk\, k\sqrt{k^2 + \left(\frac{n\pi}{a}\right)^2}\, e^{-\lambda\sqrt{k^2+(n\pi/a)^2}}. \tag{37.273}$$

We note that the cut-off functions for (37.264), (37.271), and (37.272) are the same if different values of k_z are taken into account. Making the change of variables

$$\sqrt{k^2 + \left(\frac{n\pi}{a}\right)^2} = y, \tag{37.274}$$

we find

$$\int_0^\infty dk\, k\sqrt{k^2+\left(\frac{n\pi}{a}\right)^2}e^{-\lambda\sqrt{k^2+\left(\frac{n\pi}{a}\right)^2}} = \int_{n\pi/a}^\infty dy\, y^2 e^{-\lambda y} \tag{37.275}$$

$$= \frac{d^2}{d\lambda^2}\int_{n\pi/a}^\infty dy\, e^{-\lambda y} = \frac{d^2}{d\lambda^2}\left[\frac{1}{\lambda}e^{-\lambda n\pi/a}\right]. \tag{37.276}$$

The second derivative in the final equality allows us to write the result more conveniently. This term by itself diverges like $1/\lambda$ and will not cancel the E_0 behavior. But we recall that we have an infinite sum to consider. It is found to be a geometric series that can be summed out quite simply to give

$$\sum_{n=1}^\infty e^{-\lambda n\pi/a} = e^{-\lambda\pi/a}\sum_{n=1}^\infty e^{-\lambda(n-1)\pi/a} = e^{-\lambda\pi/a}\sum_{m=0}^\infty e^{-\lambda m\pi/a} \tag{37.277}$$

$$= \frac{e^{-\lambda\pi/a}}{1-e^{-\lambda\pi/a}} = \frac{1}{e^{\lambda\pi/a}-1}. \tag{37.278}$$

Before embarking on taking the second derivative of this term as required by the relation (37.276), let us examine the behavior of this term as $\lambda \to 0$, since our purpose at the moment is to see if it reproduces the $1/\lambda^4$ behavior. From (37.277) we find

$$\sum_{n=1}^\infty e^{-\lambda n\pi/a} \underset{\lambda\to 0}{\to} \frac{a}{\pi\lambda}. \tag{37.279}$$

The second derivative in (37.276) is then

$$\frac{d^2}{d\lambda^2}\left[\frac{1}{\lambda}\frac{a}{\pi\lambda}\right] = \frac{6a}{\pi\lambda^4}. \tag{37.280}$$

Hence the most divergent part of E is found to be

$$E_{\text{divergent}} = \frac{3c\hbar L^2 a}{\pi^2}\frac{1}{\lambda^4}. \tag{37.281}$$

This, indeed, cancels E_0 as advertised.

To get the full result without using the approximation $\lambda \to 0$, we make use of the identity

$$\frac{z}{e^z-1} = \sum_{n=0}^\infty B_n\frac{z^n}{n!} \tag{37.282}$$

where B_n's are called Bernoulli's numbers. We can obtain these numbers if we multiply both sides by $(e^z - 1)$, expand e^z in an infinite power series, and compare the powers of z on both sides. A few of the values are given below:

$$B_0 = 1, \quad B_1 = -\frac{1}{2}, \quad B_2 = \frac{1}{6}, \quad B_4 = -\frac{1}{30}. \tag{37.283}$$

Using (37.282) in (37.283), one finds the following result for the first two leading terms:

$$E = \frac{3c\hbar L^2 a}{\pi^2}\frac{1}{\lambda^4} - \frac{1}{720}\frac{c\hbar L^2 \pi^3}{a^3}. \tag{37.284}$$

As expected from our previous calculation, the divergent parts of E and E_0 cancel. The next leading term in E is actually convergent.

We define the difference

$$\frac{\Delta E}{L^2} = \frac{E - E_0}{L^2}. \tag{37.285}$$

This is the only physically meaningful quantity that is independent of L^2 and it corresponds to the energy difference per unit surface area. We find

$$\frac{\Delta E}{L^2} = -\frac{1}{720}\frac{c\hbar \pi^3}{a^3}. \tag{37.286}$$

The force between the conductors is given by the change in energy per unit distance, $-\partial E/\partial a$. The force per unit surface area, F, is, therefore

$$F = -\frac{c\hbar \pi^2}{240 a^4}. \tag{37.287}$$

Thus, because of the negative sign above, there is an attractive force between the conducting planes. The value of F does not depend on the cut-off. This remarkable result is in a nutshell the Casimir effect. It has been experimentally verified. Different geometrical structures have been studied for the conductors and in each case experiment agrees with theory.

One might wonder whether the results we have derived depend on the specific form of the cut-off function we have chosen. This is found not to be so. If we take a general function $F(k)$ then as long as it is a smooth cut-off function it is found that the energy difference $(E - E_0)$ is essentially the difference between an integral and its trapezoidal approximation. This type of difference is treated through the so-called Euler–MacLaurin formalism involving standard mathematical techniques. Once again one encounters the Bernoulli numbers, leading to the same result we have just derived.

37.15 Problems

1. Two particles inside two infinite walls at $x = -a$ and $x = a$ interact with each other via the potentials

$$\text{(i)}\ V(x_1, x_2) = \lambda\delta(x_1 - x_2),$$
$$\text{(ii)}\ V(x_1, x_2) = \lambda x_1 x_2.$$

Determine the energy to the lowest order in each case.

2. Two identical spin $\frac{1}{2}$ particles are inside two infinite walls. Construct the wavefunctions for the two-particle system as a product of the space and spin wavefunctions. What is the result if the particles have spin zero or are not identical?
3. If two identical spin $\frac{1}{2}$ particles described in Problem 2 interact with each other through the spin-dependent potential $\lambda \boldsymbol{\sigma}_1 \cdot \boldsymbol{\sigma}_2$, obtain the energies of the two lowest states.
4. Consider two free particles that are identical, with common mass m. Assume that they are represented by wavefunctions $N \exp\left[-\alpha (x-a)^2\right]$ and $N \exp\left[-\alpha (x+a)^2\right]$, respectively. Determine the expectation value of the total energy with respect to individual wavefunctions. Call this energy E. Now construct normalized symmetric and unsymmetric wavefunctions for the two particles. Determine again the energy expectation values with respect to these new wavefunctions. Call them E_+ and E_-, respectively. Determine the differences $(E_\pm - E)$ and compare their magnitudes and signs.
5. Consider a system of two identical particles each with spin s. Determine the ratio of symmetric to antisymmetric spin states that one can form in this system. Show that these states together form a complete basis set for the two-particle wavefunctions.

38 Interactions of electrons and phonons in condensed matter

A condensed matter system in its most basic form can be viewed in terms of two components. One corresponds to electrons which, as discussed in Chapter 5, arrange themselves into energy levels as a consequence of Fermi statistics, with each level containing no more than two electrons with opposite spin orientations. These are the Fermi levels and the energy of the electrons in the topmost level is called the Fermi energy. Thus, in metals, where electrons can move about freely, even near absolute zero temperatures the electrons possess kinetic energy.

The second component corresponds to ions, or atomic lattices, the vibrational modes of which are described in terms of simple harmonic motion. The normal modes of these oscillations, when quantized, have particle-like properties and are called phonons. Since electrons contribute to the vibrations of the ions, the phonons interact with electrons, which in turn leads to interaction between electrons themselves through phonon exchange.

We elaborate on all of this below.

38.1 Fermi energy

38.1.1 One dimension

We repeat here the discussion in Chapter 5. Let us consider the ground state consisting of N noninteracting electrons confined in one dimension of length L. Each electron will be described by a free wavefunction

$$u(x) = \frac{1}{\sqrt{L}} e^{ikx} \chi_\lambda \tag{38.1}$$

where the χ_λ's designate the spin-up and spin-down states,

$$\chi_+ = \begin{bmatrix} 1 \\ 0 \end{bmatrix}, \quad \chi_- = \begin{bmatrix} 0 \\ 1 \end{bmatrix}, \tag{38.2}$$

which are normalized according to

$$\chi^\dagger_{\lambda_1} \chi_{\lambda_2} = \delta_{\lambda_1 \lambda_2}. \tag{38.3}$$

We assume, as we have done before, that the wavefunction satisfies the periodic boundary condition

$$u(x+L) = u(x), \tag{38.4}$$

which implies that the momentum vector can take only discrete values given by

$$k_n L = 2n\pi. \tag{38.5}$$

The energy eigenvalues will then be

$$E_n = \frac{\hbar^2 k_n^2}{2m} = \frac{2n^2\pi^2\hbar^2}{mL^2}. \tag{38.6}$$

Since the electrons are noninteracting, the ground state of the N-electron system can be built up by putting electrons into different levels. Because of the exclusion principle, however, at most one electron can be placed in each level with a given value of k_n. Since an electron has spin which can take on two values, spin up and spin down, we are allowed to put no more than two electrons in a given level as long as their spins are in opposite directions.

Thus N electrons will fill up levels with $n = 1, 2, \ldots, N/2$. The last level will consist of either one or two electrons depending on whether N is odd or even. The energy of the highest level can then be obtained by substituting $n = N/2$ in (38.6), which gives

$$E_F = \frac{N^2\pi^2\hbar^2}{2mL^2}. \tag{38.7}$$

This is the so called "Fermi energy" for the one-dimensional case. The Fermi energy is a very important concept in condensed matter systems. In metals, for example, even near absolute zero the electrons continue to have kinetic energy, with the highest value given by the Fermi energy. The so-called chemical potential is the same as Fermi energy at zero temperature.

The total energy for the N-electron system is then

$$E_{tot} = 2\sum_{n=1}^{N/2} \frac{2n^2\pi^2\hbar^2}{mL^2} \tag{38.8}$$

where the factor 2 corresponds to the two spin states. Since N is assumed to be very large, the above sum can be converted to an integral:

$$\sum_{n=1}^{N/2} n^2 \approx \int_1^{N/2} dn\, n^2 \simeq \frac{N^3}{24}. \tag{38.9}$$

Thus, the total energy of the electrons is

$$E_{tot} = \frac{N^3\pi^2\hbar^2}{6mL^2}. \tag{38.10}$$

The energy per electron is

$$E_e = \frac{E_{tot}}{N} = \frac{N^2\pi^2\hbar^2}{6mL^2}, \tag{38.11}$$

which implies that the energy of an individual electron increases as N^2.

Let us compare this result with a system consisting of N bosons. Since there is no exclusion principle preventing the bosons from occupying the same state, their ground state will consist of all N particles occupying the same state, $n = 1$. Thus the total energy of this system will be N times the ground-state energy. Hence,

$$E_{tot} = N\frac{2\pi^2\hbar^2}{mL^2} \tag{38.12}$$

and the energy of a single boson is

$$E_b = \frac{E_{tot}}{N} = \frac{\pi^2\hbar^2}{2mL^2}, \tag{38.13}$$

which remains a constant. This result is in sharp contrast to the case of electrons.

38.1.2 Three dimensions

We now go to the more realistic case of three dimensions. We consider a cube of length L, and volume, $V = L^3$, inside which the wavefunction is given by

$$\psi_{\mathbf{k}\lambda}(\mathbf{r}) = \frac{1}{\sqrt{V}}e^{i\mathbf{k}\cdot\mathbf{r}}\chi_\lambda. \tag{38.14}$$

It satisfies the periodic boundary conditions

$$k_i = \frac{2\pi n_i}{L}, \quad i = x, y, z, \quad n_i = \pm 1, \pm 2, \cdots \tag{38.15}$$

inside the cube. The χ_λ's, as before, designate the spin-up and spin down states.

We will discuss this problem in the language of second quantization that we plan to use throughout the rest of this chapter. We write the total kinetic energy, which is the unperturbed Hamiltonian, $\mathbf{H}_0$, as

$$\mathbf{H}_0 = \sum_{\mathbf{k}\lambda} \frac{\hbar^2k^2}{2m} a^\dagger_{\mathbf{k}\lambda}a_{\mathbf{k}\lambda} \tag{38.16}$$

where $a_{\mathbf{k}\lambda}$ and $a^\dagger_{\mathbf{k}\lambda}$ are the usual creation and destruction operators for the electrons with momentum $\mathbf{k}$ and spin λ. If $n_{\mathbf{k}\lambda}$ is the number operator then

$$a^\dagger_{\mathbf{k}\lambda}a_{\mathbf{k}\lambda} = n_{\mathbf{k}\lambda} \tag{38.17}$$

and

$$\mathbf{H}_0 = \sum_{\mathbf{k}\lambda} \frac{\hbar^2k^2}{2m} n_{\mathbf{k}\lambda}, \tag{38.18}$$

which implies a sum of the kinetic energy of each mode, characterized by **k** and λ, multiplied by the number of particles in that mode. If we denote by

$$|F\rangle \tag{38.19}$$

the multiparticle state corresponding to the ground state of the N electrons, then the energy eigenvalue E_0 is given by

$$E_0 = \langle F |\mathbf{H}_0| F\rangle = 2\int d^3n \frac{\hbar^2 k^2}{2m} \tag{38.20}$$

where d^3n is the number of states available and the factor 2 accounts for the two spin states for each **k**. From (38.15) we obtain d^3n as

$$d^3n = dn_x\, dn_y\, dn_z = \frac{V\, d^3k}{(2\pi)^3}. \tag{38.21}$$

Since

$$d^3k = 4\pi k^2\, dk \tag{38.22}$$

where 4π corresponds to the solid angle in the k-space, we can evaluate the integral in (38.20) to obtain the total energy of the electron system. We find

$$E_0 = \frac{V\hbar^2 k_F^5}{10\pi^2 m} \tag{38.23}$$

where the integration in k is carried out from 0 to k_F, the Fermi momentum, which is the maximum momentum value for the ground state.

We can write the above relation in terms of the total number of electrons, N. We note that

$$N = \int d^3n = \int \frac{V\, d^3k}{(2\pi)^3}, \tag{38.24}$$

which upon integration gives

$$N = \frac{Vk_F^3}{3\pi^2}. \tag{38.25}$$

Thus, E_0 can be expressed as

$$E_0 = \frac{3}{5}\left(\frac{\hbar^2 k_F^2}{2m}\right)N. \tag{38.26}$$

The energy per electron is

$$\frac{E_0}{N} = \frac{3}{5}E_F \tag{38.27}$$

where E_F is the Fermi energy, which is related to k_F by

$$E_F = \frac{\hbar^2 k_F^2}{2m}. \tag{38.28}$$

Substituting the value of k_F from relation (38.25), we obtain E_F in terms of N with $V = L^3$:

$$E_F = \frac{\pi^2 \hbar^2}{2mL^2} \left(\frac{3N}{\pi} \right)^{2/3}. \tag{38.29}$$

We can also write E_F in terms of the particle density

$$n = \frac{N}{V} \tag{38.30}$$

as

$$E_F = \frac{\hbar^2}{2m} \left(3\pi^2 n \right)^{2/3}. \tag{38.31}$$

In summary, we note that the ground state of the many-electron system at low temperatures consists of completely filled energy levels up to $E = E_F$, which is typically 5–10 eV at room temperatures. Excitations of the levels are difficult to achieve at low temperatures except near the Fermi surface: a particle inside the Fermi sea has nowhere to go since all available states are already occupied. These observations play an essential role in determining the properties of metals and other solids.

38.1.3 White dwarfs

White dwarfs are heavy stars having a mass comparable to the mass of the sun but with a radius which is 1/100 that of the sun. Such high densities imply that we can consider the white dwarfs as made of a degenerate electron gas. The particle density of electrons is of the order of 10^{27} electrons/cm^3. Plugging this number into (38.31) gives

$$E_F \approx 3 \times 10^5 \text{eV}. \tag{38.32}$$

38.1.4 Heavy nucleus

The heavy nucleus is another candidate to which the approximation of the degenerate Fermi gas can be applied. The radius of the nucleus is given by

$$R = 1.25 \times 10^{-12} A^{1/3} \text{cm} \tag{38.33}$$

where A is the number of nucleons (neutrons and protons), while the particle density is given by the usual relation

$$n = \frac{A}{(4\pi/3)\, R^3}. \tag{38.34}$$

Substituting all the numbers, we obtain

$$E_F \approx 3 \times 10^7 \text{eV}. \tag{38.35}$$

38.2 Interacting electron gas

In the previous section we were concerned with electrons that were noninteracting placed in an environment that was devoid of any other types of particles. We will now consider a somewhat more realistic situation of electrons in a metal. A metal in its simplest approximation is described as a system of free electrons in a background consisting of ions that are distributed uniformly with a continuous positive charge density. The total system is assumed to be neutral. The ions are assumed to be heavy and hence their motion is neglected. It is understood that if N is the number of electrons and V is the volume in which the whole system is confined, then our calculations will be meaningful in the so called thermodynamic limit

$$N \to \infty, \quad V \to \infty \quad \text{with} \quad n = \frac{N}{V} = \text{constant}. \tag{38.36}$$

The Coulomb potential plays an essential role in that it provides the interaction between electrons, between positive ions, and between electrons and the ions. Even though the potential is of infinite range, one introduces a screening radius μ^{-1} that makes the range finite and allows one to obtain convergent results. Thus, in our calculations we will make the change

$$\frac{1}{r} \to \frac{e^{-\mu r}}{r} \tag{38.37}$$

for the Coulomb potential. At the end of the calculation, however, we let $\mu \to 0$ consistent with the limit (38.36). We find that finite results are obtained.

We begin with the whole system confined in a cube of length L and volume $V = L^3$. The wavefunction of an electron is then given by

$$\psi_{\mathbf{k}\lambda}(\mathbf{r}) = \frac{1}{\sqrt{V}} e^{i\mathbf{k}\cdot\mathbf{r}} \chi_\lambda, \tag{38.38}$$

which is normalized such that it satisfies the periodic boundary conditions

$$k_t = \frac{2\pi n_i}{L}, \quad i = x, y, z, \quad n_i = 0, \pm 1, \pm 2, \cdots \tag{38.39}$$

inside the cube, while the χ_λ's designate the normalized spin-up and spin down states.

The Hamiltonian for the system is written as

$$H = H_{el} + H_{back} + H_{el-back} \tag{38.40}$$

where H_{el} corresponds to the electrons, interacting with each other through a screened Coulomb potential

$$H_{el} = \sum_{i=1}^{N} \frac{p_i^2}{2m} + \frac{1}{2} e^2 \sum_{i \neq j}^{N} \frac{e^{-\mu |\mathbf{r}_i - \mathbf{r}_j|}}{|\mathbf{r}_i - \mathbf{r}_j|} \tag{38.41}$$

where the summation is carried out over the N electron states. The Hamiltonian for the background, H_{back}, corresponding to the ions is written as

$$H_{back} = \frac{1}{2}e^2 \int \int d^3r\, d^3r' \frac{\rho_b(\mathbf{r})\rho_b(\mathbf{r}')e^{-\mu|\mathbf{r}-\mathbf{r}'|}}{|\mathbf{r} - \mathbf{r}'|} \tag{38.42}$$

where ρ_b is the density of the ions that are interacting with each other, also through the (repulsive) screened Coulomb potential. The interaction between the N electrons and the ions in the background is given by the attractive interaction

$$H_{el-back} = -e^2 \sum_{i=1}^{N} \int d^3r \frac{\rho_b(\mathbf{r})e^{-\mu|\mathbf{r}-\mathbf{r_i}|}}{|\mathbf{r} - \mathbf{r_i}|}. \tag{38.43}$$

We will make the approximation that the density of the ions is uniform and write

$$\rho_b(\mathbf{r}) \simeq \frac{N}{V}. \tag{38.44}$$

We will make a further approximation of replacing, for the screened potential,

$$\frac{e^{-\mu|\mathbf{r}_i-\mathbf{r}_j|}}{|\mathbf{r} - \mathbf{r}'|} \to \frac{4\pi}{\mu^2}\delta^{(3)}\left(\mathbf{r} - \mathbf{r}'\right) \tag{38.45}$$

for the purposes of simplifying the integration. It is easy to show that the integrals on both sides of (38.45) are the same. We will assume this to be generally true, as long as functions multiplying these terms are slowly varying.

Hence H_{back} is found to be

$$H_{back} = \frac{1}{2}e^2 \left(\frac{N}{V}\right)^2 \frac{4\pi}{\mu^2} \int d^3r. \tag{38.46}$$

The integral over d^3r simply gives the volume V. Thus,

$$H_{back} = \frac{1}{2}e^2 \frac{N^2}{V}\frac{4\pi}{\mu^2}. \tag{38.47}$$

For $H_{el-back}$ we, once again, make the approximation (38.45) in the integral in (38.43) to obtain

$$H_{el-back} = -e^2 \sum_{i=1}^{N} \frac{N}{V}\frac{4\pi}{\mu^2}. \tag{38.48}$$

The summation over the index i gives a factor N. Thus,

$$H_{el-back} = -e^2 \frac{N^2}{V}\frac{4\pi}{\mu^2}. \tag{38.49}$$

The sum of the two Hamiltonians is then given by

$$H_{back} + H_{el-back} = -\frac{1}{2}e^2 \frac{N^2}{V}\frac{4\pi}{\mu^2}. \tag{38.50}$$

We note that, in the language of second quantization, this is actually a c-number (i.e., proportional to a unit operator in the multiparticle Hilbert space) since N corresponds to the eigenvalue of the number operator $\mathbf{N}$.

38.2.1 Calculating H_{el}

Let us now obtain H_{el} from the formalism of second quantization. The first term of (38.41) is the kinetic energy term, which we have already discussed in the previous section. As before, we denote the unperturbed Hamiltonian corresponding to the total kinetic energy by $\mathbf{H}_0$,

$$\mathbf{H}_0 = \sum_{\mathbf{k}\lambda} \frac{\hbar^2 k^2}{2m} a^{\dagger}_{\mathbf{k}\lambda} a_{\mathbf{k}\lambda} \tag{38.51}$$

where $a_{\mathbf{k}\lambda}$ and $a^{\dagger}_{\mathbf{k}\lambda}$ are destruction and creation operators for electrons in the $\mathbf{k}, \lambda$ mode.

Our aim now is to evaluate the second term in (38.41), which corresponds to the Coulomb interaction between the electrons. We call this the perturbed Hamiltonian $\mathbf{H}'$. Let us calculate the matrix element of the screened potential. Since this is a two-body interaction, as explained in Chapter 37, we need to consider the product of the wavefunctions of the two particles designated by superscripts 1 and 2. Hence, the matrix element is given by

$$\frac{e^2}{V^2} \int\int d^3r_1 d^3r_2 \left[e^{-i\mathbf{k}_1\cdot\mathbf{r}_1} \chi^{(1)\dagger}_{\lambda_1}\right]\left[e^{-i\mathbf{k}_2\cdot\mathbf{x}_2} \chi^{(2)\dagger}_{\lambda_2}\right] \frac{e^{-\mu|\mathbf{r}_1-\mathbf{r}_2|}}{|\mathbf{r}_1-\mathbf{r}_2|} \left[e^{i\mathbf{k}_3\cdot\mathbf{r}_1} \chi^{(1)}_{\lambda_3}\right]\left[e^{i\mathbf{k}_4\cdot\mathbf{r}_2} \chi^{(2)}_{\lambda_4}\right]. \tag{38.52}$$

To simplify integration, we replace the variable $\mathbf{r}_1$ by writing

$$\mathbf{r}_1 - \mathbf{r}_2 = \mathbf{r} \tag{38.53}$$

but keep $\mathbf{r}_2 = \mathbf{r}_2$. Therefore,

$$d^3r_1 d^3r_2 = d^3r\, d^3r_2. \tag{38.54}$$

The integration can now be performed quite simply. We obtain

$$\begin{aligned}\mathbf{H}' &= \frac{e^2}{V^2}\delta_{\lambda_1\lambda_3}\delta_{\lambda_2\lambda_4} \int d^3r_2\, e^{-i(\mathbf{k}_1+\mathbf{k}_2-\mathbf{k}_3-\mathbf{k}_4)\cdot\mathbf{r}_2} \int d^3r\, e^{i(\mathbf{k}_3-\mathbf{k}_1)\cdot\mathbf{r}} \frac{e^{-\mu r}}{r} \\ &= \frac{e^2}{V}\delta_{\lambda_1\lambda_3}\delta_{\lambda_2\lambda_4}\delta_{\mathbf{k}_1+\mathbf{k}_2,\mathbf{k}_3+\mathbf{k}_4} \frac{4\pi}{(\mathbf{k}_1-\mathbf{k}_3)^2+\mu^2}.\end{aligned} \tag{38.55}$$

The Kronecker δ's imply $\lambda_1 = \lambda_3, \lambda_2 = \lambda_4$ and momentum conservation:

$$\mathbf{k}_1 + \mathbf{k}_2 = \mathbf{k}_3 + \mathbf{k}_4. \tag{38.56}$$

In deriving (38.55) we have used the following relations:

$$\chi^{(1)\dagger}_{\lambda_1}\chi^{(1)}_{\lambda_3} = \delta_{\lambda_1\lambda_3}, \quad \chi^{(2)\dagger}_{\lambda_2}\chi^{(2)}_{\lambda_4} = \delta_{\lambda_2\lambda_4}, \tag{38.57}$$

$$\frac{1}{V}\int d^3r_2\, e^{-i(\mathbf{k}_1+\mathbf{k}_2-\mathbf{k}_3-\mathbf{k}_4)\cdot\mathbf{r}_2} = \delta_{\mathbf{k}_1+\mathbf{k}_2,\mathbf{k}_3+\mathbf{k}_4}, \tag{38.58}$$

$$\int d^3r\, e^{i(\mathbf{k}_3-\mathbf{k}_1)\cdot\mathbf{r}}\frac{e^{-\mu r}}{r} = \frac{4\pi}{(\mathbf{k}_1-\mathbf{k}_3)^2+\mu^2}. \tag{38.59}$$

The integral in (38.59) has a screened potential that is, however, multiplying a rapidly oscillating function and does not satisfy the criterion for the approximation (38.45) (also, fortunately, we can integrate it exactly!).

The perturbed Hamiltonian given by the second term in (38.41) in the second quantized form is given by

$$\mathbf{H}' = \frac{e^2}{2V}\sum_{\mathbf{k}_i\lambda_i}\delta_{\lambda_1\lambda_3}\delta_{\lambda_2\lambda_4}\delta_{\mathbf{k}_1+\mathbf{k}_2,\mathbf{k}_3+\mathbf{k}_4}\frac{4\pi}{(\mathbf{k}_1-\mathbf{k}_3)^2+\mu^2}a^\dagger_{\mathbf{k}_1\lambda_1}a^\dagger_{\mathbf{k}_2\lambda_2}a_{\mathbf{k}_4\lambda_4}a_{\mathbf{k}_3\lambda_3} \tag{38.60}$$

where the index i in the summation goes over $i = 1, 2, 3, 4$. We change variables and write

$$\mathbf{k}_1-\mathbf{k}_3 = \mathbf{k}_4-\mathbf{k}_2 = \mathbf{q}, \quad \mathbf{k}_3 = \mathbf{k}, \quad \mathbf{k}_4 = \mathbf{p}. \tag{38.61}$$

We note that $\mathbf{q}$ corresponds to the momentum transfer. By summing over λ_3 and λ_4 and making use of the Kronecker δ's the interaction $\mathbf{H}$'s can be simplified to

$$\frac{e^2}{2V}\sum_{\mathbf{kpq}}\sum_{\lambda_1\lambda_2}\frac{4\pi}{q^2+\mu^2}a^\dagger_{\mathbf{k+q},\lambda_1}a^\dagger_{\mathbf{p-q},\lambda_2}a_{\mathbf{p}\lambda_2}a_{\mathbf{k}\lambda_1}. \tag{38.62}$$

This can be further simplified by separating the $q = 0$ and $q \neq 0$ parts

$$\frac{e^2}{2V}\sum_{\mathbf{kpq}}{}'\sum_{\lambda_1\lambda_2}\frac{4\pi}{q^2+\mu^2}a^\dagger_{\mathbf{k+q},\lambda_1}a^\dagger_{\mathbf{p-q},\lambda_2}a_{\mathbf{p},\lambda_2}a_{\mathbf{k}\lambda_1} \tag{38.63}$$

$$+\frac{e^2}{2V}\sum_{\mathbf{kp}}\sum_{\lambda_1\lambda_2}\frac{4\pi}{\mu^2}a^\dagger_{\mathbf{k}\lambda_1}a^\dagger_{\mathbf{p}\lambda_2}a_{\mathbf{p}\lambda_2}a_{\mathbf{k}\lambda_1}. \tag{38.64}$$

The prime in the summation sign in the first term implies that we exclude terms with $q = 0$. The second term contains only the $q = 0$ terms. To simplify the second term we now use the following anticommutator relations for the fermion operators,

$$\left\{a_{\mathbf{p}\lambda_2}, a_{\mathbf{k}\lambda_1}\right\} = 0 \quad \text{and} \quad \left\{a^\dagger_{\mathbf{p}\lambda_2}, a_{\mathbf{k}\lambda_1}\right\} = \delta_{\mathbf{kp}}\delta_{\lambda_1\lambda_2}, \tag{38.65}$$

and obtain

$$a^\dagger_{\mathbf{k}\lambda_1}a^\dagger_{\mathbf{p}\lambda_2}a_{\mathbf{p}\lambda_2}a_{\mathbf{k}\lambda_1} = -a^\dagger_{\mathbf{k}\lambda_1}a^\dagger_{\mathbf{p}\lambda_2}a_{\mathbf{k}\lambda_1}a_{\mathbf{p}\lambda_2} \tag{38.66}$$

$$a^\dagger_{\mathbf{p}\lambda_2}a_{\mathbf{k}\lambda_1} = \delta_{\mathbf{kp}}\delta_{\lambda_1\lambda_2} - a_{\mathbf{k}\lambda_1}a^\dagger_{\mathbf{p}\lambda_2}. \tag{38.67}$$

Substituting this in the second term in (38.64), we obtain

$$\frac{e^2}{2V}\frac{4\pi}{\mu^2}\sum_{\mathbf{k}\lambda_1}\sum_{\mathbf{p}\lambda_2}a^\dagger_{\mathbf{k}\lambda_1}a_{\mathbf{k}\lambda_1}\left(a^\dagger_{\mathbf{p}\lambda_2}a_{\mathbf{p}\lambda_2} - \delta_{\mathbf{kp}}\delta_{\lambda_1\lambda_2}\right) = \frac{e^2}{2V}\frac{4\pi}{\mu^2}(N^2-N) \tag{38.68}$$

where we have used the relations

$$a^{\dagger}_{\mathbf{k}\lambda_1} a_{\mathbf{k}\lambda_1} = n_{\mathbf{k}\lambda_1}, \qquad \sum_{\mathbf{k}\lambda_1} n_{\mathbf{k}\lambda_1} = \mathbf{N}, \tag{38.69}$$

$$a^{\dagger}_{\mathbf{p}\lambda_2} a_{\mathbf{p}\lambda_2} = n_{\mathbf{p}\lambda_2}, \qquad \sum_{\mathbf{p}\lambda_2} n_{\mathbf{p}\lambda_2} = \mathbf{N}, \tag{38.70}$$

and have replaced $\mathbf{N}$ by its eigenvalue N.

The first term on the right-hand side of (38.68), exactly cancels (38.50), while the second term corresponds to an energy, per particle, of

$$\frac{1}{N}\left[\frac{4\pi e^2 N}{2V\mu^2}\right] = \frac{4\pi e^2}{2V\mu^2}. \tag{38.71}$$

$V = L^3$, so $V\mu^2 = L\,(L\mu)^2$. We note that $1/\mu$ is the range of the shielded Coulomb interaction. Since we expect this range to be much smaller than L,

$$\frac{1}{\mu} \ll L, \tag{38.72}$$

in the limit $L \to \infty$ the denominator of (38.71) goes to infinity and, therefore, the energy goes to zero, the limit (38.72) is to be taken before the limit (38.36).

Hence we find that H_{back} and $H_{el-back}$ are eliminated and the total Hamiltonian written in the operator form is given by

$$\begin{aligned}\mathbf{H} &= \mathbf{H}_0 + \mathbf{H}' \\ &= \sum_{\mathbf{k}\lambda} \frac{\hbar^2 k^2}{2m} a^{\dagger}_{\mathbf{k}\lambda} a_{\mathbf{k}\lambda} + \frac{e^2}{2V} \sum_{\mathbf{kpq}}{}' \sum_{\lambda_1\lambda_2} \frac{4\pi}{q^2} a^{\dagger}_{\mathbf{k}+\mathbf{q},\lambda_1} a^{\dagger}_{\mathbf{p}-\mathbf{q},\lambda_2} a_{\mathbf{p}\lambda_2} a_{\mathbf{k}\lambda_1}\end{aligned} \tag{38.73}$$

where the sum in the momentum goes up to the Fermi momentum k_F.

The second term in the sum (38.73) for the total Hamiltonian can also be evaluated. This is the energy due to the Coulomb interaction between the electrons. If we call this E_1 then, after some lengthy calculations, one obtains

$$E_1 = -\frac{3e^2 N k_F}{4\pi}. \tag{38.74}$$

38.3 Phonons

We have considered the case of a solid that is assumed to consist of uniformly distributed ions of positive charge. These form a background to the moving electrons in the solid. Let us now consider the possibility that the ionic system undergoes small longitudinal vibrations. The individual displacements can then be described in terms of harmonic oscillators.

The collective vibrational motion involving many ions can be described in terms of normal modes, which are linear combinations of the displacements of the individual harmonic oscillators. The quantized normal modes are called phonons. They satisfy canonical commutation relations involving the field amplitudes, as we will discuss below. We note that only the longitudinal modes of the vibrations of the ion charge density are responsible for modifying the Coulomb interaction between the electrons and the background.

In Chapter 9 we studied the motion of a single harmonic oscillator based on the canonical commutation relation

$$[x, p] = i\hbar \tag{38.75}$$

where the Hamiltonian for a particle of mass m is given by

$$H = \frac{p^2}{2m} + \frac{1}{2}m\omega^2 x^2, \tag{38.76}$$

with ω as the natural frequency of the oscillation and x as the longitudinal displacement.

The following lowering and raising operators a and $a^\dagger$ were introduced in Chapter 9 to solve the harmonic oscillator problem:

$$a = \sqrt{\frac{m\omega}{2\hbar}}\left(x + i\frac{p}{m\omega}\right), \quad a^\dagger = \sqrt{\frac{m\omega}{2\hbar}}\left(x - i\frac{p}{m\omega}\right). \tag{38.77}$$

Based on the relation (38.75), these operators were found to satisfy the commutation relation

$$\left[a, a^\dagger\right] = \mathbf{1}. \tag{38.78}$$

From (38.76) and (38.77) we then obtained

$$H = \left(a^\dagger a + \frac{1}{2}\mathbf{1}\right)\hbar\omega. \tag{38.79}$$

Designating $|n\rangle$ as the eigenstate of H with eigenvalues E_n,

$$H\,|n\rangle = E_n\,|n\rangle\,, \tag{38.80}$$

we found

$$E_n = \left(n + \frac{1}{2}\right)\hbar\omega, \quad \text{for } n = 0, 1, 2, \ldots. \tag{38.81}$$

These are the energy levels of the harmonic oscillator.

To describe the collective motion of N ions undergoing oscillations, let us start with a simple model of an infinite one-dimensional chain of identical harmonic oscillators. We assume their equilibrium positions to be located equidistantly at la with $l = 0, 1, 2, \ldots$, where a is the unit distance. Let x_l be the displacement of the lth oscillator from its equilibrium position. The Hamiltonian for this system is then given by

$$H = \sum_{l=1}^{N}\frac{p_l^2}{2m} + \frac{1}{2}m\omega^2\sum_{l=1}^{N}(x_l - x_{l+1})^2 \tag{38.82}$$

where we have neglected the end effects. After moving the terms x_l^2 and x_{l+1}^2 from the second sum to the first sum, we rewrite the Hamiltonian as

$$H = \sum_{l=1}^{N} \left[\frac{p_l^2}{2m} + \frac{1}{2} m\omega^2 x_l^2 \right] - m\omega^2 \sum_{l=1}^{N} x_l x_{l+1}. \tag{38.83}$$

The first term describes a set of N independent oscillators, while the second in a sense corresponds to the interaction, or coupling, between an oscillator and its nearest neighbor. In order to obtain the energy eigenvalues of this system we will need to diagonalize this Hamiltonian and write it in terms of normal modes so that H can be expressed in terms of a new set of independent oscillators. These oscillators will correspond to quasiparticle states called phonons.

The classical equation of motion for the above Hamiltonian is given by the differential equation

$$\ddot{x}_l = -\omega^2 (2x_l - x_{l-1} - x_{l+1}). \tag{38.84}$$

The normal modes are obtained in the standard manner by writing

$$x_l = x_0 e^{ikal} e^{-i\omega_k t} \tag{38.85}$$

where k is the wave vector and ω_k the angular frequency. Expression (38.85) then corresponds to the displacement of the oscillator l at time t. We interpret this as the longitudinal vibration corresponding to sound waves of wave length $2\pi/k$ and frequency ω_k. The momentum p_l will have a functional form similar to (38.85).

Substituting (38.85) in (38.84), we find after canceling the common factors,

$$\omega_k^2 = \omega^2 [2 - e^{-ika} - e^{ika}]. \tag{38.86}$$

Hence the frequency, ω_k, of the vibrational normal mode is given by

$$\omega_k = 2\omega \sin\left(\frac{ka}{2}\right). \tag{38.87}$$

This equation relates the phonon frequency ω_k to the wavenumber k. It is referred to as a dispersion relation. The speed of phonon propagation, which is given by the group velocity $\partial\omega_k/\partial k$, is also the speed of sound in the lattice. At low energies ($ka \approx 0$) it is simply equal to ωa.

Let us now express the displacement, x_l, at $t = 0$ in the form of a Fourier transform

$$x_l = \frac{1}{\sqrt{N}} \sum_{k=1}^{N} e^{ikal} X_k, \tag{38.88}$$

with periodic boundary conditions. Its inverse is given by

$$X_k = \frac{1}{\sqrt{N}} \sum_{l=1}^{N} e^{-ikal} x_l \tag{38.89}$$

where we have used the orthogonality relation

$$\frac{1}{N}\sum_{l=1}^{N} e^{ial(k'-k)} = \delta_{k,k'}. \tag{38.90}$$

Let us now go to the quantum domain. Similar relations can be written for the momentum variable, p_l, and its Fourier transform P_k,

$$p_l = \frac{1}{\sqrt{N}}\sum_k e^{-ikal} P_k, \quad P_k = \frac{1}{\sqrt{N}}\sum_l e^{ikal} p_l. \tag{38.91}$$

Since each combination of canonical variables satisfies the commutation relation

$$[x_l, p_{l'}] = i\hbar\delta_{ll'} \tag{38.92}$$

we obtain

$$[X_k, P_{k'}] = \frac{1}{N}\sum_{l,l'} e^{-ikal} e^{ik'al'} [x_l, p_{l'}] \tag{38.93}$$

$$= \frac{i\hbar}{N}\sum_i e^{-ial(k-k')} = i\hbar\delta_{k,k'}. \tag{38.94}$$

Thus the amplitudes X_k and $P_{k'}$ themselves act as canonical operators.

To express (38.83) in terms of X_k and P_k, we note from relations (38.88) and (38.90) that

$$\sum_l x_l x_{l+m} = \frac{1}{N}\sum_{kk'} X_k X_{k'} \sum_l e^{ila(k+k')} e^{imak'} = \sum_k X_k X_{-k} e^{-iak}. \tag{38.95}$$

Thus, the second term in (38.82) is found to be

$$\frac{1}{2} m\omega^2 \sum_{l=1}^{N} (x_l - x_{l+1})^2 = \frac{1}{2} m\omega^2 \sum_k X_k X_{-k}(2 - e^{-iak} - e^{iak})$$

$$= \frac{1}{2} m \sum_k \omega_k^2 X_k X_{-k}. \tag{38.96}$$

Similarly,

$$\sum_l p_l^2 = \sum_k P_k P_{-k}. \tag{38.97}$$

The Hamiltonian for a system of N harmonic oscillators is now an operator given by

$$H = \frac{1}{2m}\sum_k P_k P_{-k} + \frac{m}{2}\sum_k \omega_k^2 X_k X_{-k}, \tag{38.98}$$

which corresponds to the expansion of the Hamiltonian in terms of normal modes. From (38.89) and (38.91) we note that $X_{-k} = X_k^\dagger$ and $P_{-k} = P_k^\dagger$.

We define the following destruction and creation operators:

$$b_k = \sqrt{\frac{m\omega_k}{2\hbar}}\left(X_k + \frac{i}{m\omega_k}P_{-k}\right), \tag{38.99}$$

$$b_k^\dagger = \sqrt{\frac{m\omega_k}{2\hbar}}\left(X_{-k} - \frac{i}{m\omega_k}P_k\right), \tag{38.100}$$

which from (38.92) and (38.94) obey the commutation relations

$$[b_k, b_{k'}^\dagger] = \delta_{kk'}, \quad [b_k, b_{k'}] = 0, \quad [b_k^\dagger, b_{k'}^\dagger] = 0. \tag{38.101}$$

These are the same type of commutation relations that the a's satisfy. The Hamiltonian is then found to be

$$H = \sum_{k=1}^{N} \hbar\omega_k \left(b_k^\dagger b_k + \frac{1}{2}\right). \tag{38.102}$$

Let

$$n_k = b_k^\dagger b_k. \tag{38.103}$$

Then

$$H = \sum_{k=1}^{N} \hbar\omega_k \left(n_k + \frac{1}{2}\right). \tag{38.104}$$

We write

$$H = E_0 + \sum_{k=1}^{N} n_k \hbar\omega_k \tag{38.105}$$

where

$$E_0 = \frac{1}{2}\sum_{k=1}^{N} \hbar\omega_k \tag{38.106}$$

is identified as the ground-state energy. The second term in (38.105) is identified as a sum corresponding to n_k elementary excitations, each with energy $\hbar\omega_k$, which, as we stated earlier, are called quasiparticles, or simply phonons.

What we have achieved by developing the above formalism is to express the total Hamiltonian in the language of second quantization, where n_k corresponds to the number of phonons of energy $\hbar\omega_k$. We can now express a state corresponding to the oscillating ionic system in a multiparticle Hilbert space by writing

$$|\psi\rangle = |n_1, n_2, \ldots n_k \ldots\rangle. \tag{38.107}$$

The operators b_k and $b_k^\dagger$ then act on this state as destruction and creation operators of these phonons. We note that (38.76) corresponds to the Hamiltonian in a single-particle Hilbert space and the commutation relations (38.75) and (38.78) represent first quantization. The expression (38.104) corresponds to the same Hamiltonian in a multiparticle Hilbert space, while the commutation relations (38.94) and (38.101) give rise to second quantization.

Just as the quantized electromagnetic field gives rise to photons that are the quanta of radiation, the vibrational modes in a crystal when quantized give rise to phonons as the quanta of sound waves. The photons are polarized in the transverse direction (in the transverse gauge); the phonons, on the other hand, are polarized in the longitudinal direction.

38.4 Electron–phonon interaction

In this section we will consider the case where the phonons generated by the ions of density $\rho_b(\mathbf{r})$ undergoing longitudinal vibrations interact with the electrons. As stated in the previous section, the Hamiltonian corresponding to the interaction of the ions and electrons is given by

$$H_{el-b} = e^2 \int d^3r' d^3r \frac{\rho_{el}(\mathbf{r}')\rho_b(\mathbf{r})e^{-\mu|\mathbf{r}_j'-r|}}{|\mathbf{r}' - \mathbf{r}|} \tag{38.108}$$

where we have assumed the electrons to be continuously distributed and described by a charge density ρ_{el}. The Coulomb interaction between the ions and the electrons is screened by a parameter μ. This is not an arbitrary quantity but a physically well-defined parameter that is inversely proportional to the lattice spacings of the ions. As we will discuss below, this situation will be different from our previous considerations in that we will find the final results to depend on this parameter.

Once again, we can replace

$$\frac{e^{-\mu|\mathbf{r}'-\mathbf{r}|}}{|\mathbf{r}' - \mathbf{r}|} \rightarrow \frac{4\pi}{\mu^2}\delta^{(3)}\left(\mathbf{r}' - \mathbf{r}\right) \tag{38.109}$$

and obtain

$$H_{el-b} = \lambda \int d^3r \rho_{el}(\mathbf{r})\rho_b(\mathbf{r}) \tag{38.110}$$

where λ is related to e and μ.

To take account of the changes in the ion density due to vibrations we write the displaced coordinates of an ion as

$$\mathbf{r}' = \mathbf{r} + \mathbf{D}(\mathbf{r}) \tag{38.111}$$

where $\mathbf{D}(\mathbf{r})$ corresponds to the displacement. Taking the x-component, we have

$$x' = x + D_x(x, y, z). \tag{38.112}$$

Therefore, the infinitesimal displacement along the x-axis is given by

$$dx' = dx + \frac{\partial D_x}{\partial x}dx = \left(1 + \frac{\partial D_x}{\partial x}\right)dx. \tag{38.113}$$

Similar results will be obtained for the displacements along the y- and z-axes. The volume element given by

$$d^3r = dx\, dy\, dz \tag{38.114}$$

changes due to these displacements so that, to first order,

$$\rho_b(\mathbf{r})\, d^3r \to \rho_b(\mathbf{r})\, d^3r = \rho_b(\mathbf{r})\,(1 + \nabla \cdot \mathbf{D})\; d^3r = [\rho_b(\mathbf{r}) + \delta\rho_b(\mathbf{r})]\, d^3r \qquad (38.115)$$

where in the last term we have written

$$\rho_b(\mathbf{r})\nabla \cdot \mathbf{D} = \delta\rho_b(\mathbf{r}). \qquad (38.116)$$

This corresponds to the changes in the density due to the vibrational fluctuations. Separating the contributions to H_{el-b} coming from the first and second terms on the right-hand side of (38.110), we write.

$$H_{el-b} = H^0_{el-b} + \lambda \int d^3\, r\rho_{el}(\mathbf{r})\delta\rho_b(\mathbf{r}) \qquad (38.117)$$

where H^0_{el-b} contains the contribution of $\rho_b(\mathbf{r})$. This contribution has already been discussed previously and is considered as an unperturbed Hamiltonian in this problem. For the second term we write

$$\delta\rho_b(\mathbf{r}) = \phi(\mathbf{r}). \qquad (38.118)$$

The electron density can be expressed in terms of the electron wavefunction $\psi(\mathbf{r})$ by the usual relation

$$\rho_{el}(\mathbf{r}) = \psi^\dagger(\mathbf{r})\psi(\mathbf{r}). \qquad (38.119)$$

We write

$$H_{el-b} = H^0_{el-b} + H_{el-phonon}. \qquad (38.120)$$

Then

$$H_{el-phonon} = \lambda \int d^3\, r\, \psi^\dagger(\mathbf{r})\psi(\mathbf{r})\phi(\mathbf{r}). \qquad (38.121)$$

This has the same form as the Hamiltonian for the fermion-scalar boson interaction.

At this stage, instead of invoking the full machinery of second quantization, we use straightforward quantum mechanics to arrive at the results that will correspond to electron–electron interaction through phonon exchange. In order to understand the consequences of the phonon interactions, let us first consider a very similar problem of photon exchange, that is, the Coulomb interaction, which we have already studied in Chapter 32.

38.4.1 Photon exchange

We consider the Coulomb problem involving the interaction between two electron charge densities described by $\rho^{(1)}_{el}(\mathbf{r})$ and $\rho^{(2)}_{el}(\mathbf{r})$. The interaction Hamiltonian corresponding to $\rho^{(1)}_{el}(\mathbf{r})$ can be written in the same form as (38.110),

$$H_{el-photon} = e \int d^3r\rho^{(1)}_{el}(\mathbf{r})\phi_e(\mathbf{r}) \qquad (38.122)$$

where $\phi_e(\mathbf{r})$ is the electrostatic potential whose source is $\rho^{(2)}_{el}(\mathbf{r})$. Thus, $\phi_e(\mathbf{r})$ satisfies the well-known equation

$$\nabla^2\phi_e(\mathbf{r}) = -4\pi e\rho^{(2)}_{el}(\mathbf{r}), \qquad (38.123)$$

whose solution can be written in the Green's function formalism as

$$\phi_e(\mathbf{r}) = \phi_0(\mathbf{r}) - 4\pi e \int d^3 r' G\left(\mathbf{r}, \mathbf{r}'\right) \rho_{el}^{(2)}(\mathbf{r}') \tag{38.124}$$

where $\phi_0(\mathbf{r})$ is the homogeneous solution and G is the Green's function that satisfies the equation

$$\nabla^2 G\left(\mathbf{r}, \mathbf{r}'\right) = \delta^{(3)}\left(\mathbf{r} - \mathbf{r}'\right). \tag{38.125}$$

We are interested in the second term of (38.124), which corresponds to the Coulomb interaction.

Equation (38.125) has been solved previously. We write it in the form of a Fourier transform:

$$G\left(\mathbf{r}, \mathbf{r}'\right) = \frac{1}{(\sqrt{2\pi})^3} \int d^3 k\, g(\mathbf{k}) e^{i\mathbf{k}\cdot(\mathbf{r}-\mathbf{r}')}. \tag{38.126}$$

We leave the Green's function in this form and insert it on the right-hand side of (38.122)

$$\int d^3 r \rho_{el}^{(1)}(\mathbf{r})\phi_e(\mathbf{r}) = -\frac{4\pi e}{(\sqrt{2\pi})^3} \int d^3 k \left[\int d^3 r \rho_{el}^{(1)}(\mathbf{r}) e^{i\mathbf{k}\cdot\mathbf{r}}\right] g(\mathbf{k}) \left[\int d^3 r' \rho_{el}^{(2)}(\mathbf{r}') e^{-i\mathbf{k}\cdot\mathbf{r}'}\right]. \tag{38.127}$$

The terms in the square brackets are the form factors, which we discussed in Chapter 20. Taking account of the fact that the charge density, $\rho(\mathbf{r})\ (= \psi^*\psi)$, is positive definite, we can designate the two form factors as $F^{(1)}(\mathbf{k})$ and $F^{(2)*}(\mathbf{k})$, respectively. Thus, the interaction Hamiltonian for Coulomb interaction can be written as

$$H_{el-photon} = -\frac{4\pi e^2}{(\sqrt{2\pi})^3} \int d^3 k\, F^{(1)}(\mathbf{k}) g(\mathbf{k}) F^{(2)*}(\mathbf{k}). \tag{38.128}$$

We can now obtain $g(\mathbf{k})$ from equations (38.125) and (38.126), taking note of the fact that

$$\delta^{(3)}\left(\mathbf{r} - \mathbf{r}'\right) = \frac{1}{(2\pi)^3} \int d^3 k\, e^{i(\mathbf{k}-\mathbf{k}')\cdot\mathbf{r}}. \tag{38.129}$$

We find

$$g(\mathbf{k}) = -\frac{1}{(\sqrt{2\pi})^3} \frac{1}{k^2}. \tag{38.130}$$

This is the Fourier transform of the Coulomb potential, which we have already considered in this and in previous chapters. Since, as we stated earlier, the Coulomb interaction is screened, we obtain, through our previous methods,

$$g(\mathbf{k}) = -\frac{1}{(\sqrt{2\pi})^3} \frac{1}{k^2 + \mu^2} \tag{38.131}$$

where μ is the screening parameter. Hence we obtain

$$H_{el-photon} = \frac{4\pi e^2}{(2\pi)^3} \int d^3 k\, F^{(1)}(\mathbf{k}) \left[\frac{1}{k^2 + \mu^2}\right] F^{(2)*}(\mathbf{k}). \tag{38.132}$$

We identify the factor inside the square brackets as $V_{photon}(ee)$, which is in fact the Fourier transform of the Coulomb potential,

$$V_{photon}(ee) = \frac{4\pi e^2}{k^2 + \mu^2}. \tag{38.133}$$

38.4.2 Phonon exchange

Let us now return to the electron–phonon interaction given by (38.121) which we write as

$$H_{el-phonon} = g \int d^3r\, \rho_{el}(\mathbf{r})\phi_{ph}(\mathbf{r}). \tag{38.134}$$

Here we point out that the attractive Coulomb interaction between the electrons and the positively charged ions generates the longitudinal vibrations in the ionic system, whose collective modes appear as phonons. Also we note that $\phi_{ph}(\mathbf{r})$ is the divergence of the displacement of the charge density of the ions. The source of this displacement is ρ_{el}. The time-dependent equation of motion of $\phi_{ph}(\mathbf{r}, \mathbf{t})$ is given by the wave equation

$$\frac{1}{v_s^2}\frac{\partial^2\phi_{ph}(\mathbf{r}, \mathbf{t})}{\partial t^2} - \nabla^2\phi_{ph}(\mathbf{r}, \mathbf{t}) = -\lambda\rho_{el}(\mathbf{r}, \mathbf{t}) \tag{38.135}$$

where v_s is the velocity of sound, and λ is a parameter that is proportional to e. If we describe the time dependence of the functions appearing in equation (38.135) by an oscillating function $\exp(-i\omega t)$ then, after differentiating and factoring out this term, the equation becomes

$$\frac{\omega^2}{v_s^2}\phi_{ph}(\mathbf{r}) + \nabla^2\phi_{ph}(\mathbf{r}) = \lambda\rho_{el}(\mathbf{r}) \tag{38.136}$$

where the functions now depend only on $\mathbf{r}$. We can write this, once again as we did for the photon exchange, in the Green's function form to obtain, ignoring the homogeneous solution,

$$\phi_{ph}(\mathbf{r}) = \lambda \int d^3r'\, G\left(\mathbf{r}, \mathbf{r}'\right) \rho_{el}(\mathbf{r}) \tag{38.137}$$

where $G\left(\mathbf{r}, \mathbf{r}'\right)$ now satisfies the equation

$$\left(\nabla^2 + \frac{\omega^2}{v_s^2}\right) G\left(\mathbf{r}, \mathbf{r}'\right) = \delta^{(3)}\left(\mathbf{r} - \mathbf{r}'\right). \tag{38.138}$$

We write

$$G\left(\mathbf{r}, \mathbf{r}'\right) = \frac{1}{(\sqrt{2\pi})^3} \int d^3k\, g(\mathbf{k}) e^{i\mathbf{k}\cdot(\mathbf{r}-\mathbf{r}')}. \tag{38.139}$$

The interaction Hamiltonian becomes

$$H_{el-phonon} = \frac{g\lambda}{(\sqrt{2\pi})^3} \int d^3k\, F(\mathbf{k}) g(\mathbf{k}) F^*(\mathbf{k}). \tag{38.140}$$

The Fourier transform $g(\mathbf{k})$ can be easily obtained from (38.138) and (38.139). We find

$$g(\mathbf{k}) = \frac{1}{(\sqrt{2\pi})^3}\left(\frac{v_s^2}{\omega^2 - v_s^2 k^2}\right). \tag{38.141}$$

We write $\omega_k = v_s k$ and insert $g(\mathbf{k})$ in (38.140). The Hamiltonian now reads

$$H_{el-phonon} = \frac{g\lambda v_s^2}{(2\pi)^3}\int d^3k\, F(\mathbf{k})\left(\frac{1}{\omega^2 - \omega_k^2}\right) F^*(\mathbf{k}). \tag{38.142}$$

We identify the factor sandwiched between the two form factor terms in (38.142) as due to phonon exchange between the two electrons,

$$\frac{g\lambda v_s^2}{\omega^2 - \omega_k^2}\theta\left(\omega_D - \omega_k\right). \tag{38.143}$$

We note that in (38.143) only the values $\omega_k \leq \omega_D$, the Debye frequency (which is the maximum frequency of vibration for the atoms that make up a crystal) appear since electrons near the Fermi surface are the only ones that participate. To obtain the interaction potential we go to the static limit, $\omega \approx 0$, as we did in the case of Coulomb and Yukawa interactions in Chapter 32. Since the Debye frequency ω_D is $\approx v_s k_F$, the presence of $\theta(\omega_D - \omega_k)$ implies that the momentum transfers $|\mathbf{k}|$ have to be less than or of the order of the Fermi momentum, k_F. As long as we are in this region, then (38.143) is a negative constant. Thus, in contrast to the Coulomb case the Fourier transform of the interaction potential between the electrons through phonon–exchange is then of the form

$$V_{phonon}(ee) = -\lambda_{eff}\,\theta(\omega_D - \omega_k) \tag{38.144}$$

with $\lambda_{eff}\left(= g\lambda v_s^2/\omega_k^2\right) > 0$. Therefore, the interaction between the electrons is attractive.

Qualitatively, this attraction arises because an electron in its motion pulls on a positively charged ion leading to vibrations and hence to the phonons. In contrast to the electron's typical time period of passage, which is $O(E_F^{-1})$, the time period for the ion to go back to its equilibrium position is $O(\omega_D^{-1})$, which is generally much larger than E_F^{-1}. Hence before the ion goes back fully to its equilibrium position a second electron comes along and pulls

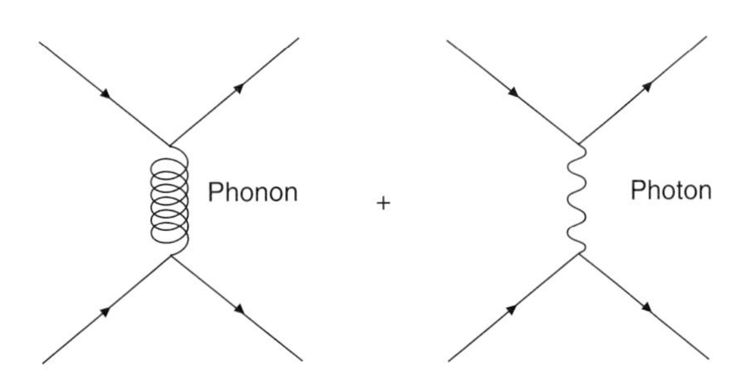

Fig. 38.1

on the ion creating, as a second-order effect, a net attraction between the first and the second electrons.

Thus an electron pair in a Fermi sea undergoes two types of interactions, as depicted in Fig. 38.1: (i) the normal Coulomb interaction, which is repulsive, and (ii) an interaction through the phonon exchange, which is attractive but which is effective only near the Fermi surface. This implies that we have

$$V_{total}(ee) = \frac{4\pi e^2}{k^2 + \mu^2} - \lambda_{eff}\,\theta(\omega_D - \omega_k). \tag{38.145}$$

It is found that in certain metals, near the Fermi surface, the sum of the above two terms is negative, which creates a net attractive force between the electrons and results in the phenomena of Cooper pairing and superconductivity.

39 Superconductivity

An electric current in a normal conductor can be thought of as a fluid made up of electrons flowing across lattices made up of heavy ions and constantly colliding with them. The kinetic energy of the electrons decreases with each collision, effectively being converted into the vibrational energy of the ions. This dissipation of energy then corresponds to electrical resistivity. It is found that the resistivity decreases as the temperature is decreased but it never completely vanishes even at absolute zero.

In a conventional superconductor, however, the electrons occur in pairs, called Cooper pairs, because of the attractive force generated by the exchange of phonons. If one looks at the energy spectrum of these pairs, there is an energy gap that is the minimum of energy needed to excite the pair. If the thermal energy (kT) of the electrons is less than the gap energy, then the Cooper pairs will act as individual entities and travel without undergoing any scattering with the ions. Therefore, there will be no resistivity. Thus, in a superconductor the resistance drops abruptly to zero below a certain temperature, called the "critical temperature." An electric current flowing in a loop of wire consisting of a superconductor then flows indefinitely with no resistance and without the help of any power source. Below, we briefly describe the mechanism that gives rise to this superconductivity.

39.1 Many-body system of half-integer spins

We consider a many-body system consisting of identical fermions that group themselves in pairs like quasiparticles where each pair consists of electrons that are degenerate in energy but have opposite linear momenta, $\mathbf{p}$ and $-\mathbf{p}$, as well as opposite spin directions. For electrons in a solid, such pairs are observed and are called Cooper pairs. The interaction between the two electrons in a pair is assumed to depend only on the distance between them. A typical matrix element of the potential, V_0, for this pairing interaction can then be written as the Fourier transform,

$$\frac{1}{2}a_l^\dagger a_{-l}^\dagger a_{-k}a_k \, \langle l, -l \, | V_0 | \, k, -k \rangle \tag{39.1}$$

where k and l each designate the combined quantum numbers for momentum and spin. For example, by k we will mean momentum $\mathbf{k}$ and spin up, while $-k$ will correspond to the same quantum numbers but with opposite signs. In (39.1) momentum conservation is explicitly taken into account.

Here $a_i^\dagger$ and a_i are the creation and destruction operators, respectively, for the electrons and satisfy the commutation relations

$$\{a_i, a_j\} = \left\{a_i^\dagger, a_j^\dagger\right\} = 0; \quad \left\{a_i, a_j^\dagger\right\} = \delta_{ij}\mathbf{1}. \tag{39.2}$$

In the case of superconductivity there is an attractive interaction between the electrons that form a pair (e.g., a Cooper pair) which is effective only near the surface of the Fermi sphere in a small region δ in the k-space as indicated below. The effective Hamiltonian for the system can be written as

$$H = \sum_{k>0} \epsilon(k)\left(a_k^\dagger a_k + a_{-k}^\dagger a_{-k}\right) - \sum_{\delta} G_{kl} a_l^\dagger a_{-l}^\dagger a_{-k} a_k. \tag{39.3}$$

The first term corresponds to noninteracting fermions. Because of the degeneracy between the states with k and $-k$ we have taken out the energy term $\epsilon(k)$ as a common factor in the first term in (39.3), which is then summed over $k > 0$. We note that the number operator, N, does not commute with H. To keep the number of particles stable we introduce the chemical potential, μ, in the first term in H and define

$$\epsilon(k) = \epsilon_k - \mu \tag{39.4}$$

where $\epsilon_k = k^2/2m$ is the unperturbed energy of an individual electron. The chemical potential is then determined from the condition that the total number of particles is given by $N = \left\langle \sum \left(a_k^\dagger a_k + a_{-k}^\dagger a_{-k}\right)\right\rangle$. In the absence of interactions, all states with $\epsilon_k < \epsilon_F$, the Fermi energy, are occupied while the states with $\epsilon_k > \epsilon_F$ are empty, so in that case $\mu = \epsilon_F$.

The second term in (39.3) is the pairing interaction, which is effective only near the Fermi surface and thus the sum runs only in the limited region designated by δ, where we have assumed the potential to be a constant. The matrix element G_{kl} is proportional to the matrix element of the potential in (39.1) and the sign of this term is chosen to be negative to indicate attraction for positive values of G_{kl}.

We will obtain below the eigenstates and eigenvalues of H through nonperturbative methods. These states are called BCS states and are fundamental to an understanding of superconductivity observed in certain solids.

Anticipating that we will be considering the system in terms of quasiparticles formed by pairs that are bound together, we make a simplifying assumption called the mean-field approximation by which, at low energies, the fluctuations of the bilinear operators $a_{-l}a_l$ from their expectation values $\langle a_{-l}a_l\rangle$ in the quasiparticle ground state (to be defined later) are minimal. This will allow us to reduce the problem from a four-body interaction to a two-body interaction within each pair. We also neglect the interaction between quasiparticles. We denote

$$\langle a_{-k}a_k\rangle = C_k \tag{39.5}$$

where we assume C_k to be real, and also equal to $\left\langle a_k^\dagger a_{-k}^\dagger\right\rangle$. We leave the precise definition of the expectation values vague at the moment. Our approximation implies that

$$\left(a_l^\dagger a_{-l}^\dagger - C_l\right)(a_{-k}a_k - C_k) \approx 0 \tag{39.6}$$

and, therefore,

$$a_l^\dagger a_{-l}^\dagger a_{-k} a_k \approx -C_l C_k + C_k a_l^\dagger a_{-l}^\dagger + C_l a_{-k} a_k. \tag{39.7}$$

We also define the quantity Δ_k as follows:

$$\Delta_k = \sum_l G_{kl} C_l. \tag{39.8}$$

The Hamiltonian in (39.3) can be expressed in the following approximate form using (39.6), (39.7), and (39.8):

$$H = \sum_{k>0} \epsilon(k) \left(a_k^\dagger a_k + a_{-k}^\dagger a_{-k} \right) + \sum_{k \in \delta} \Delta_k C_k - \sum_{k \in \delta} \Delta_k \left(a_{-k} a_k + a_k^\dagger a_{-k}^\dagger \right). \tag{39.9}$$

To obtain the eigenvalues of H we make a canonical transformation, called the Bogoliubov transformation, of the creation and destruction operators so that the new Hamiltonian will correspond to a system of noninteracting particles or quasiparticles. In essence, we want to "diagonalize" H in a manner similar to that in the two-level problems we considered earlier. To that effect we introduce the operators α_k and β_k as follows:

$$\alpha_k = u_k a_k - v_k a_{-k}^\dagger, \tag{39.10}$$

$$\beta_k = u_k a_{-k} + v_k a_k^\dagger, \tag{39.11}$$

where u_k and v_k are real quantities. We find that the operators α_i and β_i will satisfy the usual anticommutation relations that a_i's satisfy in (39.2) provided

$$u_k^2 + v_k^2 = 1. \tag{39.12}$$

An important fact to keep in mind is that the operators α_k and β_k defined in (39.10) are linear combinations of creation and destruction operators and, as a consequence, they do not conserve particle number. They are more appropriate in connection with states with quasiparticles. Let us invert the relations (39.10) and (39.11) to obtain the following, using (39.12),

$$a_k = u_k \alpha_k + v_k \beta_k^\dagger, \tag{39.13}$$

$$a_{-k} = u_k \beta_k - v_k \alpha_k^\dagger. \tag{39.14}$$

The vacuum state, $|0\rangle$, for particles, is defined by

$$a_i |0\rangle = 0. \tag{39.15}$$

We define the quasiparticle vacuum state, or quasiparticle ground state, by $|0\rangle_{quasi}$ such that

$$\alpha_i |0\rangle_{quasi} = \beta_i |0\rangle_{quasi} = 0. \tag{39.16}$$

Thus, one can consider α_i and β_i as quasiparticle destruction operators, while $\alpha_i^\dagger$ and $\beta_i^\dagger$ are the corresponding creation operators. The two vacuum states $|0\rangle$ and $|0\rangle_{quasi}$ and the relation between them, consistent with the conditions (39.13) and (39.14), are described in (39.17):

$$|0\rangle_{quasi} = \prod_k \left(u_k + v_k a_k^\dagger a_{-k}^\dagger\right)|0\rangle, \quad |0\rangle = \prod_k \left(u_k + v_k \beta_k^\dagger \alpha_k^\dagger\right)|0\rangle_{quasi}. \tag{39.17}$$

We note that the definition of $|0\rangle_{quasi}$ involves the product, $a_k^\dagger a_{-k}^\dagger$, corresponding to the creation of a pair of particles with opposite quantum numbers.

We can now give a precise expression for C_l defined in (39.5) as follows:

$$C_l = \langle a_{-l}a_l\rangle_{quasi} = \prod_{k'}\prod_k \langle 0|\left(u_{k'} + v_{k'}a_{-k'}a_{k'}\right)(a_{-l}a_l)\left(u_k + v_k a_k^\dagger a_{-k}^\dagger\right)|0\rangle = u_l v_l. \tag{39.18}$$

Hence, Δ_k, defined in (39.8) will be given by

$$\Delta_k = \sum_l G_{kl}u_l v_l. \tag{39.19}$$

The first and the third terms in (39.9) for the Hamiltonian can be expressed in terms of the new operators as follows:

$$a_k^\dagger a_k + a_{-k}^\dagger a_{-k} = 2v_k^2 + \left(u_k^2 - v_k^2\right)\left(\alpha_k^\dagger\alpha_k + \beta_k^\dagger\beta_k\right) + 2u_k v_k\left(\alpha_k^\dagger\beta_k^\dagger + \beta_k\alpha_k\right), \tag{39.20}$$

$$a_{-k}a_k + a_k^\dagger a_{-k}^\dagger = 2u_k v_k - 2u_k v_k\left(\alpha_k^\dagger\alpha_k + \beta_k^\dagger\beta_k\right) + \left(u_k^2 - v_k^2\right)\left(\alpha_k^\dagger\beta_k^\dagger + \beta_k\alpha_k\right). \tag{39.21}$$

Substituting (39.20) and (39.21) in (39.9), the Hamiltonian can be separated into three types of terms as indicated below:

$$\text{Unit operator (}c\text{ number)} \rightarrow 2\sum_k \epsilon(k)v_k^2 - \sum_{k\in\delta}\Delta_k u_k v_k. \tag{39.22}$$

$$\text{Diagonal operator} \rightarrow \left[\sum_k \epsilon(k)\left(u_k^2 - v_k^2\right) + 2\sum_{k\in\delta}\Delta_k u_k v_k\right]\left(\alpha_k^\dagger\alpha_k + \beta_k^\dagger\beta_k\right). \tag{39.23}$$

$$\text{Nondiagonal operator} \rightarrow \left[\sum_k 2\epsilon(k)u_k v_k - \sum_{k\in\delta}\Delta_k\left(u_k^2 - v_k^2\right)\right]\left(\alpha_k^\dagger\beta_k^\dagger + \beta_k\alpha_k\right). \tag{39.24}$$

Expression (39.22) is a constant term that does not contain creation or destruction operators. These types of terms are often referred to as c-numbers. In this case they correspond to the ground state of the new system, while (39.23) corresponds to the excited states of the system. The third term, given by (39.24), is the nondiagonal part that we wish to remove. To accomplish this we impose the condition

$$2\epsilon(k)u_k v_k - \Delta_k\left(u_k^2 - v_k^2\right) = 0. \tag{39.25}$$

If we take

$$x_k = \frac{v_k}{u_k} \tag{39.26}$$

then (39.25) becomes a quadratic equation in x_k. The positive solution of equation (39.25) is given by

$$x_k = \frac{E_k - \epsilon(k)}{\Delta_k} \tag{39.27}$$

where

$$E_k = \sqrt{\epsilon^2(k) + \Delta_k^2}. \tag{39.28}$$

Since $\Delta_k^2 = E_k^2 - \epsilon^2(k)$, we find

$$x_k^2 = \frac{(E_k - \epsilon(k))^2}{E_k^2 - \epsilon^2(k)} = \frac{E_k - \epsilon(k)}{E_k + \epsilon(k)}. \tag{39.29}$$

Hence,

$$u_k^2 = \frac{1}{2}\left(1 + \frac{\epsilon(k)}{E_k}\right) = \frac{1}{2}\left(1 + \frac{\epsilon(k)}{\sqrt{\epsilon^2(k) + \Delta_k^2}}\right), \tag{39.30}$$

$$v_k^2 = \frac{1}{2}\left(1 - \frac{\epsilon(k)}{E_k}\right) = \frac{1}{2}\left(1 - \frac{\epsilon(k)}{\sqrt{\epsilon^2(k) + \Delta_k^2}}\right). \tag{39.31}$$

From (39.28), (39.30), (39.31), and (39.19) we obtain the following expression for Δ_k:

$$\Delta_k = \frac{1}{2}\sum_l \frac{G_{kl}\Delta_l}{\sqrt{\epsilon^2(l) + \Delta_l^2}}. \tag{39.32}$$

This is called the "gap equation" and Δ_k is called the gap function. It is a nonlinear integral equation (when the sum is converted to an integral).

Finally, after removing the off-diagonal term, the Hamiltonian can be expressed in the following simple form:

$$H = 2\sum_{k>0} \epsilon(k) v_k^2 - \sum_{k\in\delta} \Delta_k u_k v_k + \sum_{k>0} E_k \left(\alpha_k^\dagger \alpha_k + \beta_k^\dagger \beta_k\right). \tag{39.33}$$

The ground-state energy of the quasiparticle system is given by the first two terms. The energy spectrum of the quasiparticle excited states is given by

$$E_k = \sqrt{\epsilon^2(k) + \Delta_k^2}. \tag{39.34}$$

If the pair interaction potential is further simplified to

$$G_{kl} = G \tag{39.35}$$

then Δ_k is a constant ($= \Delta$) given by

$$\Delta = G \sum_k u_k v_k. \tag{39.36}$$

where the index of summation is changed from l, which was used in (39.8) and (39.19), to k. The Hamiltonian can now be rewritten as

$$H = 2 \sum_{k>0} \epsilon(k) v_k^2 - \frac{\Delta^2}{G} + \sum_{k\in\delta} E_k \left(\alpha_k^\dagger \alpha_k + \beta_k^\dagger \beta_k \right) \tag{39.37}$$

and the relation (39.32) simplifies to

$$1 = \frac{G}{2} \sum_{k\in\delta} \frac{1}{\sqrt{\epsilon^2(k) + \Delta^2}}. \tag{39.38}$$

Let us now consider two separate cases: (1) $\Delta = 0$, and (2) $\Delta \neq 0$.

39.2 Normal states ($\Delta = 0, G \neq 0$)

The relations (39.30) and (39.31) for $\Delta = 0$ are given by

$$u_k^2 = \frac{1}{2} \left(1 + \frac{\epsilon(k)}{\sqrt{\epsilon^2(k)}} \right), \tag{39.39}$$

$$v_k^2 = \frac{1}{2} \left(1 - \frac{\epsilon(k)}{\sqrt{\epsilon^2(k)}} \right). \tag{39.40}$$

From (39.36) we note that since $G \neq 0$ we have two possibilities: (i) $u_k = 0$ or (ii) $v_k = 0$. Let us consider, specifically, the energies $\epsilon_k < \mu$, i.e., $\sqrt{\epsilon^2(k)} = -\epsilon(k)$. Therefore, from (39.39) and (39.40),

$$u_k = 0 \quad \text{and} \quad v_k = 1. \tag{39.41}$$

For this case the quasiparticle ground state obtained by putting the above values in (39.17) is given by

$$|0\rangle_{quasi} = \prod_{\epsilon_k<\mu} a_k^\dagger a_{-k}^\dagger |0\rangle, \tag{39.42}$$

which corresponds to energy levels being filled in pairs up to the Fermi level $\epsilon_k = \mu$. From (39.37), the expectation value of the energy is then given by

$$\langle 0 |H| 0\rangle_{quasi} = 2 \sum_{\epsilon_k<\mu} \epsilon(k). \tag{39.43}$$

To obtain excited states for $\Delta = 0$ we operate on $|0\rangle_{quasi}$ with $\alpha_k^\dagger$ to obtain

$$\alpha_k^\dagger |0\rangle_{quasi} = \left(u_k a_k^\dagger - v_k a_{-k} \right) |0\rangle_{quasi}, \tag{39.44}$$

which corresponds to adding a particle of energy $\epsilon(k)$ if $\epsilon_k > \mu$, therefore, $u_k = 1, v_k = 0$; or removing a particle (or creating a hole) if $\epsilon_k < \mu$, therefore, $v_k = 1, u_k = 0$. The result is that we effectively add an amount $|\epsilon(k)|$ to the ground-state energy. For this reason, the states with $\Delta = 0$ are called normal states since they are consequences of perturbation on noninteracting particles.

Let us now consider the states with $\Delta \neq 0$.

39.3 BCS states ($\Delta \neq 0$)

These are the nontrivial solutions called superconducting or BCS solutions. We note from the relation (39.37) involving Δ that, if the interaction is sufficiently attractive, that is, if G is large enough so that

$$\frac{G}{2\epsilon(k)} > 1, \tag{39.45}$$

then it will have a solution for nonzero values of Δ. The quasiparticle states for $\Delta \neq 0$ are the so-called BCS states.

39.3.1 BCS ground state

The ground-state energy, which we call E_{BCS}, is given by the first two terms of (39.33),

$$E_{BCS} = 2\sum \epsilon(k)v_k^2 - \sum \Delta_k u_k v_k \tag{39.46}$$

where u_k and v_k are given by (39.30) and (39.31), respectively. Substituting the expressions for u_k and v_k and taking $\Delta_k^2 = E_k^2 - \epsilon^2(k)$, we obtain

$$E_{BCS} = -\sum \frac{(E_k - \epsilon(k))^2}{2E_k} < 0. \tag{39.47}$$

Let us examine the dependence of E_{BCS} on u_k and v_k. We note that

$$\langle 0| a_k^\dagger a_k |0\rangle_{quasi} = v_k^2. \tag{39.48}$$

Therefore, v_k^2 is the distribution function for the quasiparticles. We find from (39.30) and (39.31) that deep inside the Fermi sea, where $\epsilon(k) = -|\epsilon(k)|$ and $|\epsilon(k)| \gg \Delta_k$, we have $|\epsilon(k)|/E_k \sim 1$ and, therefore, $v_k^2 \sim 1$ (see Fig. 39.1). This implies that in this region $E_{BCS} \sim 2\epsilon(k)$, which corresponds to a pair of particles occupying energy levels with quantum numbers $(k, -k)$. In the transition interval Δ across the Fermi surface, v_k^2 begins to drop and goes to zero while u_k^2 approaches 1. The sharp Fermi surface that we saw for $\Delta = 0$ is now smeared over the range Δ.

We note that the condition

$$N = \sum_k \langle 0| a_k^\dagger a_k + a_{-k}^\dagger a_{-k} |0\rangle_{quasi} = 2\sum_k v_k^2 \tag{39.49}$$

will determine μ, the chemical potential. It can be readily shown that if $\Delta = 0$ then $\mu = \epsilon_F$.

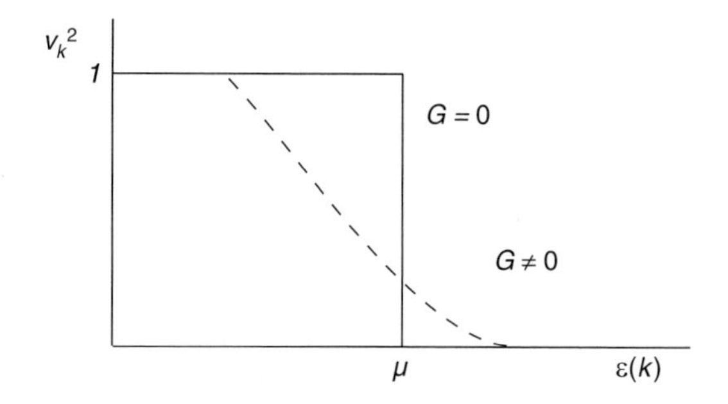

Fig. 39.1

39.3.2 Excited states and the gap function

The operators $\alpha_k^\dagger$ and $\beta_k^\dagger$ create quasiparticle excited states. Their eigenstates are given by

$$\alpha_k^\dagger |0\rangle_{quasi} \quad \text{and} \quad \beta_k^\dagger |0\rangle_{quasi}, \tag{39.50}$$

respectively. The energy of these excited states is given by

$$E_k = \sqrt{\epsilon^2(k) + \Delta_k^2}. \tag{39.51}$$

Near the Fermi surface where $\epsilon(k) \approx 0$, the excitation energy is

$$E_k = \Delta_k. \tag{39.52}$$

This is the famous "gap" between the ground state and the first excited state of the quasiparticles that describes the BCS system. The existence of this gap leads to superconductivity.

The magnitude of Δ can be estimated from relation (39.38) if we assume that the electron states near the Fermi surface are densely packed so that the summation in (39.38) can be replaced by an integral as follows:

$$\sum_{k\in\delta} \frac{1}{\sqrt{(\epsilon_k - \mu)^2 + \Delta^2}} \to \int \frac{d\epsilon_k \rho(\epsilon_k)}{\sqrt{(\epsilon_k - \mu)^2 + \Delta^2}} \tag{39.53}$$

where $\rho(\epsilon_k)$ is the energy density of states near the Fermi surface. The pair interaction operates only in the narrow region in the ϵ_k-space, which we assume to be symmetric around μ, i.e., (reinstating $\hbar$ temporarily) in the region $-\hbar\omega < (\epsilon_k - \mu) < \hbar\omega$ where ω is typically the Debye frequency. Relation (39.38) can then be written as

$$1 = G \int_0^{\hbar\omega} \frac{d\varepsilon \rho(\varepsilon + \mu)}{\sqrt{\varepsilon^2 + \Delta^2}} \simeq G\rho(\mu) \int_0^{\hbar\omega} \frac{d\varepsilon}{\sqrt{\varepsilon^2 + \Delta^2}} \tag{39.54}$$

where we have assumed that the density is essentially a constant given by its central value $\rho(\mu)$. Since

$$\int_0^{\hbar\omega} \frac{d\varepsilon}{\sqrt{\varepsilon^2 + \Delta^2}} = \sinh^{-1}\left(\frac{\hbar\omega}{\Delta}\right) \tag{39.55}$$

we obtain

$$\frac{1}{G\rho\,(\mu)} = \sinh^{-1}\left(\frac{\hbar\omega}{\Delta}\right). \tag{39.56}$$

Since it is found that $G\rho\,(\mu) \ll 1$, we conclude that

$$\Delta \approx 2\hbar\omega e^{-1/\rho(\mu)G}. \tag{39.57}$$

Thus the gap is very small, smaller than $\hbar\omega$. This gap that separates the BCS ground state from the excited states is at the basis of the observed superconductivity.

It is important to point out that $G = 0$ corresponds to an essential singularity in the solution given by (39.57). Thus, we could not have used perturbation theory, which is based on the assumption that the coupling parameter G is small.

39.4 BCS condensate in Green's function formalism

We describe below a rather elegant method to obtain the results we derived in the previous section through the Green's function technique. This formalism will also be helpful when we discuss the mechanism of spontaneous symmetry breaking in Chapter 42.

We will follow basically the same dynamics as described in the previous section, except that we will be working within the framework of (nonrelativistic) Lagrangian field theory. The Lagrangian density for free electrons has already been obtained in Chapter 25. It is given by ($\hbar = c = 1$)

$$\mathcal{L}_0 = i\psi^\dagger(x)\frac{\partial}{\partial t}\psi(x) + \psi^\dagger(x)\frac{\nabla^2}{2m}\psi(x) \tag{39.58}$$

where $x = (\mathbf{r}, t)$, and $\psi(x)$, written in the Heisenberg representation, is the electron field in the nonrelativistic limit. The Lagrangian density relevant to the problem of the pair interaction in superconductivity is

$$\mathcal{L} = \mathcal{L}_0 + \frac{g}{2}\psi^\dagger(x)\,\psi^\dagger_\beta(x)\,\psi_\beta(x)\,\psi(x) \tag{39.59}$$

where the index β corresponding to the spin is summed. The second term on the right corresponds to the four-Fermi interaction. Including the chemical potential, μ, the total Lagrangian is given by

$$\mathcal{L}(\psi, \psi^\dagger) = i\psi^\dagger\frac{\partial}{\partial t}\psi + \psi^\dagger\left(\frac{\nabla^2}{2m} + \mu\right)\psi + \frac{g}{2}\psi^\dagger(x)\,\psi^\dagger_\beta(x)\,\psi_\beta(x)\,\psi(x). \tag{39.60}$$

The four-Fermi term represents the interaction, which is assumed to be effective near the Fermi surface. When the coupling, g, becomes large, this interaction leads to the formation of a bound state between the electrons. That is, it leads to what is called a "condensate" in the electron pair.

The equation of motion for ψ_α is obtained from (39.60) through Euler–Lagrange equations in the manner described in Chapter 25, and is given as follows

$$i\frac{\partial}{\partial t}\psi_\alpha(x) + \left(\frac{\nabla^2}{2m} + \mu\right)\psi_\alpha(x) + g\psi_\beta^\dagger(x)\psi_\beta(x)\psi_\alpha(x) = 0. \tag{39.61}$$

The BCS approximation of the previous section will correspond here to replacing the bilinear operator $\psi_\beta(x)\psi_\alpha(x)$ by its vacuum expectation value,

$$g\psi_\beta(x)\psi_\alpha(x) \to g\langle 0|\psi_\beta(x)\psi_\alpha(x)|0\rangle = M\epsilon_{\alpha\beta} \tag{39.62}$$

where $\epsilon_{\alpha\beta}$ is an antisymmetric tensor to reflect the fact that $\psi_\alpha(x)$ and $\psi_\beta(x)$ anticommute. Since one can write $\psi(x) = e^{-iP.x}\psi(0)e^{iP.x}$, using the translational operator P, and since vacuum is invariant under translation, M is a constant. It is called the "order parameter," which plays a fundamental role in the formation of condensates. Equation (39.61) now reads

$$i\frac{\partial}{\partial t}\psi_\alpha(x) + \left(\frac{\nabla^2}{2m} + \mu\right)\psi_\alpha(x) + M\epsilon_{\alpha\beta}\psi_\beta^\dagger(x) = 0. \tag{39.63}$$

Equation (39.63) involves two fields, $\psi_\alpha(x)$ and $\psi_\beta^\dagger(x)$. We obtain a second equation by taking the Hermitian conjugate of (39.63), interchanging α and β, and multiplying the resulting equation by a minus sign:

$$i\frac{\partial}{\partial t}\psi_\beta^\dagger(x) - \left(\frac{\nabla^2}{2m} + \mu\right)\psi_\beta^\dagger(x) - M^*\epsilon_{\beta\alpha}\psi_\alpha(x) = 0. \tag{39.64}$$

Let us express $\psi_\alpha(x)$ and $\psi_\alpha^\dagger(x)$ in terms of their Fourier components,

$$\psi_\alpha(x) = \frac{1}{(\sqrt{2\pi})^4}\int d\omega\int d^3p\, e^{i(\mathbf{p}\cdot\mathbf{r}-\omega t)}u_\alpha(\omega,\mathbf{p}) \tag{39.65}$$

and its Hermitian conjugate

$$\psi_\alpha^\dagger(x) = \frac{1}{(\sqrt{2\pi})^4}\int d\omega\int d^3p e^{-i(\mathbf{p}\cdot\mathbf{r}-\omega t)}u_\alpha^\dagger(\omega,\mathbf{p}). \tag{39.66}$$

We make the transformation, $\omega \to -\omega$ and $\mathbf{p} \to -\mathbf{p}$ in (39.66) and interchange the limits of the integration to obtain

$$\psi_\alpha^\dagger(x) = \frac{1}{(\sqrt{2\pi})^4}\int d\omega\int d^3p\, e^{i(\mathbf{p}\cdot\mathbf{r}-\omega t)}u_\alpha^\dagger(-\omega,-\mathbf{p}). \tag{39.67}$$

We now solve for the two fields in terms of the two-level matrix

$$\Psi_\alpha = \begin{pmatrix}\psi_\alpha \\ \psi_\alpha^\dagger\end{pmatrix}, \tag{39.68}$$

which can be expressed in terms of the Fourier components (39.65) and (39.67) as

$$\begin{pmatrix} \psi_\alpha \\ \psi_\alpha^\dagger \end{pmatrix} = \frac{1}{(\sqrt{2\pi})^4} \int d\omega \int d^3p\, e^{i(\mathbf{p}\cdot\mathbf{r}-\omega t)} \begin{pmatrix} u_\alpha(\omega, \mathbf{p}) \\ u_\alpha^\dagger(-\omega, -\mathbf{p}) \end{pmatrix}. \tag{39.69}$$

Substituting (39.69) for the individual fields in the two coupled equations (39.63) and (39.64), we obtain the following matrix relation for the Fourier components $u_\alpha(\omega, \mathbf{p})$ and $u_\alpha^\dagger(-\omega, -\mathbf{p})$

$$\begin{pmatrix} \omega + \left(\mu - \frac{p^2}{2m}\right) & M \\ M^* & \omega - \left(\mu - \frac{p^2}{2m}\right) \end{pmatrix} \begin{pmatrix} u_\alpha(\omega, \mathbf{p}) \\ u_\alpha^\dagger(-\omega, -\mathbf{p}) \end{pmatrix} = 0. \tag{39.70}$$

The determinant of the 2×2 matrix above must vanish. Hence,

$$\omega^2 - \left(\mu - \frac{p^2}{2m}\right)^2 - |M|^2 = 0. \tag{39.71}$$

Taking the positive root we obtain the energy eigenvalue

$$\omega = \sqrt{\epsilon^2(p) + |M|^2} \tag{39.72}$$

where

$$\epsilon(p) = \left(\frac{p^2}{2m} - \mu\right). \tag{39.73}$$

Thus we find that because of the presence of the order parameter, M, the system has a bound state or a condensate at $\sqrt{\epsilon^2(p) + |M|^2}$. Hence, M plays the same role as the gap parameter, Δ, which we discussed in the previous section.

Let us now obtain M through the Green's function formalism for the equations (39.63) and (39.64). Since both $\psi(x)$ and $\psi^\dagger(x)$ are involved in our equations, we express the Green's function in terms of the doublet as given by (39.70) and write

$$G_{\alpha\beta}(x) = -i\left\langle 0 \left| T(\Psi_\alpha(x)\Psi_\beta^\dagger(0)) \right| 0 \right\rangle \tag{39.74}$$

where $|0\rangle$ stands for the quasiparticle vacuum. Thus $G_{\alpha\beta}(x)$ has the following matrix form:

$$G_{\alpha\beta}(x) = -i \begin{pmatrix} \left\langle 0 \left| T\left(\psi_\alpha(x)\psi_\beta^\dagger(0)\right) \right| 0 \right\rangle & \langle 0 | T(\psi_\alpha(x)\psi_\beta(0)) | 0 \rangle \\ \left\langle 0 \left| T\left(\psi_\alpha^\dagger(x)\psi_\beta^\dagger(0)\right) \right| 0 \right\rangle & \left\langle 0 \left| T\left(\psi_\alpha^\dagger(x)\psi_\beta(0)\right) \right| 0 \right\rangle \end{pmatrix}. \tag{39.75}$$

We can write this in a slightly simpler form as

$$G_{\alpha\beta}(x) = \begin{pmatrix} A_{\alpha\beta}(x) & B_{\alpha\beta}(x) \\ B_{\alpha\beta}^\dagger(x) & A_{\alpha\beta}^\dagger(x) \end{pmatrix}. \tag{39.76}$$

As usual it is best to work with the Fourier transform, which is given by

$$G_{\alpha\beta}(x) = \frac{1}{(2\pi)^4}\int d\omega \int d^3p\, e^{i(\mathbf{p}\cdot\mathbf{r}-\omega t)} G_{\alpha\beta}(\omega, \mathbf{p}). \tag{39.77}$$

Expressing the individual matrix elements on the right-hand side of (39.76) in a form similar to (39.77), we have

$$G_{\alpha\beta}(\omega, \mathbf{p}) = \begin{pmatrix} A_{\alpha\beta}(\omega, \mathbf{p}) & B_{\alpha\beta}(\omega, \mathbf{p}) \\ B^{\dagger}_{\alpha\beta}(-\omega, -\mathbf{p}) & A^{\dagger}_{\alpha\beta}(-\omega, -\mathbf{p}) \end{pmatrix}. \tag{39.78}$$

For a single-channel problem where the equation for a function ψ is given by

$$L\psi = f, \tag{39.79}$$

the Green's function is expressed as L^{-1}. If it is a coupled-channel problem, then the equation for ψ is

$$L_{ij}\psi_j = f_i \tag{39.80}$$

where $i, j = 1, 2$ and a summation over j is implied. Hence, if we write this in the matrix form we have

$$L\psi = f \tag{39.81}$$

where L is now a 2×2 matrix while ψ and f are column matrices. The Green's function then is a 2×2 matrix that is the inverse of L, i.e.,

$$G = L^{-1}. \tag{39.82}$$

We can determine G given by (39.78) by identifying its matrix elements with the inverse of the operator L, i.e., with the inverse of the matrix on the left side of (39.70),

$$\begin{pmatrix} A_{\alpha\beta}(\omega, \mathbf{p}) & B_{\alpha\beta}(\omega, \mathbf{p}) \\ B^{\dagger}_{\alpha\beta}(-\omega, -\mathbf{p}) & A^{\dagger}_{\alpha\beta}(-\omega, -\mathbf{p}) \end{pmatrix} = \begin{pmatrix} [\omega - \epsilon(p)]\,\delta_{\alpha\beta} & M\epsilon_{\alpha\beta} \\ M^*\epsilon_{\alpha\beta} & [\omega + \epsilon(p)]\,\delta_{\alpha\beta} \end{pmatrix}^{-1}. \tag{39.83}$$

Thus

$$A_{\alpha\beta}(\omega, \mathbf{p}) = \frac{[\omega + \epsilon(p)]\,\delta_{\alpha\beta}}{\omega^2 - |M|^2 - \epsilon^2(p) + i\epsilon} \tag{39.84}$$

and

$$B_{\alpha\beta}(\omega, \mathbf{p}) = \frac{-M\epsilon_{\alpha\beta}}{\omega^2 - |M|^2 - \epsilon^2(p) + i\epsilon}. \tag{39.85}$$

In the definition

$$g\langle 0 | \psi_\beta(x)\,\psi_\alpha(x) | 0\rangle = M\epsilon_{\alpha\beta} \tag{39.86}$$

we can let $x \to 0$ since M is independent of x and utilize the relations (39.75) and (39.76) to obtain

$$M\epsilon_{\alpha\beta} = g\lim_{x\to 0}\langle 0 | T(\psi_\beta(x)\,\psi_\alpha(x)) | 0\rangle = -igB_{\alpha\beta}(0). \tag{39.87}$$

Since

$$B_{\alpha\beta}(x) = \frac{1}{(2\pi)^4}\int d\omega \int d^3p\, e^{i(\mathbf{p}\cdot\mathbf{r}-\omega t)} B_{\alpha\beta}(\omega, \mathbf{p}), \tag{39.88}$$

putting $x = 0$ in (39.88) and inserting the expression (39.85) for $B_{\alpha\beta}(\omega, \mathbf{p})$ on the right-hand side we obtain

$$1 = \frac{ig}{(2\pi)^4}\int_{-\infty}^{+\infty} d\omega \int d^3p \frac{1}{\omega^2 - |M|^2 - \epsilon^2(p) + i\epsilon}. \tag{39.89}$$

Using complex integration techniques we can evaluate the integral in ω by noting that the integrand has poles at $\sqrt{\epsilon^2(p) + |M|^2} - i\epsilon'$ and $-\sqrt{\epsilon^2(p) + |M|^2} + i\epsilon'$, where ϵ' is related to ϵ. One can use Cauchy's residue theorem to evaluate this integral by completing the contour that includes the real axis and a contour along either the upper or lower half-plane. We obtain

$$1 = \frac{g}{16\pi^3}\int d^3p \frac{1}{\sqrt{\epsilon^2(p) + |M|^2}} \tag{39.90}$$

where we note that $\epsilon(p) = p^2/2m - \mu$. Henceforth we will write $\epsilon(p) = \epsilon$.

To carry out the above integration we first express d^3p in terms of the energy density of the Fermi levels, which is given by

$$\rho = \frac{dN}{VdE} \tag{39.91}$$

near the Fermi surface, where V is the volume and N is the number of states. Since

$$dN = 2\frac{d^3pV}{(2\pi)^3} \tag{39.92}$$

we have

$$d^3p = 4\pi^3\rho d\epsilon \tag{39.93}$$

where we have substituted $E = \epsilon$. Hence, (39.90) can be expressed as an integral over $d\epsilon$. We note that the interaction responsible for the formation of the condensate (i.e., the bound Cooper pair) is confined to a very small region near the Fermi surface, $-\hbar\omega/2 < \epsilon < \hbar\omega/2$, where ω is the Debye frequency. The relation (39.90) can be written, by shifting the integral, as

$$1 = \frac{g}{2}\rho \int_0^{\hbar\omega} d\epsilon \frac{1}{\sqrt{\epsilon^2 + M^2}}. \tag{39.94}$$

If we identify $(g/2)$ which appears in the Lagrangian (39.59) as G in Section 39.3, then we recover the relation (39.54) with the result

$$M = 2\hbar\omega e^{-\frac{2}{g\rho}}. \tag{39.95}$$

39.5 Meissner effect

Let $\phi_s(x)$ be a scalar wavefunction describing a Cooper pair of charge $2e$ and mass $2m$ in a superconductor. It is related to the order parameter

$$\langle 0 \left| \psi_\alpha(x) \psi_\beta(x) \right| 0 \rangle, \tag{39.96}$$

which for certain interactions can be shown to be a function of x.

According to Landau theory of phase transitions, $\phi_s(x)$ vanishes for temperatures above the critical temperature T_c and is nonzero below T_c, where we write it in a rather simplified form as

$$\phi_s(x) = \sqrt{n_s} \tag{39.97}$$

where $n_s = \left(\left| \phi_s(x) \right|^2 \right)$ is the density of the Cooper pairs, assumed to be essentially a constant. In general, $\phi_s(x)$ has a phase that we will assume to be zero.

As we have previously determined, for a particle of charge e and mass m, the electromagnetic current, is given by

$$\mathbf{j} = \frac{e}{2m} \left[\phi_s{}^*(x) \left(-i\boldsymbol{\nabla} - e\mathbf{A} \right) \phi_s + c.c. \right] \tag{39.98}$$

where $\mathbf{A}$ is the vector potential. The supercurrent, $\mathbf{j}_s$, corresponding to a Cooper pair of charge $2e$ and mass $2m$, is given by

$$\mathbf{j}_s = \frac{e}{2m} \left[\phi_s^*(x) \left(-i\boldsymbol{\nabla} - 2e\mathbf{A} \right) \phi_s + c.c. \right]. \tag{39.99}$$

Since ϕ_s in (39.97) is assumed to be a constant, the $\boldsymbol{\nabla}$ operator in (39.99) will not contribute and we obtain

$$\mathbf{j}_s = -\frac{2e^2}{m} n_s \mathbf{A}. \tag{39.100}$$

If we take the curl of both sides and use the fact that the magnetic field $\mathbf{B}$ is given by $\mathbf{B} = \boldsymbol{\nabla} \times \mathbf{A}$, we obtain

$$\boldsymbol{\nabla} \times \mathbf{j}_s = -\frac{2e^2}{m} n_s \mathbf{B}. \tag{39.101}$$

This is the well-known London equation in condensed matter physics.

We can now implement the relation

$$\boldsymbol{\nabla} \times \mathbf{B} = \frac{4\pi}{c} \mathbf{j} \tag{39.102}$$

from Maxwell's equation and obtain, from (39.101),

$$\boldsymbol{\nabla} \times \boldsymbol{\nabla} \times \mathbf{B} = -k^2 \mathbf{B} \tag{39.103}$$

where

$$k^2 = \frac{8\pi e^2 n_s}{mc}. \tag{39.104}$$

From the relation $\nabla \times \nabla \times \mathbf{B} = \nabla(\nabla \cdot \mathbf{B}) - \nabla^2\mathbf{B}$ we note that since $\nabla \cdot \mathbf{B} = 0$, according to Maxwell's equations, the left-hand side of (39.103) is $(-\nabla^2\mathbf{B})$. Therefore,

$$\nabla^2\mathbf{B} = k^2\mathbf{B}. \tag{39.105}$$

This is the wave equation for **B**.

If **B** is a function of z alone, and the superconducting surface is parallel to the x–y plane, then the above equation becomes

$$\frac{d^2\mathbf{B}}{dz^2} = k^2\mathbf{B}. \tag{39.106}$$

Since $\nabla \cdot \mathbf{B} = 0$, the z-component, B_z, of **B** is a constant. Substituting B_z(= constant) in (39.106), we find that the left-hand side will vanish and hence also the right-hand side, i.e., we have $B_z = 0$. The solution of (39.106) is then

$$\mathbf{B}(z) = \mathbf{B}_0 e^{-kz} \tag{39.107}$$

where $\mathbf{B}_0$ is a constant vector in the x–y plane. Thus, **B** decreases exponentially along the z-direction, with a penetration depth of $(1/k)$, implying that the magnetic field is expelled in a superconductor.

This phenomenon is called the Meissner effect.

Elaborating on (39.106) and (39.107), we note that if we consider the Klein–Gordon equation for a particle of mass m described by a field ϕ, then it satisfies the equation

$$-\frac{\partial^2\phi}{\partial t^2} + \nabla^2\phi = \mu^2\phi. \tag{39.108}$$

If we take the static limit, $\partial\phi/\partial t \approx 0$, and assume ϕ $(= \phi(z))$ as a function of z alone, then we obtain

$$\frac{d^2\phi}{dz^2} = \mu^2\phi. \tag{39.109}$$

The solution of (39.109), finite at $z = \infty$, is given by

$$\phi = \phi_0 e^{-\mu z}. \tag{39.110}$$

Thus, the Meissner effect corresponds to the Maxwell field becoming massive with a mass k. As we will discuss in Chapter 42, this is the nonrelativistic analog of the Higgs mechanism in which a zero-mass vector particle becomes massive.

39.5.1 Ginzburg-Landau equation

The Ginzburg–Landau equation is an equation for the wavefunction of the condensate, ϕ_s, the order parameter, which describes the macroscopic properties of a superconductor. It is based on the Landau theory of second phase transitions and minimization of the free energy of a superconducting state. The G-L equation in the presence of a magnetic field, $\mathbf{B}(= \nabla \times \mathbf{A})$ is given by

$$\frac{1}{2m}(-i\nabla - 2e\mathbf{A})^2\phi_s + a\phi_s + b\left|\phi_s\right|^2\phi_s = 0. \tag{39.111}$$

This is actually a phenomenological equation with a and b as the (temperature-dependent) parameters, but it can be related to the BCS theory. Clearly it is a nonlinear equation and therefore the solutions are quite complicated.

In the absence of any magnetic field and for a homogeneous system, so that the ∇ operator can be neglected, the equation is quite simple:

$$a\phi_s + b\left|\phi_s\right|^2 \phi_s = 0, \tag{39.112}$$

which can be written as

$$\left(a + b\left|\phi_s\right|^2\right)\phi_s = 0. \tag{39.113}$$

There is a trivial solution

$$\phi_s = 0, \tag{39.114}$$

which corresponds to a normal conductor. But there is another solution

$$\left|\phi_s\right|^2 = -\frac{a}{b}. \tag{39.115}$$

If $a < 0$ with b positive then we have a nontrivial solution corresponding to a non-zero-order parameter. This describes the superconductor state. One writes the temperature dependence of a as

$$a(T) = 0, \qquad T > T_c, \tag{39.116}$$

$$= a_0\left(T - T_c\right), \qquad T < T_c, \tag{39.117}$$

with a_0 as a positive constant. If we take b as a positive constant, then for $T > T_c$ we have a normal conductor while for $T < T_c$ we have a superconductor. Thus, T_c is considered a transition temperature for the superconducting phase.

Interestingly, if one considers the G-L equation in one dimension, then in the absence of a magnetic field, but without assuming a homogeneous medium, one can obtain an exact solution which is given by

$$\phi_s(x) = \phi_s(0)\tanh\left(\frac{x}{\sqrt{2}\xi}\right) \tag{39.118}$$

where

$$\xi = \left(\frac{1}{2ma(T)}\right)^{\frac{1}{2}}. \tag{39.119}$$

This is called the G-L coherence length.

What we have discussed above are some of the elementary aspects of the G-L equation, but it has many profound implications not only for superconductivity but for many areas in physics.

39.6 Problems

1. Start with the equation for the Cooper pair,

$$(H_0 + V)|\psi\rangle = E|\psi\rangle$$

where $|\psi\rangle$ is the pair state vector in terms of free two-particle states $|\phi_k\rangle$ with energy $2\epsilon_{k'}$(measured relative to $\epsilon_{F'}$)

$$|\psi\rangle = \sum_k a_k |\phi_k\rangle.$$

If

$$\langle\phi_{k'}| V |\phi_k\rangle = \lambda = \text{constant}$$

show that

$$a_k = -\frac{\lambda \sum_{k'} a_{k'}}{(2\epsilon_{k'} - E)}.$$

Summing both sides over k, show that one obtains the relation

$$1 = -\lambda \sum_{k'} \frac{1}{(E - 2\epsilon_{k'})}.$$

Determine the sign of λ for which there will be a solution. Express the above sum as an integral, and compare it with the result.

2. Show that the BCS Hamiltonian (39.9), apart from some c-number contributions, can be written for a specific k-value as a two-level problem.

$$\begin{bmatrix} a_k^\dagger & a_{-k} \end{bmatrix} \begin{bmatrix} \epsilon(k) & -\Delta \\ -\Delta & -\epsilon(k) \end{bmatrix} \begin{bmatrix} a_k \\ a_{-k}^\dagger \end{bmatrix}$$

where $\Delta = \Delta_k$. Diagonalize this matrix by writing

$$\begin{bmatrix} a_k \\ a_{-k}^\dagger \end{bmatrix} = \begin{bmatrix} \cos\theta & \sin\theta \\ -\sin\theta & \cos\theta \end{bmatrix} \begin{bmatrix} \alpha_k \\ \beta_k^\dagger \end{bmatrix}$$

with eigenvalues E_+, E_-. Show that one obtains

$$\cos\theta = u_k, \quad \sin\theta = v_k \quad \text{and} \quad E_+ = E_k.$$

Determine E_-.

40 Bose–Einstein condensation and superfluidity

When a collection of bosons (particles of integer spin) in a condensed matter system are cooled to temperatures near absolute zero, then as a phenomenon driven largely by Bose statistics of the particles and not by their interactions, one finds that a large fraction of the particles collapse into the lowest state, the ground state of the system. At this stage they form a single entity, a condensate, whose quantum properties can be apparent on macroscopic scales. We determine the ground-state energy and also calculate the energy of quasiparticles that arise from the multiparticle dynamics. We also briefly describe the phenomenon of superfluidity.

40.1 Many-body system of integer spins

40.1.1 Ground state and quasiparticles

Consider a system of N bosons interacting through a potential given by $V_0\left(\left|\mathbf{r}_i - \mathbf{r}_j\right|\right)$ in a coordinate representation that depends only on the magnitude of the relative separation between the particles. We write the total Hamiltonian in the second quantized form as

$$H = \sum \epsilon_k a_k^\dagger a_k + \frac{1}{2}\sum a_{k_1}^\dagger a_{k_2}^\dagger a_{k_2'} a_{k_1'} \left\langle \mathbf{k}_1\mathbf{k}_2 \left| V_0 \right| \mathbf{k}_2'\mathbf{k}_1' \right\rangle \tag{40.1}$$

where the first term corresponds to the unperturbed Hamiltonian, with ϵ_k the kinetic energy given by

$$\epsilon_k = \frac{k^2}{2m} \quad \text{where } k = |\mathbf{k}|, \tag{40.2}$$

and the second term is the interaction term.

The boson creation and destruction operators satisfy the usual commutation relations $\left[a_i^\dagger, a_j\right] = \delta_{ij}$, $\left[a_i, a_j\right] = 0 = \left[a_i^\dagger, a_j^\dagger\right]$, where i and j go over positive and negative values. We write H in the following simplified form:

$$H = \sum \epsilon_k a_k^\dagger a_k + \lambda_0 {\sum}' a_{k_1'}^\dagger a_{k_2'}^\dagger a_{k_2} a_{k_1}, \tag{40.3}$$

$$\lambda_0 = \frac{v_0}{2V}, \tag{40.4}$$

with v_0 the Fourier transform of V_0, assumed to be a constant, and V the volume of integration. The symbol $\sum'$ implies that the sum is to be carried out according to the conservation relation $\mathbf{k}_1 + \mathbf{k}_2 = \mathbf{k}_1' + \mathbf{k}_2'$.

The ground state of a collection of noninteracting, or weakly interacting, bosons is effectively a condensate of all the particles occupying their lowest energy states with $\mathbf{k}_i \approx 0$. This corresponds to Bose–Einstein condensation. Thus, the condensed particles are in a single quantum state, all with zero momentum, while the normal particles have nonzero momenta. If the interaction between the bosons is repulsive, then one will detect the presence of quasiparticles. We will discuss these properties below.

We are interested in the eigenvalues of H. We assume that the interaction is weak and the number of particles, N_0, in the condensate is close to the total number, N, of the particles. Moreover, since we will be interested in low-energy eigenvalues, the contributions coming from the terms with $\mathbf{k}_i \approx 0$ will dominate the second term in (40.3). Hence the terms with a_0 and $a_0^\dagger$ will be dominant. Their contributions to H, however, appear as a_0/V and $a_0^\dagger/V$. We will consider this problem in the thermodynamic limit $N \to \infty$ (and $N_0 \to \infty$) and $V \to \infty$ with $N/V = n$, which is assumed to be finite. Since $a_0^\dagger a_0 = N_0$, we can assume that both a_0 and $a_0^\dagger$ are $\sim N^{1/2}$ in this limit. We also note that

$$\frac{1}{V}\left[a_0, a_0^\dagger\right] = \frac{1}{V} \to 0 \quad as\ V \to \infty. \tag{40.5}$$

The leading contributions to the interaction term in (40.3) will then come from

$$\begin{aligned}&\left(a_0^\dagger\right)^2 (a_0)^2 + \left(a_0^\dagger\right)^2 a_k a_{-k} + a_k^\dagger a_{-k}^\dagger (a_0)^2 + a_0^\dagger a_k^\dagger a_k a_0 + a_k^\dagger a_0^\dagger a_0 a_k \\ &+ a_0^\dagger a_{-k}^\dagger a_{-k} a_0 + a_{-k}^\dagger a_0^\dagger a_0 a_{-k}.\end{aligned} \tag{40.6}$$

The second term in (40.3) can be written as

$$\sum{}' a_{k_1'}^\dagger a_{k_2'}^\dagger a_{k_2} a_{k_1} = N^2 + N \sum_{k\neq 0}\left[a_k^\dagger a_{-k}^\dagger + a_k a_{-k} + 2a_k^\dagger a_k + 2a_{-k}^\dagger a_{-k}\right]. \tag{40.7}$$

The Hamiltonian, H, is now given by

$$H = N^2\lambda_0 + \sum_k (\epsilon_k + 2N\lambda_0)\left[a_k^\dagger a_k + a_{-k}^\dagger a_{-k}\right] + N\lambda_0 \sum_{k\neq 0}\left[a_k^\dagger a_{-k}^\dagger + a_k a_{-k}\right]. \tag{40.8}$$

In order to obtain the energy eigenvalues we will diagonalize H, and for that purpose we make the following transformation, known as the Bogoliubov canonical transformation:

$$a_k = u_k b_k + v_k b_{-k}^\dagger, \quad a_k^\dagger = u_k b_k^\dagger + v_k b_{-k} \tag{40.9}$$

where we take u_k and v_k to be real. If we assume that $b_i, b_i^\dagger$ satisfy the same commutation relations as $a_i, a_i^\dagger$, then the following relation holds:

$$u_k^2 - v_k^2 = 1. \tag{40.10}$$

Substituting (40.9) for a_k and $a_k^\dagger$ in (40.8) and, setting the coefficient of the off-diagonal element, $\left(b_k^\dagger b_{-k}^\dagger + b_k b_{-k}\right)$, to zero, we obtain

$$(\epsilon_k + 2N\lambda_0)\, u_k v_k + N\lambda_0 \left(u_k^2 + v_k^2\right) = 0. \tag{40.11}$$

We take $x_k = v_k/u_k$, and note from (40.10) that $|x_k| < 1$. The solution of equation (40.11) is then

$$x_k = \frac{-\left(\epsilon_k + 2N\lambda_0\right) + \sqrt{\epsilon_k^2 + 4\epsilon_k N\lambda_0}}{2N\lambda_0}. \tag{40.12}$$

The relations (40.9) can now expressed in terms of x_k as

$$a_k = \frac{b_k + x_k b_{-k}^\dagger}{\sqrt{1-x_k^2}}, \quad a_k^\dagger = \frac{b_k^\dagger + x_k b_{-k}}{\sqrt{1-x_k^2}}. \tag{40.13}$$

Changing $k \to -k$ in the second relation of (40.13) and solving for b_k and $b_{-k}^\dagger$, one can invert these relations to obtain

$$b_k = \frac{a_k - x_k a_{-k}^\dagger}{\sqrt{1-x_k^2}}, \quad b_k^\dagger = \frac{a_k^\dagger - x_k a_{-k}}{\sqrt{1-x_k^2}} \tag{40.14}$$

where in the second relation of (40.13) we reverted $-k \to k$. As we will see below, b_k and $b_k^\dagger$ act like creation and destruction operators of the quasiparticles.

We note that since $\epsilon_k = 0$ for $k = 0$, the first summation in H in (40.8) will correspond to $k \neq 0$, as will the second summation:

$$H = N^2\lambda_0 + \frac{1}{2}\sum_{k\neq 0}\left(\epsilon_k + 2N\lambda_0\right)\left(a_k^\dagger a_k + a_{-k}^\dagger a_{-k}\right) + N\lambda_0 \sum_{k\neq 0}\left[a_k^\dagger a_{-k}^\dagger + a_k a_{-k}\right]. \tag{40.15}$$

Ignoring the off-diagonal terms of the type $b_k^\dagger b_{-k}^\dagger$, etc., which vanish, as already arranged, we obtain from (40.13)

$$a_k^\dagger a_k + a_{-k}^\dagger a_{-k} = \left(\frac{1+x_k^2}{1-x_k^2}\right)\left(b_k^\dagger b_k + b_{-k}^\dagger b_{-k}\right) + \left(\frac{2x_k^2}{1-x_k^2}\right) \tag{40.16}$$

where we have used the relation $\left[b_k^\dagger, b_k\right] = 1$, $\left[b_{-k}^\dagger, b_{-k}\right] = 1$ with all other commutators of b's vanishing. Similarly, we obtain

$$a_k^\dagger a_{-k}^\dagger + a_k a_{-k} = \left(\frac{2x_k}{1-x_k^2}\right)\left(b_k^\dagger b_k + b_{-k}^\dagger b_{-k}\right) + \left(\frac{2x_k}{1-x_k^2}\right) \tag{40.17}$$

where the quantities in the second term on the right-hand sides of (40.16) and (40.17) are supposed to be multiplied by a unit operator. Substituting these in (40.15) and taking account of the relation for x_k in (40.12), we find

$$H = E_0 + \frac{1}{2}\sum_{k\neq 0} \epsilon(k) \left(b_k^\dagger b_k + b_{-k}^\dagger b_{-k} \right) \tag{40.18}$$

where E_0 is the ground-state energy, and

$$\epsilon(k) = \sqrt{\epsilon_k^2 + 4\epsilon_k N\lambda_0}, \tag{40.19}$$

which is then the energy of the excited state, a quasiparticle. As indicated earlier, b_i and $b_i^\dagger$ are the destruction and creation operators for these states. The ground state, E_0, corresponds to

$$E_0 = N^2\lambda_0 - \sum_{k\neq 0} \left[\left(\frac{x_k^2}{1 - x_k^2} \right) \epsilon(k) \right]. \tag{40.20}$$

We can remove the factor $1/2$ in the second term of (40.18) by writing

$$H = E_0 + \sum_{k\neq 0} \epsilon(k)\, b_k^\dagger b_k. \tag{40.21}$$

Simplifying (40.20) further we obtain

$$E_0 = N^2\lambda_0 + \frac{1}{2}\sum_{k\neq 0} \left[\epsilon(k) - (\epsilon_k + 2N\lambda_0) \right]. \tag{40.22}$$

Let us define the quasiparticle ground state, or quasiparticle vacuum $|0\rangle_{quasi}$, as

$$b_k \, |0\rangle_{quasi} = 0. \tag{40.23}$$

Then the ground-state energy is given by

$$\langle 0 \,|H|\, 0\rangle_{quasi} = E_0 = N^2\lambda_0 + \frac{1}{2}\sum_{k\neq 0} \left[\epsilon(k) - (\epsilon_k + 2N\lambda_0) \right]. \tag{40.24}$$

The quasiparticle number distribution in the ground state is given by

$$n_k = \left\langle 0 \left| a_k^\dagger a_k \right| 0 \right\rangle_{quasi} = v_k^2 \left\langle 0 \left| b_k b_k^\dagger \right| 0 \right\rangle_{quasi} = v_k^2 \tag{40.25}$$

where we have used the canonical relation (40.9). From the relation (40.10) and (40.12) we find that for $k \to 0$,

$$v_k^2 \simeq \frac{N\lambda_0}{\sqrt{4\epsilon_k N\lambda_0}} \sim \frac{1}{k}, \tag{40.26}$$

while as $k \to \infty$,

$$v_k^2 \simeq \frac{\epsilon_k^2}{N^2\lambda_0^2} \sim \frac{1}{k^4}. \tag{40.27}$$

Finally, we note that as for the BCS theory, in Bose–Einstein condensation we also have an order parameter and a condensate wavefunction. This wavefunction is governed by a nonlinear equation called the Gross–Pitaevskii equation, which is similar to the Ginzburg–Landau equation we considered earlier.

40.2 Superfluidity

A classic example of a superfluid is the helium-4 atom ^{4}H, which has spin zero and which remains a liquid even at absolute zero.

Let us now consider $\epsilon(k)$, the energy of the quasiparticle. Since $\epsilon_k = k^2/2m$, the behavior of $\epsilon(k)$ for small k is given by

$$\epsilon(k) \approx \sqrt{4\epsilon_k N\lambda_0} = \left(\sqrt{\frac{2N\lambda_0}{m}}\right) k. \tag{40.28}$$

We note that λ_0 must be positive for the quasiparticle energy to be real, which implies that the interaction between the particles must be repulsive.

The linear behavior $\epsilon(k) \sim k$ at low energies is a characteristic feature of the quasiparticles, which is to be compared with the energy dependence of the free particles, $\epsilon_k \sim k^2$. This result is at the core of a remarkable phenomenon called superfluidity in which certain liquids, e.g., liquid helium at low temperatures, suffer no friction in going through capillaries; that is, they exhibit no viscosity. We discuss this property below.

Let us define

$$\min \frac{\epsilon(k)}{k} = v_c \tag{40.29}$$

where v_c is called the critical velocity. We find from (40.28) that the quasiparticle spectrum satisfies the relation

$$v_c > 0. \tag{40.30}$$

Note that for a noninteracting system with

$$\epsilon(k) = \frac{k^2}{2m} \tag{40.31}$$

we have $v_c = 0$ and hence no excitations occur.

To understand superfluidity we follow an argument due to Landau. Consider a liquid moving through a capillary with a constant velocity v. In a frame moving with the fluid, where the fluid is at rest, the capillary wall moves with velocity $-v$. The friction between the wall of the tube and the fluid can cause excitations in terms of the quasiparticles in the liquid by transforming the kinetic energy to internal energy. Let one of the quasiparticles generated have a momentum $\mathbf{k}$ and energy $\epsilon(k)$. Transforming back to the laboratory frame where the tube is now at rest, this fluid will have the energy

$$E' = E + [\epsilon(k) + \mathbf{k} \cdot \mathbf{v}] \tag{40.32}$$

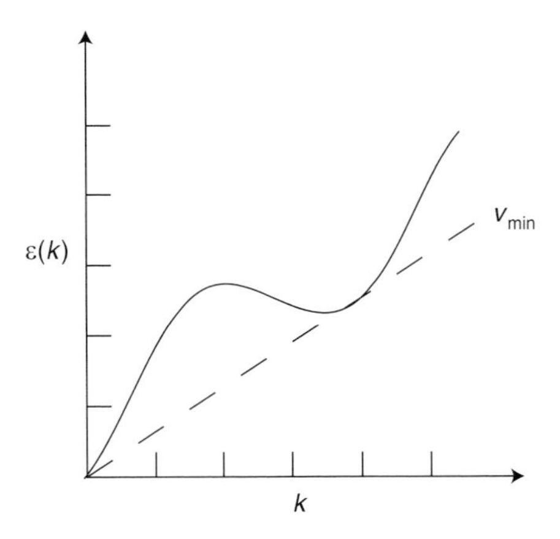

Fig. 40.1

where E' and E are final and initial energies. If no excitation is present, the energy of the fluid is E. The presence of excitations causes this energy to change by the amount $(\epsilon(k) + \mathbf{k}\cdot\mathbf{v})$. Since the energy of the flowing liquid decreases due to friction, we must have

$$\epsilon(k) + \mathbf{k}\cdot\mathbf{v} < \mathbf{0}. \tag{40.33}$$

We note that

$$\epsilon(k) + \mathbf{k}\cdot\mathbf{v} \geq \epsilon(k) - kv. \tag{40.34}$$

It follows, therefore, that if quasiparticles satisfy the property

$$v > \frac{\epsilon(k)}{k} \tag{40.35}$$

then friction will occur. However, the term on the right-hand side is the critical velocity defined in (40.29). Therefore, friction will occur when $v > v_c$.

If, however, $0 < v < v_c$ then the velocity will be in the gap where it is positive and yet below the threshold for creating quasiparticles. The fluid will then move through the tube without dissipation. In other words, the liquid will exhibit superfluidity.

The curve for $\epsilon(k)$ for liquid helium is sketched in Fig. 40.1. There are two values of k where ϵ/k is a minimum, at $k = 0$ and $k = k_1$. The minimum at $k = 0$ corresponds to low temperatures At such temperatures superfluidity occurs when

$$v < v_0 = \frac{d\epsilon(k)}{dk}\,|_{k=0}. \tag{40.36}$$

For slightly larger temperatures there is a minimum at k_1 and superfluidity occurs for

$$v < v_1 = \frac{d\epsilon(k)}{dk}\,|_{k=k_1}. \tag{40.37}$$

The quasiparticle excitations at k_1 are called rotons.

Finally, we note that for large values of k the quasiparticle energy given by (40.19) returns to the normal behavior, $\epsilon\,(k) \sim k^2$.

40.3 Problems

1. Consider two identical bosons of mass m undergoing harmonic oscillator motion each with spring constant K. Determine the energy eigenvalues of the system if the two particles interact via the following potentials:

$$\text{(i)} \qquad V(x_1, x_2) = \lambda x_1 x_2.$$

$$\text{(ii)} \qquad V(x_1, x_2) = \frac{1}{2}K(x_1 - x_2)^2.$$

2. Consider a system of N bosons each described by the wavefunction of the ground state harmonic oscillator of frequency ω,

$$u(\mathbf{r}) = \frac{1}{(\pi a^2)^{3/4}} \exp(-r^2/2a^2).$$

Assume that the bosons interact pairwise via a δ-function interaction $\lambda\delta(\mathbf{r}_i - \mathbf{r}_j)$ within each pair. Determine the kinetic, potential, and interaction energies of this system. Also obtain the total energy, E, and E divided by the total N-particle energy in the absence of interaction.

41 Lagrangian formulation of classical fields

41.1 Basic structure

In Chapter 25 we discussed the Lagrangian method as it applied to nonrelativistic problems. We extend it now to the relativistic domain in which the relativistic wavefunctions, called classical fields, replace their nonrelativistic counterparts with the Lagrangian method formulated in the four-dimensional language.

Let ϕ represent the classical field. We write the Lagrangian density $\mathcal{L}$ in terms ϕ and its derivative $\partial_\mu \phi$, and express the Lagrangian as

$$L = \int d^3x \, \mathcal{L}\left(\phi, \partial_\mu \phi\right) \tag{41.1}$$

and the action as

$$S = \int d^4x \, \mathcal{L}\left(\phi, \partial_\mu \phi\right). \tag{41.2}$$

The corresponding Euler–Lagrange equation can be written as

$$\frac{\partial \mathcal{L}}{\partial \phi} - \partial_\mu \left(\frac{\partial \mathcal{L}}{\partial\left(\partial_\mu \phi\right)} \right) = 0. \tag{41.3}$$

The Klein–Gordon (K-G) wavefunction $\phi(x)$ provides a simple example of a classical field. If we take

$$\mathcal{L} = \frac{1}{2}\left(\partial_\mu \phi \partial^\mu \phi - \mu^2 \phi^2\right) \tag{41.4}$$

then from (41.3) we obtain (with $\hbar = c = 1$)

$$\partial_\mu \phi \partial^\mu \phi + \mu^2 \phi = 0, \tag{41.5}$$

which is, indeed, the K-G equation discussed in Chapter 31. We note that this equation is consistent with the basic quantum condition obeyed by the canonical variables x_i and p_i, e.g.,

$$[x_i, p_j] = i\hbar\delta_{ij}; \quad [x_i, x_j] = 0; \quad [p_i, p_j] = 0. \tag{41.6}$$

A second quantization occurs when the field ϕ itself is quantized. This has been discussed in Chapters 37–40. In this chapter, we will discuss the classical field equations such as the K-G and Dirac equations. We will also call Maxwell's equations, which were discussed in Chapter 31, classical field equation even though the quantum conditions (41.6) were not imposed in deriving them.

41.2 Noether's theorem

One of the advantages of the Lagrangian formulation is that one can establish a relationship between symmetries of the theory and conservation laws. Consider, as an example, the following infinitesimal transformation of the field $\phi(x)$ at each space-time point, x:

$$\phi(x) \to \phi'(x) = \phi(x) + \epsilon\Delta\phi(x) \tag{41.7}$$

where $\Delta\phi(x)$ is the change in the field and ϵ is an infinitesimal parameter. We will call the theory invariant under this transformation if the action, S, and therefore, the equations of motion are invariant under (41.7). The variation in the action is given by

$$\delta S = \int d^4x \left[\frac{\partial\mathcal{L}}{\partial\phi}\delta\phi + \frac{\partial\mathcal{L}}{\partial(\partial_\mu\phi)}\delta(\partial_\mu\phi)\right]. \tag{41.8}$$

The invariance of S implies that

$$\delta S = 0. \tag{41.9}$$

From the Euler–Lagrange equations satisfied by ϕ we can write (41.8) as

$$\delta S = \int d^4x \left[\partial_\mu\frac{\partial\mathcal{L}}{\partial(\partial_\mu\phi)}\delta\phi + \frac{\partial\mathcal{L}}{\partial(\partial_\mu\phi)}\delta(\partial_\mu\phi)\right]. \tag{41.10}$$

Combining the two terms on the right we obtain

$$\delta S = \int d^4x\, \partial_\mu\left[\frac{\partial\mathcal{L}}{\partial(\partial_\mu\phi)}\delta\phi\right]. \tag{41.11}$$

From (41.7),

$$\delta\phi = \epsilon\Delta\phi. \tag{41.12}$$

Since ϵ is arbitrary, the invariance condition (41.9) implies that

$$\int d^4x\, \partial_\mu \left[\frac{\partial \mathcal{L}}{\partial\left(\partial_\mu \phi\right)} \Delta\phi \right] = 0. \tag{41.13}$$

If we define

$$J^\mu = \frac{\partial \mathcal{L}}{\partial\left(\partial_\mu \phi\right)} \Delta\phi \tag{41.14}$$

then from (41.13) we obtain

$$\partial_\mu J^\mu = 0. \tag{41.15}$$

If we designate J^μ as the "current" then according to (41.15), this current is conserved. This is the essence of Noether's theorem. We point out that this current need not be the same as the physical current we defined in other contexts.

Integrating (41.15) we obtain

$$\int d^3r\, \partial_\mu J^\mu = \int d^3r \left[\frac{\partial J^4}{\partial t} + \boldsymbol{\nabla} \cdot \mathbf{J} \right] = 0 \tag{41.16}$$

where the integral is over the infinite three-dimensional space. Therefore,

$$\int d^3r\, \frac{\partial J^4}{\partial t} = -\int d^3r \boldsymbol{\nabla} \cdot \mathbf{J}. \tag{41.17}$$

From the divergence theorem we can convert the volume integral to the surface integral at infinity. If $d\mathbf{S}$ is the element of the surface at infinity, we find

$$\int d^3r\, \boldsymbol{\nabla} \cdot \mathbf{J} = \int d\mathbf{S} \cdot \mathbf{J} = 0 \tag{41.18}$$

provided J vanishes sufficiently rapidly at infinity. If we define a "charge" as

$$Q(t) = \int d^3r\, J^4(\mathbf{r}, t) \tag{41.19}$$

then (41.17) and (41.18) give

$$\frac{dQ}{dt} = 0. \tag{41.20}$$

Thus the "charge" related to the fourth component of J^μ is conserved.

41.3 Examples

Below we outline the Lagrangian structure of some of the classical fields.

41.3.1 Klein–Gordon field

Real ϕ

We have discussed this in the previous section. The appropriate Lagrangian is

$$\mathcal{L} = \frac{1}{2}[(\partial^\mu \phi)^2 - \mu^2\phi^2] \tag{41.21}$$

where $(\partial^\mu \phi)^2 = \partial^\mu \phi \partial_\mu \phi$. It gives rise to the Euler–Lagrange equation

$$(\partial^\mu \phi)^2 + \mu^2 \phi = 0. \tag{41.22}$$

In particle physics it is customary to consider the $\mu^2\phi^2$ term in $\mathcal{L}$ to be a part of the potential following the traditional definition $L = T - V$. For example, one writes the above Lagrangian as

$$\mathcal{L} = \frac{1}{2}(\partial^\mu \phi)^2 - V(\phi) \tag{41.23}$$

where

$$V(\phi) = \frac{1}{2}\mu^2\phi^2. \tag{41.24}$$

The above designation may seem like a bookkeeping arrangement and unnecessary. In fact, the potential plays a fundamental role. For $\mu^2 > 0$ it corresponds to the mass of the scalar particle as we already know. But if $\mu^2 < 0$ it leads to a phenomenon called spontaneous symmetry breaking, which has profound consequences that are at the basis of recent advances in particle physics. We will discuss this in Chapter 42 on spontaneous symmetry breaking, where the potential is typically of the form

$$V(\phi) = \frac{1}{2}\mu^2\phi^2 + \frac{\lambda}{4}\phi^4. \tag{41.25}$$

The Euler–Lagrange equation for this case is

$$(\partial^\mu \phi)^2 + \mu^2 \phi + \lambda\phi^3 = 0. \tag{41.26}$$

Complex ϕ

The simplest Lagrangian with a complex ϕ is of the form

$$\mathcal{L} = \frac{1}{2}\partial^\mu \phi^* \partial_\mu \phi - V(\phi, \phi^*), \tag{41.27}$$

with

$$V(\phi, \phi^*) = \frac{1}{2}\mu^2\phi^*\phi. \tag{41.28}$$

The Euler–Lagrange equations lead to

$$\partial_\mu \left(\frac{\partial \mathcal{L}}{\partial\left(\partial_\mu \phi\right)} \right) - \frac{\partial \mathcal{L}}{\partial \phi} = 0 \Longrightarrow \partial_\mu \partial^\mu \phi^* + \mu^2 \phi^* = 0, \tag{41.29}$$

$$\partial_\mu \left(\frac{\partial \mathcal{L}}{\partial\left(\partial_\mu \phi^*\right)} \right) - \frac{\partial \mathcal{L}}{\partial \phi^*} = 0 \Longrightarrow \partial_\mu \partial^\mu \phi + \mu^2 \phi = 0. \tag{41.30}$$

One normally defines

$$\phi = \frac{\sigma + i\pi}{\sqrt{2}} \quad \text{and} \quad \phi^* = \frac{\sigma - i\pi}{\sqrt{2}} \tag{41.31}$$

where σ and π are scalar (Hermitian) fields. Then from (49.29) and (49.30) we deduce

$$\partial_\mu \partial^\mu \sigma + \mu^2 \sigma = 0, \tag{41.32}$$

$$\partial_\mu \partial^\mu \pi + \mu^2 \pi = 0. \tag{41.33}$$

Thus we have two degenerate scalar fields σ and π.

One could include possible additional interactions of the form given by

$$V(\phi, \phi^*) = \frac{1}{2}\mu^2\phi^*\phi + \frac{\lambda}{4}\left(\phi^*\phi\right)^2, \tag{41.34}$$

then the Lagrangian in terms of σ and π is

$$\mathcal{L} = \frac{1}{2}\left(\partial_\mu \sigma\right)^2 + \frac{1}{2}\left(\partial_\mu \pi\right)^2 - V\left(\sigma^2 + \pi^2\right) \tag{41.35}$$

where

$$V\left(\sigma^2 + \pi^2\right) = \frac{\mu^2}{2}\left(\sigma^2 + \pi^2\right) + \frac{\lambda}{4}\left(\sigma^2 + \pi^2\right)^2. \tag{41.36}$$

The equations of motion are

$$\partial_\mu \partial^\mu \sigma + \mu^2 \sigma + \lambda\sigma\left(\sigma^2 + \pi^2\right) = 0, \tag{41.37}$$

$$\partial_\mu \partial^\mu \pi + \mu^2 \pi + \lambda\pi\left(\sigma^2 + \pi^2\right) = 0. \tag{41.38}$$

Conserved current

Let us now consider the conserved current for the Lagrangian corresponding to complex scalar fields and discuss the consequences of Noether's theorem.

We return first to the Lagrangian (41.27),

$$\mathcal{L} = \frac{1}{2}\left[\partial_\mu \phi \partial^\mu \phi^* - \mu^2 \phi\phi^*\right]. \tag{41.39}$$

It is invariant under the transformation

$$\phi \to \phi' = e^{i\alpha}\phi \tag{41.40}$$

which is a continuous transformation since α can take continuous values. For an infinitesimal transformation, $\alpha = \epsilon$, we obtain

$$\phi \to \phi' = e^{i\epsilon}\phi = \phi + i\epsilon\phi. \tag{41.41}$$

Thus, from (41.21),

$$\Delta\phi = i\phi \quad \text{and} \quad \Delta\phi^* = -i\phi^*. \tag{41.42}$$

Hence, the current J^μ is given by

$$J^\mu = \frac{\partial \mathcal{L}}{\partial\left(\partial_\mu \phi\right)}\Delta\phi + \frac{\partial \mathcal{L}}{\partial\left(\partial_\mu \phi^*\right)}\Delta\phi^*, \tag{41.43}$$

which is found to be

$$J^\mu = \left(\partial^\mu \phi^*\right) i\phi + \left(\partial^\mu \phi\right)\left(-i\phi^*\right) \tag{41.44}$$

$$= i\left(\phi\partial^\mu\phi^* - \phi^*\partial^\mu\phi\right). \tag{41.45}$$

According to Noether's theorem this current is conserved. That is,

$$\partial_\mu J^\mu = 0. \tag{41.46}$$

This is the same result as we obtained for the complex scalar field discussed in Chapter 37. In this case it is in fact the physical current.

In terms of σ and π, the current J^μ can be written as

$$J^\mu = \sigma\partial^\mu\pi - \pi\partial^\mu\sigma. \tag{41.47}$$

The presence of the interaction terms does not affect the form of the current since these terms do not involve derivatives. We note that σ and π satisfy the equations given by (41.37) and (41.38) of the previous section.

41.3.2 Dirac field

The Lagrangian for the Dirac field can be written as

$$\mathcal{L} = i\overline{\psi}\gamma^{\mu}\partial_{\mu}\psi - m\overline{\psi}\psi \tag{41.48}$$

where ψ is the Dirac spinor with $\overline{\psi}$ and γ^{μ} which have been previously defined. The corresponding Euler–Lagrange equation is

$$\partial_{\mu}\left(\frac{\partial\mathcal{L}}{\partial\left(\partial_{\mu}\psi\right)}\right) - \frac{\partial\mathcal{L}}{\partial\psi} = 0, \tag{41.49}$$

which gives

$$-\partial_{\mu}i\overline{\psi}\gamma^{\mu} + m\overline{\psi} = 0. \tag{41.50}$$

After multiplying (41.50) on the right by γ^4 and using the relation $\overline{\psi}\gamma^4 = \psi^{\dagger}$, we take the Hermitian conjugate of (41.50) and obtain

$$i\gamma^{\mu}\partial_{\mu}\psi - m\psi = 0. \tag{41.51}$$

This is, of course, the Dirac equation. We will defer the discussion on the presence of interactions to the next section when we discuss Maxwell's equations.

Conserved current

The Lagrangian (41.48) is invariant under the transformation

$$\psi \to \psi' = e^{i\alpha}\psi \underset{\alpha=\epsilon}{\to} \psi + i\epsilon\psi. \tag{41.52}$$

Hence

$$\Delta\psi = i\psi, \quad \text{and} \quad \Delta\overline{\psi} = -i\overline{\psi}. \tag{41.53}$$

The current J^{μ} is then

$$J^{\mu} = \frac{\partial\mathcal{L}}{\partial\left(\partial_{\mu}\psi\right)}\Delta\psi + \frac{\partial\mathcal{L}}{\partial\left(\partial_{\mu}\overline{\psi}\right)}\Delta\overline{\psi}, \tag{41.54}$$

which reduces to

$$J^{\mu} = \overline{\psi}\gamma^{\mu}\psi. \tag{41.55}$$

This is the same expression for the current as we obtained in Chapter 33 on the Dirac equation.

41.4 Maxwell's equations and consequences of gauge invariance

The Lagrangian for Maxwell's equation is written as

$$\mathcal{L} = -\frac{1}{4}F^{\mu\nu}F_{\mu\nu} \tag{41.56}$$

where $F_{\mu\nu}$, Maxwell's tensor, is given in terms of the vector potential A_μ as follows:

$$F_{\mu\nu} = \partial_\mu A_\nu - \partial_\nu A_\mu. \tag{41.57}$$

The Euler–Lagrange equations are then given by

$$\partial_\mu \left(\frac{\partial \mathcal{L}}{\partial \left(\partial_\mu A_\nu \right)} \right) - \frac{\partial \mathcal{L}}{\partial A_\nu} = 0. \tag{41.58}$$

From (41.56) and (41.57) we find that

$$\frac{\partial \mathcal{L}}{\partial A_\nu} = 0 \tag{41.59}$$

and

$$\left(\frac{\partial \mathcal{L}}{\partial \left(\partial_\mu A_\nu \right)} \right) = \left(-\frac{1}{4} \right) \frac{\partial}{\partial \left(\partial_\mu A_\nu \right)} \left[\partial_\alpha A_\beta - \partial_\beta A_\alpha \right] \left[\partial^\alpha A^\beta - \partial^\beta A^\alpha \right] = -F^{\mu\nu}. \tag{41.60}$$

Hence (41.58) gives

$$\partial_\mu F^{\mu\nu} = 0, \tag{41.61}$$

which is Maxwell's equation.

Below we reiterate what we stated in Chapter 32, but this time in the context of the Lagrangian framework: the equation in terms of A^μ can be obtained through the definition (41.56) to give

$$\Box A^\mu + \partial^\mu (\partial_\nu A^\nu) = 0 \tag{41.62}$$

where $\Box = \partial_\nu \partial^\nu$.

We noted earlier in Chapter 31 that $F^{\mu\nu}$ and Maxwell's equations are invariant under the gauge transformation

$$A_\mu \to A'_\mu = A_\mu - g\partial_\mu \chi(x). \tag{41.63}$$

As we mentioned then, since χ depends on x, this transformation is called a "local" gauge transformation. This allows us to take advantage of the arbitrariness given by (41.63) and eliminate the second term in (41.62) by taking

$$\partial_\nu A^\nu = 0. \tag{41.64}$$

This is the Lorentz condition. Under gauge transformation (41.63) we find

$$\partial_\nu A^\nu \to \partial_\nu A'^\nu = \partial_\nu A^\nu - g\partial_\nu \partial^\nu \chi. \tag{41.65}$$

To keep condition (41.64) invariant we choose χ such that

$$\Box\chi = 0. \tag{41.66}$$

Thus, from (41.62), A_μ satisfies the equation

$$\Box A^\mu = 0 \tag{41.67}$$

provided the relations (41.62), (41.64), and (41.66) are imposed. The condition (41.64) is also called the transverse gauge, which we have considered in the past.

We note that (41.67) corresponds to the Klein–Gordon equation for each of the four components of A^μ but with zero mass. A mass term added to (41.67) will give

$$\Box A^\mu + m^2 A^\mu = 0, \tag{41.68}$$

which corresponds to the Lagrangian

$$\mathcal{L} = -\frac{1}{4}F_{\mu\nu}F^{\mu\nu} + m^2 A^\mu A_\mu. \tag{41.69}$$

But this Lagrangian is no longer invariant under gauge transformations. The presence of a mass term destroys gauge invariance. Invoking the concept of photons, we can restate this situation slightly differently and say that the photon is massless as given by (41.67) because of gauge invariance.

We will now go a step further and consider Maxwell's field interacting with an external current j^ν. The Lagrangian is then of the form

$$\mathcal{L} = -\frac{1}{4}F_{\mu\nu}F^{\mu\nu} + \frac{1}{c}j^\mu A_\mu. \tag{41.70}$$

The corresponding Maxwell's equation obtained from the Euler–Lagrange equations is

$$\partial_\mu F^{\mu\nu} = \frac{1}{c}j^\nu. \tag{41.71}$$

Let j^ν be the current due to a Dirac particle. In our discussion of the conserved currents for the Dirac field we found that

$$j^\nu = e\bar{\psi}\gamma^\nu\psi \tag{41.72}$$

where we have introduced the factor e since we want to consider the electromagnetic current. Thus (41.71) can be written as

$$\partial_\mu F^{\mu\nu} = \frac{e}{c}\bar{\psi}\gamma^\nu\psi. \tag{41.73}$$

Having discussed Maxwell's equation in the presence of a Dirac current, we will now consider the Dirac equation in the presence of an electromagnetic field, which we already know to be of the form

$$\gamma^{\mu}(i\partial_{\mu} - eA_{\mu})\psi + m\psi = 0. \tag{41.74}$$

To determine the consequences of gauge transformation we must consider (41.73) and (41.74) together and obtain the gauge transformation properties of ψ. We show below that if

$$\psi(x) \rightarrow \psi'(x) = e^{i\chi(x)}\psi(x) \tag{41.75}$$

where $\chi(x)$ is the **same** function as $\chi(x)$ in (41.63), then both (41.73) and (41.74) will be gauge-invariant equations. It is easy to confirm this for (41.73) since $F_{\mu\nu}$ that appears on the left is gauge invariant and, on the right, since both ψ and $\bar{\psi}$ appear, the contribution of $e^{i\chi(x)}$ is canceled.

Equation (41.74) under gauge transformation is given by

$$\bar{\psi}\gamma^{\mu}(i\partial_{\mu} - eA_{\mu})\psi \rightarrow \bar{\psi}'\gamma^{\mu}(i\partial_{\mu} - eA'_{\mu})\psi = e^{-i\chi}\bar{\psi}\gamma^{\mu}\left[i\partial_{\mu} - e(A_{\mu} - \frac{1}{e}\partial_{\mu}\chi)\right]e^{i\chi}\psi \tag{41.76}$$

where we have taken $g = 1/e$ in (41.63). The right-hand side above simplifies to

$$= \bar{\psi}[-\partial_{\mu}\chi + i\partial_{\mu} - eA_{\mu} + \partial_{\mu}\chi]\psi \tag{41.77}$$

$$= \bar{\psi}[i\partial_{\mu} - eA_{\mu}]\psi \tag{41.78}$$

where we have canceled $e^{-i\chi}$ with $e^{i\chi}$ that remain after taking the derivatives. Thus both (41.73) and (41.74) remain invariant under the gauge transformations of A_{μ} and ψ given, respectively, by

$$A_{\mu} \rightarrow A'_{\mu} = A_{\mu} - \frac{1}{e}\partial_{\mu}\chi, \tag{41.79}$$

$$\psi \rightarrow \psi' - e^{i\chi}\psi. \tag{41.80}$$

To make gauge invariance manifest in the field equations, we incorporate the notation D_{μ}, defined previously,

$$D_{\mu} = \partial_{\mu} - ieA_{\mu}. \tag{41.81}$$

We note that

$$D_{\mu}\psi \rightarrow D'_{\mu}\psi' = (\partial_{\mu} - ieA'_{\mu})(e^{i\chi}\psi) = e^{i\chi}(\partial_{\mu} - ieA_{\mu})\psi = e^{i\chi}D_{\mu}\psi. \tag{41.82}$$

Finally, from the above results we conclude that a gauge-invariant Lagrangian for the Dirac particle can be written as

$$\mathcal{L} = \bar{\psi}\gamma^{\mu}D_{\mu}\psi + m\bar{\psi}\psi.$$

41.4.1 Further consequences of imposing local gauge invariance

Let us now summarize the above results by considering the consequences that would arise if, starting from scratch, we asserted that our theory be locally gauge invariant. Starting with the Dirac equation, we find that while $\bar{\psi}\psi$ is invariant under $\psi(x) \to e^{i\chi(x)}\psi(x)$, the first term in the equation gives

$$i\bar{\psi}\gamma^{\mu}\partial_{\mu}\psi \to i\bar{\psi}'\gamma^{\mu}\partial_{\mu}\psi' - \bar{\psi}'\gamma^{\mu}\left(\partial_{\mu}\chi\right)\psi' \tag{41.83}$$

where there is a left-over term involving $\partial_{\mu}\chi$. The only way to cancel this term is to add a vector field, $A_{\mu}(x)$, which can be done by replacing ∂_{μ} in the Dirac equation by D_{μ} defined in (41.81). Furthermore, as we have already observed, this vector field must be massless. From these results one can say that gauge symmetry "predicts" a massless vector particle, which in QED is the photon.

A locally gauge invariant Lagrangian that combines both the Dirac and Maxwell's equations can be written as

$$\mathcal{L} = \bar{\psi}i\gamma^{\mu}D_{\mu}\psi + m\bar{\psi}\psi - \frac{1}{4}F^{\mu\nu}F_{\mu\nu}. \tag{41.84}$$

We note that the Euler–Lagrange equation for the ψ variable will give (41.74), while for the A^{μ} variables it will give (41.73). There are two important things achieved by local gauge invariance:

(i) massless vector fields;
(ii) minimal coupling $e\bar{\psi}\gamma_{\mu}\psi A^{\mu}$, which is contained in the term $\bar{\psi}\gamma^{\mu}D_{\mu}\psi$. In other words, the photon's coupling to the electron, or any other matter field, is determined by its transformation property. It is found that the above conclusions hold even under second quantization.

In the case of a complex (charged) scalar boson interacting with the Maxwell field, the Lagrangian is given by

$$\mathcal{L} = -\frac{1}{4}\left(F_{\mu\nu}\right)^2 + \left|D_{\mu}\phi\right|^2 - V\left(\phi, \phi^*\right), \tag{41.85}$$

which is invariant under

$$\phi(x) \to e^{i\alpha(x)}\phi(x), \quad A_{\mu}(x) \to A_{\mu}(x) - \frac{1}{e}\partial_{\mu}\alpha(x) \tag{41.86}$$

where V is the gauge-invariant potential.

41.4.2 Maxwell's equations with Dirac and scalar particles

It can readily be checked that the Lagrangian that is locally gauge invariant and has a Dirac particle, a charged scalar particle, and the zero-mass photon (Maxwell's equation) is given by

$$\mathcal{L} = \bar{\psi} i\gamma^{\mu} D_{\mu}\psi + \left|D_{\mu}\phi\right|^2 - \frac{1}{4}\left(F_{\mu\nu}\right)^2 + m\bar{\psi}\psi - V(\phi) \tag{41.87}$$

where

$$\left|D_{\mu}\phi\right|^2 = \left(D_{\mu}\phi\right)^* \left(D_{\mu}\phi\right) \tag{41.88}$$

and

$$V(\phi) = \frac{1}{2}\mu^2 |\phi|^2 . \tag{41.89}$$

Gauge theory

In conclusion, what we have discussed above are called local symmetries since they depend on space-time variables. We saw that such gauge symmetries generate dynamics called gauge interactions by giving rise to interaction couplings that involve vector particles. The theory underlying the above formalism is called gauge theory and one believes that all basic interactions are described by some form of gauge theory consisting of only massless particles, included among which are the vector bosons, which are often called the gauge bosons. The acquisition of mass for these particles then results from spontaneous symmetry breaking of gauge invariance, which we will discuss in the next chapter.

42 Spontaneous symmetry breaking

42.1 BCS mechanism

We consider the Lagrangian defined in Chapter 39 for the Cooper pair interaction as the BCS Lagrangian, which is given by ($\hbar = c = 1$)

$$\mathcal{L}(\psi, \psi^\dagger) = i\psi^\dagger \frac{\partial}{\partial t}\psi + \psi^\dagger \left(\frac{\nabla^2}{2m} + \mu\right)\psi + \frac{g}{2}\psi^\dagger(x)\,\psi^\dagger_\beta(x)\,\psi_\beta(x)\,\psi(x) \tag{42.1}$$

where ψ is the nonrelativistic electron field. The corresponding Euler–Lagrange equation for the field ψ_α is

$$i\frac{\partial}{\partial t}\psi_\alpha(x) + \left(\frac{\nabla^2}{2m} + \mu\right)\psi_\alpha(x) + g\psi^\dagger_\beta(x)\,\psi_\beta(x)\,\psi_\alpha(x) = 0. \tag{42.2}$$

We note that the Lagrangian and the equation of motion are invariant under the transformation

$$\psi \to e^{-i\theta}\psi, \quad \psi^\dagger \to e^{i\theta}\,\psi^\dagger \tag{42.3}$$

where θ is a real, continuous variable, independent of x. This type of transformation is designated as $U(1)$, and since θ does not depend on x, one calls it a "global" transformation. If θ were dependent on x, then we would be dealing with the so-called "local" transformation.

Now let us consider the equation

$$i\frac{\partial}{\partial t}\psi_\alpha(x) + \left(\frac{\nabla^2}{2m} + \mu\right)\psi_\alpha(x) + M\epsilon_{\alpha\beta}\psi^\dagger_\beta(x) = 0, \tag{42.4}$$

which was obtained from (42.2) after we made the substitution

$$g\psi_\beta(x)\,\psi_\alpha(x) \to g\left\langle 0\left|\psi_\beta(x)\,\psi_\alpha(x)\right|0\right\rangle = M\epsilon_{\alpha\beta} \tag{42.5}$$

where the vacuum state $|0\rangle$ is the BCS vacuum $|0\rangle_{BCS}$. Since M is just a number and not an operator, it cannot be affected by the $U(1)$ transformation, and therefore, equation (42.4) is no longer invariant under the $U(1)$ transformation (42.3). The culprit is clearly (42.5) where we replaced a composite operator by its vacuum expectation value. If we examine (42.5), we find that if the vacuum is assumed to be invariant under the $U(1)$ transformation (42.3) then we must have

$$\left\langle 0\left|\psi_\beta(x)\,\psi_\alpha(x)\right|0\right\rangle = e^{-2i\theta}\left\langle 0\left|\psi_\beta(x)\,\psi_\alpha(x)\right|0\right\rangle. \tag{42.6}$$

Since θ is an arbitrary variable, this relation would imply that $\langle 0 |\psi_\beta(x)\psi_\alpha(x)| 0\rangle$ must vanish. The only alternative available to us is to assume that vacuum does not respect the $U(1)$ symmetry. Thus the vacuum must break the $U(1)$ symmetry. This type of symmetry breaking is called "spontaneous" symmetry breaking (SSB) where the Lagrangian satisfies the symmetry but not the vacuum.

Let us consider the relation between the BCS vacuum and the ordinary vacuum,

$$|0\rangle = \prod_k \left(u_k + v_k a_k^\dagger a_{-k}^\dagger\right)|0\rangle_0, \tag{42.7}$$

where $|0\rangle$ on the left corresponds to the BCS vacuum and $|0\rangle_0$ on the right corresponds to the ordinary vacuum. By implementing the relation (42.3) for the creation operators we find that

$$U(\theta)|0\rangle = \prod_k \left(u_k + v_k e^{2i\theta} a_k^\dagger a_{-k}^\dagger\right)|0\rangle_0 \tag{42.8}$$

where $U(\theta)$ designates the $U(1)$ transformation given by (42.3). Clearly the BCS vacuum $|0\rangle$ is not invariant under $U(\theta)$. If we write

$$|\theta\rangle = U(\theta)|0\rangle \tag{42.9}$$

then $|\theta\rangle$ is a different vacuum state. Furthermore,

$$U(\theta')|\theta\rangle = |\theta + \theta'\rangle. \tag{42.10}$$

Thus, since θ' is continuous, we have an infinite number of states that are all vacuum states and they are degenerate.

We recall from our discussion on superconductivity that the presence of a nonzero vacuum expectation value (VEV) led to the formation of condensates in the BCS system. Since we have found that a nonzero VEV leads to spontaneous symmetry breaking (and vacuum degeneracy), this example illustrates the fact that the presence of condensates leads to SSB, and vice versa, i.e.,

$$\text{Condensates} \rightleftarrows SSB\,(+\text{Vacuum degeneracy}). \tag{42.11}$$

Since SSB occurs due to the formation of condensates, which are dynamical entities, the symmetry breaking such as that found in superconductivity is called "dynamical" symmetry breaking.

42.2 Ferromagnetism

Let us consider the case of ferromagnetism near the Curie temperature. The free energy density for small values of magnetization, $\mathbf{M}$, is given by

$$u(\mathbf{M}) = (\partial_i \mathbf{M})^2 + V(\mathbf{M}) \tag{42.12}$$

where $i\,(= 1, 2, 3)$ corresponds to the space coordinates and $V(\mathbf{M})$ is defined through Ginzburg–Landau theory as

$$V(\mathbf{M}) = a(T)(\mathbf{M}\cdot\mathbf{M}) + b(\mathbf{M}\cdot\mathbf{M})^2 \tag{42.13}$$

where b is a positive constant independent of temperature, T, while a is defined as

$$a = a_0\,(T - T_C) \quad \text{with } a_0 > 0. \tag{42.14}$$

Here T_C is the Curie temperature, so that a is positive when $T > T_C$ but becomes negative when T goes below T_C. The symmetry here is rotational symmetry as $u(\mathbf{M})$ is invariant under rotations.

The ground-state magnetization (the "vacuum") is obtained by taking a minimum of V,

$$\partial V/\partial M_i = 0, \tag{42.15}$$

to obtain

$$M_i(a + 2b\mathbf{M}\cdot\mathbf{M}) = 0 \tag{42.16}$$

for $i = 1, 2, 3$. For $T > T_C$, where a is positive, the solution is

$$M_i = 0, \tag{42.17}$$

which fixes both the magnitude ($|\mathbf{M}| = 0$) and direction of $\mathbf{M}$ for the ground state. We have a unique vacuum that is rotationally invariant.

However, for $T < T_C$ when $a < 0$, the minimum is given by

$$|\mathbf{M}| = (-a/2b)^{1/2}. \tag{42.18}$$

$V(\mathbf{M})$ in this case is $(a/2b)\,|\mathbf{M}|^2$, which is negative, and therefore, lower than $V(\mathbf{M}) = 0$. The value $\mathbf{M} = 0$ now corresponds to a local maximum.

Relation (42.18) fixes only the magnitude of the vacuum state but it says nothing about the direction. It can have any direction among an infinite number of possibilities, thus the vacuum state is infinitely degenerate. If we apply rotation to a vacuum state, we get another vacuum state, not the same state, i.e.,

$$U\,|0\rangle \neq |0\rangle \tag{42.19}$$

where U is the rotation operator. Once the direction of $\mathbf{M}$ is chosen, we have a single vacuum, with a specific direction, which breaks the rotational symmetry.

In summary, for $T > T_C$ all the magnetic dipoles are randomly oriented in the ground state, which is rotationally invariant, but for $T < T_C$ they are all aligned in some arbitrary direction, leading to spontaneous magnetization. This phenomenon provides an explanation for the ferromagnetism observed in certain materials.

Next we consider the classical fields.

42.3 SSB for discrete symmetry in classical field theory

We will consider classical fields, though some of the results we obtain will be expressed in the language of quantum field theory. For our first example we consider the Lagrangian density for scalar particles given as

$$\mathcal{L} = \frac{1}{2}\left(\partial_\mu \phi\right)^2 - V(\phi) \tag{42.20}$$

where, as we discussed in Chapter 41, the $\mu^2\phi^2$ term is contained in $V(\phi)$,

$$V(\phi) = \frac{1}{2}\mu^2\phi^2 + \frac{\lambda}{4}\phi^4 \tag{42.21}$$

where $\lambda > 0$. $V(\phi)$ is called the "potential" to separate it from the kinetic energy term given by the first term in (42.20). The Lagrangian (42.20) is symmetric under the discrete transformation

$$\phi \to \phi' = -\phi. \tag{42.22}$$

To find the lowest energy state we determine the minimum of the potential $V(\phi)$ through the relation

$$\frac{\partial V}{\partial \phi} = 0, \tag{42.23}$$

which gives

$$\mu^2\phi + \lambda\phi^3 = 0. \tag{42.24}$$

The solutions are found to be

$$\phi_{\text{min}} = 0, \quad \pm\sqrt{-\mu^2/\lambda}. \tag{42.25}$$

Let us consider two possibilities: (i) $\mu^2 > 0$ and (ii) $\mu^2 < 0$. The behavior of $V(\phi)$ for both these cases is shown in Fig. 42.1.

(i) For $\mu^2 > 0$, the term $\frac{1}{2}\mu^2\phi^2$ in $V(\phi)$ signifies a mass term and corresponds to the Klein–Gordon equation for a scalar particle with mass μ. The vacuum state is given by $\phi_{\text{min}} = 0$, and it is the only allowed solution. For this solution, $V(\phi_{\text{min}}) = 0$. If $|0\rangle$ designates the vacuum state, then $\phi_{\text{min}} = 0$ corresponds to

$$\langle 0|\, \phi\, |0\rangle = 0; \tag{42.26}$$

and if U designates the transformation (42.22), then since this is a unique vacuum, it does not change under U, i.e.,

$$U\,|0\rangle = |0\rangle\,. \tag{42.27}$$

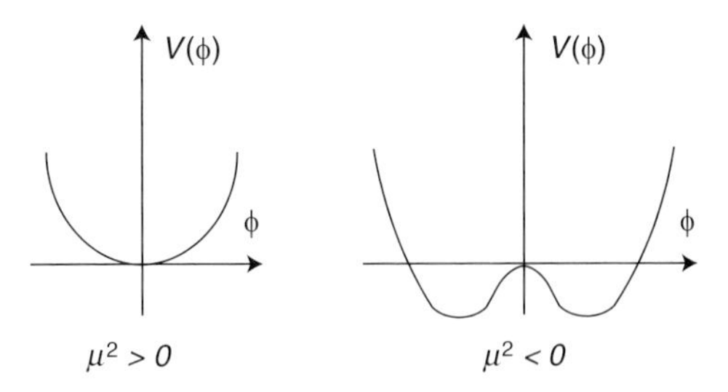

Fig. 42.1

(ii) For $\mu^2 < 0$, the term $(1/2)\,\mu^2\phi^2 = -(1/2)\,|\mu|^2\,\phi^2$ in $V(\phi)$ will have no relation to the mass of the Klein–Gordon particle. Thus in this case we must consider the scalar particle represented by the Klein–Gordon equation to have zero mass. The vacuum state will be determined by two other solutions in (42.25), $\phi_{\min} = \pm\sqrt{-\mu^2/\lambda}$, for which

$$V(\phi_{\min}) = -\frac{\mu^4}{4\lambda}, \tag{42.28}$$

which is lower than $V(\phi_{\min}) = 0$, while $\phi_{\min} = 0$ is now a local maximum. We have now a choice of two minima, $\pm\sqrt{-\mu^2/\lambda}$. Either choice will break the reflection symmetry (42.22).

If we designate the degenerate vacuum states as $|0_1\rangle$ for $\phi_{\min} = +\sqrt{-\mu^2/\lambda}$ and $|0_2\rangle$ for $\phi_{\min} = -\sqrt{-\mu^2/\lambda}$, then for both

$$\langle 0|\,\phi\,|0\rangle \neq 0. \tag{42.29}$$

Under the transformation U given by (42.22), $U\,|0_i\rangle \neq |0_i\rangle$ for $i = 1, 2$, $|0_1\rangle$ and $|0_2\rangle$ are interchanged, i.e.,

$$U\,|0_1\rangle = |0_2\rangle\,. \tag{42.30}$$

If we now select a specific vacuum state, say $|0_1\rangle$, the potential can be expressed with respect to it by writing

$$\phi(x) = v + \sigma(x) \tag{42.31}$$

with

$$v = \sqrt{m^2/\lambda}, \quad m^2 = -\mu^2 > 0 \tag{42.32}$$

and

$$\langle 0|\,\sigma\,|0\rangle = 0 \tag{42.33}$$

where v is called the vacuum expectation value of ϕ and written as

$$v = \langle 0|\,\phi\,|0\rangle\,. \tag{42.34}$$

The Lagrangian given in (42.20) can be written in terms of σ as

$$\mathcal{L} = \frac{1}{2}\left(\partial_\mu \sigma\right)^2 - V(\sigma) \tag{42.35}$$

where $V(\sigma)$ is given by

$$V(\sigma) = m^2\sigma^2 + \lambda v \sigma^3 + \frac{\lambda}{4}\sigma^4. \tag{42.36}$$

We note that the negative mass term for $V(\phi)$ in (42.21) has disappeared; instead we have a term $m^2\sigma^2$ in $V(\sigma)$, which is a positive term and, when written as $\frac{1}{2}(2m^2)$, corresponds to a mass $\sqrt{2}m$.

Thus, for $\mu^2 < 0$, the symmetry $\phi \to -\phi$ is spontaneously broken for the massless scalar theory, giving rise to a new state – a particle of mass $\sqrt{2}m$.

42.4 SSB for continuous symmetry

As our next example we consider a Lagrangian that exhibits continuous symmetry,

$$\mathcal{L} = \frac{1}{2}\left(\partial_\mu \sigma\right)^2 + \frac{1}{2}\left(\partial_\mu \pi\right)^2 - V(\sigma, \pi), \tag{42.37}$$

where

$$V(\sigma, \pi) = \frac{\mu^2}{2}\left(\sigma^2 + \pi^2\right) + \frac{\lambda}{4}\left(\sigma^2 + \pi^2\right)^2. \tag{42.38}$$

We now have two scalar particles, σ and π, both with the same mass μ if $\mu^2 > 0$, or both with zero mass if $\mu^2 < 0$.

The Lagrangian density $\mathcal{L}$ in (42.37) is invariant under the following continuous transformation, designated as the $O(2)$ transformation,

$$\begin{pmatrix} \sigma \\ \pi \end{pmatrix} \to \begin{pmatrix} \sigma' \\ \pi' \end{pmatrix} = \begin{pmatrix} \cos\alpha & \sin\alpha \\ -\sin\alpha & \cos\alpha \end{pmatrix} \begin{pmatrix} \sigma \\ \pi \end{pmatrix} \tag{42.39}$$

since it leaves $\left(\sigma^2 + \pi^2\right)$ invariant. To obtain the ground state we minimize V in (42.38) with respect to σ and π to obtain

$$\frac{\partial V}{\partial \sigma} = \sigma\left[\mu^2 + \lambda\left(\sigma^2 + \pi^2\right)\right] = 0, \tag{42.40}$$

$$\frac{\partial V}{\partial \pi} = \pi\left[\mu^2 + \lambda\left(\sigma^2 + \pi^2\right)\right] = 0. \tag{42.41}$$

For $\mu^2 > 0$, we have the trivial solution $\sigma_{\min} = \pi_{\min} = 0$ for the minimum, which corresponds to the normal vacuum $V_{\min} = 0$.

For $\mu^2 < 0$, however, we have the solution

$$\sigma^2 + \pi^2 = v^2 \tag{42.42}$$

where

$$v = \sqrt{\frac{m^2}{\lambda}}, \quad m^2 = -\mu^2 > 0. \tag{42.43}$$

Thus we have now an infinite number of vacua described by the points along the circle given by (42.42) with a radius given in (42.43). We can select a point $\sigma_{\text{min}} = v$ and $\pi_{\text{min}} = 0$, i.e.,

$$\langle 0 |\sigma| 0\rangle = v, \quad \langle 0 |\pi| 0\rangle = 0. \tag{42.44}$$

By selecting a specific "direction," however, we break the $O(2)$ symmetry. If we expand V around its minimum by writing

$$\sigma = \sigma' + v, \tag{42.45}$$

$$\pi = \pi', \tag{42.46}$$

we obtain

$$\mathcal{L} = \frac{1}{2}\left[\left(\partial_\mu \sigma'\right)^2 + \left(\partial_\mu \pi'\right)^2\right] + m^2\sigma'^2 - \lambda v \sigma'\left(\sigma'^2 + \pi'^2\right) - \frac{\lambda}{4}\left(\sigma'^2 + \pi'^2\right)^2, \tag{42.47}$$

which corresponds to a σ' particle of mass $\sqrt{2}m$, but the mass of π' is zero.

Thus for $\mu^2 < 0$, we discover that in place of zero mass particles σ and π with which we started out, we have, after SSB, a particle of mass $\sqrt{2}m$ along with a massless particle.

Let us do this problem another way, by rewriting the above Lagrangian in terms of complex scalar fields by defining

$$\phi = \frac{1}{\sqrt{2}}\left(\sigma + i\pi\right). \tag{42.48}$$

The Lagrangian density (42.37) can be re-expressed as

$$\mathcal{L} = \partial_\mu \phi^* \partial^\mu \phi - V(\phi, \phi^*) \tag{42.49}$$

where

$$V = \mu^2 \phi^* \phi + \lambda \left(\phi^* \phi\right)^2. \tag{42.50}$$

The symmetry here is the $U(1)$ symmetry

$$\phi \to \phi' = e^{-i\alpha}\phi, \tag{42.51}$$

$$\phi^* \to \phi^{*\prime} = e^{i\alpha}\phi^*. \tag{42.52}$$

For $\mu^2 < 0$, the minimum of the potential corresponds to

$$|\phi|^2 = \frac{v^2}{2} \tag{42.53}$$

where

$$v = \sqrt{\frac{m^2}{\lambda}}, \quad m^2 = -\mu^2 > 0, \tag{42.54}$$

which we can write as

$$|\langle 0 \, |\phi| \, 0\rangle| = \frac{v}{\sqrt{2}}. \tag{42.55}$$

Thus, we can only fix the magnitude of ϕ but not the phase. Since

$$\phi = \frac{1}{\sqrt{2}} (\sigma + i\pi), \tag{42.56}$$

we can select the phase by taking

$$\langle 0 \, |\sigma| \, 0\rangle = v \quad \text{and} \quad \langle 0 \, |\pi| \, 0\rangle = 0, \tag{42.57}$$

which breaks the $U(1)$ symmetry.

If we take

$$\sigma' = \sigma - v \quad \text{and} \quad \pi' = \pi \tag{42.58}$$

then the Lagrangian density in terms of σ' and π' is given by

$$\mathcal{L} = \frac{1}{2}\left[\left(\partial_\mu \sigma'\right)^2 + \left(\partial_\mu \pi'\right)^2\right] + m^2\sigma'^2 - \lambda v \sigma' \left(\sigma'^2 + \pi'^2\right) - \frac{\lambda}{4}\left(\sigma'^2 + \pi'^2\right)^2, \tag{42.59}$$

recovering our previous result that the scalar particle σ' now has mass $\sqrt{2}m$ while π' is massless.

42.5 Nambu-Goldstone bosons

In the examples of continuous symmetry that we considered above, we noted that in each case after SSB we were left with a massless scalar particle along with a massive one. This result is a consequence of a very well-known theorem for processes that exhibit SSB for continuous symmetry. These massless scalar particles are known as Nambu–Goldstone bosons.

The proof is actually quite straightforward once we take account of the fact that the mass terms appear as a coefficient of the ϕ^2 term in the potential $V(\phi)$. Thus,

$$\frac{\partial^2 V}{\partial \phi^2} = \mu^2. \tag{42.60}$$

If $V(\phi)$ is invariant under a continuous transformation of ϕ, then under an infinitesimal transformation of ϕ,

$$\phi \to \phi + \Delta\phi, \tag{42.61}$$

$V(\phi)$ will remain unchanged, i.e.,

$$V(\phi) = V(\phi + \Delta\phi). \tag{42.62}$$

Since $\Delta\phi$ is infinitesimal we can expand the right-hand side above and keep terms only up to the linear term in $\Delta\phi$. Therefore,

$$V(\phi) = V(\phi) + \Delta\phi \frac{\partial V}{\partial \phi}. \tag{42.63}$$

Hence, we must have

$$\Delta\phi \frac{\partial V}{\partial \phi} = 0. \tag{42.64}$$

Let us now take the derivative of equation (42.64) with respect to ϕ at $\phi = \phi_0$ where $V(\phi)$ has a minimum.

$$\frac{\partial \Delta\phi}{\partial \phi_0} \frac{\partial V}{\partial \phi_0} + \Delta\phi_0 \frac{\partial^2 V}{\partial \phi_0^2} = 0. \tag{42.65}$$

Because $V(\phi)$ is minimum at $\phi = \phi_0$, $\partial V / \partial \phi_0$ must vanish. Therefore,

$$\Delta\phi_0 \frac{\partial^2 V}{\partial \phi_0^2} = 0. \tag{42.66}$$

If ϕ_0 corresponds to a point where SSB takes place, then the symmetry (42.62) does not hold and $\Delta\phi_0 \neq 0$. Hence, we must have

$$\frac{\partial^2 V}{\partial \phi_0^2} = 0. \tag{42.67}$$

However, when we expand $V(\phi)$ around $\phi = \phi_0$ we obtain

$$V(\phi) = V(\phi_0) + \frac{1}{2}\left(\phi - \phi_0\right)^2 \frac{\partial^2 V}{\partial \phi_0^2} + \cdots \tag{42.68}$$

where the $\partial V / \partial \phi_0$ term is absent since ϕ_0 is a minimum. The coefficient of ϕ^2 in (42.68) is $\partial^2 V / \partial \phi_0^2$, which vanishes due to relation (42.67). Hence, from (42.60) ϕ corresponds to a zero-mass particle. Thus, SSB for continuous symmetry leads to massless scalar particles, which are the Nambu–Goldstone bosons.

42.5.1 Examples

Let us elaborate on this result by returning to the Lagrangian given by (42.37) in the previous section,

$$\mathcal{L} = \frac{1}{2}\left(\partial_\mu \sigma\right)^2 + \frac{1}{2}\left(\partial_\mu \pi\right)^2 - V(\sigma, \pi) \tag{42.69}$$

where

$$V(\sigma, \pi) = \frac{\mu^2}{2}\left(\sigma^2 + \pi^2\right) + \frac{\lambda}{4}\left(\sigma^2 + \pi^2\right)^2. \tag{42.70}$$

As discussed previously, the Lagrangian is invariant under the continuous transformation (42.39). For infinitesimal transformations,

$$V(\sigma + \Delta\sigma, \pi + \Delta\pi) = V(\sigma, \pi) + \Delta\sigma \frac{\partial V}{\partial \sigma} + \Delta\pi \frac{\partial V}{\partial \pi} \tag{42.71}$$

where σ and π transform as

$$\sigma \to \sigma' = \sigma \cos\alpha + \pi \sin\alpha = \sigma + \alpha\pi, \tag{42.72}$$

$$\pi \to \pi' = -\sigma \sin\alpha + \pi \cos\alpha = \pi - \alpha\sigma. \tag{42.73}$$

Therefore,

$$\Delta\sigma = \alpha\pi \quad \text{and} \quad \Delta\pi = -\alpha\sigma. \tag{42.74}$$

If V is invariant, then from (42.71) we have

$$\Delta\sigma \frac{\partial V}{\partial \sigma} + \Delta\pi \frac{\partial V}{\partial \pi} = 0, \tag{42.75}$$

which implies, using (42.74), that

$$\pi \frac{\partial V}{\partial \sigma} - \sigma \frac{\partial V}{\partial \pi} = 0. \tag{42.76}$$

We consider the case $\mu^2 < 0$ when SSB takes place and take the derivative of (42.75) with respect to π and σ respectively at a point where V is a minimum, i.e., where

$$\left(\frac{\partial V}{\partial \pi}\right)_{\min} = 0 = \left(\frac{\partial V}{\partial \sigma}\right)_{\min}. \tag{42.77}$$

We have already determined that for this case

$$\sigma_{\min} \neq 0 \quad \text{and} \quad \pi_{\min} = 0. \tag{42.78}$$

The derivative of (42.76) with respect to π at the minimum gives

$$\left(\frac{\partial V}{\partial \sigma}+\pi \frac{\partial^2 V}{\partial \pi\, \partial \sigma}-\sigma \frac{\partial^2 V}{\partial \pi^2}\right)_{\min}=0 \tag{42.79}$$

where the term $\partial\sigma/\partial\pi$ does not appear since the two fields are independent. The first two terms vanish due to (42.77) and (42.78), respectively; therefore, we have

$$\sigma_{\min}\left(\frac{\partial^2 V}{\partial \pi^2}\right)_{\min}=0. \tag{42.80}$$

Since $\sigma_{\min} \neq 0$ from (42.78) we obtain

$$\left(\frac{\partial^2 V}{\partial \pi^2}\right)_{\min}=0. \tag{42.81}$$

Thus from (42.60) we conclude that π is the massless Nambu–Goldstone boson.

The derivative of (42.76) with respect to σ at the minimum gives

$$\left(\pi \frac{\partial^2 V}{\partial \sigma^2}-\frac{\partial V}{\partial \pi}-\sigma \frac{\partial^2 V}{\partial \sigma\, \partial \pi}\right)_{\min}=0. \tag{42.82}$$

The first two terms vanish due to the relations (42.77) and (42.78), respectively. Hence,

$$\sigma_{\min}\left(\frac{\partial^2 V}{\partial \sigma\, \partial \pi}\right)_{\min}=0. \tag{42.83}$$

Since $\sigma_{\min} \neq 0$, we have

$$\left(\frac{\partial^2 V}{\partial \sigma\, \partial \pi}\right)_{\min}=0. \tag{42.84}$$

This does not give any new result since, from (42.70), we find

$$\left(\frac{\partial^2 V}{\partial \sigma\, \partial \pi}\right)_{\min}=2\lambda \sigma_{\min}\pi_{\min}, \tag{42.85}$$

which vanishes since $\pi_{\min}=0$.

42.6 Higgs mechanism

Let us now bring in the Maxwell field and include its interaction with a charged scalar particle described by a complex scalar field. Following our earlier discussions, the Lagrangian for a complex scalar field coupled to a Maxwell field is given by

$$\mathcal{L}=\left|D_\mu \phi\right|^2-V(\phi)-\frac{1}{4}\left(F_{\mu\nu}\right)^2 \tag{42.86}$$

where the Maxwell tensor is given by

$$F_{\mu\nu} = \partial_\mu A_\nu - \partial_\nu A_\mu \tag{42.87}$$

and

$$D_\mu = \partial_\mu - ieA_\mu, \tag{42.88}$$

$$V = \mu^2 |\phi|^2 + \lambda |\phi|^4, \tag{42.89}$$

where $|D_\mu\phi|^2 = (D_\mu\phi)^* (D_\mu\phi)$, and $(F_{\mu\nu})^2 = F_{\mu\nu}F^{\mu\nu}$. We note that the Lagrangian (42.86) is invariant under the local gauge transformation

$$\phi(x) \to e^{-i\alpha(x)}\phi(x), \quad A_\mu(x) \to A_\mu(x) - \frac{1}{e}\partial_\mu\alpha(x). \tag{42.90}$$

For $\mu^2 > 0$, this Lagrangian corresponds to a scalar particle of mass μ, and a massless photon represented by $A_\mu(x)$.

Let us digress a little and discuss what would happen if the photon had a mass. One would then have to add a mass term in the Lagrangian

$$\frac{1}{2}m^2 A_\mu(x)A^\mu(x) \tag{42.91}$$

for the photon, which would not be invariant under gauge transformation (42.90). Thus, invariance under gauge transformation plays an essential role in preventing the photon from having a mass. We also note that because of gauge invariance the zero-mass photon will have only two polarization components – the two transverse components, while if $A_\mu(x)$ were to represent a massive vector boson then it would have three degrees of freedom, including a longitudinal component.

Let us replace the photon field by a vector field, which we will describe by the same notation, $A_\mu(x)$, in a gauge-invariant Lagrangian (42.86). In writing the gauge transformation relation, the electric charge e will be replaced by another coupling constant, which we will designate as g. Thus, the new Lagrangian will look exactly like (42.86) but with an $A_\mu(x)$ that no longer represents a photon but rather an arbitrary vector particle. This Lagrangian will be invariant under the gauge transformation

$$\phi(x) \to e^{-i\alpha(x)}\phi(x), \quad A_\mu(x) \to A_\mu(x) - \frac{1}{g}\partial_\mu\alpha(x). \tag{42.92}$$

We will show below that if we take $\mu^2 < 0$ then a dramatic thing happens as a consequence of SSB that occurs. One finds that, in addition to the appearance of a massive scalar particle, which we have already discussed, $A_\mu(x)$ now acquires mass even though the Lagrangian continues to remain gauge invariant.

To elaborate on this phenomenon, let us first review the scalar sector. We write the complex ϕ as

$$\phi = \frac{1}{\sqrt{2}}(\sigma + i\pi) \tag{42.93}$$

where σ and π are real fields. We found in our earlier discussions that for $\mu^2 < 0$, there was a minimum of the potential $V(\phi)$ at

$$|\phi_{\min}|^2 = \frac{v^2}{2} \tag{42.94}$$

where

$$v = \sqrt{\frac{m^2}{\lambda}}, \quad m^2 = -\mu^2 > 0. \tag{42.95}$$

Expanding the functions about the minimum

$$\sigma = \sigma' + v\pi = \pi' \quad \text{with} \quad \phi' = \frac{1}{\sqrt{2}}\left(\sigma' + i\pi'\right) \tag{42.96}$$

one finds

$$V(\phi) \to m^2\sigma'^2 - \lambda v\sigma'\left(\sigma'^2 + \pi'^2\right) - \frac{\lambda}{4}\left(\sigma'^2 + \pi'^2\right)^2 \tag{42.97}$$

which, as we discussed earlier, corresponds to a scalar particle $\sigma'$of mass $\sqrt{2}m$ along with a massless Nambu–Goldstone boson, π'. These conclusions are familiar to us.

Now let us consider $|D_\mu\phi|^2$. If we substitute (42.96) in (42.93) to obtain ϕ, we find

$$\begin{aligned} |D_\mu\phi|^2 \to \frac{1}{2}\left[\left(\partial_\mu\sigma'\right)^2 + \left(\partial_\mu\pi'\right)^2\right] + \frac{1}{2}g^2A_\mu A^\mu\left[\sigma'^2 + \pi'^2\right] + \frac{1}{2}g^2v^2A_\mu A^\mu \\ + gA^\mu\left[\left(\partial_\mu\sigma'\right)\pi' - \left(\partial_\mu\pi'\right)\sigma'\right] - gvA^\mu\left[\partial_\mu\pi' - gA_\mu\sigma'\right]. \end{aligned} \tag{42.98}$$

We note that the third term in (42.98) is of the form (42.91) and corresponds to a vector particle of mass gv . Even though the Lagrangian (42.86) that we started with is gauge invariant, the field A_μ now has mass, thanks to SSB.

Let us now consider an important transformation in which the massless particle, π', is completely eliminated from the Lagrangian. After SSB, instead of shifting the fields according to (42.96), we do it in terms of polar coordinates:

$$\phi(x) = \frac{1}{\sqrt{2}}\left[\sigma' + v\right]\exp\left(i\pi'/v\right). \tag{42.99}$$

This is effectively a shift in the modulus of $\phi(x)$ rather than in the real part. For small values of π', we obtain, by expanding the exponential,

$$\phi(x) = \frac{1}{\sqrt{2}}\left[\sigma' + v + i\pi' + \cdots\right], \tag{42.100}$$

which is of the same form as (42.96). After this shift let us make a gauge transformation of $\phi(x)$ of the type given by (42.92),

$$\phi'(x) = \exp\left(-i\pi'/v\right)\phi(x) = \frac{1}{\sqrt{2}}\left(\sigma' + v\right). \tag{42.101}$$

We now have a single, purely real, field. The gauge transformation (42.101) for $\phi(x)$ is to be accompanied by that of $A_\mu(x)$ given by (42.92), where we take $\alpha(x) = \pi'(x)$. Hence $A_\mu(x)$ is transformed as

$$A'_\mu(x) = A_\mu(x) - \frac{1}{gv}\partial_\mu \pi' \tag{42.102}$$

and for $D_\mu \phi$ we write

$$\begin{aligned} D_\mu \phi &= \left[\partial_\mu\left(e^{i\pi'/v}\phi'\right) - ig\left(A'_\mu + \frac{1}{gv}\partial_\mu \pi'\right)\left(e^{i\pi'/v}\phi'\right)\right] \\ &= e^{i\pi'/v}\left[\partial_\mu \sigma' - igA'_\mu\left(\sigma' + v\right)\right]/\sqrt{2}. \end{aligned} \tag{42.103}$$

Hence,

$$\left|D_\mu \phi\right|^2 = \frac{1}{2}\left|\partial_\mu \sigma' - igA'_\mu\left(\sigma' + v\right)\right|^2. \tag{42.104}$$

In the expression for the Maxwell tensor, A_μ is replaced by A'_μ,

$$F_{\mu\nu} = \partial_\mu A'_\nu - \partial_\nu A'_\mu, \tag{42.105}$$

and

$$V(\phi) = \frac{\mu^2}{2}\left(\sigma' + v\right)^2 - \frac{\lambda}{4}\left(\sigma' + v\right)^4. \tag{42.106}$$

The total Lagrangian is now

$$\mathcal{L} = \frac{1}{2}\left(\partial_\mu \sigma'\right)^2 + m^2\sigma'^2 - \frac{1}{4}\left(F'_{\mu\nu}\right)^2 + \frac{1}{2}g^2v^2A'_\mu A'^\mu + \frac{1}{2}g^2A'_\mu A'^\mu\sigma'\left(\sigma' + 2v\right) - \lambda v^2\,\sigma'^4. \tag{42.107}$$

In examining this Lagrangian we notice that we still have the massive scalar boson, σ', and the massive vector boson (also called the gauge boson), A'^μ, but the massless Goldstone boson, π', has disappeared. It is still present, of course, in the definition of A'_μ in (42.102). As we mentioned earlier, unlike the massless photon, which has two components, both in the transverse directions, the massive vector particle A'_μ will have three. The third component in essence will be provided by π' through the term $\partial_\mu \pi'$ in (42.102). This is a mechanism in which one says that the massless Goldstone boson is "gauged away" or simply "eaten away." The gauge transformation (42.101) and (42.102) that was used in order to arrive at the above Lagrangian is referred to as the unitary gauge transformation.

The mechanism presented above is at the basis of the Standard Model in particle physics. The model successfully predicted, through SSB, the existence of the vector bosons W and Z as particles that accompany the massless photons. The counterpart of the heavy scalar boson is then the Higgs boson, which as yet has not been discovered experimentally. One might find the mechanism of SSB through the changing of the sign of μ^2 as somewhat artificial. It might be interesting if the actual mechanism of SSB comes about through the formation of condensates as we discussed earlier in connection with superconductivity, and if, in that case, the Higgs particle itself is one of the condensates.

43 Basic quantum electrodynamics and Feynman diagrams

In our discussion of the Lagrangian formulation we obtained the equations of motion involving the classical Dirac, Maxwell's, and Klein–Gordon fields. We will continue with these equations in this chapter, but with the fields now quantized. At the same time we will revert to the Hamiltonian formalism with the fields now expressed in the Heisenberg representation. We will confine ourselves only to the interacting Dirac and Maxwell's fields. This will be the basis of what we will call basic quantum electrodynamics (QED).

43.1 Perturbation theory

Consider a field $\Psi(x)$ which is fully interacting and satisfies the Heisenberg equation

$$\frac{\partial \Psi(x)}{\partial t} = i\left[H, \Psi(x)\right] \tag{43.1}$$

where

$$H = H(\Psi) = H_0(\Psi) + H_I(\Psi) \tag{43.2}$$

is the total Hamiltonian, which includes the free Hamiltonian H_0 as a function of the fields Ψ and the interaction Hamiltonian H_I, e.g., $\Psi(x)\gamma \cdot A(x)\Psi(x)$ for QED. Consider now another field $\Psi_{in}(x)$ that satisfies the free field equation

$$\frac{\partial \Psi_{in}(x)}{\partial t} = i\left[H_{in}^o, \Psi_{in}(x)\right] \tag{43.3}$$

where

$$H_{in}^o = H_0(\Psi_{in}), \tag{43.4}$$

which is the same as H_0 in (43.2) except that $\Psi(x)$ is replaced by $\Psi_{in}(x)$. Here the subscript "in" stands for incoming field, as will be clarified below.

We assume that $\Psi(x)$ and $\Psi_{in}(x)$ are related to each other through a unitary time evolution operator $U(t)$ as

$$\Psi(x) = U^{-1}(t)\,\Psi_{in}(x)\,U(t) \tag{43.4a}$$

such that

$$\Psi(x) \to \Psi_{in}(x) \qquad \text{as } t \to -\infty. \tag{43.5}$$

This condition in essence means that, as $t \to -\infty$, $\Psi(x)$ represents a free particle. Furthermore, it implies that

$$U(-\infty) = 1. \tag{43.6}$$

From (43.1), (43.3), and (43.4a) we also obtain

$$H = U^{-1}(t)\, H_{in} U(t) \tag{43.7}$$

where

$$H_{in} = H_0\left(\Psi_{in}\right) + H_I\left(\Psi_{in}\right). \tag{43.8}$$

Thus, H_{in} is the same operator as the total Hamiltonian, H, given by (43.2) except that the fields $\Psi(x)$ are replaced by $\Psi_{in}(x)$. Qualitatively, the point here is that Ψ is a fully interacting field, while Ψ_{in} is a free field. However, since the Hamiltonian is expressed in terms of fields, one must make the distinction about which fields appear in H.

We assume that as $t \to +\infty$, $\Psi(x)$ once again becomes a free particle, though, because of the effect of interactions during the intervening time period, it may not necessarily have the same quantum numbers, e.g., momentum, it started out with. It will satisfy the following relation:

$$\Psi(x) \to \Psi_{out}(x) \quad \text{as} \quad t \to \infty. \tag{43.9}$$

From (43.4a) and (43.9) we obtain

$$\Psi_{out}(x) = U^{-1}(\infty)\, \Psi_{in}(x)\, U(\infty). \tag{43.10}$$

Since $U(-\infty) = 1$ from (43.6), we multiply (43.10) on the left and right by $U(-\infty)$ and $U^{-1}(-\infty)$, respectively, and obtain

$$\Psi_{out}(x) = [U(-\infty)\, U^{-1}(\infty)]\Psi_{in}(x)\, [U(\infty)\, U^{-1}(-\infty)]. \tag{43.11}$$

Let us define a unitary operator $U\left(t, t'\right)$ as

$$U\left(t, t'\right) = U(t)\, U^{-1}\left(t'\right), \quad \text{with } U(t,t) = 1 \tag{43.12}$$

and write (43.11) as

$$\Psi_{out}(x) = S^{-1}\Psi_{in}(x)\, S \tag{43.13}$$

where S is defined as the S-matrix given by

$$S = U(\infty, -\infty). \tag{43.14}$$

The S-matrix then converts a free in-field into a free out-field. Our task now is to calculate S. We will accomplish this through perturbation expansion, which we outline below.

Since the presence of $U(t')$ for an arbitrary time t' does not change any of the previous equations which are all in terms of t, we will replace $U(t)$ in the previous equations by $U(t, t')$. We first consider equation (43.3) and write, using (43.4a),

$$\frac{\partial \Psi_{in}}{\partial t} = \frac{\partial}{\partial t}\left(U\Psi U^{-1}\right). \tag{43.15}$$

Therefore,

$$\dot{\Psi}_{in} = \dot{U}\Psi U^{-1} + U\dot{\Psi}U^{-1} + U\Psi\left(\dot{U}^{-1}\right) \tag{43.16}$$

where the dot represents the time-derivative. We first obtain $\dot{U}^{-1}$ through the relation

$$UU^{-1} = 1, \tag{43.17}$$

by taking the time-derivative

$$\dot{U}U^{-1} + U\dot{U}^{-1} = 0, \tag{43.18}$$

which gives

$$(\dot{U}^{-1}) = -U^{-1}\dot{U}U^{-1}. \tag{43.19}$$

Inserting this expression into the third term in (43.16) we find, with some additional modifications,

$$\dot{\Psi}_{in} = (\dot{U}U^{-1})\left(U\Psi U^{-1}\right) + \left(U\dot{\Psi}U^{-1}\right) - \left(U\Psi U^{-1}\right)\dot{U}U^{-1}. \tag{43.20}$$

From (43.4a) we obtain

$$\dot{\Psi}_{in} = (\dot{U}U^{-1})\Psi_{in} + \left(U\dot{\Psi}U^{-1}\right) - \Psi_{in}\dot{U}U^{-1}. \tag{43.21}$$

Therefore,

$$\dot{\Psi}_{in} = \left[\dot{U}U^{-1}, \Psi_{in}\right] + U\dot{\Psi}U^{-1}. \tag{43.22}$$

To obtain the second term on the right-hand side of (43.22), we note from (43.1) that

$$U\dot{\Psi}U^{-1} = iU\left[H, \Psi\right]U^{-1} = i\left[H_{in,}\Psi_{in}\right]. \tag{43.23}$$

Hence (43.22) can be written as

$$\dot{\Psi}_{in} = \left[\dot{U}U^{-1} + iH_{in}, \Psi_{in}\right]. \tag{43.24}$$

From (43.3) and (43.24) we obtain

$$\left[\left(\dot{U}U^{-1} + i\left[H_{in} - H_{in}^{o}\right]\right), \Psi_{in}\right] = 0. \tag{43.25}$$

Since this is true for any Ψ_{in}, we must have

$$\dot{U}U^{-1} + i\left[H_{in} - H_{in}^{o}\right] = 0. \tag{43.26}$$

However, from (43.8) and (43.4) we know that

$$H_{in} - H_{in}^{o} = H_I(\Psi_{in}) = H_{in}^{I}. \tag{43.27}$$

Hence,

$$\dot{U}U^{-1} = -iH_{in}^{I} \tag{43.28}$$

apart from a possible c-number, which we can add on the right-hand side. This leads to the equation

$$\dot{U}\left(t, t'\right) = -iH_{in}^{I}U\left(t, t'\right), \tag{43.29}$$

which we can solve with the condition

$$U(t,t) = 1 \tag{43.30}$$

to give

$$U(t,t') = 1 - i\int_{t'}^{t} dt_1\, H_{in}^I(t_1)\, U(t_1,t'). \tag{43.31}$$

Using this as a recursion relation we generate the following perturbation series:

$$\begin{aligned} U(t,t') &= 1 + (-i)\int_{t'}^{t} dt_1\, H_{in}^I(t_1) + (-i)^2\int_{t'}^{t} dt_1\int_{t'}^{t_1} dt_2\, H_{in}^I(t_1)\, H_{in}^I(t_2) \\ &\quad + \cdots + (-i)^n\int_{t'}^{t} dt_1\int_{t'}^{t_1} dt_2\cdots\int_{t'}^{t_{n-1}} dt_n\, H_{in}^I(t_1)\cdots H_{in}^I(t_n) + \cdots. \end{aligned} \tag{43.32}$$

This series is reminiscent of the expression we came across in treating the perturbation expansion in the interaction representation in Chapter 3. We can write this series in terms of the time-ordered product $T(H_{in}^I(t_1)\cdots H_{in}^I(t_n))$ by observing the following for a typical term in this expansion:

$$\begin{aligned} &\int_{t'}^{t} dt_1\int_{t'}^{t_1} dt_2\cdots\int_{t'}^{t_{n-1}} dt_n\, H_{in}^I(t_1)\cdots H_{in}^I(t_n) \\ &= \frac{1}{n!}\int_{t'}^{t} dt_1\int_{t'}^{t} dt_2\cdots\int_{t'}^{t} dt_n T\left(H_{in}^I(t_1)\cdots H_{in}^I(t_n) + \cdots\right). \end{aligned} \tag{43.33}$$

The S-matrix given by (43.14) can then be written as

$$\begin{aligned} S &= 1 + (-i)\int_{-\infty}^{\infty} dt_1\, T(H_{in}^I(t_1)) + \frac{(-i)^2}{2!}\int_{-\infty}^{\infty} dt_1\int_{-\infty}^{\infty} dt_2\, T(H_{in}^I(t_1)\, H_{in}^I(t_2)) \\ &\quad + \cdots + \frac{(-i)^n}{n!}\int_{-\infty}^{\infty} dt_1\int_{-\infty}^{\infty} dt_2\cdots\int_{-\infty}^{\infty} dt_n\, T(H_{in}^I(t_1)\cdots H_{in}^I(t_n)) + \cdots. \end{aligned} \tag{43.34}$$

43.2 Feynman diagrams

The T-matrix is given in terms of the S-matrix by the relation

$$\mathbf{S} = \mathbf{1} + i\,(2\pi)^4\,\delta(p_f - p_i)\mathbf{T}. \tag{43.35}$$

This definition is an extension to four dimensions of the relation between **S** and **T** given before which explicitly includes the fourth component (energy) of the momentum vector, p^μ. Since we will be considering only the transitions due to the interaction Hamiltonian that conserve energy and momentum, the following results will apply to the T-matrix with the free fields and their interaction given by

$$\text{(Dirac)}\ \boldsymbol{\psi}(x) = \sum_{s=1}^{2}\frac{1}{(\sqrt{2\pi})^3}\int d^3p\sqrt{\frac{m}{E_p}}\left[u_s(\mathbf{p})a_s(\mathbf{p})\,e^{-ip\cdot x} + v_s(\mathbf{p})b_s^\dagger(\mathbf{p})e^{ip\cdot x}\right], \tag{43.36}$$

$$\text{(Maxwell)}\ A_\mu(x) = \frac{1}{\left(\sqrt{2\pi}\right)^3} \int \frac{d^3k}{\sqrt{2\omega_k}} \left[a_\mu(\mathbf{k})\, e^{-ik\cdot x} + a_\mu^\dagger(\mathbf{k})\, e^{ik\cdot x}\right], \tag{43.37}$$

$$\text{(Interaction)}\ H_I(x) = \overline{\Psi}(x)\, \gamma_\mu \Psi_\beta(x)\ A^\mu(x). \tag{43.38}$$

Going order by order we first consider:

$T(H_I)$

Since only one space-time point is involved there is no time-ordering to be considered and so we remove the symbol T and consider the product

$$H_I(x) = \overline{\Psi}_\alpha(x)\left(\gamma_\mu\right)_{\alpha\beta} \Psi_\beta(x)\ A^\mu(x), \tag{43.39}$$

which corresponds to two fermion fields (electrons or positrons), $\Psi_\alpha(x)$ and $\Psi_\beta(x)$, and a photon field, $A^\mu(x)$, at a point x. We call this common point the "vertex." Keeping only the creation and destruction operators for each of the fields that appear in (43.39), and ignoring the γ-matrices, we have the product

$$\left(a^\dagger, b\right)\left(a, b^\dagger\right)\left(a_\gamma, a_\gamma^\dagger\right) \tag{43.40}$$

where we have ignored the dependence on spin and momentum. It is understood that these terms appear in a matrix element sandwiched between the initial and final states in the occupation number space, e.g., $\langle f|...|i\rangle$. First we discuss just the fermions in the above product. There are four possible terms

$$a^\dagger a,\ a^\dagger b^\dagger,\ ba,\ bb^\dagger \tag{43.41}$$

where a and $a^\dagger$ operators correspond to electrons, while b and $b^\dagger$ correspond to positrons.

In order to draw diagrams that each of these terms can be associated with, we make the convention that time increases from left to right. If we take a solid line as describing a fermion (electron or positron), and a wavy line as a photon, then the lines coming from the left and stopping at the vertex will be called the "incoming" particles. They will be associated with the destruction operators, e.g., a, b (and a_γ if the photons are also included). Lines coming out of the vertex going to the right are the "outgoing" particles, and correspond to $a^\dagger, b^\dagger$ (and $a_\gamma^\dagger$).

The diagrams we discussed above and the ones to follow are due to Feynman, and are called "Feynman diagrams." One can show that there is a one-to-one correspondence between the diagrams and the terms in the S-matrix expansion.

For electrons we attach an arrow to the solid line in the same direction as the direction in which it is moving, as shown in Fig. 43.1(a).

For an incoming positron, since it is described as a particle moving backward in time, we attach an arrow in the direction opposite to the direction of motion as shown in Fig. 43.1(b). Note that even though it is coming from left to right (incoming), its arrow points in the opposite direction. Similarly for the outgoing positron (Fig. 43.1(c)).

The diagrams corresponding to the individual products in (43.41) can then be drawn as shown in Fig. 43.2. From these four diagrams we note that out of the two possible arrows

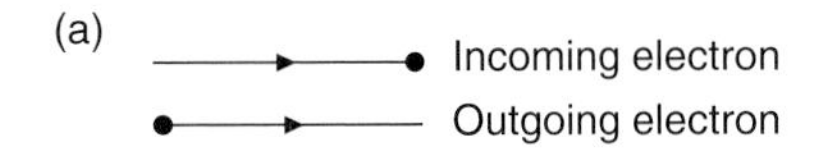

Fig. 43.1

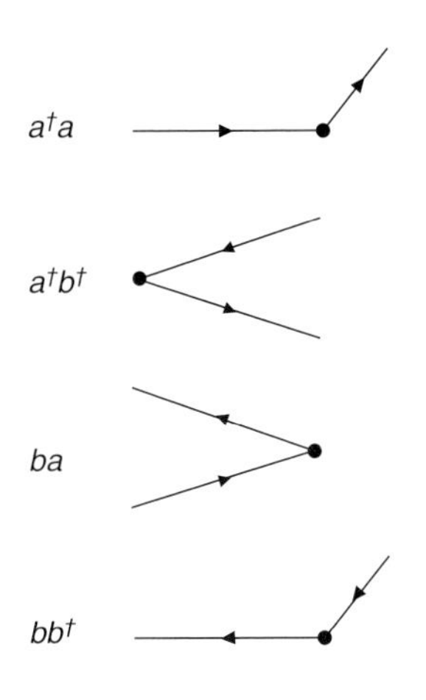

Fig. 43.2

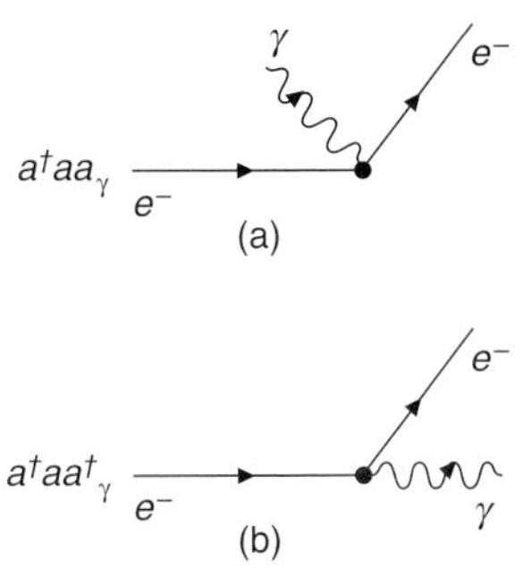

Fig. 43.3

only one points toward the vertex, while the other points away from it. Thus, as we trace a fermion line, the direction of the arrow remains continuous.

Let us now include the photons. In each of the four diagrams in Fig. 43.2 one can attach an incoming or an outgoing photon. For example, in the diagrams for $a^\dagger a$ given in Fig. 43.3(a) and (b) we have two possibilities.

If we designate the state vector in the occupation number space as

$$|\text{electrons, positrons, photons}\rangle \tag{43.42}$$

then the only nonzero matrix element that is relevant to $a^\dagger a a_\gamma^\dagger$ will correspond to

$$\left\langle 1, 0, 1 \left| a^\dagger a a_\gamma^\dagger \right| 1, 0, 0 \right\rangle. \tag{43.43}$$

There will then be eight such possibilities corresponding to the expression (43.41).

Let us pursue our discussion on the product $a^\dagger a a_\gamma^\dagger$ given by Fig. 43.3(b), which is present in the expression $\overline{\Psi}_\alpha(x)\left(\gamma_\mu\right)_{\alpha\beta}\Psi_\beta(x)\,A^\mu(x)$ and determine its contribution to the T-matrix. If we designate the initial and final states in terms of momentum eigenvalues, then the states corresponding to (43.42) that are relevant to (43.43) are

$$\begin{aligned} |\text{initial state}\rangle &= |i\rangle = |\mathbf{p}_1, 0, 0\rangle\,, \\ |\text{final state}\rangle &= |f\rangle = |\mathbf{p}_2, 0, \mathbf{k}\rangle\,. \end{aligned} \tag{43.44}$$

The operators $\Psi(x)$, $\overline{\Psi}(x)$ and $A^\mu(x)$ discussed in Chapter 37,

$$\Psi(x) = \frac{1}{(\sqrt{2\pi})^3}\sum_{s=1}^{2}\int d^3p\sqrt{\frac{m}{E_p}}\left[u_s(\mathbf{p})a_s(\mathbf{p})e^{-ip\cdot x} + v_s(\mathbf{p})b_s^\dagger(\mathbf{p})e^{ip\cdot x}\right], \tag{43.45}$$

$$\overline{\Psi}(x) = \frac{1}{(\sqrt{2\pi})^3}\sum_{s=1}^{2}\int d^3p\sqrt{\frac{m}{E_p}}\left[\overline{u}_s(\mathbf{p})a_s^\dagger(\mathbf{p})e^{ip\cdot x} + \overline{v}_s(\mathbf{p})b_s(\mathbf{p})e^{-ip\cdot x}\right], \tag{43.46}$$

$$A^\mu(x) = \frac{1}{(\sqrt{2\pi})^3}\int \frac{d^3p}{\sqrt{2\omega}}\left[a^\mu(\mathbf{p})\,\epsilon^\mu e^{-ip\cdot x} + a^{\mu\dagger}(\mathbf{p})\,\epsilon^{*\mu}e^{ip\cdot x}\right],$$

where ϵ^μ is the polarization vector. The matrix element of $\overline{\Psi}_\alpha(x)\left(\gamma_\mu\right)_{\alpha\beta}\Psi_\beta(x)\,A^\mu(x)$ corresponding to the states described by (43.44) is obtained as

$$\overline{u}_\alpha(\mathbf{p}_2)\left(\gamma_\mu\right)_{\alpha\beta}u_\beta(\mathbf{p}_1)\epsilon^\mu e^{-i(p_1-p_2-k)\cdot x} \tag{43.47}$$

and the corresponding T-matrix is then

$$\int d^4x\,H_I(x) = \int d^4x\,\overline{\Psi}(x)\,\gamma_\mu\Psi_\beta(x)\,A^\mu(x) = \overline{u}(\mathbf{p}_2)\gamma\cdot\epsilon u(\mathbf{p}_1)\delta^{(4)}(p_1-p_2-k). \tag{43.48}$$

This is then the contribution of the matrix element (43.43), expressed in momentum space and represented by Fig. 43.3(a) and (b).

43.3 $T(H_I(x_1)H_I(x_2))$ and Wick's theorem

Before we embark on calculating $T(\mathbf{H}_I(x_1)\,\mathbf{H}_I(x_2))$ and introduce Wick's theorem let us consider the time-ordered product of two scalar bosons,

$$T\left(\Phi(x_1)\Phi(x_2)\right), \tag{43.48a}$$

where $\Phi(x_i)$ are the boson fields given by

$$\begin{aligned}\Phi(x) &= \frac{1}{(\sqrt{2\pi})^3}\int \frac{d^3p}{\sqrt{2\omega_p}}[a(\mathbf{p})e^{-ip\cdot x} + a^\dagger(\mathbf{p})e^{ip\cdot x}] \\ &= b_i^{(+)} + b_i^{(-)}\end{aligned} \tag{43.49}$$

where

$$b_i^{(+)} = \frac{1}{(\sqrt{2\pi})^3}\int \frac{d^3p}{\sqrt{2\omega_p}}a(\mathbf{p})e^{-ip\cdot x} \quad \text{and} \quad b_i^{(-)} = \frac{1}{(\sqrt{2\pi})^3}\int \frac{d^3p}{\sqrt{2\omega_p}}a^\dagger(\mathbf{p})e^{ip\cdot x}. \tag{43.50}$$

Note that the superscripts (+) and (−) refer to positive and negative frequencies and correspond to destruction and creation operators (although counterintuitive, these are, nevertheless, the conventional definitions).

The time-ordered product (43.48a) can be written as

$$T(b_1b_2) = \theta\left(t_1 - t_2\right)\left[\left(b_1^{(+)} + b_1^{(-)}\right)\left(b_2^{(+)} + b_2^{(-)}\right)\right] + t_2 \leftarrow t_1 \tag{43.51}$$

where b_1 and b_2 refer to the two bosons. We express the product in square brackets in terms of the normal products we considered earlier in which the destruction operators stand to the right of all the creation operators,

$$N\left(b_1b_2\right) = b_1^{(+)}b_2^{(+)} + b_2^{(-)}b_1^{(+)} + b_1^{(-)}b_2^{(-)} + b_1^{(-)}b_2^{(+)}. \tag{43.52}$$

We obtain, in terms of N and the commutators, the following:

$$T(b_1b_2) = \theta\left(t_1 - t_2\right)\left\{N\left(b_1b_2\right) + \left[b_1^{(+)}, b_2^{(-)}\right]\right\} + \theta\left(t_1 - t_2\right)\left\{N\left(b_1b_2\right) + \left[b_2^{(+)}, b_1^{(-)}\right]\right\}. \tag{43.53}$$

We write this result as

$$T\left(b_1b_2\right) = N\left(b_1b_2\right) + C\left(b_1b_2\right) \tag{43.54}$$

where C is called the contraction term:

$$C(b_1b_2) = \theta\left(t_1 - t_2\right)\left[b_1^{(+)}, b_2^{(-)}\right] + \theta\left(t_2 - t_1\right)\left[b_2^{(+)}, b_1^{(-)}\right]. \tag{43.55}$$

This is the statement of Wick's theorem applied to the product of two scalar bosons. We note here that besides the creation and destruction operators, contained in the normal product, which give rise to the type of diagrams we have already encountered, we now have

Fig. 43.4

a new term, the contraction term $C[\Phi(x_1)\Phi(x_2)]$. It is the only term that includes both vertices, x_1 and x_2. While other terms depend only on one or the other vertex, $C[\Phi(x_1)\Phi(x_2)]$ connects the two vertices. It can be written as the vacuum expectation value,

$$C(b_1b_2) = \langle 0|T(b_1b_2)|0\rangle, \tag{43.56}$$

since

$$\langle 0|N|0\rangle = 0. \tag{43.57}$$

This is, indeed, the propagator we have considered earlier and, as the name suggests, it propagates information from one vertex to another. It contains no operators (except for the unit operator), and it is simply what one calls a c-number. The diagram representing it will be a line between the two vertices, as given by Fig. 43.4, depending on where one decides to place the vertices. These types of lines are called "internal" lines, while the lines associated with creation and destruction of particles, indicating incoming and outgoing particles, are called "external" lines.

Let us now obtain the time-ordered product of two fermions $T(\psi(x_1)\,\psi(x_2))$ which we write as $T(f_1f_2)$ given by

$$T(f_1f_2) = \theta(t_1 - t_2)f_1f_2 - \theta(t_2 - t_1)f_2f_1. \tag{43.58}$$

After writing, as in the case of scalar boson,

$$f_i = f_i^{(+)} + f_i^{(-)}, \quad i = 1, 2 \tag{43.59}$$

we obtain, this time in terms of the anticommutators, the products appearing on the right-hand side of (43.58),

$$f_1f_2 = N(f_1f_2) + \left\{f_1^{(+)}, f_2^{(-)}\right\}, \tag{43.60}$$

$$f_2f_1 = N(f_1f_2) + \left\{f_2^{(+)}, f_1^{(-)}\right\}, \tag{43.61}$$

where

$$N(f_1f_2) = f_1^{(+)}f_2^{(+)} - f_2^{(-)}f_1^{(+)} + f_1^{(-)}f_2^{(+)} + f_1^{(-)}f_2^{(-)}. \tag{43.62}$$

If we write

$$C(f_1f_2) = \theta(t_1 - t_2)\left\{f_1^{(+)}, f_2^{(-)}\right\} - \theta(t_2 - t_1)\left\{f_2^{(+)}, f_1^{(-)}\right\}, \tag{43.63}$$

which is the contraction term for two fermions, then

$$T(f_1f_2) = N(f_1f_2) + C(f_1f_2). \tag{43.64}$$

This is then the statement of Wick's theorem for the product of two fermion fields. Here once again we see the presence of the contraction term, which is the same as the propagator discussed earlier and given by the vacuum expectation value

$$C(f_1 f_2) = \langle 0 | T(f_1 f_2) | 0 \rangle . \tag{43.65}$$

Let us now discuss how to evaluate the time-ordered product. We will remove the γ-matrices, which can be inserted at the end of the calculation, and we also suppress the spin indices. We then have to evaluate

$$T\left(\overline{\psi}(x_1)\,\psi(x_1)\,\overline{\psi}(x_2)\,\psi(x_2)\;\mathbf{A}^\mu(x_1)\,\mathbf{A}^\mu(x_2)\right). \tag{43.66}$$

There are four fermion fields for which the time-ordered product can be written down. Instead of writing a general form that we will not need for our purposes, we can simplify things by considering some typical terms as we did for the single vertex. For that purpose let us keep only the creation and destruction operators that are involved at each vertex, and write for (43.66) essentially a product of two terms like (43.40), one for each vertex:

$$\left[\left(a^\dagger, b\right)\left(a, b^\dagger\right)\left(a_\gamma, a_\gamma^\dagger\right)\right]_1 \left[\left(a^\dagger, b\right)\left(a, b^\dagger\right)\left(a_\gamma, a_\gamma^\dagger\right)\right]_2 . \tag{43.67}$$

The subscripts for the square brackets above correspond to the two vertices.

As we will see below, once we know the initial and final states, assuming that they are consistent with charge conservation and other symmetry considerations, one can draw all the relevant diagrams (to second order).

Consider specifically the case where the initial state corresponds to two electrons,

$$|i\rangle = \left|e^- e^-\right\rangle, \tag{43.68}$$

which means that one can only have destruction operators $(a)_1\,(a)_2$ to give a nonzero result, one from each vertex. We will not specify the final state but we will consider as many different creation operators as are allowed through (43.67). We can discard the two disjointed terms

$$\left[a^\dagger a a_\gamma^\dagger\right]_1 \left[a^\dagger a a_\gamma^\dagger\right]_2 \tag{43.69}$$

since they are two independent diagrams irrelevant to scattering (see Fig. 43.5).

As in the case of two bosons, the only diagrams that will give a nonzero answer will be those that have the contraction term that connects the two vertices. This would happen also for the two-fermion case. Let us, therefore, consider

$$\left(a^\dagger a\right)_1 \left(a^\dagger a\right)_2 C\left[a_\gamma a_\gamma\right]. \tag{43.70}$$

One could replace any one of the a_γ's, or both, by $a_\gamma^\dagger$. This then corresponds to two electrons in the final state, with a photon propagator. The diagram is described by Fig. 43.6.

This is, indeed, the relativistic electron–electron scattering, $e^- + e^- \to e^- + e^-$, called Møller scattering. There is another diagram obtained by the interchange of the final state

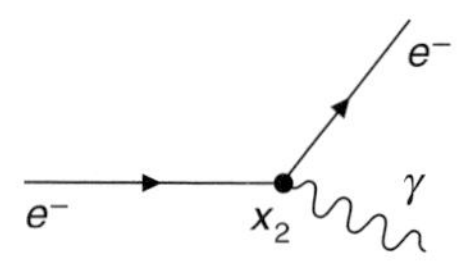

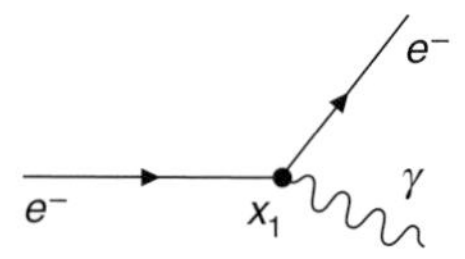

Fig. 43.5

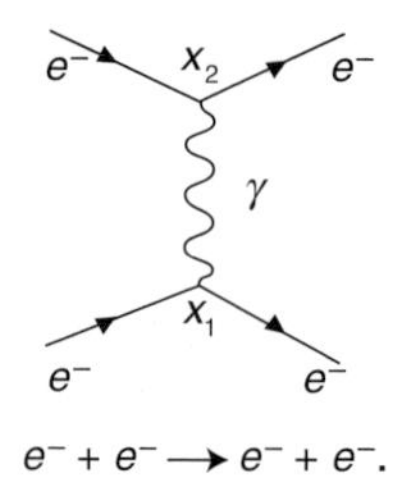

$e^- + e^- \longrightarrow e^- + e^-$.

Fig. 43.6

electrons that the full Wick's expansion contains. There are no other diagrams (to second order) when the initial state consists of two electrons.

From the above example we come to the important conclusion that to achieve a nonzero T-matrix element it is necessary, but not sufficient, to have a contraction term in the time-ordered product.

Based on this observation, let us consider the case where the initial state is an electron and a photon, i.e., an $e^-\gamma$ state:

$$|i\rangle = \left|e^-\gamma\right\rangle. \tag{43.71}$$

It is easy to see that there are only two nonzero terms. One of which corresponds to

$$\left(aa_\gamma\right)_1 \left(a^\dagger a_\gamma^\dagger\right)_2 C\left[a^\dagger a\right], \tag{43.72}$$

which gives rise to the diagram in Fig. 43.7(a). This is Compton scattering, $e^- + \gamma \rightarrow e^- + \gamma$. The other corresponds to the term

$$\left(aa_\gamma^\dagger\right)_1 \left(a^\dagger a_\gamma\right)_2 C\left[a^\dagger a\right] \tag{43.73}$$

with the diagram in Fig. 43.7(b). This is also a part of the Compton scattering with the photon lines interchanged.

Finally, let us consider the initial state that corresponds to an electron and a positron, i.e., an e^-e^+ state:

$$|i\rangle = \left|e^-e^+\right\rangle. \tag{43.74}$$

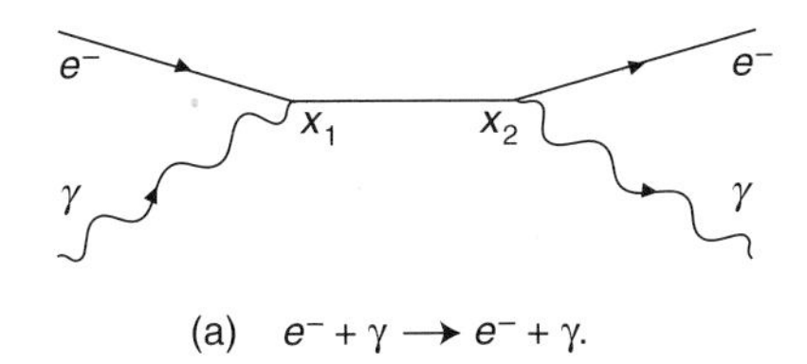

(a) $e^- + \gamma \rightarrow e^- + \gamma$.

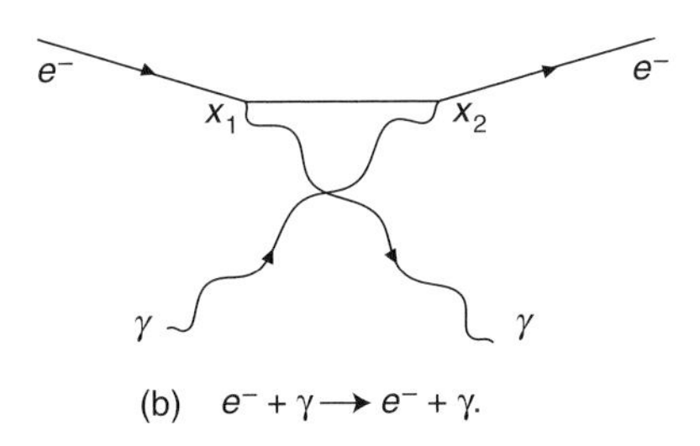

(b) $e^- + \gamma \rightarrow e^- + \gamma$.

Fig. 43.7

We now have three possibilities, the first being

$$(b^\dagger b)_1(a^\dagger a)_2 C\left[a_\gamma a_\gamma\right] \tag{43.75}$$

with the diagram in Fig. 43.8(a). This is called Bhabha scattering, $e^- + e^+ \rightarrow e^- + e^+$. The second possibility is

$$(ab)_1(a^\dagger b^\dagger)_2 C\left[a_\gamma a_\gamma\right] \tag{43.76}$$

with the diagram given in Fig. 43.8(b). This is also a part of the Bhabha scattering.

The third possibility is

$$(aa_\gamma^\dagger)_1(ba_\gamma^\dagger)_2 C\left[a_\gamma a_\gamma\right] \tag{43.77}$$

with the diagram in Fig. 43.9(a). This is called pair annihilation, $e^- + e^+ \rightarrow \gamma + \gamma$. One can also reverse the process, and write

$$(aa_\gamma^\dagger)_1(ba_\gamma^\dagger)_2 C\left[a_\gamma a_\gamma\right]. \tag{43.78}$$

Finally, we have two photons in the initial state:

$$|i\rangle = |2\gamma\rangle\,. \tag{43.79}$$

This is the reaction $\gamma + \gamma \rightarrow e^- + e^+$, which is called pair production and is shown diagrammatically in Fig. 43.9(b).

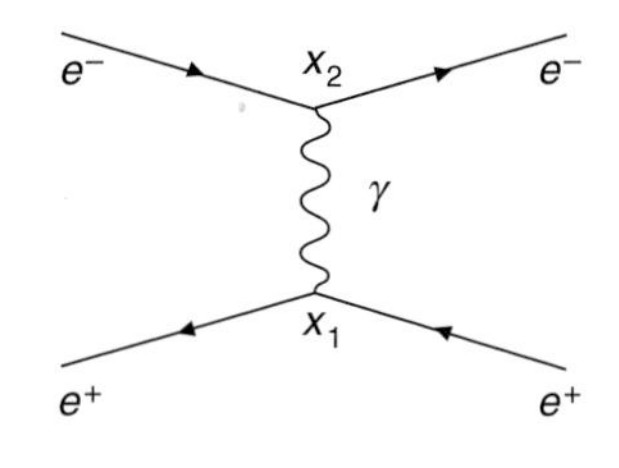

(a) $e^- + e^+ \longrightarrow e^- + e^+$.

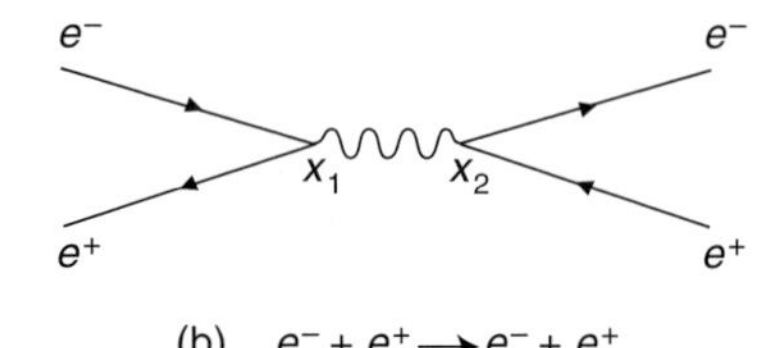

(b) $e^- + e^+ \longrightarrow e^- + e^+$.

Fig. 43.8

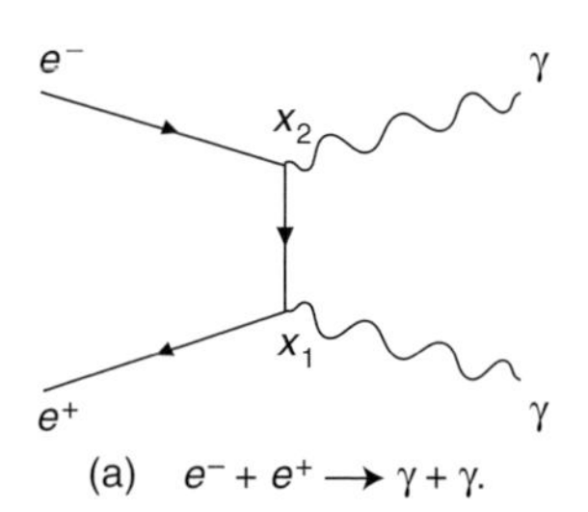

(a) $e^- + e^+ \longrightarrow \gamma + \gamma$.

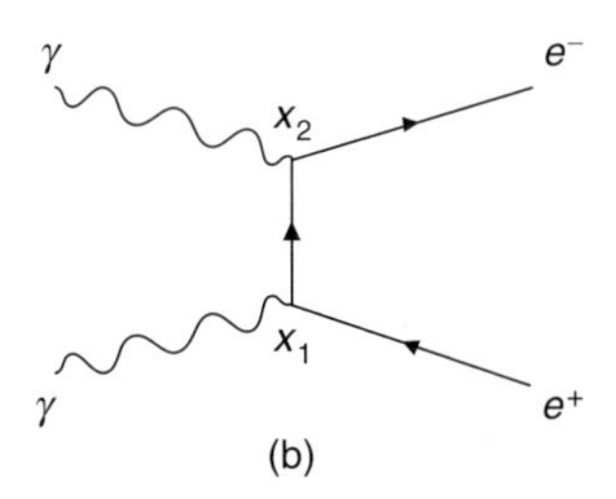

(b)

Fig. 43.9

We have now considered all possible second-order diagrams in QED.

Higher-order diagrams can similarly be constructed. For example, the process $\gamma + \gamma \to \gamma + \gamma$ corresponds to a fourth-order diagram given in Fig. 43.10. We will not be giving them here.

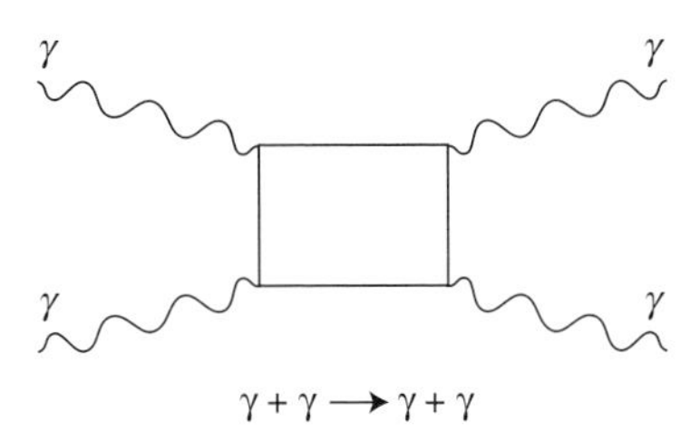

Fig. 43.10

43.4 Feynman rules

The contributions in the momentum space coming from each component of a typical Feynman diagram to M_{fi} are listed in Fig. 43.11(a) to (f).

(i) Incoming electron and positron respectively (Fig. 43.11(a)).
(ii) Outgoing electron and positron respectively (Fig 43.11(b)).
(iii) Incoming and outgoing photon (Fig 43.11(c)).
(iv) Electron propagator (Fig. 43.11(d)).
(v) Photon propagator (Fig. 43.11(e)).
(vi) Electron–photon vertex (Fig. 43.11(f)).

In addition, one must impose momentum conservation at each vertex. In calculating any loop diagrams (see Chapter 44), one must also integrate over each undetermined loop momentum, given, for example, by the integral $\int d^4p/(2\pi)^4$ if p is the undetermined loop momentum.

43.5 Cross-section for $1 + 2 \to 3 + 4$

We will consider the scattering of 2 particles → 2 particles with initial four-momenta p_1 and p_2 and final momenta p_3 and p_4. The cross sections and the related quantities follow the same definitions given previously:

$$\text{Cross-section} = \frac{\text{transition probability per unit time}}{\text{incident flux}}$$

where

$$\begin{aligned}\text{Incident flux} &= \text{no. of particles incident per unit area per unit time}\\ &= (\text{density of initial states}) \times (\text{relative velocity of incident particle}).\end{aligned}$$

If J is the incident flux then

$$J = \frac{1}{V}\,|\boldsymbol{v}_1 - \boldsymbol{v}_2| \tag{43.80}$$

(a) $u_s(\mathbf{p})$ $\bar{v}_s(\mathbf{p})$

(b) $\bar{u}_s(\mathbf{p})$ $v_s(\mathbf{p})$

(c) ε_μ ε_μ

(d) $\mathbf{p}$ $\dfrac{i(\gamma \bullet p + m)}{p^2 - m^2 + i\varepsilon}$

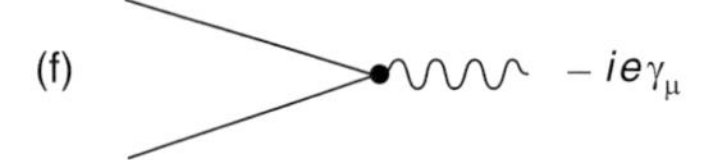

Fig. 43.11 **Feynman rules.**

where $|\boldsymbol{v}_1 - \boldsymbol{v}_2|$ represents the relative velocity between the two incident particles with velocities $\boldsymbol{v}_1$ and $\boldsymbol{v}_2$, respectively.

We now write the expression for an S-matrix element as

$$S_{fi} = \delta_{fi} + i\,(2\pi)^4\,\delta^{(4)}\left(p_f - p_i\right)T_{fi}. \tag{43.81}$$

The second term gives the probability amplitude for transitions to occur. Thus, the transition probability per unit time is given by

$$\begin{aligned}\lambda_{fi} &= \frac{1}{T}\left|i\,(2\pi)^4\,\delta^{(4)}\left(p_f - p_i\right)T_{fi}\right|^2 \\ &= \frac{1}{T}\,(2\pi)^8\,\delta^{(4)}\left(p_f - p_i\right)\delta^{(4)}\,(0)\left|T_{fi}\right|^2 \end{aligned} \tag{43.82}$$

where we have used the relation $f(x)\,\delta\,(x) = f(0)\,\delta(x)$ with $f(x)$ being $\delta(x)$ itself. Since

$$\delta^{(4)}(p) = \frac{1}{(2\pi)^4}\int d^4x\,e^{-ip\cdot x}, \tag{43.83}$$

then for the case of finite space and time

$$\delta^{(4)}\,(0) = \frac{1}{(2\pi)^4}VT, \tag{43.84}$$

which implies that

$$\lambda_{fi} = (2\pi)^4\,\delta^{(4)}\left(p_f - p_i\right)V\left|T_{fi}\right|^2. \tag{43.85}$$

The transition probability per unit time for transitions into a group of final states is given by

$$w_{fi} = (2\pi)^4 \delta^{(4)}\left(p_f - p_i\right) V \prod_{j=3}^{4} \frac{d^3 p_j V}{(2\pi)^3} \left|T_{fi}\right|^2. \tag{43.86}$$

The cross-section will be given by

$$d\sigma = \frac{w_{fi}}{J} = \frac{1}{|\mathbf{v}_1 - \mathbf{v}_2|} (2\pi)^4 \delta^{(4)}\left(p_f - p_i\right) V^4 \prod_{j=3}^{4} \frac{d^3 p_j}{(2\pi)^3} \left|T_{fi}\right|^2. \tag{43.87}$$

There are four fields involved in $2 \to 2$ scattering, each normalized in a finite volume V, and thus each contributing a factor $1/\sqrt{V}$. We take the volume factor out and write for T_{fi},

$$T_{fi} = \left(\frac{1}{\sqrt{V}}\right)^4 T'_{fi}. \tag{43.88}$$

Hence,

$$d\sigma = \frac{1}{|\mathbf{v}_1 - \mathbf{v}_2|} (2\pi)^4 \delta^{(4)}\left(p_f - p_i\right) \prod_{j=3}^{4} \frac{d^3 \mathrm{p}_j}{(2\pi)^3} \left|T'_{fi}\right|^2. \tag{43.89}$$

We make a further simplification and write

$$T'_{fi} = \beta M_{fi} \tag{43.90}$$

where the factor β is the coefficient multiplying the field operators, its form depending on whether we have fermions or bosons (which include photons). In a $2 \to 2$ reaction we have either all fermions or two fermions and two bosons, or all bosons. Thus we have the following possibilities for β:

$$\beta = \sqrt{\frac{m_1 m_2 m_3 m_4}{E_1 E_2 E_3 E_4}} \qquad \text{all fermions} \tag{43.91}$$

$$= \sqrt{\frac{m_1 m_2}{E_1 E_2 2\omega_1 2\omega_2}} \qquad \text{two fermions and two bosons} \tag{43.92}$$

$$= \sqrt{\frac{1}{2\omega_1 2\omega_2 2\omega_3 2\omega_4}} \qquad \text{all bosons.} \tag{43.93}$$

The cross-section is then

$$d\sigma = \frac{1}{|\mathbf{v}_1 - \mathbf{v}_2|} (2\pi)^4 \delta^{(4)}\left(p_f - p_i\right) |\beta|^2 \prod_{j=3}^{4} \frac{d^3 p_j}{(2\pi)^3} \left|M_{fi}\right|^2. \tag{43.94}$$

We need also to multiply the expression above by a symmetry factor S,

$$S = \prod \frac{1}{s_i!}. \tag{43.95}$$

To obtain the velocity term in the denominator we note that

$$|\boldsymbol{v}_1 - \boldsymbol{v}_2| = \left|\frac{\mathbf{p}_1}{E_1} - \frac{\mathbf{p}_2}{E_2}\right| = \frac{|\mathbf{p}_1|}{E_1 E_2}(E_1 + E_2) \tag{43.96}$$

where we have considered the center of mass system with $\mathbf{p}_2 = -\mathbf{p}_1$.

43.6 Basic two-body scattering in QED

The formalism we have developed will now enable us to calculate two-body scattering cross-sections in QED. We will not be pursuing this in any great detail, however. First, the calculations are highly labor-intensive. Second, except for the corrections that arise due to the relativistic and field-theoretic aspects of the problems, nothing overly dramatic is found at the end of the calculations.

The dramatic aspects of quantum field theory are discussed in the next chapter when we take up the subject of anomalous magnetic moment and Lamb shift. These are part of the so-called radiative corrections, which will be calculated with the help of the machinery that is already in place.

There are four basic interactions in second order QED:

(i) $e^- + e^- \to e^- + e^-$
(ii) $\gamma + e^- \to \gamma + e^-$
(iii) $e^+ + e^- \to \gamma + \gamma$
(iv) $e^- + e^+ \to e^- + e^+$

Two of these scattering processes, (i) and (ii), we have already discussed, nonrelativistically. One is Rutherford scattering calculated in the Born approximation involving the scattering of electrons off a heavy charged particle, e.g., a proton or a nucleus. A variation on this involving Dirac electrons, called Mott scattering, has also been considered. The other is Thomson scattering involving electron–photon interaction obtained through second-order time-dependent perturbation theory. We will compare the QED results with these two. Finally, we will also mention briefly two processes that are entirely field-theoretic in origin: electron–positron scattering and electron–positron annihilation (pair production).

43.7 QED vs. nonrelativistic limit: electron–electron system

43.7.1 Rutherford scattering (nonrelativistic)

We have already discussed Rutherford scattering in Chapter 20 when we considered nonrelativistic scattering due to the Coulomb potential. We discussed it again in Chapter 36 for the case of the Dirac particles. The Coulomb potential corresponding to the scattering of an electron off a nucleus, of charge Ze, say, is given by

$$V(r) = -\frac{Ze^2}{r}. \tag{43.97}$$

Let us first consider the nonrelativistic case for the scattering of a particle of charge $-e$, mass m, and momentum k off a heavy nucleus of charge Ze. The scattering amplitude in the Born approximation is given by

$$f_B(\theta) = \frac{Ze^2 m}{\hbar^2 k^2}\frac{1}{1-\cos\theta} = \frac{Ze^2 m}{2\hbar^2 k^2}\frac{1}{\sin^2(\theta/2)} \tag{43.98}$$

and the cross-section is found to be

$$\frac{d\sigma}{d\Omega} = |f_B(\theta)|^2 = \left(\frac{Ze^2 m}{2\hbar^2 k^2}\right)^2 \frac{1}{\sin^4(\theta/2)}. \tag{43.99}$$

If we write the nonrelativistic approximation

$$\hbar k = mv \tag{43.100}$$

where v is the velocity, then taking $\hbar = c = 1$, we have

$$\frac{d\sigma}{d\Omega} = \frac{Z^2 e^4}{4m^2 v^4}\frac{1}{\sin^4(\theta/2)}. \tag{43.101}$$

43.7.2 Mott scattering (Dirac)

We have already discussed Mott scattering in Chapter 36. Below we simply recapitulate what we found. For the Dirac case we obtained the T-matrix,

$$T_{fi} = \frac{iZe^2}{V}\sqrt{\frac{m^2}{E_f E_i}}\frac{\bar{u}(p_f, s_f)\gamma_4 u(p_i, s_i)}{|\mathbf{q}|^2} 2\pi\delta(E_f - E_i) \tag{43.102}$$

for the scattering of a Dirac particle with initial four-momentum p_i to final momentum p_f. The differential cross-section summed over final spins and averaged over initial spin states is found to be

$$\frac{d\sigma}{d\Omega} = \frac{4Z^2\alpha^2 m^2}{|\mathbf{q}|^4}\sum_{\text{spins}}\frac{1}{2}\left|\bar{u}(p_f, s_f)\gamma_4 u(p_i, s_i)\right|^2$$
$$= \frac{4Z^2\alpha^2 m^2}{|\mathbf{q}|^4}\frac{1}{2}\text{Tr}\left(\gamma_4\frac{\gamma\cdot p_i + m}{2m}\gamma_4\frac{\gamma\cdot p_f + m}{2m}\right) \tag{43.103}$$

where

$$\mathbf{q} = \mathbf{p}_f - \mathbf{p}_i \tag{43.104}$$

is the momentum transfer. Since energy is conserved, we take

$$\left|\mathbf{p}_f\right| = |\mathbf{p}_i| = k. \tag{43.105}$$

Therefore,

$$\left|\mathbf{q}^2\right| = \left|\mathbf{p}_f\right|^2 + |\mathbf{p}_i|^2 - 2\mathbf{p}_f \cdot \mathbf{p}_i = 2k^2\left(1 - \cos\theta\right) \tag{43.106}$$

where θ is the scattering angle. Using the trace properties we find the differential cross-section to be given by

$$\frac{d\sigma}{d\Omega} = \frac{Z^2\alpha^2 m^2}{4k^4 \sin^4\left(\theta/2\right)}\left[1 - \beta^2 \sin^2\left(\frac{\theta}{2}\right)\right]. \tag{43.107}$$

Once again, writing $\hbar k = mv$ with $\hbar = c = 1$, we obtain

$$\frac{d\sigma}{d\Omega} = \frac{Z^2\alpha^2}{4m^2 v^4 \sin^4\left(\theta/2\right)}\left[1 - \beta^2 \sin^2\left(\frac{\theta}{2}\right)\right]. \tag{43.108}$$

Thus, as we pointed out in Chapter 36, we find a correction to the Rutherford scattering formula to order $(v/c)^2$.

43.7.3 Møller scattering: $e^- + e^- \rightarrow e^- + e^-$

This process corresponds to the scattering of two electrons in a fully field-theoretic formalism. The matrix element M_{fi} for the scattering is

$$\bar{u}\left(\mathbf{p}'\right)\gamma^\mu u\left(\mathbf{p}\right)\frac{g_{\mu\nu}}{\left(q'-q\right)^2}\bar{u}\left(\mathbf{q}'\right)\gamma^\nu u\left(\mathbf{q}\right) - \bar{u}\left(\mathbf{q}'\right)\gamma^\mu u\left(\mathbf{p}\right)\frac{g_{\mu\nu}}{\left(p'-q\right)^2}\bar{u}\left(\mathbf{p}'\right)\gamma^\nu u\left(\mathbf{q}\right). \tag{43.109}$$

The two diagrams relevant for this process are contained in Fig. 43.6 with appropriate momentum designations, $e^-(p) + e^-(q) \rightarrow e^-(p') + e^-(q')$ and, with the particles interchanged, $e^-(q) + e^-(p) \rightarrow e^-(p') + e^-(q')$. The diagrams are now to be considered in the momentum space, which simply means that we remove the designations x_1 and x_2 from the vertices. The negative sign between the two terms is due to the fact that electrons satisfy Fermi statistics.

The differential cross-section for this process is given by

$$\frac{d\sigma}{d\Omega} = \frac{\alpha^2\left(2E^2 - m^2\right)^2}{4E^2\left(E^2 - m^2\right)^2}\left[\frac{4}{\sin^4\theta} - \frac{3}{\sin^2\theta} + \frac{\left(E^2 - m^2\right)^2}{\left(2E^2 - m^2\right)^2}\left(1 + \frac{4}{\sin^2\theta}\right)\right] \tag{43.110}$$

where E is the total center of mass energy of the two-electron system given by

$$4E^2 = (p+q)^2 \tag{43.111}$$

and θ is the scattering angle given by

$$\mathbf{p} \cdot \mathbf{p}' = |\mathbf{p}|\left|\mathbf{p}'\right|\cos\theta. \tag{43.112}$$

In the extreme relativistic limit, $E \gg m$, we obtain

$$\frac{d\sigma}{d\Omega} = \frac{\alpha^2}{E^2}\left(\frac{4}{\sin^4\theta} - \frac{2}{\sin^2\theta} + \frac{1}{4}\right). \tag{43.113}$$

In the nonrelativistic limit, when $E \approx m$, we obtain

$$\frac{d\sigma}{d\Omega} = \frac{\alpha^2}{m^2}\frac{1}{4v^2}\left(\frac{4}{\sin^4\theta} - \frac{3}{\sin^2\theta}\right) \tag{43.114}$$

where v is the velocity of the electron, which is given by

$$v^2 = \frac{(E^2 - m^2)}{E^2}. \tag{43.115}$$

If we write

$$\sin\theta = 2\sin\frac{\theta}{2}\cos\frac{\theta}{2} \tag{43.116}$$

for the two denominators in (43.114), and use the identity

$$1 = \left(\sin^2\frac{\theta}{2} + \cos^2\frac{\theta}{2}\right)^2 \tag{43.117}$$

for the numerator of the first term, we obtain the relation

$$\frac{d\sigma}{d\Omega} = \frac{\alpha^2}{m^2}\frac{1}{16v^4}\left(\frac{1}{\sin^4\frac{\theta}{2}} + \frac{1}{\cos^4\frac{\theta}{2}} - \frac{1}{\sin^2\frac{\theta}{2}\cos^2\frac{\theta}{2}}\right) \tag{43.118}$$

which is manifestly symmetric. In the limit $\theta \approx 0$, the first term dominates and the formula looks very much like the expression for Rutherford scattering. Here, however, m, is the reduced mass, so one must substitute $(m/2)$ for m in the above formula to obtain the Rutherford scattering result that corresponds to the scattering of an electron off a heavy nucleus.

43.8 QED vs. nonrelativistic limit: electron–photon system

43.8.1 Thomson scattering

We have already treated this problem nonrelativistically in second-order time-dependent perturbation. We found that the differential cross-section is given by

$$\frac{d\sigma}{d\Omega} = r_0^2\left|\boldsymbol{\epsilon}^{(2)}\cdot\boldsymbol{\epsilon}^{(1)}\right|^2 \tag{43.119}$$

where r_0 is the so-called Thomson radius

$$r_0 = \frac{e^2}{4\pi mc^2}. \tag{43.120}$$

43.8.2 Compton scattering: $\gamma + e^- \to \gamma + e^-$

In a fully field-theoretic treatment, the matrix element M_{fi} for this process is given as

$$\bar{u}\left(\mathbf{p}'\right)\left[\gamma\cdot\epsilon'\frac{\gamma\cdot(p+q)+m}{(p+q)^2-m^2}\gamma\cdot\epsilon+\gamma\cdot\epsilon\frac{\gamma\cdot\left(p-q'\right)+m}{\left(p-q'\right)^2-m^2}\gamma\cdot\epsilon'\right]u\left(\mathbf{p}\right). \tag{43.121}$$

The diagrams are given in Fig. 43.7 (a) and (b) (removing the designations x_1 and x_2 from the vertices). We actually did consider both the diagrams when we derived the Thomson formula. The momentum variables are indicated, which correspond to

$$\gamma(q)+e^-(p)\to\gamma(q')+e^-(p'). \tag{43.122}$$

We will consider the laboratory frame where the initial electron is at rest:

$$p^\mu = (m,0), \quad q^\mu = (k,\mathbf{k}), \tag{43.123}$$

$$p'^\mu = \left(E',\mathbf{p}'\right), \quad q'^\mu = \left(k',\mathbf{k}'\right), \tag{43.124}$$

where, because the photons have zero mass, $k = |\mathbf{k}|$ and $k' = \left|\mathbf{k}'\right|$.

The differential cross-section in the laboratory system is found to be

$$\frac{d\sigma}{d\Omega} = \frac{\alpha^2}{4m^2}\left(\frac{k'}{k}\right)^2\left(\frac{k'}{k}+\frac{k}{k'}+4\left(\epsilon'\cdot\epsilon\right)^2-2\right). \tag{43.125}$$

This is called the Klein–Nishina formula.

To obtain the relation between k' and k, we start with the energy–momentum relation

$$p^\mu + q^\mu = p'^\mu + q'^\mu, \quad \text{i.e.,} \quad q'^\mu - q^\mu = p^\mu - p'^\mu. \tag{43.126}$$

Treating the energy and three-momenta separately, the following relations are satisfied:

$$k' - k = m - E' \tag{43.127}$$

and

$$\mathbf{k}' - \mathbf{k} = -\mathbf{p}'. \tag{43.128}$$

We also note that

$$\left(q'-q\right)^2 = \left(p-p'\right)^2, \tag{43.129}$$

which gives

$$-2q'\cdot q = 2m^2 - 2E'. \tag{43.130}$$

Since

$$q' \cdot q = kk' - kk' \cos\theta \tag{43.131}$$

where θ is the scattering angle, we obtain from (43.127), (43.130), and (43.131)

$$k' = \frac{k}{1 + (k/m)(1 - \cos\theta)}. \tag{43.132}$$

Therefore, the frequency of the outgoing photon is less than that of the incoming one.

In the nonrelativistic limit, $k \to 0$, and, therefore, from (43.132), one also finds that $k' \to 0$ with

$$\frac{k'}{k} \to 1. \tag{43.133}$$

Upon substituting this relation we obtain the Thomson cross-section

$$\frac{d\sigma}{d\Omega} = \frac{\alpha^2}{m^2}\left(\epsilon \cdot \epsilon'\right)^2. \tag{43.134}$$

43.8.3 Electron–positron annihilation, pair production: $e^- + e^+ \to \gamma + \gamma$

The matrix element M_{fi} for this process is given below. The diagrams (removing the designations x_1 and x_2 from the vertices) are given in Fig. 43.9(a) and (b).

$$\bar{v}\left(\boldsymbol{p}'\right)\left[\gamma \cdot \epsilon' \frac{\gamma \cdot (p - q) + m}{(p - q)^2 - m^2} \gamma \cdot \epsilon + \gamma \cdot \epsilon \frac{\gamma \cdot \left(p - q'\right) + m}{(p - q')^2 - m^2} \gamma \cdot \epsilon'\right] u\left(\boldsymbol{p}\right). \tag{43.135}$$

The momentum variables are given by

$$\mathbf{e}^-(p) + \mathbf{e}^+(p') \to \boldsymbol{\gamma}(q) + \boldsymbol{\gamma}(q'). \tag{43.136}$$

The differential cross-section in the laboratory system is given by

$$\frac{d\sigma}{d\Omega} = \frac{\alpha^2\left(m + E'\right)}{8\left|\mathbf{p}'\right|\left(m + E' - \left|\mathbf{p}'\right|\cos\theta\right)^2}\left(\frac{k}{k'} + \frac{k'}{k} - 4\left(\epsilon \cdot \epsilon'\right)^2 + 2\right) \tag{43.137}$$

where the variables are defined similarly to the Compton scattering case,

$$p^\mu = (m, 0),\ p'^\mu = \left(E', \mathbf{p}'\right), \tag{43.138}$$

$$q^\mu = (k, \mathbf{k}),\ q'^\mu = \left(k', \mathbf{k}'\right), \tag{43.139}$$

and θ is the scattering angle $\mathbf{k}' \cdot \mathbf{p}' = k'\left|\mathbf{p}'\right|\cos\theta$. As we did for the Compton case, the relation between k' and k can be easily derived on the basis of energy momentum conservation. It is found to be

$$\frac{k'}{k} = \frac{E' - \left|\mathbf{p}'\right|\cos\theta}{m}. \tag{43.140}$$

43.8.4 Electron–positron scattering, Bhabha scattering: $e^- + e^+ \to e^- + e^+$

The matrix element T'_{fi} for this process is given below and the diagrams (removing the designations x_1 and x_2 from the vertices) in Fig. 43.8(a) and (b).

$$\bar{v}\left(\boldsymbol{p}'\right)\gamma^{\mu}v\left(\boldsymbol{p}\right)\frac{g_{\mu\nu}}{\left(q'-q\right)^2}\bar{u}\left(\boldsymbol{q}'\right)\gamma^{\nu}u\left(\boldsymbol{q}\right)-v\left(\boldsymbol{p}\right)\gamma^{\mu}u\left(\boldsymbol{q}\right)\frac{g_{\mu\nu}}{\left(p+q\right)^2}\bar{u}\left(\boldsymbol{q}'\right)\gamma^{\nu}v\left(\boldsymbol{p}'\right). \quad (43.141)$$

The differential cross-section is

$$\frac{d\sigma}{d\Omega}=\frac{\alpha^2}{8E^2}\left(\frac{1+\cos^4\theta/2}{\sin^4\theta/2}+\frac{1}{2}\left(1+\cos^2\theta\right)-2\frac{\cos^4\theta/2}{\sin^4\theta/2}\right) \quad (43.142)$$

where, as in the case of Møller scattering, E is the center-of-mass energy.

The nonrelativistic approximation is given by

$$\frac{d\sigma}{d\Omega}=\frac{\alpha^2}{m^2}\frac{1}{16v^4\sin^4\theta/2}. \quad (43.143)$$

This process is closer to Rutherford scattering in that the two particles involved in scattering are not identical as they are for the Møller case. However, here one must again replace m by $(m/2)$ to get the Rutherford result.

44 Radiative corrections

44.1 Radiative corrections and renormalization

The second-order diagrams we considered previously are often called "tree" diagrams, since they look like trees with branches. The next higher-order two-particle → two particle scattering processes in QED are given by diagrams shown in Fig. 44.1(a)–(d), which are referred to as "radiative corrections." As we will discuss below, they all involve loops where, according to the Feynman rules, one must integrate over the internal momenta. The common characteristic of these loops is that they are all divergent in the limit $k \to \infty$, called "ultraviolet" divergence, where k is an internal loop momentum.

There are also other divergent radiative corrections. These are the so-called "infrared" divergences in the limit $k \to 0$ that are present in the three diagrams Fig. 44.1(a)–(c). We will not be considering this type of divergence except to mention that they are exactly canceled by "bremsstrahlung" diagrams, Fig. 44.2(a) and (b), which involve the emission of zero-energy photons in the final state.

The treatment of the infinities in the loop diagrams and the extracting of finite quantities from them, which are then verified by experiments, form one of the remarkable success stories of QED. The basic premise of the theory is that the free-particle Hamiltonian, H_0, and the interaction, H_I, contain parameters such as charge and mass of the interacting particles

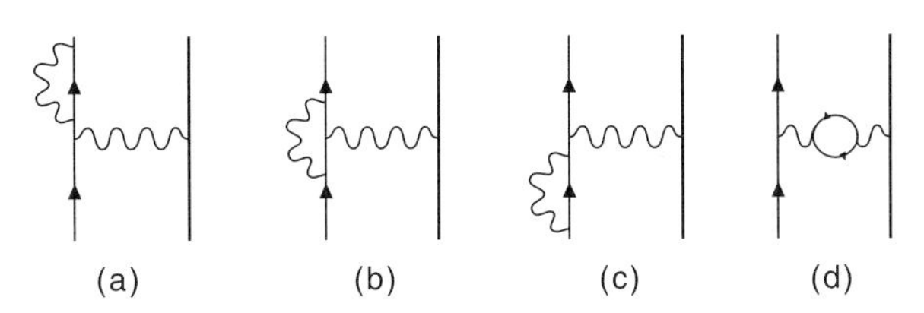

Fig. 44.1

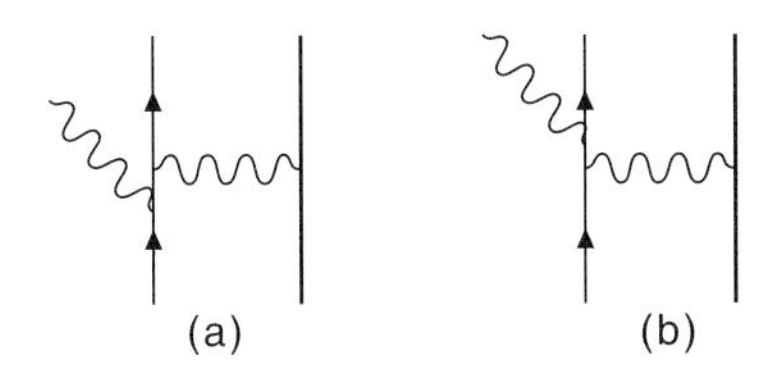

Fig. 44.2

which should be designated as "bare" quantities. These parameters appear at the lowest order of perturbation but they are not the quantities that are physically observed. Higher-order perturbative terms produce corrections to these parameters and, experimentally, what we are supposed to be observing are the final products that are "dressed-up" – "renormalized", as they are termed – through higher-order terms. These are the "physical" objects in contrast to the bare quantities and, according to QED theory, it is these that should be finite. One finds that in QED theory the various infinities can be absorbed in a consistent manner into physically well-defined observable parameters such as charge and mass. The finite quantities extracted by the theory are then subjected to experimental verification and the agreement is found to be quite dramatic. This is the essence of the renormalization program we will discuss below.

We isolate four types of loop diagrams (Fig. 44.1(a)–(d)) and consider them below.

(i) The electron self-energy diagram, due to the emission and reabsorption of a photon, appears in Figs 44.1(a) and (c). These diagrams will be a part of any higher-order diagram. The self-energy diagram contributes to the mass of the electron.
(ii) The vertex correction, also due to the emission and reabsorption of a photon, contributes to the charge and magnetic moment of the electron. This correction corresponds to Fig. 44.1(b).
(iii) The vacuum polarization diagram, Fig. 44.1(d), due to the presence of an electron loop in a photon propagator. This diagram contributes to the charge. Here the photon continues to remain massless because of gauge invariance.

To carry out the renormalization program we will make use of the Ward identity described in the Appendix and proceed in three steps. In the first step we will separate the "divergent" part from the "convergent" part when calculating the matrix elements of each of the above diagrams. In the second step we will collect all the divergent parts and describe the renormalization process through which the infinite terms are combined into two finite, physically observable, constants which by the redefinitions of these parameters will be called physical charge and mass. In the third step the left over convergent parts are discussed. These convergent parts refer to

- the anomalous magnetic moment of the electron, and
- the Lamb shift in the hydrogen atom.

We will show that our calculated results agree with the experimental results to a remarkable accuracy.

44.2 Electron self-energy

An electron propagator has been considered previously to the lowest order. Let us now consider the diagram in Fig. 44.3. This is the same diagram that appears on the external fermion legs in Fig. 44.1(a) and (c), but it is now isolated so that we can determine its contribution. It describes an electron emitting a photon and then re-absorbing it. The contribution of this diagram is of the order of α, the fine structure constant ($\simeq 1/137$). One can write it as

$$\bar{u}\,(\mathbf{p})\,\Sigma\,(\gamma\cdot p)\,u\,(\mathbf{p}) \tag{44.1}$$

where $u\,(\mathbf{p})$ is the free particle Dirac wavefunction, and $\Sigma\,(\gamma\cdot p)$ is given by the integral

$$\Sigma\,(\gamma\cdot p) = -ie_0^2\int\frac{d^4k}{(2\pi)^4}\gamma_\mu\left(\frac{\gamma\cdot p-\gamma\cdot k+m_0}{[(p-k)^2-m_0^2]k^2}\right)\gamma^\mu \tag{44.2}$$

where e_0 and m_0 are the bare charge and mass of the electron. One often writes Σ_2 for the above integral to signify that it is a second-order calculation.

If we consider the unperturbed Lagrangian given by

$$\int d^4x\,\bar{\psi}\,(x)\,(i\gamma\cdot\partial-m_0)\,\psi\,(x) \tag{44.3}$$

then we note that the term contributing to mass in the momentum space is of the form

$$\bar{u}\,(\mathbf{p})\,m_0u\,(\mathbf{p})\,. \tag{44.4}$$

Expression (44.1), therefore, adds to the bare mass term (44.2) so that the sum of the two is $\bar{u}\,(\mathbf{p})\,[m_0+\Sigma\,(\gamma\cdot p)]\,u(\mathbf{p})$.

To calculate $\Sigma\,(\gamma\cdot p)$ we use the Feynman integration technique described in the Appendix and replace the product of the two denominators in the loop integral given by (44.2) by a single denominator

$$\frac{1}{\left[(p-k)^2-m_0^2\right]k^2} = \int_0^1 dx\frac{1}{D^2} \tag{44.5}$$

where D is given by

$$D = \left[(p-k)^2-m_0^2\right]x+k^2(1-x) \tag{44.6}$$

$$= k^2-2p\cdot kx+xp^2-m_0^2x. \tag{44.7}$$

In expanding the quadratic term present in the first term of (44.6) we have used the free particle relation $p^2=m_0^2$. We next shift the variables by taking

$$\kappa = k-px \tag{44.8}$$

so that

$$D=\kappa^2-\Delta \tag{44.9}$$

where

$$\Delta = m_0^2x-x\,(1-x)\,p^2. \tag{44.10}$$

We then write

$$\Sigma\,(\gamma\cdot p) = -ie_0^2\int\frac{d^4\kappa}{(2\pi)^4}\int_0^1 dx\frac{N}{\left[\kappa^2-\Delta\right]^2} \tag{44.11}$$

where the numerator, N, is given by

$$N = \gamma_\mu \gamma \cdot p \gamma^\mu (1 - x) + m_0 \gamma_\mu \gamma^\mu. \tag{44.12}$$

This expression can be simplified further by utilizing the relations that were obtained in Chapter 35 involving the product of the γ-matrices:

$$\gamma_\mu \gamma \cdot p \gamma^\mu = \gamma_\mu \gamma^\nu p_\nu \gamma^\mu = -2\gamma^\nu p_\nu = -2\gamma \cdot p \tag{44.13}$$

and

$$\gamma_\mu \gamma^\mu = 4. \tag{44.14}$$

We then obtain

$$N = -2\gamma \cdot p(1 - x) + 4m_0. \tag{44.15}$$

Therefore,

$$\Sigma\,(\gamma \cdot p) = -ie_0^2 \int_0^1 dx\,[-2\gamma \cdot p(1 - x) + 4m_0] \int \frac{d^4\kappa}{(2\pi)^4} \frac{1}{\left[\kappa^2 - \Delta\right]^2}. \tag{44.16}$$

The integral over $d^4\kappa$ is divergent since the integrand is of the form $(d\kappa/\kappa)$, which is logarithmically divergent:

$$\int^\Lambda \frac{d\kappa}{\kappa} = \ln \Lambda \tag{44.17}$$

where the upper limit Λ is infinitely large.

In order to obtain meaningful results from the divergent integral (44.16) there are several different integration schemes available that one can follow. The scheme we will follow is called the dimensional regularization method, which is explained in detail in the Appendix. It is a technique in which certain invariance properties such as gauge invariance and Ward identity are more easily preserved. In this scheme one first carries out the integration over an arbitrary space-time dimension d by assuming the Feynman integrals to be analytic functions in the complex d-plane. The integrals are convergent for d less than 4. After integrating and achieving a finite result one takes the limit $d \to 4$. The infinity that was noted earlier in the loop integral is then reflected as a singularity which, in the cases we will discuss, is found to be a pole. We then separate this pole term from the left-over finite terms.

Thus, following the discussion in the Appendix, we have, for the integral in (44.16) in d dimensions

$$\int \frac{d^4\kappa}{(2\pi)^4} \frac{1}{\left[\kappa^2 - \Delta\right]^2} \to i \int \frac{d^dK}{(2\pi)^d} \frac{1}{\left[K^2 + \Delta\right]^2}. \tag{44.18}$$

The integral on the right-hand side has been calculated in the Appendix. One finds

$$\int \frac{d^4\kappa}{(2\pi)^4} \frac{1}{\left[\kappa^2 - \Delta\right]^2} = \frac{i}{8\pi^2 \epsilon} \tag{44.19}$$

where

$$\epsilon = 4 - d. \tag{44.20}$$

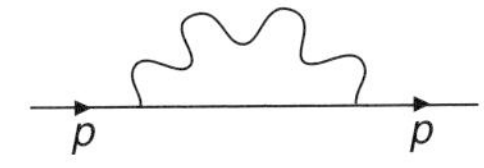

Fig. 44.3

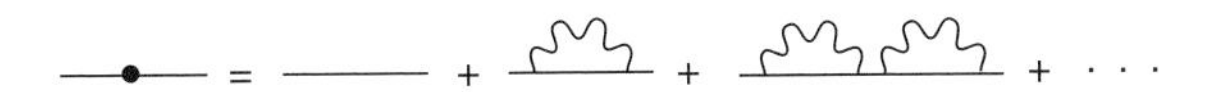

Fig. 44.4

Substituting (44.19) in (44.16) and integrating over x, we obtain

$$\Sigma(\gamma \cdot p) = e_0^2[-\gamma \cdot p + 4m_0] \frac{1}{8\pi^2\epsilon}. \tag{44.21}$$

Let us discuss the consequences of this relation to the electron self-energy question. The electron propagator is given by

$$\frac{\gamma \cdot p + m_0}{p^2 - m_0^2}, \tag{44.22}$$

which can be written as

$$\frac{1}{\gamma \cdot p - m_0}. \tag{44.23}$$

To prove this one simply needs to multiply and divide (44.22) by $(\gamma \cdot p + m_0)$ and use the relation $\gamma^\mu\gamma^\nu + \gamma^\nu\gamma^\mu = 2g^{\mu\nu}$.

Expression (44.22) is the lowest-order contribution to the electron propagator. The higher-order corrections to the propagator can be viewed as the following series of diagrams. This series can be written as shown in Fig. 44.4.

$$\frac{1}{\gamma \cdot p - m_0} + \frac{1}{\gamma \cdot p - m_0}\Sigma(\gamma \cdot p)\frac{1}{\gamma \cdot p - m_0} + \cdots, \tag{44.24}$$

which is a geometric series and sums to

$$S_F(p) = \frac{1}{\gamma \cdot p - m_0 - \Sigma(\gamma \cdot p)} \tag{44.25}$$

where the expression for $\Sigma(\gamma \cdot p)$ has already been given in (44.21).

We now designate m as the renormalized mass and write the corresponding renormalized propagator $S'_F(p)$ as

$$S'_F(p) = \frac{1}{\gamma \cdot p - m}. \tag{44.26}$$

We relate S'_F to S_F as follows:

$$Z_2 S'_F(p) = S_F(p) \tag{44.27}$$

where we have assumed S'_F to be finite, with Z_2 to contain all the infinities.

Taking the inverse of the relation (44.27) and inserting the relations (44.25) and (44.26) we find

$$\frac{\gamma \cdot p - m}{Z_2} = \gamma \cdot p - m_0 - \Sigma\left(\gamma \cdot p\right) = \gamma \cdot p - m_0 - e_0^2\left[-\gamma \cdot p + 4m_0\right]\frac{1}{8\pi^2\epsilon}. \quad (44.28)$$

By equating the coefficients of $\gamma \cdot p$ on both sides of (44.28), we obtain

$$\frac{1}{Z_2} = 1 + \frac{e_0^2}{8\pi^2\epsilon}. \quad (44.29)$$

Inverting this relation by keeping ϵ fixed, and expanding $\left(1 + e_0^2/8\pi^2\right)^{-1}$ by assuming e_0^2 to be small, we obtain the relation

$$Z_2 \simeq 1 - \frac{e_0^2}{8\pi^2\epsilon} \quad (44.30)$$

where we have kept only the first two terms in the expansion.

Equating the constant terms on both sides of (44.28) we find

$$\frac{m}{Z_2} = \left[1 + \frac{4e_0^2}{8\pi^2\epsilon}\right] m_0. \quad (44.31)$$

Multiplying both sides by Z_2 and using (44.30) we find, to the leading order in e_0^2,

$$m = \left[1 + \frac{3e_0^2}{8\pi^2\epsilon}\right] m_0. \quad (44.32)$$

Thus, if we define the relation between the renormalized mass and bare mass as

$$m = m_0 + \Sigma\left(m\right), \quad (44.33)$$

then

$$\Sigma\left(m\right) = \frac{3m_0 e_0^2}{8\pi^2\epsilon}. \quad (44.34)$$

The basic assertion in the renormalization theory is that even though $\Sigma(m)$ is divergent, the sum of this term with the unknown and possibly infinite bare mass term, m_0, is, nevertheless, finite. This is what we have called the renormalized mass, m.

From the definition given in Chapter 37, the propagator $S_F\left(p\right)$ can be expressed as a time-ordered product of fields $\psi(x)$, i.e.,

$$S_F\left(p\right) \sim \left\langle O \left| T\left(\bar{\psi}\left(x\right)\psi(x)\right)\right| O\right\rangle. \quad (44.35)$$

Similarly, we write

$$S_F'\left(p\right) \sim \left\langle O \left| T\left(\bar{\psi}'\left(x\right)\psi'(x)\right)\right| O\right\rangle \quad (44.36)$$

where $\psi'(x)$ is interpreted as a renormalized field. Equation (44.27) then corresponds to the following relation between the fields $\psi(x)$ and $\psi'(x)$:

$$\psi(x) = \sqrt{Z_2}\psi'(x), \tag{44.37}$$

with Z_2, given by (44.30), containing the divergent terms. We will complete the renormalization program in the next chapter, and at the same time determine the convergent parts of the radiative corrections.

44.3 Appendix to Chapter 44

44.3.1 Ward Identity

The matrix elements to second order in QED that involve fermions are often expressible in the form of a four-vector, M^μ, or a second-rank tensor, $T^{\mu\nu}$ associated with the external lines. They involve fermion current, $j^\mu(x) = \bar{\psi}'(x)\gamma^\mu\psi(x)$, or products of fermion currents. For example, in Moller scattering, which we considered in the previous chapter, a typical term in the matrix element is

$$\left[\bar{u}\gamma^\mu u\right]\text{[photon propagator]}\left[\bar{u}\gamma_\mu u\right] \tag{44.38}$$

which is a product of two currents.

Write $\psi'(x)$ and $\psi(x)$ as $u(\mathbf{p}')\exp[i(p'\cdot x)]$ and $u(\mathbf{p})\exp[i(p\cdot x)]$, respectively the current conservation relation, $\partial_\mu j^\mu = 0$, implies, in the momentum space, the relation

$$q_\mu j^\mu = 0 \tag{44.39}$$

where $q = p' - p$. For the matrix element M^μ this means that

$$q_\mu M^\mu = 0. \tag{44.40}$$

This is basically the statement of the Ward identity.

In scattering processes involving photons, the photon polarization vector, ϵ_μ, enters in the matrix element in the form

$$\epsilon_\mu M^\mu. \tag{44.41}$$

For example, in Compton scattering, a typical term leaving out external fields is

$$\left[\gamma^\mu\epsilon'_\mu\right]\text{[electron propagator]}\left[\gamma^\nu\epsilon_\nu\right], \tag{44.42}$$

which is of the form (44.38). Hence the Ward identity (44.40) applies to this process by replacing ϵ'_μ or ϵ_ν by an appropriate momentum vector.

In the next section we will discuss some important practical consequences of the Ward identity for two entities: the vertex diagram and the photon propagator.

44.3.2 Ward identity, vertex function, and photon propagator

The vertex diagram is described by two external electron lines and an internal photon line (propagator). The vertex term, Λ^μ, is a four-vector. It can, therefore, be expressed as a linear combination of the available independent four-vectors. These are: γ^μ and the four-vectors p^μ and p'^μ. We will write Λ^μ as

$$\Lambda^\mu = a\gamma^\mu + b\left(p^\mu + p'^\mu\right) + cq^\mu \tag{44.43}$$

where $q^\mu = p'^\mu - p^\mu$. From (44.40) we must have

$$q_\mu \Lambda^\mu = 0. \tag{44.44}$$

Therefore, for a vertex term given by $\bar{u}\left(p'\right) q_\mu \Lambda^\mu u\left(p\right)$, we will have the relation

$$\begin{aligned}\bar{u}\left(p'\right) q_\mu \Lambda^\mu u\left(p\right) &= a\left[\bar{u}\left(p'\right)\gamma \cdot p' u\left(p\right) - \bar{u}\left(p'\right)\gamma \cdot p u\left(p\right)\right] \\ &\quad + b\left[\bar{u}\left(p'\right)\left(p'^2 - p^2\right)u\left(p\right)\right] + c\left[\bar{u}(p')q^2 u\left(p\right)\right]\end{aligned} \tag{44.45}$$

where, as we stated above, $u\left(p\right)$ and $\bar{u}\left(p'\right)$ correspond to the external, free fermions, while the photon line is internal. The first term in (44.45) will vanish because $\gamma \cdot p' = \gamma \cdot p = m$; the second term vanishes as well since $p^2 = p'^2 = m^2$. The only term left is the third one since q^2, which corresponds to an internal photon, does not necessarily vanish. Following relation (44.44), the left-hand side of (44.45) vanishes. Hence,

$$c = 0. \tag{44.46}$$

The second example concerns the photon propagator involved in the Feynman diagram in Fig. 44.1d. It is of the form

$$\prod{}^{\mu\nu} = \int \gamma^\mu \,[\text{electron propagator}]\, \gamma^\nu \,[\text{electron propagator}]\,. \tag{44.47}$$

A general expression for a second-rank tensor can be written in terms of the linear combination

$$\prod{}^{\mu\nu} = ag^{\mu\nu} + bq^\mu q^\nu. \tag{44.48}$$

From (44.44) we have

$$q_\mu \prod{}^{\mu\nu} = 0. \tag{44.49}$$

Hence,

$$aq^\nu + bq^2 q^\nu = 0, \tag{44.50}$$

which gives

$$b = -\frac{a}{q^2} \tag{44.51}$$

and

$$\prod{}^{\mu\nu} = a\left(g^{\mu\nu} - \frac{q^\mu q^\nu}{q^2}\right). \tag{44.52}$$

The above relation holds whether or not the photon is connected to external electrons. If it is connected to an external fermion line, however, then its contribution to the S-matrix is of the form

$$\bar{u}(p')\gamma_\mu u(p)\prod{}^{\mu\nu} \tag{44.53}$$

where u and $\bar{u}$ represent free, external fermions. The contribution of the second term in (44.48) vanishes since

$$\bar{u}(p')\gamma_\mu q^\mu u(p) = \bar{u}(p')\left(\gamma \cdot p' - \gamma \cdot p\right)u(p) = 0. \tag{44.54}$$

Hence, we are left with only the term

$$\prod{}^{\mu\nu} = a g^{\mu\nu}. \tag{44.55}$$

Thanks to the Ward identity, the form of the photon propagator is, therefore, particularly simple.

44.3.3 Feynman integration technique

The Feynman integration result states that

$$\frac{1}{a_1 a_2 \cdots a_n} = (n-1)! \int \frac{dx_1\, dx_2 \,\cdots\, dx_n \delta\left[1 - \sum_i^n x_i\right]}{\left[a_1 x_1 + a_2 x_2 + \cdots a_n x_n\right]^n} \tag{44.56}$$

where, in our problems $a_1 a_2 ... a_n$ are functions of momenta. Since we already know the answer it will help us to devise a derivation that is simple and elegant!

First we note that we can write

$$\frac{1}{a_1} = \int_0^\infty dz_1 e^{-a_1 z_1}. \tag{44.57}$$

Taking a product of two such terms, we obtain

$$\frac{1}{a_1 a_2} = \int_0^\infty dz_1 e^{-a_1 z_1} \int_0^\infty dz_2 e^{-a_2 z_2}. \tag{44.58}$$

This is not in the form (44.56) which we want to derive. Therefore, let us take

$$z_1 = t x_1, \quad \text{and} \quad z_2 = t x_2 \tag{44.59}$$

where t is an arbitrary constant at the moment. We obtain

$$\frac{1}{a_1 a_2} = \int_0^\infty dx_1 \int_0^\infty dx_2 t^2 e^{-t(a_1 x_1 + a_2 x_2)}. \tag{44.60}$$

It is clear that we cannot leave the result in terms of an arbitrary parameter. What we need to do is to integrate over t, in which case we get a term of the form

$$\int dt e^{-t(a_1x_1+a_2x_2)}. \tag{44.61}$$

This will, indeed, give us a denominator of the form we have in (44.56). However, we cannot simply integrate over t since it will bring in an extra integration that did not exist before. Therefore, to keep the value intact, we need to put in a δ-function of t, which we take to be $\delta\left[t - t(x_1 + x_2)\right]$. Hence we write

$$\frac{1}{a_1a_2} = \int_0^\infty dx_1 \int_0^\infty dx_2 \int_0^\infty dt\, t^2 e^{-t(a_1x_1+a_2x_2)} \delta\left[t - t(x_1 + x_2)\right]. \tag{44.62}$$

Since

$$\delta\left[t - t(x_1 + x_2)\right] = \frac{1}{t}\delta\left[1 - (x_1 + x_2)\right], \tag{44.63}$$

we find

$$\frac{1}{a_1a_2} = \int_0^\infty dx_1 \int_0^\infty dx_2 \int_0^\infty dt\, t e^{-t(a_1x_1+a_2x_2)} \delta\left[1 - (x_1 + x_2)\right]. \tag{44.64}$$

We can now integrate over t to obtain

$$\frac{1}{a_1a_2} = \int_0^\infty dx_1 \int_0^\infty dx_2 \frac{\delta\left[1 - (x_1 + x_2)\right]}{(a_1x_1 + a_2x_2)^2} \tag{44.65}$$

which is, indeed, of the form (44.56) for a product of two terms.

The procedure to generalize this to n terms is now straightforward. We write

$$\frac{1}{a_1a_2\cdots a_n} = \int_0^\infty dz_1 e^{-a_1z_1} \int_0^\infty dz_2 e^{-a_2z_2} \ldots \int_0^\infty dz_n e^{-a_nz_n} \tag{44.66}$$

then make the transformation

$$z_i = tx_i \quad \text{for} \quad i = 1\cdots n \tag{44.67}$$

and then integrate over t after multiplying by $\delta\left[t - t(x_1 + x_2 + \cdots + x_n)\right]$.

44.3.4 Dimensional regularization

A typical integral that appears in the calculation of the Feynman diagrams is of the form

$$\int \frac{d^4\kappa}{(2\pi)^4} \frac{(\kappa^2)^m}{\left[\kappa^2 - \Delta + i\epsilon\right]^n}. \tag{44.68}$$

We note that

$$\kappa^2 = \kappa_4^2 - \boldsymbol{\kappa}^2. \tag{44.69}$$

The negative sign in this expression is due to the fact that we are working in the Minkowski space. The integration, however, is facilitated considerably by making a counterclockwise rotation by 90° in the κ_4-space, called a Wick rotation,

$$\kappa_4 \to iK_4, \quad \boldsymbol{\kappa} \to \mathbf{K}, \quad \text{and} \quad \kappa^2 \to -K^2. \tag{44.70}$$

This avoids any singularities in the path of integration and at the same time brings the integration into the Euclidean space. We then have

$$\int \frac{d^4\kappa}{(2\pi)^4} \frac{(\kappa^2)^m}{\left[\kappa^2 - \Delta + i\epsilon\right]^n} \to i(-)^{n+m} \int \frac{d^4K}{(2\pi)^4} \frac{(K^2)^m}{\left[K^2 + \Delta\right]^n} \tag{44.71}$$

where

$$K^2 = K_4^2 + \mathbf{K}^2 \quad \text{and} \quad d^4K = dK\, K^3 d\Omega_4. \tag{44.72}$$

On the right-hand side of the second relation, K is the magnitude in the four-dimensional Euclidean space and $d\Omega_4$ is the four-dimensional solid angle. Below we consider first the case $m = 0$:

$$\int \frac{d^4K}{(2\pi)^4} \frac{1}{\left[K^2 + \Delta\right]^n}. \tag{44.73}$$

Integrations of this type are divergent if $n \leq 2$ since the power of K involved in the numerator is $\sim K^4$. As we stated earlier, to obtain meaningful results in which one can absorb the divergences into physical parameters and, at the same time, preserve the properties that are independent of space-time, one resorts to the so-called "dimensional regularization" method. Here one integrates (44.73) in an arbitrary dimension d where the integral is convergent and then, after the integration is performed, one analytically continues the dimension from d to 4. In the process any singularities that occur are included in the result. We explain this below.

The integral in (44.73) in terms of d dimensions is given by

$$\lim_{d \to 4} \int \frac{d^dK}{(2\pi)^d} \frac{1}{\left[K^2 + \Delta\right]^n}. \tag{44.74}$$

For $n \leq 2$ and $d = 4$ this integral is divergent.

To discuss dimensional regularization, let us consider first the case with $n = 2$, and write $d^dK = dKK^{d-1}d\Omega_d$ where $d\Omega_d$ is the d-dimensional solid angle,

$$\int \frac{d^dK}{(2\pi)^d} \frac{1}{\left[K^2 + \Delta\right]^2} = \int \frac{dKK^{d-1}}{(2\pi)^d} \frac{1}{\left[K^2 + \Delta\right]^2} \int d\Omega_d. \tag{44.75}$$

To calculate this integral, we change variables in the first integral on the right-hand side,

$$\frac{K^2}{\Delta} = \frac{1-x}{x}, \tag{44.76}$$

and obtain

$$\int_0^\infty \frac{dKK^{d-1}}{(2\pi)^d} \frac{1}{\left[K^2+\Delta\right]^2} = \left(\frac{1}{2}\Delta^{\frac{d}{2}-2}\right) \frac{1}{(2\pi)^d} \int_0^1 dx\, x^{1-\frac{d}{2}} (1-x)^{\frac{d}{2}-1}. \tag{44.77}$$

To calculate (44.77) we make use of the well-known relation for the beta-function, $B(\alpha,\beta)$,

$$\int_0^1 dx\, x^{\alpha-1} (1-x)^{\beta-1} = B(\alpha,\beta) = \frac{\Gamma(\alpha)\Gamma(\beta)}{\Gamma(\alpha+\beta)} \tag{44.78}$$

where Γ's are the gamma functions. We then obtain the following by taking $\alpha = 2 - \frac{d}{2}$ and $\beta = \frac{d}{2}$,

$$\int_0^\infty \frac{dK\, K^{d-1}}{(2\pi)^d} \frac{1}{\left[K^2+\Delta\right]^2} = \left(\frac{1}{2}\Delta^{\frac{d}{2}-2}\right) \frac{1}{(2\pi)^d} \frac{\Gamma\left(2-\frac{d}{2}\right)\Gamma\left(\frac{d}{2}\right)}{\Gamma(2)}. \tag{44.79}$$

Since we wish to take the limit $d \to 4$, the argument of the first Γ-function in (44.79) vanishes in that limit. However, we note the following properties of the Γ-function:

$$\Gamma(1+\alpha) = \alpha\Gamma(\alpha), \tag{44.80}$$

$$\Gamma(1) = 1, \tag{44.81}$$

and

$$\Gamma(\alpha) \underset{\alpha\to 0}{\to} \frac{1}{\alpha} - \gamma + O(\alpha), \tag{44.82}$$

where γ is called the Euler–Mascheroni constant whose value is 0.5772. Thus $\Gamma(\alpha)$ has a pole at $\alpha = 0$ and, therefore, (44.79) has a pole at $d = 4$. The divergence we saw earlier in the integral (44.73) for $d = 4$ and $n = 2$ shows up as a pole at $d = 4$ in the complex d-plane.

The integral over the solid angle in (44.75) has no divergence. Hence in the limit $d \to 4$ one needs to calculate only the integral over the four-dimensional solid angle $d\Omega_4$. We show in the following chapter that

$$\int d\Omega_4 = 2\pi^2. \tag{44.83}$$

Hence

$$\int \frac{d^4K}{(2\pi)^4} \frac{1}{\left[K^2+\Delta\right]^2} = \left[\lim_{d\to 4} \left(\frac{1}{2}\Delta^{\frac{d}{2}-2}\right) \frac{1}{(2\pi)^d} \frac{\Gamma\left(2-\frac{d}{2}\right)\Gamma\left(\frac{d}{2}\right)}{\Gamma(2)}\right] \left(2\pi^2\right). \tag{44.84}$$

Substituting the results from (44.79)–(44.84), we obtain, to leading order in the singularity at $d = 4$,

$$\int \frac{d^4K}{(2\pi)^4} \frac{1}{\left[K^2+\Delta\right]^2} = \frac{1}{8\pi^2\epsilon} \tag{44.85}$$

where

$$\epsilon = 4 - d. \tag{44.86}$$

The following integrals will also be useful in the calculation of the radiative corrections.

$$\int \frac{d^d K}{(2\pi)^d} \frac{K^2}{\left(K^2+\Delta\right)^3} = \frac{1}{(4\pi)^{d/2}} \frac{d}{2} \frac{\Gamma\left(2-\frac{d}{2}\right)}{\Gamma(3)} \left(\frac{1}{\Delta}\right)^{2-\frac{d}{2}}, \tag{44.87}$$

$$\int \frac{d^d K}{(2\pi)^d} \frac{K^2}{\left(K^2+\Delta\right)^2} = \frac{1}{(4\pi)^{d/2}} \frac{d}{2} \frac{\Gamma\left(1-\frac{d}{2}\right)}{\Gamma(2)} \left(\frac{1}{\Delta}\right)^{1-\frac{d}{2}}. \tag{44.88}$$

On the right-hand side in (44.87) and (44.88) we have replaced the integral over $d\Omega_d$ by the integral over $d\Omega_4$ since we will be considering these integrals only in the limit $d \to 4$.

45 Anomalous magnetic moment and Lamb shift

45.1 Calculating the divergent integrals

The procedure for calculating the integrals involved in radiative corrections is now fairly clear. One needs to go through the following steps:

1. Use the Feynman integration technique to write the integral in terms of a single denominator, D.
2. Introduce a new variable, the loop momentum, κ, so that D can be expressed in terms of κ^2, and the integration can be performed over $d^4\kappa$.
3. Use a dimensional regularization scheme to do the divergent part of the integral in terms of the dimension d. After integrating, let $d \to 4$ and separate the singular (pole) term at $d = 4$ from the finite terms.

We will follow this procedure in order to calculate the loop diagrams for the vertex function and the photon propagator.

45.2 Vertex function and the magnetic moment

In the lowest order, the vertex diagram in Fig. 45.1 is written as

$$\bar{u}\left(\mathbf{p}'\right)\gamma^{\mu}u\left(\mathbf{p}\right). \tag{45.1}$$

From the Gordon decomposition relation which we derived in Chapter 35, one can write

$$\gamma^{\mu} = \left[\frac{p^{\mu}+p'^{\mu}}{2m} + \frac{i\sigma^{\mu\nu}q_{\nu}}{2m}\right] \tag{45.2}$$

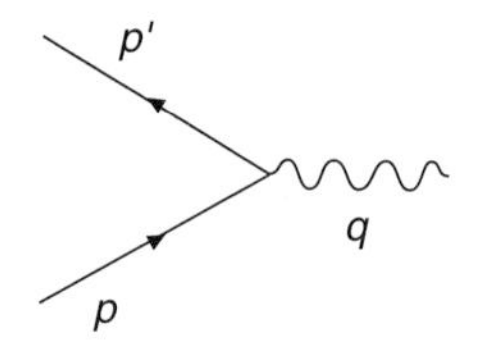

Fig. 45.1

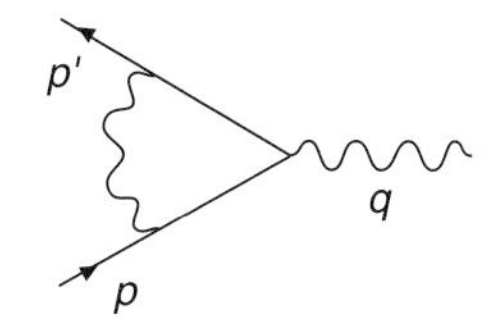

Fig. 45.2

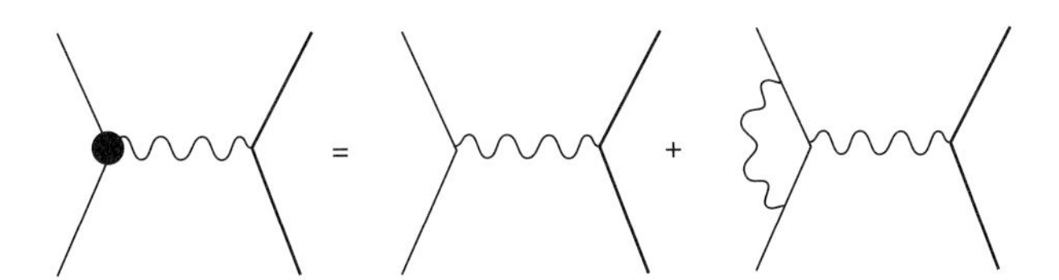

Fig. 45.3

where $q^\mu = p'^\mu - p^\mu$. The electromagnetic interaction term given by

$$\bar{\psi}(x)\,\gamma^\mu\psi(x)\,A_\mu(x) \tag{45.3}$$

contains a $\sigma^{\mu\nu}q_\nu$ term corresponding to the interaction $\boldsymbol{\mu}\cdot\mathbf{B}$ where $\mathbf{B}$ is the magnetic field and $\boldsymbol{\mu}$ is the magnetic moment, called the intrinsic magnetic moment of the electron. It is given by

$$\boldsymbol{\mu} = \frac{e}{mc}\mathbf{S} \tag{45.4}$$

where $\mathbf{S}$ is the electron spin, $\mathbf{S} = (1/2)\,\boldsymbol{\sigma}$. If one writes the above relation as

$$\boldsymbol{\mu} = \left(\frac{g}{2}\right)\frac{e}{mc}\mathbf{S} \tag{45.5}$$

where g is called the Landé factor, then (45.4) implies that $g = 2$.

We will now consider radiative corrections to the vertex diagram described by Fig. 45.2 and contained in Fig. 44.1(b). The vertex function is given by

$$\bar{u}(\mathbf{p}')\,\Lambda^\mu u(\mathbf{p}) \tag{45.6}$$

where Λ^μ is $O(\alpha)$ compared to the lowest-order term, γ^μ. The total vertex function to order α, which includes the tree diagram, can then be written (see Fig. 45.3) as

$$\bar{u}(\mathbf{p}')\,\Gamma^\mu u(\mathbf{p}) \tag{45.7}$$

where

$$\Gamma^\mu = \gamma^\mu + \Lambda^\mu. \tag{45.8}$$

We note that Γ^μ will depend on both p' and p, which we write as $\Gamma^\mu(p',p)$. To take account of the radiative corrections we write a general expression in terms of the available vectors γ^μ and $\sigma^{\mu\nu}q_\nu$,

$$\Gamma^\mu(p',p) = \gamma^\mu F_1\left(q^2\right) + i\frac{\sigma^{\mu\nu}q_\nu}{2m}F_2\left(q^2\right) \tag{45.9}$$

where $q^\mu = p'^\mu - p^\mu$, while F_1 and F_2 are the form factors (see the discussion in Section 20.6) which are in effect charge and magnetic moment distributions generated through electron–photon interactions.

We note, once again, from relations (45.2) and (45.9) that γ^μ and F_1, with $F_1(0) = 1$ to lowest order, already contain contributions from the term, $\sigma^{\mu\nu}q_\nu$, which corresponds to the intrinsic magnetic moment. The F_2 term takes account of the possibility that there may be additional contributions purely due to radiative corrections. These contributions will then be of the order α. A nonzero value of F_2 will, therefore, signal the presence of an anomalous magnetic moment.

In the limit $q^2 \to 0$, the form factors F_1 and F_2, therefore, have the following values to lowest order:

$$F_1(0) = 1, \quad F_2(0) = 0. \tag{45.10}$$

The total magnetic moment will then be

$$\boldsymbol{\mu} = [F_1(0) + F_2(0)]\frac{e}{mc}\mathbf{S} = [1 + F_2(0)]\frac{e}{mc}\mathbf{S}. \tag{45.11}$$

From (45.5) this implies that

$$\frac{g}{2} = 1 + F_2(0). \tag{45.12}$$

Thus,

$$\frac{g-2}{2} = F_2(0). \tag{45.13}$$

Any departure from $g = 2$ will point to the existence of an anomalous term. As we will show below an anomalous term arises from the vertex term described by Fig. 45.2.

45.3 Calculation of the vertex function diagram

If we write the vertex function as

$$\bar{u}(p')\,\Lambda^\mu(p',p)u(p), \tag{45.14}$$

then the diagram of Fig. 45.2 gives

$$\Lambda^\mu(p',p) = \int \frac{d^4k}{(2\pi)^4}\frac{-ig_{\alpha\beta}}{(k-p)^2 + i\epsilon}\left(-\mathrm{i}e\gamma^\alpha\right)\frac{i[\gamma\cdot(k+q) + m]}{(k+q)^2 - m^2 + i\epsilon}\gamma^\mu \times \frac{i(\gamma\cdot k + m)}{k^2 - m^2 + i\epsilon}\left(-ie\gamma^\beta\right) \tag{45.15}$$

The three factors of γ-matrices come from the three vertices in Fig. 45.2. The $g_{\alpha\beta}$ term is due to the photon propagator, and the other two factors are from the electron propagators. After simplifying the numerator we find

$$\Lambda^\mu(p',p) = 2ie^2\int \frac{d^4k}{(2\pi)^4}\frac{N^\mu}{[(k-p)^2 + i\epsilon][(k+q)^2 - m^2 + i\epsilon][k^2 - m^2 + i\epsilon]} \tag{45.16}$$

where

$$N^{\mu} = \left[\gamma \cdot k\gamma^{\mu}\gamma \cdot (k+q) + m^2\gamma^{\mu} - 2m\left(2k^{\mu} + q^{\mu}\right)\right]. \tag{45.17}$$

In deriving (45.17) we have used the properties of the products of γ-matrices given in Chapter 35, e.g., $\gamma^{\alpha}\gamma^{\mu}\gamma_{\alpha} = -2\gamma^{\mu}$.

Following the procedure outlined earlier, as a first step, we evaluate the integral by writing the product of the denominators in (45.16) as a single denominator using the Feynman integration trick,

$$\begin{aligned}
&\frac{1}{[(k-p)^2 + i\epsilon][(k+q)^2 - m^2 + i\epsilon][k^2 - m^2 + i\epsilon]} \\
&= \int_0^1 dz_1\, dz_2\, dz_3 \delta\left(z_1 + z_2 + z_3 - 1\right) \frac{2}{D^3}
\end{aligned} \tag{45.18}$$

where

$$\begin{aligned}
D &= z_1\left(k^2 - m^2\right) + z_2[(k+q)^2 - m^2] + z_3\left(k-p\right)^2 + i\epsilon \\
&= k^2 + 2k \cdot (z_2 q - z_3 p) + z_2 q^2 + z_3 p^2 - (1 - z_3)\, m^2 + i\epsilon
\end{aligned} \tag{45.19}$$

and where we have used the relation $z_1 + z_2 + z_3 = 1$. To simplify D we write

$$\kappa \equiv k + z_2 q - z_3 p \tag{45.20}$$

so that

$$D = \kappa^2 - \Delta + i\epsilon \tag{45.21}$$

where

$$\Delta \equiv -z_1 z_2 q^2 + (1 - z_3)^2\, m^2. \tag{45.22}$$

We also note that

$$\int \frac{d^4\kappa}{(2\pi)^4} \frac{\kappa^{\mu}}{D^3} = 0. \tag{45.23}$$

This follows from the fact that κ^{μ} is an odd function while D is not. Also, as explained in the Appendix,

$$\int \frac{d^4\kappa}{(2\pi)^4} \frac{\kappa^{\mu}\kappa^{\nu}}{D^3} = \int \frac{d^4\kappa}{(2\pi)^4} \frac{\frac{1}{4}g^{\mu\nu}\kappa^2}{D^3}. \tag{45.24}$$

The numerator function for N^{μ} in (45.17), after using (45.22), and (45.23) can be written as

$$\begin{aligned}
N^{\mu} &= -\frac{1}{2}\gamma^{\mu}\kappa^2 + \left(-z_2\gamma \cdot q + z_3\gamma \cdot p\right)\gamma^{\mu}\left(\left(1 - z_2\right)\gamma \cdot q + z_3\gamma \cdot p\right) \\
&\quad + m^2\gamma^{\mu} - 2m\left(\left(1 - 2z_2\right)q^{\mu} + 2z_3 p^{\mu}\right).
\end{aligned} \tag{45.25}$$

We then obtain

$$\Lambda^{\mu} = 4ie^2 \int \frac{d^4\kappa}{(2\pi)^4} \int_0^1 dz_1 dz_2 dz_3 \delta(z_1 + z_2 + z_3 - 1) \frac{1}{D^3}$$
$$\times \left[-\frac{1}{2}\gamma^{\mu}\kappa^2 + (-z_2\gamma \cdot q + z_3\gamma \cdot p)\gamma^{\mu}((1 - z_2)\gamma \cdot q + z_3\gamma \cdot p) \right.$$
$$\left. + m^2\gamma^{\mu} - 2m\left((1 - 2z_2)q^{\mu} + 2z_3 p^{\mu}\right)\right]. \tag{45.26}$$

The first term in the numerator behaves like κ^2, while $d^4\kappa$ goes like κ^4 and D^3 like κ^6. Therefore, the contribution coming from the first term is of the form $d\kappa/\kappa$, hence it is logarithmically divergent. We separate the divergent term from the rest, which are convergent, and write

$$\Lambda^{\mu} = \left(\Lambda^{\mu}\right)_{divergent} + \left(\Lambda^{\mu}\right)_{convergent}. \tag{45.27}$$

45.4 Divergent part of the vertex function

The divergent part is identified as

$$\left(\Lambda^{\mu}\right)_{divergent} = -2ie^2\gamma^{\mu} \int_0^1 dz_1 dz_2 dz_3 \delta(z_1 + z_2 + z_3 - 1) \int \frac{d^4\kappa}{(2\pi)^4} \frac{\kappa^2}{\left(\kappa^2 - \Delta\right)^3}. \tag{45.28}$$

From expression (44.87) we have in d dimensions,

$$\int \frac{d^4\kappa}{(2\pi)^4} \frac{\kappa^2}{\left(\kappa^2 - \Delta\right)^3} \to i \int \frac{d^dK}{(2\pi)^d} \frac{K^2}{\left(K^2 + \Delta\right)^3} = i \frac{1}{(4\pi)^{d/2}} \frac{d}{2} \frac{\Gamma\left(2 - \frac{d}{2}\right)}{\Gamma(3)} \left(\frac{1}{\Delta}\right)^{2-\frac{d}{2}}, \tag{45.29}$$

which for $d = 4$ gives

$$\int \frac{d^4\kappa}{(2\pi)^4} \frac{\kappa^2}{\left(\kappa^2 - \Delta\right)^3} = \frac{i}{8\pi^2\epsilon}. \tag{45.30}$$

Hence,

$$\left(\Lambda^{\mu}\right)_{divergent} = -2\mathrm{i}e^2\gamma^{\mu} \int_0^1 dz_1\, dz_2\, dz_3\, \delta(z_1 + z_2 + z_3 - 1) \frac{i}{8\pi^2\epsilon}. \tag{45.31}$$

However,

$$\int_0^1 dz_1 dz_2 dz_3 \delta(z_1 + z_2 + z_3 - 1) = \int_0^1 dz_1 \int_0^{1-z_1} dz_2 = \frac{1}{2}. \tag{45.32}$$

Therefore,

$$\left(\Lambda^{\mu}\right)_{divergent} = \frac{e^2}{8\pi^2\epsilon}\gamma^{\mu}. \tag{45.33}$$

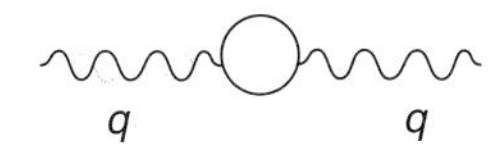

Fig. 45.4

For renormalization purposes we express, using (45.8),

$$\gamma^{\mu} + \left(\Lambda^{\mu}\right)_{divergent} = \frac{\gamma^{\mu}}{Z_1} \tag{45.34}$$

where Z_1 is the renormalization factor. Substituting the result from (45.33) in (45.34) we find

$$1 + \frac{e^2}{8\pi^2\epsilon} = \frac{1}{Z_1}. \tag{45.35}$$

Once again, keeping ϵ fixed and expanding in powers of e^2, we obtain to second order

$$Z_1 \simeq 1 - \frac{e^2}{8\pi^2\epsilon}. \tag{45.36}$$

Thus Z_1 contains the pole given by the second term. We will return to this result when we discuss the renormalization constants in other loop terms.

The remaining contributions in (45.26) are given by

$$\begin{aligned}\left(\Lambda^{\mu}\right)_{convergent} &= 4ie^2 \int \frac{d^4\kappa}{(2\pi)^4} \int_0^1 dz_1 dz_2 dz_3 \delta\left(z_1 + z_2 + z_3 - 1\right) \frac{1}{D^3} \\ &\times \left[\left(-z_2\gamma \cdot q + z_3\gamma \cdot p\right)\gamma^{\mu}\left(\left(1 - z_2\right)\gamma \cdot q + z_3\gamma \cdot p\right)\right. \\ &\left. + m^2\gamma^{\mu} - 2m\left(\left(1 - 2z_2\right)q^{\mu} + 2z_3 p^{\mu}\right)\right].\end{aligned} \tag{45.37}$$

We will return to this result in Section 45.9 along with the convergent contributions from the other loop terms.

45.5 Radiative corrections to the photon propagator

The photon propagator to the lowest order has been considered earlier. The first-order radiative correction will come from the electron loop as described by the diagram of Fig. 45.4. We write it as

$$\left(\frac{-ig_{\rho\mu}}{q^2}\right) i\Pi_2^{\mu\nu} \left(\frac{-ig_{\nu\sigma}}{q^2}\right) \tag{45.38}$$

where

$$i\Pi_2^{\mu\nu} = -\left(-ie\right)^2 \int \frac{d^4k}{(2\pi)^4} \left[\gamma^{\mu} \frac{i\left(\gamma \cdot k + m\right)}{k^2 - m^2} \gamma^{\nu} \frac{i\left(\gamma \cdot k + \gamma \cdot q + m\right)}{(k+q)^2 - m^2}\right]. \tag{45.39}$$

The factors involving the γ-matrices are due to the two vertices, while the two propagators correspond to the internal electrons in the loop. The product of the numerators inside the square bracket can be simplified by using the properties of the γ-matrices outlined in Chapter 35. We write (45.39) as

$$i\Pi_2^{\mu\nu} = -4e^2 \int \frac{d^4k}{(2\pi)^4} \frac{N^{\mu\nu}}{[k^2 - m^2][(k+q)^2 - m^2]} \tag{45.40}$$

where

$$N^{\mu\nu} = k^\mu (k+q)^\nu + k^\nu (k+q)^\mu - g^{\mu\nu} \left(k \cdot (k+q) - m^2 \right). \tag{45.41}$$

Following, once again, the Feynman integration technique, the product in the denominator can be written as

$$\frac{1}{[k^2 - m^2][(k+q)^2 - m^2]} = \int_0^1 dxdy\delta(x+y-1)\frac{1}{D^2} \tag{45.42}$$

where

$$D = x[(k+q)^2 - m^2] + (1-x)\left(k^2 - m^2\right) \tag{45.43}$$

$$= k^2 + 2xk \cdot q + xq^2 - m^2. \tag{45.44}$$

We simplify D by writing

$$\kappa = k + xq. \tag{45.45}$$

Hence,

$$D = \kappa^2 - \Delta + i\epsilon \tag{45.46}$$

where

$$\Delta = m^2 - x(1-x)q^2. \tag{45.47}$$

The numerator term, $N^{\mu\nu}$, is then given by

$$N^{\mu\nu} = \left(\kappa^\mu - xq^\mu\right)\left[\kappa^\nu + (1-x)q^\nu\right] + \left(\kappa^\nu - xq^\nu\right)\left[\kappa^\mu + (1-x)q^\mu\right]$$
$$- g^{\mu\nu}\left[(\kappa - xq)\cdot[\kappa + (1-x)q] - m^2\right]. \tag{45.48}$$

The terms linear in κ^μ, as we discussed earlier, do not contribute and hence they can be neglected. Thus, we find

$$N^{\mu\nu} = 2\kappa^\mu\kappa^\nu - g^{\mu\nu}\kappa^2 - 2x(1-x)q^\mu q^\nu + g^{\mu\nu}[m^2 + x(1-x)q^2]. \tag{45.49}$$

Hence,

$$i\Pi_2^{\mu\nu} = -4e^2 \int_0^1 dx \int \frac{d^4\kappa}{(2\pi)^4} \frac{2\kappa^\mu\kappa^\nu - g^{\mu\nu}\kappa^2 - 2x(1-x)q^\mu q^\nu + g^{\mu\nu}\left(m^2 + x(1-x)q^2\right)}{\left(\kappa^2 - \Delta\right)^2}. \tag{45.50}$$

45.6 Divergent part of the photon propagator

The first two terms in the numerator of (45.50) give divergent contributions. We, therefore, use the dimensional regularization technique to carry out the above integration. Using the relations

$$2\kappa^{\mu}\kappa^{\nu} \to \frac{2}{d}g^{\mu\nu}\kappa^2 \to -\frac{2}{d}g^{\mu\nu}K^2, \tag{45.51}$$

$$g^{\mu\nu}\kappa^2 \to -g^{\mu\nu}K^2,$$

we obtain, keeping intact the convergent terms in (45.50),

$$i\Pi_2^{\mu\nu} = -4ie^2\int_0^1 dx\int\frac{d^dK}{(2\pi)^d}\frac{\left(-\frac{2}{d}+1\right)g^{\mu\nu}K^2 - 2x(1-x)\,q^{\mu}q^{\nu} + g^{\mu\nu}\left(m^2 + x(1-x)\,q^2\right)}{\left(K^2+\Delta\right)^2}. \tag{45.52}$$

This integration is carried out in the Appendix . One finds that the above relation can be written as

$$i\Pi_2^{\mu\nu}(q) = \left(q^2 g^{\mu\nu} - q^{\mu}q^{\nu}\right)\cdot\Pi_2\left(q^2\right) \tag{45.53}$$

where

$$\Pi_2\left(q^2\right) = \frac{-8e^2}{(4\pi)^{d/2}}\int_0^1 dx\, x(1-x)\frac{\Gamma\left(2-\frac{d}{2}\right)}{\Delta^{2-d/2}}. \tag{45.54}$$

In order to separate the infinite and finite parts while preserving gauge invariance (a point which will be clear later in this section), let us write

$$\Pi_2\left(q^2\right) = \Pi_2(0) + \hat{\Pi}_2\left(q^2\right). \tag{45.55}$$

What we have done here is to make a subtraction at $q^2 = 0$, i.e., we have subtracted out the $q^2 = 0$ term, which, as we will see below, is divergent, leaving a convergent term, $\hat{\Pi}_2\left(q^2\right)$. From (45.55) we obtain, in the limit $d \to 4$,

$$\Pi_2(0) = \left(\Pi_2\left(q^2\right)\right)_{divergent} = \frac{-8e^2}{(4\pi)^2}\frac{2}{\epsilon}\int_0^1 dx\, x(1-x) \tag{45.56}$$

$$= -\frac{e^2}{6\pi^2\epsilon}. \tag{45.57}$$

The convergent part of the diagram is then

$$\hat{\Pi}_2\left(q^2\right) = \left(\Pi_2\left(q^2\right)\right)_{convergent}. \tag{45.58}$$

We will discuss the convergent part in Section 45.9.

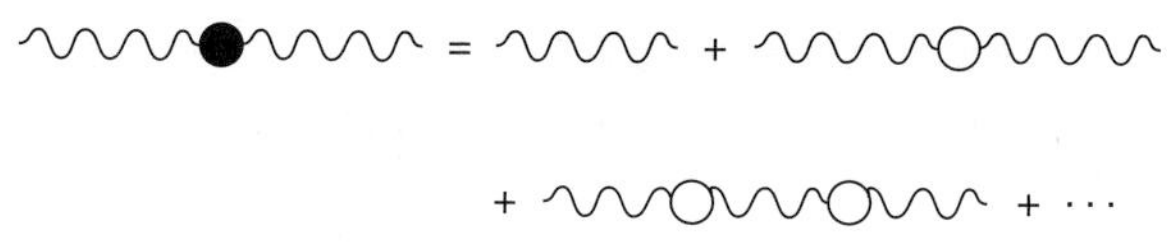

Fig. 45.5

45.7 Modification of the photon propagator and photon wavefunction

The photon propagator to lowest order is given by

$$\frac{-ig_{\mu\nu}}{q^2}. \tag{45.59}$$

The higher-order contributions to the propagator can be expressed diagrammatically as in Fig. 45.5.

If we designate the total photon propagator represented by the above series as $D_{\mu\nu}(q)$ then

$$D_{\mu\nu}(q) = \frac{-ig_{\mu\nu}}{q^2} + \left(\frac{-ig_{\mu\rho}}{q^2}\right) i\Pi_2^{\rho\sigma}(q) \left(\frac{-ig_{\sigma\nu}}{q^2}\right) + \cdots. \tag{45.60}$$

Substituting the expression for $i\Pi_2^{\rho\sigma}(q)$ given in (45.53) into (45.60) we obtain

$$D_{\mu\nu}(q) = \frac{-ig_{\mu\nu}}{q^2} + \left(\frac{-ig_{\mu\rho}}{q^2}\right) \left[\left(q^2 g^{\rho\sigma} - q^\rho q^\sigma\right) \cdot \Pi_2\left(q^2\right)\right] \left(\frac{-ig_{\sigma\nu}}{q^2}\right) + \cdots. \tag{45.61}$$

Since $D_{\mu\nu}(q)$ enters into the calculation as a matrix element,

$$\bar{u}\left(p'\right) \gamma^\mu D_{\mu\nu}(q) u\left(p\right) \quad \text{or} \quad \bar{u}\left(p'\right) \gamma^\nu D_{\mu\nu}(q) u\left(p\right) \tag{45.62}$$

where $q = p' - p$; therefore, because of the Ward identity (see the Appendix to Chapter 44), terms proportional to q^μ or q^ν in $D_{\mu\nu}(q)$ will give vanishing contributions. Thus, only the terms proportional to $g^{\mu\nu}$ need be included in the sum (45.60). Hence, we have

$$D_{\mu\nu}(q) = \frac{-ig_{\mu\nu}}{q^2} \left[1 + \Pi_2\left(q^2\right) + \left(\Pi_2\left(q^2\right)\right)^2 + \cdots\right]. \tag{45.63}$$

The geometric series in (45.63) can be easily summed to give

$$D_{\mu\nu}(q) = \frac{-ig_{\mu\nu}}{q^2[1 - \Pi_2\left(q^2\right)]}. \tag{45.64}$$

We reach a remarkable conclusion that the photon, which had zero mass, given by the location of the pole, $q^2 = 0$, remains at zero mass thanks to gauge invariance as reflected by the Ward identity. The residue of the pole in (45.64), however, is changed from the previous value of $-ig_{\mu\nu}$. We define another renormalization constant Z_3 given by

$$Z_3 = \frac{1}{[1 - \Pi_2(0)]}. \tag{45.65}$$

Thus we write

$$D_{\mu\nu}(q) = Z_3 \frac{g_{\mu\nu}}{q^2} \tag{45.66}$$

near $q^2 = 0$.

We now define the renormalized photon propagator as

$$D'_{\mu\nu}(q) = \frac{g_{\mu\nu}}{q^2} \tag{45.67}$$

and write

$$D_{\mu\nu}(q) = Z_3 D'_{\mu\nu}(q). \tag{45.68}$$

As in the case of fermion fields, if A^μ and A'^μ are bare and renormalized photon fields respectively, then the respective propagators are given by the vacuum expectation values of the time-ordered products,

$$D_{\mu\nu}(q) \sim \langle O \left| T\left(A_\mu A_\nu\right) \right| O \rangle \tag{45.69}$$

and

$$D'_{\mu\nu}(q) \sim \langle O \left| T\left(A'_\mu A'_\nu\right) \right| O \rangle. \tag{45.70}$$

Hence, one can relate the bare and renormalized photon fields as

$$A^\mu = \sqrt{Z_3} A'^\mu. \tag{45.71}$$

We have already calculated $\Pi_2(0)$, which is found from (45.57) to be

$$\Pi_2(0) = -\frac{e^2}{6\pi^2\epsilon}. \tag{45.72}$$

We can include higher-order terms of q^2 in $D_{\mu\nu}(q)$ by writing (45.65) as

$$D_{\mu\nu}(q) = \frac{-ig_{\mu\nu}}{q^2\left[1 - \Pi_2(q^2)\right]} \simeq \frac{-ig_{\mu\nu}}{q^2[1 - \Pi_2(0)][1 - \hat{\Pi}_2(q^2)]} \tag{45.73}$$

where we have approximated the denominator, using (45.55), as

$$\left[1 - \Pi_2\left(q^2\right)\right] = \left[1 - \Pi_2(0) - \hat{\Pi}_2\left(q^2\right)\right] \simeq \left[1 - \Pi_2(0)\right]\left[1 - \hat{\Pi}_2\left(q^2\right)\right]. \tag{45.74}$$

This approximation is valid to order α since the product $\Pi_2(0)\,\hat{\Pi}_2\left(q^2\right)$, which we ignored above, is of the order α^2. Using (45.74) the renormalized propagator can now be written as

$$D'_{\mu\nu}(q) = \frac{g_{\mu\nu}}{q^2[1 - \hat{\Pi}_2(q^2)]} \tag{45.75}$$

where we note that $\hat{\Pi}_2(q^2)$ is convergent.

45.8 Combination of all the divergent terms: basic renormalization

Consider now a typical coupling term in QED which involves the product of three fields with a coupling constant e_0. We write it as

$$e_0\bar{\psi}\psi A \to \frac{e_0 Z_2}{Z_1}\sqrt{Z_3}\bar{\psi}'\psi' A' = e_R\bar{\psi}'\psi' A'. \tag{45.76}$$

The right-hand side contains only the renormalized fields; therefore, we define the renormalized coupling constant, e_R, in terms of the bare coupling constant, e_0, as

$$e_R = \frac{e_0 Z_2}{Z_1}\sqrt{Z_3}. \tag{45.77}$$

Thus all the renormalization constants along with the bare coupling constant are absorbed into a single constant e_R. It is this coupling constant that is finite and experimentally measured.

From our calculations we found that

$$Z_1 = Z_2. \tag{45.78}$$

This is, actually, not an accident. One can show that this equality follows from gauge invariance. Relation (45.78) is called the Ward–Takahashi identity.

Hence,

$$e_R = e_0\sqrt{Z_3}. \tag{45.79}$$

It has been shown that the perturbation series in terms of e_R now converges, as all the infinities are taken into account and absorbed into the physical quantities.

We have thus succeeded in absorbing the infinite quantities that appeared in perturbation expansion into the bare mass and charge and redefined the resulting combination as renormalized mass and charge. Let us now turn to the leftover quantities, which are convergent.

45.9 Convergent parts of the radiative corrections

45.9.1 Anomalous magnetic moment

Let us return to the convergent part of Λ^μ, which we derived in Section 45.4:

$$\left(\Lambda^\mu(p',p)\right)_{convergent} = 4ie^2 \int \frac{d^4\kappa}{(2\pi)^4} \int_0^1 dz_1\, dz_2\, dz_3 \delta\left(z_1+z_2+z_3-1\right) \times \frac{1}{D^3}\left[\left(-z_2\gamma\cdot q + z_3\gamma\cdot p\right)\gamma^\mu\left((1-z_2)\gamma\cdot q + z_3\gamma\cdot p\right) + m^2\gamma^\mu - 2m\left((1-2z_2)q^\mu + 2z_3p^\mu\right)\right] \quad (45.80)$$

where $q = p' - p$. We will be interested in the vertex function at $q^2 = 0$.

Since we are interested in the matrix element $\bar{u}\left(p'\right)\Lambda^\mu u\left(p\right)$, we note the following:

(i) $\gamma \cdot p'$ on the left of any term inside the square bracket can be replaced by m since $\bar{u}\left(p'\right)\gamma\cdot p' = m\bar{u}\left(p'\right)$; similarly, $\gamma\cdot p$ on the right can be replaced by m.

(ii) In the absence of any factors of the γ matrices multiplying it, the term $\gamma \cdot q$ will not contribute since $\gamma\cdot q = \gamma\cdot\left(p'-p\right)$ vanishes when sandwiched between $\bar{u}\left(p'\right)$ and $u\left(p\right)$.

(iii) The terms with q^μ standing alone will also not contribute because of the Ward identity if no factors involving the γ matrices are multiplying it.

The following relations will also be very helpful in obtaining the numerator appearing in (45.80):

$$\gamma\cdot q\gamma^\mu\gamma\cdot q = 2q^\mu\gamma\cdot q - \gamma^\mu q^2 \to 0, \quad (45.81)$$

$$\gamma^\mu\gamma\cdot q \to 2p'^\mu - 2m\gamma^\mu, \quad (45.82)$$

where we have used $q_\mu = p'_\mu - p_\mu$. Utilizing the points (i)–(iii) outlined above, the numerator function, $N^{\mu\nu}$, of (45.80) can then be written in a series of simplifying steps as

$$\begin{aligned} N^\mu &= \left[\left(-z_2\gamma\cdot q + z_3\gamma\cdot p\right)\gamma^\mu\left((1-z_2)\gamma\cdot q + z_3\gamma\cdot p\right) + m^2\gamma^\mu - 4mz_3p^\mu\right] \\ &= \left[\left(-\left(z_2+z_3\right)\gamma\cdot q + mz_3\right)\gamma^\mu\left((1-z_2)\gamma\cdot q + mz_3\right) + m^2\gamma^\mu - 4mz_3p^\mu\right] \\ &= \left[2mz_3(z_3-1)p^\mu + m^2(1-2z_3-z_3^2)\gamma^\mu\right]. \end{aligned} \quad (45.83)$$

From the Gordon decomposition relation we have

$$\bar{u}\left(p'\right)\gamma^\mu u\left(p\right) = \bar{u}\left(p'\right)\left[\frac{p'^\mu + p^\mu}{2m} + \frac{i\sigma^{\mu\nu}q_\nu}{2m}\right]u\left(p\right). \quad (45.84)$$

Since $p'^\mu = p^\mu + q^\mu$, using the points noted in (iii) above, one can replace $\left(p'^\mu + p^\mu\right)$ by $2p^\mu$ in (45.84) and express p^μ in terms of γ^μ and $\sigma^{\mu\nu}q_\nu$ from the above decomposition formula. Substituting this value of p^μ in (45.83) we obtain

$$N^\mu = \left[m^2\gamma^\mu(1-4z_3+z_3^2) + \frac{i\sigma^{\mu\nu}q_\nu}{2m}\left(2m^2z_3(1-z_3)\right)\right]. \quad (45.85)$$

The coefficient of $i\sigma^{\mu\nu}q_\nu/2m$ is related to $F_2(q^2)$ through (45.9). Therefore,

$$F_2\left(q^2\right) = \frac{\alpha}{2\pi}\int_0^1 dz_1\, dz_2\, dz_3 \delta\left(z_1 + z_2 + z_3 - 1\right)\left[\frac{2m^2 z_3\left(1-z_3\right)}{m^2\left(1-z_3\right)^2 - q^2 z_1 z_2}\right] + O\left(\alpha^2\right) \tag{45.86}$$

where α is the fine structure constant ($\simeq 1/137$). At $q^2 = 0$ we find

$$\begin{aligned} F_2\left(q^2 = 0\right) &= \frac{\alpha}{2\pi}\int_0^1 dz_1\, dz_2\, dz_3 \delta\left(z_1 + z_2 + z_3 - 1\right)\frac{2m^2 z_3\left(1-z_3\right)}{m^2\left(1-z_3\right)^2} \\ &= \frac{\alpha}{\pi}\int_0^1 dz_3 \int_0^{1-z_3} dz_2 \frac{z_3}{1-z_3} = \frac{\alpha}{2\pi}. \end{aligned} \tag{45.87}$$

Hence the anomalous magnetic moment, μ_a, is given by

$$\mu_a = F_2(0) = \frac{\alpha}{2\pi} \approx 0.0011614. \tag{45.88}$$

This value is very close to the experimental value of

$$\mu_{exp} = 0.001159652209. \tag{45.89}$$

The value we obtained in (45.88) was the result of second-order calculation. Calculations have been made up to sixth order. The new value is found to be

$$\mu_{th} = 0.001159652411. \tag{45.90}$$

The theoretical value thus agrees with experiments to one part in 10^8, which is a remarkable confirmation of QED!

45.9.2 Lamb shift

As we found in our (nonrelativistic) treatment of the hydrogen atom with Coulomb potential, the radial wavefunction depends on the principal quantum number, n, and the angular momentum quantum number, l (with $l \leq (n-1)$). The energy levels, however, depend only on n. The states with different angular momenta but same n are, therefore, degenerate. Specifically, the two states $n = 2$, $l = 0$, the $2S$ state, and $n = 2$, $l = 1$ the $2P$ state have the same binding energy.

In Dirac's relativistic theory where the eigenvalue of the total angular momentum $\mathbf{J}$ ($= \mathbf{L} + \mathbf{S}$) is a good quantum number, the two $n = 2$ states mentioned above split into three states: $2S_{1/2}$ with $j = \frac{1}{2}$ and $l = j - \frac{1}{2}$; $2P_{1/2}$ with $j = \frac{1}{2}$ and $l = j + \frac{1}{2}$; $2P_{3/2}$ with $j = 3/2$ and $l = j - \frac{1}{2}$. It was found that the energy levels of the hydrogen atom depended on j in addition to n. This meant that $2S_{\frac{1}{2}}$ and $2P_{\frac{1}{2}}$ still remained degenerate.

Lamb and Retherford, in 1947, found that the $2S_{1/2}$ and $2P_{1/2}$ levels of the hydrogen atom were actually split with the $2P_{1/2}$ level more than 1000 MHz below the $2S_{1/2}$ level, in contrast to the predictions of the Dirac theory. This is called the Lamb shift. We will show below that it can be explained through the effect of radiative corrections that modify the Coulomb potential. We find that this modified term is proportional to $\delta^{(3)}(\mathbf{r})$, which will affect only the S-state and hence cause the split between the energy levels of the $2S_{1/2}$ and the $2P_{1/2}$ states.

The calculation of the corrections to the hydrogen levels due to QED is complicated because one must include the radiative corrections coming from the vertex function, and the anomalous magnetic moment, and the electron self-energy, in addition to the vacuum polarization. In other words, one must include contributions from all six diagrams of Fig. 44.1(a)–(d). We will consider only the contribution due to the photon propagator in Fig. 44.1 and Fig. 45.4 and calculate the energy shift using first-order perturbation theory.

It is convenient first to relate the photon propagator to the nonrelativistic potential, $V(r)$. As we discussed before, this potential is simply the Fourier transform of the propagator in the static limit. For example, the Fourier transform of the photon propagator

$$\frac{e^2}{q^2} \tag{45.91}$$

in the static limit, $q_4 = 0$, $q^2 = -|\mathbf{q}|^2$, is given by

$$V(r) = -\int \frac{d^3q}{(2\pi)^3} e^{i\mathbf{q}\cdot\mathbf{r}} \frac{e^2}{|\mathbf{q}|^2} = -\frac{e^2}{4\pi r} \tag{45.92}$$

where the integral is obtained by first replacing $|\mathbf{q}|^2$ by $\left(|\mathbf{q}|^2 + \mu^2\right)$ and, after integration, letting $\mu \to 0$. As expected, this potential is just the familiar Coulomb potential.

The convergent term, $\hat{\Pi}_2\left(q^2\right)$, as we derived earlier is given by

$$\left(\Pi_2\left(q^2\right)\right)_{convergent} = \hat{\Pi}_2\left(q^2\right) = \frac{-8e^2}{(4\pi)^{d/2}} \int_0^1 dx\, x(1-x) \times \Gamma\left(2-\frac{d}{2}\right)\left[\left(\frac{1}{\Delta}\right)^{2-d/2} - \left(\frac{1}{m}\right)^{2(2-d/2)}\right]. \tag{45.93}$$

The terms in the square brackets, in the limit $d \to 4$, give

$$\left[\left(\frac{1}{\Delta}\right)^{2-d/2} - \left(\frac{1}{m}\right)^{2(2-d/2)}\right] \to \left(2-\frac{d}{2}\right)\log\left(\frac{\Delta}{m^2}\right). \tag{45.94}$$

Therefore,

$$\hat{\Pi}_2\left(q^2\right) \underset{d\to 4}{=} \frac{-8e^2}{(4\pi)^{d/2}} \int_0^1 dx\, x(1-x)\,\Gamma\left(2-\frac{d}{2}\right)\left(2-\frac{d}{2}\right)\log\left(\frac{\Delta}{m^2}\right). \tag{45.95}$$

Since,

$$\left(2-\frac{d}{2}\right)\Gamma\left(2-\frac{d}{2}\right) = \Gamma\left(3-\frac{d}{2}\right) \tag{45.96}$$

we find in the limit $d \to 4$,

$$\hat{\Pi}_2\left(q^2\right) = \frac{-8e^2}{(4\pi)^2}\int_0^1 dx\, x\,(1-x)\log\left[1 - \frac{x\,(1-x)\,q^2}{m^2}\right] \tag{45.97}$$

where we have substituted the relation given by (45.47) for Δ. Since we will be interested in effects due to small values of q^2, we expand the logarithm in $\hat{\Pi}_2\left(q^2\right)$ for small q^2 and retain only the leading term, which is found to be

$$\hat{\Pi}_2\left(q^2\right) = \frac{-e^2}{2\pi^2}\left(\frac{q^2}{m^2}\right)\int_0^1 dx\, x^2\,(1-x)^2. \tag{45.98}$$

The integration then gives

$$\hat{\Pi}_2\left(q^2\right) = -\frac{\alpha}{15\pi m^2}q^2 \tag{45.99}$$

where $\alpha = e^2/4\pi$.

Thus, in summary, we write

$$\Pi_2\left(q^2\right) = \left(\Pi_2\left(q^2\right)\right)_{divergent} + \left(\Pi_2\left(q^2\right)\right)_{convergent} \tag{45.100}$$

where

$$\left(\Pi_2\left(q^2\right)\right)_{divergent} = \Pi_2\,(0) = -\frac{e^2}{6\pi^2\epsilon} \tag{45.101}$$

and

$$\left(\Pi_2\left(q^2\right)\right)_{convergent} = \hat{\Pi}_2\left(q^2\right) = -\frac{\alpha}{15\pi m^2}q^2. \tag{45.102}$$

The photon propagator modified by vacuum polarization to $O(\alpha)$ is found from Section 45.7 to be

$$\frac{e^2}{q^2[1 - \hat{\Pi}_2\left(q^2\right)]}. \tag{45.103}$$

The potential derived from it is given by

$$V\,(\mathbf{r}) = -\int \frac{d^3q}{(2\pi)^3}e^{i\mathbf{q}\cdot\mathbf{r}}\frac{e^2}{|\mathbf{q}|^2\left[1 - \hat{\Pi}_2\left(-|\mathbf{q}|^2\right)\right]}. \tag{45.104}$$

After expanding the denominator for small $|\mathbf{q}|^2$, we write

$$V\,(\mathbf{r}) = -\int \frac{d^3q}{(2\pi)^3}e^{i\mathbf{q}\cdot\mathbf{r}}\frac{e^2}{|\mathbf{q}|^2}\left[1 + \hat{\Pi}_2\left(-|\mathbf{q}|^2\right)\right] \tag{45.105}$$

where from (45.102) we have

$$\hat{\Pi}_2\left(-|\mathbf{q}|^2\right) = \frac{\alpha}{15\pi m^2}|\mathbf{q}|^2. \tag{45.106}$$

Thus, we obtain

$$V\,(\mathbf{r}) = -\int \frac{d^3q}{(2\pi)^3}e^{i\mathbf{q}\cdot\mathbf{r}}\frac{e^2}{|\mathbf{q}|^2} - \frac{\alpha e^2}{15\pi m^2}\int \frac{d^3q}{(2\pi)^3}e^{i\mathbf{q}\cdot\mathbf{r}}. \tag{45.107}$$

The integral of the first term, as expected, is the usual Coulomb potential. The second term is the familiar δ-function. Hence

$$V(\mathbf{r}) = -\frac{\alpha}{r} - \frac{4\alpha^2}{15m^2}\delta^{(3)}(\mathbf{r}). \tag{45.108}$$

The second term is the modification to the Coulomb potential due to vacuum polarization. If we denote the modified potential $V'(\mathbf{r})$ then

$$V'(\mathbf{r}) = -\frac{4\alpha^2}{15m^2}\delta^{(3)}(\mathbf{r}). \tag{45.109}$$

Using first order time-independent perturbation theory the energy shift is found to be

$$\Delta E = \int d^3r\,|\psi(\mathbf{r})|^2\,V'(\mathbf{r}) = -\frac{4\alpha^2}{15m^2}\,|\psi(0)|^2 \tag{45.110}$$

where $\psi(\mathbf{r})$ is the wavefunction of the hydrogen atom. Only the S-state which has $l = 0$ will contribute, since the wavefunctions for angular momenta $l > 0$ vanish at $r = 0$. Substituting the value of the $n = 2$ S-state wavefunction we find the energy shift due to vacuum polarization to be

$$\Delta E = -\frac{\alpha^5 m}{30\pi}. \tag{45.111}$$

This contribution is equivalent to a correction of -27.1 MHz to the energy level difference between $2S_{1/2}$ and $2P_{1/2}$. As for the contribution of the other radiative corrections, the vertex correction contributes an amount of magnitude 1010 MHz, while the anomalous magnetic moment term contributes 68 MHz. If one includes higher orders for which calculations have been made one finds that the $2S_{1/2}$ level is above $2P_{1/2}$ by 1057.864 $\pm$ 0.014 MHz. The experimental value is 1057.862 $\pm$ 0.020 MHz, which is a stunning re-confirmation of the QED theory!

45.10 Appendix to Chapter 45

45.10.1 The photon propagator integral

To obtain the integral of the first term in (45.53) we note from the Appendix to Chapter 44 that

$$\int \frac{d^dK}{(2\pi)^d}\frac{K^2}{\left(K^2+\Delta\right)^2} = \frac{1}{(4\pi)^{d/2}}\frac{d}{2}\frac{\Gamma\left(1-\frac{d}{2}\right)}{\Gamma(2)}\left(\frac{1}{\Delta}\right)^{1-\frac{d}{2}}. \tag{45.112}$$

The first term in the numerator in (45.52) is then

$$\frac{d}{2}\left(-\frac{2}{d}+1\right)\Gamma\left(1-\frac{d}{2}\right)\left(\frac{1}{\Delta}\right)^{1-\frac{d}{2}} = -\left(1-\frac{d}{2}\right)\Gamma\left(1-\frac{d}{2}\right)\left(\frac{1}{\Delta}\right)^{1-\frac{d}{2}}$$
$$= -\Delta\Gamma\left(2-\frac{d}{2}\right)\left(\frac{1}{\Delta}\right)^{2-\frac{d}{2}}. \quad (45.113)$$

The remaining two terms in the numerator in (45.52) do not have any K^2 dependence. For these terms we use (44.84),

$$\int_0^\infty \frac{d^dK}{(2\pi)^d}\frac{1}{\left[K^2+\Delta\right]^2} = \left(\frac{1}{2}\Delta^{\frac{d}{2}-2}\right)\frac{1}{(2\pi)^d}\frac{\Gamma\left(2-\frac{d}{2}\right)\Gamma\left(\frac{d}{2}\right)}{\Gamma(2)}. \quad (45.114)$$

Combining the integral from the first term in (45.52) with those from the rest of the terms we obtain, through (45.114) and (45.115),

$$i\Pi_2^{\mu\nu}(q) = -4ie^2\int_0^1 dx\frac{1}{(4\pi)^{d/2}}\frac{\Gamma\left(2-\frac{d}{2}\right)}{\Delta^{2-d/2}}\left[-\Delta g^{\mu\nu}+g^{\mu\nu}\right.$$
$$\left.\times\left[m^2+x(1-x)q^2\right]-2x(1-x)q^\mu q^\nu\right] \quad (45.115)$$

Substituting Δ and combining together the terms with coefficients $g^{\mu\nu}$ and $q^\mu q^\nu$, one can write

$$i\Pi_2^{\mu\nu}(q) = \left(q^2g^{\mu\nu}-q^\mu q^\nu\right)\cdot\Pi_2\left(q^2\right). \quad (45.116)$$

where the function $\Pi_2\left(q^2\right)$ is given by

$$\Pi_2\left(q^2\right) = \frac{-8e^2}{(4\pi)^{d/2}}\int_0^1 dx\,x(1-x)\frac{\Gamma\left(2-\frac{d}{2}\right)}{\Delta^{2-d/2}}. \quad (45.117)$$

45.10.2 Calculating $\int d\Omega_4$

First we consider the three-dimensional space and reproduce the results we are already familiar with. To facilitate going to four dimensions, we employ different notations and re-express the traditional Cartesian coordinates as well as the azimuthal and polar angles as follows:

$$(x,y,z)\to(x_1,x_2,x_3), \quad (45.118)$$

$$(\phi,\theta)\to(\theta_1,\theta_2). \quad (45.119)$$

The relation between the Cartesian coordinates and spherical coordinates is then given by

$$x = x_1 = r \sin\theta_2 \cos\theta_1, \tag{45.120}$$

$$y = x_2 = r \sin\theta_2 \sin\theta_1, \tag{45.121}$$

$$z = x_3 = r \cos\theta_2. \tag{45.122}$$

The three-dimensional volume element then transforms from the Cartesian system to the spherical system as

$$dx_1 dx_2 dx_3 = J dr d\theta_1 d\theta_2 \tag{45.123}$$

where J is the Jacobian given by

$$J = \det\left(\frac{\partial\,(x_1, x_2, x_3)}{\partial\,(r, \theta_1, \theta_2)}\right). \tag{45.124}$$

From the relations (45.120)–(45.122) we obtain

$$J = r^2 \prod_{i=1}^{2} (\sin\theta_1)^{i-1} \tag{45.125}$$

and hence (45.123) reads

$$dx_1 dx_2 dx_3 = r^2 dr \left[\prod_{i=1}^{2} (\sin\theta_i)^{i-1}\ d\theta_1 d\theta_2\right]. \tag{45.126}$$

The three-dimensional solid angle, which is the quantity in the square bracket, is then given by

$$d\Omega_3 = \prod_{i=1}^{2} (\sin\theta_i)^{i-1}\, d\theta_1 d\theta_2 = d\theta_1 \sin\theta_2 d\theta_2. \tag{45.127}$$

The integral over $d\Omega_3$ is then found to be

$$\int d\Omega_3 = \int_0^{2\pi} d\theta_1 \int_0^{\pi} \sin\theta_2 d\theta_2 = 4\pi, \tag{45.128}$$

which is, of course, a well-known result.

In a similar manner we write the transformation from Cartesian coordinates to spherical coordinates in four dimensions as follows:

$$x_1 = r \sin\theta_3 \sin\theta_2 \cos\theta_1, \tag{45.129}$$

$$x_2 = r \sin\theta_3 \sin\theta_2 \sin\theta_1, \tag{45.130}$$

$$x_3 = r \sin\theta_3 \cos\theta_2, \tag{45.131}$$

$$x_4 = r \cos\theta_3. \tag{45.132}$$

The four-dimensional volume element is

$$dx_1\, dx_2\, dx_3\, dx_4 = J\mathrm{dr}\, \mathrm{d}\theta_1 d\theta_2 d\theta_3 \tag{45.133}$$

where the Jacobian is given by

$$J = \det\left(\frac{\partial\,(x_1, x_2, x_3, x_4)}{\partial\,(r, \theta_1, \theta_2, \theta_3)}\right). \tag{45.134}$$

Substituting (45.129)–(45.132) we find

$$J = r^3 \prod_{i=1}^{3} \sin^{i-1}\theta_i. \tag{45.135}$$

Hence,

$$dx_1\, dx_2\, dx_3\, dx_4 = r^3 dr \left[\prod_{i=1}^{3} (\sin\theta_i)^{i-1}\, d\theta_1\, d\theta_2\, d\theta_3\right] \tag{45.136}$$

and the four-dimensional solid angle is

$$d\Omega_4 = \prod_{i=1}^{3} (\sin\theta_i)^{i-1}\, d\theta_1 d\theta_2 d\theta_3. \tag{45.137}$$

The integral over the solid angle is then

$$\int d\Omega_4 = \int_0^{2\pi} d\theta_1 \int_0^{\pi} \sin\theta_2 d\theta_2 \int_0^{\pi} \sin^2\theta_3 d\theta_3 = 2\pi^2. \tag{45.138}$$

Bibliography

Quantum mechanics

Abers, E. S. (2004) *Quantum Mechanics*. Upper Saddle River, NJ: Prentice Hall.

Basdevant, J. L. and Dalibard, J. (2005) *Quantum Mechanics*. Berlin/Heidelberg: Springer.

Basdevant, J. L. and Dalibard, J. (2006) *The Quantum Mechanics Solver*. Berlin/Heidelberg: Springer.

Cohen-Tanoudji, C., Diu, B., and Lalöe, F. (1977) *Quantum Mechanics*. New York: Wiley.

Constantinescu, F. and Magyari, E. (1971) *Problems in Quantum Mechanics*. Oxford: Pergamon Press.

Davydov, A. S. (1965) *Quantum Mechanics*. Elmsford, NY: Pergamon Press.

Dirac, P. A. M. (1958) *Quantum Mechanics*. London: Oxford University Press.

Gasiorowicz, S. (2003) *Quantum Physics*. New York: Wiley.

Gottfried, K. and Yan, T-M (2003) *Quantum Mechanics: Fundamentals*. New York: Springer.

Griffiths, D. (1995) *Introduction to Quantum Mechanics*. Upper Saddle River, NJ: Prentice Hall.

Kogan, V. I. and Galitskiy (1963) *Problems in Quantum Mechanics*. Englewood Cliffs, NJ: Prentice Hall.

Landau, L. and Lifschitz, E. (1958) *Quantum Mechanics*. London: Pergamon Press.

Le Bellac, M. (2006) *Quantum Physics*. Cambridge: Cambridge University Press.

Liboff, R. L. (2003) *Introductory Quantum Mechanics*. Reading, MA: Addison-Wesley.

Merzbacher, E. (1970) *Quantum Mechanics*. New York: Wiley.

Sakurai, J. J. (1994) *Modern Quantum Mechanics*. Reading, MA: Addison-Wesley.

Schiff, L. (1968) *Quantum Mechanics*. New York: McGraw-Hill.

Shankar, R. (1994) *Principles of Quantum Mechanics*. New York: Plenum.

Tamkavis, K. (2005) *Problems and Solutions in Quantum Mechanics*. Cambridge: Cambridge University Press.

Relativistic quantum mechanics and quantum field theory

Bjorken, J. D. and Drell, S. D. (1964) *Relativistic Quantum Mechanics*. New York: McGraw-Hill.

Bjorken, J. D. and Drell, S. D. (1965) *Relativistic Quantum Fields*. New York: McGraw-Hill.

Itzykson, C. and Zuber, J. B. (1980) *Quantum Field Theory*. New York: McGraw-Hill.

Jauch, J. M. and Rohrlich, F. (1976) *The Theory of Photons and Electrons*. Berlin: Springer.

Kaku, M. (1993) *Quantum Field Theory: A Modern Introduction*. New York: Oxford University Press.

Landau, L. and Lifschitz, E. (1982) *Quantum Electrodynamics*. London: Pergamon Press.
Peshkin, M. E. and Schroeder, D. V. (1995) *Introduction to Quantum Field Theory*. Reading, MA: Addison-Wesley.
Sakurai, J. J. (1967) *Advanced Quantum Mechanics*. Reading, MA: Addison-Wesley.
Srednicki, M. (2007) *Quantum Field Theory*. Cambridge: Cambridge University Press.
Zee, A. (2003) *Quantum Field Theory in a Nutshell*. Princeton, NJ: Princeton University Press.

Condensed matter physics, condensates, and related topics

Abrikosov, A. A., Gorkov, L. P., and Dzyaloshinski, L. P. (1963) *Methods of Quantum Field Theory in Statistical Physics*. New York: Dover.
Annett, J. F. (2004) *Superconductivity, Superfluids and Condensates*. Oxford: Oxford University Press.
Ashcroft, N. W. and Mermin, N. D. (1975) *Solid State Physics*. Philadelphia: Saunders College.
Fetter, A. L. and Walecka, J. D. (1971) *Quantum Theory of Many Particle Systems*. New York: McGraw-Hill.
Mahan, G. D. (1990) *Many-Particle Physics*. New York: Plenum Press.
Marder, M. P. (2000) *Condensed Matter Physics*. New York: Wiley.
Mattuck, R. D. (1967) *A Guide to Feynman Diagrams in the Many Body Problem*. New York: McGraw-Hill.
Miransky, V. A. (1993) *Dynamical Symmetry Breaking in Quantum Field Theories*. Singapore: World Scientific.
Prange, R. E. and Girvin, S. M. (1990) *The Quantum Hall Effect*. Berlin: Springer-Verlag.
Rickayzen, G. (1965) *Theory of Superconductivity*. New York: Wiley.
Stone, M. S. (1992) *Quantum Hall Effect*. Singapore: World Scientific.
Tinkham, M. (1975) *Introduction to Superconductivity*. New York: McGraw-Hill.
Tsuneto, T. (1998) *Superconductivity and Superfluidity*. Cambridge: Cambridge University Press.
Yoshioka, D. (2002) *The Quantum Hall Effect*. Berlin: Springer-Verlag.

Scattering theory

Bransden, B. H. (1983) *Atomic Collision Theory*. Reading, MA: Benjamin / Cummings.
Goldberger, M. L. and Watson, K. M. (1964) *Collision Theory*. New York: Wiley.
Joachain, C. J. (1975) *Quantum Collision Theory*. Amsterdam: North-Holland.
Newton, R. G. (2002) *Scattering Theory of Waves and Particles*. New York: Dover.

Gauge theory and elementary particles

Cheng, T.P. and Li, L. F. (1994) *Gauge Theory of Elementary Particle Physics*. New York: Oxford University Press.
Greiner, W. and Muller, B. (1993) *Gauge Theory of Weak Interactions*. Berlin: Springer.

Halzen, F. and Martin, A. D. (1984) *Quarks & Leptons: An Introductory Course in Modern Particle Physics*. New York: Wiley.
Okun, L. B. (1980) *Leptons and Quarks*. Amsterdam: North-Holland.
Quigg, C. (1983) *Gauge Theory of the Strong, Weak and Electromagnetic Interactions*. Reading, MA: Addison-Wesley.

Mathematical physics

Arfken, G. B. and Weber, H. J. (2005) *Mathematical Methods for Physicists*. Amsterdam: Academic Press.
Bronshtein, I.N., Semendyayev, K. A., Musiol, G., and Meuhling, H. (2007) *Handbook of Mathematics*. Berlin: Springer.
Butkov, E. (1966) *Mathematical Physics*. Reading, MA: Addison-Wesley.

Index